Biointeractions of Nanomaterials

Biointeractions of Nanomaterials

Edited by

Vijaykumar B. Sutariya
University of South Florida
College of Pharmacy
Tampa, Florida, USA

Yashwant Pathak
University of South Florida
College of Pharmacy
Tampa, Florida, USA

CRC Press
Taylor & Francis Group
Boca Raton London New York

CRC Press is an imprint of the
Taylor & Francis Group, an **informa** business

First published in paperback 2024

First published 2015
by CRC Press
2385 NW Executive Center Drive, Suite 320, Boca Raton FL 33431

and by CRC Press
4 Park Square, Milton Park, Abingdon, Oxon, OX14 4RN

CRC Press is an imprint of Taylor & Francis Group, LLC

© 2015, 2024 Taylor & Francis Group, LLC

This book contains information obtained from authentic and highly regarded sources. While all reasonable efforts have been made to publish reliable data and information, neither the author[s] nor the publisher can accept any legal responsibility or liability for any errors or omissions that may be made. The publishers wish to make clear that any views or opinions expressed in this book by individual editors, authors or contributors are personal to them and do not necessarily reflect the views/opinions of the publishers. The information or guidance contained in this book is intended for use by medical, scientific or health-care professionals and is provided strictly as a supplement to the medical or other professional's own judgement, their knowledge of the patient's medical history, relevant manufacturer's instructions and the appropriate best practice guidelines. Because of the rapid advances in medical science, any information or advice on dosages, procedures or diagnoses should be independently verified. The reader is strongly urged to consult the relevant national drug formulary and the drug companies' and device or material manufacturers' printed instructions, and their websites, before administering or utilizing any of the drugs, devices or materials mentioned in this book. This book does not indicate whether a particular treatment is appropriate or suitable for a particular individual. Ultimately it is the sole responsibility of the medical professional to make his or her own professional judgements, so as to advise and treat patients appropriately. The authors and publishers have also attempted to trace the copyright holders of all material reproduced in this publication and apologize to copyright holders if permission to publish in this form has not been obtained. If any copyright material has not been acknowledged please write and let us know so we may rectify in any future reprint.

Except as permitted under U.S. Copyright Law, no part of this book may be reprinted, reproduced, transmitted, or utilized in any form by any electronic, mechanical, or other means, now known or hereafter invented, including photocopying, microfilming, and recording, or in any information storage or retrieval system, without written permission from the publishers.

For permission to photocopy or use material electronically from this work, access www.copyright.com or contact the Copyright Clearance Center, Inc. (CCC), 222 Rosewood Drive, Danvers, MA 01923, 978-750-8400. For works that are not available on CCC please contact mpkbookspermissions@tandf.co.uk

Trademark notice: Product or corporate names may be trademarks or registered trademarks and are used only for identification and explanation without intent to infringe.

Publisher's Note
The publisher has gone to great lengths to ensure the quality of this reprint but points out that some imperfections in the original copies may be apparent.

Library of Congress Cataloging-in-Publication Data

Bio-interactions of nanomaterials / [edited by] Vijaykumar B. Sutariya, Yashwant Pathak.
   p. ; cm.
   Biointeractions of nanomaterials
   Includes bibliographical references and index.
   ISBN 978-1-4665-8238-5 (hardcover : alk. paper)
   I. Sutariya, Vijaykumar B., editor. II. Pathak, Yashwant, editor. III. Title: Biointeractions of nanomaterials.
   [DNLM: 1. Nanostructures--toxicity. 2. Biocompatible Materials--chemistry. 3. Nanotechnology--methods. 4. Safety Management. 5. Toxicity Tests. QT 36.5]

R857.B54
610.28--dc23
2014015951

ISBN: 978-1-4665-8238-5 (hbk)
ISBN: 978-1-03-291941-6 (pbk)
ISBN: 978-0-429-10275-2 (ebk)

DOI: 10.1201/b17191

Visit the Taylor & Francis Web site at
http://www.taylorandfrancis.com

and the CRC Press Web site at
http://www.crcpress.com

Dedicated to the loving memory of my father, Bhadabhai Chakubhai Sutariya, who passed away on April 22, 2013. He was my role model and mentor throughout my life and whatever I have achieved in life is because of his blessings. I would also like to dedicate this book to the memory of Swami Vivekananda; the world celebrated the 150th birthday of Swamijee in 2013.

Vijaykumar B. Sutariya

To the loving memories of my parents and Dr. Keshav Baliram Hedgewar, who showed the right direction; my wife Seema, who gave my life positive meaning; and my son Sarvadaman who gave a golden lining to my life.

Yashwant Pathak

Contents

Foreword ix
Preface xi
Editors xiii
Contributors xv

Chapter 1 Introduction—Biointeractions of Nanomaterials: Challenges and Solutions 1

Vijaykumar B. Sutariya, Vrinda Pathak, Ana Groshev, Mahavir B. Chougule, Sachin Naik, Deepa Patel, and Yashwant Pathak

Chapter 2 Nanoparticle Exposures in Occupational Environments 49

Li-Hao Young, Ying-Fang Wang, Ching-Hwa Chen, Chun-Wan Chen, and Perng-Jy Tsai

Chapter 3 Physicochemical Characterization–Dependent Toxicity of Nanoparticles 73

Jigar N. Shah, Ankur P. Shah, Hiral J. Shah, and Vijaykumar B. Sutariya

Chapter 4 Cytotoxicity of Stimuli-Responsive Nanomaterials: Predicting Clinical Viability through Robust Biocompatibility Profiles 103

Daniel Wehrung and Moses O. Oyewumi

Chapter 5 Biosensing Devices for Toxicity Assessment of Nanomaterials 117

Evangelia Hondroulis, Pratik Shah, Xuena Zhu, and Chen-Zhong Li

Chapter 6 Carbon Nanotubes and Pulmonary Toxicity 131

Malay K. Das and Charles Preuss

Chapter 7 Nanotoxicity of Polymeric and Solid Lipid Nanoparticles 141

Dev Prasad and Harsh Chauhan

Chapter 8 Analytical Characterization of Nanomaterials in Biological Matrices for Hazard Assessment 159

Mingsheng Xu, Daisuke Fujita, Huanxing Su, Hongzheng Chen, and Nobutaka Hanagata

Chapter 9 Nanoparticles and Human Health: A Review of Epidemiological Studies 175

Vijaykumar B. Sutariya, Ana Groshev, Vivek Dave, Hardeep Saluja, Deepak Bhatia, Prabodh Sadana, and Yashwant Pathak

Chapter 10 Toxicogenomic Approaches to Understanding the Toxicity of Nanoparticles 209

Qiwen Shi, Mahavir B. Chougule, Vijaykumar B. Sutariya, and Deepak Bhatia

Chapter 11 Nanomaterial-Based Gene and Drug Delivery: Pulmonary Toxicity Considerations 225

Mahavir B. Chougule, Rakesh K. Tekade, Peter R. Hoffmann, Deepak Bhatia, Vijaykumar B. Sutariya, and Yashwant Pathak

Chapter 12 Cardiovascular Toxicity of Nanomaterials 249

Saijie Zhu and Minghuang Hong

Chapter 13 Toxicity of Nanomaterials on the Gastrointestinal Tract 259

Jayvadan Patel and Vibha Champavat

Chapter 14 Toxicity of Nanomaterials on the Liver, Kidney, and Spleen 285

Jayvadan Patel and Anita Patel

Chapter 15 Regulatory Implications of Nanotechnology 315

Lynn L. Bergeson and Michael F. Cole

Chapter 16 Ocular Toxicity of Nanoparticles 347

Aditya Grover, Anjali Hirani, Yong Woo Lee, Vijaykumar B. Sutariya, and Yashwant Pathak

Chapter 17 Genotoxicity of Nanoparticles 353

Amaya Azqueta, Leire Arbillaga, and Adela López de Cerain

Chapter 18 Interactions of Polysaccharide-Coated Nanoparticles with Proteins 365

Christine Vauthier

Chapter 19 Models for Risk Assessments of Nanoparticles 383

Sanjay Dey, Bhaskar Mazumder, and Yaswant Pathak

Chapter 20 Immunotoxicity of Carbon Nanoparticles 425

Paulami Pal, Bhaskar Mazumder, and Yaswant Pathak

Index 443

Foreword

Nanomaterials are those in the nanometer range (10^{-9} m). These incredibly small particles can be organic or inorganic, with examples ranging from poly(lactic-*co*-glycolic acid) or gold nanoparticles to carbon nanotubes and quantum dots. These particles may be used to encapsulate drugs, recognize biological markers, or visualize body tissues among many other possibilities, all enabling their widespread application in biology, medicine, and pharmaceutics. Indeed, these nanomaterials may have beneficial effects that have not even been imagined.

The small size of these particles provides an enormous surface area, which is ideal for interactions with cells on a molecular level, but also raises the question of their biosafety. The chemical composition of the diverse nanomaterials available for biological interactions may have unforeseen consequences in living systems. Whether the good that these interactions accomplish outweighs the risk of harm will have to be addressed before nanomaterials are used on a wide scale, especially in biological systems.

This book is a collaborative effort of the editors Drs. Vijaykumar B. Sutariya and Yashwant Pathak and the numerous contributors who are leading scientists in this field. The subject matter is of prime importance in the area of nanotechnology and its applications. These contributors, knowledgeable and experienced in their field, attempt to elucidate the potential biointeractions of nanomaterials with their respective applications in efforts to answer the questions posed above. This book presents the possible biointeractions of various nanomaterials with a number of different body tissues in a multitude of applications. I would like to congratulate Drs. Vijaykumar B. Sutariya and Yashwant Pathak at the University of South Florida for editing this important and timely book.

It is my great pleasure to write a foreword and present to you *Biointeractions of Nanomaterials*. I sincerely hope you will gain as much insight as I did from these chapters.

Shyam S. Mohapatra, PhD, MBA, FAAAAI, FNAI
Distinguished USF Health Professor and Director
Division of Translational Medicine-USF Nanomedicine Research Center
Vice Chair of Research
Department of Internal Medicine
President, USF Chapter of the National Academy of Inventors

Preface

The purpose of this book is to focus on the biointeractions of nanomaterials, an area that has not been previously addressed in detail. It also covers various techniques and tests that have been developed to evaluate the toxicity of materials at the nanolevel. The interactions of nanomaterials and nanosystems within biosystems are a concern for the scientific community.

This book is targeted toward academic researchers as well as industry members who are involved in the development of nanosystems. Many graduate schools have initiated courses in nanotechnology and applications, and this book will be a great resource for students as well as professors. Additionally, this will be a useful tool for industrial scientists investigating technology to update their nanotoxicology and nanosafety understanding.

The objective of the book is to address issues related to the toxicity and safety of nanomaterials and nanosystems. It also covers the interactions of these in biological systems, and various tools and methods used to evaluate toxicity and safety issues.

The volume comprises 20 chapters written by leading scientists in the field of nanotechnology. Chapter 1 covers the challenges and solutions of biointeractions of nanomaterials. This is followed by three chapters that address the assessment and characterization of nanosystems in the bioenvironment.

The next group of chapters covers toxicity and includes biosensing devices for toxicity assessment, carbon nanotubes, and pulmonary toxicity, as well as nanotoxicity of solid lipid nanoparticles. The final group of chapters from 8 to 20 covers nanosafety concerns and solutions. Each of these chapters delves into the effects of nanoparticles on different organs and sheds light on regulatory implications of nanomaterials.

We sincerely hope this book gets an overwhelming response from the scientific community in the field of nanotechnology.

We thank and acknowledge our families, the publishers, and our contributing authors. We would also like to acknowledge Aditya Grover, Anastasia Groshev, and Anjali Hirani for their assistance in editing and obtaining copyright clearance as well as the staff of Taylor & Francis who assisted in shaping this wonderful book in the field of nanotechnology.

Vijaykumar B. Sutariya
Yashwant Pathak

Editors

Dr. Vijaykumar B. Sutariya earned his bachelor of pharmacy and master of pharmacy from L. M. College of Pharmacy, Gujarat University, Ahmedabad, India and his PhD in pharmacy from The M.S. University of Baroda, Vadodara, India. He did his postdoctoral training in the field of pharmaceutics and drug delivery at Butler University, Indianapolis, Indiana.

Dr. Sutariya is an assistant professor in the Department of Pharmaceutical Sciences at the University of South Florida (USF) College of Pharmacy. He has a joint appointment with the Department of Internal Medicine, Division of Translational Medicine at USF.

Dr. Sutariya has published more than 30 research papers in peer-reviewed journals and has presented at various national and international meetings. He is a reviewer of many international journals and an editorial board member of more than six journals related to drug delivery and pharmaceutical sciences. Dr. Sutariya's research is focused on the development of novel drug delivery systems such as nanoparticles, liposome, and thermoreversible gel. His main research focus is on brain-targeted drug delivery and ocular drug delivery. Dr. Sutariya is currently serving as a coinvestigator on two NIH grants (R01 and R15). In addition to research, Dr. Sutariya teaches various courses related to pharmaceutics in the Doctor of Pharmacy curriculum.

Dr. Yashwant Pathak completed his MS and PhD in pharmaceutical technology at Nagpur University, India and his EMBA and MS in conflict management from Sullivan University, Kentucky. He is an associate dean for faculty affairs at the College of Pharmacy, University of South Florida, Tampa, Florida. With extensive experience in academia as well as industry, he has to his credit more than 100 publications, 5 books on nanotechnology, 4 books on nutraceuticals, and several books on cultural studies, including 2 on aging studies from an Indian perspective. His areas of research include drug delivery systems and their characterization in animal models.

Contributors

Leire Arbillaga
Department of Pharmacology and Toxicology
University of Navarra
Pamplona, Spain

Amaya Azqueta
Department of Pharmacology and Toxicology
University of Navarra
Pamplona, Spain

Lynn L. Bergeson
Bergeson & Campbell, P.C.
Washington, D.C.

Deepak Bhatia
Department of Pharmaceutical Sciences
Northeast Ohio Medical University
Rootstown, Ohio

Vibha Champavat
Nootan Pharmacy College
North Gujarat, India

Harsh Chauhan
Department of Pharmacy Sciences
Creighton University
Omaha, Nebraska

Ching-Hwa Chen
Department of Environmental and
    Occupational Health, Medical College
National Cheng Kung University
Tainan, Taiwan

Chun-Wan Chen
Institute of Occupational Safety and Health
Ministry of Labor
Taipei, Taiwan

Hongzheng Chen
Department of Polymer Science and
    Engineering
Zhejiang University
Zhejiang, China

Mahavir B. Chougule
Department of Pharmaceutical Sciences
University of Hawaii
Hilo, Hawaii

Michael F. Cole
Bergeson & Campbell, P.C.
Washington, D.C.

Malay K. Das
College of Pharmacy
University of South Florida
Tampa, Florida

Vivek Dave
Wegmans School of Pharmacy
St. John Fisher College
Rochester, New York

Adela López de Cerain
Department of Pharmacology and Toxicology
University of Navarra
Pamplona, Spain

Sanjay Dey
Department of Pharmaceutical Sciences
Dibrugarh University
Dibrugarh, India

Daisuke Fujita
Advanced Key Technologies Division
National Institute for Materials Science
Ibaraki, Japan

Ana Groshev
College of Pharmacy
University of South Florida
Tampa, Florida

Aditya Grover
College of Pharmacy
University of South Florida
Tampa, Florida

Nobutaka Hanagata
Interdisciplinary Laboratory for Nanoscale
 Science and Technology
National Institute for Materials Science
Ibaraki, Japan

Anjali Hirani
School of Biomedical Engineering and
 Sciences
Virginia Tech
Blacksburg, Virginia
and
College of Pharmacy
University of South Florida
Tampa, Florida

Peter R. Hoffmann
Department of Cell and Molecular Biology
John A. Burns School of Medicine
Honolulu, Hawaii

Evangelia Hondroulis
College of Engineering and Computing
Florida International University
Miami, Florida

Minghuang Hong
Pharmaceutical Crystal Engineering Research
 Group
Shanghai Institute of Pharmaceutical Industry
Shanghai, China

Yong Woo Lee
School of Biomedical Engineering and Sciences
Virginia Tech
Blacksburg, Virginia

Chen-Zhong Li
College of Engineering and Computing
Florida International University
Miami, Florida

Bhaskar Mazumder
Department of Pharmaceutical Sciences
Dibrugarh University
Dibrugarh, India

Sachin Naik
Formulation Department
SunPharma Advanced Research Co. Ltd.
Gujarat, India

Moses O. Oyewumi
Department of Pharmaceutical Sciences
Northeast Ohio Medical University
Rootstown, Ohio

Paulami Pal
Department of Pharmaceutical
 Sciences
Dibrugarh University
Dibrugarh, India

Anita Patel
Nootan Pharmacy College
North Gujarat, India

Deepa Patel
Parul Institute of Pharmacy and Research
Gujarat, India

Jayvadan Patel
Nootan Pharmacy College
North Gujarat, India

Vrinda Pathak
College of Pharmacy
University of South Florida
Tampa, Florida

Yashwant Pathak
College of Pharmacy
University of South Florida
Tampa, Florida

Dev Prasad
School of Pharmacy
Massachusetts College of Pharmacy and
 Health Sciences
Boston, Massachusetts

Charles Preuss
Department of Molecular Pharmacology and
 Physiology
Morsani College of Medicine
University of South Florida
Tampa, Florida

Prabodh Sadana
Department of Pharmaceutical Sciences
Northeast Ohio Medical University
Rootstown, Ohio

Contributors

Hardeep Saluja
College of Pharmacy
Southwestern Oklahoma State University
Weatherford, Oklahoma

Ankur P. Shah
Pharmaceutical Technology Center
Zydus Cadila Healthcare Ltd.
Gujarat, India

Hiral J. Shah
Department of Pharmaceutics
Arihant School of Pharmacy and BRI
Gujarat, India

Jigar N. Shah
Department of Pharmaceutics
Nirma University
Ahmedabad, India

Pratik Shah
College of Engineering and Computing
Florida International University
Miami, Florida

Qiwen Shi
Department of Pharmaceutical Sciences
College of Pharmacy
Northeast Ohio Medical University
Rootstown, Ohio

Huanxing Su
State Key Laboratory of Quality Research in Chinese Medicine
and
Institute of Chinese Medical Sciences
University of Macau
Macau SAR, China
and
Interdisciplinary Laboratory for Nanoscale Science and Technology
National Institute for Materials Science
Ibaraki, Japan

Vijaykumar B. Sutariya
College of Pharmacy
University of South Florida
Tampa, Florida

Rakesh K. Tekade
Department of Pharmaceutical Sciences
University of Hawaii at Hilo
Hilo, Hawaii

Perng-Jy Tsai
Department of Environmental and Occupational Health
National Cheng Kung University
Tainan, Taiwan

Christine Vauthier
Institut Galien Paris-Sud
Université de Paris Sud Faculté de Pharmacie
Chatenay-Malabry, France

Ying-Fang Wang
Department of Environmental and Occupational Health
Medical College
National Cheng Kung University
Tainan, Taiwan

Daniel Wehrung
Department of Pharmaceutical Sciences
Northeast Ohio Medical University
Rootstown, Ohio

Mingsheng Xu
Department of Polymer Science and Engineering
Zhejiang University
Zhejiang, China

Li-Hao Young
Department of Occupational Safety and Health
School Public Health
China Medical University
Taichung, Taiwan

Saijie Zhu
College of Pharmacy
The University of Texas at Austin
Austin, Texas

Xuena Zhu
College of Engineering and Computing
Florida International University
Miami, Florida

1 Introduction—Biointeractions of Nanomaterials
Challenges and Solutions

Vijaykumar B. Sutariya, Vrinda Pathak, Ana Groshev, Mahavir B. Chougule, Sachin Naik, Deepa Patel, and Yashwant Pathak

CONTENTS

1.1 Introduction 2
    1.1.1 What Is Nanotechnology? 2
    1.1.2 Genesis of the Field 4
1.2 Nanomaterials 5
1.3 Classification of Nanomaterials 6
1.4 Application of Nanomaterials 8
    1.4.1 Applications in Medicine and Pharmacy 9
        1.4.1.1 Tissue Engineering 9
        1.4.1.2 Drug Delivery Systems 10
        1.4.1.3 Nasal Vaccination 10
        1.4.1.4 Cancer Diagnosis and Treatment 10
        1.4.1.5 Local Anesthetic Toxicity 11
        1.4.1.6 Gene Therapy and Transfection 11
        1.4.1.7 Molecular Diagnostics and Imaging 11
        1.4.1.8 Biosensor and Biolabels 12
        1.4.1.9 Antimicrobial Nanopowders and Coatings 12
        1.4.1.10 Extraction and Separation Techniques 13
        1.4.1.11 Nucleic Acid Sequence and Protein Detection 13
    1.4.2 Applications in Computer Technology 13
    1.4.3 Environmental Applications 14
        1.4.3.1 Catalysis and Elimination of Pollutants 14
        1.4.3.2 Water Remediation 15
        1.4.3.3 Sensors 15
        1.4.3.4 Fuel Cells 15
    1.4.4 Applications in Commonly Used Products 16
        1.4.4.1 Cosmetics 16
        1.4.4.2 Coatings 16
        1.4.4.3 Self-Cleaning Windows 16
        1.4.4.4 Scratch-Resistant Materials 16
        1.4.4.5 Textiles 16
        1.4.4.6 Insulation Materials 16
        1.4.4.7 Nanocomposites 16
        1.4.4.8 Paint 17

        1.4.4.9   Cutting Tools.. 17
        1.4.4.10 Lubricants 17
1.5   Nanotoxicity... 17
1.6   Biointeractions of Nanomaterials 19
    1.6.1   Interactions with the Environment 19
    1.6.2   Nanotoxicity in the Body..23
        1.6.2.1   Molecular Mechanisms of Nanomaterial Toxicity 23
        1.6.2.2   Pharmacokinetics...25
    1.6.3   Effects of Nanomaterials on Organ Systems...26
        1.6.3.1   Pulmonary System 26
        1.6.3.2   Gastrointestinal Tract..26
        1.6.3.3   Reticuloendothelial Systems 27
        1.6.3.4   Cardiovascular System..27
        1.6.3.5   Central Nervous System 27
        1.6.3.6   Integumentary System 27
1.7   Nanotoxicity: Challenges, Solutions, and the Future 27
    1.7.1   Physicochemical Characterization..28
    1.7.2   *In Vitro* Assessment...30
        1.7.2.1   DNA Synthesis and Damage 30
        1.7.2.2   Immunogenicity 31
        1.7.2.3   Oxidative Stress 31
        1.7.2.4   Cell Proliferation.. 31
        1.7.2.5   Exocytosis 32
        1.7.2.6   Cell Viability and Metabolic Activity...32
        1.7.2.7   Hemolysis...32
    1.7.3   *In Vivo* Assessment 33
        1.7.3.1   Absorption, Distribution, Metabolism, Excretion, and
                       Pharmacokinetic Studies 34
        1.7.3.2   Genotoxicity and Carcinogenic Studies..34
    1.7.4   Considerations for Preventing Nanotoxicity..35
1.8   Future Considerations..36
1.9   Summary 36
References..37

1.1 INTRODUCTION

1.1.1 What Is Nanotechnology?

Nanotechnology is the science that deals with the interactions that arise at a nanosized, molecular scale. There are several paradigms from nature, such as viruses, DNA, water molecules, and red blood cells, with sizes in the nanometer range. This chapter presents a general overview of the nanoparticles (NPs) and their biointeractions. Figure 1.1 illustrates several cases from nature and pharmaceuticals of components with nanometer dimensions. For many decades, nanotechnology has been used most frequently in the areas of engineering, electronics, and physics, and has shown remarkable developments in these fields. However, pharmaceutical and biomedical areas of application still need to be explored.

The unique field of nanotechnology represents not just one specific field but a wide range of areas from basic material sciences to personal care applications. The exciting aspect of nanotechnology is the capability to fabricate formulations by manipulating molecules and supramolecular structures for the development of devices with programmed functions. This is very promising when it is applied to the field of active pharmacological ingredient (API) delivery. The conventional form of

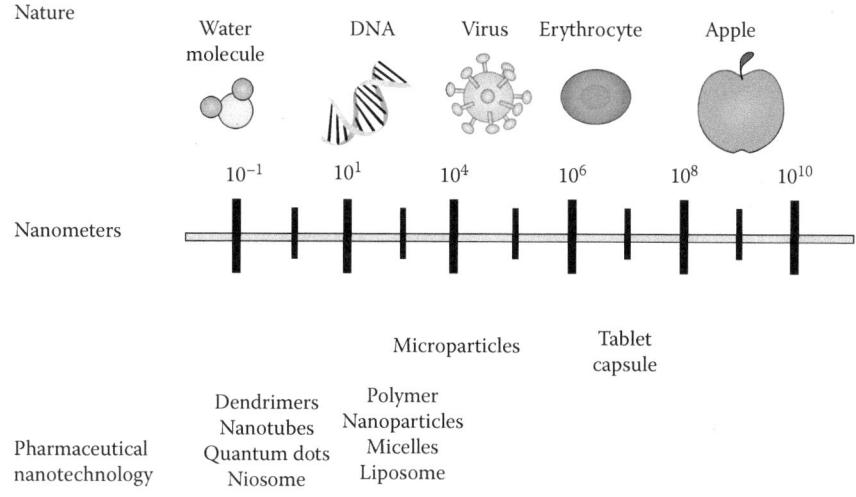

FIGURE 1.1 (**See color insert.**) Dimensions scale of nanotechnology.

novel carriers, such as liposomes, polymeric micelles, and NPs, are now known as nanovehicles. However, this is only in the terms of size. These conventional drug delivery systems would have developed to their current state regardless of the current development of nanotechnology. To fully understand the scope of nanotechnology in the drug delivery field, it may be favorable to categorize drug delivery systems by providing examples from before and after the rise of nanotechnology.

The properties of materials at the nanometer scale can be incredibly altered from those at a larger scale. As the size decreases from bulk compounds, only very small changes in the properties occur until the size of the particulates falls below 100 nm, while remarkable changes in properties can further take place. Nanostructured materials are of great interests for the development of novel properties and functions (Bhushan 2010).

Nanoparticulate drug delivery systems offer many benefits over conventional dosage forms. The advantages of nanoparticulate drug delivery systems include improved therapeutic efficacies, reductions in toxicity, improved biodistributions, and improved patient compliance. Pharmaceutical NPs contain entrapped API substances and are composed of tens or hundreds of atoms or molecules, ranging from 5 to 300 nm in size and with different morphologies, such as amorphous, crystalline, spherical, and needles, among others (Saraf 2006).

Nanosized formulations and structures can be produced by using either "bottom-up" or "top-down" fabrication methods. In "bottom-up" methods, nanoparticulate structures are developed by building up atoms or molecules in a controlled manner through the regulation of thermodynamic properties such as self-assembly, precipitation, and crystallization. On the other hand, advances in nanotechnologies can be used to fabricate nanoscale structures through size-reduction approaches. These techniques, referred to as "top-down" nanofabrication technologies, include photolithography, nanomolding, dip-pen, lithography, and nanofluidics (Figure 1.2 compares the bottom-up and top-down techniques in various manufacturing processes) (Peppas 2004, Sahoo and Labhasetwar 2003).

The reduction of size, having a crucial role in pharmacy, is essential for proper unit operations. It helps in improving the performance of dosages and by providing better formulation opportunities for drugs. Drugs with sizes in the nanometer range improve performances in various dosage forms. Nano-sized formulations provide enhancements in surface area, solubility, rate of dissolution, and oral bioavailability. It may also provide a rapid onset of therapeutic action, a reduction in doses (and frequencies), and decreased fed/fasted and patient-to-patient variabilities.

Particulate dispersions, or solid particles with a size in the nanometer range (10–1000 nm), are known as NPs, in which a drug is dissolved, entrapped, encapsulated, or attached to a NP

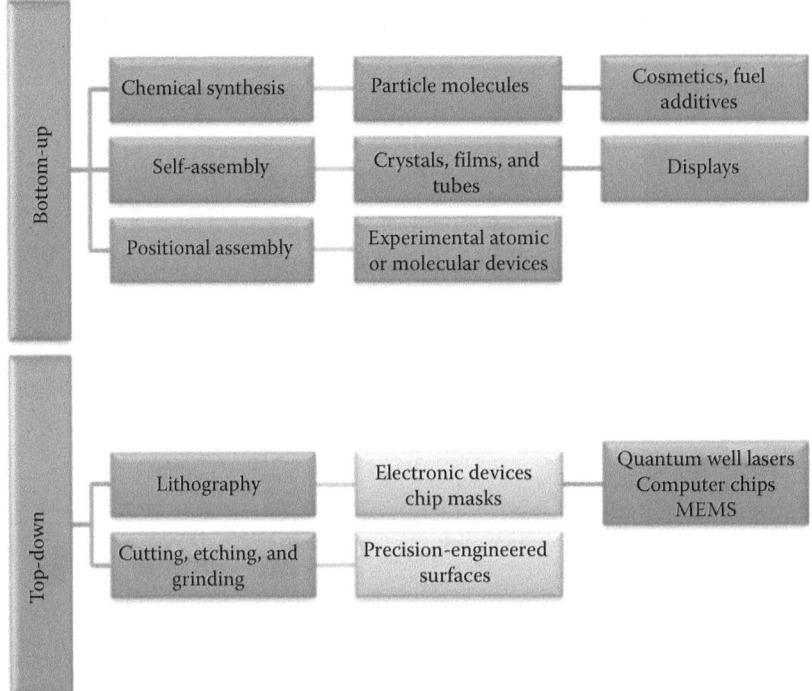

FIGURE 1.2 Bottom-up and top-down techniques in manufacturing nanoparticles.

matrix. NPs, nanospheres, or nanocapsules can be obtained in various forms, such as particles or vesicles, based on the preparation method. Major ambitions in fabricating NPs as delivery carriers are to control the particle size, surface properties, and the release of therapeutics in order to accomplish the site-specific targeting of the drug at a therapeutically optimal rate and dose regimen. NPs can be made up of several materials, including polymers, metals, and ceramics. According to their methods of manufacturing and the materials used, NPs can adopt different shapes and sizes with specific properties. Many other types of NPs are in several stages of development as drug delivery carriers, including lipid-based carriers such as liposomes, lipid emulsions, lipid–drug complexes, polymer–drug conjugates, polymer microspheres, micelles, and various ligand-targeted carriers such as immunoconjugates (Allen 2002, LaVan et al. 2003, Liu et al. 2000, Moghimi et al. 2001).

1.1.2 Genesis of the Field

Although nanotechnology as a field has developed recently, the concept has been present since much earlier. The synthesis and use of gold NPs predates the age of peer-reviewed literature. For example, artists have been utilizing colloidal gold, otherwise referred to as a gold NP solution, to create colors for pottery from the Ming dynasty and stained glass windows in medieval churches (Daniel and Astruc 2004).

The first published report on colloidal gold dates back to a celebrated, 1857 work by Faraday, although earlier unpublished experiments are likely (Faraday 1857). In 1959, Richard Feynman, an American physicist and Nobel Prize winner, envisioned the idea of manipulating particles or materials at a molecular and atomic scale in his presentation titled "There's Plenty of Room at the Bottom" to the American Physical Society's annual meeting. During his talk, he presented facts for generating nanoscale machines to manipulate, control, and image materials at the atomic level. However, the term "nanotechnology" was coined in 1974, more than a decade later, by Norio Taniguchi, a scientist at the University of Tokyo, in his work titled *On the Basic Concept of Nanotechnology*. He described extra-high precision and ultrafine dimensional structures, and also expected improvements in integrated circuits and devices of mechanical, optoelectronic, and computer memory applications (Taniguchi 1974). This is called the "top-down" approach (of carving small structures from larger ones) (Thassu et al. 2007).

The creation of the scanning tunneling microscope by Gerd Binnig and Heinrich Rohrer in 1981, from IBM Zurich Laboratories, rendered a breakthrough by allowing visualizations on a nanosized scale. Further, the invention of the atomic force microscope (AFM) in 1986 made possible the imaging of structures on an atomic scale. In 1986, another scientist, K. Eric Drexler, in his book titled *Engines of Creation*, argued about the future of nanotechnology, specifically the design of larger structures from their atomic and molecular components, known as the "bottom-up approach" (Drexler 1986). He also offered thoughts for "molecular nanotechnology," which is the self-assembly of particles into an ordered and functional structure.

Another major advance in the field of nanotechnology was established in 1985, when Harry Kroto, Robert Curl, and Richard Smalley developed a new form of carbon known as "fullerenes" (or "buckyballs"), a single molecule containing 60 carbon atoms arranged in the shape of a soccer ball. This invention led to a Nobel Prize in Chemistry in 1996. In 2000, this new area of research received recognition from the government when former President Bill Clinton launched the National Nanotechnology Initiative (NNI) to promote research and development in nanotechnology. NNI defines research and development in nanotechnology as that on the 1–100 nm range scale to create systems with novel properties that have the capacity to function on the atomic scale (Thomas and Sayre 2005). Thus, nanotechnology aims to design the formulation of structures, devices, and systems by controlling the shape and size at a nanometer range (Varshney 2012). Today, nanotechnology has progressed into an extensive field of science, with multibillion dollar investments from the public and private sectors. Along with this comes the potential to generate multitrillion dollar industries in the coming decades with an enormous potential to benefit many more applications and areas of research.

1.2 NANOMATERIALS

The history of nanomaterials (NMs) is perhaps as old as that of the universe, as nanostructures were formed in its near beginning. From the dawn of mankind, NPs were produced from fires used by early humans (Alagarasi 2011). The scientific community caught on to NMs much later.

A nanometer is one millionth of a millimeter, about 100,000 times smaller than the diameter of human hair. NMs are important because, at this scale, exclusive optical, magnetic, and electrical properties emerge, among others. These characteristics have great application potentials in electronics, medicine, and other fields. Owing to coatings or surface modifications, NMs demonstrate biocompatibility through interacting with living cells.

NMs are the foundation stones of nanoscience and nanotechnology, a large and interdisciplinary area of research and development that has been explosively developing globally in the past few years. It has the potential to revolutionize the approach in which NMs are developed, and the range and nature of functionalities that can be accessed. It has a significant commercial impact, which will continue growing in the future.

Modified NMs are resources fabricated at the nanometer scale to benefit from small sizes and novel characteristics, normally not found in their conventional, bulk counterparts. These characteristics are enhanced relative surface areas and new quantum effects. NMs encompass a higher surface area to volume ratio than their conventional forms, which can lead to superior chemical reactivities and also have an effect on their strength. Moreover, at the nanorange scale, quantum effects become much more important in determining the material's properties, leading to new optical, electrical, and magnetic characters. The range of NM commercial products available today is very broad, including sunscreens, wrinkle-free textiles, stain-resistant goods, cosmetics, electronics, paints, and varnishes (Alagarasi 2011).

1.3 CLASSIFICATION OF NANOMATERIALS

The classification of NMs is not simple. It may be appropriate to organize the types of NMs according to their chemical and physical properties. However, different types of structures synthesized using various manufacturing processes, along with different surface coatings, can obscure classifications. Other approaches to categorization may be based on NM points of origin or whether they are natural or modified (Nowack and Bucheli 2007). Given the range of NM characteristics, it may be practical to assign categories based on specific properties, such as the potential for health risks (Tervonen et al. 2009). Therefore, in order to provide a general overview of NMs, multiple categorization systems should be considered.

For instance, taking the point of origin into consideration, NMs can be generated via either natural or anthropogenic processes (Figure 1.3 demonstrates classification organization of NMs based on their origin). Naturally produced NMs can be classified into biogenic, geogenic, atmospheric, and pyrogenic categories, based on the methods and mechanisms of production by living organisms, the soil, the air, and heat, respectively. Anthropogenic NMs, or those produced as a result of human activity, can be classified into two categories—unintentionally produced and intentionally engineered NMs (Nowack and Bucheli 2007). Most often, unintentional NMs are created as a by-product of combustion processes (Nowack and Bucheli 2007). Intentionally developed NMs can be further classified into five different categories: carbon-based materials, metal-based materials, dendrimers, polymeric particles, and composites (Tuominen and Schultz 2010).

As NMs have an enormously small size, 100 nm or less in at least one dimension, descriptions of their size and shape have been attempted, such that if NMs have a nanometer size in one dimension, they are referred to as surface films; in two dimensions, fibers or strands; and in three dimensions, particles. They can be present in single, fused, aggregated, or agglomerated forms, with different shapes, such as spherical, tubular, and irregular (Figure 1.4 shows examples of NM classification based on their shape).

NMs are resources that are differentiated by an ultrafine grain size (<50 nm in size) or by a dimensionality restricted to 50 nm. NMs can be produced with different modulation dimensionalities as described by Richard W. Siegel: atomic clusters, filaments and cluster assemblies (zero), multilayers (one), ultrafine-grained over layers or buried layers (two), and nanophase materials consisting of equiaxed nanometer-sized grains (three) as shown in Figure 1.4 (Siegel and Fougere 1995).

Therefore, NMs can also be classified based on structure, morphology, and physicochemical properties. General classifications, based on the types of the NMs, take into account dendrimers, nanotubes, fullerenes, and quantum dots (QDs) (Nowack and Bucheli 2007).

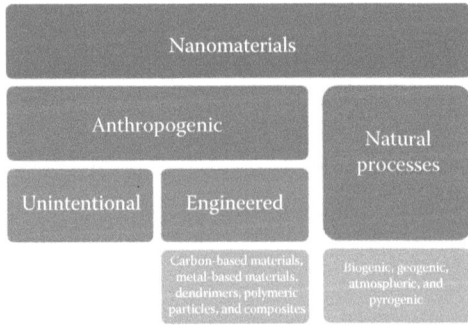

FIGURE 1.3 Classification of nanomaterials based on their origin. Natural process such as proliferation of living organisms, movement of soil, air, and the heat energy may result in the formation of nanomaterials. Anthropogenic nanomaterials are produced as a result of human activity regardless whether specific intent to do so was present or not. Depending on the process, unintentional processes or engineering design may result in nanomaterials of different shapes, sizes, and different properties.

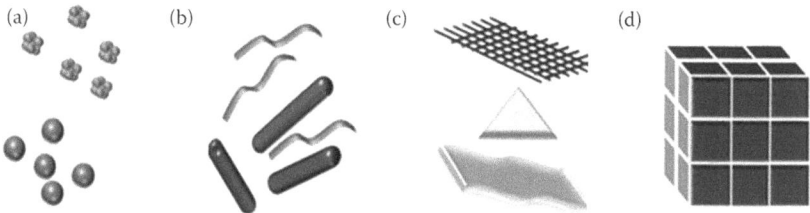

FIGURE 1.4 Classification of nanomaterials. (a) Zero-dimensional spheres and clusters. (b) One-dimensional nanofibers, wires, and rods. (c) Two-dimensional films, plates, and networks. (d) Three-dimensional nanomaterials. (Reprinted with permission from Alagarasi, A. 2011. *Introduction to Nanomaterial*. National Centre for Catalysis Research.)

NMs can be classified based on their phase composition properties, such as single-phase solids (crystalline, amorphous particles, layers, etc.), multiphase solids (matrix composites, coated particles, etc.), and multiphase systems (colloids, aerogels, ferrofluids, etc.). Based on their methods of manufacturing, NMs can be classified into three different categories: gas-phase reactions (flame synthesis, condensation, etc.), liquid-phase reactions (sol–gel, precipitation, hydrothermal processing, etc.), and mechanical procedures (ball milling, plastic deformation, etc.) (Wolfgang 2004). Based on their structural properties, NMs can be classified into two parts: (1) nanocrystalline materials, such as crystals, generally consisting of crystallite with at least one dimension in a nanometer size, and (2) nanostructured materials, such as dislocation fragments, clusters, quasicrystals, micropores, subgrains, and segregations. Nanofragmented materials, composed of dislocation fragments or subgrains whose size is less than 100 nm (Figure 1.5a), normally consist of metals and alloys subjected to megaplastic deformations. Nonporous materials mainly exhibit a high volume density of nanopores less than 100 nm situated on the conventional grain body or along their boundaries (Figure 1.5b). Nanodendrites are materials mainly consisting dendrite solidification products in the form of degenerate dendrite nanodendrites, such as dendrite cells, and become visible upon the rapid solidification

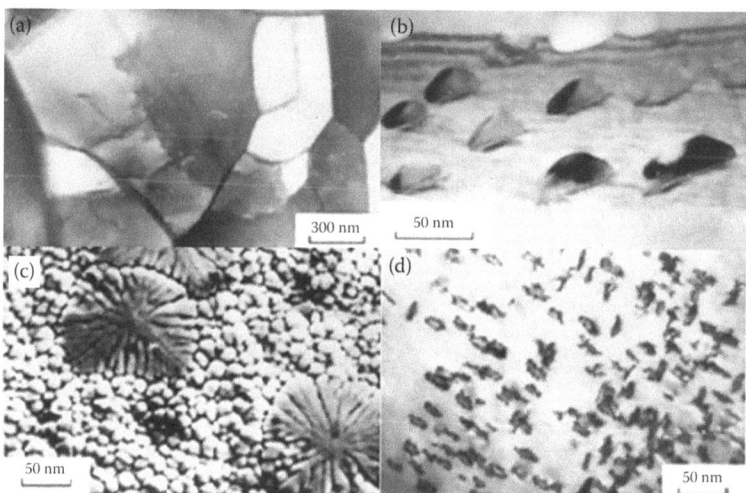

FIGURE 1.5 Nanostructured materials. (a) Structure of nanofragmented material. Melts quenched FeSi alloy, TEM. (b) Structure of nonporous materials. Nanopores are located at grain boundaries in a polycrystalline FeAl alloy produced by melt quenching, TEM. (c) Structure of nanodendrite materials. Dendrite nanocell in side grain in a FeSi produced by melt quenching are visible, SEM. (d) Structure of nanodislocation materials. Melt quench FeLi ally has high density of prismatic vacancy-type dislocation loop, TEM. (Reprinted with permission from Glezer, A. M. 2011. *Russian Metallurgy (Metally)* 4:263–269.)

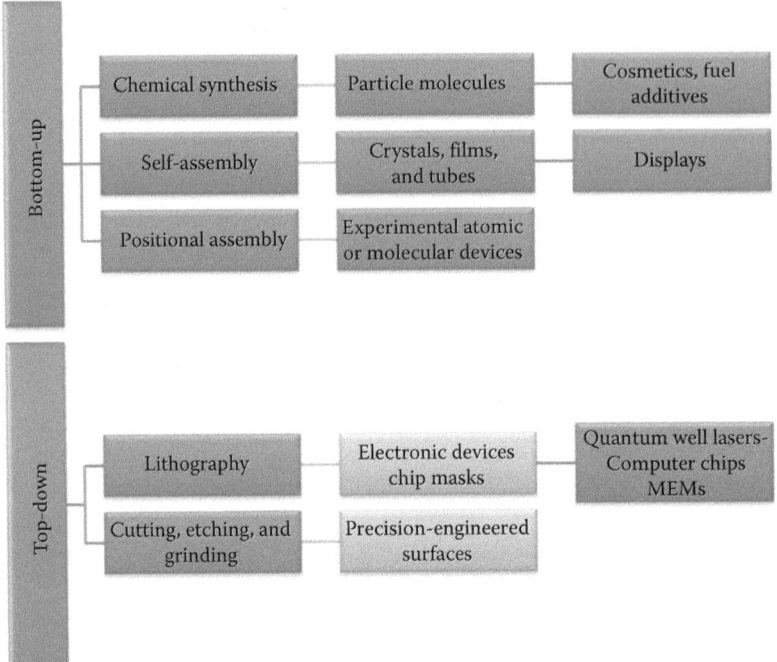

FIGURE 1.6 Summary of classification categories of the nanomaterials. NM can be classified based on their structure and state, for example, agglomeration state. More commonly, NMs are categorized based on their dimensions, morphology, and composition. (Adapted with permission from Nowack, B. and T. D. Bucheli. 2007. *Environmental Pollution* 150(1):5–22.)

of melted compounds (Figure 1.5c). Nanodislocation materials are distinguished by a high-volume fraction of nanoscale dislocation collections or a configuration of definite types (Figure 1.5d).

Nanophase materials generally contain phase transformation nanoproducts. Nanosegregations are materials that consist of grain boundaries or other element segregations with at least one dimension in the nanosized scale. Based on modern theory, nano-cluster or amorphous materials are mainly multicomponent amorphous metallic glasses with a nano-cluster structure. The clusterization of amorphous alloy is highly prominent after local plastic flow. As a result, the amorphous state of alloys manufactured by melt quenching should be considered as possessing a nanostructured shape (Glezer 2011).

NMs can be classified on the basis of their source and origin, structure, morphology, or other physicochemical properties. Figure 1.6 summarizes these various methods of classification (Buzea et al. 2007, Nowack and Bucheli 2007).

1.4 APPLICATION OF NANOMATERIALS

With applications in many areas, such as pharmacuticals, electronics, fuel cells, batteries, agriculture, the food industry, and cosmetics, NMs have certainly already established themselves in the market. A variety of NM-containing products currently exist, such as sunscreens, electronics, paints, varnishes, stain-resistant and wrinkle-free textiles, windows, bicycles, automobiles, and sports equipment such as longer-lasting tennis balls using butyl rubber and nano-clay composites. Owing to their ability to absorb ultraviolet (UV) light, numerous products exist to provide UV-blocking coatings on glass bottles. For example, nanosized titanium dioxide is widely used in sunblock creams and self-cleaning windows, and nanoscale silica is being available as a filler in a series of products, including cosmetics and dental fillings (Alagarasi 2011).

TABLE 1.1
Applications of Nanomaterials

Field of Application	Uses
Medicine and pharmacy	QDs biological imaging for medical diagnostics; early diagnosis of atherosclerosis, Alzheimer's disease, and cancer; drug delivery systems
Computer technology	Nano transistors, computer and camera display and computer memory systems
Environment	Solar cells, solar panels, fuel additives, alternate sources fuels (cellulose intoethanol for fuel), rechargeable batteries; longer, stronger, and lighter-weight wind mills to generate more electricity; advanced filters for cleaner and more purified water
Transportation	Construction of better highways, bridges rails, tunnels, parking garages, pavements in terms of performance, cost effectiveness and longevity; advanced vehicular operation to avoid accidents; aviation
Commonly used products	Baseball bats, tennis rackets, motorcycle helmets, automobile bumpers, luggage, spectacles, fabrics, food containers, sensors to alert food spoilage, cosmetic products, ceramics, tires, cleaning agents, antimicrobial/antibacterial coatings, paints, and household tools

Source: Adapted from NNI. National Nanotechnology Initiative, http://www.nano.gov/you/nanotechnology-benefits.

It is obvious that NMs trump their conventional counterparts due to their exceptional formability and better chemical, physical, and mechanical properties. Modifying material properties allows for applications as diverse as advanced ceramics, semiconductor electronics, sensors, special polymers, magnetics, and membranes. Table 1.1 summarizes the applications of NMs used in different fields.

1.4.1 Applications in Medicine and Pharmacy

1.4.1.1 Tissue Engineering

Nanotechnology has provided several elegant materials that are widely used for tissue repair and replacements and implantable devices, such as sensory aids, retina implants, structural implant materials, implant coatings, bone repairs, surgical aids, operating tools, smart instruments, tissue regeneration scaffolds, and bioresorbable materials (Table 1.2 summarizes various applications in

TABLE 1.2
Applications of Nanomaterials in Medicine (Tissue Regeneration, Growth, and Repair)

Nanosystem	Application
Tissue Regeneration, Growth, and Repair	
Nanoengineered prosthetics	Retinal, auditory, spinal, and cranial implants
Cellular manipulation	Persuasion of lost nerve tissue to grow: growth of body part
Cancer Therapy	
Carbon nanotubes	DNA mutation detection, disease protein biomarker detection
Dendrimer	Controlled release drug delivery, image contrast agents
Nanocrystals	Improved formulation for poorly soluble drugs
Nanoparticles	MRI and ultrasound image contrast agents, targeted drug delivery, permeation enhancers, reporters of apoptosis, angiogenesis
Nanoshells	Tumor-specific imaging, deep tissue thermal ablation
Nanowires	Disease protein biomarker detection, DNA mutation detection, gene expression detection
Quantum dots	Optical detection of genes and proteins in animal models and cell assays, tumor and lymph node visualization

the area of tissue regeneration, growth, and repair). For example, nanofiber scaffolds, generally used to redevelop central nervous system cells and other organs, have been shown to facilitate the regeneration of axonal tissue in hamsters with severed optic tracts (Ellis-Behnke et al. 2006).

1.4.1.2 Drug Delivery Systems

Commercially available and conventional dosage forms for drugs generally suffer from many drawbacks, such as the need for target specificity, a high rate of drug metabolism, cytotoxicity, high dose and dosing frequency requirements, and poor patient compliance, among others. Nanotechnology has facilitated drug delivery systems by improving the physical, chemical, and biological properties that can provide efficient delivery means for currently available active pharmacological ingredients (APIs). Several nanocarriers, such as polymeric NPs, polymeric micelles, liposome, niosomes, dendrimer, polymer–drug conjugates, and antibody–drug conjugates, can generally be divided into the following categories:

1. Sustained and controlled delivery systems
2. Stimuli-sensitive/environment-sensitive delivery systems
3. Functionalized systems for the delivery of bioactives
4. Multifunctional systems for the combined delivery of therapeutics, biosensing, and diagnostic
5. Site-specific, targeted drug delivery systems, including intracellular, cellular, and tissue targeting (Vasir et al. 2005)

The direct, intravenous administration of APIs may induce toxicity due to first-order drug release kinetics when compared to intravenous administration. In cases where a sustained release in required, the field of NMs offers implantable delivery systems by virtue of their size, control, and almost zero-order releases. Some novel vascular carriers, such as liposome, ethosome and transferosome, niosomes, and some implant chips, have been envisaged in recent times, which may assist in the minimization of peak plasma levels with minimal adverse reactions, allow for longer and more predictable action, decrease the frequency of dosing, and improve the levels of patient acceptance and compliance.

Furthermore, various strategies are being developed for superior, site-specific delivery using novel carriers such as polymeric NPs, liposomes, polymeric micelles, dendrimers, iron oxide, and proteins, by modifying the active and passive uptake of drugs. The targeting of drugs to tumor sites via passive delivery methods and using the improved permeation and retention (EPR) effect is thought to be a unique strategy that uses this carrier system by taking advantage of the leaky vasculature in tumors. Some surface modification techniques, using several targeted ligands via covalent binding or adsorption to the carrier system, improved their site specificity, selectivity, and formulation for active targeting. Carriers with targeted, ligand conjugations provide site specificity at various levels. In tuberculosis chemotherapy, the active targeting to lung cells is accounted to have enhanced drug bioavailability, a reduction in dosing frequency, and avoiding the nonadherence trouble that was encountered in the control of tuberculosis.

1.4.1.3 Nasal Vaccination

The use of nanosphere carriers for the delivery of vaccine is currently under development. Nasal vaccinations of antigen-coated, polystyrene nanospheres are widely used for targeting human dendritic cells (Matsusaki et al. 2005). Nanospheres had a positive effect on human dendritic cells by inducing the transcription of genes important for phagocytosis as well as an immune response.

1.4.1.4 Cancer Diagnosis and Treatment

NMs can have a great impact on cancer therapy and diagnoses. Current, commonly available cancer treatments are surgery, chemotherapy, immunotherapy, and radiotherapy. NMs provide remarkable opportunities to assist and improve available, conventional therapies, thereby allowing science to overcome the current challenges in cancer therapies. One such challenge, for instance, involves targeting and site specificity. The development of functionalized and multifunctionalized drug delivery carriers allows for active and passive targeting. One common approach is normally based on the pathophysiology of diseased sites, such as leaky vasculatures in cancer tissues (Ferrari 2005). Owing to its nanosize, NMs can modify the biodistribution and pharmacokinetic characteristics of the anticancer drug considerably as compared to the free drug. These nanoscale materials can recognize biomarkers or detect mutations in cancer cells and treat the abnormal cells by

1. Thermotherapy, including photothermal ablation therapy using silica nanoshells, carbon nanotubes (CNTs), magnetic field-induced thermotherapy using magnetic NPs, photodynamic therapy by QDs as photosensitizers, and carriers for controlled and targeted release
2. Nanostructured polymer NPs, dendrimers, and nanoshells for cancer chemotherapy
3. Radiotherapy using CNTs and dendrimers for boron neutron capture therapy

1.4.1.5 Local Anesthetic Toxicity

Local anesthetics can be very toxic, ranging from local neurotoxicity to cardiovascular collapse and coma. Aside from conventional therapies, drug-scavenging NPs have been shown to considerably enhance the survival in treated animals (Renehan et al. 2005, Weinberg et al. 2003).

1.4.1.6 Gene Therapy and Transfection

Gene therapy occurs when a normal gene is inserted in place of an abnormal, disease-causing gene by using a carrier molecule. Conventional applications of viral vectors normally produce adverse immunologic and inflammatory reactions, as well as diseases in the host. NMs have presently come forward as potential vectors of effective and promising tools in systemic gene therapy. Different polymeric NPs, such as chitosan, gelatin, poly-L-lysine, and modified silica NPs, have been researched to have an increased transfection efficiency and decreased cytotoxicity. It is well noted that NMs provide feasible options as ideal vectors in gene therapies.

Surface-functionalized NPs can be used to infuse cell membranes at a much higher level than NPs without surface functionalizations (Lewin et al. 2000). This property can be used to transport genetic material into living cells through transfection. Silica nanospheres, tagged on their outer surfaces with cationic ammonium groups, can bind anionic DNA through electrostatic interactions (Kneuer et al. 2000). Subsequently, the NPs transport the DNA into cells.

1.4.1.7 Molecular Diagnostics and Imaging

Molecular imaging is the nanoscience that deals with demonstrating, characterizing, and quantifying subcellular biological processes in intact organisms. These processes are composed of gene expression, protein–protein interactions, signal transduction, cellular metabolism, and intracellular/intercellular trafficking. Some NPs that have intrinsic diagnostic properties are QDs, iron oxide nanocrystals, and metallic NPs. They have been effectively employed in magnetic resonance, optical, ultrasonic, and nuclear imagings (Wickline and Lanza 2002). Several other applications of NPs in diagnostics include the selective labeling of cells and tissues, long-term imaging, multicolor multiplexing, the dynamic imaging of subcellular structures, fluorescence resonance energy transfer (FRET)-based analysis, and magnetic resonance imaging (MRI). FRET and MRI are two major diagnostic approaches that have been developed for molecular-level diagnostics. Conventional MRI contrast agents, such as paramagnetic and superparamagnetic materials, are now being replaced by various novel nanocarriers, such as dendrimers, QDs, CNTs, and magnetic NPs. They are established as very efficient contrast agents, offering more stable, intense, and clearer images of objects due to a high-intensity photostability and resolution, and a resistance to photobleaching. A few approved NP applications in imaging and as drug carriers are listed on Table 1.3.

In addition, different, "noninvasive" systems have been widely used for more than a quarter of a century in the field of medical imaging. For example, superparamagnetic magnetite particles

TABLE 1.3
Few Approved Nanoparticles Application in Imaging and as Drug Carriers

Compound	Use
Imaging Agents	
Endorem®—superparamagnetic iron oxide nanoparticles (available in market)	MRI agent
Gadomer®—dendrimer-based MRI agents (phase III clinical trial)	MRI agent—cardiovascular
Drug Delivery	
Abraxane®—albumin nanoparticle containing paclitaxel (available in market)	Breast cancer

covered with dextran are used as image-enhancement agents in MRI (Harisinghani et al. 2003). Intracellular imaging is also feasible via the attachment of QDs to specific molecules, permitting intracellular processes to be monitored directly.

1.4.1.8 Biosensor and Biolabels

Several analytical methods have been developed with the use of this innovative technology. Examples are the determination of different pathological proteins and physiological–biochemical indicators related to disease or disrupted metabolic conditions in the body. Several nanoenabled technologies, techniques, and analytical applications are listed in Table 1.4.

In a general way, biosensors are defined as a measurement method composed of a probe with a sensitive bioreceptor, or a biological recognition part, a physicochemical detector component, and a transducer to transduce and amplify these signals into a measurable form. A nanobiosensor, or nanosensor, is a type of biosensor that has dimensions on the nanometer scale. Applications of various nanosystems as biosensors and biolabels are listed in Table 1.5.

Nanosensors could provide devices to explore important biological processes at the cellular level *in vivo*. The fundamental functions of nanosensors are to recognize and monitor cells; to be used as biomarkers and sensors; and to act as fluorescent, biological labels (Kubik et al. 2005). Biosensors are presently used in the field of target recognition and validation; assay method development; and the determination of absorption, distribution, metabolism, excretion, and toxicity (Jain 2005).

1.4.1.9 Antimicrobial Nanopowders and Coatings

Certain nanopowders, including metal NPs, demonstrate antimicrobial activity. It has been repeatedly demonstrated that more than 90% of *Escherichia coli*, other bacteria species, and viruses are killed within a few minutes when they come in contact with nanopowder. Silver and titanium dioxide NPs (<100 nm) are assessed as coatings for surgical masks due to their antimicrobial effect (Li et al. 2006).

TABLE 1.4
Several Numbers of Nanoenabled Technologies, Techniques, and Their Analytical Applications

Technology	Techniques	Use
Bioarrays and biosensors	Nanofabrication	Nano-object detection
DNA-chips	Lab on chip nanotubes	Electrochemical detection
Protein-chips	Pill on chip nanowires	Optical detection
Glyco-chips	Nanofluidics nanoparticles	Mechanical detection
Cell-chips	Nanostructured surfaces	Electrical detection

TABLE 1.5
Application of Nanomaterials in Medicine and Pharmacy (Biosensors and Biolabels)

Nanosystem	Application
Biosensor and Biolabels	
Gold nanoparticles	For ssDNA detection; in immunohistochemistry to identify protein–protein interaction
Iron oxide nanocrystals	Monitor gene expression; detect the pathogens such as cancer, brain inflammation, arthritis, and atherosclerosis
Nanopores	Sensing single DNA molecules by nanopores
Cantilever array	Diagnosis of diabetes mellitus, for detection of bacteria, fungi, viruses; for cancer diagnosis
Carbon nanotubes	Blood glucose monitoring; sensors for DNA detection
Nanowire	Electrical detection of single viruses and biomolecules
Nanoparticle-based biodetection	Detection of pathogenic biomarkers, ultrasensitive detection of single bacteria

1.4.1.10 Extraction and Separation Techniques

Differen functionalized nanotubes are widely used as smart, nanophase extractors with molecular-identification capabilities to eliminate specific molecules from solutions (Martin and Kohli 2003). Generally, nanotube membranes can operate as channels for the specific transport of molecules and ions among solutions that are present on both sides of the membrane. Membranes containing nanotubes with internal diameters of less than 1 nm can separate small molecules on the basis of their molecular size, whereas nanotubes with bigger internal diameters of 20–60 nm can be used to separate proteins (Martin and Kohli 2003).

1.4.1.11 Nucleic Acid Sequence and Protein Detection

Targeting and identifying different diseases could be possible by detecting nucleic acid sequences that are distinctive to specific bacteria, viruses, or to definite diseases, or the abnormal concentration of certain proteins that signal the presence of different cancers and diseases (Rosi and Mirkin 2005). NM-based assay methods are presently evaluated as well as more sensitive, protein detection methods. Sequences of nucleic acids are, at present, detected by assays detecting molecular fluorophores attached to polymerase chain reaction (PCR). In spite of its high sensitivity and selectivity, PCR has major drawbacks, such as its complexity of method, susceptibility to contamination, cost, and lack of portability (Rosi and Mirkin 2005). Currently available protein detection methods, such as the enzyme-linked immunosorbent assay (ELISA), permit the detection of protein concentrations at which the disease is frequently advanced. More sensitive NM methods would transform the physical treatment of many cancers and diseases (Rosi and Mirkin 2005).

1.4.2 APPLICATIONS IN COMPUTER TECHNOLOGY

The field of microelectronic engineering continually strives toward miniaturization, where the smaller the circuit components—transistors, resistors, and capacitors—the more compact the circuit and the device can be. A reduction in size offers a few advantages, such as an increased device portability and usability, and an often, lower manufacturing cost. Also, microprocessors can run much quicker, enabling computations at far greater speeds. On the other hand, there are numerous technological impediments to these improvements, which include

1. The difficulty in the manufacture of these ultrafine precursor components
2. The dissipation of a remarkable amount of heat generated by these microprocessors due to quicker speeds
3. A short mean time of failures (poor reliability)

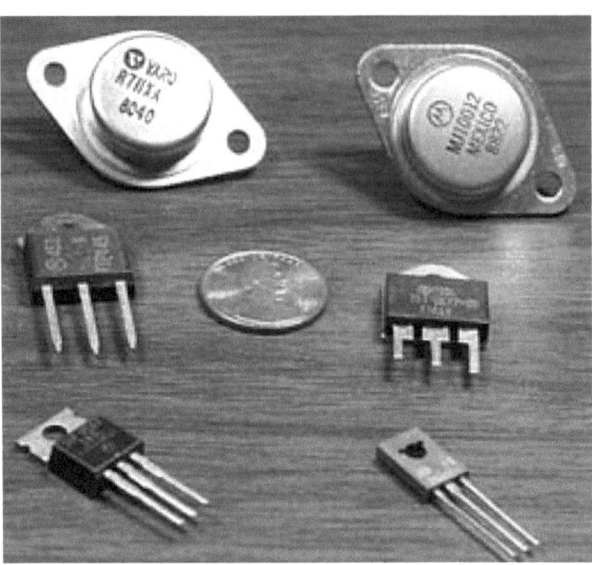

FIGURE 1.7 Examples of silicon nanowires in junctionless transistors. (Reprinted with permission from Alagarasi, A. 2011. *Introduction to Nanomaterial*. National Centre for Catalysis Research.)

The field of NM development assists the industry in breaking down these barriers by offering the manufacturers with nanocrystalline forms of starting materials; ultra-high-purity materials; materials with improved thermal conductivity; and prolonged, durable interconnections in microprocessors (Figure 1.7).

As a component of a circuit, a decrease in the size of the transistor can contribute to a decrease in the overall size. The transistor's design consists of a heavily doped source of electrons, the gate, and the drain that is *p*-doped with holes that can take up electrons (Alagarasi 2011). Conventional inversion-mode (IM) transistors suffer from a few drawbacks that limit the reduction in size and the speed of operation as a function of the material's doping concentration.

The use of nanotechnology in computer engineering has allowed for the design of junctionless transistors, which are substantially more effective and much smaller in size as compared to the IM device. In the junctionless transistor, the doping concentration is equal to that on the source and drain. The gate, controlling the current and acting as a drain, is split from the nanowire by a thin, insulating layer. If the cross section of the device is small enough, the gate can deplete the heavily doped material completely, turning the current off (Lee et al. 2009). Also, as the function of the current is controlled exclusively by the gate, the lifetime, temperature, and efficiency of the device are greatly improved.

1.4.3 Environmental Applications

1.4.3.1 Catalysis and Elimination of Pollutants

Owing to their highly reactive surface, NPs make great catalysts (Bell 2003). Aluminum powder and NPs used as a solid fuel in rocket propulsion are examples (Miller and Herr 2004, Risha et al. 2002). For comparison, bulk aluminum is largely unreactive and is extensively used in utensils. The differences in reactivities between the two forms of aluminum can be easily explained by the fact that catalysts supporting or retarding the reaction rates are dependent on surface activity, which can be very important in manipulating the rate-controlling step (Alagarasi 2011).

This catalytic, chemical activity in NMs can be used in reactions of toxic gases, such as carbon monoxide and nitrogen oxide, in automobile catalytic converters, and in power generation equipments to decrease the hazards and pollution from combustion products (Alagarasi 2011, Astruc 2008, Haruta 2002).

1.4.3.2 Water Remediation

Iron NPs that contain a small amount of palladium have been shown to transform harmful products in groundwater into less harmful end products (He and Zhao 2005). For example, the NPs are able to eliminate organic chlorine, a carcinogen, from water and soil contaminated with chlorine-based organic solvents that are normally used by dry cleaners. The NPs facilitate the chemical reactions that change the solvents to benign hydrocarbons.

1.4.3.3 Sensors

Owing to the dynamic surface of NMs, they make exceptional sensors that are susceptible to small changes in the concentration of the species (Luo et al. 2006). Applications such as the detection of anthrax and the quantification of chromium in wastewater have been demonstrated using NPs (Alagarasi 2011, Wang et al. 2004b).

1.4.3.4 Fuel Cells

Electrochemical fuel cells translate chemical energy into electricity. The electrodes are mainly responsible for the channeling of energy, making their surface area important. Another factor that plays a key role in the performance of the fuel cell is the electrocatalyst. In an ideal structure, an electrode must have a large surface area for a maximized contact with the catalyst, reactant gas, and electrolyte, thereby facilitating gas transport, supply, and good electronic conductance (Alagarasi 2011).

Microbial fuel cells (MFCs) have been used in the generation of electricity by utilizing bacteria. In MCFs, bacteria oxidize the substrate (sugar, starch, or alcohols) to generate electricity and clean water by ridding the waste water of those compounds (Figure 1.8) (Liu et al. 2004). This allows for a significant reduction in sanitation costs and the purification of domestic and industrial

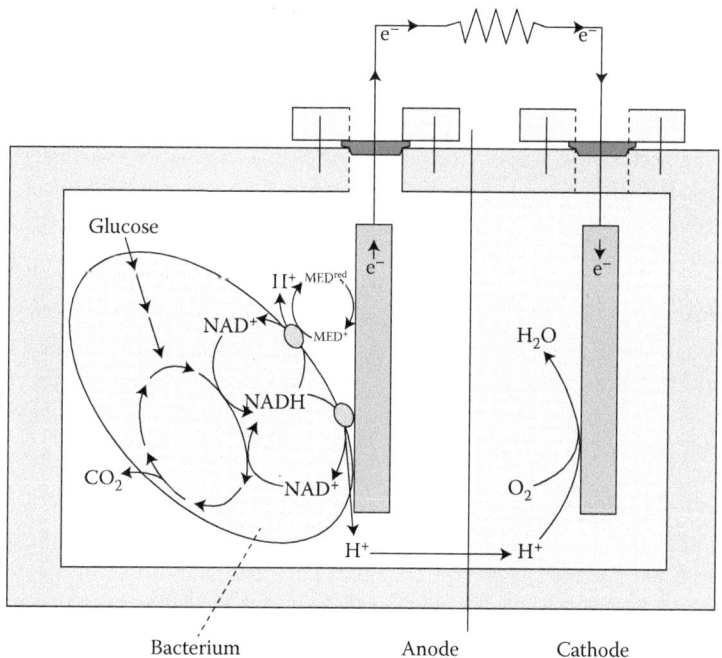

FIGURE 1.8 Schematic representation of microbial fuel cell. The bacterium in the cell either directly or indirectly transfers electron to the electrode resulting in the production of electricity in this setup. (Reprinted with permission from Rabaey, K., and W. Verstraete. 2005. *Trends in Biotechnology* 23(6):291–298.)

wastewater. The electricity-producing bacteria are kept separate from the electron acceptor by a proton exchange membrane, allowing the electrons to pass from the bacteria to the anode. The electrons are then combined with protons from oxygen to yield water.

NMs have been utilized in the improvement of MFCs to enhance surface areas, chemical stabilities, and biocompatibilities. For example, nanowires are thought to have a potential in facilitating the transfer of electrons from the organism to the electrode (Logan et al. 2006, Reguera et al. 2005).

1.4.4 Applications in Commonly Used Products

1.4.4.1 Cosmetics

It is well established that a prolonged exposure to UV light causes damage to the skin in the form of burns and an increased risk of cancer. Sunscreens and cosmetic preparations aim to decrease the risk of cancer by offering protection from sun rays. Nanosized titanium dioxide (TiO_2) and zinc oxide (ZnO) are often used in such preparations to block UV exposure without penetrating the skin and causing dermal discomfort (Nohynek et al. 2007, Nohynek et al. 2008, Schilling et al. 2010). Sunscreens containing TiO_2 and ZnO NPs have been deemed to be some of the safest sunscreen products in the market.

1.4.4.2 Coatings

NMs have been used for extremely thin coatings for decades, if not centuries. Today, thin coatings are used in a wide range of applications, including microelectronics, optoelectronic devices, architectural glass, anticounterfeit devices, and catalytically active surfaces. Structured coatings with nanosized-scale characteristics in more than one dimension assure to be a significant foundational technology for the future.

1.4.4.3 Self-Cleaning Windows

Self-cleaning windows, coated in extremely hydrophobic TiO_2, have been shown to be rather normal. The TiO_2 NPs speed up the breakdown of dirt and bacteria in the presence of water and sunlight, allowing them to be washed off the glass without any difficulty.

1.4.4.4 Scratch-Resistant Materials

Intermediate, nanosized layers, linking the hard outer layer and the substrate material, considerably improve wear- and scratch-resistant coatings. The intermediate layers are constructed to give a good bonding and graded matching of mechanical and thermal properties, leading to the improvement of adhesion properties.

1.4.4.5 Textiles

NPs have well-established applications in coating textiles, such as nylon, to provide antimicrobial qualities. In addition, adjustments in the porosity at a nanosized scale and surface roughness in different polymers and inorganic materials allow for the production of ultrahydrophobic, waterproof, and stain-resistant fabrics.

1.4.4.6 Insulation Materials

Nanocrystalline materials produced by the sol–gel method provide a foam-like structure, known as an "aerogel" (Hrubesh and Poco 1995). Aerogels are made of continuous, three-dimensional networks of particles and voids. Aerogels are porous, enormously lightweight, and have low thermal conductivity.

1.4.4.7 Nanocomposites

Materials that are made up of a combination of two or more components are known as composites. They are constructed to demonstrate the overall, best characteristics of each component, such as mechanical, biological, optical, electric, or magnetic qualities. Nanocomposites, consisting of

CNTs and polymers, allow for a better control of conductivity and are attractive for a wide range of applications, such as supercapacitors, sensors, and solar cells (Baibarac and Gomez-Romero 2006).

1.4.4.8 Paint

NPs impart improved, required, mechanical properties to composites, such as scratch-resistant paint, based on the encapsulation of NPs (Borup and Leuchtenberger 2002). The wear resistance of such coatings is estimated to be 10 times larger than that of conventional acrylic paints.

1.4.4.9 Cutting Tools

Cutting tools consisting of nanocrystalline materials, such as tungsten carbide, are much harder than conventional forms because of the enhanced microhardness property of nanosized composites as compared to that of microsized composites (Yao et al. 2002).

1.4.4.10 Lubricants

Nanospheres made up of inorganic materials may be used as lubricants, acting as nanosized ball bearings (Fleischer et al. 2003).

1.5 NANOTOXICITY

The term "nanotoxicology," coined in 2004, refers to the evaluation of the detrimental outcomes of nanostructure interactions with biological and ecological systems. It includes physicochemical determinants, routes of exposure, biodistributions, molecular determinants, genotoxicities, and regulatory aspects. Nanotoxicology has emerged as a subdiscipline of nanotechnology to address the potential environmental, health, and safety risks that come with the applications of NMs (Arora et al. 2012, Donaldson et al. 2004, Fischer and Chan 2007).

The anticipation of the toxicological hazards of nanostructure materials, due to their unique properties (e.g., chemical, electrical, and magnetic) and the potential for systemic availability and environmental occurrence, has raised concerns among many scientists, regulators, and nongovernmental agencies since the beginning of the 2000s (Colvin 2003, Santamaria 2012). During this time, multidisciplinary research programs were initiated by the National Center for Environmental Research of the United States Environmental Protection Agency, National Toxicology Program, National Institute of Environmental Health, and National Institutes of Health to initiate and promote research on the impact of NMs on human health and the environment (Santamaria 2012).

In the early 1980s, several toxicological and epidemiological studies were conducted to evaluate the respiratory toxicity and pulmonary effects of ambient, ultrafine particles present in the atmosphere as result of natural and anthropogenic activities. Enhanced inflammatory responses in the lungs of rats were found with exposure to TiO_2 and aluminum oxide (Al_2O_3) NPs as compared to larger particles of the same mass and chemical compositions (Ferin et al. 1990, Oberdorster et al. 1990). In the 1990s, sunscreen products came under evaluation for the potential dermal penetration of TiO_2 and ZnO NPs, as they were being used in dermally applied products. With a simultaneous interest in the research of NPs as drug delivery systems, potential outcomes were observed in the evaluation of the inhalation risk of engineered NMs, such as CNTs, in rodent toxicity studies (Shvedova et al. 2005, Warheit et al. 2004). This created an immense interest among toxicology communities in 2004. The significant acute inflammatory pulmonary effects were more pronounced in mice (Lam et al. 2004, Shvedova et al. 2005) as compared to rats (Warheit et al. 2004) in the intratracheal dosing of single- or multi-walled carbon nanotubes. The field of ecotoxicology was highlighted with a study that evaluated the effects of carbon fullerenes on largemouth bass and reported lipid peroxidation in the brain and gills (Oberdorster et al. 1990). Since 2000, research has also been focused on the evaluation of the toxicokinetics and toxicodynamics of NMs; the ingestion of NMs from food; and the use of NMs in medical devices, diagnostics, and therapeutics (Santamaria 2012).

The unique, physicochemical properties of NMs may play a vital role in any possible toxic effects as compared to bulk materials. The size, surface area, composition, and shape are thought to be a few origins of NM toxicity (Aillon et al. 2009, Lanone and Boczkowski 2006). The particle's size influences the distribution and elimination of NMs from the body. Size can also modify the intracellular fate of NMs by manipulating the modes of endocytosis, cellular uptake, and the efficiency of the particle's processing in the endocytic pathway (Lanone and Boczkowski 2006, Rejman et al. 2004). *In vivo* studies of TiO_2 NPs demonstrated that smaller particles (20 nm) led to a persistent inflammatory response as compared to larger particles (250 nm) in rat lungs (Buzea et al. 2007, Oberdorster et al. 1994, 2005). The particles in the nanosized range have an exponentially high surface area to volume ratio and are, hence, more reactive to their surrounding biological environment.

Most of the biological interactions of NMs take place on their surfaces. As the size decreases, the surface area drastically increases, which, in turn, leads to greater proportions of the particle's components being exposed. The small size makes it easy for NMs to translocate into organs. It can also lead to the production of reactive oxygen species (ROS), a contributor of DNA damage (Grabinski et al. 2007, Shvedova et al. 2004). Further, the surface charge determines the kinetics of the NPs within the environment in which they are subjected. The charge that the NMs carry on their surface determines their interactions within the cells. For example, cationic (positively charged) NMs are considered more toxic as compared to anionic (negatively charged) NMs (Goodman et al. 2004). Negatively charged moieties on cell membranes (phospholipid heads and other proteins) have a greater affinity toward disruption by NMs, leading to the cell penetration of these particles.

The presence of several functional groups, which can be controlled by rational design, may also contribute to cytotoxicity as well as reduce systemic toxicity. Surface coatings on NMs have been exploited for drug delivery into targeted regions. For example, NPs with a glutathione coating have been used to deliver paclitaxel into the brain to target brain cancers (Geldenhuys et al. 2011). Glutathione on the coating interacts with their specific receptors in the brain through which they permeate the blood–brain barrier. They also demonstrated the reduced toxicity of FDA-approved PEG–poly(lactic-co-glycolic) acid (PLGA) coatings on NMs.

The chemical composition of NMs, especially at the surface, can modify their interaction with the body. NMs have been made for prolonged circulation by modifying their surface with chemical functional groups for targeted drug deliveries. This functionalization of NMs can potentially alter their interaction with biological components. Such functionalization can also modify the degradation of some transition metals (e.g., QDs), which may otherwise result in the release of toxins and free radicals in the body, leading to subsequent cell death. Both nondegradable and biodegradable NMs can cause detrimental effects to cells through their intracellular accumulation and unexpected toxic degradants, respectively (Aillon et al. 2009, Garnett and Kallinteri 2006).

Shape is another important factor to be considered when studying nanotoxicity. Shapes that are spherical have lower aspect ratios, whereas shapes such as spirals and rods have higher aspect ratios (the ratio between the length and the width of an object). The shape plays a critical role in the effective clearance by altering interactions with macrophages. The internalization of NMs by macrophages was found to be modified by altering the actin-driven interactions in macrophages. The less internalization with rod-like materials relative to spherical materials was evident of possible shape-based phagocytosis of NMs in alveolar macrophages (Aillon et al. 2009, Champion and Mitragotri 2006). Similar kinds of results may be yielded by other tissues as well.

NMs with high aspect ratios are more prone to eliciting toxic effects as compared to the ones with low aspect ratios (Lippmann 1990, Poland et al. 2008). Rod- or spiral-shaped NMs, therefore, have a greater contact area with cell membranes, leading to partial endocytosis by macrophages, as the pseudopodium formed to engulf the NMs is unable to enclose them (Hoet et al. 2004) (Figure 1.5a). This damages the macrophages and also causes their hydrolases, cytokines, and oxidants to be released into the extracellular fluids, leading to further damage. Similarly, high aspect ratios have been shown to modify macrophages and the reticuloendothelial system (RES) uptake of fibrous asbestos in the lungs of rats. As a consequence, the longevity in biological systems of these long aspect ratio fibers leads to long-term carcinogenic effects (Buzea et al. 2007). Single-walled carbon nanotubes (SWCNTs) were found to be toxic in terms of acute inflammation and the onset of progressive fibrosis as compared to spherical particles (amorphous carbon black) in rat pulmonary toxicity studies (Buzea et al. 2007).

Surfactant mediums may also contribute to a potential health risk. While deliberate coatings can help reduce cytotoxicity, it has been found that certain accidental surface reactants can increase their toxicity. For example, when diesel exhausts interact with ozone, they become more toxic (Buzea et al. 2007). Properties of NMs can also change when they are subjected to different mediums. In the case of mediums, the medium itself reacts with NPs, which can sometimes lead to alterations in their physicochemical properties (Colvin 2003). NMs are sometimes easily dispersed within mediums, increasing their toxicity.

Crystallinity can also affect the toxicity of NMs. Different crystalline forms, even of the same chemical composition, exhibit different chemical properties. The classic example is that of TiO_2 crystals, in which the rutile form, which has been found to induce toxicity, is more toxic than the anatase form (Gurr et al. 2005).

1.6 BIOINTERACTIONS OF NANOMATERIALS

1.6.1 Interactions with the Environment

NMs can enter the environment either intentionally or unintentionally through wastes from industries (Figure 1.9) involved in their production and use, or from natural processes such as volcanic eruptions. Released particles can consequently deposit on land or on the surfaces of water bodies

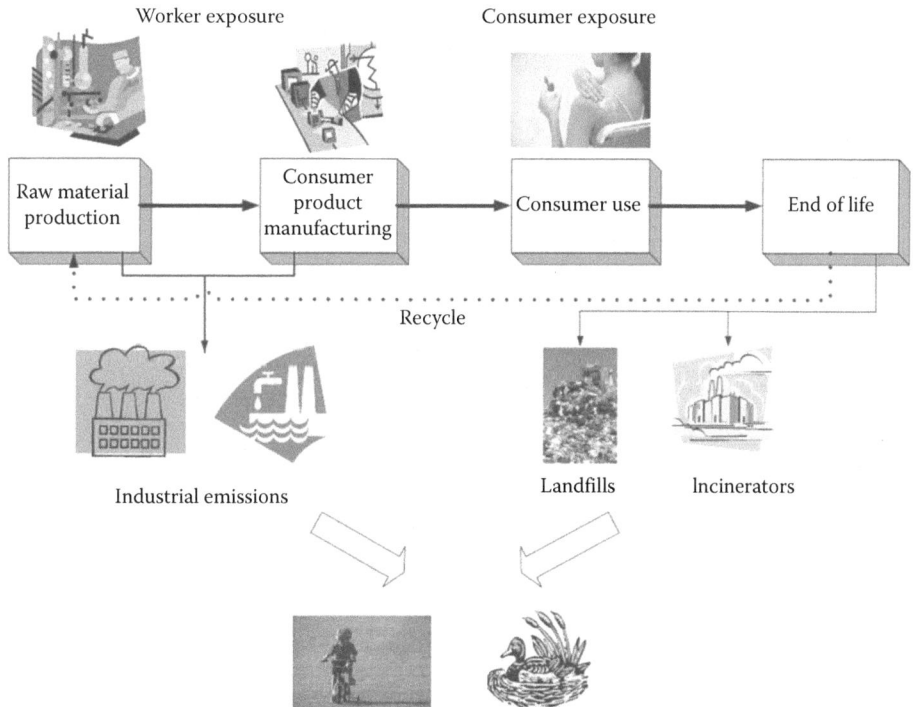

FIGURE 1.9 Transport of nanomaterials in the environment. (Reprinted with permission from Morris, J., J. Willis, and K. Gallagher. 2007. Nanotechnology White Paper. US Environmental Protection Agency. Washington, DC, www.epa.gov/osa/pdfs/nanotech/epa-nanotechnology-whitepaper-0207.pdf (February 2007).)

and interact with their specific biota. Once deposited in the soil, they can cause contamination or seep into the groundwater. NMs in solid wastes, effluents, waste water, or accidental spillages can be transported to aquatic systems by wind or rainwater.

NMs can play an important role in ecotoxicity by serving as carriers of various substances, some of which may be harmful. As the presence of environmental NMs could potentially have an effect on the bioavailability of living organisms, the persistence of NPs is being recognized as one of the key factors in environmental effects assessments. Several assays for the ecotoxicological testing of NMs have been developed, but the challenge in analyzing environmental concentrations is still dependent on reliable methods and analytical tools (Tuominen and Schultz 2010).

The toxicity of metallic NPs on bacteria has been described through various mechanisms that govern toxicity as well as the usefulness of bacterial systems to study the toxicity of manufactured NPs (Niazi and Gu 2009) (Figure 1.10). C_{60} fullerene suspensions have been found to be toxic to bacteria (Lyon et al. 2005, 2006), fathead minnows (Zhu et al. 2006), and zebra fish embryos (Usenko et al. 2007, Zhu et al. 2007). SWCNT-based NMs have been shown to be toxic to estuarine copepods, *Daphnia*, and rainbow trout (Roberts et al. 2007, Smith et al. 2007). The ZnO NPs were found to be more toxic to *Bacillus subtilis* as compared to aqueous TiO_2 and SiO_2 NP suspensions (Adams et al. 2006). The ecotoxicity of ZnO NPs was found to be significantly higher than that of TiO_2 or Al_2O_3 NPs on embryonic zebra fish experimental models (Zhu et al. 2008). A dose-dependent increase in acute toxicity was demonstrated (Zhu et al., 2006) to *Daphnia magna* in a 48-h study with water suspensions of six manufactured NMs (i.e., ZnO, TiO_2, Al_2O_3, C_{60}, SWCNTs, and multiwall carbon nanotubes (MWCNTs)), using immobilization and mortality as toxicological endpoints.

Zebra fish embryos are a useful model system for judging NM toxicity because of the similarities between the zebra fish and human genomes, early life development, and disease processes. The reduced toxicity of ZnO NMs was evaluated on fish embryos upon Fe doping in ZnO (Xia et al. 2011).

The release of NPs to the environment from its limited use and from disposable products is of particular concern. Released NMs can readily undergo transformations via biotic and abiotic processes. Understanding the fate of engineered NMs under environmental transformations will be useful in evaluating the design and development of environmentally benign NMs, as well as their use as environmental tracers in environmental sensing and contaminant remediation. This was demonstrated in a biomimetic, hydroquinone-based Fenton reaction, which provided a new method by which to characterize the expected transformations of nanoscale materials that occur under oxidative, environmental conditions (Metz et al. 2009). Current computational techniques are being used to study the interactions of NPs with biological systems (Makarucha et al. 2011). Such studies could also be used to complement experimental data on toxicity.

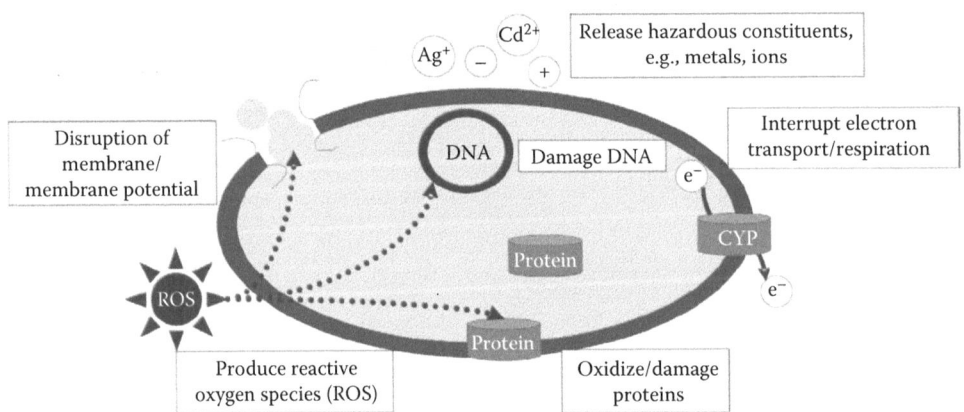

FIGURE 1.10 Biointeractions of nanomaterials. (Reprinted with permission from Klaine, S. J. et al. 2009. *Environmental Toxicology and Chemistry* 27(9):1825–1851.)

Terrestrial ecosystems are composed of soil and organisms. Soil primarily consists of air, water, organic matter, and minerals (Brady and Weil 1996). NMs move around within the pore space in soil and interact with organic matter and minerals. The fate and transport of NMs in the soil is also difficult to predict, as it depends on the physical and chemical properties of the NMs as well as the soil (Darlington et al. 2009, Doshi et al. 2008, Jaisi and Elimelech 2009, Saleh et al. 2008). These properties may lead to aggregation, adsorption, absorption, dissolution, stabilization, transport, or deposition. For example, electrostatic interactions have been observed between the negatively charged, citrate gold NP and positively charged particles in soil, leading to the attachment of NMs on soil particles. Soil solution chemistry parameters, such as ionic strength, pH, and the presence of organic matter, strongly affect the interactions of NMs with solid media; this influences the balance between the free migration of particles and the deposition of NMs (Solovitch et al. 2010). Dissolved organic matter in the soil interacts with NMs and can alter their fate, transport, and bioavailability in the soil. Owing to their small size, NMs have the capability of traveling deeper through soil pores and may get trapped within the soil matrix (Brar et al. 2010). Also, organic molecules such as humic and fulvic acids present in the soil can stabilize NMs in soil solutions and may further enhance their abilities to travel longer distances within the soil (Jaisi and Elimelech 2009). This can ultimately lead to the transport of these NMs to underground water systems. In a study by Yang et al. (2009), it was found that humic acids were adsorbed on the surface of TiO_2, Al_2O_3, and ZnO NPs, leading to a decreased zeta potential. This indicated that humic acid-coated NPs of metal oxides can be easily dispersed and suspended in solution because of enhanced electrostatic repulsions. Organic matter adsorbed on NMs also reduces their aggregation, which may influence their movement in soil solutions.

Interactions of NMs with the water between pore spaces is of great importance, as it directly affects the plant roots and hyphae, such as in the case of fungi (Navarro et al. 2008). NMs can also interact with pollutants in the soils. These pollutants can be organic, such as pesticides, or inorganic, such as metal oxides. The interaction of NMs with these pollutants can further alter the fate and behavior of NMs within the soil medium. Soil colloids with high surface areas carry absorbed minerals and other soil particles. Therefore, they play a major role in the transport of pollutants within the soil medium (Wilson et al. 2008). Only a few publications have documented the uptake of NMs by living organisms within the soil. Fabrega et al. (2009) found that the interaction of bacteria with NMs can affect the transport of NMs in soil. There is evidence to support that NMs can be transported from the soil to plants. Kurepa et al. (2010) found evidence that modified TiO_2 can enter plants cells and accumulate in certain subcellular locations. Another study (Lin and Xing 2008) observed the root uptake and toxicity of ZnO NPs in various plants. It was found that in the presence of ZnO NPs, biomass production was significantly reduced, root tips shrank, and the root epidermal and cortical cells were highly damaged. Some evidence also suggests that NMs can spread via terrestrial plants (Lin et al. 2009). The ecological impact and behavior of NMs on the entire terrestrial ecosystem remains underreported. It is imperative to assess the impact of NMs on soil and terrestrial plants as well as how much they leach to the underground water system.

In an aquatic environment, the small size and large surface area of NMs make them important binding phases for other organic and inorganic contaminants. Other properties such as high surface energy, quantum confinement, and conformational behavior are also considered important. The association and stabilization of NMs with natural organic material, as well as with other organic contaminants, is relevant in order to study their toxicological implications in aquatic ecosystems. The main plausible causes of engineered NM contamination in aquatic systems are waste water treatment plants, production facilities, industrial processes, accidents during transport, and intentional releases. Once in the environment, free NPs tend to form aggregates that can be trapped or eliminated through sedimentation (Figure 1.11). These trapped aggregates can be taken up by organisms that feed on sediment. Although this may potentially lead to distortions in the food chain, no data are currently available on this topic.

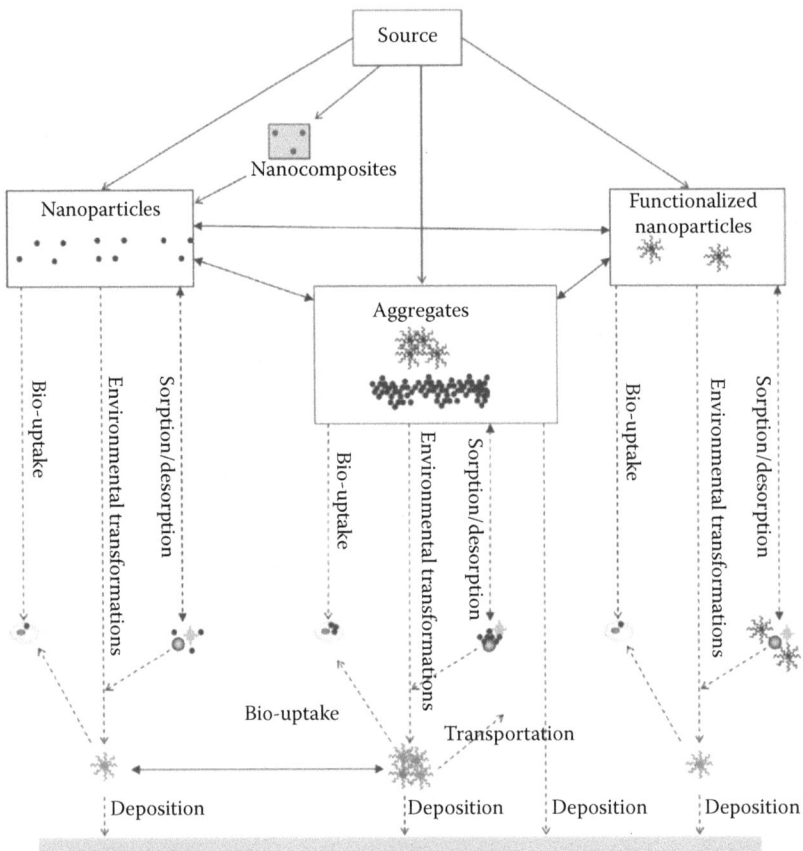

FIGURE 1.11 Fate of nanomaterials in the environment. (Reprinted with permission from Farré, M., J. Sanchís, and D. Barceló. 2011. *TrAC Trends in Analytical Chemistry* 30(3): 517–527.)

The marine environment is composed of various colloids and natural organic matter. The marine environment can be contaminated with NMs by coastal runoffs and atmospheric deposition (Figure 1.12). The coastal environment is dynamic in terms of the presence of organic matter and specific physicochemical characteristics. For example, the presence of greater amounts of organic matter near the coast can lead to a change in the temperature and salinity with depth (Yamashita et al. 2007). These characteristics may influence aggregation, and aggregates of NMs can sink to the ocean floor. It is uncertain whether NMs accumulate at the interface between cold and warm currents (Figure 1.12) or if they are recycled by biota. If they accumulate at the interface of cold and warm currents, they may pose a risk to aquatic species that feed at this zone, such as vertically migrating tuna. On the other hand, if the NMs are in the sediments, they induce risk to species living at the bottom of the sea or ocean. At the surface of the ocean, NMs may get entrapped as a microlayer due to the surface tension properties of water; this again poses a risk of toxicity to marine birds, mammals as well as other organisms living or coming in contact with the ocean surface (Simkiss 1990, Wurl and Obbard 2004).

Physicochemical properties such as pH, ionic strength, or the presence of organic ligands in the water, affects the toxicity of NMs. The results of toxicity in seawater cannot be applied to freshwater, as seawater is more alkaline, has a higher ionic strength, and has a different composition of marine bacteria that may accumulate NMs (Kennedy et al. 2004, Singaravelu et al. 2007).

Introduction—Biointeractions of Nanomaterials

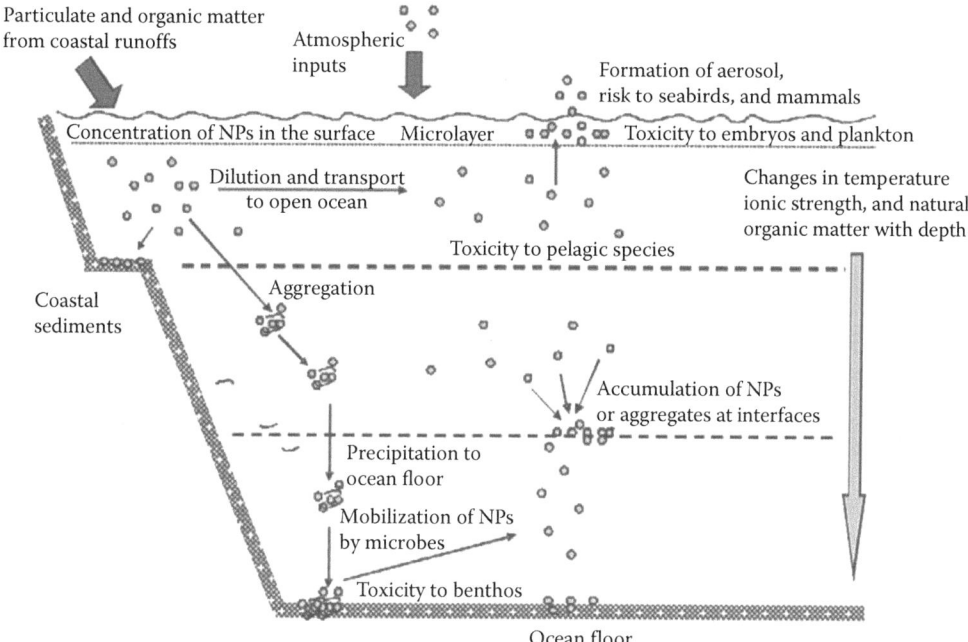

FIGURE 1.12 Schematic diagram outlining the possible fate of nanoparticles (NPs) in the marine environment and the organisms at risk of exposure. (Reprinted with permission from Klaine, S. J. et al. 2009. *Environmental Toxicology and Chemistry* 27(9): 1825–1851.)

1.6.2 Nanotoxicity in the Body

NMs effect the human body at multiple levels, broadly differentiated into molecular, cellular, and organular. The interaction of NMs with biomolecules, such as proteins and lipids, is multivariate and complex. The nano-biointeractions of NMs with the physiological environment molecules account for most of the toxicological effects induced by NMs.

At the cellular level, NMs may also cause mitochondrial injuries, enter the nucleus and damage the DNA, depolarize cell membranes, and also physically damage the membranes by forming nanosized holes. There are different methods by which NMs can interact with the cell membranes, such as via hydrophobic forces, electrostatic forces, van der Waals forces, hydrogen bonding, or receptor–ligand interactions. Once adsorbed on the surface of cells, NMs can be internalized by the cells. Sometimes, the sharp edges of NMs erode the membrane's surface, leading to perforations. The holes thus formed can act as direct entry points for NMs. Not only do they induce toxicity to the organelles inside the cell, these perforations may also lead the leakage of intracellular fluid into the surrounding medium and vice versa, thus inducing acute toxicity and possibly leading to cell death.

Cellular damage manifests itself at the organular level. As explained earlier, the production of ROS can lead to oxidative stress in biological systems. Their production is believed to be the main cause of induced toxicity in the blood, liver, spleen, kidneys, lungs, and any other organs with which they come into contact. The resultant oxidative stress can produce proinflammatory cytokines, as it is believed that ROS can affect the calcium-mediated signaling pathways within the cells.

1.6.2.1 Molecular Mechanisms of Nanomaterial Toxicity

Several different mechanisms have been proposed for the toxicity of NMs in the body (Figure 1.13). The induction of oxidative stress via free radical formation is the prime molecular mechanism of *in vivo* nanotoxicity (Lanone and Boczkowski 2006). These free radicals cause damage to biological components through the oxidation of lipids, proteins, and DNA. As a consequence of this oxidative

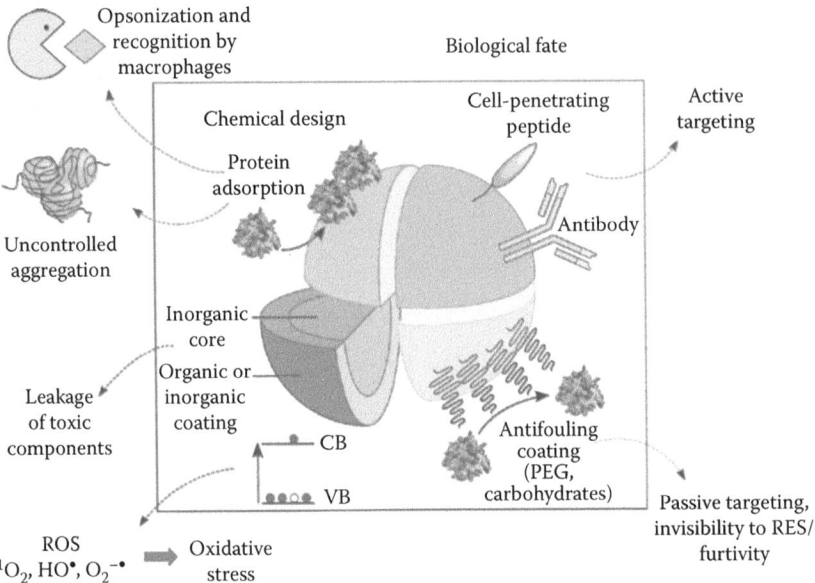

FIGURE 1.13 Biological fate of nanomaterials. The top left scenario illustrates some of the effects of the protein adsorption on the nanoparticle surface. The top right scenario represents functionalization of the surface with peptides and antibodies for uptake and cell penetration. The bottom right scenario demonstrates some of the potential coatings for blocking the surface. The bottom left scenario shows dissolution and potential ROS generation by the nanoparticle. (Reprinted with permission from Pelaz, B. et al. 2013. *Small* 9(9–10):1573–84.)

stress, the upregulation of various inflammatory factors, such as redox-sensitive transcription factors (e.g., NF-κB), activator protein-1, and kinases, may induce or enhance inflammation (Lanone and Boczkowski 2006, Rahman 2000, Rahman et al. 2005). There are several sources of free radical origins, such as phagocytic cell responses to foreign materials, insufficient amounts of antioxidants, the presence of transition metals, environmental factors, and physicochemical properties of some NMs (Lanone and Boczkowski 2006). The effect of oxidative stress may extend to organs of the RES, such as the liver, spleen, and organs of high blood flow, such as the lungs and kidneys, due to the slow clearance and high tissue accumulation of potential free radicals from NMs. Intracellular, NM interactions with cell components, such as mitochondria and nucleus, may result in the cascade of events, such as the creation of ROS, cell cycle arrest, mutagenesis, apoptosis, and nuclear DNA damage, all considered as main sources of toxicity (Aillon et al. 2009, Unfried et al. 2007). NMs may be involved in the upregulation of free radical sources in macrophages and neutrophils (Lanone and Boczkowski 2006). The immediate interaction of NMs with their surrounding environment may result in hemolysis and thrombosis. In addition, NM interactions with the immune system have been known to increase immunotoxicity (Dobrovolskaia and Mcneil 2007).

The adhesion of proteins on the surface of NMs is a normal physiological response in tackling foreign bodies. Once attached to specific proteins (opsonization), the NP–protein complex is recognized by phagocytic macrophages. Once engulfed by macrophages, it is taken to the spleen or the liver for its removal from the bloodstream (Owens and Peppas 2006). NMs can also interact with other proteins not intended for opsonization. For example, the binding of human serum albumin or apolipoproteins promotes a prolonged circulation time in the blood (Ishida et al. 2001). As mentioned earlier in the section, NM surface chemistry determines its interactions with different moieties in the body. Positively charged particles attract proteins, leading to adsorption onto their surfaces, and forming a complex known as a "protein corona." These coronae can have multiple layers: hard layers composed of proteins strongly attached to the surface of NMs, and dynamic, soft layers containing proteins, which are weakly adsorbed onto the surface (Mahmoudi et al. 2011). This NM–protein interaction can lead to the blockage of the protein's active site, mild conformational changes, or even denaturation. This can lead to a failure in the protein's ability to perform its normal biological function, such as cell signaling (Lynch and Dawson 2008, Mahmoudi et al. 2011).

Ligand-coated NPs bind to cell receptors and can be endocytosed within the cell or trigger a signaling cascade in the cell. Once inside the cell, the NMs are trapped within endolysosomal vesicles. The exact mechanism of how they escape these vesicles remains unclear. However, once that happens, they can be released into the cytosol where they can interact with other cell organelles, including the nucleus.

1.6.2.2 Pharmacokinetics

Determining the pharmacokinetics of NMs is the first crucial step in understanding its biological safety and toxicity. Pharmacokinetics is defined as the study of the mechanisms of absorption, distribution, metabolism, and excretion of a drug or its metabolite (ADME). A thorough, quantitative, pharmacokinetic analysis of NMs would reveal the target cells, tissues, or organs; the residence time; and the time and dose required to manifest toxicity. This information can be utilized to plan various focused studies that involve only the target cell, eventually helping to decipher the molecular basis of toxicity. The pharmacokinetic behavior of NMs cannot be analyzed at present due to the lack of data and the fact that any difference in the physicochemical properties might affect its pharmacokinetics.

1.6.2.2.1 Absorption

NMs can enter the human body through different routes of exposure, such as the skin, lungs, and gastrointestinal tract (GIT). As they pass through various parts of the body, they pick up different biomolecules (Mahmoudi et al. 2011) described in the previous sections (Figure 1.13). The absorption of biomolecules onto their surfaces determines their subsequent biological activities in the body (Lynch and Dawson 2008). As mentioned previously, NMs interact with proteins to form protein coronae, which determine their biodistribution and fate within the body. Protein function is altered by the conformational changes induced by adsorption onto the surfaces of NMs (Darlington et al. 2009, Lundqvist et al. 2004), which, in turn, affects their fate (Ishida et al. 2001, Paciotti et al. 2004).

1.6.2.2.2 Distribution

After absorption, NMs can be distributed to various organs, tissues, and cells. It is difficult to predict the behavior of NMs within living systems, owing to different variants within the systems. The extent of NM distribution within the body depends upon the permeability of blood vessels. Organs such as the liver, spleen, lymph nodes, and bone marrow can take up NMs. These organs contain macrophages and form the RES or mononuclear phagocyte system (MPS). This system is involved in the uptake and metabolism of foreign molecules (Saba 1970). Coatings on NMs can affect their uptake. For example, NMs coated with polyethylene glycol (PEG) resist RES uptake (Paciotti et al. 2004). It is therefore imperative to understand and study the distribution of NMs within the body.

1.6.2.2.3 Metabolism

There are limited reports on NM metabolism. There are some NMs that degrade in the tissues, such as polymer-based NMs and superparamagnetic iron oxide NMs. On the other hand, there are some that show no degradation *in vivo*, such as QDs, fullerenes, and silica NMs (Ballou et al. 2004, Khan et al. 2005, Singh et al. 2006, Yang et al. 2007). A study finding that CNTs can be degraded by neutrophil myeloperoxidase provides evidence that suggests enzymatic metabolization of NMs to a certain extent (Kagan et al. 2010). Also, coatings such as proteins used for QDs can be metabolized by proteases in the gut (Hardman 2006). It is still unclear as to how NMs may be metabolized in the body. Therefore, studies need to be conducted to address these unanswered questions.

1.6.2.2.4 Elimination

NMs can be eliminated from the body via various routes, such as exhalation; urination (via the kidneys) (Singh et al. 2006); defecation (via the biliary duct) (Hardonk et al. 1985, Renaud et al. 1989); perspiration; and through the saliva, seminal fluids, and mammary glands. Their fate of elimination, again, depends on their physicochemical properties. For example, hydroxyl functionalized SWCNT accumulate in the liver and kidneys and are excreted in the urine within 18 days (Wang et al. 2004a). On the other hand, ammonium functionalized SWCNT show neither liver uptake not fast, renal excretion (Singh et al. 2006). Contrary to that, QD are not excreted and remain intact *in vivo* (Fischer et al. 2006, Yang et al. 2007) unless they have a coating. This has been proven in the case of QDs coated with cysteine, which were excreted in mice urine (Choi et al. 2007). Studies should focus on identifying organs that could be stressed by exposure to NMs, possibly providing a molecular basis for the stress response. If there is an association found with specific organ cells and NMs characteristics (e.g., size, surface chemistry, aggregation and composition, shape), then it would be possible to establish correlations between the toxic effects of NMs and specific NM properties. Demonstrated pharmacokinetic studies of various NMs will be discussed in greater detail in the later chapters.

1.6.3 Effects of Nanomaterials on Organ Systems

There are numerous, inevitable, exposure routes by which NMs can enter the human body to elicit potential adverse effects. The specific routes are the respiratory, reticuloendothelial, cardiovascular, central nervous, and integumentary systems.

1.6.3.1 Pulmonary System

The respiratory system serves as a major entry portal for ambient particulate materials. The short-term exposure of inhaled, ultrafine carbon black, nickel, and TiO_2 particles were found to produce an enhanced inflammatory response in the rat respiratory system, as compared to fine-sized particles of similar chemical compositions (Grassian et al. 2007, Pettibone et al. 2008, Wani et al. 2011, Warheit et al. 2006). The micron-sized particles are largely trapped and cleared by the upper airway mucociliary escalator system, whereas particles less than 2.5 µm can travel to the alveoli. Inhaled NPs can become deposited in the alveolar region (Arora et al. 2012, Curtis et al. 2006, Hagens et al. 2007). The toxicity of NMs may initiate with the development of exaggerated lung responses, characterized by increased and persistent levels of pulmonary inflammation, subsequently transformed into cellular proliferation, fibro proliferative effects, and inflammatory-derived mutagenesis, which ultimately results in the development of lung tumors. As previously mentioned, various factors that are likely to influence the pulmonary toxicity of NPs are the size and number of particles, surface dose and coating, degree of aggregation, surface charges, and the method of particle synthesis (Lyon et al. 2006, Wani et al. 2011).

1.6.3.2 Gastrointestinal Tract

NMs can reach the GIT directly through the ingestion of food, water, cosmetics, drugs, and by the use of drug delivery devices, as well as after mucociliary clearance from the respiratory tract through the nasal region (Arora et al. 2012, Hagens et al. 2007). The acute toxicity of ingested nanocopper material was found to be more toxic than bulk copper material in mice. The occurrence of systemic argyria after the ingestion of colloidal nanosilver proves its secondary toxic effects after translocation from the intestinal tract (Arora et al. 2012). Reports were found for the uptake of fluorescently labeled, polystyrene NPs by intestinal lymphatic tissue (Peyer's patches) (Morishita and Peppas 2006).

1.6.3.3 Reticuloendothelial Systems

Since all of the blood exiting the GIT goes into the hepatic portal vein that directly diffuses through the liver, the RES system in the liver is exposed to all NPs absorbed from the GIT into the cardiovascular system (CVS). NPs such as carbon black and polystyrene exert their toxic effects by enhancing the secretion of proinflammatory cytokines, such as tumor necrosis factor alpha, following the stimulation of macrophages via ROS and calcium signaling (Brown et al. 2004). These proinflammatory cytokines and oxidative stresses that can potentially damage hepatocyte function and bile formation are also associated with the pathology of liver diseases (Wani et al. 2011). The direct injection of ultrafine carbon black particles into the blood induces platelet accumulation in the hepatic microvasculature of healthy mice in addition to prothrombotic changes on the endothelial surface of hepatic microvessels.

1.6.3.4 Cardiovascular System

The surface charge of NPs plays a vital role in their toxic effects on the CVS. Especially cationic, ultrafine particles, such as gold and polystyrene, have been shown to cause a lethal effect on RBCs and blood clotting while anionic particles are found to be nontoxic. The exposure to diesel exhaust particles (DEPs) was found to alter heart rates in hypertensive rats, while also inducing direct, negative effects on the heart's pacemaker activity (Hansen et al. 2007). Exposure to SWCNT has also resulted in adverse cardiovascular effects (Li et al. 2007).

1.6.3.5 Central Nervous System

The brain can be exposed to NPs by the means of two different mechanisms after inhalation; namely, trans-synaptic transport after inhalation through the olfactory epithelium and uptake through the blood–brain barrier (Jallouli et al. 2007, Lockman et al. 2004). The adverse pathologies, including hypertension and allergic encephalomyelitis, have been found to be associated with the enhanced permeation of NPs to the blood–brain barrier in experimental setups. The production of ROS (Long et al. 2006) and subsequent oxidative stress (Peters et al. 2006) by NPs has been implicated in the pathogenesis of neurodegenerative diseases, such as Parkinson's and Alzheimer's. The NP's surface charges and chemical compositions have been shown to alter blood–brain integrity and deserve considerations as to their role in brain toxicity and distribution (Wani et al. 2011).

1.6.3.6 Integumentary System

The skin is the largest primary defense organ in our body and comes into direct and indirect contact with many toxic agents. The strongly keratinized stratum conium is the rate-limiting barrier to defending against the penetration of most micron-sized particles and harmful exogenetic toxicants. Possible skin exposure to NMs can also occur during the intentional application of cosmetics and other topical drug treatments. In addition, NPs have unique scattering properties due to their small size. They may alter the optical pathway of UV photons entering the upper part of the skin's horny layer. In this way, more photons can be absorbed by the stratum conium. *In vitro* studies have shown that MWCNT initiate an irritation response in human epidermal keratinocytes by their localizing effect (Baroli et al. 2007, Zvyagin et al. 2008). QDs with diverse physicochemical properties were found to penetrate the intact stratum conium barrier and get localized within the epidermal and dermal layers (Ryman-Rasmussen et al. 2006).

1.7 NANOTOXICITY: CHALLENGES, SOLUTIONS, AND THE FUTURE

Unfortunately confined just to nanomedicine, toxicity assessments are an integral part of NM development. Even though attempts are being made to assess toxicity of widely used NMs, they are met with various challenges owing to the shortcomings of the methods used for assessment. Even though nanotoxicology is considered a well-defined field, it faces a lot of challenges even today.

1.7.1 Physicochemical Characterization

As discussed earlier, the physical and chemical properties of NMs play a very important role in their interaction with biological systems. The physicochemical properties of NPs must be scrutinized in detail to interpret any results of NP-induced toxicity by optimizing the NP design. Specific NP properties that influence cellular toxicity are still not fully understood. Hence, a thorough characterization of the NP is essential.

There are diverse approaches that are commonly used to characterize these properties, several of which are described in Table 1.6. Size (distribution) and shape determinations are typically evaluated with one or more of the following: dynamic light scattering (DLS) (Murdock et al. 2008), transmission electron microscopy (TEM), scanning electron microscopy (SEM) (Love et al. 2012, Marquis et al. 2009), and AFM. Field flow fractionation (FFF) is a chromatography-like technique in which the partition of sample species is achieved in a thin, open channel, and the particle size distribution can be calculated directly from the first principles (Bohnsack et al. 2012). Crystal structure is generally elucidated by x-ray diffraction and the surface area is determined by the Brunauer–Emmett–Teller (BET) method by nitrogen adsorption and desorption isotherms (Maurer-Jones et al. 2010).

The surface chemical composition can be examined through various techniques, such as inductively coupled–mass spectrometry (ICP-MS) and inductively coupled–optic emission spectrometry (ICP-OES), which utilize an inductively coupled plasma as the ion source and can detect metals (and some nonmetals) at concentrations below one part per trillion. ICP-MS and ICP-OES are also capable of monitoring isotopic speciations for the ions of choice (Bohnsack et al. 2012). Other techniques are secondary ion mass spectroscopy (SIMS), liquid chromatography–mass spectrometry (LC-MS), matrix-assisted laser desorption/ionization–time of flight (MALDI-TOF), and are used specifically for an elemental analysis along with mass determinations. The elemental composition of NP surfaces can be determined by Auger electron spectroscopy (AES), electron energy loss spectroscopy (EELS), and x-ray photoelectron spectroscopy (XPS). These are all high-vacuum techniques and vary in their capabilities and surface sensitivities (Love et al. 2012, Powers et al. 2012).

NP uptake by cells is a vital factor to the assessment nanotoxicity. Flow cytometry (FCM) has been used in the field of biochemistry to analyze thousands of cells in a second, which is advantageous over TEM and ICP-MS techniques (Ibuki and Toyooka 2012). Electron paramagnetic resonance (EPR)–electron spin resonance (ESR) are specialized methods to measure free radicals either directly or by "spin trapping" them with a reference molecule. They are powerful techniques to quantify oxygen ROS in toxicological evaluations of NMs. The ion abrasion SEM (IA-SEM) and focused ion beam SEM (FIB-SEM), in addition to soft x-ray tomography, are being used to elucidate three-dimensional displays of cells and tissues (Powers et al. 2012). Synchrotron radiation-induced x-ray fluorescence (SRXRF) can provide the local, electronic, and molecular structures around the atom of interest with sub-Angstrom spatial resolutions. It has also been utilized to investigate the uptake of organic mercury in zebra fish larvae (Bohnsack et al. 2012).

Along with the physical characteristics of NPs, dose characterizations are also critical for the interpretation of results. The calculation of a dosing metric is complicated because little is known about appropriate doses and how aggregation or stability influences effective dosings (Love et al. 2012). These characterization challenges are ripe for the study and application of the collective expertise of analytical chemists.

Various combinations of methods may be utilized to validate NM properties, as each method has its own disadvantages and advantages. However, chemical characterization should accompany physical characterization to assess the presence of contaminants in test samples, although exhaustive characterization is time consuming as well as very expensive. There is a need to devise a battery of standardized tests to adequately assess these properties, so that the data that are obtained are comparative and reproducible. There is also a need for a standardized reference material for NMs that can be used by toxicologists, so that data can be compared with different studies.

TABLE 1.6
Analytical Techniques to Characterize Nanoparticles

Characteristics	Analytical Methods	Parameters
Particle imaging	TEM—transmission electron microscopy	Size, shape, state of aggregation
	SEM—scanning electron microscopy	Size, shape, state of aggregation topography, elemental composition in combination with other techniques
	ESEM—environmental SEM	As SEM above
	AFM—atomic force microscopy	Size, shape, morphology, state of aggregation
	SAXS—small-angle x-ray scattering, SANS—small-angle neutron scattering	Number of bilayers, chemical group ordering
Physical property	DLS—dynamic light scatter size of particle, NTA—nanoparticle-tracking analysis, FCS—fluorescence correlation spectroscopy, FFF—field flow fractionation	Particle size distribution
	Filtration	Size fractionation
	Centrifugation	
	XRD—x-ray diffraction	Crystal structure, size
	BET—Brunauer–Emmett–Teller method	Surface area per unit mass and porosity
	SEC—size exclusion chromatography, AFFF–asymmetric field fractionation, DLS, electrophoretic method, electroacoustic technique	Size distribution ζ potential
Chemical composition	ICP-MS—inductively coupled–mass spectrometry, ICP-OES—inductively coupled–optic emission spectrometry, SIMS—secondary ion mass spectroscopy	Element composition, mass
	EDS—x-ray energy dispersive spectroscopy	Element composition
	LC-MS—liquid chromatography–mass spectrometry, MALDI-TOF—matrix-assisted laser desorption/ionization–time of flight	Fullerene structure, mass
	AES—Auger electron spectroscopy, EELS—electron energy loss spectroscopy, XPS—x-ray photoelectron spectroscopy	Surface chemistry
	NMR—nuclear magnetic resonance spectroscopy, FTIR—Fourier transformed infrared spectroscopy, Raman spectroscopy	Chemical functional groups, surface groups
Uptake	ICP-AES—inductively coupled plasma atomic emission spectroscopy, SRXRF—synchrotron radiation-induced x-ray fluorescence, NAA—neutron activation analysis, ICP-MS, TEM, flow cytometry	Cellular uptake
	IA-SEM—ion abrasion SEM, FIB-SEM—focused ion beam SEM, soft x-ray tomography	Three-dimensional imaging of nanoparticle distribution in cells and tissues
	EPR—electron paramagnetic resonance/ESR—spin resonance	Quantification of reactive oxygen species generation after cellular interaction of nanoparticles
Stability	CD—circular dichroism spectroscopy, UV-vis—UV-visible spectroscopy, DLS, ICP-AES, ICP-MS, colorimetric assays	Stability indicating changes, such as particle size distribution, dissolution, surface chemistry, degradant formation

Powers et al. (2006) have summarized five basic rules for physicochemical characterization:

1. The sample used for characterization assessments should be representative of the material.
2. The size and shape should be measured as dispersed a state as possible.
3. The most appropriate method for measurement should be applied.
4. The particles should be measured in desired amounts to ensure precision and accuracy.
5. As far as possible, the characteristics of the particles should be measured under the same conditions as its application.

1.7.2 IN VITRO ASSESSMENT

In vitro toxicity studies are valuable in the optimization of NP design. *In vitro* methods to assess NP toxicities can be classified as mechanistic or viability assays. The mechanistic assays are those that seek to assess the effects of NPs on various cellular processes, whereas viability assays are concerned solely with whether a given NP results in cell death.

Owing to their lower costs and efficacy, *in vitro* studies are the most preferred studies for the assessment of toxicity. *In vitro* assays consist of subcellular systems, such as macromolecules; organelles; cellular systems, such as individual cells, culture, barrier systems; and whole tissues, such as organs, slices, and explants. Before the administration into biological systems (mice or rabbits), it is very important to ascertain the right dosage or a safe concentration of the drug. That is where *in vivo* studies play a pivotal role. The toxicity of the drug is first tested in cell culture to assess the suitability and dosage for *in vitro* studies (Geldenhuys et al. 2011).

Although *in vitro* tests are quick and straightforward, there are some limitations, such as the correlations of the *in vitro* to the *in vivo* environments. The process of immortalization alters the cells' properties and sensitivities. Also, some cells are more sensitive than others to a certain kind to toxin as compared to other cells in a different culture. Because the cells are isolated from their natural environment, they may not be an appropriate model. For example, Lee et al. (2009) reported that 2D cell cultures that are commonly used in *in vitro* studies may not correctly reflect the actual toxicity of NPs, as they do not represent functions of 3D tissues that have complex cell-to-cell and cell-to-matrix interactions with different diffusion or transport conditions.

There may also be interferences with specific toxicology assays based on the properties of the NM. These are a result of the following unique physical and chemical properties: high surface areas that may lead to increased adsorption capacities, optical properties that may interfere with fluorescence or visible light absorption detection systems, increased catalytic activities due to enhanced surface energies, and magnetic properties that make them redox active and thus interfere with methods based on redox reactions (Kroll et al. 2009). Interferences with colorimetric assays, such as the MTT 3-(4,5 dimethylthiazol-2-yl)-2-5-diphenyl tetrazolium bromide (MTT) assay, have been reported (Doak et al. 2009, Kroll et al. 2009, Monteiro-Riviere and Inman 2006, Monteiro-Riviere et al. 2009, Pulskamp et al. 2007, Scalf and West 2006, Song et al. 2010). For example, CNTs interact with formazan crystals and make them insoluble (Wörle-Knirsch et al. 2006). Consequently, such variables can lead to conflicting results (Doak et al. 2009, Monteiro-Riviere et al. 2009, Wörle-Knirsch et al. 2006). Thus, if such interferences are suspected, additional tests would be required to confirm the findings.

1.7.2.1 DNA Synthesis and Damage

DNA synthesis assays are commonly used to assess cell proliferation or to quantify the number of cells in each stage of the cell cycle (which can subsequently reveal cell cycle arrest at a given point). The incorporation of 5-bromo-2-deoxyuridine (BrdU) into newly synthesized DNA has been frequently employed to quantify DNA synthesis in nanotoxicity assays. The genotoxicities of silver (Ag) NPs and PEG-coated cadmium selenide/zinc sulfide (CdSe/ZnS) QDs on lung epithelial cancer (A549) and skin epithelial (HSF-42) cells were assessed by utilizing the aforementioned method (Oostingh et al. 2011, Zhang et al. 2006). As damage to the DNA is highly correlated with an increased risk of cancer, it is critical to assess such damages by any NP that is likely to come in contact with humans. The comet assay (single-cell gel electrophoresis assay), which is utilized to measure the number of single-strand breaks in DNA, is the most common method to assess DNA damage. This assay has been used to assess DNA damage in cells exposed to cerium oxide (CeO_2) (Auffan et al. 2009), Ag (AshaRani et al. 2009), and SiO_2 NPs (AshaRani et al. 2009). Other methods to assess DNA damage include checking for the presence of micronuclei or other chromosomal aberrations and measuring the expression of proteins implicated in DNA repair. An increase in the expression and activation of DNA repair-related proteins was found upon cellular exposure to MWCNTs (Zhu et al. 2007).

The general DNA microarray and more specific PCR analyses are being utilized to assess the activity of functional genes involved in various cellular processes. They have been used to assess changes in gene expression upon exposure to gold (Au) nanorods (Hauck et al. 2008), SWCNTs (Nygaard et al. 2009), and SiO_2-coated CdSe/ZnS QDs (Zhang et al. 2006). PCR was utilized to study the effects of antimony trioxide (Sb_2O_3) NPs in erythroblasts (Bregoli et al. 2009) and the impact of cesium dioxide (CeO_2) NPs on the expression of genes related to oxidative stress and cell structure (Park et al. 2008).

1.7.2.2 Immunogenicity

The ability of a given NP to evoke an immune response is a vital parameter in demonstrating its toxicity on physiological systems, and it may not be explored by standard cellular toxicity studies. ELISA can accurately detect cytokine levels at picograms levels. Many investigators have studied the formation of proinflammatory cytokines (e.g., interleukin-6 and -8) following the exposure to metal oxide NPs in various cell types by this technique (Veranth et al. 2007, Schanen et al. 2009).

1.7.2.3 Oxidative Stress

An elevated amount of ROS, either due to an innate immune response, to a NP or from the ability of a specific NP (e.g., a fullerene or a metal oxide) to autocatalyze ROS formation in the cellular environment, has the potential to damage or disrupt key cellular processes (Xia et al. 2006). Generally, the presence of ROS can be directly assessed by quantifying the amount of ROS present in a given cell population or indirectly assessed by monitoring the secondary effects of prolonged oxidative stress. The spectrofluorimetry/FCM or spectrophotometry-based system can directly measure and monitor the ROS-induced formation of the fluorescent product, fluorescein, from 2,7-dihydrodichlorofluorescein diacetate (DCFDA), the superoxide-induced conversion of dihydroethidium (DHE) from the blue fluorescent form to the red fluorescent form, or the superoxide-induced conversion of nitroblue tetrazolium (NBT) to blue formazan. The DCFDA and DHE assays have experimentally shown to change ROS levels in MPMCs (Marquis et al. 2011) or human fibroblasts (AshaRani et al. 2009), which were exposed to Au or Ag NPs with different surface functionalities. The effects of ultra-small superparamagnetic iron oxide NPs (AshaRani et al. 2009) and cationic lipid-coated Fe_3O_4 NPs (Soenen et al. 2009) in human monocyte macrophages and 3T3 cells were assessed by NBT assay. The determination of lipid peroxidation or antioxidant depletion is the measurement of the secondary effects of increased cellular ROS levels. These can be done with the detection of 8-hydroxy deoxyguanosine (8-OHdG) and superoxide dismutase (SOD) activities. A green fluorescent dye, which turns red in the presence of oxidized lipids, was utilized to assess lipid peroxidation in the presence of Cd/Te QDs (Choi et al. 2007).

1.7.2.4 Cell Proliferation

The rate of cell growth is an important indicator of overall cell integrity and of the potential for NPs to interfere with proliferative processes. There are two quantitative assays commonly utilized as the standard for assessing cell proliferation: (a) cell counting by FCM or high-content image analyzers, and (b) the colony-forming efficiency (CFE) assay. The effect of SWCNTs (Mu et al. 2009) and PEG-silane modified CdSe/ZnS QDs (Zhang et al. 2006) on the proliferation of HEK293 and human lung and skin epithelial cells has been assessed by FCM. The CFE assay has been used to assess the effects of polymeric-entrapped, thiol-coated Au nanorods (Zhang et al. 2006) on murine fibroblasts and human hematopoietic progenitor cells.

1.7.2.5 Exocytosis

Changes in exocytosis may be another indicator of nanotoxicity. Carbon-fiber microelectrode amperometry has been employed to study the effects of various NPs on the secretion of small, electro-active molecules (e.g., serotonin and epinephrine). This method allows one to quantify the number of chemical messenger molecules released per vesicle, the specific release kinetics, and the frequency of vesicle fusion with a high sensitivity and time resolution. Studies in MPMCs and adrenal chromaffin cells have utilized this method to reveal the mutagenic potential of functionalized (with either positive or negative side chains) Au and Ag NP exposure (Marquis et al. 2011).

1.7.2.6 Cell Viability and Metabolic Activity

Cell viability studies are perhaps some of the most widely used assays to assess nanotoxicity, as they provide information on the mechanisms or causes of cellular toxicity and death. Any lethal consequences from NP exposure, including membrane lysis, cell cycle arrest, and apoptosis, may stop mitochondrial activity. Many different types of assays that allow for the study of toxicity are used in research. Toxicity can also be assessed by using two or more independent test systems to validate findings. The colony formation assay, or the clonogenic assay, is an *in vitro* cell-survival assay, based on the ability of a single cell to grow into a colony. It is a simple method that can be employed to avoid interference from NPs, as no dye or stain is used (Franken et al. 2006).

Assays of metabolic activity following exposure to NPs are the most common methods used to determine cell viability. The most popular test is the MTT assay in live cells. MTT [3-(4,5-dimethylthiazol-2-yl)-2,5-diphenyltetrazolium bromide] is reduced to purple formazan, which can be detected spectrophotometrically. Several similar assays (MST, MTS, XTT, WST-1) have also been employed to eliminate the possibilities of NM interference with these assays (Stone et al. 2009). Alamar blue (resazurin) is another dye that undergoes reduction by living cells to produce the fluorescent product, resorufin. It has also been extensively utilized to measure cell viability, following exposure to SiO_2-coated CdSe QDs (Sharma et al. 2009) and amino acid-functionalized Au (Ghosh et al. 2008).

1.7.2.7 Hemolysis

The risk of erythrocytic lysis is especially important for NPs that are intended to be directly introduced into the bloodstream. The assessment of hemoglobin (Hb) by spectrophotometric techniques in response to NP exposures can be a measure of both membrane disruption and extreme cellular toxicity (i.e., necrosis). This approach has been utilized to determine the median lethal dose values for functionalized Au NPs (Goodman et al. 2004). Recent studies have focused on the hemolytic potential of functionalized Au NPs while assessing their effects on ROS production in neutrophils and thrombotic capabilities (Love et al. 2012).

The assessment of the indicators of programmed cell death (i.e., apoptosis) and necrosis directly reveal a NP's ability to induce intracellular, self-destruction mechanisms and destroy cells. Such assays have been developed beyond the measurement of membrane integrity to the quantification of apoptotic protein levels and activation and DNA fragmentation. There are five main techniques generally used to determine membrane integrity: phosphatidylserine (which migrates to the extracellular surface of apoptotic cells) labeled with annexin V, propidium iodide exclusion by intact membranes (AshaRani et al. 2009), trypan blue exclusion by intact membranes (Goodman et al. 2004, Hauck et al. 2008), neutral red staining (which undergoes a color change due to protonation in intact lysosomes) (Lanone et al. 2009), and the determination of the total lactate dehydrogenase (LDH) content in extracellular mediums (Papageorgiou et al. 2007, Tkachenko et al. 2004). Another common assay looks for the exclusion of red fluorescent ethidium homodimer 1 from live cells, while measuring the uptake of calcein-AM (which fluoresces green after modification by intracellular esterases). Assessing the level of DNA fragmentation with TUNEL (terminal deoxynucleotidyl transferase deoxyuridine triphosphate nick end labeling) can be used to identify apoptosis, as demonstrated by studies on SWCNTs (Mu et al. 2009) and $Eu(OH)_3$ NPs (Patra et al. 2008). It is important to consider the NM's impedance during the selection of any of the aforementioned methods. For example, in the MTT assay, CNTs can modify the solubility of formazan through the adsorption of the reduced crystals, thereby falsely lowering the viability results (Worle-Knirsch et al. 2006). Such spontaneous reductions may also occur in graphene particles (Liao et al. 2011). The LDH assay has also failed for some NPs, including Cu (LDH was inactivated) and TiO_2 (LDH was adsorbed) (Han et al. 2011). Further advancement in this field requires the detection and quantification of sensitive toxicological markers that may be unique to nanotoxicity.

1.7.3 IN VIVO ASSESSMENT

The prediction of the safety and toxicity of nanoconstructs has been examined by the extensive testing of *in vitro* cultured cells along with *in silico* computational models. *In vivo* systems are much more complex with interdependent pathways, which are difficult to evaluate by *in vitro* analysis. However, toxicity assays in animal models can provide better correlations with human conditions.

Acute toxicity studies are performed in animal models to identify the maximum tolerated dose (MTD) and no observable effect level (NOEL) in NP dosages. In classical toxicology studies, the dose of NMs is measured by the milligrams of test items per kilogram of animal weight. However, the surface area, size, density, and surface properties of NMs are less common to take under consideration in toxicity studies. The true evaluation of nanotoxicity should be based on both the classical mg/kg exposure and the dosage based on surface area to justify the effects of nanoscale reactivities on toxicity. Acute toxicity studies normally span 14 days after a single dose or repeated dose administration, and the evaluation of organ-specific toxicity in addition to finding the right dose. At least two species, one rodent and one nonrodent species, are preferably required to conclude the results from these studies. The following parameters are monitored during the study:

- Responses to the administered dose: Following the administration of NMs, neuronal, hematological, and cardiac responses can occur and, hence, animals should be monitored for at least 30 min postadministration.
- Changes in weight: The overall health of the animal is the simplest parameter to observe for any possible toxic effects of the injected dose. The change in weight (>10%) can significantly indicate the NM's adverse effect. However, this is a preliminary observation and further investigation is required to find out the actual cause of toxicity.
- Clinical observation: The functionality of various organs systems, such as the cardiovascular, respiratory, ocular, and gastrointestinal systems, are examined to evaluate clinical changes. Imaging procedures such as ultrasound, x-ray, computed tomography (CT), and MRI are used as supportive elements.
- Clinical pathology: The plasma samples collected from the processed animals are utilized to check liver functionality by the measurement of aspartate aminotransferase (AST), alanine aminotransferase (ALT), and total bilirubin and albumin levels. Kidney functions are evaluated by assessing blood urea nitrogen and creatine levels in plasma. Cardiac function is assessed by measuring LDH and creatine phosphokinase (CPK). Amylase levels are indicators of exocrine functions.

- Gross necroscopy: Gross necroscopy can reveal valuable information about the specific toxicity of NMs. Severe intestinal bleeding was observed after the intravenous administration of G7 amine-terminated PAMAM dendrimers in CD-1 mice (Greish et al. 2012). The histology of the lung and liver was performed to evaluate the toxicity of SWCNTs (Oberdorster et al. 2005).

1.7.3.1 Absorption, Distribution, Metabolism, Excretion, and Pharmacokinetic Studies

Absorption, distribution, metabolism, excretion, and pharmacokinetic (ADME/PK) studies are vital tools to evaluate the occupational health impact and potential hazards of NMs to human beings. The pharmacokinetic and biodistribution behaviors of NMs is essential to understand, as it reflects the in-depth concentration of NMs in each organ, which can be used as the basis of phase I clinical studies in humans. Radiolabeling studies with gamma emitters, such as ^{125}I, is a popular method of *in vivo* NP quantification. A biodistribution profile of PAMAM dendrimers with differential surface charges and sizes was evaluated via ^{125}I labeling in CD-1 mice (Greish et al. 2012). ^{99}Tc was used in the scintigraphy imaging of nanoscale *N*-(hydroxypropyl) methacrylamide copolymers in mice (Line et al. 2005). The limitation of this method is the stability of the nanoconstruct–radiolabel conjugates, which needs to be evaluated before beginning the imaging study. The application of ICP-MS can eliminate the aforementioned limitations and provide acute quantifications of NMs in parts per million levels. The biodistribution profile of gold nanorods and spheres in tumor-bearing mice was evaluated and analyzed by ICP-MS (Arnida et al. 2011).

1.7.3.2 Genotoxicity and Carcinogenic Studies

The mutagenic, teratogenic, and carcinogenic potential of NMs are being evaluated in these studies. Widely accepted *in vivo* genotoxicity tests are the metaphase chromosomal analysis and the bone marrow micronucleus test. In 1975, Schmid developed the mouse bone marrow micronucleus test as an alternative to cytogenic studies on mammalian bone marrow cells (Hayashi et al. 1983). The damage to the chromosome or mitotic apparatus (resulting in micronucleus formation), as a consequence of NMs exposure in animal bone marrow cells, is being detected by the micronucleus test. Micronuclei (MNi) are formed because of acentric fragments of chromosomes or due to chromosomal breakages or mitotic spindle apparatus damages (Sarto et al. 1987). FCM has been used to differentiate bone marrow cells from abnormal, peripheral RBCs (Sarto et al. 1987). Teratogenic studies are not regularly required as part of toxicity studies. Such studies follow the International Conference on Harmonization (ICH) of Technical Requirements for Registration of Pharmaceuticals for Human Use guidelines (Baber 1994). ICH guidelines involve segments of test procedure standardizations for fertility and reproductive performance, embryo–fetal development, perinatal, and postnatal analyses on maternal and newborn cases. *In vivo* carcinogenic studies involve long-term observation for the development of tumors following the administration of either single- or repeated-dose NMs. It requires a large number of animals and usually extends up to 30 months.

Apart from the aforementioned studies, chronic and subchronic studies are essential to assess the toxic effects of NMs, which are nonbiodegradable and consist of metal oxides, such as gold, carbon, and silica particles. In such cases, acute and subacute observations are not sufficient to evaluate the safety of NPs due to their long residence times in the body.

The major challenges, or the most discouraging aspects, of *in vivo* tests are its length, expense, and ethical issues. Even though *in vivo* tests may counter most of the limitations faced by *in vitro* studies, there are still many challenges that impede their use for the assessment of nanotoxicity. First, there is difficulty in determining the actual dosages of NMs in the environment to which animal and humans are exposed, owing to their small size and quantity. Sometimes, if the concentration of the known dose is high, it may lead to agglomeration. Vehicles are used for dose delivery in order to avoid agglomeration. It is another challenge to ensure that the vehicle used for delivery does not interfere with the NMs and does not induce any toxicity of its own. For example, some studies found that phosphate-buffered saline, commonly used as a vehicle for dose delivery, is a poor dispersion agent (Sager et al. 2007, Sayes et al. 2007). Once delivered into the body, NMs may interact with other components of the biological system, such as proteins, different salt concentrations, and variable pH values, to form unstable suspensions, thereby negatively affecting their biodistribution and activity (Buford et al. 2007). Hence, the results thus obtained may be conflicting.

1.7.4 Considerations for Preventing Nanotoxicity

As it has been highlighted, the importance of the tight control over NM parameters is crucial for the prevention and control of nanotoxicity. One of the most prominent factors that increases the variability of the properties and activity of NMs is agglomeration. The phenomenon of agglomeration involves the adhesion of particles to each other, mainly because of van der Waal's forces, which dominate at the nanoscale level due to the increased surface area to volume ratio (Powers et al. 2007). NMs agglomerate after their synthesis in both the dry and suspension forms. The challenge for synthetic chemists is to prevent agglomeration, as it can lead to changes in physical and chemical properties. The major properties affected are size, size distribution, surface-to-volume ratio and, hence, surface reactivity. Since these parameters play a major role in the toxicity of NPs and are altered due to agglomeration, it is prudent to account for these changes in the study design (Borm et al. 2006, Teeguarden et al. 2007).

Agglomeration is influenced by several intrinsic and extrinsic factors, such as the composition of NMs and their concentration, size, surface coating, zeta potential, and temperature, among others (Teeguarden et al. 2007). It is well known that NPs can pass through biological barriers due to their small size. Agglomeration can alter their biological responses due to a decrease in the total available surface area, leading to an underestimation of toxic potential, especially in the case of drug delivery and safety and toxicity assessments (Sager et al. 2007). There are different methods available to deagglomerate NMs. Sonication is the most preferable and widely used method because it disperses NPs in a liquid by cavitation and does not have much of an effect on the properties of the particles. However, the attained deagglomeration is incomplete, as the particles do not reach their primary size and display the tendency to reagglomerate over time (Murdock et al. 2008). Another important method for preventing the agglomeration of NMs is surface modification. The particles can be coated with polymers or dispersed in ionic or nonionic surfactants (Farah et al. 2008, Sager et al. 2007, Skebo et al. 2007, Wick et al. 2007). While surface modification allows the particles to be stabilized and avoids agglomeration, it also raises concerns that they may shield or influence the effects of NMs on biological systems (Warheit 2008, Derfus et al. 2004, Warheit et al. 2005). The stability of such surface coatings inside a biological environment is another critical issue.

At the initial synthesis stage, the scientist may need to consider specific, physical parameters. However, in order to control the risk of NMs at all lifecycle points, interdisciplinary collaborations may be required. Additionally, owing to the increase in the production of NMs, the chances of their release into the environment and their subsequent effects on ecosystems are becoming important issues that need to be addressed. To do that, it is first necessary to assess the fate and behavior of NMs in the environment. It is still unclear how, at what concentrations, and in what forms, the NMs will be released into the environment. The answers to these questions will guide the formulation of regulatory guidelines that will protect the ecosystem and will also permit the full industrialization of the benefits that nanotechnology offers. It is important to focus current research efforts on the release, behavior (reaction to changes in environments), and fate (aggregation, adsorption, etc.) of NMs. A lifecycle assessment of the release of NMs into the environment is, therefore, imperative as the implementation of effective and protective regulatory policies (Navarro et al. 2008).

1.8 FUTURE CONSIDERATIONS

The study of NM toxicity is currently in the initial phases of development. The environmental testing of NMs requires the development of testing guidelines to allow the comparison and interpretation of data from clinical as well as environmental studies, and the close cooperation in interdisciplinary research. Unlike microorganisms and biomolecules, NMs can be engineered in a laboratory. Their physicochemical properties can be modified to enable systematic studies. Monte Carlo simulations have been used to model the effects of NP size and ligand densities on cellular uptake and tumor targeting. This would help to improve NP designs for optimal tumor accumulations in diagnostic and drug delivery applications (Buford et al. 2007, Borm et al. 2006, pp. 68–70, Powers et al. 2007). Such stimulation tools can be developed by conducting studies in a systematic manner and with a properly selected biological system. On that basis, researchers could create a database that would facilitate finding commonalities in experimental results. The outcomes of these studies can also be entered into a database, which can further help researchers to use computer simulation programs to identify appropriate nanostructure designs for a specific application.

There are several regulatory agencies that are looking into the toxicity caused by NMs. NNI was previously mentioned in the chapter; it has a section devoted to identifying the potential exposure, possible toxicity, and the need for personal protective equipments when working with nanoscale materials. Several other U.S. agencies (the National Institute for Occupational Safety and Health, the National Science Foundation Nanoscale Science and Engineering technology, and the Environmental Protection Agency) are working to assess, support, and monitor the impact of nanotechnology on health and the environment. The NIH–National Toxicology Program funds research on the toxicity of NMs.

The development of communication activities to enable technical information to be summarized, critiqued, and ultimately synthesized for various interested parties, including decision makers and consumers, would also be beneficial in tackling toxicity. Applied methods (sample preparation, experimental setup, and toxicity analysis) in current and future studies should be fully documented to enhance the transparency and comparability of obtained data. Finally, a global understanding of nanotechnology-specific risks is essential. If the global research community can take cognizance of these issues, then we can surely look forward to the advent of safe nanotechnologies.

1.9 SUMMARY

The field of nanotechnology takes root well before the era of peer-reviewed literature when colloidal gold was used to coat pottery during the Ming dynasty and breathtaking, stained-glass windows of seventeenth-century cathedrals. Nanotechnology rose to the interest of science through the famous work of Faraday in 1857 about making a "beautiful ruby fluid" (Faraday 1857). Conceptual explanations of nanotechnology received more publicity in the 1959 presentation by Richard Feyman, a celebrated Nobel Prize winner. However, it was the invention of electron microscopy in 1981 that caused a burst in the growth of nanotechnology. Furthermore, in 2000, the National Nanotechnology Initiative helped the field develop into the booming, multitrillion dollar industry that we have today, with many applications from conductors in computer technology to cancer treatments in medicine.

The exciting capabilities of nanotechnology are bringing about its infiltration into almost every aspect of life. Nanoparticles allow for the design and administration of delivery systems with targeted and sustained release capacities and decreased off-target activities and toxicities in the fields of medicine and pharmacy. New application venues have been discovered in computer technologies where almost every circuitry component can be redesigned with the use of nanotechnology, yielding faster and more efficient operations in compact designs. Nanotechnology allows for the remediation and recycling of resources and energy in environmental sustainability efforts. Other uses of nanotechnology have been applied in coatings, cosmetics, paints, and lubricants, among others.

The potential health and environmental risks have become more of a concern with the growing, widespread use of nanotechnology. Indeed, for the field to move forward, a thorough understanding of nanotoxicology is required. One of the major challenges of nanotoxicity assessments is the vast number of vehicles of exposure and the various pharmacokinetics highly sensitive to even small changes in the physicochemical properties of nanomaterials. Additionally, highly reactive nanomaterials may experience a change in properties upon interacting with their environments, such as acquiring or shedding coating, reacting with present compounds, and agglomerating together. Present research methods allow for the detailed characterization of nanomaterials and their action. However, multiple tests may be required to account for the multiple facets of each nanomaterial and its interactions with the environment. As the field of nanotechnology continues to develop, a careful scrutiny of all the properties of nanomaterials is required. In the future, new protocols will be necessary to improve our capacity to predict the toxicity of nanomaterials.

REFERENCES

Adams, L. K., D. Y. Lyon, and P. J. J. Alvarez. 2006. Comparative eco-toxicity of nanoscale TiO_2, SiO_2, and ZnO water suspensions. *Water Research* 40(19):3527–3532. doi: 10.1016/j.watres.2006.08.004.

Aillon, K. L., Y. M. Xie, N. El-Gendy, C. J. Berkland, and M. L. Forrest. 2009. Effects of nanomaterial physicochemical properties on *in vivo* toxicity. *Advanced Drug Delivery Reviews* 61(6):457–466. doi: 10.1016/j.addr.2009.03.010.

Alagarasi, A. 2011. *Introduction to Nanomaterial*. Chennai, India: National Centre for Catalysis Research.

Allen, T. M. 2002. Ligand-targeted therapeutics in anticancer therapy. *Nature Reviews Cancer* 2(10):750–763. doi: 10.1038/Nrc903.

Arnida, M. M. Janat-Amsbury, A. Ray, C. M. Peterson, and H. Ghandehari. 2011. Geometry and surface characteristics of gold nanoparticles influence their biodistribution and uptake by macrophages. *European Journal of Pharmaceutics and Biopharmaceutics* 77(3):417–423. doi: 10.1016/j.ejpb.2010.11.010.

Arora, S., J. M. Rajwade, and K. M. Paknikar. 2012. Nanotoxicology and *in vitro* studies: The need of the hour. *Toxicology and Applied Pharmacology* 258(2):151–165. doi: 10.1016/j.taap.2011.11.010.

AshaRani, P. V., G. L. K. Mun, M. P. Hande, and S. Valiyaveettil. 2009. Cytotoxicity and genotoxicity of silver nanoparticles in human cells. *ACS Nano* 3(2):279–290. doi: 10.1021/Nn800596w.

Astruc, D. 2008. *Nanoparticles and Catalysis*: Wiley Online Library.

Auffan, M., J. Rose, T. Orsiere, M. De Meo, A. Thill, O. Zeyons, O. Proux, A. Masion, P. Chaurand, O. Spalla, A. Botta, M. R. Wiesner, and J. Y. Bottero. 2009. CeO_2 nanoparticles induce DNA damage towards human dermal fibroblasts *in vitro*. *Nanotoxicology* 3(2):161–171. doi: 10.1080/17435390902788086.

Baber, N. 1994. International conference on harmonisation of technical requirements for registration of pharmaceuticals for human use (ICH). *British Journal of Clinical Pharmacology* 37(5):401.

Baibarac, M., and P. Gomez-Romero. 2006. Nanocomposites based on conducting polymers and carbon nanotubes: From fancy materials to functional applications. *Journal of Nanoscience and Nanotechnology* 6(2):289–302. doi: 10.1166/Jnn.2006.002.

Ballou, B., B. C. Lagerholm, L. A. Ernst, M. P. Bruchez, and A. S. Waggoner. 2004. Noninvasive imaging of quantum dots in mice. *Bioconjugate Chemistry* 15(1):79–86. doi: 10.1021/bc034153y.

Baroli, B., M. G. Ennas, F. Loffredo, M. Isola, R. Pinna, and M. A. Lopez-Quintela. 2007. Penetration of metallic nanoparticles in human full-thickness skin. *Journal of Investigative Dermatology* 127(7):1701–1712. doi: 10.1038/sj.jid.5700733.

Bell, A. T. 2003. The impact of nanoscience on heterogeneous catalysis. *Science* 299(5613):1688–1691.

Bhushan, B. 2010. Introduction to nanotechnology. In: B. Bhushan (ed.), *Springer Handbook of Nanotechnology*. 1–13. Berlin Heidelberg: Springer.

Bohnsack, J. P., S. Assemi, J. D. Miller, and D. Y. Furgeson. 2012. The primacy of physicochemical characterization of nanomaterials for reliable toxicity assessment: A review of the zebrafish nanotoxicology model. In: J. Reineke (ed.), *Nanotoxicity*, 261–316. New York: Springer.

Borm, P., F. C. Klaessig, T. D. Landry, B. Moudgil, J. Pauluhn, K. Thomas, R. Trottier, and S. Wood. 2006. Research strategies for safety evaluation of nanomaterials, part V: Role of dissolution in biological fate and effects of nanoscale particles. *Toxicological Sciences* 90(1):23–32.

Borup, B., and W. Leuchtenberger. 2002. Soft silanes for scratch-proof surfaces. *Materials World* 10(3):20–21.

Brady, N. C., and R. R. Weil. 1996. *The Nature and Properties of Soils*. New Jersey: Prentice-Hall Inc.

Brar, S. K., M. Verma, R. D. Tyagi, and R. Y. Surampalli. 2010. Engineered nanoparticles in wastewater and wastewater sludge–evidence and impacts. *Waste Management* 30(3):504–520. doi: 10.1016/j.wasman.2009.10.012 S0956-053X(09)00460-7 [pii].

Bregoli, L., F. Chiarini, A. Gambarelli, G. Sighinolfi, A. M. Gatti, P. Santi, A. M. Martelli, and L. Cocco. 2009. Toxicity of antimony trioxide nanoparticles on human hematopoietic progenitor cells and comparison to cell lines. *Toxicology* 262(2):121–129. doi: 10.1016/j.tox.2009.05.017.

Brown, D. M., K. Donaldson, P. J. Borm, R. P. Schins, M. Dehnhardt, P. Gilmour, L. A. Jimenez, and V. Stone. 2004. Calcium and ROS-mediated activation of transcription factors and TNF-alpha cytokine gene expression in macrophages exposed to ultrafine particles. *American Journal of Physiology—Lung Cellular and Molecular Physiology* 286(2):L344–L353. doi: 10.1152/ajplung.00139.2003.

Buford, M. C., R. F. Hamilton Jr., and A. Holian. 2007. A comparison of dispersing media for various engineered carbon nanoparticles. *Particle and Fibre Toxicology* 4:6. doi: 10.1186/1743-8977-4-6.

Buzea, C., I. I. Pacheco, and K. Robbie. 2007. Nanomaterials and nanoparticles: Sources and toxicity. *Biointerphases* 2(4):MR17–MR71. doi: 10.1116/1.2815690.

Champion, J. A., and S. Mitragotri. 2006. Role of target geometry in phagocytosis. *Proceedings of the National Academy of Sciences of the United States of America* 103(13):4930–4934. doi: 10.1073/pnas.0600997103.

Choi, A. O., S. J. Cho, J. Desbarats, J. Lovric, and D. Maysinger. 2007. Quantum dot-induced cell death involves Fas upregulation and lipid peroxidation in human neuroblastoma cells. *Journal of Nanobiotechnology* 5:1. doi: 10.1186/1477-3155-5-1.

Choi, H. S., W. Liu, P. Misra, E. Tanaka, J. P. Zimmer, B. Itty Ipe, M. G. Bawendi, and J. V. Frangioni. 2007. Renal clearance of quantum dots. *Nature Biotechnology* 25(10):1165–1170. doi: nbt1340 [pii] 10.1038/nbt1340.

Colvin, V. L. 2003. The potential environmental impact of engineered nanomaterials. *Nature Biotechnology* 21(10):1166–1170. doi: 10.1038/Nbt875.

Curtis, J., M. Greenberg, J. Kester, S. Phillips, and G. Krieger. 2006. Nanotechnology and nanotoxicology: A primer for clinicians. *Toxicology Review* 25(4):245–260.

Daniel, M. C., and D. Astruc. 2004. Gold nanoparticles: Assembly, supramolecular chemistry, quantum-size-related properties, and applications toward biology, catalysis, and nanotechnology. *Chemical Reviews* 104(1):293–346. doi: 10.1021/Cr030698+.

Darlington, T. K., A. M. Neigh, M. T. Spencer, O. T. Nguyen, and S. J. Oldenburg. 2009. Nanoparticle characteristics affecting environmental fate and transport through soil. *Environmental Toxicology & Chemistry* 28(6):1191–1199. doi: 10.1897/08-341.1 08-341 [pii].

Derfus, A. M., W. C. W. Chan, and S. N. Bhatia. 2004. Probing the cytotoxicity of semiconductor quantum dots. *Nano Letters* 4(1):11–18.

Doak, S. H., S. M. Griffiths, B. Manshian, N. Singh, P. M. Williams, A. P. Brown, and G. J. S. Jenkins. 2009. Confounding experimental considerations in nanogenotoxicology. *Mutagenesis* 24(4):285–293.

Dobrovolskaia, M. A., and S. E. Mcneil. 2007. Immunological properties of engineered nanomaterials. *Nature Nanotechnology* 2(8):469–478. doi: 10.1038/nnano.2007.223.

Donaldson, K., V. Stone, C. L. Tran, W. Kreyling, and P. J. A. Borm. 2004. Nanotoxicology. *Occupational and Environmental Medicine* 61(9):727–728. doi: 10.1136/oem.2004.013243.

Doshi, R., W. Braida, C. Christodoulatos, M. Wazne, and G. O'Connor. 2008. Nano-aluminum: Transport through sand columns and environmental effects on plants and soil communities. *Environmental Research* 106(3):296–303. doi: S0013-9351(07)00096-5 [pii] 10.1016/j.envres.2007.04.006.

Drexler, K. E. 1986. *Engines of Creation*. 1st ed. Garden City, NY: Anchor Press/Doubleday.

Ellis-Behnke, R. G., Y. X. Liang, S. W. You, D. K. C. Tay, S. G. Zhang, K. F. So, and G. E. Schneider. 2006. Nano neuro knitting: Peptide nanofiber scaffold for brain repair and axon regeneration with functional return of vision. *Proceedings of the National Academy of Sciences of the United States of America* 103(19):7530–7530. doi: 10.1073/pnas.0602514103.

Fabrega, J., J. C. Renshaw, and J. R. Lead. 2009. Interactions of silver nanoparticles with *Pseudomonas putida* biofilms. *Environmental Science & Technology* 43(23):9004–9009. doi: 10.1021/es901706j.

Faraday, M. 1857. The bakerian lecture: Experimental relations of gold (and other metals) to light. *Philosophical Transactions of the Royal Society of London* 147:145–181. doi: 10.2307/108616.

Farah, A. A., R. A. Alvarez-Puebla, and H. Fenniri. 2008. Chemically stable silver nanoparticle-crosslinked polymer microspheres. *Journal of Colloid and Interface Science* 319(2):572–576.

Farré, M., J. Sanchís, and D. Barceló. 2011. Analysis and assessment of the occurrence, the fate and the behavior of nanomaterials in the environment. *TrAC Trends in Analytical Chemistry* 30(3):517–527.

Ferin, J., G. Oberdorster, D. P. Penney, S. C. Soderholm, R. Gelein, and H. C. Piper. 1990. Increased pulmonary toxicity of ultrafine particles. 1. Particle clearance, translocation, morphology. *Journal of Aerosol Science* 21(3):381–384. doi: 10.1016/0021-8502(90)90064-5.

Ferrari, M. 2005. Cancer nanotechnology: Opportunities and challenges. *Nature Reviews Cancer* 5(3):161–171. doi: 10.1038/Nrc1566.

Fischer, H. C., and W. C. W. Chan. 2007. Nanotoxicity: The growing need for *in vivo* study. *Current Opinion in Biotechnology* 18(6):565–571. doi: 10.1016/j.copbio.2007.11.008.

Fischer, H. C., L. Liu, K. S. Pang, and W. C. W. Chan. 2006. Pharmacokinetics of nanoscale quantum dots: *In vivo* distribution, sequestration, and clearance in the rat. *Advanced Functional Materials* 16(10):1299–1305. doi: 10.1002/adfm.200500529.

Fleischer, N., M. Genut, I. Rapoport, and R. Tenne. 2003. New nanotechnology solid lubricants for superior dry lubrication, *Proceedings of the 10th European Space Mechanisms and Tribology Symposium*, pp.65–66.

Franken, N. A. P., H. M. Rodermond, J. Stap, J. Haveman, and C. Van Bree. 2006. Clonogenic assay of cells *in vitro*. *Nature Protocols* 1(5):2315–2319.

Garnett, M. C., and P. Kallinteri. 2006. Nanomedicines and nanotoxicology: Some physiological principles. *Occupational Medicine* 56(5):307–311.

Geldenhuys, W., T. Mbimba, T. Bui, K. Harrison, and V. Sutariya. 2011. Brain-targeted delivery of paclitaxel using glutathione-coated nanoparticles for brain cancers. *Journal of Drug Targeting* 19(9):837–845. doi: 10.3109/1061186X.2011.589435.

Ghosh, P. S., C. K. Kim, G. Han, N. S. Forbes, and V. M. Rotello. 2008. Efficient gene delivery vectors by tuning the surface charge density of amino acid-functionalized gold nanoparticles. *ACS Nano* 2(11):2213–2218. doi: 10.1021/nn800507t.

Glezer, A. M. 2011. Structural classification of nanomaterials. *Russian Metallurgy (Metally)* 2011(4):263–269.

Goodman, C. M., C. D. McCusker, T. Yilmaz, and V. M. Rotello. 2004. Toxicity of gold nanoparticles functionalized with cationic and anionic side chains. *Bioconjugate Chemistry* 15(4):897–900. doi: 10.1021/bc049951i.

Grabinski, C., S. Hussain, K. Lafdi, L. Braydich-Stolle, J. Schlager. 2007. Effect of particle dimension on biocompatibility of carbon nanomaterials. *Carbon* 45(14):2828–2835.

Grassian, V. H., P. T. O'Shaughnessy, A. Adamcakova-Dodd, J. M. Pettibone, and P. S. Thorne. 2007. Inhalation exposure study of titanium dioxide nanoparticles with a primary particle size of 2 to 5 nm. *Environmental Health Perspectives* 115(3):397–402. doi: 10.1298/Ehp.9469.

Greish, K., G. Thiagarajan, H. Herd, R. Price, H. Bauer, D. Hubbard, A. Burckle, S. Sadekar, T. Yu, A. Anwar, A. Ray, and H. Ghandehari. 2012. Size and surface charge significantly influence the toxicity of silica and dendritic nanoparticles. *Nanotoxicology* 6(7):713–723. doi: 10.3109/17435390.2011.604442.

Gurr, J.-R., A. S. S. Wang, C.-H. Chen, and K.-Y. Jan. 2005. Ultrafine titanium dioxide particles in the absence of photoactivation can induce oxidative damage to human bronchial epithelial cells. *Toxicology* 213(1–2):66–73. doi: http://dx.doi.org/10.1016/j.tox.2005.05.007.

Hagens, W. I., A. G. Oomen, W. H. de Jong, F. R. Cassee, and A. J. A. M. Sips. 2007. What do we (need to) know about the kinetic properties of nanoparticles in the body? *Regulatory Toxicology and Pharmacology* 49(3):217–229. doi: 10.1016/j.yrtph.2007.07.006.

Han, X. L., R. Gelein, N. Corson, P. Wade-Mercer, J. K. Jiang, P. Biswas, J. N. Finkelstein, A. Elder, and G Oberdorster. 2011. Validation of an LDH assay for assessing nanoparticle toxicity. *Toxicology* 287(1–3):99–104. doi: 10.1016/j.tox.2011.06.011.

Hansen, C. S., M. Sheykhzade, P. Moller, J. K. Folkmann, O. Amtorp, T. Jonassen, and S. Loft. 2007. Diesel exhaust particles induce endothelial dysfunction in apoE-/- mice. *Toxicology and Applied Pharmacology* 219(1):24–32. doi: 10.1016/j.taap.2006.10.032.

Hardman, R. 2006. A toxicologic review of quantum dots: Toxicity depends on physicochemical and environmental factors. *Environmental Health Perspectives* 114(2):165–172.

Hardonk, M. J., G. Harms, and J. Koudstaal. 1985. Zonal heterogeneity of rat hepatocytes in the *in vivo* uptake of 17 nm colloidal gold granules. *Histochemistry* 83(5):473–477.

Harisinghani, M. G., J. Barentsz, P. F. Hahn, W. M. Deserno, S. Tabatabaei, C. H. van de Kaa, J. de la Rosette, and R. Weissleder. 2003. Noninvasive detection of clinically occult lymph-node metastases in prostate cancer. *New England Journal of Medicine* 348(25):2491–2499. doi: 10.1056/Nejmoa022749.

Haruta, M. 2002. Catalysis of gold nanoparticles deposited on metal oxides. *Cattech* 6(3):102–115.

Hauck, T. S., A. A. Ghazani, and W. C. W. Chan. 2008. Assessing the effect of surface chemistry on gold nanorod uptake, toxicity, and gene expression in mammalian cells. *Small* 4(1):153–159. doi: 10.1002/smll.200700217.

Hayashi, M., T. Sofuni, and M. Ishidate Jr. 1983. An application of acridine orange fluorescent staining to the micronucleus test. *Mutation Research Letters* 120(4):241–247.

He, F., and D. Y. Zhao. 2005. Preparation and characterization of a new class of starch-stabilized bimetallic nanoparticles for degradation of chlorinated hydrocarbons in water. *Environmental Science & Technology* 39(9):3314–3320. doi: 10.1021/Es048743y.

Hoet, P. H., I. Bruske-Hohlfeld, and O. V. Salata. 2004. Nanoparticles—Known and unknown health risks. *Journal of Nanobiotechnology* 2(1):12. doi: 1477-3155-2-12 [pii] 10.1186/1477-3155-2-12.

Hrubesh, L. W., and J. F. Poco. 1995. Thin aerogel films for optical, thermal, acoustic and electronic applications. *Journal of Non-Crystalline Solids* 188(1–2):46–53. doi: 10.1016/0022-3093(95)00028-3.

Ibuki, Y., and T. Toyooka. 2012. Nanoparticle uptake measured by flow cytometry. In: J. Reineke (ed.), *Nanotoxicity*, 157–166. New York: Springer.

Ishida, T., H. Harashima, and H. Kiwada. 2001. Interactions of liposomes with cells *in vitro* and *in vivo*: Opsonins and receptors. *Current Drug Metabolism* 2(4):397–409.

Jain, K. K. 2005. The role of nanobiotechnology in drug discovery. *Drug Discovery Today* 10(21):1435–1442. doi: 10.1016/S1359-6446(05)03573-7.

Jaisi, D. P., and M. Elimelech. 2009. Single-walled carbon nanotubes exhibit limited transport in soil columns. *Environmental Science & Technology* 43(24):9161–9166. doi: 10.1021/es901927y.

Jallouli, Y., A. Paillard, J. Chang, E. Sevin, and D. Betbeder. 2007. Influence of surface charge and inner composition of porous nanoparticles to cross blood-brain barrier *in vitro*. *International Journal of Pharmaceutics* 344(1–2):103–109. doi: 10.1016/j.ijpharm.2007.06.023.

Kagan, V. E., N. V. Konduru, W. Feng, B. L. Allen, J. Conroy, Y. Volkov, Vlasova II, N. A. Belikova et al. 2010. Carbon nanotubes degraded by neutrophil myeloperoxidase induce less pulmonary inflammation. *Nature Nanotechnology* 5(5):354–359.

Kennedy, C. B., S. D. Scott, and F. G. Ferris. 2004. Hydrothermal phase stabilization of 2-line ferrihydrite by bacteria. *Chemical Geology* 212(3–4):269–277. doi: http://dx.doi.org/10.1016/j.chemgeo.2004.08.017.

Khan, M. K., S. S. Nigavekar, L. D. Minc, M. S. Kariapper, B. M. Nair, W. G. Lesniak, and L. P. Balogh. 2005. In vivo biodistribution of dendrimers and dendrimer nanocomposites—Implications for cancer imaging and therapy. *Technology in Cancer Research and Treatment* 4(6):603–613. doi: d=3022&c=4191&p=13201&do=detail [pii].

Klaine, S. J., P. J. J. Alvarez, G. E. Batley, T. F. Fernandes, R. D. Handy, D. Y. Lyon, S. Mahendra, M. J. McLaughlin, and J. R. Lead. 2009. Nanomaterials in the environment: Behavior, fate, bioavailability, and effects. *Environmental Toxicology and Chemistry* 27(9):1825–1851.

Kneuer, C., M. Sameti, U. Bakowsky, T. Schiestel, H. Schirra, H. Schmidt, and C. M. Lehr. 2000. A nonviral DNA delivery system based on surface modified silica-nanoparticles can efficiently transfect cells *in vitro*. *Bioconjugate Chemistry* 11(6):926–932. doi: 10.1021/Bc0000637.

Kroll, A., M. H. Pillukat, D. Hahn, and J. Schnekenburger. 2009. Current *in vitro* methods in nanoparticle risk assessment: Limitations and challenges. *European Journal of Pharmaceutics and Biopharmaceutics* 72(2):370–377.

Kubik, T., K. Bogunia-Kubik, and M. Sugisaka. 2005. Nanotechnology on duty in medical applications. *Current Pharmaceutical Biotechnology* 6(1):17–33.

Kurepa, J., T. Paunesku, S. Vogt, H. Arora, B. M. Rabatic, J. Lu, M. B. Wanzer, G. E. Woloschak, and J. A. Smalle. 2010. Uptake and distribution of ultrasmall anatase TiO_2 Alizarin red S nanoconjugates in *Arabidopsis thaliana*. *Nano Lett* 10(7):2296–2302. doi: 10.1021/nl903518f.

Lam, C. W., J. T. James, R. McCluskey, and R. L. Hunter. 2004. Pulmonary toxicity of single-wall carbon nanotubes in mice 7 and 90 days after intratracheal instillation. *Toxicological Sciences* 77(1):126–134. doi: 10.1093/toxsci/kfg243.

Lanone, S., and J. Boczkowski. 2006. Biomedical applications and potential health risks of nanomaterials: Molecular mechanisms. *Current Molecular Medicine* 6(6):651–663. doi: 10.2174/156652406778195026.

Lanone, S., F. Rogerieux, J. Geys, A. Dupont, E. Maillot-Marechal, J. Boczkowski, G. Lacroix, and P. Hoet. 2009. Comparative toxicity of 24 manufactured nanoparticles in human alveolar epithelial and macrophage cell lines. *Particle and Fibre Toxicology* 6(1):14.

LaVan, D. A., T. McGuire, and R. Langer. 2003. Small-scale systems for *in vivo* drug delivery. *Nature Biotechnology* 21(10):1184–1191. doi: 10.1038/Nbt876.

Lee, C.-W., A. Afzalian, N. D. Akhavan, R. Yan, I. Ferain, and J.-P. Colinge. 2009. Junctionless multigate field-effect transistor. *Applied Physics Letters* 94(5):053511–053511-2.

Lee, J., G. D. Lilly, R. C. Doty, P. Podsiadlo, and N. A. Kotov. 2009. *In vitro* toxicity testing of nanoparticles in 3D cell culture. *Small* 5(10):1213–1221.

Lewin, M., N. Carlesso, C. H. Tung, X. W. Tang, D. Cory, D. T. Scadden, and R. Weissleder. 2000. Tat peptide-derivatized magnetic nanoparticles allow *in vivo* tracking and recovery of progenitor cells. *Nature Biotechnology* 18(4):410–414.

Li, Y., P. Leung, L. Yao, Q. W. Song, and E. Newton. 2006. Antimicrobial effect of surgical masks coated with nanoparticles. *Journal of Hospital Infection* 62(1):58–63. doi: 10.1016/i.jhin.2005.04.015.

Li, Z., T. Hulderman, R. Salmen, R. Chapman, S. S. Leonard, S. H. Young, A. Shvedova, M. I. Luster, and P. P. Simeonova. 2007. Cardiovascular effects of pulmonary exposure to single-wall carbon nanotubes. *Environmental Health Perspectives* 115(3):377–382. doi: 10.1289/ehp.9688.

Liao, K. H., Y. S. Lin, C. W. Macosko, and C. L. Haynes. 2011. Cytotoxicity of graphene oxide and graphene in human erythrocytes and skin fibroblasts. *ACS Applied Materials & Interfaces* 3(7):2607–2615. doi: 10.1021/Am200428v.

Lin, D., and B. Xing. 2008. Root uptake and phytotoxicity of ZnO nanoparticles. *Environmental Science & Technology* 42(15):5580–5585.

Lin, S., J. Reppert, Q. Hu, J. S. Hudson, M. L. Reid, T. A. Ratnikova, A. M. Rao, H. Luo, and P. C. Ke. 2009. Uptake, translocation, and transmission of carbon nanomaterials in rice plants. *Small* 5(10):1128–1132. doi: 10.1002/smll.200801556.

Line, B. R., A. Mitra, A. Nan, and H. Ghandehari. 2005. Targeting tumor angiogenesis: Comparison of peptide and polymer-peptide conjugates. *Journal of Nuclear Medicine* 46(9):1552–1560.

Lippmann, M. 1990. Effects of fiber characteristics on lung deposition, retention, and disease. *Environmental Health Perspectives* 88:311–317.

Liu, M. J., K. Kono, and J. M. J. Frechet. 2000. Water-soluble dendritic unimolecular micelles: Their potential as drug delivery agents. *Journal of Controlled Release* 65(1–2):121–131. doi: 10.1016/S0168-3659(99)00245-X.

Liu, H., R. Ramnarayanan, and B. E. Logan. 2004. Production of electricity during wastewater treatment using a single chamber microbial fuel cell. *Environmental Science & Technology* 38(7):2281–2285.

Lockman, P. R., J. M. Koziara, R. J. Mumper, and D. D. Allen. 2004. Nanoparticle surface charges alter blood-brain barrier integrity and permeability. *Journal of Drug Targeting* 12(9–10):635–641. doi: 10.1080/10611860400015936.

Logan, B. E., B. Hamelers, R. Rozendal, U. Schröder, J. Keller, S. Freguia, P. Aelterman, W. Verstraete, and K. Rabaey. 2006. Microbial fuel cells: Methodology and technology. *Environmental Science & Technology* 40(17):5181–5192. doi: 10.1021/es0605016.

Long, T. C., N. Saleh, R. D. Tilton, G. V. Lowry, and B. Veronesi. 2006. Titanium dioxide (P25) produces reactive oxygen species in immortalized brain microglia (BV2): Implications for nanoparticle neurotoxicity. *Environmental Science & Technology* 40(14):4346–4352.

Love, S. A., M. A. Maurer-Jones, J. W. Thompson, Y. S. Lin, and C. L. Haynes. 2012. Assessing nanoparticle toxicity. *Annual Review of Analytical Chemistry* 5(5):181–205. doi: 10.1146/annurev-anchem-062011-143134.

Lundqvist, M., I. Sethson, and B. H. Jonsson. 2004. Protein adsorption onto silica nanoparticles: Conformational changes depend on the particles' curvature and the protein stability. *Langmuir* 20(24):10639–10647. doi: 10.1021/la0484725.

Luo, X., A. Morrin, A. J. Killard, and M. R. Smyth. 2006. Application of nanoparticles in electrochemical sensors and biosensors. *Electroanalysis* 18(4):319–326.

Lynch, I., and K. A. Dawson. 2008. Protein-nanoparticle interactions. *Nano Today* 3(1–2):40–47. doi: http://dx.doi.org/10.1016/S1748-0132(08)70014-8.

Lyon, D. Y., L. K. Adams, J. C. Falkner, and P. J. J. Alvarez. 2006. Antibacterial activity of fullerene water suspensions: Effects of preparation method and particle size. *Environmental Science & Technology* 40(14):4360–4366. doi: 10.1021/Es0603655.

Lyon, D. Y., J. D. Fortner, C. M. Sayes, V. L. Colvin, and J. B. Hughes. 2005. Bacterial cell association and antimicrobial activity of a C-60 water suspension. *Environmental Toxicology and Chemistry* 24(11):2757–2762. doi: 10.1897/04-649r.1.

Mahmoudi, M., I. Lynch, M. R. Ejtehadi, M. P. Monopoli, F. B. Bombelli, and S. Laurent. 2011. Protein-nanoparticle interactions: Opportunities and challenges. *Chemical Reviews* 111(9):5610–5637. doi: 10.1021/cr100440 g.

Makarucha, A. J., N. Todorova, and I. Yarovsky. 2011. Nanomaterials in biological environment: A review of computer modelling studies. *European Biophysics Journal with Biophysics Letters* 40(2):103–115. doi: 10.1007/s00249-010-0651-6.

Marquis, B. J., Z. Liu, K. L. Braun, and C. L. Haynes. 2011. Investigation of noble metal nanoparticle zeta-potential effects on single-cell exocytosis function *in vitro* with carbon-fiber microelectrode amperometry. *Analyst* 136(17):3478–3486. doi: 10.1039/c0an00785d.

Marquis, B. J., M. A. Maurer-Jones, K. L. Braun, and C. L. Haynes. 2009. Amperometric assessment of functional changes in nanoparticle-exposed immune cells: Varying Au nanoparticle exposure time and concentration. *Analyst* 134(11):2293–2300. doi: 10.1039/B913967b.

Martin, C. R., and P. Kohli. 2003. The emerging field of nanotube biotechnology. *Nature Reviews Drug Discovery* 2(1):29–37. doi: 10.1038/Nrd988.

Matsusaki, M., K. Larsson, T. Akagi, M. Lindstedt, M. Akashi, and C. A. K. Borrebaeck. 2005. Nanosphere induced gene expression in human dendritic cells. *Nano Letters* 5(11):2168–2173. doi: 10.1021/Nl050541s.

Maurer-Jones, M. A., Y. S. Lin, and C. L. Haynes. 2010. Functional assessment of metal oxide nanoparticle toxicity in immune cells. *ACS Nano* 4(6):3363–3373. doi: 10.1021/Nn9018834.

Metz, K. M., A. N. Mangham, M. J. Bierman, S. Jin, R. J. Hamers, and J. A. Pedersen. 2009. Engineered nanomaterial transformation under oxidative environmental conditions: Development of an *in vitro* biomimetic assay. *Environmental Science & Technology* 43(5):1598–1604. doi: 10.1021/Es802217y.

Miller, T. F., and J. D. Herr. 2004. Green rocket propulsion by reaction of Al and Mg powders and water. *AIAA paper* 4037:2004.

Moghimi, S. M., A. C. Hunter, and J. C. Murray. 2001. Long-circulating and target-specific nanoparticles: Theory to practice. *Pharmacological Reviews* 53(2):283–318.

Monteiro-Riviere, N. A., and A. O. Inman. 2006. Challenges for assessing carbon nanomaterial toxicity to the skin. *Carbon* 44(6):1070–1078.

Monteiro-Riviere, N. A., A. O. Inman, and L. W. Zhang. 2009. Limitations and relative utility of screening assays to assess engineered nanoparticle toxicity in a human cell line. *Toxicology and Applied Pharmacology* 234(2):222–235.

Morishita, M., and N. A. Peppas. 2006. Is the oral route possible for peptide and protein drug delivery? *Drug Discovery Today* 11(19):905–910.

Morris, J., J. Willis, and K. Gallagher. 2007. Nanotechnology White Paper. US Environmental Protection Agency. Washington, DC, www.epa.gov/osa/pdfs/nanotech/epa-nanotechnology-whitepaper-0207.pdf (February 2007).

Mu, Q., G. Du, T. Chen, B. Zhang, and B. Yan. 2009. Suppression of human bone morphogenetic protein signaling by carboxylated single-walled carbon nanotubes. *ACS Nano* 3(5):1139–1144. doi: 10.1021/nn900252j.

Murdock, R. C., L. Braydich-Stolle, A. M. Schrand, J. J. Schlager, and S. M. Hussain. 2008. Characterization of nanomaterial dispersion in solution prior to *in vitro* exposure using dynamic light scattering technique. *Toxicological Sciences* 101(2):239–253. doi: 10.1093/toxsci/kfm240.

Navarro, E., A. Baun, R. Behra, N. B. Hartmann, J. Filser, A. J. Miao, A. Quigg, P. H. Santschi, and L. Sigg. 2008. Environmental behavior and ecotoxicity of engineered nanoparticles to algae, plants, and fungi. *Ecotoxicology* 17(5):372–386. doi: 10.1007/s10646-008-0214-0.

Niazi, J. H., and M. B. Gu. 2009. Toxicity of metallic nanoparticles in microorganisms-a review. In Y. J. Kim, U. Platt, M. B. Gu, and H. Iwahashi (eds.), *Atmospheric and Biological Environmental Monitoring*, 193–206. Heidelberg, Germany: Springer.

Nohynek, G. J., E. K. Dufour, and M. S. Roberts. 2008. Nanotechnology, cosmetics and the skin: Is there a health risk? *Skin Pharmacology and Physiology* 21(3):136–149.

Nohynek, G. J., J. Lademann, C. Ribaud, and M. S. Roberts. 2007. Grey goo on the skin? Nanotechnology, cosmetic and sunscreen safety. *CRC Critical Reviews in Toxicology* 37(3):251–277.

Nowack, B., and T. D. Bucheli. 2007. Occurrence, behavior and effects of nanoparticles in the environment. *Environmental Pollution* 150(1):5–22.

Nygaard, U. C., J. S. Hansen, M. Samuelsen, T. Alberg, C. D. Marioara, and M. Lovik. 2009. Single-walled and multi-walled carbon nanotubes promote allergic immune responses in mice. *Toxicological Sciences: An Official Journal of the Society of Toxicology* 109(1):113–123. doi: 10.1093/toxsci/kfp057.

Oberdorster, G., J. Ferin, G. Finkelstein, P. Wade, and N. Corson. 1990. Increased pulmonary toxicity of ultrafine particles .2. Lung lavage studies. *Journal of Aerosol Science* 21(3):384–387. doi: 10.1016/0021-8502(90)90065-6.

Oberdorster, G., J. Ferin, and B. E. Lehnert. 1994. Correlation between particle-size, *in-vivo* particle persistence, and lung injury. *Environmental Health Perspectives* 102:173–179. doi: 10.2307/3432080.

Oberdorster, G., E. Oberdorster, and J. Oberdorster. 2005. Nanotoxicology: An emerging discipline evolving from studies of ultrafine particles. *Environmental Health Perspectives* 113(7):823–839.

Oostingh, G. J., E. Casals, P. Italiani, R. Colognato, R. Stritzinger, J. Ponti, T. Pfaller et al. 2011. Problems and challenges in the development and validation of human cell-based assays to determine nanoparticle-induced immunomodulatory effects. *Particle and Fibre Toxicology* 8(1):8. doi: 10.1186/1743-8977-8-8.

Owens, D. E. 3rd, and N. A. Peppas. 2006. Opsonization, biodistribution, and pharmacokinetics of polymeric nanoparticles. *International Journal of Pharmaceutics* 307(1):93–102. doi: S0378-5173(05)00668-X [pii] 10.1016/j.ijpharm.2005.10.010.

Paciotti, G. F., L. Myer, D. Weinreich, D. Goia, N. Pavel, R. E. McLaughlin, and L. Tamarkin. 2004. Colloidal gold: A novel nanoparticle vector for tumor directed drug delivery. *Drug Delivery* 11(3):169–183. doi: 10.1080/10717540490433895 2PLY5CG1FR8GVUVF [pii].

Papageorgiou, I., C. Brown, R. Schins, S. Singh, R. Newson, S. Davis, J. Fisher, E. Ingham, and C. P. Case. 2007. The effect of nano- and micron-sized particles of cobalt-chromium alloy on human fibroblasts in vitro. *Biomaterials* 28(19):2946–2958. doi: 10.1016/j.biomaterials.2007.02.034.

Park, E. J., J. Choi, Y. K. Park, and K. Park. 2008. Oxidative stress induced by cerium oxide nanoparticles in cultured BEAS-2B cells. *Toxicology* 245(1–2):90–100. doi: 10.1016/j.tox.2007.12.022.

Patra, C. R., R. Bhattacharya, S. Patra, N. E. Vlahakis, A. Gabashvili, Y. Koltypin, A. Gedanken, P. Mukherjee, and D. Mukhopadhyay. 2008. Pro-angiogenic properties of europium(III) hydroxide nanorods. *Advanced Materials* 20(4):753-+. doi: 10.1002/adma.200701611.

Pelaz, B.z, G. Charron, C. Pfeiffer, Y. Zhao, J. M. de la Fuente, X.-J. Liang, W. J. Parak, and P. del Pino. 2013. Interfacing engineered nanoparticles with biological systems: Anticipating adverse nano–bio interactions. *Small.* 9(9–10):1573–84. doi: 10.1002/smll.201201229. Epub 2012 Oct 30.

Peppas, N. A. 2004. Intelligent therapeutics: Biomimetic systems and nanotechnology in drug delivery. *Advanced Drug Delivery Reviews* 56(11):1529–1531. doi: 10.1016/j.addr.2004.07.001.

Peters, A., B. Veronesi, L. Calderon-Garciduenas, P. Gehr, L. C. Chen, M. Geiser, W. Reed, B. Rothen-Rutishauser, S. Schurch, and H. Schulz. 2006. Translocation and potential neurological effects of fine and ultrafine particles a critical update. *Particle and Fibre Toxicology* 3:13. doi: 10.1186/1743-8977-3-13.

Pettibone, J. M., A. Adamcakova-Dodd, P. S. Thorne, P. T. O'Shaughnessy, J. A. Weydert, and V. H. Grassian. 2008. Inflammatory response of mice following inhalation exposure to iron and copper nanoparticles. *Nanotoxicology* 2(4):189–204. doi: 10.1080/17435390802398291.

Poland, C. A., R. Duffin, I. Kinloch, A. Maynard, W. A. Wallace, A. Seaton, V. Stone, S. Brown, W. Macnee, and K. Donaldson. 2008. Carbon nanotubes introduced into the abdominal cavity of mice show asbestos-like pathogenicity in a pilot study. *Nature Nanotechnology* 3(7):423–428. doi: 10.1038/nnano.2008.111 nnano.2008.111 [pii].

Powers, K. W., S. C. Brown, V. B. Krishna, S. C. Wasdo, B. M. Moudgil, and S. M. Roberts. 2006. Research strategies for safety evaluation of nanomaterials. Part VI. Characterization of nanoscale particles for toxicological evaluation. *Toxicological Sciences* 90(2):296–303.

Powers, K. W., P. L. Carpinone, and K. N. Siebein. 2012. Characterization of nanomaterials for toxicological studies. *Methods in Molecular Biology* 926:13–32. doi: 10.1007/978-1-62703-002-1_2.

Powers, K. W., M. Palazuelos, B. M. Moudgil, and S. M. Roberts. 2007. Characterization of the size, shape, and state of dispersion of nanoparticles for toxicological studies. *Nanotoxicology* 1(1):42–51. doi: 10.1080/17435390701314902.

Pulskamp, K., S. Diabaté, and H. F. Krug. 2007. Carbon nanotubes show no sign of acute toxicity but induce intracellular reactive oxygen species in dependence on contaminants. *Toxicology Letters* 168(1):58.

Rabaey, K., and W. Verstraete. 2005. Microbial fuel cells: Novel biotechnology for energy generation. *Trends in Biotechnology* 23(6):291–298.

Rahman, I. 2000. Regulation of nuclear factor-kappa B, activator protein-1, and glutathione levels by tumor necrosis factor-alpha and dexamethasone in alveolar epithelial cells. *Biochemical Pharmacology* 60(8):1041–1049. doi: 10.1016/S0006-2952(00)00392-0.

Rahman, I., S. K. Biswas, L. A. Jimenez, M. Torres, and H. J. Forman. 2005. Glutathione, stress responses, and redox signaling in lung inflammation. *Antioxidants & Redox Signaling* 7(1–2):42–59. doi: 10.1089/ars.2005.7.42.

Reguera, G., K. D. McCarthy, T. Mehta, J. S. Nicoll, M. T. Tuominen, and D. R. Lovley. 2005. Extracellular electron transfer via microbial nanowires. *Nature* 435(7045):1098–1101. doi: 10.1038/nature03661.

Rejman, J., V. Oberle, I. S. Zuhorn, and D. Hoekstra. 2004. Size-dependent internalization of particles via the pathways of clathrin- and caveolae-mediated endocytosis. *Biochemical Journal* 377:159–169. doi: 10.1042/Bj20031253.

Renaud, G., R. L. Hamilton, and R. J. Havel. 1989. Hepatic metabolism of colloidal gold-low-density lipoprotein complexes in the rat: Evidence for bulk excretion of lysosomal contents into bile. *Hepatology* 9(3):380–392. doi: S0270913989000649 [pii].

Renehan, E. M., F. K. Enneking, M. Varshney, R. Partch, D. M. Dennis, and T. E. Morey. 2005. Scavenging nanoparticles: An emerging treatment for local anesthetic toxicity. *Regional Anesthesia and Pain Medicine* 30(4):380–384. doi: 10.1016/j.rapm.2005.04.004.

Risha, G., E. Boyer, R. Wehrman, and K. Kuo. 2002. Performance comparison of HTPB-based solid fuels containing nano-sized energetic powder in a cylindrical hybrid rocket motor. Paper read at 38th AIAA/ASME/SAE/ASEE Joint Propulsion Conference and Exhibit, Indianapolis, IN.

Roberts, A. P., A. S. Mount, B. Seda, J. Souther, R. Qiao, S. J. Lin, P. C. Ke, A. M. Rao, and S. J. Klaine. 2007. In vivo biomodification of lipid-coated carbon nanotubes by *Daphnia magna*. *Environmental Science & Technology* 41(8):3025–3029. doi: 10.1021/Es062572a.

Rosi, N. L., and C. A. Mirkin. 2005. Nanostructures in biodiagnostics. *Chemical Reviews* 105(4):1547–1562. doi: 10.1021/Cr030067f.

Ryman-Rasmussen, J. P., J. E. Riviere, and N. A. Monteiro-Riviere. 2006. Penetration of intact skin by quantum dots with diverse physicochemical properties. *Toxicological Sciences* 91(1):159–165. doi: 10.1093/toxsci/kfj122.Saba, T. M. 1970. Physiology and physiopathology of the reticuloendothelial system. *Archives of Internal Medicine* 126(6):1031–1052.

Sager, T. M., D. W. Porter, V. A. Robinson, W. G. Lindsley, D. E. Schwegler-Berry, and V. Castranova. 2007. Improved method to disperse nanoparticles for *in vitro* and *in vivo* investigation of toxicity. *Nanotoxicology* 1(2):118–129.

Sahoo, S. K., and V. Labhasetwar. 2003. Nanotech approaches to drug delivery and imaging. *Drug Discovery Today* 8(24):1112–1120.

Saleh, N., H. J. Kim, T. Phenrat, K. Matyjaszewski, R. D. Tilton, and G. V. Lowry. 2008. Ionic strength and composition affect the mobility of surface-modified Fe0 nanoparticles in water-saturated sand columns. *Environmental Science & Technology* 42(9):3349–3355.

Santamaria, A. 2012. Historical overview of nanotechnology and nanotoxicology. *Methods in Molecular Biology* 926:1–12. doi: 10.1007/978-1-62703-002-1_1.

Saraf, S. 2006. Recent developments in pharmaceutical nanotechnology. *The Pharmaceutical Magazine*, 1–3.

Sarto, F., S. Finotto, L. Giacomelli, D. Mazzotti, R. Tomanin, and A. G. Levis. 1987. The micronucleus assay in exfoliated cells of the human buccal mucosa. *Mutagenesis* 2(1):11–17. doi: 10.1093/mutage/2.1.11.

Sayes, C. M., K. L. Reed, and D. B. Warheit. 2007. Assessing toxicity of fine and nanoparticles: Comparing *in vitro* measurements to *in vivo* pulmonary toxicity profiles. *Toxicological Sciences: An Official Journal of the Society of Toxicology* 97(1):163–180. doi: 10.1093/toxsci/kfm018.

Scalf, J., and P. West. 2006. Part I: Introduction to nanoparticle characterization with AFM. Application Note-Pacific Nanotechnologies, www.nanoparticles.org/pdf/scalf-west.pdf (last accessed November 23, 2009).

Schanen, B. C., A. S. Karakoti, S. Seal, D. R. Drake, 3rd, W. L. Warren, and W. T. Self. 2009. Exposure to titanium dioxide nanomaterials provokes inflammation of an *in vitro* human immune construct. *ACS Nano* 3(9):2523–2532. doi: 10.1021/nn900403 h.

Schilling, K., B. Bradford, D. Castelli, E. Dufour, J. F. Nash, J. Pape, S. Schulte, I. Tooley, J. van den Bosch, and F. Schellauf. 2010. Human safety review of "nano" titanium dioxide and zinc oxide. *Photochemical & Photobiological Sciences* 9(4):495–509.

Sharma, H., S. N. Sharma, U. Kumar, V. N. Singh, B. R. Mehta, G. Singh, S. M. Shivaprasad, and R. Kakkar. 2009. Formation of water-soluble and biocompatible TOPO-capped CdSe quantum dots with efficient photoluminescence. *The Journal of Materials Science: Materials in Medicine* 20 Suppl 1:S123–S130. doi: 10.1007/s10856-008-3494-2.

Shvedova, A. A., E. Kisin, N. Keshava, A. R. Murray, O. Gorelik, and S. Arepalli. 2004. Exposure of human bronchial cells to carbon nanotubes caused oxidative stress and cytotoxicity. In: Galaris D. (ed.), *Proceedings Meeting of the Society for Free Radical Research European Section*. Philadelphia: Taylor & Francis Group.

Shvedova, A. A., E. R. Kisin, R. Mercer, A. R. Murray, V. J. Johnson, A. I. Potapovich, Y. Y. Tyurina et al. 2005. Unusual inflammatory and fibrogenic pulmonary responses to single-walled carbon nanotubes in mice. *American Journal of Physiology—Lung Cellular and Molecular Physiology* 289(5):L698–L708. doi: 10.1152/ajplung.00084.2005.

Siegel, R. W., and G. E. Fougere. 1995. Mechanical properties of nanophase metals. *Nanostructured Materials* 6(1):205–216.

Simkiss, K. 1990. Surface effects in ecotoxicology. *Functional Ecology* 4(3):303–308. doi: 10.2307/2389590.

Singaravelu, G., J. S. Arockiamary, V. G. Kumar, and K. Govindaraju. 2007. A novel extracellular synthesis of monodisperse gold nanoparticles using marine alga, *Sargassum wightii* Greville. *Colloids and Surfaces B: Biointerfaces* 57(1):97–101. doi: S0927-7765(07)00025-2 [pii] 10.1016/j.colsurfb.2007.01.010.

Singh, R., D. Pantarotto, L. Lacerda, G. Pastorin, C. Klumpp, M. Prato, A. Bianco, and K. Kostarelos. 2006. Tissue biodistribution and blood clearance rates of intravenously administered carbon nanotube radiotracers. *Proceedings of the National Academy of Sciences of the United States of America* 103(9):3357–3362. doi: 0509009103 [pii] 10.1073/pnas.0509009103.

Skebo, J. E., C. M. Grabinski, A. M. Schrand, J. J. Schlager, and S. M. Hussain. 2007. Assessment of metal nanoparticle agglomeration, uptake, and interaction using high-illuminating system. *International Journal of Toxicology* 26(2):135–141.

Smith, C. J., B. J. Shaw, and R. D. Handy. 2007. Toxicity of single walled carbon nanotubes to rainbow trout (*Oncorhynchus mykiss*): Respiratory toxicity, organ pathologies, and other physiological effects. *Aquatic Toxicology* 82(2):94–109. doi: 10.1016/j.aquatox.2007.02.003.

Soenen, S. J. H., A. R. Brisson, and M. De Cuyper. 2009. Addressing the problem of cationic lipid-mediated toxicity: The magnetoliposome model. *Biomaterials* 30(22):3691–3701. doi: 10.1016/j.biomaterials.2009.03.040.

Solovitch, N., J. Labille, J. Rose, P. Chaurand, D. Borschneck, M. R. Wiesner, and J. Y. Bottero. 2010. Concurrent aggregation and deposition of TiO_2 nanoparticles in a sandy porous media. *Environmental Science & Technology* 44(13):4897–4902. doi: 10.1021/es1000819.

Song, M.-M., W.-J. Song, H. Bi, J. Wang, W.-L. Wu, J. Sun, and M. Yu. 2010. Cytotoxicity and cellular uptake of iron nanowires. *Biomaterials* 31(7):1509–1517.

Stone, V., H. Johnston, and R. P. Schins. 2009. Development of *in vitro* systems for nanotoxicology: Methodological considerations. *Critical Reviews in Toxicology* 39(7):613–626. doi: 10.1080/10408440903120975.

Taniguchi, N. 1974. On the basic concept of nanotechnology. Paper read at Proc. Intl. Conf. Prod. Eng. Tokyo, Part II, Japan Society of Precision Engineering.

Teeguarden, J. G., P. M. Hinderliter, G. Orr, B. D. Thrall, and J. G. Pounds. 2007. Particokinetics *in vitro*: Dosimetry considerations for *in vitro* nanoparticle toxicity assessments. *Toxicological Sciences* 95(2):300–312.

Tervonen, T., I. Linkov, J. R. Figueira, J. Steevens, M. Chappell, and M. Merad. 2009. Risk-based classification system of nanomaterials. *Journal of Nanoparticle Research* 11(4):757–766. doi: 10.1007/s11051-008-9546-1.

Thassu, D., Y. Pathak, and M. Deleers (eds.). 2007. Nanoparticulate drug-delivery systems: An overview. In: *Nanoparticulate Drug Delivery Systems*, 1–32. New York: Informa.

Thomas, K., and P. Sayre. 2005. Research strategies for safety evaluation of nanomaterials, Part I: Evaluating the human health implications of exposure to nanoscale materials. *Toxicological Sciences* 87(2):316–321. doi: kfi270 [pii] 10.1093/toxsci/kfi270.

Tkachenko, A. G., H. Xie, Y. L. Liu, D. Coleman, J. Ryan, W. R. Glomm, M. K. Shipton, S. Franzen, and D. L. Feldheim. 2004. Cellular trajectories of peptide-modified gold particle complexes: Comparison of nuclear localization signals and peptide transduction domains. *Bioconjugate Chemistry* 15(3):482–490. doi: 10.1021/Bc034189q.

Tuominen, M., and E. Schultz. 2010. Environmental aspects related to nanomaterials—A literature survey. The Finnish Environment 1 26: ISBN 978-952-11-3813-3 (PDF); ISSN 1796–1637 (online).

Unfried, K., C. Albrecht, L. O. Klotz, A. Von Mikecz, S. Grether-Beck, and R. P. F. Schins. 2007. Cellular responses to nanoparticles: Target structures and mechanisms. *Nanotoxicology* 1(1):52–71. doi: 10.1080/00222930701314932.

Usenko, C. Y., S. L. Harper, and R. L. Tanguay. 2007. *In vivo* evaluation of carbon fullerene toxicity using embryonic zebrafish. *Carbon* 45(9):1891–1898. doi: 10.1016/j.carbon.2007.04.021.

Varshney, M. 2012. "Nanotechnology" current status in pharmaceutical science: A review. *IJTA*. 6:14–24.

Vasir, J. K., M. K. Reddy, and V. D. Labhasetwar. 2005. Nanosystems in drug targeting: Opportunities and challenges. *Current Nanoscience* 1(1):47–64. doi: 10.2174/1573413052953110.

Veranth, J. M., E. G. Kaser, M. M. Veranth, M. Koch, and G. S. Yost. 2007. Cytokine responses of human lung cells (BEAS-2B) treated with micron-sized and nanoparticles of metal oxides compared to soil dusts. *Particle and Fibre Toxicology* 4:2. doi: 10.1186/1743-8977-4-2.

Wang, H., J. Wang, X. Deng, H. Sun, Z. Shi, Z. Gu, Y. Liu, and Y. Zhao. 2004a. Biodistribution of carbon single-wall carbon nanotubes in mice. *Journal of Nanoscience and Nanotechnology* 4(8):1019–1024.

Wang, L., L. Wang, T. Xia, L. Dong, G. Bian, and H. Chen. 2004b. Direct fluorescence quantification of chromium (VI) in wastewater with organic nanoparticles sensor. *Analytical Sciences* 20(7):1013–1017.

Wani, M. Y., M. A. Hashim, F. Nabi, and M. A. Malik. 2011. Nanotoxicity: Dimensional and morphological concerns. *Advances in Physical Chemistry* 2011, 1–15 (Article ID 450912) doi: 10.1155/2011/450912.

Warheit, D. B. 2008. How meaningful are the results of nanotoxicity studies in the absence of adequate material characterization? *Toxicological Sciences* 101(2):183–185.

Warheit, D. B., W. J. Brock, K. P. Lee, T. R. Webb, and K. L. Reed. 2005. Comparative pulmonary toxicity inhalation and instillation studies with different TiO_2 particle formulations: Impact of surface treatments on particle toxicity. *Toxicological Sciences* 88(2):514–524.

Warheit, D. B., B. R. Laurence, K. L. Reed, D. H. Roach, G. A. M. Reynolds, and T. R. Webb. 2004. Comparative pulmonary toxicity assessment of single-wall carbon nanotubes in rats. *Toxicological Sciences* 77(1):117–125. doi: 10.1093/toxsci/kfg228.

Warheit, D. B., T. R. Webb, C. M. Sayes, V. L. Colvin, and K. L. Reed. 2006. Pulmonary instillation studies with nanoscale TiO_2 rods and dots in rats: Toxicity is not dependent upon particle size and surface area. *Toxicological Sciences* 91(1):227–236. doi: 10.1093/toxsci/kfj140.

Weinberg, G., R. Ripper, D. L. Feinstein, and W. Hoffman. 2003. Lipid emulsion infusion rescues dogs from bupivacaine-induced cardiac toxicity. *Regional Anesthesia and Pain Medicine* 28(3):198–202. doi: 10.1053/rapm.2003.50041.

Wick, P., P. Manser, L. K. Limbach, U. Dettlaff-Weglikowska, F. Krumeich, S. Roth, W. J. Stark, and A. Bruinink. 2007. The degree and kind of agglomeration affect carbon nanotube cytotoxicity. *Toxicology Letters* 168(2):121–131.

Wickline, S. A., and G. M. Lanza. 2002. Molecular imaging, targeted therapeutics, and nanoscience. *Journal of Cellular Biochemistry* 87(Suppl 39), 90–97. doi: 10.1002/Jcb.10422.

Wilson, M. A., N. H. Tran, A. S. Milev, G. S. Kannangara, H. Volk, and G. Q. Lu. 2008. Nanomaterials in soils. *Geoderma* 146(1):291–302.

Wolfgang, L. 2004. *Industrial Application of Nanomaterials Chances and Risks*. Dusseldorf, Germany: VDI Technologiezentrum.

Worle-Knirsch, J. M., K. Pulskamp, and H. F. Krug. 2006. Oops they did it again! Carbon nanotubes hoax scientists in viability assays. *Nano Letters* 6(6):1261–1268. doi: 10.1021/Nl060177c.

Wurl, O., and J. P. Obbard. 2004. A review of pollutants in the sea-surface microlayer (SML): A unique habitat for marine organisms. *Marine Pollution Bulletin* 48(11–12):1016–1030. doi: http://dx.doi.org/10.1016/j.marpolbul.2004.03.016.

Xia, T. A., Y. Zhao, T. Sager, S. George, S. Pokhrel, N. Li, D. Schoenfeld et al. 2011. Decreased dissolution of ZnO by iron doping yields nanoparticles with reduced toxicity in the rodent lung and zebrafish embryos. *ACS Nano* 5(2):1223–1235. doi: 10.1021/Nn1028482.

Xia, T., M. Kovochich, J. Brant, M. Hotze, J. Sempf, T. Oberley, C. Sioutas, J. I. Yeh, M. R. Wiesner, and A. E. Nel. 2006. Comparison of the abilities of ambient and manufactured nanoparticles to induce cellular toxicity according to an oxidative stress paradigm. *Nano Letters* 6(8):1794–1807. doi: 10.1021/Nl061025k.

Yamashita, Y., A. Tsukasaki, T. Nishida, and E. Tanoue. 2007. Vertical and horizontal distribution of fluorescent dissolved organic matter in the Southern Ocean. *Marine Chemistry* 106(3–4):498–509. doi: http://dx.doi.org/10.1016/j.marchem.2007.05.004.

Yang, K., D. Lin, and B. Xing. 2009. Interactions of humic acid with nanosized inorganic oxides. *Langmuir* 25(6):3571–3576. doi: 10.1021/la803701b 10.1021/la803701b [pii].

Yang, R. S., L. W. Chang, J. P. Wu, M. H. Tsai, H. J. Wang, Y. C. Kuo, T. K. Yeh, C. S. Yang, and P. Lin. 2007. Persistent tissue kinetics and redistribution of nanoparticles, quantum dot 705, in mice: ICP-MS quantitative assessment. *Environmental Health Perspectives* 115(9):1339–1343. doi: 10.1289/ehp.10290.

Yao, Z., J. J. Stiglich, and T. S. Sudarshan. 2002. Nano-grained tungsten carbide–cobalt (WC/Co). Working paper. Fairfax: Materials Modification, Print.

Zhang, T., J. L. Stilwell, D. Gerion, L. Ding, O. Elboudwarej, P. A. Cooke, J. W. Gray, A. P. Alivisatos, and F. F. Chen. 2006. Cellular effect of high doses of silica-coated quantum dot profiled with high throughput gene expression analysis and high content cellomics measurements. *Nano Letters* 6(4):800–808. doi: 10.1021/nl0603350.

Zhu, L., D. W. Chang, L. M. Dai, and Y. L. Hong. 2007. DNA damage induced by multiwalled carbon nanotubes in mouse embryonic stem cells. *Nano Letters* 7(12):3592–3597. doi: 10.1021/Nl071303v.

Zhu, S. Q., E. Oberdorster, and M. L. Haasch. 2006. Toxicity of an engineered nanoparticle (fullerene, C-60) in two aquatic species, Daphnia and fathead minnow. *Marine Environmental Research* 62:S5–S9. doi: 10.1016/j.marenvres.2006.04.059.

Zhu, X. S., L. Zhu, Y. Li, Z. H. Duan, W. Chen, and P. J. J. Alvarez. 2007. Developmental toxicity in zebrafish (*Danio rerio*) embryos after exposure to manufactured nanomaterials: Buckminsterfullerene aggregates (nC(60)) and fullerol. *Environmental Toxicology and Chemistry* 26(5):976–979. doi: 10.1897/06-583.1.

Zhu, X. S., L. Zhu, Z. H. Duan, R. Q. Qi, Y. Li, and Y. P. Lang. 2008. Comparative toxicity of several metal oxide nanoparticle aqueous suspensions to Zebrafish (*Danio rerio*) early developmental stage. *Journal of*

Environmental Science and Health Part a—Toxic/Hazardous Substances & Environmental Engineering 43(3):278–284. doi: 10.1080/10934520701792779.

Zvyagin, A. V., X. Zhao, A. Gierden, W. Sanchez, J. A. Ross, and M. S. Roberts. 2008. Imaging of zinc oxide nanoparticle penetration in human skin *in vitro* and *in vivo*. *Journal of Biomedical Optics* 13(6):064031. doi: 10.1117/1.3041492.

2 Nanoparticle Exposures in Occupational Environments

Li-Hao Young, Ying-Fang Wang, Ching-Hwa Chen, Chun-Wan Chen, and Perng-Jy Tsai

CONTENTS

2.1 Introduction 49
2.2 Routes for Occupational Nanoparticle Exposures 50
    2.2.1 Inhalation 50
    2.2.2 Dermal 52
    2.2.3 Ingestion 52
2.3 Exposure Assessment 52
    2.3.1 Exposure Metrics 52
    2.3.2 Measurement Instruments 53
        2.3.2.1 Time-Integrated Measurements 54
        2.3.2.2 Time-Resolved Measurements 54
2.4 Occupational Nanoparticle Exposures in Various Industrial and Developmental Laboratories 55
2.5 Hierarchy and Prudent Practices of Exposure Control and Management 56
    2.5.1 Engineering Control 56
        2.5.1.1 Elimination or Substitution 60
        2.5.1.2 Isolation and Containment 61
        2.5.1.3 Ventilation 61
    2.5.2 Administrative Controls 62
        2.5.2.1 Employee Training 62
        2.5.2.2 Transfer, Labeling, and Storage 63
        2.5.2.3 Housekeeping 63
        2.5.2.4 Personal Hygiene 64
    2.5.3 Personal Protective Equipment 64
        2.5.3.1 Respirator 64
        2.5.3.2 Protective Clothes, Gloves, and Other Equipment 64
    2.5.4 Fire and Explosion Control 65
    2.5.5 Management of Nanomaterial Spills and Waste 65
    2.5.6 Occupational Health Surveillance 66
2.6 A Qualitative Risk-Based Control and Management Strategy: Control Banding 67
2.7 Conclusions 68
References 68

2.1 INTRODUCTION

Nanotechnology, as defined by the Interagency Subcommittee on Nanoscale Science, Engineering, and Technology (NSET) of the U.S. Federal Office of Science and Technology Policy, is "research and technology development at the atomic, molecular or macromolecular levels, in the length scale of approximately 1–100 nm range, to provide a fundamental understanding of phenomena and materials at the nanoscale level and to create and use structures, devices and systems that have novel properties and functions because of their small and/or intermediate size." ISO/TC 146/SC 2/WG1 N 320 defines the nanoparticle as "a particle with a nominal diameter smaller than about 100 nm," and nanoaerosol as "an aerosol comprised of or consisting of nanoparticles and nanostructured particles." An ultrafine particle is defined as "a particle sized about 100 nm in diameter or less," and this terminology is widely used by toxicologists. In principle, sources of airborne nanoparticles are divided into two groups: naturally occurring and anthropogenic. Naturally occurring nanoparticles can be emitted from volcano eruptions and forest fires, among other sources. For anthropogenic nanoparticles, they could be further divided into engineered nanoparticles and fugitive nanoparticles. For engineered nanoparticles, they are associated with the use of nanotechnology for the use or manufacture of nanoparticles, such as carbon blacks, fumed silica, titanium dioxide, metallic oxides, fullerenes, carbon nanotubes, and quantum dots, among others. Process fugitive nanoparticles are particles produced unintentionally during an intentional operation. The major sources of these particles include the processes associated with the combustion, welding, metal processing, diesel engine operations, and so on. In principle, more complex chemical compositions, irregular shapes, and polydispersed size distributions can be seen in process fugitive nanoparticles than those of engineered nanoparticles (Oberdörster et al., 2005, Kreyling et al., 2006). A workplace associated with the emission of anthropometric nanoparticles, therefore, is likely to have nanoparticle exposures to related workers. In addition, it should be noted that nanomaterials are widely used in medical imaging, drug delivery, cosmetic, electronic, and other scopes. Therefore, exposures might occur for workers involved in scientific research in the laboratory or developing production processes in pilot manufacturing plants.

Exposures to nanoparticles might cause serious inflammation after the deposition of nanoparticles in the respiratory tract because of their large particle numbers and surface areas, and complexed chemical compositions, sizes, shapes, and electric charges (Oberdörster, 2001; Donaldson et al., 2002; Kreyling et al., 2002). Recent toxicological studies have suggested that they can easily penetrate cells or tissue and may result in many irreversible health effects, such as chronic pulmonary inflammation, epithelial cell hyperplasia, cardiovascular disease, and lung tumors (Dockery et al., 1993; Oberdörster G, 2000; Kreyling et al., 2002; Oberdörster E, 2004). Considering that the aforementioned preliminary toxicological data have been found with nanomaterials, a more prudent practice to assess occupational nanoparticle exposures and resultant health risks, and eventually to initiate appropriate exposure controls and risk management strategies, would be necessary.

2.2 ROUTES FOR OCCUPATIONAL NANOPARTICLE EXPOSURES

Like many other agents, three main exposure routes are associated with occupational nanoparticle exposures, including inhalation, dermal contact, and ingestion.

2.2.1 INHALATION

In principle, inhalation is considered as the main exposure route in most workplaces. Inhaled particles may deposit in various regions of the respiratory system affected by five main, distinct deposition mechanisms of sedimentation: inertial impaction, interception, diffusion, sedimentation, and electrostatic attraction. The human respiratory system can be divided into three main regions: the head airways (from the nose and mouth to the larynx), the tracheobronchials (from the larynx to the terminal bronchioles), and the alveoli (Figure 2.1).

For micrometer particles at the head airways, inertial impaction is the main deposition mechanism. For nose breathers, therefore, most deposition occurs at the head airway. If a person breathes through the mouth, the main deposition site becomes the larynx. However, the most dominant

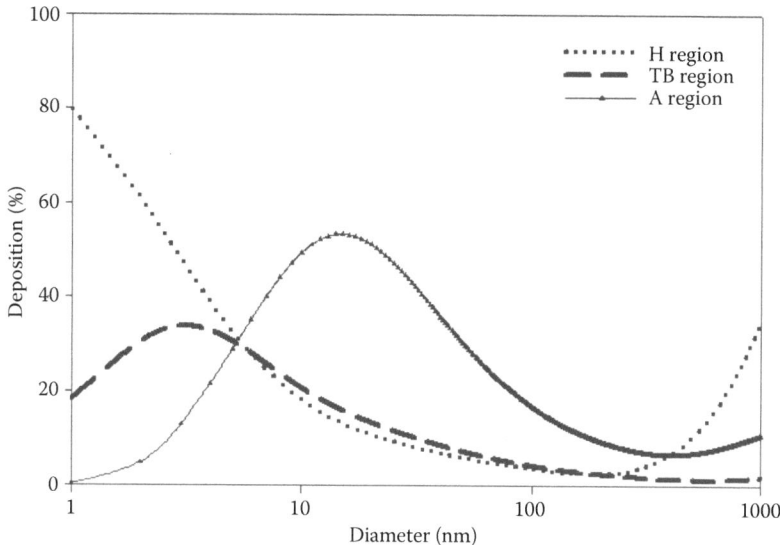

FIGURE 2.1 Calculated deposition curves of nanoparticles for the head airway (H), tracheobronchial (TB), and alveolar (A) regions of the respiratory tract using the LUDEP model software.

mechanism for this region is the diffusion of nanoparticles. In particular, when the nanoparticles are smaller in size, an increased effect would result with particle diffusions. The head airway does not have cilia. Therefore, the pharyngeal lymphatic ring located in the pharynx might provide the most defense against the deposition of foreign matter in this region.

The entire tracheobroncial region (except for a part of the larynx) is coated with cilia. These flickers move in a coordinated manner with a frequency of 10–20 times per second to transport mucus and deposited particles upwards to the larynx. It takes ~0.5, 2, and 5 h for deposited particles to be transported from the broad, medium broad, and fine airways to the larynx, respectively.

In the lowermost airways (alveolar region), there is very limited mixture between just inhaled air and residual air (i.e., air that remains from previous breaths). The deposition of particles with diameters between 100 nm and 1 µm in this region is governed by the transfer of particles from newly inhaled air to residual air. Almost all the particles will deposit at this region through diffusion and sedimentation for small and large particles, respectively. The mechanisms for clearing particles from the alveolar region are not completely known yet. The main mechanisms might be associated with the devouring of the particles by phagocytes.

The International Commission on Radiological Protection (ICRP) developed a model to predict particle depositions at different regions of the human respiratory tract (ICRP, 1994). The model covers various breathing characteristics governed by the involved workloads, and a wide range of particle sizes from 1 nm to 100 µm. It is interesting to note that most micrometer-sized particles deposit at the head airways. Therefore, the nose is considered as an effective prefilter for protecting against particles larger than a couple of micrometers, which makes the human lungs fairly well protected from coarse particles depositing at the more vulnerable lower respiratory tract. But for nano-sized particles with diameters <10 nm, they will deposit at the head airways (up to 80%), whereas those in the 10–100 nm range will mainly deposit at the alveolar region (up to 50% for 20 nm particles). For 1 nm particles, however, 90% deposit at the head region, with 10% at the tracheobronchial region, and nearly nothing at the alveolar region. Figure 2.2 shows the regional deposition as a function of particle size for adults engaged in light work estimated according to the ICRP Task Group lung model based on Hinds' parameterization, gender, and activity-weighted average deposition efficiencies (Hinds, 1999).

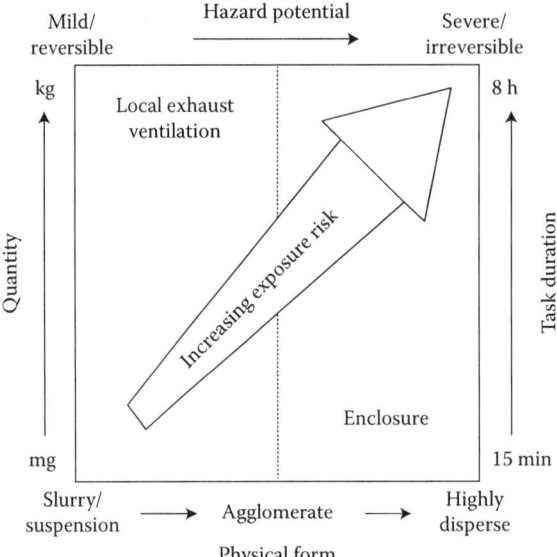

FIGURE 2.2 Major factors influencing exposure risk and hence selection of control measure. (From NIOSH, 2012. *General Safe Practices for Working with Engineered Nanomaterials in Research Laboratories.* Cincinnati, OH: U.S. DHHS, CDC, NIOSH. DHHS (NIOSH) Pub. No. 2012-147. With permission.)

2.2.2 Dermal

Dermal exposure to nanoparticles is considered a potential exposure route for workers associated with the handling of nano-sized powders. At present, many manufacturing processes in nano-industries involve direct or indirect skin contact and result in the penetration of nanoparticles through intact skin (Lademann et al., 1999; Pflücker et al., 2001; Tinkle et al., 2003; Ryman-Rasmussen et al., 2006). Monteiro-Riviere also indicated that skin could be permeable to quantum dot nanoparticles (Monteiro-Riviere, 2009). However, it is unknown whether skin penetration would result in adverse health effects.

2.2.3 Ingestion

Very little work has been done to investigate ingestion exposures. Behrens et al. (2001) has indicated that smaller particles could be transferred across the intestinal wall more easily than larger ones. The most probable ingestion exposure route for nanoparticles could be due to hand-to-mouth contact, and then the transfer to other organs via ingestion in the intestinal tract.

2.3 EXPOSURE ASSESSMENT

2.3.1 Exposure Metrics

The current metric for setting occupational exposure limits (OELs) for micro-sized particles is based on its mass concentration. However, the appropriateness of the above metric for nanoparticles has been questioned because nanoparticles feature high particle counts and large surface area per unit masses. Kreyling et al. (2006) reported that the nanoparticle is less than 10% of PM2.5 concentrations in terms of mass but contributes more than 90% of its particle number concentration. To date, both exposure metrics of the total surface area and total number concentrations of nanoparticles have shown good correlations with their resultant health effects. For example, McCawley et al. (2001) found that the particle number concentration was a more appropriate metric for lung disease. Several toxicological studies have indicated that there is a better dose–response relationship while using ultrafine and fine particle surface areas as exposure metrics (Oberdörster, 2000; Trans et al., 2000; Brown et al., 2002). Therefore, the change in exposure metrics for nanoparticles from a mass basis to both alternative metrics of particle number and surface area concentrations have been proposed (Kreyling et al., 2006; Maynard and Aitken, 2007). However, at this moment, no scientific consensus can be referred to for the selection of an appropriate exposure metric (Maynard and Aitken, 2007; Paik et al., 2008). Therefore, a multimetric exposure assessment appears to be reasonable, given the present situation.

2.3.2 Measurement Instruments

In principle, the sampling equipment for assessing occupational exposures should be personal, or at least portable, from an easy-to-use point of view. In addition, the instrument should also be battery-powered, robust, and relatively inexpensive in order to lower the cost. There are various devices that can be used to measure the mass, surface area, or the number concentrations of particles in workplaces. However, it should also be noted that not any developed instrument could be used for simultaneously measuring the nanoparticle exposures of the above metrics. Since no health-related sampling convention has been established for nanoparticles, the developed device could be used for the specific aims that have been set for the given study. Table 2.1 shows a summary of the sampling devices used in workplaces for nanoparticle exposure and assessment studies (ISO, 2007; Kuhlbusch et al., 2011; CJ Tsai et al., 2012). A detailed description of instruments is presented in the following subsections.

TABLE 2.1
Instruments and Techniques for Monitoring Nanoparticle Exposures

Metrics	Devices	Sampling Types	Size Range	Remarks
Mass	Personal nanoparticle sampler (PENS)	Time-integrated	<100 nm and >4 μm	Collection of both mass concentrations of respirable particles and nanoparticles simultaneously
	Nano-MOUDI	Time-integrated	0.01–18 μm	Particle size distribution by mass, chemical, and morphology identification
	Low-pressure cascade impactor	Time-integrated	>20 nm	Particle size distribution by mass, chemical and morphology identification
Number	CPC	Time-resolved	0.01–1 μm	Real-time number concentration
	SMPS	Time-resolved	2.5–1000 nm	Real-time size-selective (mobility diameter) detection of number concentration
	ELPI	Time-resolved	6 nm–10 μm	Real-time size-selective (aerodynamic diameter) detection of active surface area concentration. Size-selected samples may be further analyzed off-line
	FMPS	Time-resolved	5.6–560 nm	Same as SMPS
	P-Trake	Time-resolved	0.02–1 μm	Same as CPC
Surface area	NSAM	Time-resolved	0.01–1 μm	Deposited surface area concentration of TB and AL regions
	EAD/MEAD	Time-resolved	0.01–1 μm	Aerosol length (EAD), lung-deposited surface area (MEAD) of H, TB, and AL regions
	AeroTrak 9000	Time-resolved	0.01–1 μm	Same as NASM
	LQ1-DC	Time-resolved	0.01–1 μm	Active surface area

2.3.2.1 Time-Integrated Measurements

- *Personal nanoparticle sampler (PENS)*: A novel, active, PENS consisting of a respirable cyclone and a micro-orifice impactor, with a cutoff aerodynamic diameter of 4 μm and 100 nm, which respectively enables the collection of the respirable particulate mass and nanoparticles simultaneously, was developed to meet the critical demand for the personal sampling of engineered nanomaterials in workplaces (Tsai et al., 2012).
- *Nano-MOUDI*: The Nano-MOUDI impactor collects particles ranging from >18 to 0.01 μm on 13 impaction stages. It can be used to characterize particle size distributions of collected particles by mass. The morphology of the particles deposited on each impaction stage can be further analyzed by using scanning electron microscope (SEM) or transmission electron microscopy (TEM). Their crystallography, metal contents, and chemical composition can be determined through further sample analysis techniques, such as x-ray diffraction (XRD), atomic absorption spectrophotometer (AA), and gas chromatography/mass spectrometry (GC/MS).

2.3.2.2 Time-Resolved Measurements

- *Codensation particle counter (CPC)*: The CPC is a real-time instrument used for measuring particle number concentrations. Air is drawn into a saturator tube filled with supersaturated isopropyl alcohol vapor. The vapor is condensed on air-containing aerosols, and the aerosols grow until they can be sensed by a photodetector as they pass through a laser beam. The CPC measures particles ranging from a few nanometers to a micrometer, with measurable number concentrations ranging from 0 to 100,000 particles/cm^3.
- *Scanning mobility particle sizer (SMPS)*: The instrument measures the particle size distributions and number concentrations. The SMPS is composed of a differential mobility analyzer (DMA) and a CPC installed in series. The DMA first selects a particle size interval of the sampled aerosol based on its electrical mobility in the scanning electric field, and then the CPC counts the particles exiting into the classifier. It should be noted that the particles must be previously neutralized at the DMA inlet using radioactive sources, generating numerous positive and negative charges in order to reach a state of charge equilibrium. The SMPS measures particles ranging from a few nanometers to a micrometer, with measurable number concentrations ranging from 10 to 760 nm.
- *Nanoparticle surface area monitor (NSAM)*: Recently, it has been found that the response function curves of an electrical aerosol detector (EAD) (TSI Inc., Model 3070A, St. Paul, MN, USA) matched with trancheobronchial (TB) and alveolar (A) depositions while the ion trap voltages were set at 100 V and 200 V, respectively (Fissan et al., 2007). Based on the above observations, the EAD was modified and a TSI NSAM (TSI Inc., Model 3550, St. Paul, MN, USA) was built by adjusting its built-in, ion trap voltage. For an NSAM, sampled particles are charged by a corona effect-induced ion diffusion at their surfaces. The number of charges carried by a particle is related to its surface area (particle size). The charged particles, after passing through the ion trap, are then collected on an electrical filter, allowing the determination of the surface area concentration of collected particles. The preset ion trap voltages of 100 and 200 V allow the concentrations of nanoparticle surface area deposited in the tracheobronchial or alveolar region to be estimated, respectively. The instrument measures the total surface area for particle diameters ranging from 10 to 1000 nm. This is good for measuring workplace exposure to nanoparticles and for inhalation toxicology and epidemiology studies, but it does not provide the particle size distribution and total surface area of the collected particles.
- *EAD and MEAD*: For EAD, sampled particles are first passed through a small cyclone to remove particles with a diameter larger than 1.0 μm. The sampled aerosol flow (i.e., 2.5 lpm) is then split into two: the portion of 1.5 lpm is directly introduced into the aerosol charging chamber, and the other portion of 1.0 lpm is used as the carrier for unipolar ions,

generated in the corona discharging chamber, after the particles and vapor contaminants are removed by high-efficiency particulate air (HEPA) and active carbon filters. The two split flows are mixed in the aerosol charging chamber. Particles exited from the charging chamber pass through an ion trap, with the voltage set at 20 V, before the electrical charges carried by particles are measured in an aerosol electrometer of a Faraday cage-type downstream of the ion trap.

A modified EAD (MEAD), which consists of a unipolar charger to electrically charge sampled particles, an ion trap to remove excess ions, and an aerosol electrometer to meter the resultant current, is used in this study with its ion trap controlled by an external, high-voltage power supply (Stanford Research System Inc., Model PS325/2500 V-25 W, Sunnyvale, CA, USA). The ion trap in the MEAD now serves as a size selector, by changing its voltage and allowing different fractions of charged aerosols to pass through. The total current of passed aerosols is then measured as the output of the MEAD. Owing to the charging method used in EADs, a difference in the output of the EAD signal is expected when test particles are in the same size distributions with different materials.

Li et al. (2009a) evaluated the performance of EADs with the same ion trap voltage settings as those in NSAMs. Poly- and monodisperse nanoparticles of Ag, NaCl, and oleic acid were generated and used as test aerosols. For the EAD TB curves, the variation of EAD TB line slopes is about 15% when varying the dielectric constant of particle materials from infinite to 2.5, and the dependence of size on TB correlation curves is negligible (less than 5%). The variation of the slope on EAD A correlation lines is generally within 10%. Much of the variation is a result of the size, not from particle materials. In spite of detectable particle material and size effects on the EAD TB and A correlation curves, it was concluded that the effects were generally minor at a relatively low concentration (i.e., EAD readout less than 1 pA). However, the particle material effect is more pronounced for polydispersed particle testing. The slope variation due to the particle size in the correlation line is estimated to be about 5% (Li et al., 2009b). The charging efficiencies and the charge distribution of particles passing through the EAD charger were further characterized using monodispersed Ag and PSL particles. It was found that the effect of the particle material on NSAM readouts is generally minor at low aerosol concentrations. The material effect is, however, more pronounced at high aerosol concentrations. To reproduce the measured, intrinsic charging efficiency and charge distribution of particles in the EAD charger, the birth-and-death particle charging model with two Nit (ion concentration times aerosol residence time) values were proposed. Modeling results using the proposed model are in very good agreement with the experimental charge distribution of particles with a diameter less than 1 μm. Based on the information above, it was concluded that MEAD could be modified as a sizer to indicate particle concentrations and size distributions (Li et al., 2009a). However, the above instrument has never been validated in the field, especially for nanoparticles with different material characteristics.

2.4 OCCUPATIONAL NANOPARTICLE EXPOSURES IN VARIOUS INDUSTRIAL AND DEVELOPMENTAL LABORATORIES

Table 2.1 provides a summary of nanoparticle exposure concentrations from various industry workplaces or development laboratories. The workplace levels, outdoor ambient levels, or the same activity levels were measured and listed for comparison. Many monitored activities were related to the end phase in a commercial-scale production (i.e., bagging and packaging of the product) (Wake, 2001; Kuhlbusch et al., 2004; Fujitani et al., 2008; Demou et al., 2008; Peters et al., 2009; Wang et al., 2010). Others were monitored for research-scale activities only, focusing on welding and grinding (Wake, 2001; Zimmer and Biswas, 2001; Maynard, 2003; Stephenson et al., 2003; Möhlmann, 2005; Cheng et al., 2008; Dash and D'Arcy, 2008; Evans et al., 2008; Elihn and Berg,

2009; Methner et al., 2010); harvesting the product, dryer and reactor maintenance, and cleaning (Bello et al., 2008; Han et al., 2008; Methner, 2008, 2010; Yeganeh et al., 2008; Park et al., 2009); activities using flame spray pyrolysis (Wegner et al., 2011); and various activities in manufacturing (Bello et al., 2009; Methner et al., 2010; Wang et al., 2011). From Table 2.2, it can be seen that nanoparticle concentrations found in the ambient air and workplace atmospheres are on the orders of $\sim 10^3$–10^4 and $\sim 10^3$–10^6, respectively.

2.5 HIERARCHY AND PRUDENT PRACTICES OF EXPOSURE CONTROL AND MANAGEMENT

Despite the many scientific uncertainties and the lack of regulatory guidelines, a prudent and preventive approach should be adopted to minimize personal exposures to and fire/explosion risks from nanomaterials. This statement is in accordance with the Precautionary Principle, which stipulates a precautionary approach to avoiding or diminishing a harm that is scientifically plausible but uncertain (COMEST, 2005). The most effective means by which to prevent or reduce harm are anticipating potential hazards early in the developmental stage of a process and then incorporating health and safety considerations into the design, implementation, and operation phases. This prevention-through-design (PtD) strategy aims to incorporate safe practices and measures into the life cycle of the process at the earliest stage possible (NIOSH, 2011a). Once past the design stage, it is often very difficult, as well as costly, to modify the process, equipment, or workplace for health and safety purposes. The present strategy follows the traditional, industrial hygiene (IH) hierarchy of control that emphasizes hazard reductions as close to the source as possible. The three main categories of control include engineering controls, administrative controls, and the use of personal protective equipment (PPE) (Plog et al., 2012). It is important to note that engineering controls are considered more effective, and, thus, preferred over the other control measures, as they are typically implemented close to the emission source and less dependent on the workers' involvement or behavior. In addition, the selection of control measures should be chosen such that they are proportional to the risk of exposure. The major factors that will influence the risk of exposure are the physical form, quantity, and the duration of the task. For example, exposure risks increase when working with easily dispersed, large quantities of hazardous nanomaterials over long task durations, as illustrated in Figure 2.3 (NIOSH, 2012).

With respect to the physical form, the National Research Council (NRC) of the National Academy of Sciences (NRC, 2011) recommended that preferences should be given to handling materials of lower risk, in the order of

- Solid materials with embedded nanostructures
- Solid nanomaterials with nanostructures fixed to the material's surface
- Nanoparticles suspended in liquids
- Dry dispersible engineered nanoparticles, nanoparticle agglomerates, or nanoparticle aggregates

Specific control measures include (1) elimination or substitution; (2) isolation or containment; (3) ventilation; (4) employee training; (5) labeling and storage; (6) work procedures; (7) cleaning and maintenance; (8) personal hygiene; (9) task duration; and (10) respirator, clothing, and goggles. Figure 2.4 shows the details of control categories and their relative priority described in detailed in the following sections.

2.5.1 Engineering Control

Engineering controls broadly include eliminating or substituting hazardous nanomaterials, and physically changing some aspects of the process or workplace to reduce emissions and, thus,

TABLE 2.2
Summary of Nanoparticle Exposure Number Concentrations from Various Development Laboratory and Manufacturing Workplaces

Industry/Exposure Material	Activity/Process	Number Concentration Background	Number Concentration Workplace	Reference
Carbon black	Bagging	694–3836 × 10^3	4–50 × 10^3	Wake (2001)
Nickel powder	Bagging	3–16 × 10^3	4–212 × 10^3	Wake (2001)
Precious metal	Sieving	19–62 × 10^3	23–71 × 10^3	Wake (2001)
Titanium dioxide	Bagging	10–58 × 10^3	4–17 × 10^3	Wake (2001)
Zinc refining	Sintering	20–23 × 10^3	12–14 × 10^3	Wake (2001)
	Casting	20–23 × 10^3	56–100 × 10^3	Wake (2001)
Plasma coating	Wire coating	2–8 × 10^3	3–905 × 10^3	Wake (2001)
Galvanizing	Galvanizing	15–37 × 10^3	10–683 × 10^3	Wake (2001)
Steel foundry	Molding	13–73 × 10^3	118–>500 × 10^3	Wake (2001)
Welding	MIG	10–19 × 10^3	117–>500 × 10^3	Wake (2001)
Plastic welding	Welding	1–5 × 10^3	111–3766 × 10^3	Wake (2001)
Hand soldering	Tinning	2–11 × 10^3	12–>500 × 10^3	Wake (2001)
Experimental apparatus	Welding processes	—	3 × 10^6	Zimmer and Biswas (2001)
Hardwood	High-speed grinding		0.8 × 10^6	Maynard (2003)
Aluminum	High-speed grinding		0.1 × 10^6	
Industrial maintenance and metal fabrication workplace environment	Welding	—	2 × 10^5	Stephenson et al. (2003)
Carbon black	Packaging/bag filling	10^4–10^5	10^4–10^5	Kuhlbusch et al. (2004)
Silica	Smelting	—	1 × 10^5	Möhlmann (2005)
—	Metal grinding	—	1.3 × 10^5	Möhlmann (2005)
—	Soft soldering	—	4 × 10^5	Möhlmann (2005)
—	Plasma cutting	—	5 × 10^5	Möhlmann (2005)
Bakery	Bakery	—	6.4 × 10^5	Möhlmann (2005)
Airport	Airport terminal	—	7 × 10^5	Möhlmann (2005)
—	Hard soldering	—	0.54–35 × 10^5	Möhlmann (2005)
Welding	Welding	—	1–400 × 10^5	Möhlmann (2005)
Parking lot		—	0.3–21.7 × 10^4	Ramachandran et al. (2005)
Carbon black	Reactor/pelletizing	10^4–10^5	10^4–10^5	Kuhlbusch and Fissan (2006)
Fullerenes	Scooping, brushing, sweeping	4.6 × 10^3/ 1.3–6.3 × 10^4	4.8 × 10^3/ 0.54–6.3 × 10^4	Yeganeh et al. (2008)
Fullerenes	Bagging	1.2–1.6 × 10^4	>1.6 × 10^4	Fujitani et al. (2008)
MWCNT	Recovering CNT; blending composites	1.0 × 10^4	>4 × 10^4	Han et al. (2008)
CNT	Removal and detaching of CNTs	0.35–0.4 × 10^4	0.3–0.4 × 10^4	Bello et al. (2008)
Metal-based	Reactor maintenance and cleaning, powder handling and packing, workplace cleaning	0.85 × 10^4	5.9 × 10^4	Demou et al. (2008)
Metal	Reactor cleanout	1.2 × 10^4	2.9 × 10^4	Methner (2008)
Automotive gray iron foundry	Welding	—	2 × 10^4–4 × 10^6	Evans et al. (2008)

continued

TABLE 2.2 (continued)
Summary of Nanoparticle Exposure Number Concentrations from Various Development Laboratory and Manufacturing Workplaces

Industry/Exposure Material	Activity/Process	Number Concentration Background	Number Concentration Workplace	Reference
Iron foundry	Welding	—	2×10^4–3×10^5	Cheng et al. (2008)
Automotive plant	Welding	—	2×10^5	Dash and D'Arcy (2008)
Iron foundry	Molding and pouring		2.07×10^4–2.82×10^5	Cheng et al. (2008)
CNT	Dry cutting CNT-carbon laminate	4.8×10^3	2.94×10^5	Bello et al. (2009)
Lithium titanate Me-oxide	Bagging, milling, powder sifting, loading dock	$\sim 1.5 \times 10^5$	$\sim 1.5 \times 10^5$	Peters et al. (2009)
Seven industrial plants	Welding	—	3×10^4–1×10^5	Elihn and Berg (2009)
Silver (commercial production facility)	Reactor (opening the hatch)	7.9×10^6	18.92×10^6	Park et al. (2009)
	Dryer (opening the door)	5.11×10^6	6.45×10^6	
	Grinder (opening the hatch)	4.03×10^6	7.08×10^6	
Titanium dioxide production facility			1.2×10^4–1.7×10^4	Plitzko (2009)
CNT (base-carbon)	Wet cutting	—	9.4×10^4	Bello et al. (2009)
CNT (base-alumina)	Dry cutting	—	1.48×10^5	Bello et al. (2009)
CNT (alumina)	Dry cutting	—	0.38×10^5	Bello et al. (2009)
CNT (base-carbon)	Dry cutting	—	2.83×10^5	Bello et al. (2009)
CNT (carbon)	Dry cutting	—	2.94×10^5	Bello et al. (2009)
Carbon black manufacturing factory	Packaging	3.46×10^3	25.7×10^3	Wang et al. (2010)
	Warehouse	18.6×10^3	42.1×10^3	
	Pelletizing	—	13.7×10^3	
	Outdoor background	3.41×10^3	—	
Carbon nanofiber (research and development laboratory)	Weighting, mixing		4000	Methner et al. (2010)
	Wet sawing		5000	
MWCNT (research and development laboratory)	Opening growth chamber (no exhaust)		4.2×10^4	Methner et al. (2010)
	Opening growth chamber (with exhaust)		300	
Carbon nanopearls (research and development laboratory)	Opening furnace		160	Methner et al. (2010)
	Transferring material to vial		1940	
Fullerenes and MWCNT (research and development laboratory)	Weighing raw MWCNT in hood		1480–1580	Methner et al. (2010)
	Weighing functionalized MWCNT		680	
	Sonication of raw MWCNT		2200–2800	

TABLE 2.2 (continued)
Summary of Nanoparticle Exposure Number Concentrations from Various Development Laboratory and Manufacturing Workplaces

Industry/Exposure Material	Activity/Process	Number Concentration Background	Number Concentration Workplace	Reference
	Sonication of functionalized MWCNT		730	
Aluminum (research and development laboratory) (pilot scale)	Cleaning/brush down of plasma torch		6700	Methner et al. (2010)
	Cleaning/brush down of filter chamber		10,450	
	Cleaning/brush down of cyclone		15,580	
Silica-iron nanomaterial manufacturer (pilot scale)	Inside spray enclosure		79,700	Methner et al. (2010)
	Outside spray enclosure		2300	
	Inside fiberizer		6360	
Titanium dioxide manufacturer	Manual loading/unloading trays inside booth		15,500	Methner et al. (2010)
	Dumping into mixing tank using focused LEV		3500	
	Spray dryer drum changeout		1,44,800	
	Shop vacuum (no HEPA)		80,700	
Manganese, silver, cobalt, and iron oxide	Passivator tumbler exhaust port		5970	Methner et al. (2010)
	Opening reactor no LEV		85,900	
	Reactor cleanout- manganese no LEV		16,900	
	Reactor cleanout- manganese with LEV		100	
	Reactor cleanout- silver no LEV		6100	
	Reactor cleanout- silver with LEV		0	
	Reactor cleanout- cobalt no LEV		12,900	
	Reactor cleanout- cobalt with LEV		1900	
Nylon 6 nanofiber manufacturer	Propane forklift exhaust		45,000	Methner et al. (2010)
	Electric arc welding		84,600	
Die casting plant	Machining		1.22×10^5	Park et al. (2010)
	Trim		1.65×10^5	
	Die cast		2.83×10^5	
Fastener manufacturing factory	Forming	—	2.13×10^5	Wang et al. (2011)
	Threading	—	1.42×10^5	
	Heat treating	—	3.47×10^5	
	Ambient	0.126×10^5	—	
FSP pilot plants (FePO$_4$)	Flame spray pyrolysis	–	6.9×10^4–8.4×10^4	Wegner et al. (2011)

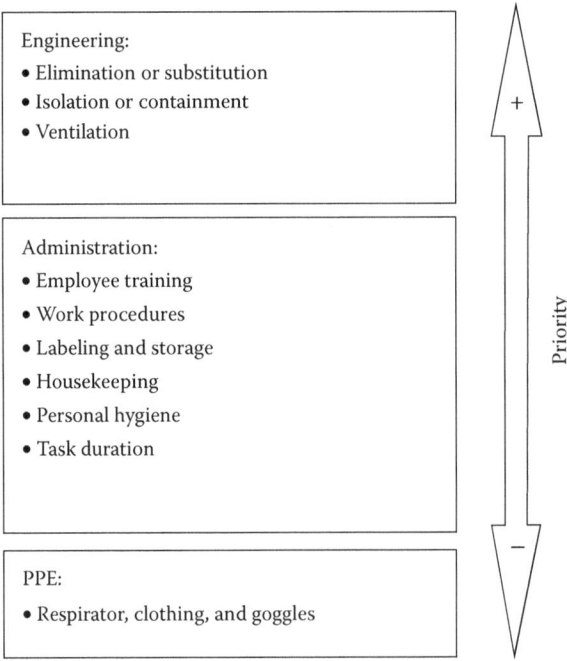

FIGURE 2.3 The relative priority of different control categories and specific measures.

> CAUTION
> **Nanomaterial sample**
> **Consisting of** [Technical description here]
> **In case of** Container breakage
> **Contact:** [Point of contact]
> at [Contact's telephone number]

FIGURE 2.4 Recommended package labeling for transfer and storage of nanomaterials.

employee exposures. As described previously, engineering controls are most cost effective when they are considered at the early design stage, implemented close to the emission sources, and generally independent of the workers' involvement.

2.5.1.1 Elimination or Substitution

As they are synthesized and used for their unique properties, it is often not feasible to eliminate or substitute the nanomaterials themselves. However, substitution opportunities do exist for process modifications. For example, substituting dry production processes (i.e., nanomaterials in dry powder form) with wet ones (i.e., in liquid suspension), optimizing the operating conditions to a less "energetic" process, and substituting for nontoxic or less hazardous alternative solvents. When the above measures are applicable, elimination and substitution are among the most effective means of exposure control. For more information and successful examples of alternative assessment methodologies, process changes, or chemical substitutions, the readers are referred to the EPA's Design for the Environment (DfE; http://www.epa.gov/dfe/alternative_assessments.html) and the Toxics Use Reduction Institute at the University of Massachusetts Lowell (http://www.turi.org).

2.5.1.2 Isolation and Containment

Isolation and containment involve the physical separation of a potential nanomaterial emission source (e.g., a process, task, or equipment) from workers to minimize the risk of exposure. Separation can be done by either locating or enclosing the emission source in an area separated from the workers, or by isolating the workers in a controlled booth or room for remote operations or observations. As shown earlier in Figure 2.1, operations involving easily dispersed, dry, or highly toxic nanomaterials need to be conducted in enclosures. For example, dry nanomaterials, should be handled inside a glove box. A number of nanomaterials such as carbon black, silica fumes, titanium oxide, and metal oxides are synthesized inside closed circuit reactors, which provide workers with an adequate isolation from the direct exposure to nanomaterials. However, other safe procedures are required when workers enter the enclosure for maintenance. It is noted that isolation and containment are commonly coupled with ventilation in order to vent and remove the suspended nanomaterials.

2.5.1.3 Ventilation

There are two types of ventilation available for the control of airborne nanomaterials: dilution (or general) ventilation (DV) and local exhaust ventilation (LEV). Typically provided by a building's heating, ventilation, and air-conditioning systems, DV is the dilution of contaminated air with uncontaminated air for the control of airborne hazards. Unlike LEV, the supply and exhaust of air in DV occurs across a relatively large area, room, or building. As a result, DV has the following limitations: (1) The contaminant quantity must be small enough, such that the airflow rate needed for dilution is practical; (2) the contaminant concentration must be low enough, such that workers' exposure is below the threshold limit; (3) the contaminant toxicity must be low; and (4) the generation of contaminant should be relatively uniform within the area or room. As such, DV alone is not as effective or satisfactory for the control of health hazards as is LEV (ACGIH, 2010). Nevertheless, a well-designed DV can supplement and enhance the effectiveness of LEV. OSHA laboratory standard specifics of 4–12 room air changes per hour (ACH) is normally adequate for DV in laboratories where LEV is used as a primary method of control. To prevent the migration of nanomaterials into nearby rooms or areas, the workplace should be kept under negative pressure. However, ACGIH notes that the ACH is a poor basis for ventilation criteria in workplaces where hazards, heat, or odor controls are required.

A LEV system is characterized by its enclosing or exterior hood that intends to capture and remove contaminants in proximity or as close as possible to the emission source. Compared to exterior hoods, in particular, the enclosing hoods are designed to "contain" the contaminants inside the enclosure, operating under a negative pressure differential with respect to the worker's breathing zone. As a result, enclosures are more protective and should be considered whenever possible for working with nanomaterials. Examples of such enclosures include laboratory chemical hoods, glove boxes, biological safety cabinets (BSCs), and powder handling enclosures. For more details on the operation of the aforementioned enclosing hoods, readers are referred to the guidance provided by the NRC (NRC, 2011) and NIOSH (NIOSH, 2012). A brief summary of these two guidance documents is provided as follows.

Laboratory chemical hoods are the most common types of LEV systems in laboratory settings. They consist of an exhaust fan that draws the air into the hood, a moveable sash, airflow guiding slots, and a work surface. It is important to note that, although they look alike, the negative-pressure chemical hood should not be mistaken with the positive-pressure laminar flow, clean bench. Clean benches are those that blow HEPA filter-treated air into the operator's face with the intention of protecting the product instead of the worker, and, thus, should not be used to control exposures to nanomaterials or hazardous materials. In the case of chemical hoods, the entry of air into the hood needs to be smooth, uniform, and with an appropriate face velocity to ensure its proper performance. Depending on the supply of air distribution, required airflow ventilation rates are recommended in the range of 60–100 cubic feet per minute per square feet (cfm/ft^2) of open hood face (ACGIH, 2010). Higher ventilation rates are selected for hoods with poor air distributions, such as disturbance from traffic past hoods, or nearby ceiling-grille/diffuser-supplied air. Studies have shown that the escape of nanomaterials from chemical hoods can occur when the hood face velocity is below 80 feet per minute (fpm) or above 120 fpm (Tsai et al., 2009). A low velocity allows the supposedly exhaust air to exit the hood, whereas a high velocity causes a strong turbulent wake in between the worker and the hood face, both of which can carry or pull nanomaterials out of the hood. Conventional, constant-flow hoods, in particular, are prone to such problems if the sash is not positioned at a proper height. Better alternatives include the by-pass hood and the constant-velocity hood. Recently, a novel air–curtain chemical hood was developed under the support of the Taiwan Institute of Occupational Safety and Health (IOSH) (Huang et al., 2007) and shown to be effective in containing airborne nanoparticles inside the hood due to the downward air–curtain from the double-layer sash (Tsai et al., 2010). Finally, owing to the unknown hazards from nanoparticles, the exhaust air of the LEV should be filtered or scrubbed, whenever feasible, before venting outdoors and should not be recirculated indoors.

In addition to chemical hoods, glove boxes, BSCs, and powder handling enclosures could be effective at containing airborne nanomaterials when appropriate practices and operations are followed. A glove box enclosure provides a better means for isolation between the workers and nanomaterials and, hence, better protection. However, extra cautious steps must be given when transferring nanomaterials into, out of, and when cleaning the glove box. BSCs are designed specifically for the handling of microorganisms (e.g., virus and bacteria), which are in a range of sizes similar to nanomaterials. With a sophisticated HEPA-filtered airflow design, the most widely used and readily available Class II BSCs provide protection to not only the worker but also to the product and the environment. However, the complex airflow patterns inside the cabinet may cause the loss of nanomaterials. Powder handling enclosures are typically small, ventilated enclosures for handling small quantities of dry powders. Unlike chemical hoods, the relatively low airflow rate and velocity in the powder handling enclosure reduces the potential for the loss or escape of nanomaterials.

2.5.2 Administrative Controls

Administrative controls are a critical and integral part of control strategies, as they are crucial to the awareness, practices, and procedures adopted by workers to minimize exposure and complement engineering controls. On the other hand, a designated work area should be given to nanomaterial-related operations and processes. The designated area can be a laboratory, a specific area, or an enclosure with its entry posted with signs of potential hazards, PPE requirements, and access only to authorized personnel. To avoid the migration of nanomaterials outside the designated area, the use of walk-off pads, the creation of a buffer zone, and the installation of decontamination facilities are advisable. In addition, the number of nanomaterial workers and the duration of work should be reduced as much as is practical.

2.5.2.1 Employee Training

Employee training is arguably the most important element of administrative controls. It aims to communicate and educate workers, as well as their employers, on the presence of potential hazards, safe work procedures, good personal hygiene, regular housekeeping, the proper disposal of nanomaterials, and the use of PPE. Among them, hazard communication is the foundation of successful administrative controls, because both the workers and employers need to know exactly why precautious practices are required for working with nanomaterials, such that they are fully involved in the nanomaterial-related safety and health issues. Furthermore, the workers need to know the procedures to perform their work (e.g., transfer, label, storage, cleaning, and disposal) efficiently and in a way that minimizes exposure. As nanomaterial health and safety sciences are evolving rapidly, there is a need to regularly update training programs when new information is available.

2.5.2.2 Transfer, Labeling, and Storage

Nanomaterials that are classified as hazardous, or other suspected hazardous materials, should be packaged, marked, and labeled in accordance with applicable federal, state, and local standard procedures for its transfer and transport (e.g., 49 CFR Part 172 and Part 173). For those materials that are not classified as hazardous or are suspected to be hazardous, they may still pose health and safety risks if they are accidentally released during their transfer and transport. Therefore, the NRC recommends that shipments of any nanomaterials, regardless of whether they are hazardous or not, should be packaged using the equivalent of a Department of Transportation-certified packaging group I (PG I; great danger) container that is labeled properly (such as Figure 2.5). If the nanomaterials are in the form of dry, dispersible powders, the following line of text should be added to the label: "Nanomaterials can exhibit unusual reactivity and toxicity. Avoid breathing dust, ingestion, and skin contact" (NRC, 2011). For its transfer and storage, the container should be unbreakable and able to be sealed tightly. When appropriate, secondary containments or outer packages should be used.

2.5.2.3 Housekeeping

Good housekeeping should be practiced regularly where nanomaterials are handled, to avoid the accumulation and/or resuspension of nanomaterials. At a minimum of once every day or at the end of work shift, all working surfaces that are potentially contaminated with nanomaterials should be cleaned using a vacuum equipped with a HEPA and/or wet wiping methods (e.g., moistened disposable wipes). The vacuum cleaner should be labeled "For Use with Nanomaterials Only" and should be used solely for this purpose (Ellenbecker and Tsai, 2008), and it should be explosion-proof in the case of potential explosion hazard (Ostiguy, 2009). Never use compressed air or dry sweeping methods, as these can cause the resuspension of dry nanomaterials.

		\multicolumn{4}{c}{Exposure probability score}
		Extremely unlikely (0–25)	Less likely (26–50)	Likely (51–75)	Probable (76–100)
Hazard severity score	Very High (76–100)	RL 3	RL 3	RL 4	RL 4
	High (51–75)	RL 2	RL 2	RL 3	RL 4
	Medium (51–75)	RL 1	RL 1	RL 2	RL 3
	Low (76–100)	RL 1	RL 1	RL 1	RL 2

FIGURE 2.5 The four levels of risk in a 4 × 4 matrix, as a function of hazard severity score and exposure probability score.

2.5.2.4 Personal Hygiene

Good personal hygiene should also be practiced regularly where nanomaterials are handled to avoid the migration of nanomaterials out of the workplace. In addition to the required PPE described in the following section, hand-washing and showering facilities should be available and easily accessible for the decontamination of workers, particularly before drinking, eating, smoking, and leaving the workplace. To prevent carrying nanomaterials back home, it is necessary to provide locker rooms and laundry services for the storage and decontamination of work clothes. Do not eat, drink, or smoke in the areas where nanomaterials are handled.

2.5.3 PERSONAL PROTECTIVE EQUIPMENT

Exposure controls by means of PPE should be considered on a precautionary basis, despite the limited information on the effectiveness of PPE for the protection against nanoparticles. PPE include respirators, clothing, gloves, glasses, and other garments designed to protect the workers by means of filtration, shielding, or a clean air supply. PPE should be considered as a last resort, when engineering and administrative controls are unfeasible or unsatisfactory, to provide sufficient protection to the workers. Specific PPE requirements should conform to the OSHA regulation 29 CFR Part 1910 Subpart I.

2.5.3.1 Respirator

An evaluation of the airborne exposure to nanomaterials by IH professionals is necessary to determine whether or not respirators are required. As a general guideline, activities that involve agitating dry nanomaterials (e.g., cleaning, scooping, and pouring) or handling nanomaterials without effective control measures are at a high risk of inhalation exposure, and, thus, require respiratory protection. When respirators are required, the IH professionals are responsible for developing and implementing a respiratory protection program (RPP) that includes the following key elements: respiratory selection procedures, medical evaluation and fit testing, procedures and schedules for respirator cleaning and maintenance, employee training, and program effectiveness evaluations. For more information about the RPP, readers are referred to the NIOSH publications *Approaches to Safe Nanotechnology* (NIOSH, 2009a) and *Respirator Selection Logic* (Bollinger, 2004). A brief summary is provided below.

The respirator and, when applicable, cartridge should be selected according to the actual and acceptable exposure levels. Thus, the desired protection factor will increase when the acceptable exposure level is low and/or the actual exposure level is high. Ellenbecker and Tsai (2008) recommended that, if it is required, the respirator should be at minimum a half-mask, P100 cartridge-type respirator that has been properly fitted for the workers. In addition, disposable respirators with at least an N95 rating are acceptable for workers not required to have respirators. However, Ostiguy (2009) recommended against the use of negative-pressure respirators because a loose fit will result in the penetration of nanomaterials into the respirator. This highlights that employee training on the proper use of respirators is crucial in ensuring the effectiveness of the respirator.

Recently, the NIOSH has proposed recommended exposure limits (RELs) for airborne carbon nanotubes and nanofibers (CNT and CNF), and nano titanium dioxide (TiO_2) (NIOSH, 2010a, 2011b). With respect to 8-h, time-weighted average (TWA) mass concentrations, the REL is 7 µg/m^3 of elemental carbon for respirable CNT and CNF, whereas it is 0.3 mg/m^3 for nano-TiO_2. In workplaces where the estimated exposure level exceeds the above REL, an RPP may be necessary. OSHA shows that the assigned protection factor (APF) for fit-tested half/dust mask and elastomeric half mask is 10, loose-fitting or hood-powered air-purifying respirator is 25, and elastomeric full facepiece is 50 (OSHA, 2009a).

2.5.3.2 Protective Clothes, Gloves, and Other Equipment

The NIOSH has recommended that protective clothing normally required for wet chemistry laboratory would be appropriate for the handling of nanomaterials until new information is available, including close-toed shoes, long pants without cuffs, long-sleeved shirts, and laboratory coats.

Gauntlet-type or polymer (e.g., nitrile) gloves with extended sleeves are required when handling nanomaterials in their dry or liquid forms. It is important to note that the gloves should also be resistant to the liquid chemicals that hold the nanomaterials. In addition, the gloves should be changed frequently, as the penetration of and exposures to nanomaterials are not known to have good warning properties. Unlike nanomaterials, a skin burn sensation, such as from the exposure to strong acid, is a good warning sign that exposure had occurred.

Spectacle-type safety glasses, with side shields, face shields, chemical splash goggles, and other safety eyewear, should be worn appropriate to the type and level of hazard. Note that safety glasses and face shields do not provide sufficient protection against airborne, dry nanomaterials.

2.5.4 Fire and Explosion Control

Airborne particles can cause fires and are explosion hazards when their concentrations are high enough in the presence of oxygen and an ignition source. Nanomaterials have a substantially larger surface-to-volume ratio than that of a same quantity of larger particles, and may, thus, present a lower ignition energy, higher reactivity, combustion/catalytic potential, and combustion rate (NIOSH, 2009a; Ostiguy, 2010). Therefore, relatively inert materials could become highly reactive and combustible at the nanoscale. Many organic compounds, metal oxides, and some nonmetallic inorganic compounds are potentially combustible, while aluminum, magnesium, zirconium, and lithium are known to have high explosive potentials (Ostiguy, 2010). Recent studies show that the minimum ignition energy of some nanomaterials is lower than the same material at a microscale. In particular, metallic nanomaterials may display pyrophoric characteristics (e.g., aluminum begins to burn on contact with air). However, several common nanomaterials, such as aluminum, iron, zinc, copper, and several carbon nanomaterials, have been shown to have explosive characteristics (maximum explosion pressure and rate of pressure rise) similar to conventional, microscale powders (HSE, 2010; Steinkrauss et al., 2010). As a result, the authors concluded that nanomaterials may be more prone to ignition, but, once ignited, the explosion violence is no more severe than microscale powders. In general, the potential and severity of explosions increases with an increasing amount of combustible nanomaterials. As a result, the risk of explosions at typical, laboratory-scale research is smaller than that at pilot plants or full-scale production plants (NIOSH, 2012). According to Ellenbecker and Tsai (2008), laboratory scale refers to working with substances in which the containers used for reactions, transfers, and other handling of substances are designed to be easily and safely manipulated by one person. Storage of nanomaterials, however, potentially presents a case where the quantity may be large and thus requires special attention to fire and explosion hazards. Nevertheless, any nanomaterial-related activities should be carried out in ways that minimize the release and airborne resuspension and in an environment that is isolated from an ignition source, which could be an electrical, thermal, electrostatic, mechanical, or chemical source.

2.5.5 Management of Nanomaterial Spills and Waste

Cleanup procedures and kits should be developed and prepared in the case of nanomaterial spills. In the case of significant spills (e.g., more than a few grams), it is prudent to activate emergency response procedures. The key elements in the cleanup procedures include PPE (for minimizing the inhalation and dermal exposures), containment, restricted access, and the removal of nanomaterial spills. In addition, considerations should be given to potential complications due to the reactions and compatibility between nanomaterials and cleaning materials; for example, inside the vacuum cleaner filter and canister. Nanomaterial spills should be promptly contained by barricade tapes or other barriers, walk-off mats placed at the entry points, and by reducing the amount of air currents passing over the surface of the spill area. Dry nanomaterials can be cleaned up using HEPA-equipped vacuums and/or wet cleaning methods (e.g., moistened disposable wipes or humidification of dry nanomaterials). Liquid spills containing nanomaterials can be cleaned up using absorbent mats and liquid traps. These methods aim to minimize the personal exposure and to reduce the airborne resuspension and transport of nanomaterials outside the spill area. A typical nanomaterial spill kit includes the following items (NIOSH, 2012): (1) barricade tape, (2) nitrile or other chemically impervious gloves, (3) elastomeric respirator with appropriate filters, (4) absorbent materials, (5) wipes, (6) sealable plastic bags, (7) walk-off mats, (8) HEPA-equipped vacuums, and (9) a spray bottle with deionized water or other appropriate liquid. Any waste resulting from the cleanup should be managed and disposed in accordance with the laboratory's hazardous waste procedures.

Similarly, as only a little is known about the environmental risks, it would be prudent to utilize all available knowledge and resources (e.g., IH professionals and applicable regulations) to manage and dispose wastes containing nanomaterials in a safe and environmentally friendly manner. Until new information is available, it is appropriate to consider nanomaterials, and any other materials or objects that come into contact with them, as hazardous wastes. In Switzerland, nanomaterials are considered as hazardous substances, owing to their explosive and pyrogenic properties (Steinkrauss et al., 2010). In the United States, EPA regulations place the burden of determining a waste hazardous or not, and in what hazard classification, on the waste generator. This responsibility, in practice, has commonly been carried out by IH professionals or their waste disposal firms. Nevertheless, there are some guidelines provided by the Department of Energy (DOE), British Standards Institution, and NRC, given as follows. In the process of a hazard evaluation, it should be noted that nanomaterials often have different reactivities than bulk materials, and, thus, one shall not rely solely upon the bulk material properties for nanomaterial properties. For liquid wastes, the hazards from both the liquid and the nanomaterials should be considered. The DOE recommends collecting nanomaterial wastes in a sealable plastic bag or container under appropriate ventilation controls. When it is full, double-bag the waste and label it with "contains nanomaterials" and any other particular hazard notes. In addition, notify the hazardous waste handlers that nanomaterials are in the wastes.

A comprehensive list of PPE and materials for use in work and cleanup with information on brand/supplier names, specifications, and costs has been researched and compiled by Adeleye et al. (2011). It is therefore possible to evaluate the economic implications of implementing specific sets of control measures described earlier.

2.5.6 Occupational Health Surveillance

Occupational health surveillance involves the ongoing systematic collection, analysis, and dissemination of exposure and health data on groups of workers for the purpose of preventing illnesses and injury (Halperin, 1996). It broadly includes both hazardous and medical surveillance. As an integral part of an effective safety and health program, the NIOSH has continually recommended implementing occupational health surveillance when workers are exposed to potentially hazardous materials. Hazard surveillance involves identifying potentially hazardous workplace practices or exposures, and assessing the extent to which they can be linked to workers, the effectiveness of controls, and the reliability of exposure measures (NIOSH, 2009b). At present, in the absence of adequate health information, hazard surveillance with a preventive focus is particularly relevant to the production and handling of nanomaterials in order to establish prudent measures for controlling its exposure.

Medical surveillance, on the other hand, involves monitoring actual health events or testing changes in a biologic function of "exposed" workers. One of its major purposes is to detect early signs of work-related illness and disease before medical treatment, also known as medical screening. As a result, medical surveillance should be considered as a second line of defense, behind engineering and administrative controls and the use of PPE. Other benefits from medical surveillance include providing baseline information with regard to work accommodation, acclimatization, and potential adverse health effects, which were not known earlier because of limited knowledge. OSHA has a number of standards that include the requirements for medical surveillance (OSHA, 2009b), while the NIOSH also has recommended the medical surveillance of workers exposed to certain occupational hazards (NIOSH, 2009b). It is noted that none of the above standards and recommendations has specifically addressed nanomaterials. Only recently, the NIOSH has provided an interim guidance for medical screening and hazard surveillance for workers potentially exposed to engineered nanoparticles, with their recommendations as follows:

- Take prudent measures to control exposures to engineered nanoparticles
- Conduct hazard surveillances as the basis for implementing controls
- Continue the use of established medical surveillance approaches

Therefore, it is prudent to conduct medical surveillances of workers exposed to nanomaterials composed of chemicals for which there are existing OSHA standards or NIOSH recommendations (NIOSH, 2009b).

2.6 A QUALITATIVE RISK-BASED CONTROL AND MANAGEMENT STRATEGY: CONTROL BANDING

Control banding (CB) is a qualitative risk assessment strategy that determines workplace risks and subsequently offers control measures appropriate for the determined level of risk. The original CB strategy was developed in the 1980s by IH professionals in the pharmaceutical industry to manage risks from exposures to a large number of new chemical compounds without firm toxicological and exposure data (Zalk and Nelson, 2008; NIOSH, 2009c). The uncertainties faced by the pharmaceutical industry resemble those encountered at the early developmental stage of nanotechnology. CB is based on the assumptions that the level of risk is a function of the severity of hazard and the probability of exposure, and there are a limited number of control measures available, despite the numerous hazards that exist. Typically, the hazards and exposures are stratified into 2–5 different levels, also known as "bands." The two bands are then combined together, resulting into risk-stratified control bands, in which the degree of control increases with an increasing risk. In essence, in a context of uncertainty, CB intends to focus limited resources on exposure controls and describes how strictly a risk needs to be managed (NIOSH, 2010b). With that in mind, CB strategies in recent years have gained international attention for its relative robust, ease-of-use framework and solution-oriented approach.

Currently, there are several CB tools available to advise on the control of nanomaterial exposures, such as CB Nanotool (Paik et al., 2008; Zalk et al., 2009), Precautionary Matrix for Synthetic Nanomaterials (http://www.nanotechnologie.admin.ch; Höck et al., 2008), ANSES's CB tool for nanomaterials (http://www.anses.fr; ANSES, 2010; Riediker et al., 2012), Stoffenmanager Nano (http://nano.stoffenmanager.nl; van Duuren-Stuurman et al., 2012), Guidance on Working Safely with Nanomaterials and Nanoproducts (Cornelissen et al., 2011), and GoodNanoGuide (http://goodnanoguide.org). A commentary article by Brouwer (2012) has compared and described the similarities and differences among several CB tools. Below, we introduce the CB Nanotool developed and evaluated by Paik et al. (2008) and Zalk et al. (2009), and implemented at the Lawrence Livermore National Laboratory as the required risk assessment approach for all work with nanomaterials. This brief introduction is intended to serve as guidance on developing control and management strategies.

In the CB Nanotool, the control band of a specific operation is based on the overall risk level (RL) determined for that operation. The RL is the result of a combination of the hazard severity score and the exposure probability score. The factors that determine the severity score are selected to take into account the physicochemical (40 points) and toxicological (30 points) characteristics of the nanomaterial, and the toxicological characteristics (30 points) of their parent material. They include the nanomaterial's surface chemistry, shape, diameter, solubility, the nanomaterial's as well as their parent material's dermal toxicity, carcinogenicity, mutagenicity, asthmagen, reproductive toxicity, and OEL. The factors that determine the probability score include the estimated amount of chemical use (25 points), the dustiness/mistiness (30 points) of the nanomaterial, the number of workers with similar exposures, and the frequency and duration of operation (45 points). If a factor is determined as "unknown," 75% of the highest point value would be assigned for that factor. The two overall scores each have a maximum point of 100, stratified across four levels. The hazard severity score and exposure probability score result in a 4 × 4 matrix with its intersections forming bands of RL, as shown in Figure 2.5.

The four levels of risk are then linked to the four corresponding control bands as follows:

RL 1: General ventilation
RL 2: Fume hoods or local exhaust ventilation
RL 3: Containment
RL 4: Seek specialist advice

As shown above, the CB Nanotool needs rather comprehensive information for hazard banding, and thus requires the involvement of IH professionals. A spreadsheet of CB Nanotool 2.0, along with instructions for scoring and a Nanomaterial Information Form for data collection is available at http://www.controlbanding.net.

In light of numerous uncertainties, the CB Nanotool and other CB tools provide a relatively simple yet efficient instrument that can facilitate risk management. As with any qualitative and quantitative risk assessment methods, there are limitations to the CB approach. For example, the selected hazard and exposure factors, and the scoring of these factors, only reflect what is known at the time of the development of the model. In addition, there is currently very limited toxicological data from which to determine the appropriate control levels. As a result, the CB should be implemented with some degree of caution, and continually be refined when new information on nanomaterial toxicity and exposure data and more validation studies become available. Finally, it is important to recognize that CB is *not* a replacement for IH professionals, nor does it eliminate the need for quantitative exposure assessments.

2.7 CONCLUSIONS

The field of nanotechnology is rapidly evolving and expanding and, thus, the number of workers that could potentially be exposed to a wide variety of nanomaterials will increase. The unique physiochemical properties of some of these nanomaterials are shown to pose safety and health risks that are not anticipated from their bulk materials. Until new information on their toxicity and exposure become available, it is at present necessary to take precautionary measures to minimize workers' exposures to hazardous nanomaterials. Prudent practices and preventive measures should be implemented according to the traditional IH hierarchy of control in the following order whenever possible: PtD, engineering controls, administrative controls, the use of PPEs, and occupational health surveillance.

In the face of numerous uncertainties, the CB strategy emerges as a simple yet efficient approach to directing limited resources to areas where it is needed the most, that is, exposure control. It offers desired control advices and solutions based on the level of risk determined by bands of hazard severities and exposure probabilities. It is important to recognize that CB is intended as a complement to IH professions, not a replacement, and it is particularly useful when limited knowledge is known about the hazard and exposure. With that being said, the effectiveness of CB builds upon up-to-date knowledge as well as sound IH principles. As a result, the CB should be implemented with some degree of caution, and continually be refined when new information becomes available.

REFERENCES

ACGIH (American Conference of Industrial Hygienists). 2010. *Industrial Ventilation: A Manual of Recommended Practice*, 27th ed., Cincinnati, OH: ACGIH, USA.
Adeleye, A., Huang, D., Layton, Z., Paladugu, S., Twining, J. 2011. CERNS: A condensed environmental health & safety reference for nanotechnology startups. Master's thesis, University of California-Santa Barbara, CA, USA.

ANSES (French Agency for Environmental and Occupational Health Safety). 2010. *Development of a Control Banding Tool for Nano Materials*. France: ANSES. Request no. 2008-SA-0407.

Behrens, A., Brandner, S., Genoud, N., Aguzzi, A. 2001. Normal neurogenesis and scrapie pathogenesis in neural grafts lacking the prion protein homologue Doppel. *EMBO Rep.* 2:347–352.

Bello, D., Hart, A.J., Ahn, K., Hallock, M., Yamamoto, N., Garcia, E.J., Ellenbecker, M.J., Wardle, B.L. 2008. Particle exposure levels during CVD growth and subsequent handling of vertically-aligned carbon nanotube films. *Carbon*. 46:974–981.

Bello, D., Wardie, B.L., Yamamoto, N., deVilloria, R.G., Garcia, E.J., Hart, A.J., Ahn, K., Ellenbecker, M.J., Hallock, M. 2009. Exposure to nanoscale particles and fibers during machining of hybrid advanced composites containing carbon nanotubes. *J. Nanopart. Res.* 5:231–249.

Bollinger, N. 2004. *NIOSH Respirator Selection Logic*. Cincinnati, OH: U.S. DHHS, CDC, NIOSH. DHHS (NIOSH) Publication No. 2005-100.

Brouwer, D.H. 2012. Control banding approaches for nanomaterials. *Ann. Occup. Hyg.* 56:506–514.

Brown, J.S., Zeman, K.L., Bennett, W.D. 2002. Ultrafine particle deposition and clearance in the healthy and obstructed lung. *Am. J. Resp. Crit. Care Med.* 166:1240–1247.

Cheng, Y.H., Chao, Y.C., Wu, C.H., Tsai, C.J., Uang, S.N., Shih, T.S. 2008. Measurements of ultrafine particle concentrations and size distribution in an iron foundry. *J. Hazard. Mater.* 158:124–130.

COMEST (World Commission on the Ethics of Scientific Knowledge and Technology). 2005. *The Precautionary Principle*. Paris, France: United Nations Educational, Scientific and Cultural Organization.

Cornelissen, R., Jongeneelen, F., van Broekhuizen, P., van Broekhuizen, F. 2011. *Guidance Working Safely with Nanomaterials and Nanoproducts*. Amsterdam, Netherlands: IVAM.

Dasch, J., D'Arcy, J. 2008. Physical and chemical characterization of airborne particles from welding operations in automotive plants. *J. Occup. Environ. Hyg.* 5:444–454.

Demou, E., Peter, P., Hellweg, S. 2008. Exposure to manufactured nanostructured particles in an industrial pilot plant. *Ann. Occup. Hyg.* 52:695–706.

Dockery, D.W., Pope, C.A., Xu, X., Spengler, J.D., Ware, J.H., Fay, M.E. 1993. An association between air pollution and mortality in six U.S. cities. *N. Engl. J. Med.* 329:1753–1759.

Donaldson, K., Brown, D., Clouter, A., Duffin, R., MacNee, W., Renwick, L., Tran, L., Stone, V. 2002. The pulmonary toxicology of ultrafine particles. *J. Aerosol. Med.* 15:213–220.

Ellenbecker, M., Tsai, S.-J.C. 2008. *Interim Best Practices for Working with Nanoparticles*. Lowell, MA: Center for High-Rate Nanomanufacturing, University of Massachusetts-Lowell (rev.1).

Elihn, K., Berg, P. 2009. Ultrafine particle characteristics in seven industrial plants. *Ann. Occup. Hyg.* 53:475–484.

Evans, D.E., Heitbrink, W.A., Slavin, T.J., Peters, T.M. 2008. Ultrafine and Respirable particles in an automotive grey iron foundry. *Ann. Occup. Hyg.* 52:9–21.

Fissan, H., Trampe, A., Neunman, S., Pui, D.Y.H., Shin, W.G. 2007. Rationale and principle of an instrument measuring lung deposition area. *J. Nanopart. Res.* 9:53–59.

Fujitani, Y., Kobayashi, T., Arashidani, K., Kunugita, N., Suemura, K. 2008. Measurement of physical properties of aerosols in a fullerene factory for inhalation exposure assessment. *J. Occup. Environ. Hyg.* 5:380–389.

Han, J.H., Lee, E.J., Lee, J.H., So, K.P., Lee, Y.H., Bae, G.N., Lee, S.-B., Ji, J.H., Cho, M.H., Yu, I.J. 2008. Monitoring multiwalled carbon nanotube exposure in carbon nanotube research facility. *Inhal. Toxicol.* 20:741–749.

Halperin, W.E. 1996. The role of surveillance in the hierarchy of prevention. *Am. J. Ind. Med.* 29:321–323.

Hinds, W.C. 1999. Condensation and evaporation, in: *Aerosol Technology Properties, Behavior, and Measurement of Airborne Particles*. 2nd ed., New York, USA: John Wiley and Sons, Inc, 278–303.

Höck, J., Hofmann, H., Krug, H., Lorenz, C. 2008. *Guidelines on the Precautionary Matrix for Synthetic Nanomaterials*. Berne, Switzerland: Federal Office of Public Health and Federal Office for the Environment, 2011 Version 2.1.

HSE (Health and Safety Executive). 2010. *Fire and Explosion Properties of Nanopowders*. UK: Health and Safety Executive. Research Report RR782.

Huang, R.F., Wu, Y.D., Chen, H.D., Chen, C.C., Chen, C.W., Chang, C.P., Shih, T.S. 2007. Development and evaluation of an air-curtain fume cabinet with considerations of its aerodynamics. *Ann. Occup. Hyg.* 51:189–206.

ICRP (International Commission on Radiological Protection) 1994. *Human Respiratory Tract Model for Radiological Protection*, Publication 66, Annals of ICRP. London, UK: Oxford, Pergamon.

ISO. 2007. *Workplace Atmospheres-Ultrafine, Nanoparticle and Nano-Structured Aerosols-Inhalation Exposure Characterization and Assessment*. ISO/TR 27628. Geneva, Switzerland: International Organization for Standardization.

Kreyling, W.G., Semmler, M., Erbe, F., Mayer, P., Takenaka, S., Schulz, H. 2002. Translocation of ultrafine insoluble iridium particles from lung epithelium to extrapulmonary organs is size dependent but very low. *J. Toxicol. Environ. Health.* 65:1513–1530.

Kreyling, W.G., Semmler, M., Moller, W. 2006. Health implications of nanoparticles. *J. Nanopart. Res.* 8:543–562.

Kuhlbusch, T.A.J., Asbach, C., Fissan, H., Göhler, D., Stintz, M. 2011. Nanoparticle exposure at nanotechnology workplaces: A review. *Part. Fibre Toxicol.* 8:22.

Kuhlbusch, T.A.J., Neumann, S., Fissan, H. 2004. Number size distribution, mass concentration, and particle composition of PM1, PM2.5, and PM10 in bag filling areas of carbon black production. *J. Occup. Environ. Hyg.* 1:660–671.

Kuhlbusch, T.A.J., Neumann, S., Fissan, H. 2006. Particle characteristics in the reactor and pelletizing areas of carbon black production. *J. Occup. Environ. Hyg.* 3:558–567.

Lademann, J., Weigmann, H.J., Rickmeyer, C., Barthelmes, H., Schaefer, H., Mueller, G. et al. 1999. Penetration of titanium dioxide microparticles in a sunscreen formulation into the horny layer and the follicular orifice. *Skin Pharmacol. Appl. Skin Physiol.* 12:247–256.

Li, L., Chen, D.R., Tsai, P.J. 2009a. Use of an electrical aerosol detector (EAD) for nanoparticle size distribution measurement. *J. Nanopart. Res.* 11:111–120.

Li, L., Chen, D.R., Tsai, P.J. 2009b. Evaluation of an electrical aerosol detector (EAD) for the aerosol integral parameter measurement. *J. Electrostatics* 67:765–773.

Maynard, A.D. 2003. Estimating aerosol surface area for number and mass concentration measurements. *Ann. Occup. Hyg.* 42:123–144.

Maynard, A.D., Aitken, R.J. 2007. Assessing exposure to airborne nanomaterials: Current abilities and future requirements. *Nanotoxicology* 1:26–41.

McCawley, M.A., Kent, M.S., Berakis, M.T. 2001. Ultrafine beryllium number concentration as a possible metric for chronic beryllium disease risk. *Appl. Occup. Environ. Hyg.* 16:631–638.

Methner, M., Hodson, L., Dames, A., Geraci, C. 2010. Nanoparticle emission assessment technique (NEAT) Part B. *J. Occup. Environ. Hyg.* 7:163–176.

Methner, M.M. 2008. Effectiveness of local exhaust ventilation (LEV) in controlling engineered nanomaterial emissions during reactor cleanout operations. *J. Occup. Environ. Hyg.* 5:D63–D69.

Möhlmann, C. 2005. Vorkommen ultrafeiner Aerosole an Ar beitsplätzen. *Gefahrstoffe-Reinhaltung der Luft.* 65: 469–471.

Monteiro-Riviere, N.A., Ryman-Rasmussen, J.P. 2009. Mechanisms of quantum dot nanoparticle cellular uptake. *Toxicol. Sci.* 110:138–155.

NIOSH (National Institute for Occupational Safety and Health). 2009a. *Approaches to Safe Nanotechnology: Managing the Health and Safety Concerns Associated with Engineered Nanomaterials.* Cincinnati, OH: U.S. Department of Health and Human Services (DHHS), Centers for Disease Control and Prevention (CDC), NIOSH. DHHS (NIOSH) Publication No. 200–25.

NIOSH. 2009b. *Current Intelligence Bulletin 60: Interim Guidance for Medical Screening and Hazard Surveillance for Workers Potentially Exposed to Engineered Nanoparticles.* Cincinnati, OH: U.S. DHHS, CDC, NIOSH. DHHS (NIOSH) Publication No. 2009–116.

NIOSH. 2009c. *Qualitative Risk Characterization and Management of Occupational Hazards: Control Banding (CB).* Cincinnati, OH: U.S. DHHS, CDC, NIOSH. DHHS (NIOSH) Publication No. 200–52.

NIOSH. 2010a. (Draft) *NIOSH Current Intelligence Bulletin: Occupational Exposure to Carbon Nanotubes and Carbon Nanofibers.* Cincinnati, OH: U.S. DHHS, CDC, NIOSH. http://www.cdc.gov/niosh/topics/ptd (accessed March 1, 2013).

NIOSH. 2010b. *Control Banding.* Cincinnati, OH: U.S. DHHS, CDC, NIOSH. http://www.cdc.gov/niosh/topics/ctrlbanding (accessed March 1, 2013).

NIOSH. 2011a. *Prevention through Design.* Cincinnati, OH: U.S. DHHS, CDC, NIOSH. http://www.cdc.gov/niosh/topics/ptd (accessed March 1, 2013).

NIOSH, 2011b. *Current Intelligence Bulletin 63: Occupational Exposure to Titanium Dioxide.* Cincinnati, OH: U.S. DHHS, CDC, NIOSH. DHHS (NIOSH) Publication No. 2011–160.

NIOSH, 2012. *General Safe Practices for Working with Engineered Nanomaterials in Research Laboratories.* Cincinnati, OH: U.S. DHHS, CDC, NIOSH. DHHS (NIOSH) Pub. No. 2012–147.

NRC (National Research Council). 2011. *Prudent Practices in the Laboratory: Handling and Management of Chemical Hazards, Updated Version.* Washington, DC: The National Academies Press.

Oberdörster, E. 2004. Manufactured nanomaterials (Fullerenes, C60) induce oxidative stress in the brain of juvenile largemouth bass. *Environ. Health Perspect.* 112:1058–1062.

Oberdörster, G. 2000. Toxicology of ultrafine particles: *In vivo* studies. *Philos. Trans. R. Soc. Lond. A.* 358:2719–2740.

Oberdörster, G. 2001. Pulmonary effects of inhaled ultrafine particles. *Int. Arch. Occup. Environ. Health* 74: 1–8.

Oberdörster, G., Oberdörster, E., Oberdörster, J. 2005. Nanotoxicology: An emerging discipline evolving from studies of ultrafine particles. *Environ. Health Perspect.* 113:823–839.

OSHA (Occupational Safety and Health Administration). 2009a. *Screening and Surveillance: A Guide to OSHA Standards.* Washington, DC: OSHA, Department of Labor. OSHA 3162–12R.

OSHA. 2009b. *Assigned Protection Factors for the Revised Respiratory Protection Standard.* Washington, DC: OSHA, Department of Labor. OSHA 3352-02.

Ostiguy, C., Roberge, B., Ménard, L., Endo, C.-A. 2009. *Best Practices Guide to Synthetic Nanoparticle Risk Management.* Montréal, Québec, Canada: IRSST. Report R-599.

Ostiguy, C., Roberge, B., Woods, C., Soucy, B. 2010. *Engineered Nanoparticles: Current Knowledge about Occupational Health and Safety Risks and Prevention Measures.* Montréal, Québec, Canada: IRSST. Report R-656.

Paik, S.Y., Zalk, D.M., Swuste, P. 2008. Application of a pilot control banding tool for risk level assessment and control of nanoparticle exposures. *Ann. Occup. Hyg.* 52:419–428.

Park, J., Kwak, B.K., Bae, E., Lee, J., Kim, Y., Choi, K., Yi, J. 2009. Characterization of exposure to silver nanoparticles in a manufacturing facility. *J. Nanopart. Res.* 11:1705–1712.

Park, J.Y., Ramachandran, G., Raynor, P.C., Olson, Jr., G.M. 2010. Determination of particle concentration rankings by spatial mapping of particle surface area, number, and mass concentrations in a restaurant and a die casting plant. *J. Occup. Environ. Hyg.* 7:466–476.

Peters, T., Elzey, S., Johnson, R., Park, H., Grassian, V., Maher, T., O'Shaughnessy, P. 2009. Airborne monitoring to distinguish engineered nanomaterials from incidental particles for environmental health and safety. *J. Occup. Environ. Hyg.* 6:73–81.

Pflücker, F., Wendel, V., Hohenberg, H., Gartner, E., Will, T., Pfeiffer, S., Wepf, R., Gers-Barlag, H. 2001. The human stratum corneum layer: An effective barrier against dermal uptake of different forms of topically applied micronised titanium dioxide. *Skin Pharmacol. Appl. Skin Physiol.* 14:92–97.

Plitzko, S. 2009. Workplace exposure to engineered nanoparticles. *Inhal. Toxicol.* 21:25–29.

Plog, B.A., Quinlan, P.J., Villarreal, J. 2012. *Fundamentals of Industrial Hygiene*, 6th ed., USA: National Safety Council.

Ramachandran, G., Paulsen, D., Watts, W., Kittelson, D. 2005. Mass, surface area and number metrics in diesel occupational exposure assessment. *J. Environ. Monit.* 728–735.

Riediker, M., Ostiguy, C., Triolet, J., Troisfontaine, P., Vernez, D., Bourdel, G., Thieriet, N., Cadène, A. 2012. Development of a control banding tool for nanomaterials. *J. Nanomaterials.* 1:1–8.

Ryman-Rasmussen, J.P., Riviere, J.E., Monteiro-Riviere, N.A. 2006. Penetration of intact skin by quantum dots with diverse physicochemical properties. *Toxicol. Sci.* 91:159–165.

Steinkrauss, M., Fierz, H., Lerena, P., Suter, G. 2010. *Fire and Explosion Properties of Synthetic Nanomaterials.* Berne, Switzerland: Federal Office for the Environment (FOEN). Environmental studies no. 1011.

Stephenson, D., Seshadri, G., Veranth, J.M. 2003. Workplace exposure to submicron particle mass and number concentrations from manual arc welding of carbon steel. *AIHA J.* 64:516–521.

Tinkle, S.S., Antonini, J.M., Rich, B.A., Roberts, J.R., Salmen, R., DePree, K., Adkins, E.J. 2003. Skin as a route of exposure and sensitization in chronic beryllium disease. *Environ. Health Perspect.* 111:1202–1208.

Tran, C.L., Buchanan, D., Cullen, R.T., Searl, A., Jones, A.D., Donaldson, K. 2000. Inhalation of poorly soluble particles. II. Influence of particle surface area on inflammation and clearance. *Inhal. Toxicol.* 12:1113–1126.

Tsai, C.J., Liu, C.N., Hung, S.M., Chen, S.C., Uang, S.N., Cheng, Y.S., Zhou, Y. 2012. Novel active personal nanoparticle sampler for the exposure assessment of nanoparticles in workplaces. *Environ. Sci. Technol.* 46:4546–4552.

Tsai, S.J., Ada, E., Isaacs, J., Ellenbecker, M.J. 2009. Airborne nanoparticle exposures associated with the manual handling of nanoaluminia and nanosilver in fume hoods. *J. Nanopart. Res.* 11:147–161.

Tsai, S.J., Huang, R.F., Ellenbecker, M.J. 2010. Airborne nanoparticle exposures while using constant-flow, constant-velocity, and air-curtain-isolated fume hoods. *Ann. Occup. Hyg.* 54:78–87.

Van Duuren-Stuurman, B., Vink, S.R., Verbist, K.J.M., Heussen, H.G.A., Brouwer, D.H., Kroese, D.E.D., Tielemans, E., Fransman, W. 2012. Stoffenmanager nano version 1.0: A web-based tool for risk prioritization of airborne manufactured nano objects. *Ann. Occup. Hyg.* 56:525–541.

Wang, Y.F., Tsai, P.J., Chen, C.W., Chen, D.R., Hsu, D.J. 2010. Using a modified electrical aerosol detector to predict nanoparticle exposure to different regions of the respiratory tract for workers in carbon black manufacturing industry. *Environ. Sci. Technol.* 44:6767–6774.

Wang, Y.F., Tsai, P.J., Chen, C.W., Chen, D.R., Dai, Y.T. 2011. Size distributions and exposure concentrations of nanoparticles associated with the emissions of oil mists from fastener manufacturing processes. *J. Hazard. Mater.* 198:182–187.

Wake, D. 2001. Ultrafine particles in the workplace. HSL Report number ECO/00/18.

Wegner, K., Schimmoeller, B., Thiebaut, B., Fernandez, C.N., Rao, T. 2011. Pilot plants for industrial nanoparticle production by flame spray pyrolysis. *KONA Powder Part. J.* 29:251–263.

Yeganeh, B., Kull, C.M., Hull, M.S., Marr, L.C. 2008. Characterization of airborne particles during production of carbonaceous nanomaterials. *Environ. Sci. Technol.* 42:4600–4606.

Zalk, D.M., Nelson, D.I. 2008. History and evolution of control banding: A review. *J. Occup. Environ. Hyg.* 5:330–346.

Zalk, D.M., Paik, S.Y., Swuste, P. 2009. Evaluating the control banding nanotool: A qualitative risk assessment method for controlling nanoparticle exposures. *J. Nanoparticle Res.* 11:1685–1704.

Zimmer, A.T., Biswas, P. 2001. Characterization of the aerosols resulting from arc welding processes. *J. Aerosol Sci.* 32:993–1008.

3 Physicochemical Characterization–Dependent Toxicity of Nanoparticles

Jigar N. Shah, Ankur P. Shah, Hiral J. Shah, and Vijaykumar B. Sutariya

CONTENTS

3.1 Nanotechnology: Introduction 73
    3.1.1 Advantages of Nanosystems 75
    3.1.2 Limitations of Nanosystems 76
3.2 Nanoparticles: Importance in Today's Era 76
    3.2.1 Nanomedicine: Global Scenario 77
    3.2.2 Nanomedicine: Marketed Products 78
3.3 Nanosystems: Different Types and Applications 78
3.4 Nanoparticles: Physicochemical Characteristics (Properties) and Their Multidimensional Issues 80
3.5 Nanoparticles: Physicochemical Characterization-Dependent Toxicity 84
    3.5.1 Structural Properties-Dependent Toxicity 86
        3.5.1.1 Particle Shape and Aspect Ratio 86
        3.5.1.2 Structural Arrangement (Crystalline Form) 86
        3.5.1.3 Surface Chemistry and Charge 87
    3.5.2 Dose-Dependent Toxicity 87
    3.5.3 Concentration and Drug–Loading-Dependent Toxicity 87
    3.5.4 Size, Size Distribution and Surface Area-Dependent Toxicity 88
    3.5.5 Material Composition-Dependent Toxicity 91
    3.5.6 Bio-Persistence-Dependent Toxicity 91
    3.5.7 Cellular Uptake and Binding-Dependent Toxicity 94
    3.5.8 Surface Coatings-Dependent Nanotoxicity 95
3.6 Nanoparticles Toxicity Comparison Studies 95
3.7 Conclusion 95
3.8 Future Perspectives 96
References 97

3.1 NANOTECHNOLOGY: INTRODUCTION

Nanotechnology is a multidomain system that includes a vast and diverse array of devices from engineering, physics, chemistry, and biology. The growing field of nanotechnology, along with rapid advances in science and technology, is undergoing an explosive advancement and has opened up new opportunities for further development of medical science by identifying new and effective medical treatments for human health care. The combination of scientific and engineering principles is involved in the design, synthesis, development, characterization, and application of materials and devices in a complete, functional, and organizational manner to establish one of the various dimensions in a nanometer scale (i.e., 1 billionth of a meter) (Emerich and Thanos 2003; Sahoo and Labhasetwar 2003; Williams 2004; Matsudai and Hunt 2005; Moghimi et al. 2005; Shaffer 2005; Emerich 2005; Chan 2006; Cheng et al. 2006; Yoshikawa et al. 2006). The applications of nanotechnology in medicine and physiology employ materials and devices designed to interact with the body at subcellular levels with a high degree of specificity. This can be potentially translated into targeted cellular and tissue-specific clinical applications designed to achieve maximal therapeutic efficacies with minimal side effects (Sahoo et al. 2007).

The word "nanoparticle" in drug targeting was conceptualized in the late 1960s and early 1970s by Paul Ehrlich, a physician with a great interest in bacteriology and immunology. This field was given special attention in the years between 1970 and the early 1980s. Further developments focused on especially interesting improvements, such as nanoparticles for the delivery of drugs across the blood–brain barrier (BBB) and PEGylated (i.e., poly(ethylene glycol)) nanoparticles with a prolonged blood circulation time. The first commercial nanoparticle product containing a drug (Abraxane™, human serum albumin nanoparticles containing paclitaxel) appeared on the market in the beginning of 2005 (Kreuter 2007).

Nanoparticles are particulate dispersions or solid particles with a size range of 1–1000 nm (Figure 3.1). The 1–100 nm scale is of a particular interest for biological interfaces. For example, particles less than 12 nm in diameter may cross the blood–brain barrier (Oberdörster et al. 2004; Sarin et al. 2008; Sonavane et al. 2008), and objects of 30 nm or less can be endocytosed by cells (Conner and Schmid 2003). Based on the methods of preparation, different nanoparticulate systems such as nanoparticles, nanospheres, or nanocapsules can be obtained. Nanocapsules are systems in which the drug is confined to a cavity surrounded by a unique, polymeric membrane, while nanospheres are matrix systems in which the drug is physically and uniformly dispersed. In recent years, biodegradable, polymeric nanoparticles, particularly those coated with hydrophilic polymer such as PEG, have been used as potential drug delivery devices because of their ability to circulate for a prolonged period of time; target a particular organ; as carriers of DNA in gene therapy; and their ability to deliver proteins, peptides, and genes (Langer 2002; Bhadra et al. 2002; Kommareddy et al. 2005; Lee and Kim 2005).

The growth of nanotechnology has led to many nanoparticle-containing consumer products appearing in our daily lives. The current markets suggest that there are more than a thousand products or product lines available worldwide based on nanotechnology principles (Rejeski 2009). Moreover, the presence of nanoparticles in these products has also become more widespread. Therefore, it has become necessary to evaluate the safety of nanoparticles in terms of their exposure to their makers and users. Each and every living organism on earth has been exposed to nanometer-sized foreign particles; we inhale them with every breath, consume them with every drink, and continuously encounter nanometer-sized entities. The majority of nanoparticles do not cause ill effects and go unnoticed but, occasionally, an intruder will cause appreciable harm to the organism.

The nanoparticle's effect on human health and environmental conditions remains unclear (Colvin 2003; Maynard et al. 2006; Nel et al. 2006; Helmus 2007). This issue has become highlighted due to a large number of scientific reports and published studies. It becomes necessary to know the mechanisms

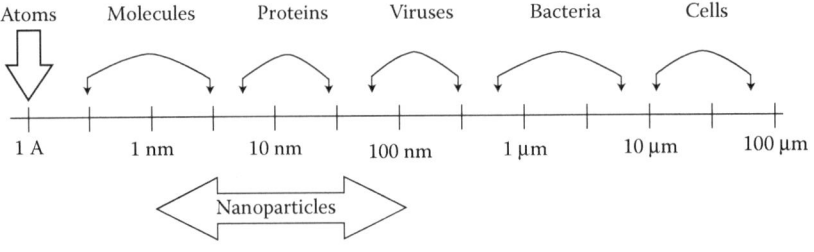

FIGURE 3.1 Scale showing size of nanoparticles compared to biological components and microorganisms.

and reasons behind the possible interactions between different types of cells and nanoparticles as functions of its size, shape, and surface chemistry (Lewinski et al. 2008). Studies also mention that it is difficult to derive any conclusions due to the variability of various nanoparticle parameters, such as physicochemical properties, cell target, dosing parameters, and the biochemical assays used.

Recent studies have confirmed that the physicochemical characteristics of nanoparticles, such as particle size, shape, surface area, solubility, chemical composition, and dispersion factor, play very significant role in determining their biological responses (Oberdörster et al. 2005; Nel et al. 2006; Powers et al. 2006). For example, small-sized nanoparticles can enter the cell and mitochondria through various pathways, subsequently inducing oxidative stress and cell death via apoptosis (Xia et al. 2007). The relatively larger surface areas of nanoparticles can produce a large number of reactive oxygen species (ROS), which can damage DNA (Karakoti et al. 2006; Hussain et al. 2009). Slightly or completely soluble nanoparticles might release toxic or nontoxic ions that undergo chemical reactions to form ROS (Brunner et al. 2006). The entry of nanoparticles into cells is dependent on the nature of the (charged) functional groups that coat their surfaces (Zhang and Monteiro-Riviere 2009). Given this information, it is very important for researchers to determine how these physicochemical characteristics of nanoparticles affect their biological responses. For example, nanoparticles that have different abilities to generate reactive species (RS) have a greater biological effect than the dependence on surface area (Sayes et al. 2006a,b,c).

"Nanotoxicity" is the study of the potential toxic impacts of different nanoparticles and nanomaterials on biological and ecological systems. Nanotoxicity studies arose from diverse fields, such as molecular toxicology, material science, molecular biology, analytical chemistry, and engineering. The basic aim of the nanotoxicity field is to build design rules for the development of safe nanoparticles. Therefore, systematic and scientific studies are essential and should be based on well-characterized physicochemical properties and their effects on cellular-viability and -function in relevant model systems. Risk assessment strives to determine risks based on the possibility of exposure and the hazards of the potentially toxic substance—in this case, nanoparticles—to make regulatory decisions (Sara et al. 2012).

3.1.1 Advantages of Nanosystems

The two basic properties of nanoparticles, like their small size and biodegradability, are important advantages for effective drug delivery (Singh and Lillard 2009). Nanoparticles have the following advantages:

1. Because of their small size, they can extravasate through the endothelium in inflammatory sites, epithelium (e.g., intestinal tract and liver), tumors, or penetrate microcapillaries. In general, the nanosize of these particles allows for its efficient uptake by a variety of cell types and its selective drug accumulation at target sites (Desai et al. 1997; Panyam and Labhasetwar 2003; Panyam et al. 2003b).
2. They protect therapeutic agents against enzymatic degradation (i.e., nucleases and proteases) (Ge et al. 2002).
3. They increase the aqueous solubility and bioavailability of the drug (Parveen et al. 2012).
4. They provide a targeted delivery of the drug by attaching targeting ligands to their surface or by the use of magnetic guidance (Mohanraj and Chen 2006).
5. They decrease the toxic side effects of the drug (Parveen et al. 2012).
6. They allow the development of rapid formulations (Parveen et al. 2012).
7. Their controlled-release and particle-degradation characteristics can be readily modulated by the choice of matrix constituents. Drug-loading capacities are relatively high and drugs can be incorporated into the systems without any chemical reactions; this is an important factor for preserving the drug's activity (Mohanraj and Chen 2006).
8. They offer appropriate forms for all routes of administration (Parveen et al. 2012).

3.1.2 Limitations of Nanosystems

The following are limitations to the use of nanosystems:

1. Their small size and large surface area can lead to problems in formulation stabilities and shelf life due to particle–particle aggregations, making the physical handling of nanoparticles difficult in liquid and dry forms, and also resulting in limited drug loading and burst releases (Mohanraj and Chen 2006).
2. Possible difficulties in drug incorporation and the achievement of desired releases (Jong and Borm 2008).
3. Problems regarding biocompatibility, biodistribution, and targeting (Jong and Borm 2008);
4. A number of different classes of nanoparticles, with different physicochemical properties, account for the adverse biological responses and produce safety and toxicological issues (Parveen et al. 2012).
5. Possible adverse effects of residual materials after drug delivery (Jong and Borm 2008).

3.2 NANOPARTICLES: IMPORTANCE IN TODAY'S ERA

The current era has witnessed an unprecedented growth in the research and applications of nanotechnology. There is strong optimism that nanotechnology will bring significant advances in the diagnosis and treatment of diseases. Nanotechnology has widespread applications in medicine, including drug delivery, both *in vivo* and *in vitro* diagnostics, nutraceuticals, and the production of improved biocompatible materials. Before the discovery of nanotechnology, healthcare was challenged by three interlocking crises that made healthcare systems unsustainable:

- Rising costs
- Changing demographics
- Quality

Table 3.1 shows the scenario before the discovery of nanoparticles, about how life-threatening diseases responded to therapy. For example, in the case of Alzheimer's disease and oncology, the percentage responding to therapy was 30 and 25, respectively. This indicates a need of specialized therapy, which can cure the diseases up to 90%.

TABLE 3.1
Outcome of the Old Model of R&D: Patient Response Rates to Major Drug Therapies

Category of Disease	% Who Respond to Therapy
Analgesic for pain (COX-2 inhibitors)	80
Asthma	60
Cardiac arrythmias	60
Schizophrenia	60
Migraine (acute)	52
Migraine (prophylaxis)	50
Rheumatoid arthritis	50
Osteoporosis	48
HCV	47
Alzheimer's disease	30
Oncology	25

Physicochemical Characterization–Dependent Toxicity of Nanoparticles

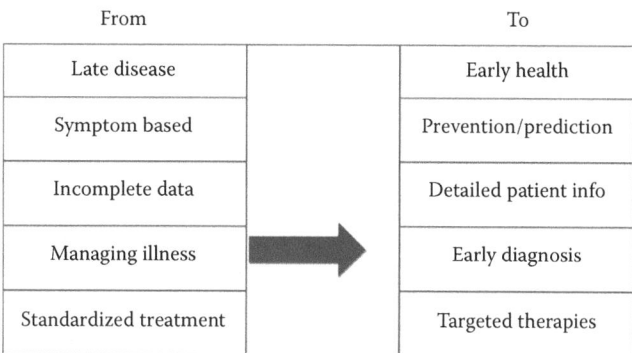

FIGURE 3.2 Shift of current healthcare system from late disease treatment to early health diagnosis.

There are number of factors responsible for the high costs and quality issues, such as inconsistencies in the delivery of healthcare and variable outcomes from region-to-region and hospital-to-hospital. There is strong need to develop a new business model which will cover:

- Personalized medicine and targeted therapy
- From blockbuster to biche buster

Figure 3.2 shows the current scenario of medical science and the health care system, which is moving from late disease treatment to early health diagnosis. Nanomedicine represents a huge promise for healthcare and is part of personalized medicine.

- Conventional medicine is reactive to tissue-level problems occurring at the symptomatic level. Nanomedicine can diagnose and treat problems at the molecular level and inside single cells prior to the development of traditional symptoms.
- Conventional medicine is not readily available in abundance for humanity because it is sophisticated, expensive, and labor-intensive. Nanomedicine can be much more preventive, comparatively inexpensive because it will minimize the use of expensive human experts, and can be more readily mass produced and distributed.
- There is a potential for greatly improved "directed therapies" for treating cancer and cardiovascular diseases using new nano-drug/gene delivery systems.
- Tiny, implantable devices can monitor health.
- Tiny, implantable devices with nanobiosensors can treat chronic diseases (diabetes, cardiovascular issues, arthritis, Parkinson's, Alzheimer's) with fewer side effects.
- New point-of-care and home healthcare devices can develop.

3.2.1 Nanomedicine: Global Scenario

According to a survey by BCC Research (www.bccresearch.com), the total, global sales of nanomedicine reached $72.8 billion in 2011 and is expected to increase to $130.9 billion in 2016, a five-year compound annual growth rate (CAGR) of 12.5%. The market for nanomedicine can be broken down into six main segments based on therapeutic applications: cardiovascular, anti-inflammatories, anti-infectives, central nervous system (CNS) products, anticancers, and other applications.

The cardiovascular segment of the market, worth $4 billion in 2011, is expected to increase at a CAGR (compound annual growth rate) of 16.5% to reach $8.6 billion in 2016. The anti-inflammatories segment of the market, worth $7.3 billion in 2011, is expected to increase at a CAGR of 15.2% to reach $14.8 billion in 2016. The anti-infectives segment was valued at $9.3 billion in

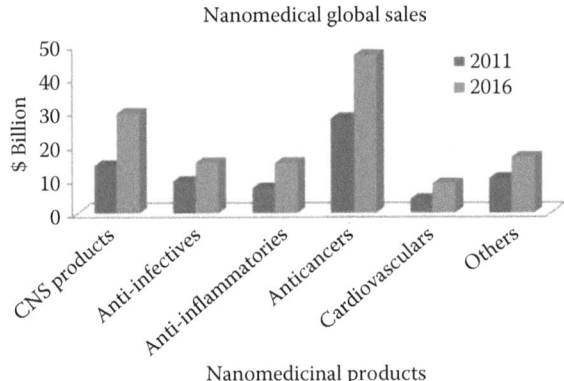

FIGURE 3.3 Nanomedical global sales by therapeutics area, 2011 and 2016 ($billions). (Data adapted from www.bccresearch.com.)

2011 and, by 2016, that value should reach $14.8 billion, at a CAGR of 9.7%. The CNS products segment was valued at $14 billion in 2011 and, by 2016, that value should reach $29.5 billion, at a CAGR of 16.1%. The largest segment, comprised of anticancers, was worth $28 billion in 2011, and after increasing at a CAGR of 10.8%, should be worth $46.7 billion in 2016. The segment made up of other applications was worth $10.2 billion in 2011, and, in 2016, should be worth $16.5 billion, a CAGR of 10.1% (Figure 3.3).

Nanomedicine has a widely established market, where a large number of nanomedicine products has made a significant contribution to the healthcare system. In the field of nanomedicine research, the United States accounts for one-third of all publications and half of patent filings. A comparison between Europe as a whole and the United States shows that, while Europe is at the forefront of research, the United States leads in the number of patent filings. The strong patenting activity of U.S. scientists and companies indicates a more advanced commercialization status than elsewhere.

3.2.2 Nanomedicine: Marketed Products

There are a number of nano-delivery products already marketed globally, like polymeric nanoparticles, liposomes, PEGylated liposomes, nanocrystals, micellar nanoparticles, solid-lipid nanoparticles, virosomes, and the like for various diseases such as cancer chemotherapy, bacterial and fungal infections, osteoarthritis, renal diseases, multiple sclerosis, enzyme replacement therapy, and hepatitis B. Forty-four marketed nano-delivery products are mentioned in Table 3.2.

Table 3.3 shows various marketed nanoparticle products used for the *in vivo* imaging of liver tumors and abdominal structures; *in vitro* diagnostics in the diagnosis of pregnancy, ovulation, and HIV; and for immunodiagnostics and biomaterials for dental and bone defects and others.

3.3 NANOSYSTEMS: DIFFERENT TYPES AND APPLICATIONS

Nanoparticles are divided into a number of classes based on their size, important characteristics, and applications. First, there are nanoclusters that are defined as semi-crystalline nanostructures with at least one dimension between 1 and 10 nm and a narrow size distribution. Then, there are the nanopowders, "an agglomeration of noncrystalline nanostructural subunits with at least one dimension less than 100 nm." There are also nanocrystals, single crystalline nanomaterials with at least one dimension less than 100 nm. Some other types of nanoparticles are mentioned (brief description) in Table 3.4.

TABLE 3.2
List of 44 Marketed Nano-Delivery Products

Product	Generic	Formulation	Indication	Manufacturer
Abraxane	Paclitaxel	Polymeric nanoparticles	Cancer chemotherapy	Celgene
Abelcet	Amphotericin B	Liposomal formulation	Fungal infections	Elan/Alkermes, Enzon, Cephalon
Adagen	Adenosine deaminase	PEGylation	Enzyme replacement therapy	Enzon, Sigma-Tau
AmBisome	Amphotericin B	Liposomal formulation	Oral and perioral infections	Astellas/Gilead Sciences
Amphotec	Amphotericin B	Liposomal formulation	Oral and perioral infections	Three Rivers Pharmaceuticals/ALZA
Avinza	Morphine sulfate	Nanocrystal formulation	Moderate-to-severe pain	Elan/Alkermes, Pfizer
Copaxone	Glatiramer acetate	Copolymer of L-glutamic acid, L-alanine, L-tyrosine, and L-lysine)	Multiple sclerosis	Teva Pharmaceuticals
Curosurf	Poractant alfa	Liposome	Neonatal respiratory distress	Chiesi Farmaceutici SpA
DaunoXome	Daunorubicin	PEGylated liposome formulation	Cancer chemotherapy	Gilead Sciences
DepoCyt	Cytarabine	Sustained-release liposomes	Cancer chemotherapy	SkyePharma/Enzon
Depodur	Morphine sulfate	Liposome	Pacira Pharmaceuticals	Pacira Pharmaceuticals
Diprivan	Propofol	Liposomes	Induction of anesthesia	AstraZeneca
Doxil/Caelyx	Doxorubicin	PEGylated liposome formulation	Cancer chemotherapy	ALZA/OrthoBiotech/ Schering Plough
Elestrin	Elestrin Estradiol gel	Phosphate nanoparticles	Menopausal symptoms	BioSante
Elyzol	Metronidazole	Dental gel	Parodontitis Camurus	Camurus
Emend	Aprepitant	Nanocrystal formulation	Antiemetic	Merck & Co+ Elan/ Alkermes
Epaxal	Hepatitis A vaccine	Virosome technology	Prevention of Hepatitis A infection	Berna Biotech
Episil	Bioadhesive barrier	Fluidcrystal	Oral pain	Sinclair/Teva
Estrasorb	Estradiol gel	Micellar nanoparticles	Menopausal symptoms	Novavax/Espirit Pharma
Focalin XR	Dexmethylphen idate hcl	Nanocrystals	ADHD	Novartis Élan/Alkermes
Fosrenol	Lanthanum carbonate	Inorganic nanoparticles	End-stage renal disease	Shire
Genexal PM	Paclitaxel	Polymeric micelles	Cancers	Samyang
Indaflex	Indomethacin	Solid/lipid nanoparticles	Osteoarthritis	AlphaRx
Inflexal V	Subunit influenza vaccine	Virosome	Influenza prophylaxis	Crucell
Invega Sustenna	Paliperidone	Nanocrystal	Antipsychotic	Janssen
Macugen	Pegaptanib	Pegylated anti-VEGF aptamer	Age-related macular degeneration	OSI Pharmaceuticals/ Pfizer

continued

TABLE 3.2 (continued)
List of 44 Marketed Nano-Delivery Products

Product	Generic	Formulation	Indication	Manufacturer
Myocet	Doxorubicin citrate complex	Liposome encapsulated	Cancer chemotherapy	Cephalon/Zeneus Pharma/Sopherion Therapeutics
Megace ES	Megestrol acetate	Nanocrystal formulation	Cancer therapy	Elan/ Alkermes + Par + Bristol-Myers Squibb
MuGard	Hydrogel mouth rinse	Nanogel	Head and neck cancers	Access Pharma
Naprelan	Naproxen	Nanocrystal formulation	Arthritis, gout	Elan/Alkermes
Nanoxel	Paclitaxel	Polymeric nanoparticles	Cancer chemotherapy	Dabur Pharma
Neulasta	Filgrastim	Pegylation	Neutropenia	Amgen
Oncospar	Oncospar PEG-Lasparaginase	Pegylation	Cancers	Enzon/Schering-Plough
Pegasys	Peginterferon alfa 2a	Pegylation	Hepatitis B, hepatitis C	Roche/Nektar
PegIntron	Peginterferon alfa 2b	Pegylation	Chronic hepatitis C	Schering-Plough
Rapamune	Sirolimus	Nanocrystal formulation	Immunosuppression	Wyeth Élan/Alkermes
Renagel	Sevelamer HCl	Poly (allylamine) resin	Hyperphosphatemia in hemodialysis	Genzyme
Salinum	Potassium, magnesium, chlorine	Oral liquid	Xerostomia	
Somavert	Pegvisomant	Polymer protein conjugate	Acromegaly	Pfizer
Ritalin LA	Methylphenidate HCl	Pulsatile release Nanocrystal formulation	ADHD	Elan/Novartis
Survanta	Beractant	Liposome encapsulated	Neonatal respiratory distress	Abbott
Tricor	Fenofibrate	Nanocrystal formulation	Lipid reduction	Abbott Élan/Alkermes
Triglide	Fenofibrate	Nanocrystal formulation	Lipid reduction	SkyePharma/First Horizon Pharmaceuticals/Sciele Pharma
Verelan/ Verelan PM	Verapamil	Elan's SODAS Multiparticulate technology	Hypertension	Elan/Alkermes Schwarz

Source: Data from www.bccresearch.com

3.4 NANOPARTICLES: PHYSICOCHEMICAL CHARACTERISTICS (PROPERTIES) AND THEIR MULTIDIMENSIONAL ISSUES

Nanoparticles have specific physical, chemical, and biological properties. These unique properties make them ideal molecules for diagnostics and therapeutics in modern medicine. Because of these desirable properties, nanoparticles have wide applications in the industrial, medical, and cosmetic fields. Nanoparticles have the ability to penetrate lung or dermal barriers, and translocate within and damage living organisms. This is due to their small size, which makes them able to cross

TABLE 3.3
Fifteen Marketed Imaging/Diagnostic and Biomaterial

Product	Composition	Indication	Company
In Vivo Imaging			
Resovist	Iron nanoparticles	Liver tumors	Schering, Berlin
Feridex/Endorem	Iron nanoparticles	Liver tumors	Advanced Magnetics, Guerbet
Gastromark/Lumirem	Iron nanoparticles	Imaging abdominal structures	Advanced Magnetics, Guerbet
In Vitro Diagnostics			
Lateral flow tests	Colloidal gold	Pregnancy, ovulation, HIV, etc.	British Biocell, Amersham/GE, Nymox
Clinical cell separation	Magnetic nanoparticles	Immunodiagnostics	Dynal/InVitrogen, Miltenyl Biotec, Immunicon
Biomaterials			
Ceram X duo	Nanoparticle composite	Dental filling material	Dentspley
Filtek Supreme	Nanoparticle composite	Dental filling material	3M Espe
Mondial	Nanoparticle-containing dental prosthesis	Dental restoration	Heraeus Kulzer
Premise	Nanoparticle composite	Dental repair	Sybron Dental Specialities
Tetric Evoceram	Nanoparticle composite	Dental repair	Ivoclar Vivadent
Ostim	Nano-hydroxy apatite	Bone defects	Osartis
Perossal	Nano-hydroxy apatite	Bone defects	Aap implantate
Vitoss	Nano-hydroxy apatite	Bone defects	Orthovita
Acticoat	Silver nanoparticles	Antimicrobial wound care	Nucryst
Active Implants			
Pacemaker	Fractal electrodes	Heart failure	Biotronik

Source: Data from Company Websites, *Nature Biotechnology*, October 2006. With permission.

physiological barriers and traverse within the circulatory systems of humans and animals, reaching most tissues and organs and adversely affecting cellular processes and causing disease. However, the toxicity of each of these materials depends greatly upon the structural arrangement of its many atoms. Variations in the shape, size, and structural chemistry of nanoparticles yield a large number of distinct materials with highly different physical, chemical, and toxicological properties. For example, asbestos, a toxic nanomaterial, causes lung cancer and other diseases. It exists in various forms with slight variations in shape and chemistry and gives significantly varying toxicity.

The important properties of nanoparticles are shape (including aspect ratios where appropriate), size, and the morphological substructure of the particle. Nanoparticles are available in different forms, such as aerosols (solid/liquid phase in air), suspensions (solid in liquids), or emulsions (two liquid phases). In the presence of chemical agents (surfactants), the surface and interfacial properties may be modified and such agents can indirectly stabilize against particle coagulation or aggregation by conserving the particle charge and modifying the outermost layer of the particle. Nanoparticles have intrinsic toxicity profiles. The physicochemical characteristics of nanoparticles that might increase its toxicity potential include:

- Particle size and size distribution
- Particle shape
- Agglomeration/aggregation
- Crystal structure (crystallinity)

TABLE 3.4
Types of Nanosystems with Their Sizes, Characteristics, and Application

Type of Nanosystem	Characteristics	Applications
Carbon nanotubes Size: 0.5–3 nm and 20–1000 nm length	• Third allotropic crystalline form of carbon sheets either single layer (single-walled nanotube, SWNT) or multiple layer (multi-walled nanotube, MWNT) • These crystals have remarkable strength and unique electrical properties (conducting, semiconducting, or insulating)	• Functionalization enhanced solubility, penetration to cell cytoplasm and to nucleus, as carrier for gene delivery, peptide delivery
Nanocrystals Quantum dots Size: 2–9.5 nm	• Broad UV excitation and high photostability • Bright fluorescence • Semiconducting properties • Narrow emission	• Long-term multiple color imaging of liver cell • DNA hybridization • Immunoassay • Receptor-mediated endocytosis • Labeling of breast cancer marker HeR_2 surface of cancer cells
Iron oxide nanocrystal Size: 4–5 nm; Hydrodynamic radius: 15–25 nm	• Superparamagnetism	• Magnetic resonance imaging (disease detection such as cancer, arthritis, and atherosclerosis) • Intracellular monitoring
Dendrimer Size: <10 nm	• Highly branched, nearly monodisperse polymer system produced by controlled polymerization • Three main parts—core, branch, and surface	• Long circulatory, controlled delivery of bioactives • Targeted delivery of bioactives to macrophages • Liver targeting
Silica nanoparticles Size: 10 nm–50 μm	• Silanized and coated with oligonucleotide. Observable by fluorescence method	• Efficient nucleic acid hybridization • Detection of DNA • Nanobiosensor for trace analysis
Polymeric micelles Size: 10–100 nm	• Amphiphilic block copolymer micelles • High drug entrapment, payload, and biostability	• Long circulatory, target specific active and passive drug delivery • Diagnostic value
Liposomes Size: 50–100 nm	• Phospholipid vesicles • Biocompatible • Good entrapment efficiency • Offer easy surface functionalization	• Long circulatory time • Offer passive and active delivery of gene, protein, peptide, and various other bioactives
Metallic nanoparticles Size: <100 nm	• Gold and silver colloids • Very small size resulting in high surface area available for functionalization • Better stability	• Drug and gene delivery • Highly sensitive diagnostic assays • Thermal ablation and radiotherapy enhancement
Polymeric nanoparticles Size: 10–1000 nm	• Biocompatible and biodegradable • Offer complete drug protection	• Excellent carrier for controlled and sustained delivery of drugs • Stealth and surface-modified nanoparticles can be used for active and passive delivery of bioactives
Nanoshells	• Nanoshells typically have a silicon core that is sealed in an outer metallic core • By manipulating the ratio of wall to core, the shells can be precisely tuned to scatter or absorb very specific wavelengths of light	• Gold-encased nanoshells have been used to convert light into heat, enabling the destruction of tumors by selective binding to malignant cells

Source: Data adapted from Nahar, M. et al. 2006. *Crit Rev Ther Drug Carrier Syst* 23(4): 259–318.

Physicochemical Characterization–Dependent Toxicity of Nanoparticles

- Surface chemistry, charge, and area
- Stability over time and dissolution
- Dosing metric
- Cellular binding and uptake
- Biopersistance solubility
- Surface coatings
- Material composition

In principle, a large number of particles could overload the body's phagocytes, thereby triggering stress reactions that can lead to inflammation and weaken the body's defense against other pathogens. In addition to questions about what happens if non- or slowly degradable nanoparticles accumulate in body organs, another concern is their potential interaction or interference with biological processes inside the body. Due to the importance of this class of particles, the term nanotoxicology has been coined to establish the relationship between nanoparticle physicochemical properties (e.g., size, surface properties, and crystal phase) and their toxic potential. It is the biokinetic evaluation of engineered nanostructures and nanodevices (Oberdörster et al. 2005; Garnett and Kallinteri 2006). Due to an intensive expansion in nanotechnology, there is a need for detailed investigations in this area as has been widely used in the pharmaceutical industry, medicine, and engineering technology in the last two decades (Curtis et al. 2006; Kurath and Maasen 2006). Particle toxicologies and the consequent adverse health effects of asbestos fibers, carbon nanotubes (CNTs), coal dust, and many others, serve as a historical reference point for the development of nanotoxicologic concepts (Oberdörster et al. 2005; Kurath and Maasen 2006). Figure 3.4 shows multidimensional issues affecting nanotoxicity.

The physicochemical properties of nanoparticles must be examined in detail to create design rules and to begin interpreting any results due to nanoparticle-induced toxicity. Because the field of

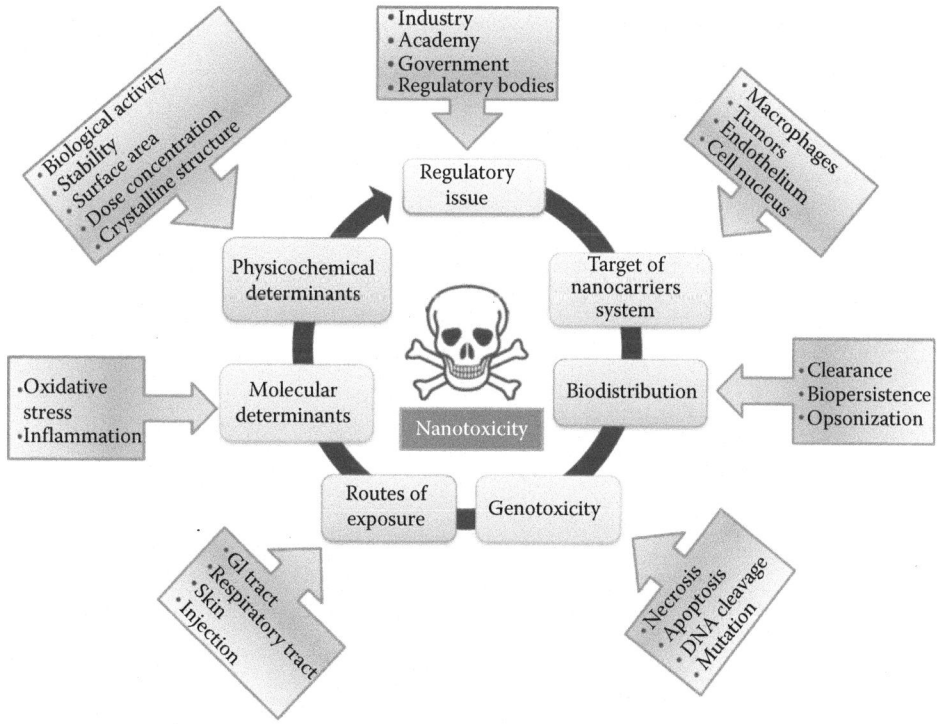

FIGURE 3.4 (See color insert.) Multidimensional issues affecting nanotoxicity.

nanotoxicity is relatively new and the specific nanoparticle properties that influence cellular toxicity are still not fully understood, a thorough characterization of the nanoparticle is essential.

3.5 NANOPARTICLES: PHYSICOCHEMICAL CHARACTERIZATION-DEPENDENT TOXICITY

Nanotoxicology is the study of nanoparticle toxicities which are unusual and not seen in larger particles. They have greater surface-area-to-volume ratios, higher its chemical reactivities and biological activities. Because of their large surface area, nanoparticles will, on exposure to tissues and fluids, immediately adsorb onto the surface of the macromolecules they encounter. This may, for instance, affect the regulatory mechanisms of enzymes and other proteins. Human skin, lungs, and the gastrointestinal tract are in constant contact with the external environment. These are the most-favorable points of entry for natural or anthropogenic nanoparticles. Injections and implants are other possible routes of exposure, primarily limited to engineered materials. Some nanoparticles, depending on their composition and size, can produce irreversible damage to cells by oxidative stress and/or organelle injury (Buzea et al. 2007).

Several diseases with unknown causes, including autoimmune diseases, Crohn's, Alzheimer's, and Parkinson's, appear to be correlated with nanoparticle exposures. Conversely, the toxic properties of some nanoparticles may be beneficial, as they are thereby able to fight disease at a cellular level, and could be used as medical treatments by, for example, targeting and destroying cancerous cells. Cristin Buzea et al. (2007) summarize the possible adverse health effects associated with the inhalation, ingestion, and contact with nanoparticles as shown in Figure 3.5. They emphasize that nanotoxicity depends on various physicochemical properties, like particle size, shape, aggregation, chemical composition, crystalline structure, and surface functionalization. In addition, the nanotox-

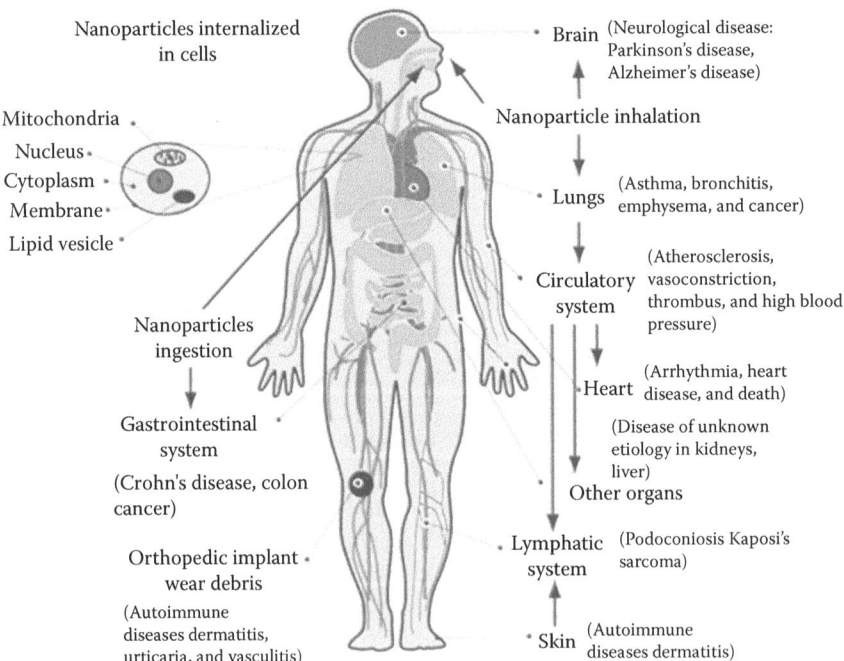

FIGURE 3.5 Schematics of human body with pathways of exposure to nanoparticles, affected organs, and associated diseases from epidemiological, *in vivo* and *in vitro* studies. (Reprinted with permission from Buzea, C., Blandino, I.I.P., Robbi, K. 2007. *Biointerphases* 2(4): MR17–MR172.)

icity to an organism is also determined by the individual's genetic complement, which provides the biochemical toolbox by which it can adapt to- and fight toxic substances (Buzea et al. 2007).

Figure 3.5 shows the most extreme adverse health effects produced by nanoparticles in different parts of body tissues and organs; there is a high need to increase the awareness of potential toxicities of some nanoparticles. Diseases associated with inhaled nanoparticles are asthma, bronchitis, emphysema, lung cancer, and neurodegenerative diseases, such as Parkinson's and Alzheimer's. Nanoparticles in the gastrointestinal tract have been linked to Crohn's disease and colon cancer. Nanoparticles that enter the circulatory system are related to the occurrence of arteriosclerosis, blood clots, arrhythmia, heart diseases, and ultimately cardiac death. Translocation to other organs, such as the liver, and spleen, may lead to diseases of these organs as well. The exposure to some nanoparticles is associated with the occurrence of autoimmune diseases such as, systemic lupus erythematosus, scleroderma, and rheumatoid arthritis (Buzea et al. 2007).

There are a number of factors responsible for the toxicity of nanoparticles: (1) The toxicity of bulk materials, for example, heavy metals whose toxicity is easy to quantify (Curtis et al. 2006); (2) the electrical properties of nanoparticles, as they can create and/or scavenge ROS and free radicals (Oberdörster et al. 2005; Curtis et al. 2006; Lanone and Boczkowski 2006; Berube et al. 2007; Duffin et al. 2007; Medina et al. 2007); (3) the size of nanoparticles may be linked to the toxicity of nanoparticles as mentioned in ultrafine particle studies (Curtis et al. 2006). The formation of ultrafine particle agglomerates is also responsible for toxicity. They pass through biological barriers such as the skin, vascular endothelium, and the blood–brain barrier, therefore affecting the absorption, distribution, and excretion of these particles (Oberdörster et al. 2005; Moghimi et al. 2005; Curtis et al. 2006; Garnett and Kallinteri 2006; Lanone and Boczkowski 2006; Hagens et al. 2007); (4) shape is one of the factors that determines toxicity, for example, CNTs (Oberdörster et al. 2005; Moghimi et al. 2005; Curtis et al. 2006; Lanone and Boczkowski 2006; Badea et al. 2007; Wagner et al. 2007; Warheit et al. 2007); and (5) nanoparticles can also trigger immune responses. However, studies are currently under way to identify their role in possible allergic reactions (Moghimi et al. 2005; Curtis et al. 2006; Lanone and Boczkowski 2006).

From the understanding of the toxicological properties of fibrous particles, it is believed that the most important parameters in determining the adverse health effects of nanoparticles are the dose, dimension, and durability (Oberdörster 2002). However, recent studies show different correlations between the various physicochemical properties of nanoparticles (Table 3.5) and the associated health effects, raising some uncertainties as to which are the most important parameters in deciding their toxicity: mass, number, size, bulk or surface chemistry, aggregation, or all of them.

TABLE 3.5
List of Physicochemical Characteristics of Nanoparticles Induced Nanotoxicity

S. No.	Physicochemical Characteristics of Nanoparticles Involved in Toxicity
1	Structural properties of nanoparticles a. Particle shape and aspect ratio b. Structural arrangement (crystalline form) c. Surface chemistry and charge
2	Dose
3	Concentration
4	Size, size distribution and surface area of nanoparticles
5	Material composition of formulation
6	Biopersistence solubility
7	Cellular uptake and binding
8	Surface coatings/modifications

3.5.1 Structural Properties-Dependent Toxicity

Nanoparticle structural properties are important in understanding their relation to nanotoxic behaviors. Structural properties include the various attributes of a nanosystem, such as shape, aspect ratio, surface morphology, structural arrangement, spatial distribution, density, and geometric features. Out of these attributes, the key attributes such as shape, aspect ratio, structural arrangement, surface chemistry, and surface charge, which are responsible for various toxicities (by different mechanisms) are discussed here. The development of electron microscopy improved the accessibility and feasibility of determining these attributes at the nanometer scale to identify their roles in specific toxicities.

3.5.1.1 Particle Shape and Aspect Ratio

It was found that the particle aspect ratio is directly proportional to its toxicity. For example, lung cancer was associated with asbestos fibers (size >10 μm), mesothelioma with fibers (size >5 μm), and asbestosis with fibers (size >2 μm). All of these fibers had a minimum thickness of about 150 nm (Lippmann 1990). Long fibers (longer than 20 μm for humans) will not be effectively cleared from the respiratory tract due to the inability of macrophages to phagocytize them (Hoet et al. 2004). The biopersistence of these long aspect ratio fibers leads to long-term carcinogenic effects (Oberdörster 2002).

The toxicity of long aspect fibers is closely related to their bio-durability, depending on its dissolution and mechanical properties (breaking). Longer fibers that break perpendicularly to their long axis become shorter and can be removed by macrophages. Asbestos fibers break longitudinally, resulting in more fibers with smaller diameters, being harder to clear (Hoet et al. 2004). If clearance from the lung is slow, the longer these fibers will stay in the lung, the higher the probability of an adverse response. Fibers that are sufficiently soluble in lung fluid can disappear in a matter of months, while insoluble fibers are likely to remain in the lungs indefinitely. Even short, insoluble fibers that are efficiently phagocytized by alveolar macrophages may induce biochemical reactions (the release of cytokines, ROS, and other mediators).

Long aspect ratio engineered nanoparticles, such as CNTs, have recently attracted a lot of attention due to their possible negative health effects (Warheit et al. 2004; Cui et al. 2005; Monteiro-Riviere et al. 2005; Muller et al. 2005) (Table 3.6), as suggested by their morphological similarities with asbestos. CNTs can be single-walled (SWCNTs) or multi-walled (MWCNTs), with varying diameters and lengths and closed capped sections or open ends (Dai 2002). Due to their hydrophobicity and tendency to aggregate, they are harmful to living cells in culture (Monteiro-Riviere et al. 2005; Cui et al. 2005). For many applications, CNTs are oxidized to create hydroxyl and carboxyl groups, especially at their ends, making them more readily dispersed in aqueous solutions (Bottini et al. 2006). SWCNTs were reported to produce significant pulmonary toxicity as compared to spherical particles (amorphous carbon black) (Lam et al. 2004; Bottini et al. 2006). For comparison purposes, equal doses of carbon black or silica nanoparticles did not induce granulomas, nor thickening of the alveolar wall, causing only weak inflammation and limited damage. The enhanced toxicity was attributed to its physicochemical properties and fibrous nature. CNTs are either not eliminated from the lungs or are very slowly eliminated; 81% are found in rat lungs 60 days after exposure (Muller et al. 2005).

3.5.1.2 Structural Arrangement (Crystalline Form)

There is quite a distinction between composition and chemistry. Although particles may have similar compositions, they may have different chemical or crystalline structures. The toxicity of a material depends on the type of its crystalline nature (Gurr et al. 2005). For example, rutile and anatase, both allotropes of titanium dioxide (i.e., polymorphs with the same chemical composition but different crystalline structure), possess different chemical and physical properties. Table 3.7 shows different crystalline forms of various types of nanoparticles with their toxicities and the mechanism involved.

TABLE 3.6
Studies on Toxicity Due to Particle Shape and Aspect Ratio

Type of Nanoparticle	Toxicity/Organ Affected	Mechanism Involved/Results	Reference
ZnO nanorods and nanospheres	Cytotoxicity	The nanorods and nanospheres had similar cytotoxicities after exposure for 12 h; after 24 h, however, the nanorod ZnO nanoparticles were more toxic than the nanosphere ZnO nanoparticles with values of EC50 for ZnO-R8 and ZnO-S6 of 8.5 and 12.1 µg mL^{-1}, respectively.	Hsiao and Huang (2011)
Denritic TiO$_2$ nanoparticles	Cytotoxicity/Mouse macrophage cell line	Dendritic TiO$_2$ nanoparticles induce the highest degree of cytotoxicity toward the mouse macrophage cell line (J774A.1), followed by spindle- and sphere-shaped nanoparticles.	Yamamoto et al. (2004)
Nanotubes	Multifocal granulomas lesions New mechanism of pulmonary toxicity	On a dose per mass basis the nanotubes were more toxic than quartz particles although the mass dose was very high and mechanical blockage of some airways was noted.	Warheit et al. (2004)
Single-walled CNTs (fiber shaped)	Carcinogenic effects/Lung granulomas after intrathecal administration	Nanotubes which may be of a few nanometers in diameter but with a length that could be several micrometers.	Lam et al. (2004)

3.5.1.3 Surface Chemistry and Charge

Surface chemistry is a critical parameter in determining the toxicity of nanoparticles and is especially relevant to molecular cell chemistry and oxidative stress. Based on their chemistry and charge (zeta potential), nanoparticles can show different cellular uptakes, subcellular localizations, and the ability to catalyze the production of ROS (Xia et al. 2006). These toxicity effects are described in Table 3.8.

3.5.2 DOSE-DEPENDENT TOXICITY

A dose is defined as the *amount* or *quantity* of substance that will reach a biological system. The dose is directly related to exposure, or the concentration of the substance in the relevant medium (air, food, water), multiplied by the duration of contact. Various toxicity effects, well studied by scientists and based on doses, are discussed in Table 3.9 with organ specificities and the mechanisms by which it is caused.

3.5.3 CONCENTRATION AND DRUG–LOADING-DEPENDENT TOXICITY

The resultant effects of different concentrations on nanoparticle toxicities are very contradictory. Research shows that certain concentrations of materials are not as toxic as were observed by other researchers. This may be due to differences in the aggregation properties of nanoparticles in air and water, resulting in inherent discrepancies between inhalation studies and instillation or *in vitro* experiments. The aggregation may depend on the zeta potential, material type, size, and many others factors.

The aggregation of nanoparticles is one of the most important and essential parameters in determining its toxicity, because large, aggregated particles more effectively cleared by macrophages as compared to smaller particles, leading to the reduced toxicity of nanoparticle aggregates larger than

TABLE 3.7
Studies on Toxicity Due to Different Structural Arrangement

Type of Nanoparticle	Toxicity/Organ Affected	Mechanism Involved/Results	Reference
TiO_2 nanoparticle	Cytotoxicity and genotoxicity	The anatase crystal form may be more toxic because of its greater oxidizing potential, which would generate a greater amount of reactive species.	Jiang et al. (2008)
Carbon nanotubes MWCNTs	Apoptosis human T cells	The cytotoxicity of pristine versus oxidized MWCNTs was compared and found that oxidized MWCNTs induced greater human T cell apoptosis.	Bottini et al. (2006)
TiO_2 nanoparticle	Necrosis	Anatase form of TiO_2 nanoparticles causes greater toxic response compared to the rutile form, especially in the viable cells. It caused increased levels of various inflammatory indicators like LDH and IL-8, which may be because of its greater oxidizing potential in generating ROS.	Braydich-Stolle et al. (2009) Jiang et al. (2008) Wu et al. (2010)
TiO_2 nanoparticles	Cytotoxicity/dermal fibroblasts and human lung epithelial cells	Photoactivation of anatase TiO_2 will increase cytotoxicity but concentrations over 100 mg/mL will be significant enough to cause any ill effects.	Sayes et al. (2006a,b,c) Warheit et al. (2006)
Rutile nano TiO_2	Pulmonary inflammatory response from 20 nm TiO_2	Rutile nanoparticles of 20 nm TiO_2 causes pulmonary inflammatory response in THP-1 and A549 cell line.	Chen et al. (2006a)

100–200 nm (Takenaka et al. 2001; Oberdörster et al. 2005). At high concentrations, nanoparticles would allow particle aggregation (Churg et al. 1998; Gurr et al. 2005) and, therefore, reduced toxic effects as compared to lower concentrations (Takenaka et al. 2001) as observed in Table 3.10. The majority of aggregates have a size >100 nm, that is, a size that seems to be a threshold for many of the adverse health effects of small particles. Therefore, nanoparticles at high concentrations will lead to the formation of aggregates that may not be as toxic as lower concentrations of the same nanoparticles.

3.5.4 Size, Size Distribution and Surface Area-Dependent Toxicity

Smaller nanoparticles (<100 nm) cause more adverse respiratory health effects (inflammation) than larger particles of the same material (Ferin et al. 1992; Oberdörster et al. 1994, 2005; Wilson et al. 2002; Donaldson and Stone 2003; Gurr et al. 2005) as mentioned in Table 3.11. This was proven by the inhalation and instillation (Oberdörster et al. 2005) of 20 and 250 nm diameter TiO_2 particles in the rat (with the same crystalline structure). The results showed that smaller particles led to a persistently high inflammatory reaction in the lungs as compared to larger sized particles. In the postexposure period (up to 1 year), it was observed that the smaller particles had, (1) a significantly prolonged retention, (2) an increased translocation to the pulmonary interstitium and pulmonary

TABLE 3.8
Studies on Toxicity Due to Surface Chemistry and Charge

Type of Nanoparticle	Toxicity/Organ Affected	Mechanism Involved/Results	Reference
Carbon nanotubes SWCNTs	Mesothelial cells	Well-dispersed SWCNTs are less toxic than agglomerated SWCNTs.	Wick et al. (2007)
Carbon nanotubes, SWCNTS and MWCNTS	Genotoxicity/guinea-pig alveolar and mouse macrophages	Cause chromosomal breakage and changes in chromosome number; purified SWCNTs are more toxic than MWCNTs.	Di-Giorgio et al. (2011) Jia et al. (2005)
Carbon nanotubes	Dermal toxicity	*In vitro* studies using a human keratinocyte cell line showed that carbon nanotube exposure resulted in accelerated oxidative stress and cellular toxicity, which may be interpreted as potential for dermal toxicity.	Shvedova et al. (2003)
Carbon nanotubes SWCNTs	Cytotoxic/rat kidney cells	High toxicity to normal rat kidney cells of two SWCNTs was due to their carboxylic acid functionalization.	Wang et al. (2011)
Protein absorption to nanorods	Uptake of the nanomaterial via receptor-mediated endocytosis	Protein adsorption to the surface of the nanorods flips their charge immediately to similar negative value of the serum proteins in the original media. Therefore, nanoparticles that had a positive effective surface charge upon preparation are no longer cationic in the cellular media. Protein adsorption to the nanoparticle surface can mediate the uptake of the nanomaterial via receptor-mediated endocytosis.	Alkilany et al. (2009) Conner and Schmid (2003)
Gold nanoparticles	—	Goodman et al. found that cationic gold nanospheres (2 nm in diameter) are toxic (at certain doses). Interestingly, the same nanoparticles with a negatively charged surface found to be not toxic at the same concentration and in the same cell line. This observation was explained by the ability of the cationic nanoparticles to interact with the negatively charged cellular membrane and the resultant membrane disruption.	Goodman et al. (2004)

persistence, (3) greater epithelial effects (such as type II cell proliferation), and (4) the functional impairment of alveolar macrophages (Oberdörster et al. 1994).

Reductions in size to the nano level cause an enormous increase in the surface-to-volume ratio. Relatively more molecules of the chemical are present on the surface, enhancing the intrinsic toxicity (Donaldson et al. 2004). This may be one of the reasons why nanoparticles are generally more

TABLE 3.9
Studies on Dose-Dependent Toxicity

Type of Nanoparticle	Toxicity/Organ Affected	Mechanism Involved/Results	Reference
TiO_2 nanoparticle	Cytotoxicity and Genotoxicity	A dose- and time-dependent increase in oxidative stress has been observed for TiO_2 nanoparticles; the anatase form generates the greatest amount.	Hackenberg et al. (2010) Barillet et al. (2010)
Carbon nanotubes SWCNT	Changes in cell morphology/keratinocyte and bronchial epithelial cells	High doses of SWCNT results in ROS generation, lipid peroxidation, oxidative stress, mitochondrial dysfunction, and changes in cell morphology.	Shvedova et al. (2003) Sayes et al. (2006a,b,c) Muller et al. (2005)
Carbon nanotubes	Chronic lung inflammation/rodents	Intratracheal administration of high doses of nanotubes demonstrated foreing body granuloma formation and interstitial fibrosis	Warheit et al. (2004)
Silica nanoparticles	Dose-dependent cytotoxicity	15 and 46 nm of silica nanoparticles showed similar dose-dependent cytotoxicity *in vitro*. Also increase in toxicity both at increasing doses and at increasing exposure time.	Lin et al. (2006)
Silica nanoparticles	Membrane damage	High dose causes reduction in cell viability/cell proliferation and by lactate dehyrogenase release from the cells indicating membrane damage.	Chang et al. (2007)
Silver nanoparticles	Biliary hyperplasia Liver	Adverse effects reported in AgNP oral dosing studies were mild and were only evident at doses of 125 mg/kg and above. A lowest observed adverse effects level (LOAEL) of 125 mg/kg in one 90 d study using 60 nm AgNP in 0.5% CMC corresponded to elevated cholesterol and cholestatic enzymes (alkaline phosphatase) and was accompanied by biliary hyperplasia. The same group found similar cholestatic enzyme effects and slight hemoconcentration at 300 mg/kg with the same material in a 28 d study in rats. This suggests that the biliary system may be a target for Ag accumulation or metabolism.	Kim et al. (2010)
Gold nanoparticles	Hemolysis Hemolysis in a select species of mouse	The toxicity of orally ingested AuNP at therapeutic or biologically relevant doses is low, with only one study showing adverse effects suggestive of hemolysis in one species (mouse) at doses of 1100 μg/kg (1.1 mg/kg). Nevertheless, given the demonstrated ability of small gold nanoparticle to enter cells and the known toxicity of solubilized gold therapeutic agents, hemolysis should be kept in mind.	Zhang et al. (2010c) Ronconi et al. (2006) Hillyer and Albrecht (2001)

TABLE 3.10
Studies Concentration and Drug-Loading-Dependent Toxicity

Type of Nanoparticle	Toxicity/Organ Affected	Mechanism Involved/Results	Reference
Nanoparticles	—	High concentration of nanoparticles would promote particle aggregation and therefore reduce toxic effects compared to lower concentrations; aggregation of nanoparticles is essential in determining their toxicity, due to a more effective macrophage clearance for larger particles compared to smaller ones	Gurr et al. (2005) Takenaka et al. (2001) Churg et al. (1998)
Gold nanoparticles	Reduce viability and reproductive performance	Naked nanoparticle cause biomolecule interactions and alteration of downstream process	Vecchio et al. (2012)
Starch-coated silver nanoparticle	Cytotoxicity and genotoxicity	Reduce ATP content of cell, damage to mitochondria, increased production of ROS	Asharani et al. (2009)

toxic than larger particles of the same, insoluble material when compared by mass. Surface area was, therefore, a driver for inflammation from these materials; the differences in severity of the response disappeared when the dose was expressed as the surface area. These examples emphasize the importance of particle size, and, by implication, the surface area that is presented to the biological system to induce particle toxicity (SCENIHR/002/05). Indeed, smaller nanoparticles have higher surface areas and particle numbers per unit mass compared to larger particles. A larger surface area leads to an increased reactivity (Roduner 2006) and is an increased source of ROS, as demonstrated by *in vitro* experiments (Donaldson and Stone 2003).

3.5.5 MATERIAL COMPOSITION-DEPENDENT TOXICITY

The type and composition of the material used to prepare nanoparticles is also responsible for its nanotoxicity. The examples are metal-, metal oxide-, and carbon-based nanoparticles. These materials are representative of each of the three most common and widely studied nanoparticle classes. Generally, cells exposed to metal (e.g., Ag) nanoparticles show increased indications of cellular stress and functional changes. The cellular toxicity of TiO_2 nanoparticles is due to its crystallinity, oxidizing potential, and aggregation properties. For carbon based nanoparticles, investigators generally agree that they are toxic and adversely affect a variety of cells due to factors such as metal impurities, particulate states, structural differences, and the surface properties of CNTs, which greatly influence their apparent cytotoxicity (Table 3.12). To advance the field, both through material characterizations of nanoparticles prior to toxicity studies and standardized, reliable methods to assess the cytotoxicity of materials are needed (Sara et al. 2012).

3.5.6 BIO-PERSISTENCE-DEPENDENT TOXICITY

The dose of nanoparticles retained in the respiratory tract is expressed as the initial number of deposited particles minus the number of particles subjected to clearance, which can occur by both physicochemical and physiological processes. Clearances are believed to occur by the following major mechanisms: (1) through the mucociliary escalator in the nose and tracheobronchial region, (2) phagocytosis by alveolar macrophages, (3) dissolution, and (4) translocation. Nanoparticles that cannot be cleared by any of these processes are considered to be biopersistent and are predicted to accumulate during chronic exposures (Sanchez et al. 2009) (Table 3.13).

TABLE 3.11
Studies on Toxicity Due to Size, Size Distribution, and Surface Area

Type of Nanoparticle	Toxicity/Organ Affected	Mechanism Involved/Results	Reference
Gold nanopartciles	Cytotoxic/necrosis cells	Oxidative stress, mitochondrial damage	Pan et al. (2009)
Gold nanoparticles—titanium dioxide	Chemical reactivity	Oxidation with dioxygen	Turner et al. (2008)
TiO$_2$ nanoparticles	Lung tumor Lung	While comparing the effect of inhaled TiO$_2$ nanoparticles with different sizes, it was observed that low-dose exposure to 20 nm diameter particles resulted in greater lung tumor incidence than high-dose exposure of 300 nm diameter particles, which correlates the effect of surface area on toxicity.	Oberdörster et al. (2005) Hoet et al. (2004)
TiO$_2$ nanoparticles	Respiratory health effects Lung inflammation	Smaller particles led to a persistently high inflammatory reaction in the lungs compared to larger size particles: (1) a significantly prolonged retention, (2) increased translocation to the pulmonary interstitium and pulmonary persistence of nanoparticles, (3) greater epithelial effects (such as type II cell proliferation), (4) impairment of alveolar macrophages function.	Oberdörste et al. (1994)
Polystyrene nanoparticles	Preferential uptake of smaller nanoparticle—causing accumulation in liver and spleen	It has been shown that polystyrene nanoparticles are preferentially taken up across M cells. Size influences absorption, as shown by greater absorption of smaller (50 nm) polystyrene particles compared to larger (100 nm) particles. The largest particles in this study (300 nm) were not absorbed. Additionally, larger particles remained within the submucosa or GALT of the intestine and colon, while smaller particles entered the bloodstream and accumulated in the liver and spleen.	Des Rieux et al. (2007) Jani et al. (1990)
TiO$_2$ nanoparticles	Inflammation and altered macrophage chemotactic responses in rat lungs	Ultrafine TiO particles (29 nm) increased inflammation and altered macrophage chemotactic responses in rat lungs, when compared to TiO particles that were 250 nm.	Renwick et al. (2004)
TiO$_2$ nanoparticles	Oxidative damage in a human bronchial epithelial cell line	TiO$_2$ particles induce oxidative damage in a human bronchial epithelial cell line in a size-dependent manner.	Gurr et al. (2005)

TABLE 3.12
Studies on Toxicity Due to Type of Material Used

Type of Nanoparticle	Toxicity/Organ Affected	Mechanism Involved/Results	Reference
Carbon nanotubes; ZnO and TiO$_2$ NPs	Genotoxicity	CNTs also cause more DNA damage and genotoxicity than the spherical ZnO and TiO$_2$ NPs.	Yang et al. (2009)
ZnO nano and TiO$_2$ nano	Cytotoxicity	Nano-ZnO might also release zinc ions into the cell culture medium, potentially resulting in greater cell damage. Nanoparticles with positive charge usually penetrate cells more readily than do negatively charged nanoparticles, suggesting that nano ZnO has greater potential to enter cells and subsequently harm them relative to nano-TiO$_2$.	Hussain et al. (2009) Harush-Frenkel et al. (2008) Osaka et al. (2009)
Cadmium quantum dots	Cytotoxicity	Toxicity to be due to release of highly toxic free Cd^{2+} ions.	Derfus et al. (2004) Kirchner et al. (2005)
PEG modified gold nanorods	Cytotoxicity	Cytotoxicity due to presence of very less amount of stabilizer CTAB.	Niidome et al. (2006)
Silver nanoparticles	Argyria and argyrosis/skin and occular tissues	The adverse effects classically associated with excessive Ag ingestion in humans are argyria and argyrosis.	Varner et al. (2010) Brandt et al. (2005)
TiO$_2$ nanoparticles	Myocardial lesions Heart	TiO$_2$ nanoparticles are able to induce high LDH, creatine kinase (CK), alpha-hydroxybutyrate dehydrogenase (HBDH), and aspartate aminotransferase (AST) activities used as markers of myocardial lesions, irrespective of the form of TiO$_2$, anatase or a mixture of anatase and rutile, or the route of exposure, intraperitoneal injections in mice, or oral gavage in rats.	Ivo et al. (2012).
TiO$_2$ nanoparticles	Hepatocyte apoptosis Liver	Li et al. analyzing the interactions between anatase TiO$_2$ nanoparticles and the DNA extracted by the liver of mice intraperitoneally exposed. They reported that nanoanatase TiO$_2$ could be inserted into DNA base pairs, bind to DNA nucleotide, and alter the secondary structure of DNA. Moreover, nanoparticles could cause liver DNA cleavage and hepatocyte apoptosis.	Ivo et al. (2012).

TABLE 3.13
Studies on Toxicity Due to Biopersistence of Nanoparticles

Type of Nanoparticle	Toxicity/Organ Affected	Mechanism Involved/Results	Reference
Silver nanoparticles	Argyria	The median onset of argyria in cases stemming from colloidal silver was 20 months, while in a case arising from soluble Ag ($AgNO_3$), the time of onset was only 8 months. This is consistent with experimental evidence from animal studies that highly soluble Ag has higher bioavailability. Tissue distribution and bioavailability of ingested AgNP are low. There is indirect evidence that AgNP may have lower bioavailability than ionic Ag. In a 28-day study of 14 nm PVP-coated AgNP administered at 12.6 mg/kg, 63% of the daily dose was eliminated in the feces, compared to 49% of the daily dose of ionic Ag in the form of silver acetate (AgOAc).	Bowden et al. (2011) Loeschner et al. (2011)

3.5.7 Cellular Uptake and Binding-Dependent Toxicity

Mechanisms for the cellular uptake of nanoparticles have important implications for nanoparticulate drug delivery and toxicity studies. Many types of nanoparticles are tested for their potential use in medical products, particularly in imaging and gene and drug deliveries. For these applications, cellular uptake is usually a prerequisite and is governed by experimental procedures in addition to the size and surface characteristics, such as hydrophobicity and charge. Although positive charge appears to improve the efficacy of imaging, gene transfer, and drug delivery, a higher cytotoxicity from such constructs has been reported. Table 3.14 gives information about various mechanisms that are involved in the uptake of nanoparticles by cells and cause toxicities such as

TABLE 3.14
Studies Cellular Uptake and Binding-Dependent Toxicity

Type of Nanoparticle	Toxicity/Organ Affected	Mechanism Involved/Results	Reference
TiO_2 nanoparticle	Apoptosis	Increased levels of signaling molecules such as caspase-3.	Hussain et al. (2010); Ge et al. (2011)
Multi-walled carbon nanotubes	Cell necrosis	The cellular uptake of MWCNTs increases cell death (Alamar blue assay), increases membrane damage (LDH assay), and induces the release of cytokines (e.g., tumor necrosis factor α, IL-12, IL-10, IL-6, IL-1β, and IL-8) in BEAS-2B cells.	Tsukahara and Hainu (2011)
TiO_2; SiO_2; MWCNTs (<8, >50, 20–30) nanoparticles	Cellular uptake	Agglomerated TiO_2 nanoparticles form particle clusters and thus causes accumulation of higher amount of adsorbed protein compared to larger size particles, such as SiO_2 nanoparticles.	Sohaebuddin et al. (2010)

necrosis and apoptosis. For example, cationic surface charges for most nanoparticles correlate to higher cellular uptakes and greater cytotoxicities in nonphagocytic cells. Cationic nanoparticles appear to cause plasma-membrane disruptions to a great extent and anionic nanoparticles, apoptosis (Frohlich 2012).

3.5.8 Surface Coatings-Dependent Nanotoxicity

As the particle surface is in direct contact with body cells and biological media, these plays a very critical role in inducing toxicity. Materials, such as surfactants, in reactions with nanoparticles can change its physicochemical properties, react chemically (Oberdörster et al. 2005; Yin et al. 2005; Gupta and Gupta 2005), and affect their cytotoxicity. Surface coatings can render noxious particles, nontoxic, and less toxic particles, highly toxic. The presence of oxygen, ozone (Risom et al. 2005), oxygen radicals (Sayes et al. 2004), and transition metals (Donaldson and Stone 2003) on nanoparticle surfaces lead to the creation ROS and the induction of inflammation. For example, the specific cytotoxicity of silica is strongly associated with the occurrence of surface radicals and ROS (Hoet et al. 2004) (Table 3.15). Diesel exhaust particles in interactions with ozone can cause increased inflammation in the lungs of rats compared to diesel particles alone (Risom et al. 2005). Spherical gold nanoparticles with various surface coatings are not toxic to human cells, despite the fact that they are internalized (Goodman et al. 2004; Connor et al. 2005). CdSe quantum dots can be rendered nontoxic when appropriately coated (Derfus et al. 2004).

3.6 NANOPARTICLES TOXICITY COMPARISON STUDIES

Several studies have investigated the toxicity of nanoparticles by comparing their toxic effects as related to their physicochemical properties (Warheit et al. 2004; Lam et al. 2004; Soto et al. 2005; Muller et al. 2005; Braydich-Stolle et al. 2005; Hussain et al. 2009). These studies concluded that CNTs are extremely toxic and produce more damage to the lungs than carbon black or silica (Muller et al. 2005). Certain varieties of CNT aggregates, and some carbon blacks, were shown to be as cytotoxic as asbestos (Soto et al. 2005). Silver nanoparticle aggregates were found to be more toxic than asbestos, while TiO_2, aluminum, iron oxide, and zirconium oxide were found to be less toxic (Soto et al. 2005). Table 3.16 shows different types of nanomaterials, their sizes, and relative cytotoxicity indices (RCI) on macrophage cells (Soto et al. 2005). Table 3.17 shows the possible mechanisms and pathways by which nanotoxicity can take place for different types of cationic nanoparticles.

3.7 CONCLUSION

Considering several studies on the applications of nanotechnology versus the limited amount of studies related to the toxicities associated with their physicochemical properties, it is evident that there is a serious risk posed by it, attributed to the associated toxicities and due to the widespread commercialization of this novel technology. This demands for an urgent intervention of the regulatory bodies toward the identification of these risks and methods to mitigate and control them, thereby making this promising technology of the modern era free from any threats to human use. Immediate attention is envisaged to establish legislative mechanisms in regards to the application of nanotechnology to govern its safe use for mankind, thereby bridging the existing gaps between enormous commercial interests and the community's expectations for regulatory safeguards and protections.

Exhaustive studies are needed for the characterization of critical nanoparticles aimed at changing and controlling various physicochemical properties, such as size, shape, concentration, crystalline structure, and surface chemistry, and their impact on toxicities in humans. Apart from the focus on LD50 determinations, research should be aimed at determining the ED 50 (dose required to produce a therapeutic response in 50% of the population) to ascertain the effective therapeutic dose of

TABLE 3.15
Studies Surface Coating/Surface Modification on Nanoparticles

Type of Nanoparticle	Toxicity/Organ Affected	Mechanism Involved/Results	Reference
Quantum dots	Cytotoxicity/plasma membrane, mitochondria, nucleus	"Naked" quantum dots induce ROS resulting in damage to plasma membranes, mitochondria, and nucleus.	Lovric et al. (2005)
Cadmium telluride quantum dots	Lipid peroxidation of cell	Nonmodified quantum dots induced toxicity due to surface bioactive coating. Surface modification with N-acetylcysteine, reduced toxicity.	Choi et al. (2007)
TiO_2 nanoparticles	Lung toxicity	Warheit et al. found that only noncoated particles resulted in BAL alteration and lung histopathological changes compared to their coated counterparts, suggesting a possible influencing role of surface chemistry in nano-TiO_2-induced lung toxicity. In this regard, several studies have investigated the influence of particle chemical surface properties, surface coatings, and functionalization on pulmonary responses showing conflicting results.	Warheit et al. (2004)
Silica-coated coated rutile TiO_2 nanoparticles	Inflammatory reactions	Significant inflammatory reactions after acute and subacute exposure to coated nanoparticles, particularly silica-coated rutile TiO_2 NPs, rutile TiO_2 NPs surface modified with unspecified amount of zirconium, silicon, aluminum, and coated with polyalcohols and rutile Fe-doped TiO_2 NPs were reported.	Ivo et al. (2012)

nanoparticles to help in providing more accurate dosages. Studies on the toxicological implications of functionalized nanoparticles through surface modifications may help in deriving approaches to reduce the toxic potential of nanoparticles and, thus, manipulate or control the uptake and toxicity of these nanoparticles.

3.8 FUTURE PERSPECTIVES

Nano regulation is still undergoing major changes to encompass environmental, health, pharmaceutical, and safety issues. With appropriate efforts in focusing research to identify the associated toxicological risks, stringent regulatory interventions, improved corporate standards, and a will to safeguard the interest of society, the nanotechnology platform is on the verge of being a blessing to mankind.

TABLE 3.16
Nanomaterials, Their Size and Relative Cytotoxicity Index on Macrophage Cell

Material	Mean Aggregate Size (μm)	Mean Particle Size (nm)	RCI (at 5 μg/mL)	RCI (at 10 μg/mL)
Ag (silver)	1	30	1.5	0.8
Ag (silver)	0.4	30	1.8	0.1
Al_2O_3	0.7	50	0.7	0.4
Fe_2O_3	0.7	50	0.9	0.1
ZrO_2	0.7	20	0.7	0.6
TiO_2	1	Fibers 5–15 nm dia.	0.3	0.05
TiO_2	2.5	20 nm	0.4	0.1
Si_3N_4	1	60	0.4	0.06
Asbestos chrysolite	7	Fibers 20 nm dia. up to 500 aspect ratio	1	1
Carbon black	0.5	20	0.8	0.6
Single-walled carbon nanotubes	10	100 nm dia.	1.1	0.9
Multi-walled carbon nanotubes	2	15 nm dia.	0.9	0.8

Source: Data from Soto, K.F. et al. 2005. *J Nanoparticle Res* 7: 145–169.

TABLE 3.17
Possible Mechanisms of Nanotoxicity Caused by Cationic Nanoparticles

Membrane Damage/Leakage/Thinning	Cationic Nanoparticles
Protein binding/unfolding responses/loss of function/fibrillation	TiO_2, carbon nanoparticles
DNA cleavage and mutation	Ag nanoparticles
Mitochondrial damage: electron transfer/ATP/apoptosis	Ag and gold nanoparticles
Lysosomal damage: proton pump activity/lysis/frustrated phagocytosis	Ag, gold nanoparticles and carbon nanotubes (CNTs)
Inflammation: signaling cascade/cytokines/chemokines/adhesion	Metal oxide nanoparticles (e.g., TiO_2) and CNTs
Fibrogenesis and tissue remodeling injury	CNTs
Blood platelet, vascular endothelial, and clotting abnormalities	SiO_2
Oxidative stress injury, radical production, GSH (glutathione) depletion, lipid peroxidation, membrane oxidation, protein oxidation.	CNTs, metal oxide nanoparticles, cationic nanoparticles

Source: Data from Ramakrishna, D. and Pragna, R. 2011. *J. Int. Fed. Clin. Chem. Lab. Med* 22:1.

REFERENCES

Alkilany, A.M., Nagaria, P.K., Hexel, C.R., Shaw, T.J., Murphy, C.J., Wyatt, M.D. 2009. Cellular uptake and cytotoxicity of gold nanorods: Molecular origin of cytotoxicity and surface effects. *Small* 5:701–708.

Asharani, P.V., LowKah, M.G., Hande, M.P., Valiyaveettil, S. 2009. Cytotoxicity and genotoxicity of silver nanoparticles in human cells. *Am Chem Soc Nanotechnol* 3:279–290.

Badea, I., Wettig, S., Verrall, R., Foldvari, M. 2007. Topical non-invasive gene delivery using gemini nanoparticles in interferon-gamma-deficient mice. *Eur J Pharm Biopharm* 65(3):414–422.

Barillet, S., Simon-Deckers, A., Herlin-Boime, N., Mayne-L' H., Reynaud, C. et al. 2010. Toxicological consequences of TiO$_2$, SiC nanoparticles and multiwalled carbon nanotubes exposure in several mammalian cell types: An in-vitro study. *J Nanoparticle Res* 12:61–73.

BeruBe, K., Balharry, D., Sexton, K., Koshy, L., Jones, T. 2007. Combustion derived nanoparticles: Mechanisms of pulmonary toxicity. *Clin Exp Pharmacol Physiol* 34(10):1044–1050.

Bhadra, D., Bhadra, S., Jain, P., Jain, N.K. 2002. Pegnology: A review of PEG-ylated systems. *Pharmazie* 57:5–29.

Bottini, M., Bruckner, S., Nika, K., Bottini, N., Bellucci, S., Magrini, A. et al. 2006. Multi-walled carbon nanotubes induce T lymphocyte apoptosis. *Toxicol Lett* 160(2):121–126.

Bowden, L.P., Royer, M.C. et al. 2011. Rapid onset of argyria induced by a silver-containing dietary supplement. *J Cutaneous Pathol* 38(10):832–835.

Brandt, D., Park, B. et al. 2005. Argyria secondary to ingestion of homemade silver solution. *J Am Acad Dermatol* 53(2):S105-1–S105-7.

Braydich-Stolle, L., Hussain, S., Schlager, J., Hofmann, M.C. 2005. *In vitro* cytotoxicity of nanoparticles in mammalian germ-line stem cells. *Toxicol Sci* 88(2):412–419.

Braydich-Stolle, L.K., Schaeublin, N.M., Murdock, R.C., Jiang, J., Biswas, P. et al. 2009. Crystal structure mediates mode of cell death in TiO$_2$ nanotoxicity. *J Nanoparticle Res* 11:1361–1374.

Brunner, T.J., Wick, P., Manser, P., Spohn, P., Grass, R.N., Limbach, L.K. et al. 2006. *In vitro* cytotoxicity of oxide nanoparticles: Comparison to asbestos, silica, and effects of particle solubility. *Environ Sci Technol* 44:4373–4381.

Buzea, C., Blandino, I.I.P., Robbi, K. 2007. Nanomaterials and nanoparticles: Sources and toxicity. *Biointerphases* 2(4):MR 17–MR 172.

Chan, W.C. 2006. Bionanotechnology progress and advances. *Biol Blood Marrow Transplant* 12:87–91.

Chang, J.S, Chang, K.L.B, Hwang, D.F. et al. 2007. *In vitro* toxicity of silica nanoparticles at high concentrations strongly depends on the metabolic activity type of the cell line. *Environ Sci Technol* 41:2064–2080.

Chen, H.W., Su, S.F., Chein, C.T., Lin, W.H., Yu, S.L., Chou, C.C., Chen, J.J.W., Yang, P.C. 2006a. Titanium dioxide nanoparticles induce emphysema-like lung injury in mice. *J Fed Am Soc Exp Biol* 20:2393–2395.

Cheng, M.M., Cuda, G., Bunimovich, Y.L. et al. 2006. Nanotechnologies for biomolecular detection and medical diagnostics. *Curr Opin Chem Biol* 10:11–19.

Choi, A.O., Cho, S.J., Desbarats, J. et al. 2007. Quantum dot-induced cell death involves Fas upregulation and lipid peroxidation in human neuroblastoma cells. *J Nanobiotechnol* 12:1.

Churg, A., Stevend, B., Wright, J.L. 1998. Comparison of the uptake of fine and ultrafine TiO$_2$ in a tracheal explant system. *Am J Physiol* 274:L81–L86.

Colvin, V.L. 2003. The potential environmental impact of engineered nanomaterials. *Nat Biotechnol* 21:1166–1170. doi: 10.1038/nbt875.

Conner, S.D., Schmid, S.L. 2003. Regulated portals of entry into the cell. *Nature* 422:37–44. doi: 10.1038/nature01451ER.

Connor, E.E., Mwamuka, J., Gole, A., Murphy, C.J., Wyatt, M.D. 2005. Gold nanoparticles are taken up by human cells but do not cause acute cytotoxicity. *Small* 1:325–327.

Cui, D., Tian, F., Ozkan, C.S., Wang, M., Gao, H. 2005. Effect of single wall carbon nanotubes on human HEK293 cells. *Toxicol Lett* 155:73–85.

Curtis, J., Greenberg, M., Kester, J., Phillips, S., Krieger, G. 2006. Nanotechnology and nanotoxicology: A primer for clinicians. *Toxicol Sci* 25(4): 245–260.

Dai, H. 2002. Carbon nanotubes: Opportunities and challenges. *Surf Sci* 500: 218–241.

Derfus, A.M., Chan, W.C.W., Bhatia, S.N. 2004. Probing the cytotoxicity of semiconductor quantum dots. *Nano Lett* 4:11–18.

Des Rieux, A., Fievez, V. et al. 2007. An improved *in vitro* model of human intestinal follicle-associated epithelium to study nanoparticle transport by M cells. *Eur J Pharmaceut Sci* 30(5):380–391.

Desai, M.P., Labhasetwar, V., Walter, E., Levy, R.J., Amidon, G.L. 1997. The mechanism of uptake of biodegradable microparticles in Caco-2 cells is size dependent. *Pharm Res* 14, 1568–1573.

Di-Giorgio, M.L., Di Bucchianico, S., Ragnelli, A.M., Aimola, P., Santucci, S., Poma, A. 2011. Effects of single and multiwalled carbon nanotubes on macrophages: Cytotoxicity and genotoxicity and electron microscopy. *Mutation Res: Gene Toxicol Environ Mutagen* 722:20–31.

Donaldson, K., Stone, V. 2003. Current hypotheses on the mechanisms of toxicity of ultrafine particles. *Ann Ist Super Sanita* 39:405–410 and references therein.

Donaldson, K., Stone, V., Tran, C.L., Kreyling, W., Borm, P.J.A. 2004. Nanotoxicology. *Occup Environ Med* 61:727–728.

Duffin, R., Mills, N.L., Donaldson, K. 2007. Nanoparticles—A thoracic toxicology perspective. *Yonsei Med J* 48(4):561–572.

Emerich, D.F. 2005. Nanomedicine—Prospective therapeutic and diagnostic applications. *Expert Opin Biol Ther* 5:1–5.

Emerich, D.F., Thanos, C.G. 2003. Nanotechnology and medicine. *Expert Opin Biol Ther* 3:655–663.

Ferin, J., Oberdörster, G., Penney, D.P. 1992. Pulmonary retention of ultrafine and fone particles in rats. *Am J Respir Cell Mol Biol* 6:535–552.

Frohlich, E. 2012. The roll of surface charge in cellular uptake and cytotoxicity of medical nanoparticles. *Int J Nanomed* 7:5577–5591.

Garnett, M.C., Kallinteri, P. 2006. Nanomedicines and nanotoxicology: Some physiological principles. *Occup Med (Lond)* 56(5):307–311.

Ge, H., Hu, Y., Jiang, X. et al. 2002. Preparation, characterization, and drug release behaviors of drug nimodipine-loaded poly (epsilon-caprolactone)-poly(ethyleneoxide)-poly(epsilon-caprolactone) amphiphilic triblock copolymer micelles. *J Pharm Sci* 91:1463–1473.

Ge, Y., Bruno, M., Wallace, K., Winnik, W., Prasad, R.Y. 2011. Proteome profiling reveals potential toxicity and detoxification pathways following exposure of BEAS-2B cells to engineered nanoparticle titanium dioxide. *Proteomics* 11:2406–2422.

Goodman, C.M., McCusker, C.D., Yilmaz, T., Rotello, V. 2004. Toxicity of gold nanoparticles functionalized with cationic and anionic side chains. *Bioconjugate Chem* 15:897–900.

Gupta, A.K., Gupta, M. 2005. Cytotoxicity suppression and cellular uptake enhancement of surface modified magnetic nanoparticles. *Biomaterials* 26:1565–1573.

Gurr, J. R., Wang, A.S.S., Chen, C.H., Jan, K.Y. 2005. Ultrafine titanium dioxide particles in the absence of photoactivation can induce oxidative damage to human bronchial epithelial cells. *Toxicology* 213:66–73.

Hackenberg, S., Friehs, G., Froelich, K., Ginzkey, C., Koehler, C. et al. 2010. Intracellular distribution, gen and cytotoxic effects of nanosized titanium dioxide particles in the anatase crystal phase on human nasal mucosa cells. *Toxicol Lett* 195:9–14.

Hagens, W.I., Oomen, A.G., De Jong, W.H., Cassee, F.R., Sips, A.J. 2007. What do we (need to) know about the kinetic properties of nanoparticles in the body? *Regul Toxicol Pharmacol* 49(3):217–219.

Harush-Frenkel, O., Rozentur, E., Benita, S., Altschuler, Y. 2008. Surface charge of nanoparticles determines their endocytic and transcytotic pathway in polarized MDCK cells. *Biomacromolecules* 9:435–443.

Helmus, M. 2007. The need for rules and regulations. *Nat Nanotechnol* 2:333–334. doi:10.1038/nnano.2007.165.

Hillyer, J.F., Albrecht, R.M. 2001. Gastrointestinal persorption and tissue distribution of differently sized colloidal gold nanoparticles. *J Pharmaceut Sci* 90:1927–1936.

Hoet, P.H.M., Bruske-Hohlfeld, I., Salata, O.V. 2004. Nanoparticles—Known and unknown health risks. *J Nanobiotechnol* 2:12–27.

Hsiao, L.H., Huang, Y.J. 2011. Effects of various physicochemical characteristics on the toxicities of ZnO and TiO_2 nanoparticles toward human lung epithelial cells. *Sci Total Environ* 409:1219–1228.

Hussain, S., Boland, S., Baeza-Squiban, A., Hamel, R., Thomassen, L.C.J., Martens, J.A. et al. 2009. Oxidative stress and proinflammatory effects of carbon black and titanium dioxide nanoparticles: Role of particle surface area and internalized amount. *Toxicology* 260:142–149.

Hussain, S., Thomassen, L.C.J, Ferecatu, I., Borot, M.-C., Andreau, K. et al. 2010. Carbon black and titanium dioxide elicit distinct apoptotic pathways in bronchial epithelial cells. *Particle Fibre Toxicol* 7:10.

Jani, P., Halbert, G.W. et al. 1990. Nanoparticle uptake by the rat gastrointestinal mucosa—Quantitation and particle size dependency. *J Pharmacy Pharmacol* 42(12):821–826.

Jia, G., Wang, H.F., Yan, L., Wang, X., Pei, R.J. et al. 2005. Cytotoxicity of carbon nanotube, multiwall nanotube and fullerene. *Environ Sci Technol* 39:1378–1383.

Jiang, J., Oberdorster, G., Elder, A., Gelein, R., Mercer, P., Biswas, P. 2008. Does nanoparticle activity depend upon size and crystal phase? *Nanotoxicology* 2:33–42.

Jong, W.H.D., Borm, P.J. 2008. Drug delivery and nanoparticles: Applications and hazards. *Int J Nanomed* 3(2):133–149.

Karakoti, A.S., Hench, L.L., Seal, S. 2006. The potential toxicity of nanomaterials—The role of surfaces. *JOM* 58:77–82.

Kim, Y.S., Song, M.Y. et al. 2010. Subchronic oral toxicity of silver nanoparticles. *Particle Fibre Toxicol* 7:1–11.

Kirchner, C., Liedl, T., Kudera, S. et al. 2005. Cytotoxicity of colloidal CdSe and CdSe/ZnS nanoparticles. *Nano Letters* 5:331–338.

Kommareddy, S., Tiwari, S.B., Amiji, M.M. 2005. Long-circulating polymeric nanovectors for tumor-selective gene delivery. *Technol Cancer Res Treat* 4:615–625.

Kreuter, J. 2007. Nanoparticles—A historical perspective. *Int J Pharm* 331:1–10.

Kurath, M., Maasen, S. 2006. Toxicology as a nanoscience?—Disciplinary identities reconsidered. *Part Fibre Toxicol* 3(6) doi: 10.1186/1743-8977-3-6.

Lam, C.W., James, J.T., McCluskey, R., Hunter, R.L. 2004. Pulmonary toxicity of single-wall carbon nanotubes in mice 7 and 90 days after intratracheal instillation. *Toxicol Sci* 77:126–134.

Langer, R. 2002. Biomaterials in drug delivery and tissue engineering: One laboratory's experience. *Acc Chem Res* 33:94–101.

Lanone, S., Boczkowski, J. 2006. Biomedical applications and potential health risks of nanomaterials: Molecular mechanisms. *Curr Mol Med* 6(6):651–663.

Lee, M., Kim, S.W. 2005. Polyethylene glycol-conjugated copolymers for plasmid DNA delivery. *Pharm Res* 22:1–10.

Lewinski, N., Colvin, V., Drezek, R. 2008. Cytotoxicity of nanoparticles. *Small* 4:26–49. doi:10.1002/smll.2007 00595.

Lin, W., Huang, Y.W., Zhou, X.D. et al. 2006. *In vitro* toxicity of silica nanoparticles in human lung cancer cells. *Toxicol Appl Pharmacol* 217:252–259.

Lippmann, M. 1990. Effects of fiber characteristics on lung deposition, retention, and disease. *Environ Health Perspec* 88:311–317.

Loeschner, K., Hadrup, N. et al. 2011. Distribution of silver in rats following 28 days of repeated oral exposure to silver nanoparticles or silver acetate. *Particle Fiber Technol* 8. doi: 10.1186/1743-8977-8-18.

Lovric, J., Cho, S.J., Winnik, F.M. et al. 2005. Unmodified cadmium telluride quantum dots induce reactive oxygen species formation leading to multiple organelle damage and cell death. *Chem Biol* 12:1159–1161.

Matsudai, M., Hunt, G. 2005. Nanotechnology and public health. *Nippon Koshu Eisei Zasshi* 52:923–927.

Maynard, A.D., Aitken, R.J., Butz, T. et al. 2006. Safe handling of nanotechnology. *Nature* 444:267–269. doi: 10.1038/444267a

Medina, C., Santos-Martinez, M.J., Radomski, A., Corrigan, O.I., Radomski, M.W. 2007. Nanoparticles: Pharmacological and toxicological significance. *Br J Pharmacol* 150(5):552–558.

Moghimi, S.M., Hunter, A.C., Murray, J.C. 2005. Nanomedicine: Current status and future prospects. *FASEB J* 19:311–330.

Mohanraj, V.J., Chen, Y. 2006. Nanoparticles—A review. *Trop J Pharm Res* 5(1):561–573.

Monteiro-Riviere, N.A., Nemanich, R.J., Inman, A.O., Wang, Y.Y., Riviere, J. E. 2005. Multi-walled carbon nanotube interactions with human epidermal keratinocytes. *Toxicol Lett* 155:377–384.

Muller, J., Huaux, F., Moreau, N., Misson, P., Heiliea, J.F., Delos, M., Arras, M., Fonseca, A., Nagyb, J.B., Lison, D. 2005. Respiratory toxicity of multi-wall carbon nanotubes. *Toxicol Appl Pharmacol* 207:221–231.

Nahar, M., Dutta, T., Murugesan, S. et al. 2006. Functional polymeric nanoparticles: An efficient and promising tool for active delivery of bioactives. *Crit Rev Ther Drug Carrier Syst* 23(4):259–318.

Nel, A., Xia, T., Madler, L., Li, N. 2006. Toxic potential of materials at the nanolevel. *Science* 311:622–627. doi: 10.1126/science.1114397.

Niidome, T., Yamagata, M., Okamoto, Y. et al. 2006. PEG-modified gold nanorods with a stealth character for *in vivo* applications. *J Control Release* 114:343–347.

Oberdörster, G. 2002. Toxicokinetics and effects of fibrous and nonfibrous particles. *Inhalation Toxicol* 14:29–56.

Oberdörster, G., Ferin, J., Lehnert, B.E. 1994. Correlation between particle size, *in vivo* particle persistence, and lung injury. *Environ Health Perspect* 102(5):173–179.

Oberdörster, G., Oberdorster, E., Oberdorster, J. 2005. Nanotoxicology: An emerging discipline evolving from studies of ultrafine particles. *Environ Health Perspect* 113(7):823–839.

Oberdörster, G., Sharp, Z., Atudorei, V. et al. 2004. Translocation of inhaled ultrafine particles to the brain. *Inhal Toxicol* 16:437–445. doi: 10.1080/08958370490439597.

Osaka, T., Nakanishi, T., Shanmugam, S., Takahama, S., Zhang, H. 2009. Effect of surface charge of magnetite nanoparticles on their internalization into breast cancer and umbilical vein endothelial cells. *Colloids Surf B: Biointerfaces* 71:325–330.

Pan, Y., Leifert, A., Ruau, D. et al. 2009. Gold nanoparticles of diameter 1.4 mm trigger necrosis by oxidative stress and mitochondrial damage. *Small* 5(18):2067–2076.

Panyam, J., Labhasetwar, V. 2003. Biodegradable nanoparticles for drug and gene delivery to cells and tissue. *Adv. Drug Delivery Rev* 55:329–347.

Panyam, J., Sahoo, S.K., Prabha, S., Bargar, T., Labhasetwar, V., 2003b. Fluorescence and electron microscopy probes for cellular and tissue uptake of poly(D,L-lactide-coglycolide) nanoparticles. *Int J Pharm* 262, 1–11.

Parveen, S., Misra, R., Sahoo, S.K. 2012. Nanoparticles: A boon to drug delivery, therapeutics, diagnostics and imaging. *Nanomedicine: NBM* 8:147–166. doi: 10.1016/j.nano.2011.05.016.

Powers, K., Brown, S., Krishna, V., Wasdo, S., Moudgil, B., Roberts, S. 2006. Research strategies for safety evaluation of nanomaterials. Part VI. Characterization of nanoscale particles for toxicological evaluation. *Toxicol Sci* 90:296–303.

Ramakrishna, D. and Pragna, R. 2011. Nanoparticles: Is toxicity a concern? *J. Int. Fed. Clin. Chem. Lab. Med* 22:1.

Rejeski, D. 2009. Nanotechnology and consumer products. Project on Emerging Nanotechnologies: WoodrowWilson International Center for Scholars.

Renwick, L.C., Brown, D., Clouter, A., Donaldson, K. 2004. Increased inflammation and altered macrophage chemotactic responses caused by two ultrafine particle types. *Occup Environ Med* 61:442–446.

Risom, L., Moller, P., Loft, S. 2005. Oxidative stress-induced DNA damage by particulate air pollution. *Mutat Res* 592:119–137.

Roduner, E. 2006. Size matters: Why nanomaterials are different. *Chem Soc Rev* 35:583–592.

Ronconi, L., Marzano, C. et al. 2006. Gold (III) dithiocarbamate derivatives for the treatment of cancer: Solution chemistry, DNA binding, and hemolytic properties. *J Med Chem* 49:1648–1657.

Sahoo, S.K., Labhasetwar, V. 2003. Nanotech approaches to drug delivery and imaging. *Drug Discov Today* 8:1112–1120.

Sahoo, S.K., Parveen, S., Panda, J.J. 2007. The present and future of nanotechnology in human health care. *Nanomedicine: Nanotechnol Biol Med* 3:20–31.

Sanchez, V.C., Pietruska, J.R., Miselis, N.R., Hurt, R.H., Kane, A.B. 2009. Biopersistence and potential adverse health impacts of fibrous nanomaterials: What have we learned from asbestos? *Wiley Interdiscip Rev Nanomed Nanobiotechnol* 1(5):511–529.

Sara, A.L., Melissa, A.M., John, W.T., Yu-Shen, L., Christy, L.H. 2012. Assessing nanoparticle toxicity. *Annu Rev Anal Chem* 5:181–205.

Sarin, H., Kanevsky, A.S., Wu, H.T. et al. 2008. Effective transvascular delivery of nanoparticles across the blood–brain tumor barrier into malignant glioma cells. *J Transl Med* 6:1–15. doi: 10.1186/1479-5876-6-80.

Sayes, C.M., Fortner, J.D., Guo, W., Lyon, D., Boyd, A.M., Ausman, K.D., Tao, Y.J. et al. 2004. The differential cytotoxicity of water-soluble fullerenes *Nano Letters* 4:1881–1887.

Sayes, C.M., Gobin, A.M., Ausman, K.D. et al. 2006a. Functionalization density dependence of single-walled carbon nanotubes cytotoxicity in-vitro. *Toxicol Lett* 161:135–142.

Sayes, C.M., Wahi, R., Kurian, P.A., Liu Y.P., West, J.L., Ausman, K.D. et al. 2006b. A cytotoxicity and inflammatory response study with human dermal fibroblasts and human lung epithelial cells. *Toxicol Sci* 92:174–185.

Sayes, M.C., Wahi, R., Kurian, P.A., Liu, Y., West, J.L., Ausman, K.D. et al. 2006c. Correlating nanoscale titania structure with toxicity: A cytotoxicity and inflammatory response study with human dermal fibroblasts and human lung epithelial cells. *Toxicol Sci* 92:174–85.

Scientific committee on emerging and newly identified health risks (SCENIHR) modified opinion (after public consultation) on the appropriateness of existing methodologies to assess the potential risks associated with engineered and adventitious products of nanotechnologies 2006. http://ec.europa.eu/health/ph_risk/documents/synth_report.pdf

Shaffer, C. 2005. Nanomedicine transforms drug delivery. *Drug Discov Today* 10:1581–1582.

Shvedova, A.A., Castranova, V., Kisin, E.R. ct al. 2003. Exposure to carbon nanotube material: Assessment of nanotube cytotoxicity using human keratinocyte cells. *J Toxicol, Environ Health* 66, no. Part A: 1909–1926.

Singh, R., Lillard, J.W. Jr. 2009. Nanoparticle-based targeted drug delivery. *Exp Mol Pathol* 86:215–223.

Sohaebuddin, S. K., Thevenot, P. T., Baker, D., Eaton, J. W., Tang, L. 2010. Nanomaterial cytotoxicity is composition, size and cell type dependent. *Part Fibre Toxicol* 7:22.

Sonavane, G., Tomoda, K., Makino, K. 2008. Biodistribution of colloidal gold nanoparticles after intravenous administration: Effect of particle size. *Colloids Surf B* 66:274–280. doi: 10.1016/j.colsurfb.2008.07.004.

Soto, K.F., Carrasco, A., Powell, T.G., Garza, K.M., Murr, L.E. 2005. Comparative *in vitro* cytotoxicity assessment of some manufactured nanoparticulate materials characterized by transmission electron microscopy. *J Nanoparticle Res* 7:145–169.

Takenaka, S., Karg, E., Roth, C., Schulz, H., Ziesenis, A., Heinzmann, U., Schramel, P., Heyder, J. 2001. Pulmonary and systemic distribution of inhaled ultrafine silver particles in rats. *Environ Health Persp* 109(Suppl. 4):547–551.

Tsukahara, T., Haniu, H. 2011. Cellular cytotoxic response induced by highly purified multi-wall carbon nanotube in human lung cells. *Mol Cell Biochem* 352:57–63.

Turner, M., Golovko, V.B., Vaughan, O.P.H. et al. 2008. Selective oxidation with dioxygen by gold nanoparticle catalysts derived from 55-atom clusters. *Nature* 454:U31-981.

Varner, K.E., El-badawy, A., Feldhake, D., Venkatapathy, R. 2010. *State-of-the-Science Review: Everything Nanosilver and More. U.S. Environmental Protection Agency*. Washington, DC, EPA/600/R-10/084.

Vecchio, G., Antonio, G., Virgilio, B., Gabriele, M., Stefenia, S., Roberto, C., Pier, P.P. 2012. Concentration-dependent, size-independent toxicity of citrate capped AuNPs in *Drosophila melanogaster*. *PLoS ONE* 7(1):e29980. doi: 10.1371/journal.pone.0029980.

Wagner, A.J., Bleckmann, C.A., Murdock, R.C., Schrand, A.M., Schlager, J.J., Hussain, S.M. 2007. Cellular interaction of different forms of aluminum nanoparticles in rat alveolar macrophages. *J Phys Chem* 111(25):7353–7359.

Wang, R.H., Mikoryak, C., Li, S.Y., Bushdiecker, D., Musselman, I.H. et al. 2011. Cytotoxicity screening of single-walled carbon nanotubes: Detection and removal of cytotoxic contaminants from carboxylated carbon nanotubes. *Mol Pharmacol* 8:1351–1361.

Warheit, D.B., Laurence, B.R., Reed, K.L., Roach, D.H., Reynolds, G.A.M., Webb, T.R. 2004. Comparative toxicity assessment of single wall carbon nanotubes in rats. *Toxicol Sci* 77:117–125.

Warheit, D.B., Webb, T.R., Reed, K.L., Frerichs, S., Sayes, C.M. 2007. Pulmonary toxicity study in rats with three forms of ultrafine-TiO_2 particles: Differential responses related to surface properties. *Toxicology* 230(1):90–104.

Wick, P., Manser, P., Limbach, L.K., Dettlaff, W.U., Krumeich, F. et al. 2007. The degree and kind of agglomeration affect carbon nanotube cytotoxicity. *Toxicol Lett* 168:121–131.

Williams, D. 2004. Nanotechnology: A new look. *Med Device Technol* 15:9–10.

Wilson, M.R., Lightbody, J.H., Donaldson, K., Sales, J., Stone, V. 2002. Interactions between ultrafine particles and transition metals *in vivo* and *in vitro*. *Toxicol Appl Pharmacol* 184:172–179.

Wu, J., Sun, J., Xue, Y. 2010. Involvement of JNK and P53 activation in G2/M cell cycle arrest and apoptosis induced by titanium dioxide nanoparticles in neuron cells. *Toxicol Lett* 199:269–276.

www.bccresearch.com (accessed on March, 2013)

Xia, T., Kovochich, M., Brant, J., Hotze, M., Sempf, J., Oberley, T., Sioutas, C., Yeh, J.I., Wiesner, M.R., Nel, A.E. 2006. Comparison of the abilities of ambient and manufactured nanoparticles to induce cellular toxicity according to an oxidative stress paradigm. *Nano Letters* 6:1794–1807.

Xia, T., Kovochich, M., Nel, A.E. 2007. Impairment of mitochondrial function by particulate matter (PM) and their toxic components: Implications for PM-induced cardiovascular and lung disease. *Front Biosci* 12:1238–1246.

Yamamoto, A., Honma, R., Sumita, M., Hanawa, T. 2004. Cytotoxicity evaluation of ceramic. *J Biomed Mater Res* 68A:244–256.

Yang, H., Liu, C., Yang, D.F., Zhang, H.S., Xi, Z.G. 2009. Comparative study of cytotoxicity, oxidative stress and genotoxicity induced by four typical nanomaterials: The role of particle size, shape and composition. *J Appl Toxicol* 29:69–78.

Yin, H., Too, H.P., Chow, G.M. 2005. The effect of particle size and surface coating on the cytotoxicity of nickel ferrite. *Biomaterials* 26:5818–5826.

Yoshikawa, T., Tsutsumi, Y., Nakagawa, S. 2006. Development of nanomedicine using intracellular DDS. *Nippon Rinsho* 64:247–252.

Zhang, L.W., Monteiro-Riviere, N.A. 2009. Mechanisms of quantum dot nanoparticle cellular uptake. *Toxicol Sci* 110:138–155.

Zhang, X.D., Wu, H.Y. et al. 2010c. Toxicologic effects of gold nanoparticles *in vivo* by different administration routes. *Int J Nanomed* 5:771–781.

4 Cytotoxicity of Stimuli-Responsive Nanomaterials
Predicting Clinical Viability through Robust Biocompatibility Profiles

Daniel Wehrung and Moses O. Oyewumi

CONTENTS

4.1 Introduction 103
4.2 Photoresponsive Materials 104
4.3 pH-Responsive Systems 106
4.4 Enzyme-Responsive Systems 107
4.5 Factors Influencing Biocompatibility of Stimuli-Responsive Nanomaterials 107
4.6 Representative Methods of Biocompatibility Assessment for Stimuli-Responsive Nanomaterials 108
4.7 Perspectives on the Field of Stimuli-Responsive Biomaterials 109
Acknowledgments 112
Abbreviations 112
References 113

4.1 INTRODUCTION

Recent advances in synthetic methods of polymerization such as controlled-radical polymerization, encompassing atom transfer radical polymerization, reversible addition–fragmentation chain transfer, and nitroxide-mediated polymerization have allowed more control over polymerization reactions and as such, allowed the development of increasingly sophisticated polymers (Sauer et al., 2001; Checcot et al., 2003; Rakhmatullina et al., 2007). These and other advancements in synthetic techniques have facilitated the development of stimuli-responsive or "smart" drug delivery systems. Stimuli-responsive delivery systems respond to some form of stimuli, be it external or internal. External stimuli are those that are applied from outside the body and include light, magnetic fields, and ultrasound; while internal (also known as proximal) stimuli are those found in the microenvironment of a specific part of the body (e.g., diseased tissue) or a specific cellular compartment (e.g., lysosome). Internal stimuli include pH, redox potential, local temperature, and enzyme overexpression. Regardless of the form of stimuli used to activate the system, the overarching principle remains the same; when the material is activated by its stimulus, it undergoes a physical/chemical change (hydrophobic–hydrophilic balance, oxidation state, secondary structure), which results in the release of entrapped therapeutic payloads. This allows for temporal and spatial control of drug release from the system, which in turn increases drug efficacy, decreases side effects, and enhances the therapeutic window.

The potential benefits that these systems offer have not gone unnoticed, and there has been a rapid increase in the number of publications centered on stimuli-responsive drug delivery systems in the last 15 years (Schmers et al., 2010). However, to date the majority of these publications have focused on the unique chemistry of activation, and have not thoroughly investigated the biocompatibility of the novel material. While obvious proof of concept studies are critical to furthering the field of stimuli-responsive drug delivery systems, these studies need to be accompanied by a thorough investigation into the biocompatibility of the novel material if they are to translate to a clinically relevant solution for drug delivery. While many reports have been made regarding the design and potentials of stimuli-responsive drug delivery systems, the issues of biocompatibility and safety concerns have not been properly addressed. For this reason, this chapter will focus on presenting key points on stimuli-responsive biomaterials and nanocarriers while assessing the need for biocompatibility testing of new stimuli-responsive materials as well as the safety concerns of stimulus that are used to trigger the system.

4.2 PHOTORESPONSIVE MATERIALS

The use of light as the activating stimulus has many potential advantages over other forms of stimuli. The intrinsic properties of light allow for precise temporal and spatial control of system activation due to the inherent ability to tune such systems to respond according to a multitude of parameters, including wavelength, intensity, duration, and location of exposure. Additionally, since photoresponsive materials utilize an externally applied light source for activation, these systems have the potential to function as generic platforms, which can be used to treat a variety of diseases through manipulation of the therapeutic payload. Photoresponsive materials have been developed for numerous applications including selective drug delivery (Lu et al., 2008), photothermal ablation therapy (Boca et al., 2011), and micropatterning of implantable devices (Sun et al., 2012). For those systems aimed at providing selective drug delivery, a variety of approaches have been investigated, including nanoimpellers, polymer scission, isomer switching, micelle/nanoparticle (NP) disruption, and even pulsed (on–off) release—capable of turning drug release "on" or "off" based on the wavelength of light used to irradiate the system.

Lu et al. (2008) developed a system for selective drug delivery using a photo-activated nanoimpeller that controlled the release of the anticancer drug camptothecin from mesoporous silica NPs. In this system, azobenzene was linked to organosilane, which was subsequently attached to mesoporous silica NPs such that it covered the surface of the pores. Drug molecules loaded into these pores could then be selectively released inside cancer cells upon photoactivation due to the switching of azobenzene between the *cis* and *trans* isomers when irradiated with visible light. The biocompatibility of this system was investigated following 72 h incubation by performing a cell count after staining the cells with propidium iodide and Hoechst 33342. At the concentrations tested (10 and 100 µg/mL), the cellular viabilities were near 100% for both cell lines (PANC-1 and SW480), which demonstrated *in vitro* biocompatibility of the system. This system utilized visible light (violet, 413 nm) for activation of the chromophore. Irradiation of the cells with 413 nm light was found to not inhibit cellular proliferation even at the maximum exposure time of 10 min.

Fomina et al. (2011) took a different approach in the development of a photo-activated NP-based system for selective drug delivery. In their system, irradiation with ultraviolet (UV) or near-infrared (NIR) light resulted in the photolysis of the protecting group 4-bromo-7-hydroxycoumarin (Bhc). This triggered a cascade of rearrangement reactions and cyclizations that ultimately resulted in complete breakdown of the polymer and release of entrapped drug. Using the 3-(4,5-dimethyl-thiazol-2-yl)-2,5-diphenyltetrazolium bromide (MTT) assay, the authors were able to claim biocompatibility of both the intact NPs as well as the scission products over a concentration range of 1.17–300 µg/mL. The authors also stated that the dose of NIR light required to activate the system was below the previously published limits for safe irradiation of living systems, but did not provide any experimental evidence to validate this claim.

The biocompatibility of micelles constructed from a novel comb-like poly(ethylene glycol) (CPEG) derivative functionalized with the photoresponsive 2-diazo-1,2-naphthoquinone (DNQ) motif was described by Chen et al. (2011). These micelles rupture and release their therapeutic payload in response to irradiation with UV light (365 nm). This disruption in the micelle's structure is caused by the DNQ motif, which upon irradiation switches from hydrophobic DNQ to hydrophilic 3-indenecarboxylic acid. In order to evaluate the biocompatibility of this new material, the authors utilized the MTT assay. Following a 24 h incubation of the CPEG-g-DNQ micelles with both HepG2 and HUVEC cell lines, cellular viabilities remained above 90% throughout the entire concentration range tested (10–200 mg/L). The observed biocompatibility through the cell-based test was attributed to the incorporation of CPEG within the polymer's structure. A major concern in the study is that only one time point was used, coupled with the limitation that the by-products of micelle disruption and the effects of irradiating cells with 365 nm light were not reported.

Another method utilizing a photoresponsive system was developed by Boca et al. and utilizes chitosan-coated silver nanotriangles (Chit-AgNTs) as photothermal transducers in order to induce local hyperthermia in tumor tissues following irradiation (photothermal ablation therapy) (Boca et al., 2011). This system utilized NIR light (800 nm) in order to induce local hyperthermia, and has the additional benefit of utilizing chitosan, a generally regarded as safe (GRAS) excipient (Keefe, 2011). In this work, biocompatibility claims were based on assessment of the Chit-AgNTs in both human cancerous cells (NCI-H460) and benign human cell (human embryonic kidney) on the basis of changes in cellular morphology as well as counting viable cells following propidium iodide/Hoechst double staining. Both these methods of assessment were conducted following 24 h incubation of the Chit-AgNTs with the cells. Analysis of the cellular morphology showed that the Chit-AgNTs treated cells showed little deviation from untreated cells, with only 3.5% of the cells displaying an abnormal, rounded morphology. Further analysis using the double staining procedure showed that over the concentration range tested (0.17–1.71 µg/mL), the Chit-AgNTs reduced the viability of normal human embryonic kidney cells by less than 5%. Interestingly, the Chit-AgNTs were significantly more cytotoxic to cancerous cells; showing cellular viabilities of 85% and 75% at the 1.37 and 1.71 µg/mL concentrations, respectively.

Despite the intensive efforts in the development of new photoresponsive materials, investigation into the biological effects of the light required to activate such systems has been limited. To date, most photoresponsive materials rely on UV light to activate the system due to the exquisite sensitivity of UV-responsive chromophores. However, this poses a serious problem to the clinical relevance of such systems due to UV light's genotoxicity and lack of tissue penetration (McMillan et al., 2008). In order to circumvent the well-established limitations associated with UV light, as well as to take advantage of some of the unique and beneficial characteristics of NIR light (e.g., tissue transparency and biologically benign) (Zhao et al., 2007), many research groups have shifted their focus toward the discovery/development of chromophores that respond efficiently to NIR light. Another reason for the recent emergence of NIR light as a trigger comes as a result of recent advances in laser technology; specifically the development of high-energy femtosecond pulsed lasers. These new lasers are capable of activating NIR-responsive chromophores; a large step toward the clinical application of NIR chromophores since to date no known NIR chromophore exists that can be activated via a continuous wave laser (Zhao, 2007).

While the general trend in the literature is to focus on the photochemistry of these new materials, some researchers have investigated the effects of NIR light on biological systems. One group has reported that the safe dose of NIR light from a femtosecond pulsed laser is 2.5–4 nJ/laser pulse (Wantanabe et al., 2004). This work has been often cited in many reports on new NIR-responsive materials in lieu of conducting biocompatibility assessment of new nanomaterials and NIR light (dose, wavelength, pulse duration, and frequency) required to activate the specific novel material being studied. Nevertheless, Chen et al. (2006) described a photoresponsive DNA–gold nanorod conjugate that responds to NIR light. In the work, Chen et al. used the trypan blue exclusion assay in order to assess cellular viability of HeLa cells following varying doses of 800 nm NIR light. Their findings indicated that 800 nm light, dosed at or below 79 µJ/pulse, did not significantly affect cellular viability (each pulse lasted 130 femtoseconds and the pulses were delivered at 1 Hz over a period of 1 min). When the dose was increased to 88 μJ/pulse, a significant reduction in cellular viability was observed. This reduction in cellular viability at higher doses of NIR light was not a problem for the system developed by Chen et al. because their DNA–gold nanorod conjugate responded sufficiently to NIR light dosed at the 79 μJ/pulse level—which trigger the release of DNA from the conjugate and resulted in the successful transfection of green fluorescence protein (GFP) into HeLa cells.

4.3 pH-RESPONSIVE SYSTEMS

While light is an attractive stimulus for activation, it is not the only stimulus that has been investigated for the development of stimuli-responsive systems. Indeed, the use of an internal stimulus sidesteps any possible biocompatibility complications that could arise through the application of an external stimulus. Much effort has been put forth into the development of systems that respond to changes in pH. The rationale behind this approach is twofold; first, it allows for the selective delivery of therapeutic agents to diseased tissues such as solid tumors, which exhibit a markedly reduced pH (5.5–6.5) compared to healthy tissues (pH 7.4) due to the core of the tumor tissue being exposed to hypoxic/anoxic conditions (Brown and Wilson, 2004). This requires the core of the tumor to rely extensively on glycolysis and lactic acid fermentation for energy production, thus making the tumor microenvironment acidic (Brown and Wilson, 2004). Second, the use of pH as the stimulus allows systems to be engineered to release the drug once they are inside intracellular compartments that exhibit a difference in pH—namely, the lysosome in which the pH is maintained around 5.5 (Liberman and Marks, 2009). The main drawback to pH-responsive systems, as well as any system that relies on an internal stimulus, is that this approach allows for a large degree of variability at both the intra- and interpatient levels.

Perhaps one of the most promising new materials for pH-responsive systems was that described by Sankaranarayanan et al., which utilized the novel copolymer poly([2,2′-(propane-2,2-diylbis(oxy)) bis(ethane-2,1-diyl) diacrylate]-co-[hexane-1,6-diyl diacrylate]-4,4′-trimethylene dipiperidine) (poly-β-aminoester ketal-2) (Sankaranarayanan et al., 2010). This new copolymer utilizes two pH-responsive motifs in order to change its solubility as well as selectively rupture upon incubation in low pH (5–6) environments while maintaining its stability at physiological pH. In order to evaluate the biocompatibility of this new copolymer, the authors employed the MTT assay. Following a 20 h incubation of the copolymer with RAW 264.7 cells, it was found that the copolymer did not induce cytotoxic effects at concentrations of 3.7 μg/mL or lower. However, cytotoxicity was seen at the highest concentration tested: 11.11 μg/mL. No further tests were conducted in order to evaluate the biocompatibility of the copolymer at different pH values or the products of the copolymer's breakdown.

Another approach that utilizes a pH-responsive material has been investigated for NP-mediated drug delivery to tumor tissues through the shedding of the NP's stealth (PEG) layer. The idea behind this system is that once the outer PEG layer is shed, the positively charged core of the NP is exposed, resulting in increased uptake of the drug-loaded NPs through the electrostatic interaction between the NP and the plasma membrane (e.g., nonspecific adsorptive pinocytosis). This approach was investigated by Poon et al. who utilized the pH-sensitive nature of the iminobiotin–neutravidin bond to link PEG to a poly-L-lysine core (Poon et al., 2011). When tested *in vivo*, these NPs were able to show stronger persistence in the tumor when utilized in two different tumor models. This supports the authors' hypothesis that pH-responsive materials are a potentially viable means of selectively targeting solid tumors, regardless of the type of cancer being treated. The biocompatibility of this system was assessed through subjecting these NPs to the MTT assay. Using uncoated latex beads as a reference, HeLa cells were incubated for 48 h with varying concentrations of the novel NPs or latex beads. The results showed that the NPs prepared with a pH-sheddable PEG layer were not cytotoxic; exhibiting levels of cellular viability similar to those seen for the latex beads. The promising *in vivo* results obtained by this group warrant further investigation into the biocompatibility of this system as a novel pH-responsive system for drug delivery.

4.4 ENZYME-RESPONSIVE SYSTEMS

Like their pH-responsive counterparts, systems that respond to specific enzymes exhibit great potential due to the disease-specific nature in which the expression of particular enzymes is upregulated. In order for cancerous cells to continue growing, promote angiogenesis, and metastasize to other tissues, extensive remodeling of the extracellular matrix is required (Forsyth et al., 1999). This means that various enzymes, such as the matrix metalloproteases (MMPs), must be secreted in exceedingly large quantities. This overexpression of MMPs makes them a potential target for use in a stimuli-responsive drug delivery system, which can be designed such that part of the system must be cleaved by the MMPs in order to allow drug release. This is exactly the strategy that was employed by Tauro et al. who linked platinum to a short six amino acid peptide sequence that was a known substrate for MMPs (Tauro et al., 2008). These platinum–peptide complexes were then loaded into hydrogels such that the rate of platinum release was dependent on MMP cleavage of the peptide. After demonstrating successful MMP-controlled platinum release *in vitro*, the peptide complexes were evaluated for biocompatibility against malignant glioma cells (HTB-14) using the MTT assay. Following 24 h incubation, the intact peptide complex did not show any cytotoxicity against the HTB-14 cells at any of the concentrations tested (0.5–200 µM). Likewise, the products of peptide cleavage by MMPs did not show any cytotoxic effects when studied under the same conditions as the intact peptides. Further studies were conducted to evaluate the biocompatibility of the peptide-loaded hydrogels. These studies, evaluated using the MTT assay following 48 and 96 h incubations, also showed no cytotoxic effects induced by the hydrogel or the peptides contained within. Follow-up studies were conducted in which exogenous MMPs (MMP-2 and MMP-9) were added to the cell culture media. These studies demonstrated that the platinum–peptide complexes contained within the hydrogel showed no cytotoxicity at 24 or 48 h unless exogenous MMP-2 and MMP-9 were added to the media. Taken together, the data indicate that the system is biocompatible until the peptide is cleaved and the cytotoxic platinum component is released.

A similar approach was utilized by Pedersen et al. in the development of a liposomal formulation composed of a retinoid–phospholipid prodrug (Pedersen et al., 2010). This prodrug was able to form liposomes due to the incorporation of the phospholipid component. The cytotoxic retinoid portion was only released once it was cleaved off of the phospholipid by secretory phospholipase A_2 IIA ($sPLA_2$). Using the MTT assay, the biocompatibility of the prodrug was assessed using two cell lines: HT-29 (human colon carcinoma) and Colo-205 (human colon adenocarcinoma). The HT-29 cells do not secrete $sPLA_2$, while the Colo-205 cells do secrete this enzyme. This allowed the investigators to evaluate the cytotoxicity of their prodrug in the presence and absence of the activating enzyme in quantities that closely mimic the natural expression level of $sPLA_2$ seen clinically in colon adenocarcinoma. Further control over the system was demonstrated through the addition of exogenous $sPLA_2$ (5 nM). Following 24 and 48 h incubations, the prodrug was not able to induce significant cytotoxicity in the absence of the $sPLA_2$ enzyme (HT-29 cells). However, when $sPLA_2$ was present, the cytotoxicity of the prodrug increased over 20-fold; with IC_{50} values below 10 µM for HT-29 cells when exogenous $sPLA_2$ was coincubated, and below 20 µM for Colo-205 cells (no exogenous $sPLA_2$ added). This data indicated that the prodrug formulation was biocompatible until cleaved by the activating enzyme $sPLA_2$, and that cells that naturally secrete this enzyme (Colo-205 cells) are able to cleave the prodrug into its active form within 24 h.

4.5 FACTORS INFLUENCING BIOCOMPATIBILITY OF STIMULI-RESPONSIVE NANOMATERIALS

Many factors affect the body's response to the introduction of foreign materials, including morphology, size, presence of complement activation groups (e.g., hydroxyl groups), and the surface chemistry of the foreign material. The surface chemistry (hydrophilic/hydrophobic nature and surface charge) is possibly the most important parameter in determining how the body will react to the foreign material due to its influence on the adsorption of proteins onto the material's surface; a process that occurs almost instantaneously (Franz et al., 2011). The general rule is that the more hydrophobic a material, the faster it will be recognized and removed from the body due to increased hydrophobic interactions with phagocytic cells as well as various proteins (including, but not limited to, opsonins) (Carstensen et al., 1992). Since some stimuli-responsive materials undergo extensive remodeling following exposure to their stimulus, the surface chemistry of these materials has the potential to change drastically. This concept also applies to biodegradable materials, whose surface chemistry can change as a function of time. Thus, additional studies are warranted to elicit any changes that occur after stimulus-induced activation (or degradation in the case of biodegradable materials) in order to assess deviations from the initial surface chemistry, and any effects these changes have on the material's biocompatibility.

One of the best-studied methods of altering the surface chemistry to favor biocompatibility is through the addition of PEG to the surface of the material (known as PEGylation). PEG is a hydrophilic molecule that serves as a steric boundary, effectively preventing the adsorption of proteins onto the surface of the material and imparting "stealth" characteristics (Muller et al., 1992; Owens and Peppas, 2006). The incorporation of PEG into polymer formulations to form PEG copolymers can be a particularly advantageous technique to increase the biocompatibility of stimuli-responsive and biodegradable polymers. As these PEG copolymers degrade or undergo activation, the copolymer continues to display PEG molecules at its surface, thereby allowing the PEG molecules to continue preventing protein adsorption (Owens and Peppas, 2006). Another strategy that has been suggested is the precoating of the material with a hydrophilic protein such as human serum albumin (HSA). This approach was found to reduce the extent of neutrophil activation caused by foreign material–neutrophil interactions (Nimeri et al., 2002). In this approach, the selection of an appropriate protein for precoating is crucial as precoating with some proteins has been shown to exacerbate neutrophil activation. This phenomenon has been demonstrated for formulations that used fibrinogen or immunoglobulin G as the precoating protein (Nimeri et al., 2002). Since this technique involves coating only the surface of the material with the protein, it is unlikely to have significant beneficial effects on the biocompatibility of the poststimulus material. However, it remains an established method for increasing the biocompatibility of the prestimulus material.

4.6 REPRESENTATIVE METHODS OF BIOCOMPATIBILITY ASSESSMENT FOR STIMULI-RESPONSIVE NANOMATERIALS

Stimuli-responsive materials pose several distinct challenges to biocompatibility testing due to their inherent design to undergo drastic changes in response to a particular stimulus. Therefore, it is essential that both the pre- and poststimulus materials be evaluated for biocompatibility; as the biocompatibility profile of the prestimulus material gives little to no indication as to the biocompatibility of that material following stimulus exposure. As such, the regimen of tests conducted should be based on the unique properties of the material in question as well as the intended application. When applicable, the stimulus itself should also be evaluated for biocompatibility. Furthermore, this is not intended to be an exhaustive list of all the methods used to establish biocompatibility. Instead, the intent here is to highlight some of the established methods used to evaluate various critical aspects of biocompatibility testing. As the field of stimuli-responsive materials advances and more complex materials are created, additional parameters will need to be evaluated in order to obtain a complete biocompatibility profile. Additional information and testing guidelines (such as appropriate cell line selection) can be found in the International Organization for Standardization's (ISO) 10993 document.

- *Blood compatibility*: Blood compatibility is a critical parameter, as most nanocarriers are designed to selectively accumulate at their site of action through passive targeting. Passive targeting requires that the nanocarriers remain in circulation, and therefore in contact with the blood, for as long as possible in order to achieve the maximum amount of accumulation at the target site. Investigation into the stability/retention time of the nanocarrier in circulation *in vivo*, the degree of platelet aggregation, and the induction of red blood cell lysis should all be evaluated in order to establish an accurate profile of how the material/nanocarrier interacts with blood (Wehrung et al., 2012a,b,c, 2013). Blood stability/retention time studies can also be used to obtain biodistribution data.

- *Macrophage activation*: Activation of macrophages can be assessed *in vitro* through monitoring reactive oxygen species (ROS) production or nitrite production (Wehrung et al., 2012a,b,c, 2013).
- *Cytotoxicity*: Cytotoxicity is most commonly assessed through the MTT assay (Wehrung et al., 2012c, 2013). Alternative assays such as the lactate dehydrogenase (LDH) assay can also be employed in order to differentiate apoptosis from necrosis (Mrakovcic et al., 2013).
- *Inflammatory response*: Measurement of cytokine release via enzyme-linked immunosorbent assay (ELISA) is a simple and reliable way to assess an inflammatory response *in vitro* or *in vivo* (Wehrung et al., 2013). Additionally, the *in vitro* macrophage activation assays noted earlier have also been used to investigate a material's potential to generate an inflammatory response. Histological examination of tissues following hematoxylin and eosin (H&E) staining (Wehrung et al., 2013) or the cage implant system are also established methods (Koschwanez and Reichert, 2007).
- *Tissue-specific toxicity*: Histological examination of specific tissues coupled with monitoring changes in serum biochemical markers such as alanine aminotransferase (ALT) and creatinine have been used for identifying tissue-specific toxicity (Wehrung et al., 2013).
- *Carcinogenicity/mutagenicity*: The Ames test as well as the Comet assay are the preferred methods for investigation into a novel material's carcinogenicity/mutagenicity (Onuki et al., 2008; Tice et al., 2000).
- *Nanocarrier stability in biological conditions*: Particle size measured via dynamic light scattering techniques following incubation in biological fluids (e.g., serum, plasma, tissue extracts) or in simulate biological conditions (e.g., fetal bovine serum, PBS, Dulbecco's Modified Eagle Medium) is a simple and effective way to ensure the stability of nanocarriers once exposed to a biological environment (Wehrung et al., 2012a,b,c). The adsorption of proteins as well as the high salt concentrations found in biological fluids can quickly change a nanocarrier's zeta potential, thereby causing a loss of particle stability and allowing particle agglomeration (Ortis-Gil et al., 2013). Particle agglomeration can cause drastic changes in the biodistribution/pharmacokinetic profile of the nanocarriers, and in some cases allow large enough aggregates to form that occlusion of the microvasculature becomes a serious safety concern.

4.7 PERSPECTIVES ON THE FIELD OF STIMULI-RESPONSIVE BIOMATERIALS

The ideal properties of any stimuli-responsive system are (i) biocompatibility of the system (material before and after stimulus exposure, as well as the stimulus required for activation), (ii) biodegradability, (iii) efficient and rapid response to the stimulus, (iv) high drug-loading efficiency, and (v) minimal premature drug leakage (e.g., drug release before system activation). While the majority of publications evaluate their novel systems for most of the aforementioned criteria, biocompatibility is frequently not evaluated or only minimally evaluated (Tables 4.1 and 4.2). This trend is a grievous oversight, since biocompatibility is absolutely essential if a stimuli-responsive material is to translate into a clinically relevant solution. The majority of published works rely solely on *in vitro* testing—primarily though the MTT assay—and use a prohibitively limited number of time points and concentrations in order to evaluate biocompatibility. More detailed studies are required in order to ensure that these new materials are in fact biocompatible. Additionally, many investigators do not investigate the biocompatibility of the poststimulus material(s), which may not be biocompatible

TABLE 4.1
Examples of Biocompatibility Assessment Methods for Delivery Systems that Respond to Internal Stimuli

Biomaterial(s)	Stimulus Source	Biocompatibility Assessment Method(s)	Cell Line(s)	Time Point(s)	Additional Comments	Reference
PMAA nanohydrogels	Redox potential and pH	CCK-8 assay	HEK-293 and U251MG	24 and 48 h	Biocompatibility of nanohydrogels was assessed over the concentration range of 0.1–100 µg/mL	Pan et al. (2012)
HPHSEP-*star*-PEP	Redox potential	MTT assay and AO/EB staining	NIH-3T3	24 h	Biocompatibility of NPs was assessed over the concentration range of 0.5–1000 µg/mL	Liu et al. (2011)
Synthetic elastin hydrogel[a]	Salt concentration	H&E staining	M-1, HEK-293, and HT-1080	3 and 6 days	*In vivo* studies in male guinea pigs, studied up to 96 days postimplantation	Mithieus et al. (2004)
Poly-L-lysine–neutravidin–PEG	pH	MTT assay	HeLa	48 h	Biocompatibility of NPs was assessed over concentration range of $1 \times 10^6 – 1 \times 10^{12}$ # particles/mL	Poon et al. (2011)
P(NIPAM-*co*-MAA)	pH	MTT assay	SKOV3	24 and 48 h	Biocompatibility assessed over the concentration range of 12.5–200 µg/mL	Dai et al. (2012)
Poly-β-aminoester ketal-2	pH	MTT assay	RAW 264.7	20 h	Biocompatible up to 3.7 µg/mL, at 11.1 µg/mL toxicity was observed	Sankaranarayanan et al. (2010)
Nanodiamonds	pH	MTT assay and cellular morphology	HeLa, A549, and K562	24 h	Biocompatibility of nanodiamonds was assessed over the concentration range of 0.1–100 µg/mL in serum-free media	Li et al. (2010)
Nanodiamond–cisplatin	pH	MTT assay	HeLa	24 h	Biocompatibility of nanodiamond-cisplatin was assessed based on comparison to free cisplatin at 37.5 and 300 µM	Guan et al. (2010)
PCL-*b*-PEEP	Temperature	MTT assay	HEK-293	72 h	Biocompatibility of polymer was assessed over the concentration range of 0.078–10 mg/mL	Wang et al. (2009)
PNIPAM, PVCL, and PVCL-graft-$C_{11}EO_{42}$	Temperature	MTT assay and LDH assay	Caco-2 and Calu-3	3 and 12 h	Biocompatibility of polymers was assessed over the concentration range of 0–10 mg/mL	Vihola et al. (2005)
Elastase–alginate	Enzyme (elastase)	MTT assay and IL-8 production	Calu-3	24 and 4 h	Biocompatibility of microparticles was assessed at 1 mg/mL	Sivadas et al. (2011)
PEGDA–MMP-sensitive peptide–platinum	Enzyme (MMP-2 and MMP-9)	MTT assay	U-87 MG	48 and 96 h	Biocompatibility of hydrogels and peptides alone was assessed over the concentration range of 0.5–110 µM	Tauro et al. (2008)
Retinoid–phospholipid prodrug/liposome	Enzyme (sPLA$_2$)	MTT assay	HT-29 and Colo-205	24 h	Biocompatibility of liposomes was assessed over the concentration range of 0.1–110 µM	Pedersen et al. (2010)

[a] Indicates *in vivo* studies were conducted along with *in vitro* studies.

TABLE 4.2
Examples of Biocompatibility Assessment Methods for Delivery Systems that are Responsive to External Stimuli

Biomaterial(s)	Stimulus Source	Biocompatibility Assessment Method(s)	Cell Line(s)	Time Point(s)	Additional Comments	Reference
LAMS–azobenzene nanoimpeller	Light (413 nm)	Propicium iodide and Hoechst staining	SW480 and PANC-1	72 h	Biocompatibility of NPs was assessed at 10 and 100 µg/mL using latex beads as a reference	Lu et al. (2008)
BhcP and ONBP	Light (740 nm)	MTT assay	RAW 264.7	24 h	Biocompatibility of NPs before and after irradiation was assessed over the concentration range of 1.17–300 µg/mL	Fomina et al. (2011)
αα-Cyclodextran/azobenzene	Light (365 and 435 nm)	None	None	None	Biocompatibility was claimed based on the use of GRAS reagents	Chen et al. (2011)
PEG–anthracene crosslinker	Light (365 nm)	MTT assay	HCE	3 and 7 days	Biocompatibility assessed by incubating a photogel insert with cells	Wells et al. (2011)
Alkoxylphenacyl-based polycarbonates	Light (280 nm)	MTT assay, nitrite production, in vivo studies	RAW 264.7	1–3 days, 14 days	Assessed in vivo by S.C. and I.V. administration to Balb/c mice	Wehrung et al. (2013)
Coumarin polyesters	Light (254 and 350 nm)	MTT assay, ROS production, nitrite production	RAW 264.7	24 h	Biocompatibility assessed before and after irradiation over the concentration range 5–50 µg/mL	Maddipatla et al. (2013)
O-Nitrobenzyl alcohol and bromo-coumarin	Light (350 and 740 nm)	MTT assay	RAW 264.7	24 h	Biocompatibility of NPs before and after irradiation was assessed over the concentration range of 0.13–100 µg/mL and compared to PLGA	Lux et al. (2012)
Chit-AgNTs	Light (800 nm)	Cell morphology, propidium iodide and Hoechst staining	HEK and NCI-H460	24 h	Biocompatibility of nanotriangles was assessed over the concentration range of 0.3–1.8 µg/mL	Boca et al. (2011)
Azobenzene–DNA capped mesoporous silica	Light (365 and 450 nm)	MTT assay	CCRF–CEM	2 h	Biocompatibility of NPs was assessed at 100 µg/mL	Yuan et al. (2012)
CPEG-g-DNQ	Light (365 nm)	MTT assay	HepG2 and HUVEC	24 h	Biocompatibility of micelles was assessed over the concentration range of 10–200 µg/mL	Chen et al. (2011)
Fluorene and boron-dipyrromethene polymer	Light (400–800 nm)	MTT assay	A498	24 h	Biocompatibility of NPs was assessed over the concentration range of 5–20 µM	Chong et al. (2012)
4-Hydroxyazobenzene and pyridine-based azophenols and azopyrimidine	Light (355 nm)	Colony count	BL21(DE3)	24 h	Biocompatibility of compounds was assessed over the concentration range of 0.1–30 mM	Garcia-Amoros et al. (2012)
No material, NIR light	Light (800 nm)	Trypan blue exclusion assay	HeLa	N/A	Biocompatibility of NIR light was assessed over the energy range of 48–108 µJ/pulse for 1 min, and at 79 µJ/pulse for 1–7 min	Chen et al. (2006)
No material, UV light	Light (315–400 nm)	Flow cytometry and comet assay	HaCaT	N/A	Biocompatibility of UV light was assessed over the dose range of 25–300 kJ/m²	McMillan et al. (2008)
Fe_2O_4/P(NIPPAAm-co-AAc)	Magnetic	Cell morphology, MTS assay and live–dead staining	L929	48 h	Biocompatibility of nanohydrogels assessed using extracts incubated with cells at a concentration of 5 mg/mL	Chou et al. (2013)

even when the intact/prestimulus material is. In order to adequately demonstrate biocompatibility, further testing needs to be done on the novel materials, their poststimulus product(s), as well as the stimulus required for activation of the system (when applicable). The potential results of this oversight may come in the form of reduced efficacy and increased side effects *in vivo* or the complete failure of the system. Hence, it is absolutely critical in the development of new stimuli-responsive materials that the biocompatibility of the entire system is thoroughly evaluated early on in the research efforts if the system is to ever become a clinically viable option.

While it is not possible to state a specific regimen of tests that are required to establish biocompatibility due to variations in the intended applications of novel stimuli-responsive materials, it is clear that certain criteria for establishing biocompatibility are ubiquitous, and can be applied to different types of stimuli-responsive materials. The current literature is lacking due to the reliance on *in vitro* testing without any accompanying tests being conducted *in vivo*. It is our opinion that evaluation of the following parameters should be considered the minimum requirements for establishing biocompatibility: cytotoxicity, hemocompatibility, immune response, and acute toxicity. Cytotoxicity needs to be evaluated and can serve as an excellent way to screen materials *in vitro* for further testing. The MTT assay provides satisfactory results regarding cytotoxicity provided the investigators utilize a wide range of concentrations and perform the test using appropriate cell lines. Additional testing on effects on red blood cells (RBCs) is also necessary if the material is intended to be used in applications in which the system will be exposed to blood (e.g., the vast majority of materials). On this front, both the induction of RBC lysis as well as coagulation studies should be assessed using *in vitro* and *in vivo* methods. Researchers also need to complete studies that evaluate the host's immune response to the novel material. Immune response tests can be carried out *in vitro* or *in vivo* with a preference for *in vivo* testing. Finally, acute toxicity needs to be evaluated *in vivo* within a dose range that is a reasonable expectation as to what the system will be dosed at clinically. Plasma levels of cytokines can be measured during this test in order to maximize the amount of information obtained, as well as to meet the aforementioned requirement of monitoring the host's immune response. Histological evaluation conducted postmortem is perhaps the simplest way to assess acute toxicity in a tissue-specific manner, and should be coupled with monitoring plasma biochemical markers in order to obtain the clearest possible picture of the novel material's acute toxicity profile. Regardless of the methods used and specific tests performed, investigators should reference the International Organization for Standardization (ISO) 10993 in order to determine the appropriate means by which their novel material should be evaluated for biocompatibility based on its intended application.

ACKNOWLEDGMENTS

The support by NEOMED research fund is gratefully appreciated. DW is a recipient of 2013-2014 AFPE predoctoral fellowship. The authors do not have personal and/or financial conflict of interest.

ABBREVIATIONS

AO/EB staining	acridine orange/ethidium bromide double staining
BhcP	4-bromo-7-hydroxycoumarin polymer
CCK-8 assay	cell counting kit-8 assay
Chit-AgNTs	chitosan-coated silver nanotriangles
CPEG-*g*-DNQ	comb-like polyethylene glycol-2-diazo-1,2-naphthoquinone
Fe_2O_4/P(NIPPAAm-*co*-AAc)	iron oxide/poly(*N*-isopropylacrylamide-*co*-acrylic acid)
Fe_2O_4-loaded lecithin/PLGA	iron oxide-loaded lecithin/poly(lactic-*co*-glycolic acid)

GRAS	generally regarded as safe
H&E staining	hematoxylin and eosin staining
HPHSEP-*star*-PEP	hyperbranched multiarm copolyphosphates
LAMS	light-activated mesoporous silica nanoparticles
LDH assay	lactate dehydrogenase assay
MMP-2	matrix metalloprotease 2
MMP-9	matrix metalloprotease 9
MTT assay	3-(4,5-dimethylthiazol-2-yl)-2,5-diphenyltetrazolium bromide assay
ONBP	*o*-nitrobenzyl polymer
PCL-*b*-PEEP	poly(ε-caprolactone)-*b*-poly(ethylethylene phosphate)
PEG–anthracene crosslinker	polyethylene glycol–anthracene crosslinker
PEGDA–MMP-sensitive peptide–platinum	poly(ethylene glycol) diacrylate hydrogel–matrix metalloprotease-sensitive peptide–platinum
PLGA	poly(lactic-*co*-glycolic acid)
PMAA nanohydrogels	poly(methacrylic acid)-based nanohydrogels
P(NIPAM-*co*-MAA)	poly[(*N*-isopropylacrylamide)-*co*-(methacrylic acid)]
PNIPAM, PVCL, and PVCL-graft-$C_{11}EO_{42}$	poly(*N*-isopropylacrylamide), poly(*N*-vinylcaprolactam), and poly(*N*-vinylcaprolactam) grafted poly(ethylene oxide)
Poly-β-aminoester ketal-2	poly([2,2'-(propane-2,2-diylbis(oxy))bis(ethane-2,1-diyl) diacrylate]-*co*-[hexane-1,6-diyl diacrylate-] 4,4'-trimethylene dipiperidine)
Poly-L-lysine–neutravidin–PEG	Poly-L-lysine–neutravidin–poly(ethylene glycol)
$sPLA_2$	secretory phospholipase A_2 IIA

REFERENCES

Boca SC, Potara M, Gabudean AM, Juhem A, Baldeck PL, Astilean S. Chitosan-coated triangular silver nanoparticles as a novel class of biocompatible, highly effective photothermal transducers for *in vitro* cancer cell therapy. *Cancer Lett.* 2011, 311:131–140.

Brown JM, Wilson WR. Exploiting tumor hypoxia in cancer treatment. *Nat. Rev. Cancer.* 2004, 4:437–447.

Carstensen H, Muller RH, Muller BW. Particle size, surface hydrophobicity and interaction with serum of parenteral fat emulsions and model drug carriers as parameters related to RES uptake. *Clin. Nutr.* 1992, 11:289–297.

Checcot F, Lecommandoux S, Klok HA, Gnanou Y. From supramolecular polymersomes to stimuli-responsive nano-capsules based on poly(diene-b-peptide) diblock copolymers. *Eur. Phys. J. E. Soft Matter.* 2003, 10(1):25–35.

Chen CC, Lin YP, Wang CW, Tzeng HC, Wu CH, Chen YC, Chen CP, Chen LC, Wu YC. DNA-gold nanorod conjugates for remote control of localized gene expression by near infrared irradiation. *J. Am. Chem. Soc.* 2006, 128:3709–3715.

Chen CJ, Liu GY, She YT, Zhu CS, Pang SP, Liu XS, Ji J. Biocompatible micelles based on comb-like PEG derivates: Formation, characterization, and photo-responsiveness. *Macromol. Rapid Commun.* 2011, 32:1077–1081.

Chen X, Gooding J, Zou G, Su W, Zhang Q. Polydiacetylene vesicles containing $\alpha\alpha$-cyclodextrin and azobenzene as photocontrolled nanocarriers. *Chem. Phys. Chem.* 2011, 12:2714–2718.

Chong H, Nie C, Zhu C, Yang Q, Liu L, Lv F, Wang S. Conjugated polymer nanoparticles for light-activated anticancer and antibacterial activity with imaging capability. *Langmuir.* 2012, 28:2091–2098.

Chou FY, Lai JY, Shih CM, Tsai MC, Lue SJ. *In vitro* biocompatibility of magnetic thermo-responsive nanohydrogel particles of poly(N-isopropylacrylamide-co-acrylic acid) with Fe_3O_4 cores: Effect of particle size and chemical composition. *Colloid. Surface. B.* 2013, 104:66–74.

Dai Y, Ma P, Cheng Z, Kang X, Zhang X, Hou Z, Li C, Yang D, Zhai X, Lin J. Up-conversion cell imaging and pH-induced thermally controlled drug release from $NaYF_4$:Yb^{3+}/Er^{3+} at hydrogel core-shell hybrid microspheres. *ACS Nano.* 2012, 6(4):3327–3338.

Fomina N, McFearin CL, Sermsakdi M, Morachis JM, Almutairi A. Low power biologically benign NIR light triggers polymer disaasembly. *Macromolecules*. 2011, 44:8590–8597.

Forsyth PA, Wong H, Laing TD, Rewcastle NB, Morris DG, Muzik H et al. Gelatinase-a (MMP-2), gelatinase-B (MMP-9) and membrane type matrix metalloproteinase-1 (MT1-Mmp) are involved in different aspects of the pathophysiology of malignant gliomas. *Br. J. Cancer*. 1999, 79(11–12):1828–35.

Franz S, Rammelt S, Scharnweber D, Simon JC. Immune response to implants—A review of the implications for the design of immunomodulatory biomaterials. *Biomaterials*. 2011, 32:6692–6709.

Garcia-Amoros J, Diaz-Lobo M, Nonell S, Velasco D. Fastest thermal isomerization of an azobenzene for nanosecond photoswitching applications under physiological conditions. *Angew. Chem. Int. Ed.* 2012, 51:12820–12823.

Guan B, Zou F, Zhi J. Nanodiamond as the pH-responsive vehicle for an anticancer drug. *Small*. 2010, 6(14):1514–1519.

Keefe DM. 2011. FDA Agency Response Letter GRAS Notice No. GRN 000397, http://www.fda.gov/Food/IngredientsPackagingLabeling/GRAS/NoticeInventory/ucm287638.htm.

Koschwanez HE, Reichert WM. *In vitro, in vivo* and post explantation testing of glucose-detecting biosensors: Current methods and recommendations. *Biomaterials*. 2007, 28(25):3687–36703.

Liberman M, Marks AD. Relationship between cell biology and biochemistry. *Basic Medical Biochemistry A Clinical Approach,* 3rd edition. Wolters Kluwer Health, Philadelphia, PA, 2009.

Liu J, Pang Y, Huang W, Zhu Z, Zhu X, Zhou Y, Yan D. Redox-responsive polyphosphate nanosized assemblies: A smart drug delivery platform for cancer therapy. *Biomacromolecules*. 2011, 12:2407–2415.

Li J, Zhu Y, Li W, Zhang X, Peng Y, Huang Q. Nanodiamonds as intracellular transporters of chemotherapeutic drug. *Biomaterials*. 2010, 31:8410–8418.

Lu J, Choi E, Tamanoi F, Zink JI. Light-activated nanoimpeller-controlled drug release in cancer cells. *Small*. 2008, 4(4):421–426.

Lux CG, McFearin CL, Joshi-Barr S, Sankaranarayanan J, Fomina N, Almutairi A. Single UV or near IR triggering event leads to polymer degradation into small molecules. *ACS Macro. Lett*. 2012, 1:922–926.

McMillan TJ, Leatherman E, Ridley A, Shorrocks J, Tobi SE, Whiteside JR. Cellular effects of long wavelength UV light (UVA) in mammalian cells. *J. Pharm. Pharmacol.* 2008, 60:969–976.

Mithieus SM, Rasko JEJ, Weiss AS. Synthetic elastin hydrogels derived from massive elastic assemblies of self-organized human protein monomers. *Biomaterials*. 2004, 25:4921–4927.

Mrakovcic M, Absenger M, Reidl R, Smole C, Roblegg E, Frohlich LF, Frohlich E. Assessment of long-term effects of nanoparticles in a microcarrier cell culture system. *PLoS ONE*. 2013, 8(2):e56791. doi:10.1371/journal.pone.0056791.

Muller RH, Wallis KH, Troster SD, Kreuter J. *In vitro* characterization of poly(methyl-methacrylate) nanoparticles and correlation to their *in vivo* fate. *J. Control. Release*. 1992, 20:237–246.

Maddipatla MVSN, Wehrung D, Tang C, Fan W, Oyewumi, MO, Miyoshi T, Joy A. Photoresponsive coumarin polyesters that show crosslinking and polymer chain scission properties. *Macromolecules*. 2013, 46: 5133–5140.

Nimeri G, Ohman L, Elwing H, Wettero J, Bengtsson T. The influence of plasma proteins and platelets on oxygen radical production and F-actin distribution in neutrophils adhering to polymer surfaces. *Biomaterials*. 2002, 23:185–1795.

Onuki Y, Bhardwaj U, Papadimitrakopoulos F, Burgess DJ. A review of the biocompatibility of implantable devices: Current challenges to overcome foreign body response. *J. Diabetes Sci. Technol*. 2008, (2)6:1003–1015.

Ortis-Gil G, Natte K, Thiermann R, Girod M, Rades S, Kalbe H, Thunemann AF, Maskos M, Osterle W. On the role of surface composition and curvature on biointerface formation and colloidal stability of nanoparticles in a protein-rich model system. *Colloid. Surface. B*. 2013, 108:110–119.

Owens DE, Peppas NA. Opsonization, biodistribution and pharmacokinetics of polymeric nanoparticles. *Int. J. Pharm*. 2006, 307:93–102.

Pan YJ, Chen YY, Wang DR, Wei C, Guo J, Lu DR, Chu CC, Wang CC. Redox/pH dual stimuli-responsive biodegradable nanohydrogels with varying responses to dithiothreitol and glutathione for controlled drug release. *Biomaterials*. 2012, 33:6570–6579.

Pedersen PJ, Adolph SK, Subramanian AK, Arouri A, Andresen TL, Mouritsen OG, Madsen R, Madsen MW, Peters GH, Clausen MD. Liposomal formulation of retinoids designed for enzyme triggered release. *J. Med. Chem*. 2010, 53:3782–3792.

Poon Z, Chang D, Zhao X, Hammond PT. Layer-by-layer nanoparticles with a pH-sheddable layer for *in vivo* targeting of tumor hypoxia. *J. Am. Chem. Soc*. 2011, 5(6):4284–4292.

Rakhmatullina E, Braun T, Chami M, Malinova V, Meier W. Self-organization behavior of methacrylate-based amphiphilic di and triblock copolymers. *Langmuir.* 2007, 23(24):12371–12379.

Sankaranarayanan J, Mahmoud EA, Kim G, Morachis JM, Almutairi A. Multiresponse strategies to modulate burst degradation and release from nanoparticles. *J. Am. Chem. Soc.* 2010, 4(10):5930–5936.

Sauer M, Haefele T, Graff A, Nardin C, Meier W. Ion-carrier controlled precipitation of calcium phosphate in giant ABA triblock copolymer vesicles. *Chem Commun.* 2001, 2452–2453.

Schmers JM, Fustin CA, Gohy JF. Light-responsive block copolymers. *Macromolecules. Rapid Commun.* 2010, 31:1588–1607.

Sivadas N, Cryan SA. Inhalable bioresponsive microparticles for targeted drug delivery in the lungs. *J. Pharm. Pharmacol.* 2011, 63:369–375.

Sun S, Chamsaz EA, Joy A 2012. Photoinduced polymer chain scission of alkoxylphenacyl based polycarbonates. *ACS Macro. Lett.* 1:1184–1188.

Tauro JR, Lee BS, Lateef SS, Gemeinhart RA. Matrix metalloprotease selective peptide substrates cleavage within hydrogel matrices for cancer chemotherapy activation. *Peptides.* 2008, 29:1965–1973.

Tice RR, Argurell E, Anderson D, Burlinson B, Hartmann A, Kobayashi H et al. Single cell gel/Comet assay: Guidelines for *in vitro* and *in vivo* genetic toxicology testing. *Environ. Mol. Mutagen.* 2000, 35:206–221.

Vihola H, Laukkanen A, Valtola L, Tenhu H, Hirvonen J. Cytotoxicity of thermosensitive polymers poly(N-isopropylacrylamide), poly(N-vinylcaprolactam) and amphiphilically modified poly(N-vinylcaprolactam). *Biomaterials.* 2005, 26:3055–3064.

Wang YC, Li Y, Yang XZ, Yuan YY, Yan LF, Wang J. Tunable thermosensitivity of biodegradable polymer micelles of poly(ϵ-caprolactone) and polyphosphoester block copolymers. *Macromolecules.* 2009, 42:3026–3032.

Wantanabe W, Arkawa N, Matsunaga S, Higashi T, Fukui K, Isobe K, Itoh K. Femtosecond laser disruption of subcellular organelles in a living cell. *Opt. Express.* 2004, 12(18):4203–4213.

Wehrung D, Geldenhuys WJ, Oyewumi MO. Effects of gelucire content on stability, macrophage interaction and blood circulation of nanoparticles engineered from nanoemulsions. *Colloid. Surface. B.* 2012a, 94:259–265.

Wehrung D, Oyewumi MO. Antitumor effect of novel gallium compounds and efficacy of nanoparticle-mediated gallium delivery in lung cancer. *J. Biomed. Nanotechnol.* 2012b, 8:1–11.

Wehrung D, Geldenhuys WJ, Bi L, Oyewumi MO. Biocompatibility, efficacy and biodistribution of Gelucire-stabilized nanoparticles engineered for docetaxel delivery. *J. Nanosci. Nanotechnol.* 2012c, 12(3):2901–2911.

Wehrung D, Sun S, Chamsaz EA, Joy A, Oyewumi MO. Biocompatibility and *in vivo* tolerability of a new class of photoresponsive alkoxylphenacyl-based polycarbonates. *J. Pharm. Sci.* 2013, 102(5):1650–1660.

Wells LA, Brook MA, Sheardown H. Generic anthracene-based hydrogel crosslinkers for photo-controllable drug delivery. *Macromol. Biosci.* 2011, 11:988–998.

Yuan Q, Zhang Y, Chen T, Lu D, Zhao Z, Zhang X, Li Z, Yan CH, Tan W. Photo-manipulated drug release from mesoporous nanocontainers controlled by azobenezene-modified nucleic acid. *ACS Nano.* 2012, 6(7):6337–6344.

Zhao Y. Rational design of light controllable polymer micelles. *Chem. Rec.* 2007, 7:286–294.

5 Biosensing Devices for Toxicity Assessment of Nanomaterials

Evangelia Hondroulis, Pratik Shah, Xuena Zhu, and Chen-Zhong Li

CONTENTS

5.1 Introduction 117
    5.1.1 Biosensors 117
    5.1.2 Nanotoxicity 119
5.2 Cellular-Based Biosensors for Nanotoxicity 120
5.3 Techniques and Devices for Nanotoxicity Testing 120
    5.3.1 Carbon Fiber Microelectrode 121
    5.3.2 Atomic Force Microscopy 121
5.4 Biosensors for Nanotoxicity Biomarker Detection 122
    5.4.1 Background 122
    5.4.2 Common Methods for Nanotoxicity Assessment 122
        5.4.2.1 Cell Viability/Proliferation Assay 122
        5.4.2.2 Direct/Indirect Intracellular ROS Measurement 123
        5.4.2.3 Assays on the Genomic Level 123
    5.4.3 Biosensing Approaches for Inflammatory Biomarkers Detection 123
    5.4.4 Paper-Based Biosensor for ROS-Induced DNA Oxidative Damage
            Biomarkers Detection 124
5.5 Conclusion 124
References 125

5.1 INTRODUCTION

5.1.1 BIOSENSORS

Living cells are associated with electrical characteristics and are thus responsive to, and even generate, electric fields and currents. Knowledge of these electrical properties of cells has led to the development of the field of bioelectronics. Bioelectronics is the application of electronics to biology and medicine and can be broken down into two categories. Physically interfacing electronic devices with biological systems have led to technologies such as the cardiac pacemaker, implantable electrical bone growth simulators, deep brain simulators, and electrical nerve simulation (Nowak et al. 2011). The other aspect of bioelectronics is electronics for both the detection and characterization of biological materials, such as on the cellular and subcellular level. This can be seen in the example of cell-based biosensors that use live cells as sensing elements to monitor the physiological changes induced by internal aberrations or external stimuli (Asphahani and Zhang 2007).

Biosensors are becoming valuable tools for analyzing various physical, chemical, and biological processes. Since 1956, when Professor Leland C. Clark Jr. first published a paper on the oxygen electrode (Clark 1956), researchers have incorporated and enhanced biosensing technologies in fields such as health care, the food industry, and environmental monitoring. The attraction to biosensors stems from their accurate, precise, and reproducible measurements in a cheap, small, and portable manner.

A biosensor is commonly defined as a device that detects, records, and transmits information regarding a physiological change or process. Biologically derived recognition entities (enzymes, antibodies, microorganisms, cell receptors, cells, etc.) are coupled to a transducer that detects the biological reaction and converts it into a signal, which can be physicochemical, optical, electrochemical, thermometric, or magnetic (Figure 5.1).

Biosensing technology has spread throughout many disciplines due to its great specificity, sensitivity, and diversity in uses. Molecular and enzymatic biosensors were among the first to be introduced in the late 1960s (Updike and Hicks 1967; Guilbault and Montalvo 1969) with thermal, optical, and electrochemical biosensors following shortly thereafter (Mosbach and Danielsson 1974; Clemens et al. 1976; Weaver et al. 1976; Volkl et al. 1980).

Planar microelectrode biosensors, used to monitor cellular behavior, were first introduced by Thomas et al. in 1972 to monitor the electrical activity of contracting embryonic, chick heart cells (Thomas et al. 1972). Since then, microelectrode biosensors have been used to study cell cultures *in vitro* under different conditions. For instance, Gross et al. used a microelectrode biosensor to monitor and eventually stimulate neuronal cell activity *in vitro* from the brain and spine (Gross et al. 1977, 1982, 1993). Other uses include monitoring metabolism (McConnell et al. 1992), fluorescent probes and reporter genes (Zysk and Baumbach 1998), and electrophysiology (DeBusschere and Kovacs 2001).

Whole-cell impedance-based biosensors, pioneered by Giaever and Keese (1984) were developed to monitor the proliferation and motion of a population of anchorage-dependent cell cultures. By monitoring whole-cell activity, one can monitor changes in membrane receptors, channels, and enzymes that may be expressed by the cell. Morphological changes can also be monitored using electrical impedance sensing (EIS) biosensors, since cellular membranes exhibit dielectric properties (Pancrazio et al. 1999). EIS biosensors are especially beneficial for monitoring the behavior of the whole cell because they provide information about the total physiological responses of cells to external stimuli. Biosensors that incorporate whole cells can have an advantage over other biosensors for certain applications because they can provide functional information without damaging the cells. Most current biosensors are used to detect enzymes, DNA/RNA (deoxyribonucleic acid), and immunological components, converting the biological phenomena into electrical signals (Katz and Willner 2003; Song et al. 2006; Luong et al. 2008) and allowing for specifically targeted results.

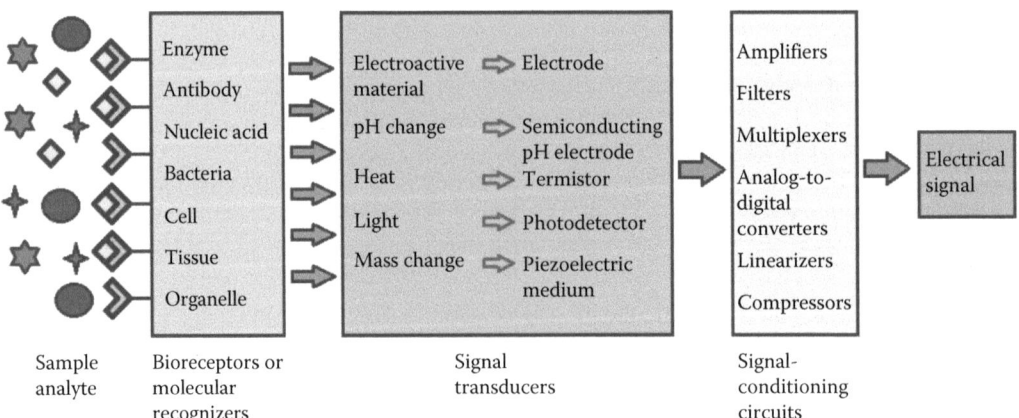

FIGURE 5.1 Schematic of the biosensor.

5.1.2 Nanotoxicity

Biomedical engineering, drug delivery, environmental health, pharmaceutical industries, and even electronics and communication technologies, all incorporate nanotechnology, leading to greater potentials for advancements in current research. For example, in the health-care field, nanomaterials are being considered in the development of new drugs and new therapies for disease control and improving the quality of life (Bianco et al. 2005; Slowing et al. 2007; Faraji and Wipf 2009). More recently, nanomaterials have been used in tissue engineering and medical imaging, leading to improved diagnostics and new therapeutic treatments (Harrison and Atala 2007; Kim et al. 2008; Shi 2009). However, due to their novel stature, nanoscale materials (including nanotubes, nanowires, nanowhiskers, fullerenes or buckyballs, and quantum dots) have to be tested for unintended hazards for human health and the environment (Oberdorster et al. 2005; Kreyling et al. 2006).

To ensure the compatibility of nanomaterials for medical applications and for the safety of the environment, testing for toxicological parameters is a necessary first step in nanotechnology research. The bulk material properties of metals change when they are in the nanoscale, and they may pose certain threats to biological systems that their bulk counterparts may not. To date, a number of studies are addressing nanotoxicity (Soto et al. 2005; Hillegass et al. 2010; Soenen et al. 2011); however, there is a variability of methods, materials, and cell lines used (Lewinski et al. 2008), leading to the need for a standard testing method, or methods, in nanotoxicity testing, which is becoming increasingly important to validate these novel techniques.

The use of nanomaterials in biomedical sciences has placed nanomaterials directly in contact with biological materials, and, thus, it is necessary to observe their interaction closely.

Another risk to be considered is the emission of hazardous air pollutants associated with the use and manufacture of nanomaterials that contain particulate matter on the order of 1–100 nm in size. Any material in the respirable size range, <100 nm in diameter, may have toxic effects on lung fibroblasts after inhalation (Mossman et al. 2007). In particular, nanoparticles with sizes <20 nm affect the alveolar region of the lung (Elder et al. 2009).

Methods such as the mitochondrial reduction of tetrazolium salts into an insoluble dye (the MTT test) and enzyme lactate dehydrogenase (LDH) release tests are traditional *in vitro* biological methods used in the current nanotoxicity studies. These measure cellular viability and proliferation and, thus, are used as markers for cell viability. They consist of procedures that provide a general sense of cytotoxicity, as they show results only at a final time point (Hussain et al. 2005). As a result, the kinetic model (absorption, distribution, metabolism, and excretion) of the nanoparticle uptake is not usually observed with these conventional methods. Following biological exposure, the particles may transport across cell membranes, especially into the mitochondria, causing internal damage that may affect cell behavior and, over time, may lead to cell death (Wilhelm et al. 2002).

In this chapter, the various types of biosensors used to detect nanotoxicity will be explored. We will explore biosensing methods measuring nanotoxicity toward cell monolayers, single cells, and individual components of the cell. The integration of biomolecules with nanotechnology has great future perspectives in the rapidly developing fields of environmental (pollution control and monitoring) and biomedical research, drug delivery, electronics, and communication technologies. With the increasing number of nanomaterial applications, assessing their toxicity should be the first important step toward creating safety guidelines for their handling and disposal. Studies of the biological effects of nanoscale materials that might answer these questions have lagged behind other aspects of nanotechnology development. Biosensing technology is shown to be sensitive enough to measure the micromotions of a cell and, is therefore able to monitor the progression of the cytotoxicity with a rapid, real-time, and multisample analysis, creating a versatile, noninvasive tool that is able to provide quantitative information with respect to alterations in cellular function under various nanomaterial exposures.

5.2 CELLULAR-BASED BIOSENSORS FOR NANOTOXICITY

Whole-cell EIS-based sensors, pioneered by Giaever and Keese, were the first demonstration of a system capable of monitoring the proliferation and motion of a population of anchorage-dependent cell cultures *in vitro* (Giaever and Keese 1984). Giaever and Keese cultured human lung fibroblast cells on modified cell culture dishes consisting of a large reference electrode (2 cm^2) and four smaller electrodes (3×10^{-4} cm^2). They applied an alternating current (AC) voltage, through a resistor, to a single small electrode in the dish, resulting in a constant current source, which enabled the impedance to be determined by the measurement of the resulting voltage. They were able to observe the effects of cell proliferation (impedance increase) as well as the micromotion of the cells (fluctuations in observed impedance).

Giaever and Keese then used their biosensor to examine the effects of different proteins on cell adhesion, spreading, and motility (Giaever and Keese 1986), to create a mathematical model of cell motion (Giaever and Keese 1989), and the use of this impedance method in cell-based sensor applications (Giaever and Keese 1991, 1992, 1993). Connelly et al. modified Giaever and Keese's biosensor design by adding a glass ring around the electrode area, to contain the cell culture media, and inserting a permeable, cellulose nitrate membrane to separate the culture dish into two sides, each with two measurement electrodes, creating a control and a test electrode (Connolly et al. 1990).

EIS technology is shown to be sensitive enough to measure the micromotion of a cell and, therefore, be able to monitor the progression of cytotoxicity with rapid, real-time, and multisample analysis, creating a versatile, noninvasive tool that is able to provide quantitative information with respect to alterations in cellular function under various nanomaterial exposures. Thus, EIS biosensors are ideal for detecting toxic nanomaterials in industrial products, chemical substances, environmental samples (e.g., air, soil, and water), or biological systems (e.g., bacteria, viruses, or tissue components), as they are able to monitor the progression of the cytotoxicity in real time, demonstrating the kinetic effects of the nanomaterials toward whole cells.

Chip-based biosensors show a promising future for monitoring cellular nanotoxicity as they allow rapid, real-time, and multisample analysis creating a versatile, noninvasive tool that is able to provide quantitative information with respect to alterations in cellular function under various nanomaterial exposures. A different chip-based approach to evaluate nanotoxicity was experimented by Kim et al. (2011). They fabricated a chip using a lithography technique where gold was the sensing electrode and was modified with RGD-MAP-C to enhance cell (SH-SY5Y) adhesion on the chip. Silica nanoparticles of various sizes and surface chemistries were examined to understand the effects of induced nanotoxicity on SH-SY5Y cells by studying the cell viability at different concentrations of nanoparticles ranging from 50 to 400 μg/mL and at various time points. Electrochemical measurements of nanotoxicity were recorded using differential pulse voltametry and were compared to absorption- and fluorescence-based techniques to evaluate the benefits of electrochemical measurements in assessing nanotoxicity.

5.3 TECHNIQUES AND DEVICES FOR NANOTOXICITY TESTING

Nanoparticles with many novel properties are used in various applications and come in contact with complex and dynamic biological systems. It is challenging to characterize nanoparticles throughout their biological interaction and to quantify the uptake rate and localization inside cells. Cells under nanotoxicity may undergo necrosis or repairable, oxidative DNA damage, recovering from it eventually or resulting in apoptosis. Nanotoxicity may alter cell differentiation, proliferation, morphology, or cell–cell communication. Traditional methods for evaluating nanotoxicity consist of bulk analysis, where the resultant is assumed to be the average of the whole cell population. Cell heterogeneity has been recently studied by researchers, suggesting that the cells in a subpopulation may exhibit different behavior from the general population (Irish et al. 2006; Graf and Stadtfeld 2008). This means that the cell population study in aggregation may hinder some very important cell mechanisms, understating that is only visible at single-cell level. Single-cell nanotoxicity studies are believed to provide more realistic cell behavior under nanoparticle interactions.

5.3.1 Carbon Fiber Microelectrode

Carbon fiber microelectrodes have been widely used for single-cell analysis due to their ability to detect diffusion-limited current at very high scan rates, allowing quick measurements of fast temporal events of the cells (Bard 2001). Carbon fiber microelectrodes have high sensitivities due to their very small tip (5–10 μm) and low noise for very sensitive detections of changes in cell behavior. The carbon fiber microelectrode amperometry technique has the potential to reveal the biophysics of exocytosis and, thus, can be an important tool in understanding cellular communication under the influence of nanoparticles.

Marquis et al. (2008) conducted carbon fiber microelectrode amperometry to characterize serotonin exocytosis from murine peritoneal mast cells cocultured with fibroblasts in the presence of Au nanoparticles. The interaction between various concentrations of serum-coated Au nanoparticles sized between 12 and 46 nm after coculturing with mast cells for 48 h, suggested an altered exocytosis mechanism. The study reported decreases in granule transport and fusion events, and increased intracellular matrix expansions and a higher number of serotonin exocytosis per granule. A further expansion of these studies showed the effects on cell viability when the nanoparticle exposure was extended for 48 and 72 h (Marquis et al. 2009). Love and Haynes (2010) carried out a study to evaluate the effect of citrate-reduced noble nanoparticles, Au (28 nm), and Ag (61 nm), on neuroendocrine cells. An inductively coupled plasma–atomic emission spectroscopy (ICP–AES) measurement was carried out for the uptake quantification; cells were lysed and the total metal content was a measure of the nanoparticle uptake. The rate of uptake for 1 nM of Ag and Au nanoparticles was found to be different for each type after 24 h of exposure, 3.4×10^4 versus 7.5×10^5 nanoparticles per cell, respectively, suggesting a higher internalization of the Au nanoparticles. The differences in the rates of nanoparticle internalization were assumed to be affected by several factors such as size, surface charge, and functionalization. Transmission electron microscopy (TEM) was used for the localization of nanoparticle assessment and to verify the internalization of nanoparticles in cellular granules and not only on the cell membrane. Carbon fiber microelectrode amperometry revealed the changing exocytosis behavior of chromaffin cells in this study. Metal oxides are commonly used in consumer products. The carbon fiber microelectrode amperometry technique has been used to understand the nanotoxicity of metal oxide (nonporous SiO_2, porous SiO_2, and nonporous TiO_2) nanoparticles on immune cells by Maurer-Jones et al. (2010). The results revealed functional changes in chemical messenger secretions from mast cell granules. Nanoparticle surface properties are known to play a major role in deciding their interactions with biological systems. Marques et al. (2011) executed a study on noble nanoparticles, Au (~26.5 nm), and Ag (~33.3 nm), with different zeta potentials (surface charge), by modifying the nanoparticle surface using cationic or anionic thiols. It was noted that positive surface-charged nanoparticles (Au+ and Ag+) were more susceptible to internalization by the mast cells compared to their negatively charged counterparts (Au– and Ag–). Carbon fiber microelectrode amperometry was further utilized by Love et al. (2012) to evaluate the changes in cellular communication in neuroendocrine cells after size-dependent Ag nanoparticle- and surface-functionalized Au nanoparticle exposure for 24 h. The study revealed that even if the Ag nanoparticles (15–60 nm) did not alter cell viability, they showed size-dependent cellular uptake and an increase in the speed of exocytosis-release kinetics. On the other hand, polyethylene glycol (PEG)-functionalized Au nanoparticles did not change cell viability. However, they decreased the number of molecules released from each vesicle.

5.3.2 Atomic Force Microscopy

Atomic force microscopy is a powerful, force-sensitive technique and has been successfully applied in single-cell studies to gather the information on cell structure, topography, membrane nanostructures, and the mechanics (e.g., adhesion force, elasticity) of mammalian cells at a nanoscale resolution under physiological or near-physiological conditions. An atomic force microscope can be used to study the mechanics of cells under the influence of nanoparticles. Wu et al. (2012) applied atomic force microscopy to reveal insights on the toxic effects of diesel exhaust particles on vascular endothelial cells at the single-cell level to understand the biophysical properties of the cells. Atomic force microscopy was utilized in two different strategies in this experiment: (1) to measure the mechanical properties of the cell, such as Young's modulus and adhesion force in the growth medium and (2) for topography and membrane visualization, cells were fixed and imaged.

5.4 BIOSENSORS FOR NANOTOXICITY BIOMARKER DETECTION

5.4.1 Background

The rapid growth of the nanotechnology industry has led to a large-scale production and application of engineered nanomaterials, which are used not only in medicine and industry, but also in various consumer products such as food products, textiles, sunscreens, and cosmetics (Pujalte et al. 2011). While, on the other hand, the increased utilization of nanomaterials could affect human health and the environment due to increased exposures (Colvin 2003; Chow et al. 2005; Owen and Depledge 2005). The current knowledge is limited to the potential health effects caused by nanomaterials; however, it shows that they may cause adverse effects at the routes of exposure such as the skin, gastrointestinal tract, and lungs (Chow et al. 2005; Xia et al. 2009). Furthermore, some nanomaterials made of certain metals may have genotoxic or carcinogenic effects. One of the most discussed mechanisms behind the health effects induced by nanomaterials is their ability to enhance the generation of reactive oxygen species (ROS), causing oxidative stress, DNA damage (MacNee and Donaldson 2003; Valko et al. 2005, 2006; Durocher et al. 2009; Jia et al. 2009; Jones and Grainger 2009; Landsiedel et al. 2009; Moller et al. 2010; Singh et al. 2009), and unregulated cell signaling, which eventually leads to changes in cell motility, apoptosis, and carcinogenesis (Kasai et al. 2001; Kasai 2002; Song et al. 2012). Therefore, there is a great need for setting up reliable methods to assess the potential toxicity of the nanomaterials with short deadlines and reasonable costs, ensuring their compatibility for medical applications and for the safety of the environment. While most reliable methods for toxicity evaluation rely on costly, *in vivo* experiments, *in vitro* assays present a promising screening method. Cell cultures offer a useful prescreening method to assess the toxicity of various external agents on cells.

5.4.2 Common Methods for Nanotoxicity Assessment

An assortment of assays is used to assess the toxic effects of nanomaterials *in vitro*. These assays can be generally classified into three groups based on their objects of measurement, that is, cell viability/proliferation, direct/indirect intracellular ROS levels, and genomic markers.

5.4.2.1 Cell Viability/Proliferation Assay

Traditional *in vitro* cell viability/proliferation or cytotoxicity assays, for the measurement of cellular viability and proliferation, are used in the current nanotoxicity studies. These assays include the Alamar Blue assay, which incorporates a fluorometric/colorimetric growth indicator based on the detection of metabolic activity (Fahmy and Cormier 2009; Jones and Grainger 2009; Poonam et al. 2011); the Trypan Blue assay, where cells with an intact membrane are able to exclude the Trypan Blue dye (Karlsson et al. 2008; Bhattacharya et al. 2009; Jones and Grainger 2009; Hillegass et al. 2010; Poonam et al. 2011); the Neutral Red assay, based on the ability of viable cells to incorporate and bind the supravital dye, neutral red, in the lysosomes (Pujalte et al. 2011; Saquib et al. 2012); formazan-based assays (MTT, MTS, and WST), which are used for the detection of various stages in the apoptosis process of cells (Lai et al. 2008; Jones and Grainger 2009; Napierska et al. 2009;

Hillegass et al. 2010; Pujalte et al. 2011; Akhtar et al. 2012; Saquib et al. 2012) and the clonogenic assay, based on the ability of a single cell to grow into a colony (Hillegass et al. 2010; Poonam et al. 2011).

5.4.2.2 Direct/Indirect Intracellular ROS Measurement

The effects of ROS on cell metabolism typically involve the mechanisms in the apoptosis process. Direct/indirect intracellular ROS measurement assays include the glutathione (GSH) assay, a luminescent-based assay for the detection and quantification of GSH in cells (Jones and Grainger 2009; Fahmy and Cormier 2009; Pan et al. 2009; Poonam et al. 2011; Pujalte et al. 2011; Saquib et al. 2012; Akhtar et al. 2012); lipid peroxidation measurement, which measures increased concentrations of end products of lipid peroxidation, indicating increased oxidative damages in the cells (Li et al. 2008; Fahmy and Cormier 2009; Jones and Grainger 2009; Hillegass et al. 2010; Poonam et al. 2011); 2-, 7-dichlorofluorescein (DCFH) assay, which detects intracellular DCFH oxidation due to the presence of hydrogen peroxides (Bhattacharya et al. 2009; Jones and Grainger 2009; Fahmy and Cormier 2009; Pan et al. 2009; Pujalte et al. 2011; Poonam et al. 2011; Akhtar et al. 2012); and electroparamagnetic resonance (EPR) assay to directly measure free radicals in cells and tissues (Bhattacharya et al. 2009; Poonam et al. 2011).

5.4.2.3 Assays on the Genomic Level

Assays on the genomic level measure any damage that occurs to the DNA of the cells. Examples of this type of assay include the comet assay, also known as single-cell gel electrophoresis assay, a technique for the detection of DNA damage at the level of the individual eukaryotic cell (Lai et al. 2008; Mroz et al. 2008; Bhattacharya et al. 2009; Jones and Grainger 2009; Hillegass et al. 2010; Poonam et al. 2011; Mei et al. 2012; Saquib et al. 2012); and DNA damage biomarker assay, as it is well known that excessive generation of ROS can oxidize cellular biomolecules and the resulting free radicals also lead to oxidative modifications in DNA, including strand breaks and base oxidations (Bhattacharya et al. 2009; Hillegass et al. 2010; Song et al. 2012).

5.4.3 BIOSENSING APPROACHES FOR INFLAMMATORY BIOMARKERS DETECTION

Most current *in vitro* cytotoxicity assays rely on the evaluation of cell viability as discussed previously. However, the cellular stress and inflammatory responses can also be used to evaluate cellular toxicity by the detection of specific biomarkers (Kroll et al. 2009). Cytokines are of particular interest as one group of such biomarkers, and their release by cells has been studied as a marker of a cellular immune response (Pfaller et al. 2009). Cytokines regulate the growth and function of immune cells during inflammation and in later immune responses (Pfaller et al. 2009). Therefore, the concentration of cytokines in the medium, typically below 10 ng/mL, will increase in response to inflammation.

Immunoassays are often used for the *in vitro* detection of secreted cytokines. Their principles are based on capturing cytokines by a specific antibody and then measuring their levels. A sandwich immunoassay format has been used for the detection of interleukin 6 (IL-6), interleukin 8 (IL-8), or monocyte chemotactic protein-1 (MCP-1) (Ida et al. 1992; Kajikawa et al. 1996; Battaglia et al. 2005; Liu et al. 2005; Yang et al. 2005; Tan et al. 2008; Kemmler et al. 2009; Pfafflin and Schleicher 2009). In particular, enzyme-linked immunosorbent assay (ELISA) is a simple and sensitive method for monitoring the effects of nanoparticles on immune cells by assessing the levels of cytokines that are released into the cell culture supernatant after the addition of nanoparticles to a cell culture. The ELISA method was first described in 1971 (Lequin 2005), and enables the simple and accurate quantification of inflammatory markers in cell culture supernatants through antibodies and enzymatic detection reactions. ELISA results have been reported for nanoparticles of different compositions and origins, for example, for titanium dioxide (Tao and Kobzik 2002), iron oxide (Wottrich et al. 2004), zinc oxide, carbon black (Monteiro-Riviere and Inman 2006; Duffin et al. 2007), carbon nanotubes (Davoren et al. 2007), fullerenes, silica (Rao et al. 2004), and quantum dots (Ryman-Rasmussen et al. 2007). The most commonly tested human and murine inflammatory markers are the chemokine IL-8, followed by tumor necrosis factor-a (TNF-a), and IL-6.

However, ELISA tests are usually time consuming and require multiple operations. Therefore, the real-time detection of cytokines has been investigated with surface-immobilized immunoassay detection on optical sensing instruments, such as the surface plasmon resonance (SPR) or waveguide-grating sensors (Battaglia et al. 2005; Kemmler et al. 2009; Yang et al. 2005). These techniques are sensitive to changes in the refractive index at the liquid–solid interface and, thus, allow the adsorption of monitoring biomolecules at the sensor surface with high sensitivity. Combined with immunoassays, they allow the sensitive detection of cytokines.

5.4.4 Paper-Based Biosensor for ROS-Induced DNA Oxidative Damage Biomarkers Detection

As mentioned in the previous section, one of the most discussed mechanisms behind the health effects induced by nanomaterials is their ability to enhance ROS generation. An excessive generation of ROS can oxidize cellular biomolecules (i.e., DNA) (Valko et al. 2006), leading to oxidative modifications in DNA. 8-Hydroxyguanine and its nucleoside, 8-hydroxy-2′-deoxyguanosine (8-OHdG), are the most studied DNA damage products due to the relative ease of their measurement and premutagenic potential (Pulido and Parrish 2003; Risom et al. 2005; Valko et al. 2005, 2006). Elevated 8-OHdG levels have been noted in numerous tumors (Risom et al. 2005; Valko et al. 2006) and, are thus widely used as a biomarker for oxidative stress and carcinogenesis (Valavanidis et al. 2009).

A novel lateral flow immunoassay (LFIA) has been developed (Zhu et al. 2013) to measure the concentration of 8-OHdG and, thus, reveals the nanotoxicity at the genomic level. LFIA, also known as the immunochromatographic test strip, has been widely used as an in-field and point-of-care diagnostic tool for testing cancer biomarkers (Lin et al. 2008; Liu et al. 2009; Oh et al. 2009; Zeng et al. 2009; Kawde et al. 2010; Yang et al. 2011), proteins (Mao et al. 2008), drugs (Pattarawarapan et al. 2007; Tang et al. 2009; Wang et al. 2011), hormones (Lu et al. 2005), and metabolites in biomedical (Liu et al. 2008; Mao et al. 2008, 2009; Dungchai et al. 2010), food, and environmental settings.

The principle of the immunostrip is mainly based on a competitive-type immunoreaction in the lateral flow strip. In a typical assay, a sample solution containing a desired concentration of 8-OHdG is applied to the sample application pad. The sample moves along the strip, due to capillary force, and is finally captured by specific antibodies through immunoreactions. The accumulation of gold nanoparticles on the test zone induces a characteristic red band that is visible to the naked eye. This color change indicates the colorimetric detection of 8-OHdG in a sample and the color intensity is inversely proportional to the concentration. The LFIA strip provides a simple approach for a rapid nanotoxicity assessment.

5.5 CONCLUSION

With the increasing number of nanomaterial applications, assessing their toxicity should be the first important step toward creating safety guidelines for their handling and disposal. Studies of the biological effects of nanoscale materials that might answer these questions have lagged behind other aspects of nanotechnology development. In this chapter, we have discussed various biosensing methods for monitoring nanotoxicity, including techniques for single-cell nanotoxicity testing, such as carbon fiber microelectrodes and atomic force microscopy, chip-based sensors for nanotoxicity, and biosensors for nanotoxicity biomarker detections, such as biosensing approaches for inflammatory biomarker detections and paper-based biosensors for ROS-induced DNA oxidative damage biomarker detections.

REFERENCES

Akhtar, M. J., M. Ahamed, M. Fareed, S. A. Alrokayan, and S. Kumar. 2012. Protective effect of sulphoraphane against oxidative stress mediated toxicity induced by CuO nanoparticles in mouse embryonic fibroblasts BALB 3T3. *J Toxicol Sci* 37 (1):139–148.

Asphahani, F. and M. Zhang. 2007. Cellular impedance biosensors for drug screening and toxin detection. *Analyst* 132 (9):835–841.

Bard, A. J. and L. R. Faulkner. 2001. *Electrochemical Methods: Fundamentals and Applications*. 2nd ed. John Wiley & Sons: New York.

Battaglia, T. M., J. F. Masson, M. R. Sierks, S. P. Beaudoin, J. Rogers, K. N. Foster, G. A. Holloway, and K. S. Booksh. 2005. Quantification of cytokines involved in wound healing using surface plasmon resonance. *Anal Chem* 77 (21):7016–7023.

Bhattacharya, K., M. Davoren, J. Boertz, R. P. Schins, E. Hoffmann, and E. Dopp. 2009. Titanium dioxide nanoparticles induce oxidative stress and DNA-adduct formation but not DNA-breakage in human lung cells. *Part Fibre Toxicol* 6:17.

Bianco, A., K. Kostarelos, and M. Prato. 2005. Applications of carbon nanotubes in drug delivery. *Curr Opin Chem Biol* 9 (6):674–679.

Chow, J. C., J. G. Watson, N. Savage, C. J. Solomon, Y. S. Cheng, P. H. McMurry, L. M. Corey et al. 2005. Nanoparticles and the environment. *J Air Waste Manag Assoc* 55 (10):1411–1417.

Clark, L. C. Jr. 1956. Monitor and control of blood and tissue oxygen tensions. *ASAIO J* 2 (1):41–48.

Clemens, A. H., P. H. Chang, and R. W. Myers. 1976. Development of an automatic system of insulin infusion controlled by blood sugar, its system for the determination of glucose and control algorithms. *J Annu Diabetol Hotel Dieu May* :269–278.

Colvin, V. L. 2003. The potential environmental impact of engineered nanomaterials. *Nat Biotechnol* 21 (10):1166–1170.

Connolly, P., P. Clark, A. S. G. Curtis, J. A. T. Dow, and C. D. W. Wilkinson. 1990. An extracellular microelectrode array for monitoring electrogenic cells in culture. *Biosens Bioelectron* 5 (3):223–234.

Davoren, M., E. Herzog, A. Casey, B. Cottineau, G. Chambers, H. J. Byrne, and F. M. Lyng. 2007. In vitro toxicity evaluation of single walled carbon nanotubes on human A549 lung cells. *Toxicol in Vitro* 21 (3): 438–448.

DeBusschere, B. D. and G. T. A. Kovacs. 2001. Portable cell-based biosensor system using integrated CMOS cell-cartridges. *Biosens Bioelectron* 16 (7–8):543–556.

Duffin, R., L. Tran, D. Brown, V. Stone, and K. Donaldson. 2007. Proinflammogenic effects of low-toxicity and metal nanoparticles *in vivo* and *in vitro*: Highlighting the role of particle surface area and surface reactivity. *Inhal Toxicol* 19 (10):849–856.

Dungchai, W., O. Chailapakul, and C. S. Henry. 2010. Use of multiple colorimetric indicators for paper-based microfluidic devices. *Anal Chim Acta* 674 (2):227–233.

Durocher, S., A. Rezaee, C. Hamm, C. Rangan, S. Mittler, and B. Mutus. 2009. Disulfide-linked, gold nanoparticle based reagent for detecting small molecular weight thiols. *J Am Chem Soc* 131 (7):2475–2477.

Elder, A., S. Vidyasagar, and L. DeLouise. 2009. Physicochemical factors that affect metal and metal oxide nanoparticle passage across epithelial barriers. *Wiley Interdiscip Rev—Nanomed Nanobiotechnol* 1 (4): 434–450.

Fahmy, B. and S. A. Cormier. 2009. Copper oxide nanoparticles induce oxidative stress and cytotoxicity in airway epithelial cells. *Toxicol in Vitro* 23 (7):1365–1371.

Faraji, A. H. and P. Wipf. 2009. Nanoparticles in cellular drug delivery. *Bioorg Med Chem* 17 (8):2950–2962.

Giaever, I. and C. R. Keese. 1984. Monitoring fibroblast behavior in tissue-culture with an applied electric-field. *Proc Nat Acad Sci USA—Biol Sci* 81 (12):3761–3764.

Giaever, I. and C. R. Keese. 1986. Use of electric fields to monitor the dynamical aspect of cell behavior in tissue culture. *IEEE Trans Biomed Eng* 33 (2):242–247.

Giaever, I. and C. R. Keese. 1989. Fractal motion of mammalian-cells. *Physica D* 38 (1–3):128–133.

Giaever, I. and C. R. Keese. 1991. Micromotion of mammalian cells measured electrically. *Proc Natl Acad Sci USA* 88 (17):7896–7900.

Giaever, I. and C. R. Keese. 1992. Toxic—Cells can tell. *Chemtech* 22 (2):116–125.

Giaever, I. and C. R. Keese. 1993. A morphological biosensor for mammalian-cells. *Nature* 366 (6455):591–592.

Graf, T. and M. Stadtfeld. 2008. Heterogeneity of embryonic and adult stem cells. *Cell Stem Cell* 3 (5):480–483.

Gross, G. W., B. K. Rhoades, D. L. Reust, and F. U. Schwalm. 1993. Stimulation of monolayer networks in culture through thin-film indium–tin oxide recording electrodes. *J Neurosci Methods* 50 (2):131–143.

Gross, G. W., E. Rieske, G. W. Kreutzberg, and A. Meyer. 1977. A new fixed-array multi-microelectrode system designed for long-term monitoring of extracellular single unit neuronal activity *in vitro*. *Neurosci Lett* 6 (2–3):101–105.

Gross, G. W., A. N. Williams, and J. H. Lucas. 1982. Recording of spontaneous activity with photoetched microelectrode surfaces from mouse spinal neurons in culture. *J Neurosci Methods* 5 (1–2):13–22.

Guilbault, G. G. and J. G. Montalvo, Jr. 1969. A urea-specific enzyme electrode. *J Am Chem Soc* 91 (8): 2164–2165.

Harrison, B. S. and A. Atala. 2007. Carbon nanotube applications for tissue engineering. *Biomaterials* 28 (2): 344–353.

Hillegass, J. M., A. Shukla, S. A. Lathrop, M. B. MacPherson, N. K. Fukagawa, and B. T. Mossman. 2010. Assessing nanotoxicity in cells *in vitro*. *Wiley Interdiscip Rev—Nanomed Nanobiotechnol* 2 (3):219–231.

Hussain, S. M., K. L. Hess, J. M. Gearhart, K. T. Geiss, and J. J. Schlager. 2005. *In vitro* toxicity of nanoparticles in BRL 3A rat liver cells. *Toxicol in Vitro* 19 (7):975–983.

Ida, N., S. Sakurai, K. Hosoi, and T. Kunitomo. 1992. A highly sensitive enzyme-linked-immunosorbent-assay for the measurement of interleukin-8 in biological-fluids. *J Immunol Methods* 156 (1):27–38.

Irish, J. M., N. Kotecha, and G. P. Nolan. 2006. Mapping normal and cancer cell signalling networks: Towards single-cell proteomics. *Nat Rev Cancer* 6 (2):146–155.

Jia, H. Y., Y. Liu, X. J. Zhang, L. Han, L. B. Du, Q. Tian, and Y. C. Xu. 2009. Potential oxidative stress of gold nanoparticles by induced-NO releasing in serum. *J Am Chem Soc* 131 (1):40–41.

Jones, C. F. and D. W. Grainger. 2009. *In vitro* assessments of nanomaterial toxicity. *Adv Drug Deliv Rev* 61 (6):438–456.

Kajikawa, O., R. B. Goodman, M. C. Johnson, K. Konishi, and T. R. Martin. 1996. Sensitive and specific immunoassays to detect rabbit IL-8 and MCP-1: Cytokines that mediate leukocyte recruitment to the lungs. *J Immunol Methods* 197 (1–2):19–29.

Karlsson, H. L., P. Cronholm, J. Gustafsson, and L. Moller. 2008. Copper oxide nanoparticles are highly toxic: A comparison between metal oxide nanoparticles and carbon nanotubes. *Chem Res Toxicol* 21 (9):1726–1732.

Kasai, H. 2002. Chemistry-based studies on oxidative DNA damage: Formation, repair, and mutagenesis. *Free Radic Biol Med* 33 (4):450–456.

Kasai, H., N. Iwamoto-Tanaka, T. Miyamoto, K. Kawanami, S. Kawanami, R. Kido, and M. Ikeda. 2001. Life style and urinary 8-hydroxydeoxyguanosine, a marker of oxidative DNA damage: Effects of exercise, working conditions, meat intake, body mass index, and smoking. *Jpn J Cancer Res* 92 (1):9–15.

Katz, E. and I. Willner. 2003. Probing biomolecular interactions at conductive and semiconductive surfaces by impedance spectroscopy: Routes to impedimetric immunosensors, DNA-sensors, and enzyme biosensors. *Electroanalysis* 15 (11):913–947.

Kawde, A.-N., X. Mao, H. Xu, Q. Zeng, Y. He, and G. Liu. 2010. Moving enzyme-linked immunosorbent assay to the point-of-care dry reagent strip biosensors. *Am J Biomed Sci* 2 (1):23–32.

Kemmler, M., B. Koger, G. Sulz, U. Sauer, E. Schleicher, C. Preininger, and A. Brandenburg. 2009. Compact point-of-care system for clinical diagnostics. *Sens Actuat B—Chemical* 139 (1):44–51.

Kim, J., J. E. Lee, S. H. Lee, J. H. Yu, J. H. Lee, T. G. Park, and T. Hyeon. 2008. Designed fabrication of a multifunctional polymer nanomedical platform for simultaneous cancer-targeted imaging and magnetically guided drug delivery. *Adv Mater* 20 (3):478–483.

Kim, T. H., S. R. Kang, B. K. Oh, and J. W. Choi. 2011. Cell chip for detection of silica nanoparticle-induced cytotoxicity. *Sensor Lett* 9 (2):861–865.

Kreyling, W. G., M. Semmler-Behnke, and W. Moller. 2006. Health implications of nanoparticles. *J Nanoparticle Res* 8 (5):543–562.

Kroll, A., M. H. Pillukat, D. Hahn, and J. Schnekenburger. 2009. Current *in vitro* methods in nanoparticle risk assessment: Limitations and challenges. *Eur J Pharm Biopharm* 72 (2):370–377.

Lai, J. C., M. B. Lai, S. Jandhyam, V. V. Dukhande, A. Bhushan, C. K. Daniels, and S. W. Leung. 2008. Exposure to titanium dioxide and other metallic oxide nanoparticles induces cytotoxicity on human neural cells and fibroblasts. *Int J Nanomed* 3 (4):533–545.

Landsiedel, R., M. D. Kapp, M. Schulz, K. Wiench, and F. Oesch. 2009. Genotoxicity investigations on nanomaterials: Methods, preparation and characterization of test material, potential artifacts and limitations—Many questions, some answers. *Mutat Res* 681 (2–3):241–258.

Lequin, R. M. 2005. Enzyme immunoassay (EIA)/enzyme-linked immunosorbent assay (ELISA). *Clin Chem* 51 (12):2415–2418.

Lewinski, N., V. Colvin, and R. Drezek. 2008. Cytotoxicity of nanoparticles. *Small* 4 (1):26–49.

Li, S. Q., R. R. Zhu, H. Zhu, M. Xue, X. Y. Sun, S. D. Yao, and S. L. Wang. 2008. Nanotoxicity of TiO(2) nanoparticles to erythrocyte *in vitro*. *Food Chem Toxicol* 46 (12):3626–3631.

Lin, Y. Y., J. Wang, G. D. Liu, H. Wu, C. M. Wai, and Y. H. Lin. 2008. A nanoparticle label/immunochromatographic electrochemical biosensor for rapid and sensitive detection of prostate-specific antigen. *Biosens Bioelectron* 23 (11):1659–1665.

Liu, G., X. Mao, J. A. Phillips, H. Xu, W. Tan, and L. Zeng. 2009. Aptamer–nanoparticle strip biosensor for sensitive detection of cancer cells. *Anal Chem* 81 (24):10013–10018.

Liu, G., J. Wang, R. Barry, C. Petersen, C. Timchalk, P. L. Gassman, and Y. Lin. 2008. Nanoparticle-based electrochemical immunosensor for the detection of phosphorylated acetylcholinesterase: An exposure biomarker of organophosphate pesticides and nerve agents. *Chemistry* 14 (32):9951–9959.

Liu, M. Y., A. M. Xydakis, R. C. Hoogeveen, P. H. Jones, E. O. B. Smith, K. W. Nelson, and C. M. Ballantyne. 2005. Multiplexed analysis of biomarkers related to obesity and the metabolic syndrome in human plasma, using the luminex-100 system. *Clin Chem* 51 (7):1102–1109.

Love, S. A. and C. L. Haynes. 2010. Assessment of functional changes in nanoparticle-exposed neuroendocrine cells with amperometry: Exploring the generalizability of nanoparticle–vesicle matrix interactions. *Anal Bioanal Chem* 398 (2):677–688.

Love, S. A., Z. Liu, and C. L. Haynes. 2012. Examining changes in cellular communication in neuroendocrine cells after noble metal nanoparticle exposure. *Analyst* 137 (13):3004–3010.

Lu, F., K. H. Wang, and Y. H. Lin. 2005. Rapid, quantitative and sensitive immunochromatographic assay based on stripping voltammetric detection of a metal ion label. *Analyst* 130 (11):1513–1517.

Luong, J. H. T., K. B. Male, and J. D. Glennon. 2008. Biosensor technology: Technology push versus market pull. *Biotechnol Adv* 26 (5):492–500.

MacNee, W. and K. Donaldson. 2003. Mechanism of lung injury caused by PM10 and ultrafine particles with special reference to COPD. *Eur Respir J Suppl* 40:47s–51s.

Mao, X., M. Baloda, A. S. Gurung, Y. H. Lin, and G. D. Liu. 2008. Multiplex electrochemical immunoassay using gold nanoparticle probes and immunochromatographic strips. *Electrochem Commun* 10 (10):1636–1640.

Mao, X., Y. Q. Ma, A. G. Zhang, L. R. Zhang, L. W. Zeng, and G. D. Liu. 2009. Disposable nucleic acid biosensors based on gold nanoparticle probes and lateral flow strip. *Anal Chem* 81 (4):1660–1668.

Marquis, B. J., Z. Liu, K. L. Braun, and C. L. Haynes. 2011. Investigation of noble metal nanoparticle zeta-potential effects on single-cell exocytosis function *in vitro* with carbon-fiber microelectrode amperometry. *Analyst* 136 (17):3478–3486.

Marquis, B. J., M. A. Maurer-Jones, K. L. Braun, and C. L. Haynes. 2009. Amperometric assessment of functional changes in nanoparticle-exposed immune cells: Varying Au nanoparticle exposure time and concentration. *Analyst* 134 (11):2293–2300.

Marquis, B. J., A. D. McFarland, K. L. Braun, and C. L. Haynes. 2008. Dynamic measurement of altered chemical messenger secretion after cellular uptake of nanoparticles using carbon-fiber microelectrode amperometry. *Anal Chem* 80 (9):3431–3437.

Maurer-Jones, M. A., Y. S. Lin, and C. L. Haynes. 2010. Functional assessment of metal oxide nanoparticle toxicity in immune cells. *ACS Nano* 4 (6):3363–3373.

Mcconnell, H. M., J. C. Owicki, J. W. Parce, D. L. Miller, G. T. Baxter, H. G. Wada, and S. Pitchford. 1992. The cytosensor microphysiometer—Biological applications of silicon technology. *Science* 257 (5078):1906–1912.

Mei, N., Y. Zhang, Y. Chen, X. Guo, W. Ding, S. F. Ali, A. S. Biris, P. Rice, M. M. Moore, and T. Chen. 2012. Silver nanoparticle-induced mutations and oxidative stress in mouse lymphoma cells. *Environ Mol Mutagen* 53 (6):409–419.

Moller, P., N. R. Jacobsen, J. K. Folkmann, P. H. Danielsen, L. Mikkelsen, J. G. Hemmingsen, L. K. Vesterdal, L. Forchhammer, H. Wallin, and S. Loft. 2010. Role of oxidative damage in toxicity of particulates. *Free Radic Res* 44 (1):1–46.

Monteiro-Riviere, N. A. and A. O. Inman. 2006. Challenges for assessing carbon nanomaterial toxicity to the skin. *Carbon* 44 (6):1070–1078.

Mosbach, K. and B. Danielsson. 1974. An enzyme thermistor. *Biochim Biophys Acta* 364 (1):140–145.

Mossman, B. T., P. J. Borm, V. Castranova, D. L. Costa, K. Donaldson, and S. R. Kleeberger. 2007. Mechanisms of action of inhaled fibers, particles and nanoparticles in lung and cardiovascular diseases. *Part Fibre Toxicol* 4:4.

Mroz, R. M., R. P. Schins, H. Li, L. A. Jimenez, E. M. Drost, A. Holownia, W. MacNee, and K. Donaldson. 2008. Nanoparticle-driven DNA damage mimics irradiation-related carcinogenesis pathways. *Eur Respir J* 31 (2):241–251.

Napierska, D., L. C. Thomassen, V. Rabolli, D. Lison, L. Gonzalez, M. Kirsch-Volders, J. A. Martens, and P. H. Hoet. 2009. Size-dependent cytotoxicity of monodisperse silica nanoparticles in human endothelial cells. *Small* 5 (7):846–853.

Nowak, K., E. Mix, J. Gimsa, U. Strauss, K. K. Sriperumbudur, R. Benecke, and U. Gimsa. 2011. Optimizing a rodent model of Parkinson's disease for exploring the effects and mechanisms of deep brain stimulation. *Parkinsons Dis* 2011:414682.

Oberdorster, G., A. Maynard, K. Donaldson, V. Castranova, J. Fitzpatrick, K. Ausman, J. Carter et al. 2005. Principles for characterizing the potential human health effects from exposure to nanomaterials: Elements of a screening strategy. *Part Fibre Toxicol* 2:8.

Oh, S. W., Y. M. Kim, H. J. Kim, S. J. Kim, J. S. Cho, and E. Y. Choi. 2009. Point-of-care fluorescence immunoassay for prostate specific antigen. *Clinica Chim Acta* 406 (1–2):18–22.

Owen, R. and M. Depledge. 2005. Nanotechnology and the environment: Risks and rewards. *Mar Pollut Bull* 50 (6):609–612.

Pan, Y., A. Leifert, D. Ruau, S. Neuss, J. Bornemann, G. Schmid, W. Brandau, U. Simon, and W. Jahnen-Dechent. 2009. Gold nanoparticles of diameter 1.4 nm trigger necrosis by oxidative stress and mitochondrial damage. *Small* 5 (18):2067–2076.

Pancrazio, J. J., J. P. Whelan, D. A. Borkholder, W. Ma, and D. A. Stenger. 1999. Development and application of cell-based biosensors. *Ann Biomed Eng* 27 (6):697–711.

Pattarawarapan, M., S. Nangola, T. R. Cressey, and C. Tayapiwatana. 2007. Development of a one-step immunochromatographic strip test for the rapid detection of nevirapine (NVP), a commonly used antiretroviral drug for the treatment of HIV/AIDS. *Talanta* 71 (1):462–470.

Pfafflin, A. and E. Schleicher. 2009. Inflammation markers in point-of-care testing (POCT). *Anal Bioanal Chem* 393 (5):1473–1480.

Pfaller, T., V. Puntes, E. Casals, A. Duschl, and G. J. Oostingh. 2009. *In vitro* investigation of immunomodulatory effects caused by engineered inorganic nanoparticles—The impact of experimental design and cell choice. *Nanotoxicology* 3 (1):46–59.

Poonam, T. and S. Mahant. 2011. *In vitro* methods for nanotoxicity assessment: Advantages and applications. *Arch Appl Sci Res* 3 (2):389–403.

Pujalte, I., I. Passagne, B. Brouillaud, M. Treguer, E. Durand, C. Ohayon-Courtes, and B. L'Azou. 2011. Cytotoxicity and oxidative stress induced by different metallic nanoparticles on human kidney cells. *Part Fibre Toxicol* 8:10.

Pulido, M. D. and A. R. Parrish. 2003. Metal-induced apoptosis: Mechanisms. *Mutat Res* 533 (1–2):227–241.

Rao, K. M., D. W. Porter, T. Meighan, and V. Castranova. 2004. The sources of inflammatory mediators in the lung after silica exposure. *Environ Health Perspect* 112 (17):1679–1686.

Risom, L., P. Moller, and S. Loft. 2005. Oxidative stress-induced DNA damage by particulate air pollution. *Mutat Res* 592 (1–2):119–137.

Ryman-Rasmussen, J. P., J. E. Riviere, and N. A. Monteiro-Riviere. 2007. Surface coatings determine cytotoxicity and irritation potential of quantum dot nanoparticles in epidermal keratinocytes. *J Invest Dermatol* 127 (1):143–153.

Saquib, Q., A. A. Al-Khedhairy, M. A. Siddiqui, F. M. Abou-Tarboush, A. Azam, and J. Musarrat. 2012. Titanium dioxide nanoparticles induced cytotoxicity, oxidative stress and DNA damage in human amnion epithelial (WISH) cells. *Toxicol in Vitro* 26 (2):351–361.

Shi, D. L. 2009. Integrated multifunctional nanosystems for medical diagnosis and treatment. *Adv Funct Mater* 19 (21):3356–3373.

Singh, N., B. Manshian, G. J. Jenkins, S. M. Griffiths, P. M. Williams, T. G. Maffeis, C. J. Wright, and S. H. Doak. 2009. Nanogenotoxicology: The DNA damaging potential of engineered nanomaterials. *Biomaterials* 30 (23–24):3891–3914.

Slowing, I. I., B. G. Trewyn, S. Giri, and V. S. Y. Lin. 2007. Mesoporous silica nanoparticles for drug delivery and biosensing applications. *Adv Funct Mater* 17 (8):1225–1236.

Soenen, S. J., P. Rivera-Gil, J. M. Montenegro, W. J. Parak, S. C. De Smedt, and K. Braeckmans. 2011. Cellular toxicity of inorganic nanoparticles: Common aspects and guidelines for improved nanotoxicity evaluation. *Nano Today* 6 (5):446–465.

Song, M. F., Y. S. Li, H. Kasai, and K. Kawai. 2012. Metal nanoparticle-induced micronuclei and oxidative DNA damage in mice. *J Clin Biochem Nutr* 50 (3):211–216.

Song, S., H. Xu, and C. H. Fan. 2006. Potential diagnostic applications of biosensors: Current and future directions. *Int J Nanomed* 1 (4):433–440.

Soto, K. F., A. Carrasco, T. G. Powell, K. M. Garza, and L. E. Murr. 2005. Comparative *in vitro* cytotoxicity assessment of some manufactured nanoparticulate materials characterized by transmission electron microscopy. *J Nanoparticle Res* 7 (2–3):145–169.

Takhar, P. and S. Mahant. 2011. *In vitro* methods for nanotoxicity assessment: Advantages and applications. *Arch Appl Sci Res* 3 (2):389–403.

Tan, W., L. Sabet, Y. Li, T. Yu, P. R. Klokkevold, D. T. Wong, and C. M. Ho. 2008. Optical protein sensor for detecting cancer markers in saliva. *Biosens Bioelectron* 24 (2):266–271.

Tang, D., J. C. Sauceda, Z. Lin, S. Ott, E. Basova, I. Goryacheva, S. Biselli, J. Lin, R. Niessner, and D. Knopp. 2009. Magnetic nanogold microspheres-based lateral-flow immunodipstick for rapid detection of aflatoxin B2 in food. *Biosens Bioelectron* 25 (2):514–518.

Tao, F. and L. Kobzik. 2002. Lung macrophage–epithelial cell interactions amplify particle-mediated cytokine release. *Am J Respir Cell Mol Biol* 26 (4):499–505.

Thomas, C. A. Jr., P. A. Springer, G. E. Loeb, Y. Berwald-Netter, and L. M. Okun. 1972. A miniature microelectrode array to monitor the bioelectric activity of cultured cells. *Exp Cell Res* 74 (1):61–66.

Updike, S. J. and G. P. Hicks. 1967. The enzyme electrode. *Nature* 214 (5092):986–988.

Valavanidis, A., T. Vlachogianni, and C. Fiotakis. 2009. 8-Hydroxy-2′-deoxyguanosine (8-OHdG): A critical biomarker of oxidative stress and carcinogenesis. *J Environ Sci Health C Environ Carcinog Ecotoxicol Rev* 27 (2):120–139.

Valko, M., H. Morris, and M. T. Cronin. 2005. Metals, toxicity and oxidative stress. *Curr Med Chem* 12 (10):1161–1208.

Valko, M., C. J. Rhodes, J. Moncol, M. Izakovic, and M. Mazur. 2006. Free radicals, metals and antioxidants in oxidative stress-induced cancer. *Chem Biol Interact* 160 (1):1–40.

Volkl, K. P., N. Opitz, and D. W. Lubbers. 1980. Continuous measurement of concentrations of alcohol using a fluorescence–photometric enzymatic method. *Fresenius Zeitschrift Fur Analytische Chemie* 301 (2):162–163.

Wang, L., W. Ma, W. Chen, L. Liu, Y. Zhu, L. Xu, H. Kuang, and C. Xu. 2011. An aptamer-based chromatographic strip assay for sensitive toxin semi-quantitative detection. *Biosens Bioelectron* 26 (6):3059–3062.

Weaver, J. C., C. L. Cooney, S. P. Fulton, P. Schuler, and S. R. Tannenbaum. 1976. Experiments and calculations concerning a thermal enzyme probe. *Biochim Biophys Acta* 452 (2):285–291.

Wilhelm, C., F. Gazeau, J. Roger, J. N. Pons, and J. C. Bacri. 2002. Interaction of anionic superparamagnetic nanoparticles with cells: Kinetic analyses of membrane adsorption and subsequent internalization. *Langmuir* 18 (21):8148–8155.

Wottrich, R., S. Diabate, and H. F. Krug. 2004. Biological effects of ultrafine model particles in human macrophages and epithelial cells in mono- and co-culture. *Int J Hyg Environ Health* 207 (4):353–361.

Wu, Y., T. Yu, T. A. Gilbertson, A. Zhou, H. Xu, and K. T. Nguyen. 2012. Biophysical assessment of single cell cytotoxicity: Diesel exhaust particle-treated human aortic endothelial cells. *PLoS One* 7 (5):e36885.

Xia, T., N. Li, and A. E. Nel. 2009. Potential health impact of nanoparticles. *Annu Rev Public Health* 30:137–150.

Yang, C. Y., E. Brooks, Y. Li, P. Denny, C. M. Ho, F. X. Qi, W. Y. Shi et al. 2005. Detection of picomolar levels of interleukin-8 in human saliva by SPR. *Lab Chip* 5 (10):1017–1023.

Yang, Q. H., X. Q. Gong, T. Song, J. M. Yang, S. J. Zhu, Y. H. Li, Y. Cui, Y. X. Li, B. B. Zhang, and J. Chang. 2011. Quantum dot-based immunochromatography test strip for rapid, quantitative and sensitive detection of alpha fetoprotein. *Biosens Bioelectron* 30 (1):145–150.

Zeng, Q., X. Mao, H. Xu, S. Wang, and G. Liu. 2009. Quantitative immunochromatographic strip biosensor for the detection of carcinoembryonic antigen tumor biomarker in human plasma. *Am J Biomed Sci*: (1): 70–79.

Zhu, X., E. Hondroulis, W. Liu, and C. Z. Li. 2013. Biosensing approaches for rapid genotoxicity and cytotoxicity assays upon nanomaterial exposure. *Small* 9: 1821–1830.

Zysk, J. R., and W. R. Baumbach. 1998. Homogeneous pharmacologic and cell-based screens provide diverse strategies in drug discovery: Somatostatin antagonists as a case study. *Comb Chem High Throughput Screen* 1 (4):171–183.

6 Carbon Nanotubes and Pulmonary Toxicity

Malay K. Das and Charles Preuss

CONTENTS

6.1 Introduction 131
6.2 Potential Applications.. 132
6.3 CNT-Mediated Toxicity... 132
6.4 Pulmonary Toxicity of SWCNTs... 134
6.5 Pulmonary Toxicity of MWCNTs 135
6.6 Metallic Contamination of CNTs and Pulmonary Toxicity 135
6.7 Pulmonary Toxicity of Functionalized CNTs.. 136
6.8 Conclusion 137
References.. 137

6.1 INTRODUCTION

Carbon nanotubes (CNTs), a distinct molecular form of carbon atoms that was discovered in the late 1980s (Kelly and Kim, 2007), were exhaustively discovered by Sumio Iijima in 1991 (Iijima, 1991). Later on in the year 2000, President Bill Clinton established the National Nanotechnology Initiative to lead the United States into the next industrial revolution (White House, 2000). Nanomaterials are the building blocks of this new industry. One of the major objectives of the initiative was to develop materials that are 10 times stronger than steel but a fraction of the weight for making all kinds of land, sea, air, and space vehicles lighter and more fuel efficient. This statement specifically implicates CNTs, a novel and lightweight material with the strongest tensile strength of all synthetic fibers (Ball, 1999). The presidential initiative directed NASA to search for applications of carbon nanotubes and other nanomaterial in aerospace.

CNTs are allotropes of carbon with a thin cylindrical nanostructure (their diameter is about 10,000 times smaller than a human hair). Nanotubes have been constructed with a length-to-diameter ratio of up to 132,000,000:1 (Wang et al., 2009), significantly larger than any other material. These cylindrical carbon molecules have unusual properties, which are valuable for nanotechnology, electronics, optics, and other fields of material science and technology. In particular, owing to their extraordinary thermal conductivity and mechanical and electrical properties, CNTs find applications as additives to various structural materials. For instance, nanotubes form a tiny portion of the material(s) in some (primarily carbon fiber) baseball bats, golf clubs, or car parts (Gullapalli and Wong, 2011).

CNTs can be produced by the deposition of carbon atoms vaporized from graphite by electric arcs or by laser onto metal particles. More recently, they have been produced by chemical vapor deposition (CVD). High-pressure carbon monoxide conversion (HiPco™, Rice University, TX) is a CVD process and is a more advanced method that uses CO as a carbon source; up to 97% of the carbon in the HiPco product ends up in CNTs (Bronikowski et al., 2001). An individual CNT molecule is about 1 nm in diameter and several microns long (Ajayan and Ebbesen, 1997). Microscopically, individual CNT fibers aggregate into bundles or ropes, which in turn agglomerate loosely into small clumps.

Nanotubes are categorized as single-walled carbon nanotubes (SWCNTs) and multiwalled carbon nanotubes (MWCNTs). Individual nanotubes naturally align themselves into "ropes" held together by van der Waals forces, more specifically, *pi*-stacking. Applied quantum chemistry, specifically orbital hybridization, best describes chemical bonding in nanotubes. The chemical bonding of nanotubes is composed entirely of sp^2 bonds, similar to those of graphite. These bonds, which are stronger than the sp^3 bonds found in alkanes and diamonds, provide nanotubes with their unique strength (Gullapalli and Wong, 2011).

SWCNTs are tubes of graphite that are normally capped at the ends. They have a single cylindrical wall. The structure of a SWCNT can be visualized as a layer of graphite a single atom thick, called graphene, which is rolled into a seamless cylinder. Most SWCNTs typically have a diameter of close to 1 nm. The tube length, however, can be many thousands of times longer. SWCNTs are more pliable, yet harder to make, than MWCNTs. They can be twisted, flattened, and bent into small circles or around sharp bends without breaking. The diameters of MWCNTs are typically in the range of 5 to 50 nm. The interlayer distance in MWCNTs is close to the distance between graphene layers in graphite. MWCNTs are easier to produce in high-volume quantities than SWCNTs. However, the structure of MWCNTs is less well understood because of its greater complexity and variety (Mintmire et al., 1992; Dekker, 1999; Martel et al., 2001).

Double-walled carbon nanotubes (DWCNTs) form a special class of nanotubes because their morphology and properties are similar to those of SWCNT but their resistance to chemicals is significantly improved. This is especially important when functionalization is required (this means grafting of chemical functions at the surface of the nanotubes) to add new properties to the CNT. In the case of SWCNTs, covalent functionalizations will break some C=C double bonds, leaving "holes" in the structure on the nanotube and, thus, modifying both its mechanical and electrical properties. In the case of DWCNTs, only the outer wall is modified. DWCNT synthesis on the gram scale was first proposed in 2003 (Flahaut et al., 2003) by the CCVD technique from the selective reduction of oxide solutions in methane and hydrogen.

6.2 POTENTIAL APPLICATIONS

The strength and flexibility of CNTs potentiate their use in controlling other nanoscale structures, which suggests they will have an important role in nanotechnology engineering. The highest tensile strength of an individual MWCNT has been tested to be 63 GPa (Yu et al., 2000). Over the years, new applications have taken advantage of their unique electrical properties, extraordinary strength, and efficiency in heat conduction. CNTs exhibit unique structural, electromagnetic, chemical, mechanical, and electrical properties, and have been considered for use in numerous technological applications (Table 6.1).

The current use and application of nanotubes has mostly been limited to the use of bulk nanotubes, which is a mass of rather unorganized fragments of nanotubes. Bulk nanotube materials may never achieve a tensile strength similar to that of individual tubes, but such composites may, nevertheless, yield strengths sufficient for many applications. Bulk CNTs have already been used as composite fibers in polymers to improve the mechanical, thermal, and electrical properties of the bulk product tips for atomic force microscope (AFM) probes (www.nanoscience.com); and scaffolding for bone growth in tissue engineering (Haddon et al., 2006). MWCNT AFM tips offer a much greater scanning life than standard probes, and combine abilities to achieve high resolution when measuring high aspect ratio features.

6.3 CNT-MEDIATED TOXICITY

People could be exposed to CNTs through accidental exposures by coming in contact with the aerosol form of CNTs during the production or exposure as a result of biomedical use. CNTs are in the nanometer size range and, hence, easily enter into the lungs via the respiratory tract through

TABLE 6.1
Applications of CNTs

CNTs Property	Applications	References
1. Structural	Textiles—CNT can make waterproof and/or tear-resistant fabrics Body armor—Combat jackets that use CNT fibers to stop bullets and to monitor the condition of the wearer Polyethylene—Adding CNT to polyethylene can increase the polymer's elastic modulus by 30% Synthetic muscles—Owing to their high contraction/extension ratio given an electric current, CNTs are ideal for synthetic muscle High tensile strength fibers—Fibers produced with polyvinyl alcohol required 600 J/g to break	www.mit.edu, Ali et al. (2009), Alan et al. (2003)
2. Electromagnetic	Artificial muscles—CNTs have sufficient contractility to make them candidates to replace muscle tissue Magnets—MWCNT coated with magnetite can generate strong magnetic fields. Recent advances show that MWCNT decorated with magnemite nanoparticles can be oriented in a magnetic field and enhance the electrical properties of the composite material in the direction of the field Superconductor Ultracapacitors Transistors	www.newscientist.com, Rina Tannenbaum et al. (2010, 2011), Tang et al. (2001)
3. Chemical	Desalination—Water molecules can be separated from salt by forcing them through networks of CNTs, which require far lower pressures than conventional reverse osmosis methods Air pollution filter—CNT membranes can filter carbon dioxide from power plant emissions Biotech container—CNT can be filled with biological molecules, aiding biotechnology	Ken et al. (2011), www.sciencedaily.com
4. Mechanical	Oscillator—Oscillators based on CNT have achieved higher speeds than other technologies (>50 GHz) Nanotube membrane—CNTs as filters in membranes have a high specific surface area and high flux that results in fast flow rates for gases and liquids. Liquids flow up to five orders of magnitude faster than predicted by classical fluid dynamics	Sholl and Johnson (2006), Mainak et al. (2005)
5. Medical	Biological sensor Kanzius cancer therapy—SWCNTs are inserted around cancerous cells, and then excited with radio waves, which cause them to heat up and kill the surrounding cells Materials for bone cell proliferation Bone formation Multifunctional drug-delivery systems with intended diagnosis and targeting purposes	O'Connell et al. (2002), Cherukuri et al. (2004), Haddon et al. (2006), Shi et al. (2007), Balaji et al. (2008), Dalton (2005)

the inhalation of air. After entering the lungs, they distribute rapidly in the central nervous system, peripheral nervous system, lymph, blood, heart, spleen, kidney, bone marrow, and liver (Singh et al., 2006, Kayat et al., 2011). Parameters such as the structure (SWCNTs or MWCNTs), size distribution, surface area, surface chemistry, surface charge, concentration, dose, and agglomeration state, as well as the purity of the samples, have a considerable impact on the reactivity and toxicity of CNTs (Jelena et al., 2007). The toxicity of CNTs has been attributed mainly due to their fiber-shaped, nano-sized structure (Monteiller et al., 2007, Donaldson et al., 2006). The nanoparticulate nanotubes are better dispersed in the lung tissues, resulting in higher toxicities. The needle-like fiber shape of CNTs is similar to asbestos fibers. The asbestos-like, length-dependent, pathogenic behaviors of CNTs induce pulmonary toxicity (Craig et al., 2008). Depending upon the size and physical structure of nanosized particles, they are deposited in different regions of the respiratory tract (Oberdorster et al., 2005) and cause pulmonary toxicity. These include pulmonary inflammation, pulmonary fibrosis, the induced accumulation of neutrophils and eosinophils, mechanical blockages, and increases in various cytotoxicity/inflammatory markers (bronchoalveolar lavage cells, polymorphonuclear leukocytes lactate dehydrogenase (LDH), tumor necrosis factor-α (TNF-α), interleukin-1β (IL-1β), and mucin) in the lungs (Muller et al., 2005, Han et al., 2008).

6.4 PULMONARY TOXICITY OF SWCNTs

SWCNTs can be about 1–4 nm in diameter and several hundred nanometers in length, making them structurally similar to the thread-like filaments of pulmonary basement membranes (type IV collagen) (Shvedova et al., 2005, Kuhn, 1995). SWCNTs are quickly incorporated into the pulmonary interstitial space with rare incorporations by alveolar macrophages; SWCNTs were commonly found in submicron groups and were rarely found as a single structure (Mercer et al., 2008). Table 6.2 shows the major toxicities caused by SWCNTs in animals and human. Cytotoxicity in human alveolar carcinoma epithelial cells (Casey et al., 2008) and oxidative stress in mouse lung epithelial cells (Jacobsen et al., 2008) can be caused by SWCNT exposure. The aspiration or inhalation of SWCNTs in mice caused a rapid granulomatous response and progressive interstitial fibrosis (Shvedova et al., 2008a). In addition, SWCNTs caused a greater increase of connective tissue thickness to the alveolar interstitial space of mice when compared with MWCNTs, suggesting that SWCNTs are more fibrogenic in mice lungs (Mercer et al., 2011). CNT-induced fibrosis is thought to be caused by fibroblast proliferation, collagen production, and elevations of matrix metalloproteinase-9 as demonstrated in human lung fibroblast cell cultures treated with SWCNTs (Wang et al., 2010). CNTs are able to induce an allergic immune response in mice using a subcutaneous injection and intranasal models (Nygaard et al., 2009). The pharyngeal aspiration of SWCNTs in mice significantly decreased the clearance of *Listeria monocytogenes* exposure from the lungs of the mice. In addition, the SWCNT-treated mice had a decreased lung macrophage phagocytic activity as well as less nitric oxide production. This suggests that SWCNT-treated mice are more prone to pulmonary infections (Shvedova et al., 2008b). A mouse lung epithelial cell line exposed to SWCNTs was positive for a Comet assay (single-cell gel elecrophoresis), which measures genotoxicity, that is, DNA damage, as measured by strand breaks or formamidopyrimidine DNA glycosylase sites (Jacobsen et al., 2008). Chinese hamster lung fibroblast cell cultures exposed to SWCNTs were positive for genotoxicity in the Comet assay and micronucleus test (Kisin et al., 2011). Single intragastric administrations of SWCNT to rats caused genotoxicity in their liver and lungs as measured by the formation of 8-oxo-7,8-dihydro-2′-deoxyguanosine, which is a measure of oxidatively

TABLE 6.2
Pulmonary Toxicity of SWCNTs

Species	Toxicity	References
Human	Lung cytotoxicity	Casey et al. (2008)
Mouse	Lung epithelial cells oxidative stress and genotoxicity	Jacobsen et al. (2008)
Mouse	Lung interstitial fibrosis and K-ras gene mutations	Shvedova et al. (2008a)
Mouse	Lung fibrosis	Mercer et al. (2011)
Human	Lung cell culture fibrosis	Wang et al. (2010)
Mouse	Intranasal immune response	Nygaard et al. (2009)
Mouse	Decreased clearance of lung bacteria	Shvedova et al. (2008b)
Hamster	Genotoxicity of lung fibroblast cells	Kisin et al. (2011)

generated DNA damage. Intratracheal installations of SWCNTs to mice caused genotoxicity in their bronchoalveolar cells as measured by the Comet assay (Jacobsen et al., 2009). The inhalation of SWCNTs in mice for 4 days caused significant K-ras gene mutations. This gene mutation is associated with lung cancer (Shvedova et al., 2008a).

6.5 PULMONARY TOXICITY OF MWCNTs

MWCNTs can have a mean length of 4 μm and a diameter of 50 nm, which makes them more fiber like; they are similar in structure to collagen fibers of the alveolar interstitium (Porter et al., 2010, Mercer and Crapo, 1990). An acute pulmonary response in mice to MWCNTs was investigated and a single treatment with MWCNTs was found to cause lung cytotoxicity and inflammation in the mice when administered by oropharyngeal aspiration (Han et al., 2010). When mice aspirated MWCNTs, they had transient inflammation and pulmonary damage, as shown by a rapid granulomatous response and progressive interstitial fibrosis (Mercer et al., 2011). Rats were intratracheally exposed to MWCNTs, which caused lung cytotoxicity and inflammation as shown by collagen-rich granulomas and an overproduction of TNF-α (Muller et al., 2005). Mice, when given MWCNTs by oropharyngeal aspiration, developed lung cytotoxicity and inflammation as measured by granuloma formation and increased inflammatory cell infiltration (Wang et al., 2011). Rats were treated with MWCNTs via intratracheal instillation and they developed lung cytotoxicity and inflammation as measured by fibrosis formation and increased inflammatory cell infiltration (Aiso et al., 2010). Mice were exposed to MWCNTs by either inhalation or intratracheal instillation. Inhalation led to the proliferation and thickening of the alveoli, and intratracheal instillation led to the inflammation of the bronchi and alveolar damage. The differences in toxicity observed by the two exposure routes are probably due to the MWCNTs' aggregation size and distribution in the mouse lungs (Li et al., 2007). Rats were exposed to MWCNTs via inhalation, which caused pulmonary granulomatous and neutrophilic inflammation (Ma-Hock et al., 2009). Human lung epithelial cells exposed to MWCNTs caused genotoxicity as measured by the Comet assay (Karlsson et al., 2008). The major pulmonary toxicities are listed in Table 6.3.

6.6 METALLIC CONTAMINATION OF CNTs AND PULMONARY TOXICITY

Arsenic (As), iron (Fe), and nickel (Ni) are some examples of metal catalysts used in the synthesis of CNTs and may contribute from 25% to 40% of the CNTs' weight. It is impossible to entirely remove catalyst metal contaminants in CNTs without destroying the structural entity. These metal contaminants significantly contribute to oxidative stress, indicated by the formation of free radicals and the accumulation of peroxidative products, the depletion of total antioxidant reserves, and a loss of cell viability (Shvedova et al., 2003, Firme and Bandaru, 2010). The zymosan-stimulated

TABLE 6.3
Pulmonary Toxicity of MWCNTs

Species	Toxicity	References
Mouse	Lung cytotoxicity and inflammation	Han et al. (2010)
Mouse	Lung cytotoxicity and inflammation	Mercer et al. (2011)
Rat	Lung cytotoxicity and inflammation	Muller et al. (2005)
Mouse	Lung cytotoxicity and inflammation	Wang et al. (2011)
Rat	Lung cytotoxicity and inflammation	Aiso et al. (2010)
Mouse	Lung cytotoxicity and inflammation	Li et al. (2007)
Rat	Lung inflammation	Ma-Hock et al. (2009)
Human	Genotoxicity of lung epithelial cells	Karlsson et al. (2008)

mouse leukemic macrophage cell line (RAW 264.7) was treated with nonpurified SWCNT (26% Fe by weight) or purified SWCNT (0.23% Fe by weight) and the nonpurified SWCNT generated more hydroxyl radicals than did the purified SWCNT (Kagan et al., 2006). 26 weight% iron-rich SWCNTs resulted in a significant loss of intracellular low-molecular-weight thiols (GSH) and the accumulation of lipid hydroperoxides in mouse macrophages. The human alveolar epithelial cell line (A549) and rat alveolar macrophage cell line (NR8383) were treated with nonpurified SWCNT and purified (acid treated) SWCNT. The metal contaminants in these CNTs were mainly cobalt (Co) and nickel (Ni) with the purified SWCNT having a 2.5 weight% metal content versus the nonpurified SWCNT having an 8 weight% metal content. Both the human alveolar epithelial cell line and the rat alveolar macrophage cell line had increased intracellular reactive oxygen species and decreased mitochondrial membrane potential when treated with the nonpurified SWCNT (Pulskamp et al., 2007). Death due to the intratracheal instillation of SWCNTs at high doses in mice has been reported. The deaths were preceded by lethargy, inactivity, and loss of body weight (Lam et al., 2004). These studies suggest that metal contamination in CNTs (from their synthesis) may importantly contribute to their pulmonary toxicity.

6.7 PULMONARY TOXICITY OF FUNCTIONALIZED CNTs

Surface functionalization renders the CNTs more interactive with the physiological systems, resulting in higher toxicities. In a study, mice were treated with SWCNTs or acid-functionalized SWCNTs (SWCNT-AF) via oropharyngeal aspiration at a dose of 10 or 40 μg. Acid functionalization reduced the carbon content by about 20% as well as reduced the content of cobalt (Co), chromium (Cr), iron (Fe), manganese (Mn), and nickel (Ni) by 24–33%. The SWCNT-AF-treated mice demonstrated greater pulmonary toxicity than SWCNT-treated mice as measured by an increased number of pulmonary neutrophils (Tong et al., 2009). Human lung tumor cell lines (H596) were treated with nonfunctionalized MWCNTs and acid-functionalized MWCNTs (MWCNT-AT). The acid treatment resulted in adding carbonyl, carboxyl, and/or hydroxyl functional groups to the CNT. The MWCNT-AT-treated H596 cells demonstrated greater cytotoxicity than did the MWCNT-treated H596 cells as determined by the MTT [3-(4,5-dimethylthiazol-2-yl)-2,5-diphenyltetrazolium bromide] assay (Magrez et al., 2006). Human lung epithelial cells (A549) were treated with nonfunctionalized MWCNTs and highly functionalized MWCNTs (hf-MWCNT-NH2). The water-soluble hf-MWCNT-NH2 caused cytotoxicity in A549 cells at the doses of >100 μg/mL as determined by the calcein/propidium iodide assay, whereas MWCNT did not cause cytotoxicity in this assay (Coccini et al., 2010). In another study, A549 cells were treated with nonfunctionalized MWCNT and hydroxyl-functionalized MWCNT-OH at low concentrations (1–40 μg) for 2, 4, and 24 h, which attempts to mimic human environmental exposure. MWCNT damaged the A549 cell membranes at lower concentrations than did the MWCNT-OH, which might be due to differences in aggregation and water solubilities. Also, MWCNT-OH induced more apoptosis in the A549 cells than did MWCNTs (Ursini et al., 2012). In 2004, Kam et al. examined endocytosis (intracellular localization) of SWCNTs and SWCNTs–biotin–streptavidin conjugates within human promyelocytic leukemia (HL60) cells and human T (Jurkat) cells. They found that unfunctionalized SWCNTs exhibited little toxicity, but the SWCNTs–biotin–streptavidin complexes caused extensive cell death. The surface charge plays an important role in the structure–activity relationships that determine the profibrogenic potential of f-CNTs in the lung. Carboxylated (COOH)-, polyethylene glycol (PEG)-, amine (NH2)-, sidewall amine (sw-NH2)-, and polyetherimide (PEI)-modified MWCNTs were successfully established from raw or as-prepared (AP-) MWCNTs, and comprehensively characterized by transmission electron microscopy (TEM), x-ray photoelectron spectroscopy (XPS), Fourier transform infrared spectroscopy (FTIR), and dynamic light scattering (DLS) to obtain information about morphology, length, degree of functionalization, hydrodynamic size, and surface charge. Cellular screening in BEAS-2B and THP-1 cells showed that, compared to AP-MWCNTs, anionic functionalization (COOH and PEG) decreased the production of profibrogenic cytokines and growth factors (including IL-1β, TGF-β1, and PDGF-AA), while neutral and weak cationic functionalization (NH2 and sw-NH2) showed intermediary effects. In contrast, the strongly cationic PEI-functionalized tubes induced robust biological effects. These differences could be attributed to differences in cellular uptake and NLRP3 inflammasome activation, which depends on the propensity toward lysosomal damage and cathepsin-B release in macrophages (Li et al., 2013). The strong cationic PEI-MWCNTs induced significant lung fibrosis, while carboxylation significantly decreased the extent of pulmonary fibrosis. These studies suggest that functionalized CNT can exert different toxicological effects than its corresponding nonfunctionalized or pristine CNTs. The biocompatibility of CNTs sometimes depends on the functionalization density. It has been found that the functionalized SWNT-phenyl-SO3H and SWNT-phenyl-(COOH)2 covalently bound sidewall functional groups, are less cytotoxic than the functionalized SWNTs in 1% Pluronic F108, which is stabilized in a micellar solution without covalent functionalization (Says et al., 2006).

6.8 CONCLUSION

CNTs are cylindrical structures that have useful properties in the manufacturing of electronics, optics, as well as medical devices, for example, drug-delivery systems. They have the potential for numerous biological and medical applications. The horizon of the biomedical applications of CNTs is likely to widen in the future to include multifunctional drug-delivery systems with intended diagnosis and targeting purposes. However, the safety of these CNTs is yet to be proven, a prerequisite for categorizing them as "generally regarded as safe." SWCNTs and MWCNTs are the two main types of CNTs. CNTs have a similar shape to asbestos fibers, which has led to the concern of CNT-mediated pulmonary toxicity. The available literature suggests that SWCNTs may exert greater toxic responses as compared to MWCNTs; surface functionalizations increase the cytotoxicity of CNTs on different cell lines. Therefore, it is recommended that the pulmonary toxicity of the CNTs must be resolved before taking up any venture in the pharmaceutical and medical fields.

REFERENCES

Aiso, S., Yamazaki, K., Umed, Y., Asakura, M., Takaya, M., Toya, T., Koda, S., Nagano, K., Arito, H., and Fukushima, S., 2010. Pulmonary toxicity of intratracheally instilled multiwall carbon nanotubes in male Fischer 344 rats. *Ind Health* 48(6):783–95.

Ajayan, P. M. and Ebbesen, T. W., 1997. Nanometre-size tubes of carbon. *Rep Prog Phys* 60:1025–62.

Alan, B. D., Steve, C., Edgar, M., Joselito, M. R., Von, H. E., John, P. F., Jonathan, N. C., Bog, G. K., and Ray, H. B., 2003. Super-tough carbon-nanotube fibres. *Nature* 423:703. doi:10.1038/423703a

Ali, E. A., Jiyoung, O. H., Mikhail, E. K., Alexander, A. K., Shaoli, F., Alexandre et al., 2009. Giant-stroke, superelastic carbon nanotube aerogel muscles. *Science* 323(5921):1575–8. DOI: 10.1126/science.1168312.

Balaji, S., Xinfeng, S., Walboomers, X. F., Hongbing, L., Vincent, C., Lon, J. W., Antonios, G. M., and John, A. J., 2008. *In vivo* biocompatibility of ultra-short single-walled carbon nanotube/biodegradable polymer nanocomposites for bone tissue engineering. *Bone* 43(2):362–70.

Ball, P., 1999. Focus carbon nanotubes. *Nat Sci* update. http://www.nature.com/nsu/991202/991202-1.html.

Bronikowski, M. J., Willis, P. A., Colbert, D. T., Smith, K. A., and Smalley, R. E., 2001. Gas-phase production of carbon single-walled nanotubes from carbon monoxide via the HiPco process: A parametric study. *J Vac Sci Technol A* 19: 1800–5.

Casey, A., Herzog, E., Lyng, F. M., Byrne, H. J., Chambers, G., and Davoren, M., 2008. Single walled carbon nanotubes induce indirect cytotoxicity by medium depletion in A549 lung cells. *Toxicol Lett* 179:78–84.

Cherukuri, P., Bachilo, S. M., Litovsky, S. H., and Weisman, R. B., 2004. Near-infrared fluorescence microscopy of single-walled carbon nanotubes in phagocytic cells. *J Am Chem Soc* 126:15638–9.

Coccini, T., Roda, E., Sarigiannis, D. A., Mustarelli, P., Quartarone, E., Profumo, A., and Manzo, L., 2010. Effects of water-soluble functionalized multi-walled carbon nanotubes examined by different cytotoxicity methods in human astrocyte D384 and lung A549 cells. *Toxicology* 269(1):41–53.

Craig, A. P., Rodger, D., Ian, K., Andrew, M., William, A. H., Wallace, A. S., Vicki, S., Simon, B., William, M., and Donaldson, K., 2008. Carbon nanotubes introduced into the abdominal cavity of mice show asbestos-like pathogenicity in a pilot study. *Nat Nanotechnol* 3: 423–8.

Dalton, A., 2005. Nanotubes may heal broken bones. http://www.wired.com/medtech/health/news/2005/08/68512.

David, S. S. and Karl, J. H., 2006. Making high-flux membranes with carbon nanotubes. *Science* 312 (5776):1003–4. DOI: 10.1126/science.1127261.

Dekker, C., 1999. Carbon nanotubes as molecular quantum wires. *Phys Today* 52(5):22–8. doi:10.1063/1.882658.

Donaldson, K., Aitken, R., Tran, L., Stone, V., Duffin, R., Forrest, G. et al., 2006. Carbon nanotubes: A review of their properties in relation to pulmonary toxicology and workplace safety. *Toxicol Sci* 92:5–22.

Firme, C. P. and Bandaru, P. R., 2010. Toxicity issues in the application of carbon nanotubes to biological systems. *Nanomedicine: Nanotechnol Biol Med* 6:245–56.

Flahaut, E., Bacsa, R., Peigney, A., and Laurent, C., 2003. Gram-scale CCVD synthesis of double-walled carbon nanotubes. *Chem Commun* 12(12):1442–3. doi:10.1039/b301514a. PMID 12841282.

Gullapalli, S. and Wong, M. S., 2011. Nanotechnology: A guide to nano-objects. *Chem Eng Prog* 107(5):28–32.

Haddon, R. C., Zanello, L. P., Zhao, B., and Hu, H., 2006. Bone cell proliferation on carbon nanotubes. *Nano Lett* 6(3):562–7. doi:10.1021/nl051861e. PMID 16522063.

Han, S. G., Andrews, R., Gairola, C. G., and Bhalla, D. K., 2008. Acute pulmonary effects of combined exposure to carbon nanotubes and ozone in mice. *Inhal Toxicol* 20:391–8.

Han, S. G., Andrews, R., and Gairola, C. G., 2010. Acute pulmonary response of mice to multi-wall carbon nanotubes. *Inhal Toxicol* 22:340–47.

http://www.sciencedaily.com/releases/2011/03/110314140632.htm. New desalination process developed using carbon nanotubes.

http://www.wired.com/medtech/health/news/2005/08/68512.

Iijima, S., 1991. Helical microtubules of graphitic carbon. *Nature* 354:56–8.

Jacobsen, N. R., Pojana, G., White, P., Møller, P., Cohn, C. A., Korsholm, K. S., Vogel, U., Marcomini, A., Loft, S., and Wallin, H., 2008. Genotoxicity, cytotoxicity, and reactive oxygen species induced by single-walled carbon nanotubes and C_{60} fullerenes in the FE1-Muta™ mouse lung epithelial cells. *Environ Mol Mutagen* 49:476–87.

Jacobsen, N. R., Moller, P., Jensen, K. A., Vogel, U., Ladefoged, O., Loft, S., and Wallin, H., 2009. Lung inflammation and genotoxicity following pulmonary exposure to nanoparticles in ApoE$^{-/-}$ mice. *Part Fibre Toxicol* 6:2.

Jelena, K., Szwarc, H., and Moussa, F., 2007. Toxicity studies of carbon nanotubes. *Adv Exp Med Biol* 620:181–204.

Kagan, V. E., Tyurina, Y. Y., Tyurin, V. A., Konduru, N. V., Potapovich, A. I., Osipov, A. N. et al., 2006. Direct and indirect effects of single walled carbon nanotubes on RAW. *Toxicol Lett* 165:88–100.

Kam, N. S., Jessop, T. C., Wender, P. A., and Dai, H., 2004. Nanotube molecular transporters: Internalization of carbon nanotube–protein conjugates into mammalian cells. *J Am Chem Soc* 126:6850–1.

Karlsson, H. L., Cronholm, P., Gustafsson, J., and Moller, L., 2008. Copper oxide nanoparticles are highly toxic: A comparison between metal oxide nanoparticles and carbon nanotubes. *Chem Res Toxicol* 21:1726–32.

Kayat, J., Gajbhiye, V., Tekade, R. K., and Jain, N. K., 2011. Pulmonary toxicity of carbon nanotubes: A systematic report. *Nanomedicine: Nanotechnol Biol Med* 7:40–9.

Kelly, Y. and Kim, M. A., 2007. Nanotechnology platforms and physiological challenges for cancer therapeutics. *Nanomed Nanotechnol Biol Med* 3:103–10.

Ken, G., Sae-Khow, O., and Mitra, S., 2011. Water desalination using carbon-nanotube-enhanced membrane distillation. *ACS Appl Mater Interfaces* 3(2):110. DOI: 10.1021/am100981s.

Kisin, E. R., Murray, A. R., Sargent, L., Lowry, D., Chirila, M., Siegrist, K. J. et al., 2011. Genotoxicity of carbon nanofibers: Are they potentially more or less dangerous than carbon nanotubes or asbestos? *Toxicol Appl Pharmacol* 252:1–10.

Kuhn, C., 1995. Basement membrane (type IV) collagen. *Matrix Biol* 14:439–45.

Lam, C. W., James, J. T., McCluskey, R., and Hunter, R. L., 2004. Pulmonary toxicity of single-wall carbon nanotubes in mice 7 and 90 days after intratracheal instillation. *Toxicol Sci* 77:126–34.

Li, J. G., Li, W. X., Xu, J. Y., Cai, X. Q., Liu, R. L., Li, Y. J., Zhao, Q. F., and Li, Q. N., 2007. Comparative study of pathological lesions induced by multiwalled carbon nanotubes in lungs of mice by intratracheal instillation and inhalation. *Environ Toxicol* 22:415–21.

Li, R., Wang, X., Ji, Z., Sun, B., Zhang, H., Chang, C. H. et al., 2013. The surface charge and cellular processing of covalently functionalized multiwall carbon nanotubes determine pulmonary toxicity. *ACS Nano* 7(3):2352–2368.

Magrez, A., Kasas, S., Salicio, V., Pasquier, N., Seo, J. W., Celio, M., Catsicas, S., Schwaller, B., and Forró, L., 2006. Cellular toxicity of carbon-based nanomaterials. *Nano Lett.* 6:1121–5.

Ma-Hock, L., Treumann, S., Strauss, V., Brill, S., Luizi, F., Mertler, M., Wiench, K., Garner, A. O., Ravenzwaay, B., and Landsiedle, R., 2009. Inhalation toxicity of multiwall carbon nanotubes in rats exposed for 3 months. *Toxicol Sci* 112:468–81.

Majumder, M., Chopra, N., Andrews, R., and Hinds, B. J., 2005. Nanoscale hydrodynamics: Enhanced flow in carbon nanotubes. *Nature* 438(3): 44. doi:10.1038/438044a.

Martel, R., Derycke, V., Lavoie, C., Appenzeller, J., Chan, K., Tersoff, J., and Avouris, P. H., 2001. Ambipolar electrical transport in semiconducting single-wall carbon nanotubes. *Phys Rev Lett* 87(25):256805. doi:10.1103/PhysRevLett.87.256805. PMID 11736597.

Mercer, R. R. and Crapo, J. D., 1990. Spatial distribution of collagen and elastin fibers in the lungs. *J Appl Physiol* 69:756–65.

Mercer, R. R., Scabilloni, J. F., Wang, L., Kisin, E., Murray, A. R., Schwegler-Berry, D. S. A. A., and Castranova, V., 2008. Alteration of deposition pattern and pulmonary response as a result of improved dispersion of aspirated single-walled carbon nanotubes in a mouse model. *Am J Physiol Lung Cell Mol Physiol* 294:L87–97.

Mercer, R. R., Hubbs, A. F., Scabilloni, J. F., Wang, L., Battelli, L. A., Friend, S., Castranova, V., and Porter, D. W., 2011. Pulmonary fibrotic response to aspiration of multi-walled carbon nanotubes. *Part Fibre Toxicol* 8:21.

Mintmire, J. W., Dunlap, B. I., and White, C. T., 1992. Are fullerene tubules metallic? *Phys Rev Lett* 68(5):631–4. doi:10.1103/PhysRevLett.68.631. PMID 10045950.

MIT Institute for Soldier Nanotechnologies. Web: www.mit.edu

Monteiller, C., Macnee, W., Faur, S., Jones, A., Miller, B., and Donaldson, K., 2007. The proinflammatory effect of low toxicity, low solubility particles on epithelial cells *in vitro*, the role of surface area. *Occup Environ Med* 64:609–15.

Muller, J., Huauxa, F., Moreaub, N., Missona, P., Heiliera, J. F., Delosc, M., et al., 2005. Respiratory toxicity of multi-wall carbon nanotubes. *Toxicol Appl Pharmacol* 207:221–31.

Nanotube Tips. NanoScience Instruments (www.nanoscience.com).

Nygaard, U. C., Hansen, J. S., Samuelsen, M., Alberg, T., Marioara, C. D., and Lovik, M., 2009. Single-walled and multi-walled carbon nanotubes promote allergic immune responses in mice. *Toxicol Sci* 109:113–23.

Oberdorster, G., Oberdorster, E., and Oberdorster, J., 2005. Nanotoxicology: An emerging discipline evolving from studies of ultrafine particles. *Environ Health Perspect* 113:823–39.

O'Connell, M. J., Bachilo, S. M., Huffman, C. B., Moore, V. C., Strano, M. S., Haroz, E. H. et al., 2002. Band gap fluorescence from individual single-walled carbon nanotubes. *Science* 297:593–6.

Porter, D. W., Hubbs, A. F., Mercer, R. R., Wu, N., Wolfarth, M. G., Sriram, K. et al., 2010. Mouse pulmonary dose- and time course-responses induced by exposure to multi-walled carbon nanotubes. *Toxicology* 269:136–47.

Pulskamp, K., Diabate, S., and Krug, H. F., 2007. Carbon nanotubes show no sign of acute toxicity but induce intracellular reactive oxygen species in dependence on contaminants. *Toxicol Lett* 168:58–74.

Rina Tannenbaum, I. T., Kim, Nunnery, G. A., Jacob, K. I., Schwartz, J., and Liu, X., 2010. Synthesis, characterization, and alignment of magnetic carbon nanotubes tethered with magnemite nanoparticles. *J Phys Chem C* 114(15):6944–51. http://dx.doi.org/10.1021/jp9118925

Rina Tannenbaum, I. T., and Kim Jacob, K. I., 2011. Anisotropic conductivity of magnetic carbon nanotubes embedded in epoxy matrices. *Carbon* 49(1):54–61. http://dx.doi.org/10.1016/j.carbon.2010.08.041

Says, C. M., Liang, F., Hudson, J. L., Mendez, J., Guo, W., Beach, J. M. et al., 2006. Functionalization density dependence of single wall carbon nanotubes cytotoxicity *in vitro*. *Toxicol Lett* 161:135–42.

Shi, X., Sitharaman, B., Pham, Q. P., Liang, F., Wu, K. W., Billups, W.E., Wilson, L. J., and Mikosa, A.G., 2007. Fabrication of porous ultra-short single-walled carbon nanotube nanocomposite scaffolds for bone tissue engineering. *Biomaterials* 28(28):4078–4090.

Sholl, D. S. and Johnson, J. K., 2006. Making high-flux membranes with carbon nanotubes. *Science* 312(5776):1003–4. DOI: 10.1126/science.1127261.

Shvedova, A. A., Castranova, V., Kisin, K. R., Schwegler-Berry, D., Murray, A. R., and Gandelsman, V. Z., 2003. Exposure to carbon nanotube material: Assessment of nanotube cytotoxicity using human keratinocyte cells. *J Toxicol Environ Health A* 66:1909–26.

Shvedova, A. A., Kisin, E. R., Mercer, R. R., Murray, A. R., Johnson, V. J., Potapovich, A. I. et al., 2005. Unusual inflammatory and fibrogenic pulmonary responses to single-walled carbon nanotubes in mice. *Am J Physiol Lung Cell Mol Physiol* 289:L698–708.

Shvedova, A. A., Castranova, V., Kisin, E. R., Schwegler-Berry, D., Murray, A. R., Gandelsman, V. Z., Maynard, A., and Baron, P., 2003. Exposure to carbon nanotube material: Assessment of nanotube cytotoxicity using human keratinocyte cells. *J Toxicol Environ Health* 66:1909–26.

Shvedova, A. A., Kisin, E., Murray, A. R., Johnson, V. J., Gorelik, O., Arepalli, S. et al., 2008a. Inhalation vs. aspiration of single-walled carbon nanotubes in C57BL/6 mice: Inflammation, fibrosis, oxidative stress, and mutagenesis. *Am J Physiol Lung Cell Mol Physiol* 295:L552–65.

Shvedova, A. A., Fabisiak, J. P., Kisin, E. R., Murray, A. R., Roberts, J. R., Antonini, J. A. et al., 2008b. Sequential exposure to carbon nanotubes and bacteria enhances pulmonary inflammation and infectivity. *Am J Respir Cell Mol Biol* 38:579–90.

Singh, R., Pantarotto, D., Lacerda, L., Pastorin, G., Klumpp, C., Prato, M. et al., 2006. Tissue biodistribution and blood clearance rates of intravenously administered carbon nanotube radiotracers. *Proc Natl Acad Sci USA* 103:3357–62.

Tang, Z. K., Zhang, L., Wang, N., Zhang, X. X., Wen, G. H., Li, G. D., Wang, J. N., Chan, C. T., and Sheng, P., 2001. Superconductivity in 4 angstrom single-walled carbon nanotubes. *Science* 292(5526):2462–5. DOI: 10.1126/science.1060470.

Tong, H., McGee, J. K., Saxena, R. K., Kodavanti, U. P., Devlin, R. B., Gilmour, M. I. et al., 2009. Influence of acid functionalization on the cardiopulmonary toxicity of carbon nanotubes and carbon black particles in mice. *Toxicol Appl Pharmacol* 239:224–32.

Ursini, C. L., Cavallo, D., Fresegna, A. M., Ciervo, A., Maiello, R., Buresti, G., Casciardi, S., Tombolini, F., Bellucci, S., and Iavicoli, S., 2012, Comparative cyto-genotoxicity assessment of functionalized and pristine multiwalled carbon nanotubes on human lung epithelial cells. *Toxicol in Vitro* 26:831–40.

Wang, L., Mercer, R. R., Rojanasakul, Y., Qiu, A., Lu, Y., Scabilloni, J. F., Wu, N., and Castranova, V., 2010. Direct fibrogenic effects of dispersed single-walled carbon nanotubes on human lung fibroblasts. *J Toxicol Environ Health A* 73:410–22.

Wang, X., Katwa, P., Podila, R., Chen, P., Ke, P. C., Rao, A. M., Walters, D. M., Wingard, C. J., and Brown, J. M., 2011. Multi-walled carbon nanotube instillation impairs pulmonary function in C57BL/6 mice. *Part Fibre Technol* 8(24):1–13.

Wang, X., Li, Q., Xie, J., Jin, Z., Wang, J., Li, Y., Jiang, K., and Fan, S., 2009. Fabrication of ultralong and electrically uniform single-walled carbon nanotubes on clean substrates. *Nano Lett* 9(9):3137–41. doi:10.1021/nl901260b. PMID 19650638.

Web address: www.newscientist.com/section/tech

White House. 2000. National Nanotechnology Initiative: Leading to the next industrial revolution. The White House, Office of the Press Secretary, Washington, DC. http://clinton4.nara.gov/textonly/WH/New/html/20000121_4.html.

Yu, M.-F., Lourie, O., Dyer, M. J., Moloni, K., Kelly, T. F., and Ruoff, R. S., 2000. Strength and breaking mechanism of multiwalled carbon nanotubes under tensile load. *Science* 287(5453):637–40. doi:10.1126/science.287.5453.637. PMID 10649994.

7 Nanotoxicity of Polymeric and Solid Lipid Nanoparticles

Dev Prasad and Harsh Chauhan

CONTENTS

7.1 Pharmaceutical Nanoparticles 141
7.2 Preparation and Characterization Techniques for Polymeric and Solid Lipid Nanoparticles 142
7.3 Nanotoxicity of Polymeric and Solid Lipid Nanoparticles 142
    7.3.1 Nanotoxicity Potential of PNs and SLNs 143
    7.3.2 Potential Mechanism of Nanotoxicity 145
        7.3.2.1 Size of Nanoparticles 145
        7.3.2.2 Interaction of PNs and SLNs with Biological Systems 148
        7.3.2.3 Physicochemical Characteristics of Polymers/Lipids Utilized in Nanoparticles 149
        7.3.2.4 Concentration of Polymers/Lipids Utilized in the Nanoparticles 150
        7.3.2.5 Conversion of Polymers/Lipids 151
        7.3.2.6 Degradation of Polymeric/Lipid Nanoparticles 151
        7.3.2.7 Route of Administration of Polymeric/Lipid Nanoparticles 151
        7.3.2.8 Biocompatibility of Polymeric/Lipid Nanoparticles 152
        7.3.2.9 Excipients/Residual Solvents Used during the Preparation of Polymeric/Lipid Nanoparticles 152
7.4 Methods for Nanotoxicity Assessment 152
    7.4.1 Physicochemical Characterization 152
    7.4.2 *In Vitro* Cell Culture Techniques 152
        7.4.2.1 Cell Viability Assay 152
        7.4.2.2 Cell Uptake Studies of Nanoparticles 153
        7.4.2.3 Assays for Alteration in Gene Expression 153
7.5 Conclusion 153
References 154

7.1 PHARMACEUTICAL NANOPARTICLES

Nanoparticles, as the name suggest, are the particles in a size ranging between 1 and 1000 nm. The utilization of nanoparticles, especially in the field of medicine and pharmaceutical sciences, provides the flexibility to alter fundamental physical properties of compounds such as solubility, diffusivity, half-life of drug in blood circulation, drug release characteristics, and immunogenicity. A drug can be dissolved, entrapped, encapsulated, or attached to a nanoparticle matrix [1–3]. Over the years, a number of nanoparticulate systems have been developed for the treatment and diagnosis of cancer, diabetes, pain, asthma, allergy, and infections among many other diseases and conditions [4,5]. Further, the discovery of new chemical entities for better treatment and control of a wide spectrum of diseases has necessitated the use of these carrier systems for faster and more efficient delivery. In the case of diagnostic applications, nanoparticles allow detection on the molecular level; they help recognize abnormalities such as fragments of viruses, precancerous cells, and disease markers that cannot be detected by conventional methods. Nanoparticulate imaging contrast agents have also been shown to improve the sensitivity and specificity of magnetic resonance imaging [6]. Overall, nanoparticles offers numerous advantages such as better solubility and drug dissolution rate, improvement in the absorption profile and bioavailability of drug, allowing site-specific targeting of drugs, allowing controlled drug release, providing effective and/or easier routes of administration, reducing therapeutic toxicity, prolonging the effect of drug in target tissue, improving the stability of the drugs against chemical and enzymatic degradation, and subsequently lowering healthcare costs.

Polymeric nanoparticles (PNs) are prepared using naturally occurring, chemically modified, or synthesized polymers. Polymer-based nanoparticles include PNs, micelles, and dendrimers. In these nanoparticles, drug can either be physically entrapped within the polymer or covalently bound to the polymer matrix depending on the preparation method. Therefore, the compounds formed may have a structure like that of capsules (PNs), amphiphilic core/shell (polymeric micelles), or hyperbranched macromolecules (dendrimers) [7].

Solid lipid nanoparticle (SLN) is a type of nanoparticle drug delivery system that uses various lipids to incorporate drug molecules. SLNs can incorporate both hydrophilic and lipophilic drugs [2,8]. SLNs are an ideal drug delivery system without using organic solvents and their ability to encapsulate the drug within its lipid matrix, allowing for sustained drug release.

Overall, pharmaceutical nanoparticles offer number of advantages and appear to be the future of drug industry. Owing to the significant increase in their usage, it is very important to closely look at the toxicity associated with the use of nanoparticles. This chapter discusses PNs and SLNs and focuses on the toxicities associated with the use of these nanoparticles.

7.2 PREPARATION AND CHARACTERIZATION TECHNIQUES FOR POLYMERIC AND SOLID LIPID NANOPARTICLES

PNs and SLNs can be prepared by using various techniques. Figure 7.1 shows different preparation and characterization techniques for these systems. These preparation and characterization techniques are similar to pharmaceutical nanoparticles and involve solution preparations such as nanosuspensions or emulsion utilizing polymers/lipids. These solution preparations can be characterized for particle size, zeta potential, stability, and so on. In most of the cases, the solution preparation is converted to solid state by removing water/solvent using freeze drying or solvent evaporation. Furthermore, these systems are characterized for stability in solid state. Techniques such as PCS, Cryo-FESEM, AFM, p-XRD, DSC, and TGA were used extensively for the characterization of nanoparticle systems. Tables 7.1 and 7.2 list some examples of PNs and SLNs prepared using various techniques.

7.3 NANOTOXICITY OF POLYMERIC AND SOLID LIPID NANOPARTICLES

Since the last decade, the use of nanoparticles has been explored considerably as a novel drug delivery system for almost all the pathological conditions. However, less emphasis has been given to study the interaction of nanoparticles with the biological systems. Nanotoxicology is a field of research that primarily focuses on the interactions of nanostructures with biological systems, particularly exploring the effects on the physical and chemical properties (e.g., size, shape, surface chemistry, composition, and aggregation) of nanostructures with the induction of toxic biological responses. Recently, Keck and Müller [37] classified nanoparticles into four different classes (I–IV) from absolutely "no" risk to "high" risk. The classification is based on the physical attribute (the nanoparticles size (>/< 100 nm)) and the chemical behavior (such as the size-related differences in interaction with human cells, and on biodegradability/nonbiodegradability in the body) of the nanoparticles. By superimposing additional criterions such as biocompatibility (B) and nonbiocompatibility (NB)

Nanotoxicity of Polymeric and Solid Lipid Nanoparticles

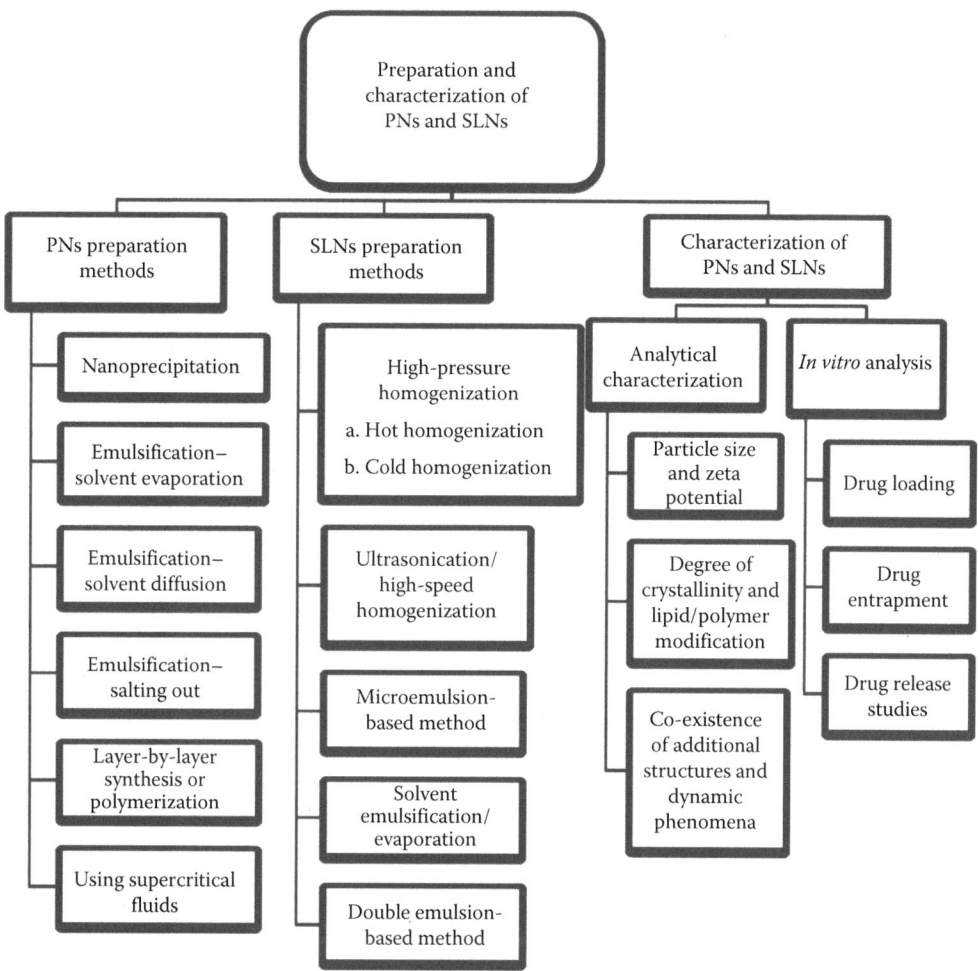

FIGURE 7.1 (**See color insert.**) Preparation and characterization techniques of PNs and SLNs.

of the nanoparticle surface, the classification is further extended into eight total classes from I-B (best tolerated) to IV-NB (highest potential risk). Nanoparticles such as nanoemulsions, liposomes, and drug nanocrystals in the size range from 100 to 1000 nm are classified into low-risk class I. Nanoparticles possessing medium risk either due to their size or nonbiodegradability are classified into class II and class III. The highest toxicity risks possess nanoparticles of class IV. They are <100 nm and can therefore access all cells. On top they are nonbiodegradable [37].

7.3.1 Nanotoxicity Potential of PNs and SLNs

Pharmaceutical PNs and SLNs can vary in size ranges from 50 to 1000 nm and thus can be considered to possess low to high risk in terms of toxicity assessment [37]. Polymers used in preparing PNs can be biodegradable or nonbiodegradable, whereas lipids used in the SLN are usually a mixture of lipids and can show polymorphism. Further, the preparation of nanoparticles requires the drug and polymers/lipids to go through various unit operations, including melting, mixing, interaction with water, solvents, and so on [2]. The transformation/degradation of these inert polymers/lipids can leads to toxicity. Polymer and lipids used in the preparation of nanoparticles are considered to be safe although their toxicity has not been studied extensively. SLNs are usually made up of physiological

TABLE 7.1
Preparation Techniques of PNs with Examples

Method	Drug	Use	Polymer
Nanoprecipitation [9–12]	Paclitaxel	Anticancer	TPGS-functionalized PLGA
	Loperamide	Opioid agonist	PLGA–PEG–PLGA triblock copolymer
	Mitoxantrone	Anticancer	Dextran, PLA, and DPPE
	Prednisolone	Glucocorticoid	Eudragit S100
Emulsification–solvent evaporation [13–16]	Latanoprost acid	Antihypertensive	PLA–PEG
	Amylin	Pancreatic hormone	Poly-caprolactone
	Ketoprofen	NSAID	Eudragit E100 and Eudragit RS
	Levofloxacin	Quinolone antibiotic	PCL and PLGA
Emulsification–solvent diffusion [17–19]	Triamcinolone acetonide	Endotoxin-induced uveitis	PLGA
	Daunorubicin	Anticancer	PLGA and PDLLA
	5-Fluorouracil	Anticancer	N-succinyl-chitosan
	α-Tocopherol acetate	Antioxidant	Poly-(lactide)
Emulsification–salting out [20–22]	Ibuprofen	NSAID	Eudragit L100-55 and PVAL
	Celecoxib	NSAID	PLGA
	Savoxepine	Atypical neuroleptic	PLA
Using supercritical fluids [23–25]	Dexamethasone	Glucocorticoid	PCL with mesoporous MCM-41 silica
	Plasmid DNA		PLGA
	Retinyl palmitate	Antioxidant	PLLA
Layer-by-layer synthesis or polymerization [26,27]	Steroidal aromatase inhibitor, exemestane in cPCL synthesized by ring opening polymerization of caprolactone		
	Tryptophan hydroxylase-2 gene in magnetic PMMA/PEI shell		

TABLE 7.2
Preparation Techniques of SLN with Examples

Preparation Techniques	Drug	Use	Lipids
High-pressure homogenization [28–30]	Triamcinolone	Corticosteroid	Glyceryl palmitostearate
A. Hot homogenization method	Acetonide acetate	Ca channel blocker	Glyceryl tripalmitate
	Nitredipine	Antioxidant	Cetyl palmitate
	Lutein	Immunosuppressant	Glyceryl monostearate
	Cyclosporin-A	Antitumor	Cetyl palmitate
B. Cold homogenization method	Vitamin E	Sunscreen	Acetyl palmitate
	Vinorelbin-bitartrate	Antitumor	Glyceryl monostearate
Microemulsion technique [31–33]	Curcuminoides	Curcumin	Stearic acid
	Ketoprofen	NSAID	Glyceryl monostearate
	Quercetin	Alzheimer's	Bees wax
	Topotecan	Cytotoxic	Carnauba wax
	Tetrandine	Antitumor	Compitrol 888 ATO
Double emulsion technique [34–36]	Ceramide	Topical	Tripalmitin
	Insulin	Diabetes	Stearic acid
	5-Flurouracil	Chemotherapy	Palmitic acid
			Glyceryl monostearate
Ultrasonication technique	Indomethacin	Analgesic	Glyceryl behenate
			Tribehenate

lipids, which are biodegradable with simple natural processes such as enzyme digestion [38]. The biodegradability of SLNs is one of the most important advantages in making them an outstanding drug delivery vehicle. These lipids and their degradation products are considered nontoxic to human body cells. However, owing to their nanosize range, their biotoxicity is an important issue, as the human body reacts very differently to nanoparticles as compared with larger particles of the same material. It has been reported that lipid carriers prepared with several lipids and emulsifying agents did not exhibit any cytotoxicity up to the concentration of 2.5% lipids [39]. Furthermore, it has been shown that higher concentration of 10% lipids led to a viability of 80% human granulocytes in culture [39]. Several examples of nanotoxicity of inorganic nanoparticles have recently been documented. In this chapter, we have drawn parallel to the toxicity reported for other nanoparticles that can be associated with PNs and SLNs along with citing specific examples for PNs and SLNs.

7.3.2 Potential Mechanism of Nanotoxicity

Table 7.3 summarizes reported works on cytotoxicity of various nanoparticles and the mechanisms of toxicity along with potentially useful design methods. As of today, most of the research on nanotoxicity has been done on the eukaryotic cells [40]. The common mechanisms of action that can lead to nanotoxicity include cell wall destruction, oxidative stress, protein denaturation, interaction of ions with cellular materials, and so on [41]. The chemical reactivity and biological activity of nanosized particles are often greater than larger-sized particles and thus increase the potential of toxicity. Nanoparticles are also much more mobile and have greater access to cellular material. Cationic and ionic polymers are commonly used in the pharmaceutical industry and they have the potential to significantly interact with the cellular material and cause toxicity. The above-mentioned mechanism of toxicity can also be associated with SLNs if the particle size is less than 100 nm although no specific studies have been done.

7.3.2.1 Size of Nanoparticles

Nanoparticles having sizes less than 100 nm have greater potential for toxicity as size can significantly affect particle properties. Smaller-sized particles usually have greater solubility and the adhesiveness to surfaces/membranes. Pharmaceutical advantages such as increasing solubility and drug targeting to specific sites justify the use of these smaller-sized particles but there is a second important size limit, the 100 nm. Larger particles can only enter the cell by phagocytosis and can only be taken up by macrophages. Therefore, these particles possess a lower toxicity risk. Particles below 100 nm can be internalized by any cell through endocytosis; thus, these particles may have a higher toxicity risk. There are many examples of PNs with sizes less than 100 nm, for example, Zou et al. [59] developed the curcumin PN using PLGA PEG PLGA triblock polymer, while Woo et al. [60] used diblock copolymer, polyethylene glycol–poly-L-lactic acid (mPEG–PLA), monovalent metal salt of a biodegradable polyester (D,L-PLACOONa), and calcium chloride to prepare PNs of size less than 100 nm. Furthermore, Fang et al. [61] developed an SLN of size less than 50 nm for a novel chemotherapeutic agent (PK-L4), an analog of amsacrine. In another example, Jeon et al. [62] prepared a surface-modified SLN containing retinyl palmitate by the hot-melt method using Gelucire 50/13 and Precirol ATO5 with dicetyl phosphate as surface modifier. The particle size of these SLNs was less than 50 nm. Table 7.4 showed the applications, concerns, and biological/mechanistic studies of inhaled nanoparticles of particle size less than 100 nm and greater than 500 nm [63]. The study focuses on inhaled nanoparticles and is important to mention as several PNs and SLNs are designed to deliver drugs through the pulmonary route of administration. Although the particle size between 100 and 500 nm is not represented in this table, it can be assumed that most of the parameters discussed in the table holds true for this size range as well.

Since PNs and SLNs are formulated in nanosize ranges, it becomes critical to evaluate the nanosize-related toxicity reported with different types of nanoparticles, such as inorganic and metallic nanoparticles. Ferin et al. and Oberdorster et al. found that when rats were exposed to

TABLE 7.3
Summary of Potential Mechanism of Nanotoxicity Based on Different Nanomaterial

Nanomaterials	Possible Mechanisms/Routes of Toxicity	Suggested Mitigation Strategies
Organic Nanomaterials		
MWCNT and SWCNT [42–44]	Frustrated phagocytosis: cell wall disruption and enzymes release, generation of ROS, and inflammation due to aggregation of CNT	Surface functionalization with antioxidants, complete purification to eliminate residual metal ions; coating with polymer matrix to prevent release of toxic metal ions; surface functionalization.
Fullerenes	Generation of ROS and resulting oxidative stress	Modulation of cationic charge density and increasing specific cellular interaction through functionalization with appropriate receptors
Cationic polymer nanospheres and dendrimers [45–47]	Endosomal cellular uptake	
Metals		
Gold nanoparticles [48]	Denaturing of protein	Coating of nanoparticles with amphipathic agents to modulate charge, size, dispersion and hydrophobicity
Silver nanoparticles [49]	Generation of ROS, Alteration of transport through membrane	Capping agents to prevent release of Ag+ ions
Metals Oxides		
TiO_2 [50,51]	Oxidative stress due to generation of ROS and photoactivity; cell death and fibrillation due to interference in macrophage cell membrane functions	Capping NMs with; coating with anti-oxidants (ascorbate, glutathione, alpha-tocopherol (vitamin E)), enzymatic scavengers of activated oxygen such as superoxide dismutase, surfactants, polymers or complexing ligands
ZnO [52]	High dissolution of ZnO nanoparticles under physiological condition and release of toxic cations	
Al_2O_3 [42,53]	Induce proinflammatory response	
SiO_2 [54–56]	Protein denaturation; oxidative stress due to ROS generation	Control of redox properties
CeO_2 [52]	Inducing protein aggregation	Surface passivation
Other Nanoparticles		
Co/Ni ferrite Nanoparticles, magnetic metallic nanoparticles [57]	Low cell viability due to the release of toxic cations	Encapsulation with polymer, Capping with phosphonic and hydroxamic acids; positively charged tetraheptylammonium (for Ni nanoparticles)
CdSe [58]	High cell mortality due to easy dissolution and release of ions	

Source: Adapted from P. Somasundaran et al., *KONA Powder and Particle Journal*, 28, 2010, 38–49. With permission.

TABLE 7.4
Applications, Concerns, and Biological/Mechanistic Studies of Inhaled Nanoparticles of Particle Size Less than 100 nm and Greater than 500 nm

	Nanoparticles (<100 nm)	Larger Particles (>500 nm)
General Characteristics		
Ratio: Particle number/mass or surface area/mass	High	Low
Agglomeration/aggregation in air and/or liquids	Likely (depends on medium)	Less likely
Deposition in respiratory tract protein/lipid adsorption in vitro	Diffusion dominates Yes, important for biokinetics	Sedimentation, impaction, and interception dominate
Translocation to Secondary Target Organs		
Clearance	Yes	Less important
Mucociliary	Probably yes	Generally not
Alveolar macrophages	Poor	Efficient
Epithelial cells	Yes	Efficient
Lymphatic circulation	Yes	Mainly under overload
Blood circulation	Yes	Under overload
Sensory neurons (uptake and transport)	Yes	Under overload
Protein/lipid adsorption in vivo	Yes	No
Cell entry/uptake	Yes (caveolae, clathrin, lipid rafts, diffusion)	Some Primarily phagocytic cells
Mitochondria	Yes	No
Nucleus	Yes (<40 nm)	No
Direct Effects (Chemistry and Dose Dependent)		
At secondary target organs	Yes	No
At portal of entry (respiratory tract)	Yes	Yes
Inflammation	Yes	Yes
Oxidative stress	Yes	Yes
Activation of signaling pathways	Yes	Yes
Primary genotoxicity	Some	No
Carcinogenicity	Yes	Yes

Source. Adapted from G. Oberdörster, *Journal of Internal Medicine*, 267, 2010, 89–105 With permission.

nanometer-scale particles, a significantly greater inflammatory response in the lungs occurs in comparison to the exposure to the larger particles with an identical chemical composition (TiO_2 and Al_2O_3) [64–66]. The results were based on the same mass ratio. It was concluded that this greater inflammatory response was due to an increased rate of interstitialization of 20–30 nm nanometer-scale particles compared with 200–500 nm in the lungs. Another study by Oberdürster et al. [67] also showed that inflammatory responses following inhalation in rats depended on particle size; however, when normalized by surface area, size-invariant dose–response function was observed.

Some reports of PN toxicity has been reported recently. In this study, the influence of the size of poly (lactic-*co*-glycolic acid) (PLGA) and titanium dioxide (TiO_2) nanoparticles on cytotoxicity was determined. Nanoparticles of three different sizes were studied in two cell lines—RAW264.7 cells and BEAS-2B. Size-dependent cytotoxic effects were observed after incubation with PLGA or TiO_2 nanoparticles for 24 h. Even though PLGA nanoparticles did not activate significantly fatal toxicity, it was found that it led to tumor necrosis factor alpha (TNF-a) release. It was also observed that the larger particle size (200 nm) PLGA did not trigger any negative response. Relatively more toxic TiO_2 nanoparticles triggered cell death and other responses such as ROS generation, mitochondrial depolarization, plasma membrane damage, intracellular calcium concentration increase, and size-dependent TNF-a release. The authors concluded that the cytotoxic effects could be due to the size-dependent interaction between nanoparticles and biomolecules, as smaller particles tend to adsorb more biomolecules [68].

In another study, the cationic branch-like macromolecules polyamidoamine (PAMAM) dendrimers used as drug delivery systems for gene-based therapies such as RNA interference were studied. The study reveals that the effect of PAMAM dendrimers on their interactions with cellular signal transduction pathways such as the epidermal growth factor receptor (EGFR) is an important area to examine the toxicity of these nanoparticles. The EGFR is an important signaling cascade that regulates cell growth, differentiation, migration, survival, and apoptosis. In this study, the impact of naked unmodified Superfect (SF), a commercially available generation 6 PAMAM dendrimer, on the EGFR tyrosine kinase-extracellular-regulated kinase 1/2 (ERK1/2) signaling pathway in human embryonic kidney (HEK 293) cells was investigated. It was found that the SF PAMAM dendrimer delivery system, at doses routinely recommended for transfection of nucleic acids, can stimulate the EGFR–ERK1/2 signal transduction pathway in human cells via an oxidative-stress-dependent mechanism. It is a well-known fact that EGFR is an important signaling pathway for important life processes such as cell growth, differentiation, and migration as well as mediating several pathologies, most notably cancer. These findings have important implications for the safe use of PAMAM dendrimers [69].

7.3.2.2 Interaction of PNs and SLNs with Biological Systems

It is very important to study the interaction of PNs and SLNs with a biological system as it can be the major route for nanotoxicity of these systems. Only few studies have documented the interactions of PNs with the biological systems. A study of enhanced transfection by an antioxidative polymeric gene carrier that reduces polyplex-mediated cellular oxidative stress looked into the role of transfection-induced cellular ROS in nanoparticle-mediated oxidative stress and showed that toxicity might be a concern. The authors studied the cellular stress during transfection for polycation-based nonviral gene delivery. It was found that polyplexes can induce cellular ROS production even at subtoxic concentrations. The study emphasizes that the effect on cellular ROS stresses should be considered while designing polycation-based nonviral gene delivery systems, especially for clinical applications where a substantial amount of polyplex may be administered [70]. Efficient transfection and low cytotoxicity are prerequisite properties for nonviral gene carriers to reach clinical settings. Since polycation-based gene delivery systems rely mostly on the formation of nanosized polyplexes with nucleic acid drugs such as plasmid DNA (pDNA) and siRNA, it is likely that the polyplex will induce oxidative stress after cellular uptake. Polyethylenimines (PEIs) are among the most extensively studied cationic polymers for nonviral gene delivery because of their ability to form nanosized particles with nucleic acid drugs and to mediate consistent transfection in several cell types [71]. However, like most polycations used in nonviral transfection protocols, the relatively high cytotoxicity of PEI is considered a major drawback. The interactions of PEI with cellular membranes may reduce membrane fluidity by facilitating the formation of huge clusters on the surface of the membrane, leading to necrotic cell death. In addition, it was recently reported that PEI and its copolymer can induce oxidative stress responses in epithelial cells and macrophages. Grandinetti et al. [72] also showed that PEI/DNA polyplexes localized to mitochondria after transfection. The interaction of PEI polyplex with mitochondria seems to be related with the interference of mitochondrial function and loss of mitochondrial membrane potential. The mitochondrial damages by polyplexes could stimulate the additional production of ROS due to the uncoupling of electron transport machineries, leading to significant increase in cellular oxidative stress [73]. Subtle alterations in the expression of cellular signal transporters induced by ROS may either directly or indirectly affect the reporter gene expression.

7.3.2.3 Physicochemical Characteristics of Polymers/Lipids Utilized in Nanoparticles

Table 7.5 lists different physicochemical parameters important for characterizing toxicity of nanoparticles. It is very important to extensively characterize the composition, shape, surface area, surface charge, and so on along with the size of PNs and SLNs. During nanoparticles development, the focus is to develop particles that contain at least one dimension smaller than 100 nm with high reproducibility and monodispersity. These nanoparticles could have a number of possible causes of toxicity: (1) nanoparticles have been demonstrated to have electronic, optical, and magnetic properties that are related to their physical dimensions, and the breakdown of these nanoparticles could lead to a unique toxic effect that is difficult to predict [74]; (2) nanoparticles surfaces are involved in many catalytic and oxidative reactions [75]. If these reactions induce cytotoxicity, then the toxicity could be greater than a similar bulk material because the surface-area-to-volume ratio for nanoscale material is much greater and (3) some nanoparticles contain metals or compounds with known toxicity and, thus, the breakdown of these materials could elicit similar toxic responses to the components themselves.

Ideally, FDA (Food and Drug Administration)-approved or GRAS (generally regarded as safe)-listed polymers are utilized for the preparation of PNs and SLNs for drug delivery systems. Table 7.6 lists various polymers and lipids considered to be safe and commonly used in preparing nanoparticles.

Further, there is a consistent and increasing use of new polymers/lipids, which enhances the efficiency of these nanoparticles but the safety of these materials remains a concern as very little time and effort are devoted in this direction.

TABLE 7.5
Lists of Different Physicochemical Parameters Important for Characterizing Toxicity of Nanoparticles

Interaction of Nanoparticles with Biological Medium

What does the material look like?
- Particle size/size distribution
- Agglomeration state/aggregation
- Shape

What is the material made of?
- Overall composition (including chemical composition and crystal structure)
- Surface composition
- Purity (including levels of impurities)

What factors affect how a material interacts with its surroundings?
- Surface area
- Surface chemistry, including reactivity, hydrophobicity
- Surface charge

Overarching considerations to take into account when characterizing engineered nanomaterials in toxicity studies:
- Stability—how do material properties change with time (dynamic stability), storage, handling, preparation, delivery, and so on? Include solubility, and the rate of material release through dissolution
- Context/media—how do material properties change in different media; that is, from the bulk material to dispersions to material in various biological matrices? ("as administered" characterization is considered to be particularly important)
- Where possible, materials should be characterized sufficiently to interpret the response to the amount of material against a range of potentially relevant dose metrics, including mass, surface area, and number concentration

Source: D.R. Boverhof, R.M. David, *Analytical and Bioanalytical Chemistry*, 396, 2010, 953–961.

TABLE 7.6
List of Various Polymers and Lipids Considered to be Safe and Commonly Used in Preparing Nanoparticles

Polymeric Nanoparticles		Solid Lipid Nanoparticles	
Polymer Categories	**Lipids**	**Oils**	**Emulsifiers/Coemulsifiers**
Poly (2-hydroxy ethyl methacrylate)	Tricaprin	Clove oil	Soybean lecithin
Poly (N-vinyl pyrrolidone)	Trilaurin	Coconut oil	Egg lecithin
Poly (methyl methacrylate)	Trimyristin	Corn oil	Phosphatidylcholine
Poly (vinyl alcohol)	Tripalmitin	Cottonseed oil	Poloxamer
Poly (acrylic acid)	Tristearin	Fractionated coconut oil	Poloxamine
Polyacrylamide	Hydrogenated coco-glycerides	Hydrogenated castor oil	Tyloxapol
Poly (ethylene-co-vinyl acetate)	Witepsol	Hydrogenated palm oil	Polysorbate
Poly (ethylene glycol)	Glyceryl monostearate	Palm kernel oil peg-6 esters	Sodium cholate
Poly (methacrylic acid).	Glyceryl behenate	Hydrogenated soybean oil	Sodium glycocholate
Polylactides (PLA)	Glyceryl palmitostearate	Glyceryl ricinoleate	Taurocholic acid sodium salt
Polyglycolides (PGA)	Cetyl palmitate	Glyceryl stearate	Taurodeoxycholic acid sodium salt
Poly (lactide-co-glycolides) (PLGA)	Stearic acid	Cetyl esters wax	Butanol
Polyanhydrides	Palmitic acid	Ethyl oleate	Butyric acid
Polyorthoesters	Decanoic acid	Canola oil	
	Behenic acid	Caprylic/capric mono/diglycerides	
	Acidan N12	Caprylic/capric triglyceride	
	Glyceryl trioleate		
	Glyceryl palmitostearate		

7.3.2.4 Concentration of Polymers/Lipids Utilized in the Nanoparticles

Polymeric and lipid concentrations in PNs and SLNs usually varies according to the drug loading and in many cases, high drug loading results in adding high concentrations of polymers and lipids for the optimized drug delivery. Although it is safe to presume that low polymer and lipid contents may not cause any toxicity, however, increasing the concentration significantly can raise the possibility of toxicity by PNs and SLNs. Excessive lipid intake may result in fat-overload syndrome characterized by hyperlipidemia, hepatosplenomegaly, and gastrointestinal disorders [77,78]. For the purpose of drug formulation using SLNs, the quantity of lipid administered into human is significantly lower. Therefore, concern for lipid overdose will be low as long as the formulation is given within the recommended therapeutic dose range of the active ingredient. This possibility of concentration-dependent toxicity of PNs and SLNs can be highlighted by the studies carried out on inorganic nanoparticulate formulations of TiO_2. Nurkiewicz et al. [79] reported that compared with fine particles, nanoparticle inhalation produced significantly greater microvascular dysfunction at a similar mass pulmonary deposition. This study was based on the comparison of microvascular reactivity of rats exposed to five doses of nano-TiO_2 (4–38 µg depositions) with five doses of fine TiO_2 (8–90 µg depositions). The results found that 67 µg fine TiO_2 produced a similar effect as 10 µg nano-TiO_2, that is more than six times of fine particle compared with nanoparticle deposition is needed to produce equivalent levels of microvascular dysfunction, oxidative stress, and nitric oxide (NO) quenching. The study also determined the changes produced on nanoparticle exposures in the reactive species and endogenous NO production. In the study, fine TiO_2 (primary particle diameter ~1 μm) and TiO_2 nanoparticles (primary particle diameter ~21 nm) were exposed to rats via aerosol inhalation. The results showed that nanoparticle exposure led to significant increase in the microvascular oxidative stress by around 60% and nitrosative stress by around fourfold. These reactive stresses corresponded with a decline in NO production in a dose-dependent manner. Furthermore, radical scavenging resulted in a partially restored NO production and normal microvascular activity. These results showed that nanoparticle exposure leads to an increase in local reactive species because of decrease in NO bioavailability and microvascular dysfunction.

In another study, Scholer et al. [80] studied the behavior of murine macrophages in the presence of different concentrations of SLN. The SLN consisting of stearic acid or dimethyl-dioctadecyl ammonium bromide were found to be cytotoxic at the concentration of 0.01%, whereas SLNs consisting of triglycerides, cetylpalmitate, or paraffin lipids were found to be safe at the same concentration. The authors concluded that decreased production in IL-6 was the possible reason for this type of toxicity. In a separate study containing high doses of compritol and cetyl palmitate, it was suggested that the toxicity is dependent on lipid matrix as well as administered doses. The results were based on histopathology and it was found that high-dose compritol containing SLN formulation led to the accumulation of lipids in the liver and spleen of mice and resulted in pathological alterations.

7.3.2.5 Conversion of Polymers/Lipids

Polymers and lipids have a significant tendency to convert into different polymorphic forms. These conversions can easily take place in biological fluids where various enzymes/surfactants are presents as well as there can be significant changes in pH. The positive charges on cationic lipids present in SLNs can promote nonspecific binding to circulating blood cells such as erythrocytes, lymphocytes, as well as endothelial cells. Likewise, the presence of unprotected surface negative charges on lipid molecules such as phosphatidylglycerol and phosphatidylserine serve as binding sites for plasma opsonin, which favors uptake by macrophages. No systemic studies have been carried out to study these changes/conversions in polymeric and SLNs. Many studies in inorganic nanoparticles have shown that ultrafine TiO_2 particles caused more inflammation in rat lungs than exposure to fine TiO_2 [81,82]. Warheit et al. in their studies reasoned that the differences in toxic potencies seem to be a result of their unique size, surface area/activity, and/or crystal properties [83–85]. Since different polymorphic forms of polymers and lipids can result in changes in crystal properties, it can be assumed that toxicity of polymers can be associated with these aspects of PNs and SLNs.

7.3.2.6 Degradation of Polymeric/Lipid Nanoparticles

Once distributed and sequestered in cells, the core nanoparticle metabolic processing mechanism is still not fully addressed. Degradation of lipids during sterilization is an important issue in SLNs as it can cause potential toxicity. Polymer-based nanoparticles and super-paramagnetic iron oxide nanoparticles for magnetic resonance imaging contrast agents are shown to degrade, but quantum dots (QDs), fullerenes, and silica nanoparticles are examples of nanoparticles without clear indication of degradation *in vivo* [86–88]. For example, Fischer et al. [89] and Ballou et al. [90] show that the core ZnS-capped CdSe QDs remain intact and fluorescent *in vivo* after 1 month; however, neither study analyzed the metabolism of the organic coating on the nanoparticles' surface. The breakdown of the nanoparticles could elicit unique molecular responses that are not predictable and, thus, the understanding and cataloging of what, when, and how much nanostructures degrade are extremely important.

7.3.2.7 Route of Administration of Polymeric/Lipid Nanoparticles

PNs and SLNs can be delivered using various administration routes. In some cases, a nanoparticle with similar composition could be delivered using different routes of administration. Since the concentration of nanoparticles in blood depends on the route of administration, it can contribute to potential nanotoxicity as it determines the access of these nanoparticles to different cells and organs of the body. For example, polymeric and lipid nanoparticles with a size range above 100 nm can be nontoxic if applied dermally compared with the administration in an invasive way like intravenously as injection. In the latter case, they could potentially irritate the immune system after uptake by macrophages. Nanotoxicity should be assessed if there is an increased risk due to a special route of pharmaceutical administration [37].

7.3.2.8 Biocompatibility of Polymeric/Lipid Nanoparticles

Biocompatibility of polymers and lipid excipients is another essential parameter to be taken into account while assessing PN and SLN toxicity for drug delivery systems in various applications. Usually, polymers and lipids utilized in PNs and SLNs are inert and do not generate any undesired response from the body.

7.3.2.9 Excipients/Residual Solvents Used during the Preparation of Polymeric/Lipid Nanoparticles

Organic solvents are commonly used in the preparation of PNs and can add to the toxicity of the formulation. Care should be taken to remove all the residual organic solvent as they have a strong potential of toxicity. Residual solvents are not a toxicity concern for SLN but high concentrations of surfactants/emulsifiers/preservatives present in SLNs can cause nanotoxicity. Studies on SLN containing sterylamine and different triglycerides suggested that the toxicity of the SLN is dependent on the composition and method of purification used. In the study, dialysis was found to be the most efficient method to remove excess surfactant thus reducing the toxicity [91]. The positively charged surfactant used in the preparation of SLNs can interact with the negatively charged cellular membrane resulting in possible toxicity.

7.4 METHODS FOR NANOTOXICITY ASSESSMENT

Although there is no standard protocol for nanotoxicity testing currently, it will suffice to mention that the key elements of a toxicity screening strategy should include physicochemical characterization of nanoparticles and various *in vitro* assays to establish the nanotoxicity of nanoparticles.

7.4.1 Physicochemical Characterization

The adequate physicochemical characterization of nanomaterials prior to undertaking experiments for *in vitro* toxicity assessments is extremely important. The major physicochemical characterization includes (1) size, including surface area, size distribution, chemical composition (purity crystallinity, electronic properties, etc.); (2) solubility; (3) shape and aggregation; and (4) surface structure including surface reactivity and particle reactivity in solution [85,92].

7.4.2 In Vitro Cell Culture Techniques

In vitro cell culture techniques could be used for the characterization of nanoparticle uptake and localization, biodistribution and qualitative analysis of nanotoxicity. Further, various cell culture assays for cytotoxicity (altered metabolism, decreased growth, lytic or apoptotic cell death), proliferation, genotoxicity, and altered gene expression can provide important assessment about the nanotoxicity [93].

7.4.2.1 Cell Viability Assay

Nanocarriers can be evaluated in terms of their potential toxicity to the cells by the use of various cell viability assays. These assays are basically indicators of cellular damage. Several standard methods for quantification of cell viability are available, which differ in sensitivity, reliability, and ease of use. These assays are based on three basic parameters. The first assay type is based on the measurement of cellular metabolic activity. An early indication of cellular damage is a reduction in metabolic activity. The MTS colorimetric assay is a commonly used assay of this category. In this assay, living (metabolically active) cells reduce tetrazolium salts (3-(4,5-dimethylthiazol-2-yl)-5-(3-carboxymethoxyphenyl)-2-(4-sulfophenyl)-2H-tetrazolium, inner salt; MTS) and the electron-coupling reagent, phenazine methosulfate to formazan, which is soluble in tissue culture medium. The absorbance of the formazan product formed can then be determined at 492 nm. Since the production of formazan is proportional to the number of living cells, the intensity of the color produced is a good indicator of cell viability [94]. Another set of cytotoxicity assays is based on the measurement of membrane integrity. The cell membrane forms a functional barrier around the cell, and traffic into and out of the cell is highly regulated by transporters, receptors, and secretion pathways. When cells are damaged, they become "leaky" and this forms the basis for the second type of assay. Lactate dehydrogenase (LDH) is a stable cytoplasmic enzyme that is present in most cells. This enzyme cannot be measured extracellularly unless cell damage has occurred. Thus, membrane integrity can be determined by measuring LDH in the extracellular medium. The amount of enzyme activity correlates to the proportion of damaged cells [95]. Cell membrane integrity can also be assessed by the uptake or exclusion of dyes, such as trypan blue or eosin. These dyes are normally excluded from the inside of healthy cells; however, if the cell membrane has been compromised, they freely cross the membrane and stain intracellular components. The third type of assay is the direct measure of cell number, since dead cells normally detach from a culture plate, and are washed away in the medium. Cell number can be measured by direct cell counting, or by the measurement of total cell protein or DNA, which are proportional to the number of cells [96].

7.4.2.2 Cell Uptake Studies of Nanoparticles

Cell uptake study can be done by using the following techniques:

1. Flow cytometry (including FACS)
2. Epi-fluorescence microscopy
3. Confocal microscopy

To study cellular uptake, appropriate cell lines such as the human ovarian carcinoma cell line and the human epidermal carcinoma cell line are selected and cellular accumulation is measured. As all these methods are fluorescence-based methods, suitable dye is attached to the targeting moiety and its uptake is studied [97].

7.4.2.3 Assays for Alteration in Gene Expression

Gene expression assays, that is, gene profiling, is an important tool for screening different environmental particles, including nanoparticles. Techniques used to assess gene expression include Northern blot analysis, ribonuclease protection assays, quantitative real-time polymerase chain reaction, polymerase chain reaction arrays, and microarrays.

7.5 CONCLUSION

PNs and SLNs are widely used in the pharmaceutical industry as drug delivery systems. They are usually regarded as safe and very few studies have been done to determine the nanotoxicity of these systems. Since PNs and SLNs are parallel to different nanoparticulate systems (like inorganic nanoparticles showing nanotoxicity) and possess similar properties like particle size, it is required that the nanotoxicity of these systems should be studied in detail to fully exploit the advantages offered by PNs and SLNs.

REFERENCES

1. L. Zhang, F.X. Gu, J.M. Chan, A.Z. Wang, R.S. Langer, O.C. Farokhzad, Nanoparticles in medicine: Therapeutic applications and developments, *Clinical Pharmacology and Therapeutics*, 83, 2007, 761–769.
2. W. Mehnert, K. Mäder, Solid lipid nanoparticles: Production, characterization and applications, *Advanced Drug Delivery Reviews*, 47, 2001, 165–196.
3. M. Hamidi, A. Azadi, P. Rafiei, Hydrogel nanoparticles in drug delivery, *Advanced Drug Delivery Reviews*, 60, 2008, 1638–1649.
4. K. Cho, X. Wang, S. Nie, Z. Chen, D.M. Shin, Therapeutic nanoparticles for drug delivery in cancer, *Clinical Cancer Research*, 14, 2008, 1310–1316.
5. C. Pinto Reis, R.J. Neufeld, A.J. Ribeiro, F. Veiga, Nanoencapsulation I. Methods for preparation of drug-loaded polymeric nanoparticles, *Nanomedicine: Nanotechnology, Biology and Medicine*, 2, 2006, 8–21.
6. M. Rawat, D. Singh, S. Saraf, nanocarriers: Promising vehicle for bioactive drugs, *Biological & Pharmaceutical Bulletin*, 29, 2006, 1790–1798.
7. C. Vauthier, K. Bouchemal, Methods for the preparation and manufacture of polymeric nanoparticles, *Pharmaceutical Research*, 26, 2009, 1025–1058.
8. A. Grana, A. Limpach, H. Chauhan, Formulation considerations and applications of solid lipid nanoparticles, *American Pharmaceutical Review*, 16, 2013.
9. G. Wang, B. Yu, Y. Wu, B. Huang, Y. Yuan, C.S. Liu, Controlled preparation and antitumor efficacy of vitamin E TPGS-functionalized PLGA nanoparticles for delivery of paclitaxel, *International Journal of Pharmaceutics*, 446, 2013, 24–33.
10. Y.-C. Chen, W.-Y. Hsieh, W.-F. Lee, D.-T. Zeng, Effects of surface modification of PLGA-PEG-PLGA nanoparticles on loperamide delivery efficiency across the blood–brain barrier, *Journal of Biomaterials Applications*, 27, 2013, 909–922.
11. H. Wang, S. Han, J. Sun, T. Fan, C. Tian, Y. Wu, Amphiphilic dextran derivatives nanoparticles for the delivery of mitoxantrone, *Journal of Applied Polymer Science*, 126, 2012, E35–E43.
12. S.J. Kshirsagar, M.R. Bhalekar, J.N. Patel, S.K. Mohapatra, N.S. Shewale, Preparation and characterization of nanocapsules for colon-targeted drug delivery system, *Pharmaceutical Development and Technology*, 17, 2012, 607–613.
13. A. Giarmoukakis, G. Labiris, H. Sideroudi, Z. Tsimali, N. Koutsospyrou, K. Avgoustakis, V. Kozobolis, Biodegradable nanoparticles for controlled subconjunctival delivery of latanoprost acid: *In vitro* and *in vivo* evaluation. *Preliminary Results, Experimental Eye Research*, 112, 2013, 29–36.
14. L.H. Guerreiro, D. Da Silva, E. Ricci-Junior, W. Girard-Dias, C.M. Mascarenhas, M. Sola-Penna, K. Miranda, L.M.T. Lima, Polymeric particles for the controlled release of human amylin, *Colloids and Surfaces B: Biointerfaces*, 94, 2012, 101–106.
15. L.T.M. Hoa, N.T. Chi, L.H. Nguyen, D.M. Chien, Preparation and characterisation of nanoparticles containing ketoprofen and acrylic polymers prepared by emulsion solvent evaporation method, *Journal of Experimental Nanoscience*, 7, 2012, 189–197.
16. W.S. Cheow, M.W. Chang, K. Hadinoto, Antibacterial efficacy of inhalable levofloxacin-loaded polymeric nanoparticles against *E. coli* biofilm cells: The effect of antibiotic release profile, *Pharmaceutical Research*, 27, 2010, 1597–1609.
17. J. Liu, Z. Qiu, S. Wang, L. Zhou, S. Zhang, A modified double-emulsion method for the preparation of daunorubicin-loaded polymeric nanoparticle with enhanced *in vitro* anti-tumor activity, *Biomedical Materials*, 5, 2010, 065002.
18. C. Yan, D. Chen, J. Gu, J. Qin, Nanoparticles of 5-fluorouracil (5-FU) loaded *N*-succinyl-chitosan (Suc-Chi) for cancer chemotherapy: Preparation, characterization—In-vitro drug release and anti-tumour activity, *Journal of Pharmacy and Pharmacology*, 58, 2006, 1177–1181.
19. J. Anais, N. Razzouq, M. Carvalho, C. Fernandez, A. Astier, M. Paul, H. Fessi, A. Lorino, Development of α-tocopherol acetate nanoparticles: Influence of preparative processes, *Drug Development and Industrial Pharmacy*, 35, 2009, 216–223.
20. S.A. Galindo-Rodríguez, F. Puel, S. Briançon, E. Allémann, E. Doelker, H. Fessi, Comparative scale-up of three methods for producing ibuprofen-loaded nanoparticles, *European Journal of Pharmaceutical Sciences*, 25, 2005, 357–367.
21. P.A. McCarron, R.F. Donnelly, W. Marouf, Celecoxib-loaded poly (D,L-lactide-*co*-glycolide) nanoparticles prepared using a novel and controllable combination of diffusion and emulsification steps as part of the salting-out procedure, *Journal of Microencapsulation*, 23, 2006, 480–498.

22. E. Allémann, J.-C. Leroux, R. Gurny, E. Doelker, *in vitro* extended-release properties of drug-loaded poly (DL-lactic acid) nanoparticles produced by a salting-out procedure, *Pharmaceutical Research*, 10, 1993, 1732–1737.
23. M. de Matos, A. Piedade, C. Alvarez-Lorenzo, A. Concheiro, M. Braga, H. de Sousa, SCF-assisted processing of dexamethasone-loaded poly (ε-caprolactone)/MCM-41 materials for biomedical applications, in: *Bioengineering (ENBENG)*, 2012 IEEE 2nd Portuguese Meeting in, IEEE, 2012, pp. 1–4.
24. A.S. Mayo, B.K. Ambati, U.B. Kompella, Gene delivery nanoparticles fabricated by supercritical fluid extraction of emulsions, *International Journal of Pharmaceutics*, 387, 2010, 278–285.
25. A. Sane, J. Limtrakul, Formation of retinyl palmitate-loaded poly (l-lactide) nanoparticles using rapid expansion of supercritical solutions into liquid solvents (RESOLV), *The Journal of Supercritical Fluids*, 51, 2009, 230–237.
26. A. Kumar, K. Sawant, Encapsulation of exemestane in polycaprolactone nanoparticles: Optimization, characterization, and release kinetics, *Cancer Nanotechnology*, 4, 2013, 57–71.
27. T. Tencomnao, K. Klangthong, N. Pimpha, S. Chaleawlert-umpon, S. Saesoo, N. Woramongkolchai, N. Saengkrit, Acceleration of gene transfection efficiency in neuroblastoma cells through polyethyleneimine/poly(methyl methacrylate) core-shell magnetic nanoparticles, *International Journal of Nanomedicine*, 7, 2012, 2783.
28. A.A. Date, M.D. Joshi, V.B. Patravale, Parasitic diseases: Liposomes and polymeric nanoparticles versus lipid nanoparticles, *Advanced Drug Delivery Reviews*, 59, 2007, 505–521.
29. Y. Li, L. Dong, A. Jia, X. Chang, H. Xue, Preparation and characterization of solid lipid nanoparticles loaded traditional chinese medicine, *International Journal of Biological Macromolecules*, 38, 2006, 296–299.
30. R.H. Muller, C.M. Keck, Challenges and solutions for the delivery of biotech drugs—A review of drug nanocrystal technology and lipid nanoparticles, *Journal of Biotechnology*, 113, 2004, 151–170.
31. Q.-B. Han, M.-L. Li, S.-H. Li, Y.-K. Mou, Z.-W. Lin, H.-D. Sun, Ent-kaurane diterpenoids from *Isodon rubescens* var. *lushanensis*, *Chemical and Pharmaceutical Bulletin*, 51, 2003, 790–793.
32. K. Jores, W. Mehnert, M. Drechsler, H. Bunjes, C. Johann, K. Mäder, Investigations on the structure of solid lipid nanoparticles (SLN) and oil-loaded solid lipid nanoparticles by photon correlation spectroscopy, field-flow fractionation and transmission electron microscopy, *Journal of Controlled Release*, 95, 2004, 217–227.
33. V. Venkateswarlu, K. Manjunath, Preparation, characterization and *in vitro* release kinetics of clozapine solid lipid nanoparticles, *Journal of Controlled Release*, 95, 2004, 627–638.
34. S. Hatziantoniou, G. Deli, Y. Nikas, C. Demetzos, G.T. Papaioannou, Scanning electron microscopy study on nanoemulsions and solid lipid nanoparticles containing high amounts of ceramides, *Micron*, 38, 2007, 819–823.
35. J. Liu, T. Gong, H. Fu, C. Wang, X. Wang, Q. Chen, Q. Zhang, Q. He, Z. Zhang, Solid lipid nanoparticles for pulmonary delivery of insulin, *International Journal of Pharmaceutics*, 356, 2008, 333–344.
36. B. Du, Y. Yan, Y. Li, S. Wang, Z. Zhang, Preparation and passive target of 5-fluorouracil solid lipid nanoparticles, *Pharmaceutical Development and Technology*, 15, 2010, 346–353.
37. C.M. Keck, R.H. Müller, Nanotoxicological classification system (NCS)—A guide for the risk-benefit assessment of nanoparticulate drug delivery systems, *European Journal of Pharmaceutics and Biopharmaceutics*, 84, 2013, 445–448.
38. J. Pardeike, A. Hommoss, R.H. Müller, Lipid nanoparticles (SLN, NLC) in cosmetic and pharmaceutical dermal products, *International Journal of Pharmaceutics*, 366, 2009, 170–184.
39. R. Müller, S. Maaβen, H. Weyhers, F. Specht, J. Lucks, Cytotoxicity of magnetite-loaded polylactide, polylactide/glycolide particles and solid lipid nanoparticles, *International Journal of Pharmaceutics*, 138, 1996, 85–94.
40. P. Somasundaran, X. Fang, S. Ponnurangam, B. Li, Nanoparticles: Characteristics, mechanisms and modulation of biotoxicity, *KONA Powder and Particle Journal*, 28, 2010, 38–49.
41. R.H. Müller, E.B. Souto, T. Göppert, S. Gohla, Production of biofunctionalized solid lipid nanoparticles for site-specific drug delivery. In: *Nanotechnologies for the Life Sciences*, Challa S.S.R. Kumar (ed.), Wiley-VCH Verlag GmbH & Co. KGaA: Weinheim, 2007.
42. S. Dey, V. Bakthavatchalu, M.T. Tseng, P. Wu, R.L. Florence, E.A. Grulke, R.A. Yokel, S.K. Dhar, H.-S. Yang, Y. Chen, Interactions between SIRT1 and AP-1 reveal a mechanistic insight into the growth promoting properties of alumina (Al_2O_3) nanoparticles in mouse skin epithelial cells, *Carcinogenesis*, 29, 2008, 1920–1929.
43. M. Monthioux, B. Smith, B. Burteaux, A. Claye, J. Fischer, D. Luzzi, Sensitivity of single-wall carbon nanotubes to chemical processing: An electron microscopy investigation, *Carbon*, 39, 2001, 1251–1272.

44. L. Ding, J. Stilwell, T. Zhang, O. Elboudwarej, H. Jiang, J.P. Selegue, P.A. Cooke, J.W. Gray, F.F. Chen, Molecular characterization of the cytotoxic mechanism of multiwall carbon nanotubes and nano-onions on human skin fibroblast, *Nano Letters*, 5, 2005, 2448–2464.
45. W. Tansey, S. Ke, X.-Y. Cao, M.J. Pasuelo, S. Wallace, C. Li, Synthesis and characterization of branched poly (L-glutamic acid) as a biodegradable drug carrier, *Journal of Controlled Release*, 94, 2004, 39–51.
46. S. Hong, A.U. Bielinska, A. Mecke, B. Keszler, J.L. Beals, X. Shi, L. Balogh, B.G. Orr, J.R. Baker, M.M. Banaszak Holl, Interaction of poly (amidoamine) dendrimers with supported lipid bilayers and cells: Hole formation and the relation to transport, *Bioconjugate Chemistry*, 15, 2004, 774–782.
47. T. Xia, M. Kovochich, J. Brant, M. Hotze, J. Sempf, T. Oberley, C. Sioutas, J.I. Yeh, M.R. Wiesner, A.E. Nel, Comparison of the abilities of ambient and manufactured nanoparticles to induce cellular toxicity according to an oxidative stress paradigm, *Nano Letters*, 6, 2006, 1794–1807.
48. T.S. Hauck, A.A. Ghazani, W.C. Chan, Assessing the effect of surface chemistry on gold nanorod uptake, toxicity, and gene expression in mammalian cells, *Small*, 4, 2008, 153–159.
49. E. Navarro, A. Baun, R. Behra, N.B. Hartmann, J. Filser, A.-J. Miao, A. Quigg, P.H. Santschi, L. Sigg, Environmental behavior and ecotoxicity of engineered nanoparticles to algae, plants, and fungi, *Ecotoxicology*, 17, 2008, 372–386.
50. H.L. Karlsson, P. Cronholm, J. Gustafsson, L. Mo¨ller, Copper oxide nanoparticles are highly toxic: A comparison between metal oxide nanoparticles and carbon nanotubes, *Chemical Research in Toxicology*, 21, 2008, 1726–1732.
51. W.F. Vevers, A.N. Jha, Genotoxic and cytotoxic potential of titanium dioxide (TiO_2) nanoparticles on fish cells in vitro, *Ecotoxicology*, 17, 2008, 410–420.
52. T. Xia, M. Kovochich, M. Liong, L. Ma¨dler, B. Gilbert, H. Shi, J.I. Yeh, J.I. Zink, A.E. Nel, Comparison of the mechanism of toxicity of zinc oxide and cerium oxide nanoparticles based on dissolution and oxidative stress properties, *ACS Nano*, 2, 2008, 2121–2134.
53. E. Oesterling, N. Chopra, V. Gavalas, X. Arzuaga, E.J. Lim, R. Sultana, D.A. Butterfield, L. Bachas, B. Hennig, Alumina nanoparticles induce expression of endothelial cell adhesion molecules, *Toxicology Letters*, 178, 2008, 160–166.
54. I. Slowing, B.G. Trewyn, V.S.-Y. Lin, Effect of surface functionalization of MCM-41-type mesoporous silica nanoparticles on the endocytosis by human cancer cells, *Journal of the American Chemical Society*, 128, 2006, 14792–14793.
55. K. Fujiwara, H. Suematsu, E. Kiyomiya, M. Aoki, M. Sato, N. Moritoki, Size-dependent toxicity of silica nano-particles to Chlorella kessleri, *Journal of Environmental Science and Health, Part A*, 43, 2008, 1167–1173.
56. S.C. Brown, M. Kamal, N. Nasreen, A. Baumuratov, P. Sharma, V.B. Antony, B.M. Moudgil, Influence of shape, adhesion and simulated lung mechanics on amorphous silica nanoparticle toxicity, *Advanced Powder Technology*, 18, 2007, 69–80.
57. D. Guo, C. Wu, X. Li, H. Jiang, X. Wang, B. Chen, In vitro cellular uptake and cytotoxic effect of functionalized nickel nanoparticles on leukemia cancer cells, *Journal of Nanoscience and Nanotechnology*, 8, 2008, 2301–2307.
58. C. Kirchner, T. Liedl, S. Kudera, T. Pellegrino, A. Muñoz Javier, H.E. Gaub, S. Stölzle, N. Fertig, W.J. Parak, Cytotoxicity of colloidal CdSe and CdSe/ZnS nanoparticles, *Nano Letters*, 5, 2005, 331–338.
59. P. Zou, L. Helson, A. Maitra, S.T. Stern, S.E. McNeil, Polymeric curcumin nanoparticle pharmacokinetics and metabolism in bile duct cannulated rats, *Molecular Pharmaceutics*, 10, 2013, 1977–1987.
60. H.N. Woo, H.K. Chung, E.J. Ju, J. Jung, H.-W. Kang, S.-W. Lee, M.-H. Seo, J.S. Lee, J.S. Lee, H.J. Park, Preclinical evaluation of injectable sirolimus formulated with polymeric nanoparticle for cancer therapy, *International Journal of Nanomedicine*, 7, 2012, 2197.
61. Y.-P. Fang, P.-C. Wu, Y.-B. Huang, C.-C. Tzeng, Y.-L. Chen, Y.-H. Hung, M.-J. Tsai, Y.-H. Tsai, Modification of polyethylene glycol onto solid lipid nanoparticles encapsulating a novel chemotherapeutic agent (PK-L4) to enhance solubility for injection delivery, *International Journal of Nanomedicine*, 7, 2012, 4995.
62. H.S. Jeon, J.E. Seo, M.S. Kim, M.H. Kang, D.H. Oh, S.O. Jeon, J. Seong Hoon, Y.W. Choi, S. Lee, A retinyl palmitate-loaded solid lipid nanoparticle system: Effect of surface modification with dicetyl phosphate on skin permeation in vitro and anti-wrinkle effect in vivo, *International Journal of Pharmaceutics*, 452, 2013, 311–320.
63. A.D. Maynard, D.B. Warheit, M.A. Philbert, The new toxicology of sophisticated materials: Nanotoxicology and beyond, *Toxicological Sciences*, 120, 2011, S109–S129.
64. G. Oberdörster, Safety assessment for nanotechnology and nanomedicine: Concepts of nanotoxicology, *Journal of Internal Medicine*, 267, 2010, 89–105.

65. J. Ferin, G. Oberdörster, D.P. Penney, S.C. Soderholm, R. Gelein, H.C. Piper, Increased pulmonary toxicity of ultrafine particles? I. Particle clearance, translocation, morphology, *Journal of Aerosol Science*, 21, 1990, 381–384.
66. G. Oberdörster, J. Ferin, G. Finkelstein, P. Wade, N. Corson, Increased pulmonary toxicity of ultrafine particles? II. Lung lavage studies, *Journal of Aerosol Science*, 21, 1990, 384–387.
67. G. Oberdürster, Toxicology of ultrafine particles: *In vivo* studies, *Philosophical Transactions of the Royal Society of London. Series A: Mathematical, Physical and Engineering Sciences*, 358, 2000, 2719–2740.
68. S. Xiong, S. George, H. Yu, R. Damoiseaux, B. France, K. Ng, J.-C. Loo, Size influences the cytotoxicity of poly (lactic-co-glycolic acid) (PLGA) and titanium dioxide (TiO_2) nanoparticles, *Archives of Toxicology*, 87, 2013, 1075–1086.
69. S. Akhtar, B. Chandrasekhar, S. Attur, M.H.M. Yousif, I.F. Benter, On the nanotoxicity of PAMAM dendrimers: Superfect® stimulates the EGFR–ERK1/2 signal transduction pathway via an oxidative stress-dependent mechanism in HEK 293 cells, *International Journal of Pharmaceutics*, 448, 2013, 239–246.
70. M.S. Lee, N.W. Kim, K. Lee, H. Kim, J.H. Jeong, Enhanced transfection by antioxidative polymeric gene carrier that reduces polyplex-mediated cellular oxidative stress, *Pharmaceutical Research*, 2013, 1–10.
71. T.G. Park, J.H. Jeong, S.W. Kim, Current status of polymeric gene delivery systems, *Advanced Drug Delivery Reviews*, 58, 2006, 467–486.
72. G. Grandinetti, N.P. Ingle, T.M. Reineke, Interaction of poly (ethylenimine)–DNA polyplexes with mitochondria: Implications for a mechanism of cytotoxicity, *Molecular Pharmaceutics*, 8, 2011, 1709–1719.
73. N. Li, C. Sioutas, A. Cho, D. Schmitz, C. Misra, J. Sempf, M. Wang, T. Oberley, J. Froines, A. Nel, Ultrafine particulate pollutants induce oxidative stress and mitochondrial damage, *Environmental Health Perspectives*, 111, 2003, 455.
74. P. Borm, F.C. Klaessig, T.D. Landry, B. Moudgil, J. Pauluhn, K. Thomas, R. Trottier, S. Wood, Research strategies for safety evaluation of nanomaterials, part V: Role of dissolution in biological fate and effects of nanoscale particles, *Toxicological Sciences*, 90, 2006, 23–32.
75. A. Nel, T. Xia, L. Mädler, N. Li, Toxic potential of materials at the nanolevel, *Science*, 311, 2006, 622–627.
76. D.R. Boverhof, R.M. David, Nanomaterial characterization: Considerations and needs for hazard assessment and safety evaluation, *Analytical and Bioanalytical Chemistry*, 396, 2010, 953–961.
77. R. Zini, P. Riant, J. Barré, J.-P. Tillement, disease-induced variations in plasma protein levels, *Clinical Pharmacokinetics*, 19, 1990, 218–229.
78. J. McElnay, P. D'Arcy, Protein binding displacement interactions and their clinical importance, *Drugs*, 25, 1983, 495–513.
79. T.R. Nurkiewicz, D.W. Porter, A.F. Hubbs, J.L. Cumpston, B.T. Chen, D.G. Frazer, V. Castranova, Nanoparticle inhalation augments particle-dependent systemic microvascular dysfunction, *Particle and Fibre Toxicology*, 5, 2008, 1–12.
80. N. Schöler, H. Hahn, R. Müller, O. Liesenfeld, Effect of lipid matrix and size of solid lipid nanoparticles (SLN) on the viability and cytokine production of macrophages, *International Journal of Pharmaceutics*, 231, 2002, 167–176.
81. E. Bermudez, J.B. Mangum, B.A. Wong, B. Asgharian, P.M. Hext, D.B. Warheit, J.I. Everitt, Pulmonary responses of mice, rats, and hamsters to subchronic inhalation of ultrafine titanium dioxide particles, *Toxicological Sciences*, 77, 2004, 347–357.
82. J. Ferin, G. Oberdorster, D. Penney, Pulmonary retention of ultrafine and fine particles in rats, *American Journal of Respiratory Cell and Molecular Biology*, 6, 1992, 535–542.
83. D.B. Warheit, W.J. Brock, K.P. Lee, T.R. Webb, K.L. Reed, Comparative pulmonary toxicity inhalation and instillation studies with different TiO_2 particle formulations: Impact of surface treatments on particle toxicity, *Toxicological Sciences*, 88, 2005, 514–524.
84. D.B. Warheit, T.R. Webb, K.L. Reed, S. Frerichs, C.M. Sayes, Pulmonary toxicity study in rats with three forms of ultrafine-TiO_2 particles: Differential responses related to surface properties, *Toxicology*, 230, 2007, 90–104.
85. D.B. Warheit, How meaningful are the results of nanotoxicity studies in the absence of adequate material characterization? *Toxicological Sciences*, 101, 2008, 183–185.
86. R.S. Yang, L.W. Chang, J.P. Wu, M.H. Tsai, H.J. Wang, Y.C. Kuo, T.K. Yeh, C.S. Yang, P. Lin, Persistent tissue kinetics and redistribution of nanoparticles, quantum dot 705, in mice: ICP-MS quantitative assessment, *Environmental Health Perspectives*, 115, 2007, 1339–1343.
87. R. Singh, D. Pantarotto, L. Lacerda, G. Pastorin, C. Klumpp, M. Prato, A. Bianco, K. Kostarelos, Tissue biodistribution and blood clearance rates of intravenously administered carbon nanotube

radiotracers, *Proceedings of the National Academy of Sciences of the United States of America*, 103, 2006, 3357–3362.
88. H.C. Fischer, W.C.W. Chan, nanotoxicity: The growing need for *in vivo* study, *Current Opinion in Biotechnology*, 18, 2007, 565–571.
89. H.C. Fischer, L. Liu, K.S. Pang, W.C.W. Chan, Pharmacokinetics of nanoscale quantum dots: *In vivo* distribution, sequestration, and clearance in the rat, *Advanced Functional Materials*, 16, 2006, 1299–1305.
90. B. Ballou, B.C. Lagerholm, L.A. Ernst, M.P. Bruchez, A.S. Waggoner, noninvasive imaging of quantum dots in mice, *Bioconjugate Chemistry*, 15, 2004, 79–86.
91. A. Heydenreich, R. Westmeier, N. Pedersen, H. Poulsen, H. Kristensen, Preparation and purification of cationic solid lipid nanospheres—Effects on particle size, physical stability and cell toxicity, *International Journal of Pharmaceutics*, 254, 2003, 83–87.
92. R.C. Murdock, L. Braydich-Stolle, A.M. Schrand, J.J. Schlager, S.M. Hussain, Characterization of nanomaterial dispersion in solution prior to *in vitro* exposure using dynamic light scattering technique, *Toxicological Sciences*, 101, 2008, 239–253.
93. C.F. Jones, D.W. Grainger, *In vitro* assessments of nanomaterial toxicity, *Advanced Drug Delivery Reviews*, 61, 2009, 438–456.
94. W. Lin, Y.-W. Huang, X.-D. Zhou, Y. Ma, *In vitro* toxicity of silica nanoparticles in human lung cancer cells, *Toxicology and Applied Pharmacology*, 217, 2006, 252–259.
95. L. Dong, K.L. Joseph, C.M. Witkowski, M.M. Craig, Cytotoxicity of single-walled carbon nanotubes suspended in various surfactants, *Nanotechnology*, 19, 2008, 255702.
96. R.P. Schins, R. Duffin, D. Höhr, A.M. Knaapen, T. Shi, C. Weishaupt, V. Stone, K. Donaldson, P.J. Borm, Surface modification of quartz inhibits toxicity, particle uptake, and oxidative DNA damage in human lung epithelial cells, *Chemical Research in Toxicology*, 15, 2002, 1166–1173.
97. S.-J. Chiu, S. Liu, D. Perrotti, G. Marcucci, R.J. Lee, Efficient delivery of a Bcl-2-specific antisense oligodeoxyribonucleotide (G3139) via transferrin receptor-targeted liposomes, *Journal of Controlled Release*, 112, 2006, 199–207.

8 Analytical Characterization of Nanomaterials in Biological Matrices for Hazard Assessment

Mingsheng Xu, Daisuke Fujita, Huanxing Su, Hongzheng Chen, and Nobutaka Hanagata

CONTENTS

8.1 Introduction 159
8.2 Characterization of Primary NMs in Dry State 160
    8.2.1 Terminology and Definitions for NMs 160
    8.2.2 Physicochemical Characterization of Primary NMs 163
    8.2.3 Characterization in Biological Matrices 166
    8.2.4 Characterization of NMs in Biological Matrices by Electron Microscopy 166
    8.2.5 Characterization of NMs in Biological Matrices by Raman Spectroscopy 167
    8.2.6 Characterization of Dissolution of NMs in a Biological Matrix 168
8.3 Conclusion 170
Acknowledgments 170
References 171

8.1 INTRODUCTION

Novel nanomaterials (NMs) are playing key roles in nanotechnology innovations. As a consequence of their small size, NMs exhibit unique physicochemical properties and biological effects as compared to their respective bulk materials. NMs are increasingly used in a wide variety of industrial processes and consumer products. Although the obvious beneficial effects of NMs are well recognized, the increased presence of NMs raises concerns about the adverse effects on the environment, health, and society (so-called NanoEHS). There is growing evidence to suggest that complicated interaction processes may occur between NMs and biological systems [1,2] and to show that NMs, including carbon nanotubes, fullerenes, quantum dots such as CdS, oxide nanoparticles (NPs) such as ZnO, and CuO and TiO_2, exhibit various toxic effects on biological systems [3–6].

In spite of numerous publications on the investigations of the potential impacts of NMs, the progress is very slow. There are, at present, no standard methods for nanotoxicology, which might possibly lead to opposing results for the same NM. For example, in general, the concerns about the toxicity of NMs have been primarily based on the assumption that small-sized NMs are more toxic than large ones, and there are many reports showing that toxicity increases with a decrease in the particle size of NMs [7]. However, there are also findings that the cytotoxicity of NPs, such as SiO_2, is not a function of their primary size [3]. Many studies have attributed the toxic effects of metal-based NPs, such as ZnO, primarily to released metal cations [9], but other studies showed that the toxicity of ZnO NPs is independent of the amount of soluble Zn ions in the cell culture media [10]. At present, researchers have not been able to establish a single physicochemical parameter, for instance, size, shape, surface [7], chemical composition [8], ion [9], or conduction [3], which best describes the toxicity of NMs. Instead, a variety of physicochemical parameters have been suggested to contribute to the biological behaviors of NMs. Thus, a full characterization of NMs [11] is essential in order to correlate the physicochemical parameters with toxicity, and the identification of the main physicochemical factors that govern the toxic effects of NMs is important for the safe design and synthesis of NMs. This also allows researchers to establish predictive paradigms for the potential risks of NMs [3,12]. Knowledge about the interactions of NMs with biological surroundings is still in its infancy. This constitutes the key challenge and issue when assessing the toxicity of NMs.

In the field of materials science, the characterization methods for NM properties are well developed and established, including the use of advanced scanning probe microscopy [13–15] and electron microscopy [1,3,16,17]. For nanotoxicity studies, there are suggestions for the minimum analytical characterization of NMs for hazard assessments in biological matrices, including size distribution, shape, chemical composition, surface properties, agglomeration/aggregation, and solubility [11]. Different protocols for dispersing the same type of NM, for example, ultrasonication, shaking, and vortexing, or various solvents, will lead to various states (charge, agglomeration, aggregation, or dissolution) in a suspension. This suggests that performing characterizations in biological environments is of more importance than the measurement of primary NMs for hazard evaluations. Therefore, most of the current techniques become less powerful for measurements when NMs are in biological environments due to the dynamical interaction of the NMs with their surrounding matrix [1,2,18]. Because most of the toxic analyses of NMs have been on the basis of the primary characteristics of NMs measured prior to the dispersion of NMs in biological matrices, it is not surprising that there are controversial conclusions for the same type NM and the same cell lines regarding toxic effects and their correlation with physiochemical parameters.

8.2 CHARACTERIZATION OF PRIMARY NMs IN DRY STATE

8.2.1 Terminology and Definitions for NMs

Many terms and labels have been used with the prefix "nano" in practice, as well as in science and engineering. This causes misunderstandings in the real meaning of such terms. However, no internationally harmonized definition of "nanomaterial" exists yet, even though a wide range of definitions have indeed been discussed and proposed [19]. Therefore, it is very important to establish precise and unambiguous definitions that can explain the NMs as part of standardization for nanotoxicology. We propose a nomenclature for NMs collected from the existing literature.

The term "material" is often used intuitively without a legislative definition. According to Lovestam et al. [19], the term "material" refers to a single or closely bound ensemble of substances, at least one of which is in a condensed phase, where the constituents of the substance are atoms and molecules. The term "condensed phase" is generally used in thermodynamics for phases where a strong interaction between the constituents (i.e., atoms and molecules) exists; therefore, either the solid or liquid phases of a substance.

The ISO/TS 80004 series of standards, from the International Organization for Standardization (ISO) [20,21], describe vocabulary for nanotechnology and its applications. These were largely motivated by health, safety, and environment concerns of nanotechnology.

- ISO/TS 80004-1:2010 Nanotechnologies—Vocabulary—Part 1: Core terms. This document lists a number of core terms relevant to nanoscale materials, several of which are related to NMs, such as nanoscale, NM, nano-object, nanostructure, nanostructured material, manufactured NM, engineered NM, nanoscale phenomenon, and nanoscale property.

The definition of "nanoscale" encompasses the size ranging from approximately 1 nm to 100 nm. The "approximately" is assumed to be applicable for both the lower and upper limits of the definition, and size can refer to all three dimensions [19]. The "nano-object" is defined as a material with one, two, or three external dimensions in the nanoscale. The term "nanostructure" refers to a composition of interrelated constituent parts, in which one or more of those parts is/are a nanoscale region. The "nanostructured material" is defined as a material having an internal or surface nanostructure. This definition suggests that the dimensions of a nanostructured material can be larger than 100 nm; there exists overlapping between nano-objects and nanostructured materials: nano-objects can be nanostructured. The "nanoscale property" refers to a characteristic of a nano-object or nanoscale region. The "nanoscale phenomenon" refers to an effect attributable to nano-objects or nanoscale regions. The "nanomaterial" refers to a material with any external dimensions in the nanoscale or having an internal structure or surface structure in the nanoscale. Thus, the nanomaterial is defined as the sum of nano-objects and nanostructured materials. The "manufactured nanomaterial" refers to a NM intentionally produced for commercial purposes to have specific properties or a specific composition. And lastly, the "engineered nanomaterial" refers to a NM designed for a specific purpose or function. Lovestam et al. [19] have a suggestion on the definition of "nanomaterials." However, we believe that it is more important to clearly define the term "nanoscale," because all other definitions related to NMs are based upon this. The current definition of "nanoscale" in the ISO/TS 80004-1:2010 document does not make the term "size" clear. As associated to a material, we suggest that the "size" is the size of an original, individual material, or a named "primary" material, for example, a particle, wire, or tube, rather than an agglomerate and aggregate.

- ISO/TS 27687:2008 Nanotechnologies—Terminology and definitions for some types of nano-objects, i.e. nanoparticle, nanoplate, nanofiber, nanotube, nanorod, nanowire, and quantum dot. This document is revised as ISO/TS 80004-2. "Nanoparticle" is defined as a nano-object with all three external dimensions in the nanoscale. The "nanoplate" is defined as a nano-object with one external dimension in the nanoscale and the two other, significantly larger, external dimensions. There is no definition of "nanosheet" in the ISO-related document at this time. The size range of the "nanosheet" may be different from that of the "nanoplate." With regard to the thickness of graphene (0.334 nm), we propose the definition of "nanosheet" to be a nano-object with one external dimension in the range from 0.3 to 30 nm and the two other, significantly larger, external dimensions. Accordingly, to avoid conflictions with "nanosheet," we suggest a change of the lower size limit of "nanoplate" (approximately 1 nm) to 30 nm. Hence, the proposed definition of "nanoplate" is a nano-object with one external dimension in the range from 30 to 100 nm and the two other, significantly larger, external dimensions. Thus, nanosheets and nanoplates are two-dimensional NMs. The "nanofiber" is defined as a nano-object with two similar external dimensions in the nanoscale and a third, significantly larger dimension. "Nanotube" refers to a hollow nanofiber, "nanorod" refers to a solid nanofiber, "nanowire" refers to an electrically conducting or semiconducting nanofiber, and "quantum dot" refers to a crystalline NP that exhibits size-dependent properties due to quantum confinement effects on the electronic states.
- ISO/TS 80004-3:2010 Nanotechnologies—Vocabulary—Part 3: Carbon nano-objects. This document defines terms such as graphene, fullerene, and carbon nanotubes. "Graphene" is defined as a single layer of carbon atoms with each atom bound to three neighbors in a honeycomb structure. This differs from bilayer, trilayer, and few-layer graphene sheets. The chemical community often regards graphene oxide (GO) and reduced graphene oxide (rGO) as graphene, which causes confusions and is actually incorrect. More important is that these graphene-based materials have distinguishing properties.

- ISO/TS 80004-4:2011 Nanotechnologies—Vocabulary—Part 4: Nanostructured materials. This document gives terms and definitions for materials in the field of nanotechnologies where one or more components are nanoscale regions and the materials exhibit properties attributable to the presence of those nanoscale regions. A nanostructured material has an internal or surface structure with a significant fraction of features, grains, voids, or precipitates in the nanoscale. Articles that contain nano-objects or nanostructured materials are not necessarily nanostructured materials themselves.
- ISO/TS 80004-5:2011 Nanotechnologies—Vocabulary—Part 5: Nano/bio interface. This document lists terms and definitions related to the interface between NMs and biology with the intention of facilitating communications among those who have an interest in the application or use of nanotechnologies in biology or biotechnology, or the use of biological principles or matter in nanotechnology.
- ISO/TS 80004-7:2011 Nanotechnologies—Vocabulary—Part 7: Diagnostics and therapeutics for healthcare. This document is applicable to the use of nanotechnologies in medical diagnostics and therapeutics, and exploitation of material features at the nanoscale for diagnostic or therapeutic purposes in relation to human diseases.
- ISO/TR 13121:2011, Nanomaterial risk evaluation. This document describes a process for identifying, evaluating, addressing, and making decisions about, and communicating the potential risks of, developing and using manufactured NMs. This is done in order to protect the health and safety of the public, consumers, workers, and the environment; offering guidance on the information needed to make sound risk evaluations and risk management decisions, as well as how to manage in the face of incomplete or uncertain information. It includes methods to update assumptions, decisions, and practices and suggests methods that organizations can use to be transparent and accountable in how they manage NMs. It also describes a process of organizing, documenting, and communicating about these.

Owing to their high specific surface area, NMs have a higher free surface energy as compared to their bulk counterparts. Thus, they often have the tendency to agglomerate or aggregate to reduce their free energy. ISO/TS 27687 also gives definitions for particles clustered in agglomerates and aggregates. The "agglomerate" is defined as a collection of weakly bound particles or aggregates, or mixtures, of the two where the resulting external surface area is similar to the sum of the surface areas of the individual components. "Aggregate" refers to a particle composed of strongly bonded or fused particles where the resulting external surface area may be significantly smaller than the sum of the calculated surface areas of the individual components. Here, the term "particle" refers to a minute piece of matter with defined physical boundaries (ISO 14644-6:2007, ISO/TS 27687). We can replace "particle" with "nanomaterial" for the definitions of NMs clustered in agglomerates and aggregates.

NMs are of interest because of their unique optical, magnetic, electrical, and other properties at this scale. These unique properties have the potential for great impacts in electronics, medicine, and other fields. Two principal factors that cause the properties of NMs to differ significantly from other materials are increased relative surface area and quantum confinement effects. NMs have a much greater surface-area-to-volume ratio than their conventional forms. Accordingly, this translates to a very high surface reactivity with the surrounding surface, ideal for catalysis or sensor applications. Quantum confinement effects can become much more important in determining the material's properties and characteristics, leading to novel optical, electrical, and magnetic behaviors relative to bulk materials without a change in the chemical composition. According to the quantum confinement theory, electrons in the conduction band and holes in the valence band are spatially confined by the potential barrier of the surface. Owing to the confinement of both electrons and holes, the lowest energy optical transition from the valence to the conduction band will increase in energy, effectively increasing the band gap. As a result, the absorption energy of quantum dots will shift to higher frequencies with the decreasing diameter of the dots. This is readily observed from the reflected colors of quantum dots with varying diameters, shifting from blue to red with an increase in size. However, for metallic NPs with diameters ca. >2 nm, the operating principle is different from semiconductor quantum dots since there is no band gap between valence and conduction bands and the energy states form a continuum analogous to bulk metals. For these metallic NPs, another phenomenon known as surface plasmon resonance (SPR) is active for these structures, involving specific scattering interactions between the impinging light and the nanostructures. For the smallest of metallic NPs with dimensions ca. <2 nm, the surface plasmon absorption disappears. Quantum confinement effects also cause a change in the other optical properties of metallic NPs, because the spacing between intraband energy levels increases with decreasing particle size. This change will lead to changes in light emission characteristics. The reason that this is mentioned is because there is no direct, material-independent relationship between the size and novel effects. Thus, case-by-case studies are necessary for every NM, and we emphasize that more attention is needed toward the novel nanoscale properties when the potential impact of NMs is evaluated [3].

8.2.2 Physicochemical Characterization of Primary NMs

The characterization of NMs in their dry states is the first step for the assessment of their potential hazards. The characterizations include chemical composition and impurities; geometric properties such as size distribution, shape, surface morphology, and surface area; crystalline structure; and optical, electronic, and magnetic properties. The chemical composition, crystalline structure, shape, surface morphology, and surface area of NMs can be characterized by conventional techniques developed in materials science, including x-ray photoelectron spectroscopy (XPS), x-ray diffraction (XRD), scanning electron microscope (SEM), transmission electron microscopy (TEM), and atomic force microscopy (AFM). Here, we do not address these techniques in detail and pay no attention to standardized characterization techniques, such as crystalline structure by XRD, morphology by SEM, and TEM and AFM, which is discussed elsewhere. We only emphasize what one should pay attention to when a physicochemical property is under investigation for hazard assessment by using common and mature techniques.

Equipped with energy dispersive spectrometer (EDS) microanalysis system, SEM/EDS, and TEM/EDS can locally obtain the elemental composition of NMs, that is, their relative proportions (atomic% for specimen). EDS can be used as a semiquantitative mode to determine chemical composition by peak-height ratio relative to a standard. In an EDS spectrum, there may be energy peak overlaps among different elements, particularly those corresponding to x-rays generated by emission from different energy-level shells (K, L, and M) in different elements. EDS cannot detect the lightest elements, typically below the atomic number of Na, by detectors equipped with a Be window. Despite the high spatial resolution, the sensitivity of the EDS technique is lower than that of XPS and the detection limit is about 0.1 wt%, which prevents the detection of trace impurities. Thus, for the testing of low concentration elements (less than 1 wt%), the accuracy and precision is about 10%. Electron backscatter diffraction (EBSD), an accessory system of SEM, provides quantitative microstructural information about the crystallographic nature, such as grain size, grain boundary character, grain orientation, texture, and the phase identity of a specimen. In the case of the measurement of poor-conduction specimens by SEM and TEM, a conducting layer (2–5 nm thick) of Au, Pt, or C is needed to solve the problem of the charge-up effect for the imaging.

XPS utilizes photoionization (using soft x-rays with a photon energy of 200–2000 eV to examine core levels) and the analysis of the kinetic energy distribution of emitted photoelectrons to study the composition and electronic states of the surface region of a sample. XPS is a quantitative spectroscopic technique that measures the elemental composition, empirical formula, chemical state, and electronic state of the elements that exist at a surface, generally from the topmost 1–12 nm of a material. While XPS is a surface-sensitive technique, a depth profile of the sample in terms of XPS quantities can be obtained by combining a sequence of ion gun etching cycles interleaved with XPS measurements from the current surface. The detection limit is approximately 0.1–1.0 at% (0.1 at% = 1 part per thousand = 1000 ppm), and the minimum analysis area ranges from 1 to 200 µm. Importantly, XPS allows for the determination of the valence band structure of a specimen. We recently studied the cytotoxicity of SiO_2, Al_2O_3, CeO_2, ZnO, CuO, Al-doped ZnO, and Al-doped CuO NPs [3]. We found that the cytotoxicity in the A549 and NIH3T3 cell lines was associated with the electronic properties of the nano-oxides by measuring the density of states (DOS) near Fermi level of the nano-oxides with core-level XPS. Although there is other evidence showing the influence of electronic properties of metal oxides [12] and carbon nanotubes [22] on the toxicological potential [23] at the cellular and whole animal levels, the effect of the electronic characteristics of NMs on biological systems has not yet gained great attention. It is known that most of the intracellular and *in vivo* toxicities from NMs arise from the production of excess reactive oxygen species (ROS) [24,25]. ROS and other radicals are involved in a variety of biological phenomena, such as mutation, carcinogenesis, degenerative diseases, inflammation, aging, and development [26]. ROS are well recognized for playing a dual role as deleterious and beneficial species. It is also known that many important biological processes involve redox reactions. For example, the process of cell respiration also depends heavily on the reduction of NAD^+ to NADH and the reverse reaction (the oxidation of NADH to NAD^+). All cellular membranes are especially vulnerable to oxidation due to their high concentrations of unsaturated fatty acid. The redox potential is defined as the ratio between the oxidant and reductant, for example, ROS and scavengers [26]. The redox state of a biological system is kept within a narrow range under normal conditions—similar to the manner in which the biological system regulates its pH parameter. Under pathological conditions, the redox state can be changed toward lower (redosis) or higher (oxidosis) values. Although redox is a thermodynamic parameter, reducing power is not and, therefore, can be calculated in biological systems to supply valuable information concerning cellular responses to oxidative stress. This parameter represents and encompasses the overall capability of the cell, biological fluid, or tissue to donate electrons (oxidation potential) and the overall concentration of the reducing equivalents responsible for this ability. Therefore, *electrons* play an important role in biological redox reactions, and all the factors that influence electron concentrations, electron transfers, and electron transport may have effects on biological redox reactions. The links connecting the electronic properties of NMs to toxic effects need more investigations.

Belonging to the same family as scanning probe microscopy, AFM is one of the most powerful tools in nanotechnology. It enables us to quickly obtain three-dimensional, topographic images of various materials and structures with a resolution at the nanoscale or near the atomic level. AFM is extensively used in a wide range of disciplines, such as materials science, solid-state physics, electronics, and life science [15,27], for academic research as well as for industrial fabrication and inspection. AFM can be operated in ambient air and in liquid; it works under extreme circumstances, such as variable temperature, high magnetic field, and ultrahigh vacuum. Furthermore, AFM has evolved into one of the most important tools in nanotechnology with its unmatched capabilities of monitoring biological events in their native environments [28]. AFM forms images by sliding a sharp tip over a sample surface through a raster scanner and subsequently recording relevant signals. Because each tip has its own three-dimensional shape with a finite size, the acquired image does not reflect the actual shape of the specimen but a dilation or convolution of the sample topography, dependent on the shape of the tip [13]. This issue will become significant when the dimensions of NMs are comparable to the radius of curvature of the tip's apex (normally about 5–10 nm) and, thus, the tip effect should be taken into account if accurate size information is critical for the study. An ultrasharp AFM tip can be used as a simple way to mitigate the tip effect [13]. In spite of the tip effect, AFM allows for the study of electrical [29], magnetic [30], and other properties of NMs.

One of the origins of toxicities in NMs is assumed to be the small size feature of NMs. Thus, the availability of techniques for the measurement of size (size distribution, agglomerates/aggregates) is crucial for nanotoxicity studies. Several advanced techniques are available as different techniques based on different measurement principles. However, these may yield different results when measuring the very same object, and the measured results are normally not comparable [31,32].

The methods for the measurement of size can be divided into single NM methods, such as SEM, TEM, and AFM, and ensemble methods, such as dynamic light scattering (DLS) [11]. The advanced single NM methods can, simultaneously, visualize and more accurately measure the size of the original NMs if measurement parameters are more optimal than other methods [13]. However, for polydisperse NMs, statistical significance is a problem since the number of individual NMs in an acquired high-resolution image is very limited. This issue can be addressed by acquiring multiple images and using automated image recognition software for the determination of size. DLS is a conventional method for measuring the size distribution in a suspension. DLS measures the Brownian motion of NPs and relates this movement to an equivalent hydrodynamic diameter, with the motion of smaller particles being overestimated. In reality, DLS measures the time-dependent fluctuations in scattering intensity caused by the constructive and destructive interference resulting from the relative Brownian movements of the NPs within a sample [33]. This is an indirect method for the calculation of the size by means of a physical model. The calculated equivalent hydrodynamic diameters are not compared with the particle sizes measured by direct methods, such as SEM, TEM, and AFM. The DLS measurement requires information of the dispersant's refractive index in which the NMs are dispersed. This is a big problem for cell culture media because there is little information on the refractive index of these media. Operated in accordance with ISO 13321 and ISO 22412, DLS cannot distinguish between different types of particles in a suspension, and becomes significantly less reliable for providing a wide particle size distribution because of the sixth power dependence of the light scattering intensity on the sizes of the scattering particles. Instead, the method produces a single average value and a number (the polydispersity index) indicating the polydispersity of the particle population [19]. Thus, DLS is more suitable only for the measurement of monodisperse particles in low-viscosity suspension, but not for polydisperse particles. To get a better size distribution characterization of polydisperse particles in a suspension, NP tracking analysis (NTA) (size distribution between 20 and 1000 nm, relying on the refractive index of particles and the viscosity of solvent) [32] or disk centrifugation (size distribution between about 5 nm and about 75 μm) [34] is preferable. NTA is a method for visualizing and analyzing particles in liquids that relates the rate of Brownian motion to particle size. The rate of movement is related only to the viscosity of the liquid, the temperature, and size of the particle, and is not influenced by the particle density or refractive index. The NTA system visualizes particles as point scatters moving under Brownian motion, but this is not a resolved image and, thus, no morphology information can be provided, as this scatter is isotropic. NTA counts particles by simultaneously video tracking many individual particles and relating the Brownian movement to a particle size according to the formula derived from the Stokes–Einstein equation [35]. However, there are problems with this particle-tracking approach. To obtain an accurate individual radius measurement, an accurate measurement of a particle's mean-squared movement must be made. As each measured movement is a random variable, this necessitates measuring a large number. Thus, ideally, a particle needs to be tracked over many frames [36]. NTA can be time consuming and requires some operational skills for the adjustment of all software settings. Recently, Walker [36] reported an alternative data-processing method to recover a particle size distribution from NP tracking data. NTA has an advantage over DLS, particularly for polydisperse samples, in that it does not, in principle, suffer from an *intensity weighting* effect, as large and small particles will usually be imaged to different regions of the camera detector array. Disk centrifugation, also called differential centrifugal sedimentation (DCS), applies rotation speeds of up to 45,000g. The centrifuge, in the form of a disk and filled with a density gradient fluid, is orientated in a vertical direction for analytical purposes [38]. The size of the particles is calculated following Stokes' law, which requires knowledge of the particle's properties such as the density of particles [39]. External or internal calibration standards are required to determine particle size [40]. DSL, NTA, and DCS are all indirect and ensemble methods; among the analytical techniques [37,40], electron microscopy is presently the most accurate technique to characterize the geometric properties of NMs.

8.2.3 Characterization in Biological Matrices

At present, to analyze NMs in the environment [41–43] and cosmetics [44], a generally used scheme includes isolation, purification, and concentration steps (if necessary) before the identification of the NMs contained in the original matrices [37]. There are many methods for separating NMs from the complex surroundings. These techniques include ultrafiltration, nanofiltration, dialysis, field-flow fractionation (FFF), and size exclusion chromatography (SEC) [37,40]. Once extracted from the complex matrices, the NMs can be characterized by the conventional analytical techniques addressed above. However, the state of characterized NMs that are isolated by the schemed processing is certainly different from that in complex matrices, and the analysis is extremely susceptible to artifacts; even chemical compositions may be altered. Such *ex situ* analysis provides less information to understand the biological behaviors of NMs for nanotoxicology.

The characterization of NMs in a biological matrix is more difficult and challenging as compared to the characterization of primary NMs, because the surrounding environment can have a substantial impact on the behaviors of NMs. The physicochemical properties of NMs in biological environments tend to change with time. The interaction of NMs with biological surroundings is especially important for the hazard assessments of NMs. It is well known that in a biological environment, NMs can adsorb proteins via physical or chemical interactions, forming a protein corona. The formation of the protein corona dynamically alters the size and surface composition of the NM [34]. This gives a biological identity of NMs distinct from the original ones. The protein corona's behavior is believed to be very important for the physiological response, including signaling, kinetics, transport, uptake, transport, accumulation, and toxicity [2,45]. It has been suggested that the NM itself does not constitute the effective unit for interaction with biological systems but, rather, it is the combination of the NM with its specific protein corona that reacts with the biological system. Compared to the intensive study of the protein corona, there are a few reports on the interaction of NMs in the biological environment with the ions in the biological system [1,46]. The effective surface charge of a NM will be changed due to the consequences of the interactions with the biological matrix. Thus, deliberate attempts at surface modifications in NMs may not show effects as expected. Owing to these matrix effects, physicochemical properties, such as size, agglomeration/aggregation states, surface charge and coatings, dissolution, and other related properties, may change in different solvents, test media, and biological environments. Characterizations in a biological matrix should provide information on any changes in the NM's characteristics in order to make correlations with its biological outcome.

8.2.4 Characterization of NMs in Biological Matrices by Electron Microscopy

By TEM and XPS, we found that ZnO and CuO NPs can selectively adsorb ions in biological environments, forming a bio-nanomaterial complex, also called an ion corona [1]. Owing to the high energy of the electron beam used in the TEM study (200 keV), we did not detect signals from key components of serum proteins, such as N, by the EDS technique. However, we did observe the disappearance of certain matters during the TEM observation because biomolecules can be easily damaged by high-energy electron beams, and detected an N signal by XPS. This is the first clear observation of the adsorption of various ions, not biological molecules, by NMs, after being dispersed in biological matrices. Without an elemental identification, this kind of ion corona may be mistakenly considered as a protein corona. Walczyk et al. [45] studied the proteins surrounding modified polystyrene and silica particles by TEM operated 80 keV as well as other relevant techniques. We believe that proteins and ions are mingled together when surrounding NMs in biological environments, and that the identification of the constituents of nanomaterial-bio complexes will significantly improve our understanding of what is selectively acquired by NMs in biological environments; how those acquired constituents influence the reactivity, transport, transformation, bioavailability, and toxicity of NMs; and how the NMs interact with living cells.

Confocal fluorescence microscopy is one of the most common ways to monitor the uptake and localization of fluorescently labeled NMs. However, not all NMs can be easily fluorescently labeled. Furthermore, the label may be released from the NMs into cells and, thus, the distribution of fluorescence within the cells does not necessarily represent the presence or subcellular distribution of the NMs. It is also possible that the label may change the biological behaviors of NMs.

TEM is frequently used to investigate the uptake, aggregation/agglomeration, and location of NMs in cells [4] as a function of the size, shape, surface modification, or surface charge. For instance, Gratton et al. [47] reported the internalization of monodisperse hydrogel particles into HeLa cells as a function of size, shape, and surface charge. It was found that HeLa cells can internalize nonspherical particles with dimensions as large as 3 μm by using different mechanisms of endocytosis.

By TEM, Belade et al. [48] showed TiO_2 and carbon black NPs were widely and rapidly accumulated in 16HBE bronchial epithelial cells and MRC5 fibroblasts. Moreover, the NMs accumulated chiefly as aggregates in cytosolic vesicles and were absent from the mitochondria or nuclei. Despite similar accumulation patterns, TiO_2 aggregates had a higher size than carbon black aggregates. Interestingly, they found that the intracellular NM accumulation was dissociated from the observed cytotoxicity. This is common; the uptake of NMs into cells does not always cause toxic effects. In spite of the high spatial resolution of conventional electron microscopy, the sample preparation for electron microscopy studies is very complicated, including fixation, drying, sectioning, and staining [4,49].

An alternative option is cryo-TEM and cryo-SEM, where rapid freezing of the sample allows for the analysis in a state similar to the hydrated state in the original solution [16,50]. The biological sample is spread on an electron microscopy grid and is preserved in a frozen-hydrated state by rapid freezing, usually plunge freezing in liquid ethane near the temperature of liquid nitrogen [51]. By maintaining specimens at the liquid nitrogen temperature or colder, they can be introduced into the high vacuum for measurements at variable temperatures. Thus, plunge freezing, cryo-transfer, and imaging have the potential to produce a representative view of NMs in biological matrices. Cryo-SEM can be applied to investigate the fine structures of bulk samples. Win et al. [52] studied the effects of particle sizes and surface modifications on cellular uptake by cryo-SEM. The acquired image clearly indicates that some NPs were found throughout the endoplasm and around the nucleus of Caco-2 cells, and some were adsorbed on the cell membrane. In spite of the requirement of a number of sample preparation steps and time-consuming measurements, the cryo-SEM and cryo-TEM paves the way for a high-resolution image of aqueous specimens.

8.2.5 Characterization of NMs in Biological Matrices by Raman Spectroscopy

Raman spectroscopy is well established for chemically identifying materials by sensing the vibration of chemical bonds, rather than performing elemental analyses. Raman spectroscopy is a laser-based optical, label-free technique that excites vibrations of molecular bonds in a material. It relies on the inelastic scattering, or Raman scattering, of a laser. The laser light interacts with molecular vibrations, phonons, or other excitations in the system, resulting in the energy of the laser photons being shifted up or down. The shift in energy gives information about the vibrational modes in the system. Raman spectrum serves as a *molecular fingerprint* of a material, yielding information on molecular bonds, conformations, and intermolecular interactions.

The Raman spectroscopy technique has been employed for biomedicine applications [53]. Each biomolecule has a unique "fingerprint" of Raman peaks at well-defined frequencies, as they contain a variety of molecular bonds (e.g., C–H, C=C, O–H, aromatic ring). Frequency shifts are recorded in wavenumbers (cm^{-1}), and biological molecules have vibrations in the range of about 600–3000 cm^{-1}. Thus, probing the change in the compositional and structural characteristics of a biological system by the Raman spectroscopy technique can provide information on the state of the biological system under investigation. Through enhancements with noble metal NPs, such as Au and Ag, surface-enhanced Raman scattering (SERS) [54] has been applied to the detection of biomolecules such as DNA [55], DNA/RNA mononucleotides [56], and proteins [57], to the labeling of cells [58,59] and tissues [60], and to the monitoring of apoptotic processes [61]. Furthermore, the development of the high-speed, confocal, Raman microscope system [14,62,63] allows for the transient imaging of live cells and the tracking of NMs within the cells.

By detecting the Raman features of NMs in biological matrices, it is possible to investigate the behaviors of NMs, such as graphene [14]; carbon nanotubes [62–64]; polymer NMs [65]; and Au, Ag, and SiO_2 NPs, due to their high Raman signals, resulting from high Raman cross sections, in biological systems. It is possible for Raman spectroscopy to identify the proteins and ions adsorbed by NMs without isolation in biological environments, despite a few report so far. The combined system of Raman spectroscopy with AFM would provide other functions for the analysis of NMs in a biological matrix. Monitoring the changes in the Raman signatures of cells or subcellular compartments can provide information on cell metabolism due to interactions with NMs.

Liu et al. [64] performed *ex vivo* investigations of the long-term fate of polyethyleneglycol-functionalized single-walled carbon nanotubes (SWCNTs) intravenously injected into mice. Raman spectroscopy and Raman mapping were used to probe the blood circulation behavior and biodistribution of the functionalized SWCNTs in various organs of the mice by exploiting the characteristic resonant Raman spectrum of SWCNTs. The results showed the presence of the functionalized SWCNTs in the intestine, feces, kidney, and bladder of mice, suggesting the excretion and clearance of SWCNTs from mice via the biliary and renal pathways. No toxic effects from the functionalized SWCNTs to mice were observed in necropsy, histology, and blood chemistry measurements despite the long retention time, which is mostly due to the surface functionalization [66].

Andersson et al. [67] investigated the uptake and distribution of TiO_2 NPs (Figure 8.1) with differences in size, shape, and crystal structure in the A549 cell line by TEM and Raman spectroscopy. NP retention was found in the vicinity of organelles and the nucleus. The uptake of TiO_2 NPs (Figure 8.2) was found to be slow, endosomal-particle uptake behavior, and strongly dependent on the hard, agglomeration size rather than the primary particle size, which quantitatively agreed with the measured intracellular oxidative stress.

A major advantage of Raman spectroscopy for the imaging and characterization of NMs in biological tissues lies in the possibility of using wavelengths in the near-infrared range for excitation. The long penetration depth and the low autofluorescence in this spectral region allows for the Raman imaging of whole, small animals [53]. Zavaleta et al. [68] reported the imaging of SWCNTs in tumors in nude mice at a depth of 2 mm in the animal by monitoring the G-band peak (~1593 cm^{-1}) of SWCNTs. *In vivo* imaging enables the observation of the same animal over time. One shortcoming of this technique is the poor spatial resolution (about 350 nm) focused by an ×100 optical lens with numerical aperture of 0.9 [62], for finding NMs by the low-resolution optical microscope system and for detections by the large laser spot size. The Raman spectroscopy technique enables one to obtain the spatially resolved chemical fingerprinting of both NMs and biological components in living cells and other biological organisms, which makes it possible to classify both the distribution of NMs and their impact on biological systems. One disadvantage of the Raman technique is the relatively low spatial resolution; the system normally uses optical microscope to observe a sample.

8.2.6 Characterization of Dissolution of NMs in a Biological Matrix

The release of metal cations from NMs in the cell culture medium and the role of the metal cations in cytotoxicity are still unclear. Many studies have attributed the toxic effects of metal-based NPs primarily to released metal cations [9,69]. For example, Zn^{2+} ions were assumed to release from ZnO particles, as they were dispersed into cell culture medium or outside cells [70–72]. Consequently, the toxic effect of ZnO NPs was mainly attributed to the dissolved Zn^{2+} ions in the cell culture medium [69,70]. By contrast, it was reported that the majority of ZnO NPs did not dissolve in bronchial epithelial growth medium (BEGM) [73]. Furthermore, it was clearly shown that the toxicity of ZnO NPs to human colon-derived RKO cells was independent of the amount of soluble Zn^{2+} ions in the cell culture medium [10]. In the case of CuO NPs, there are also contradicting reports of the contribution

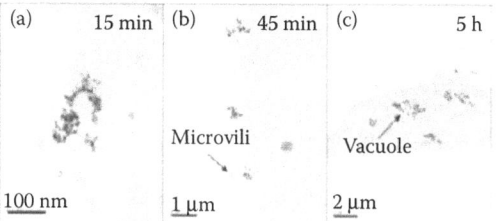

FIGURE 8.1 TEM images show time dependence of TiO$_2$ uptake (10 μg/mL) by A549 cell lines following (a) 15 min, (b) 45 min, and (c) 5 h incubation. (Reprinted with permission from Andersson, P. A., Lejon, C., Ekstrand-Hammarstrom, B. et al. 2011. *Small* 7: 514–523.)

of released Cu^{2+} to the observed toxicity [74–76]. The adverse biological effects of CuO NMs [12] and nano-sized Ag [77] were explained by their solubility, based on data analysis by measurements using inductively coupled plasma–mass spectrometry (ICP–MS) relevant techniques or simply on the comparable toxicity to Cu salts [76,78] such as CuSO$_4$. By contrast, it was reported that the cytotoxic effects related to the released copper fraction were found to be significantly lower than the effects related to CuO particles [79,80]. There is no directly analytical technique to characterizing the dissolution of NMs in a cell culture medium. At present, the "free released ion concentration" is commonly determined by using ICP-based techniques [81,82] to measure the supernatant collected

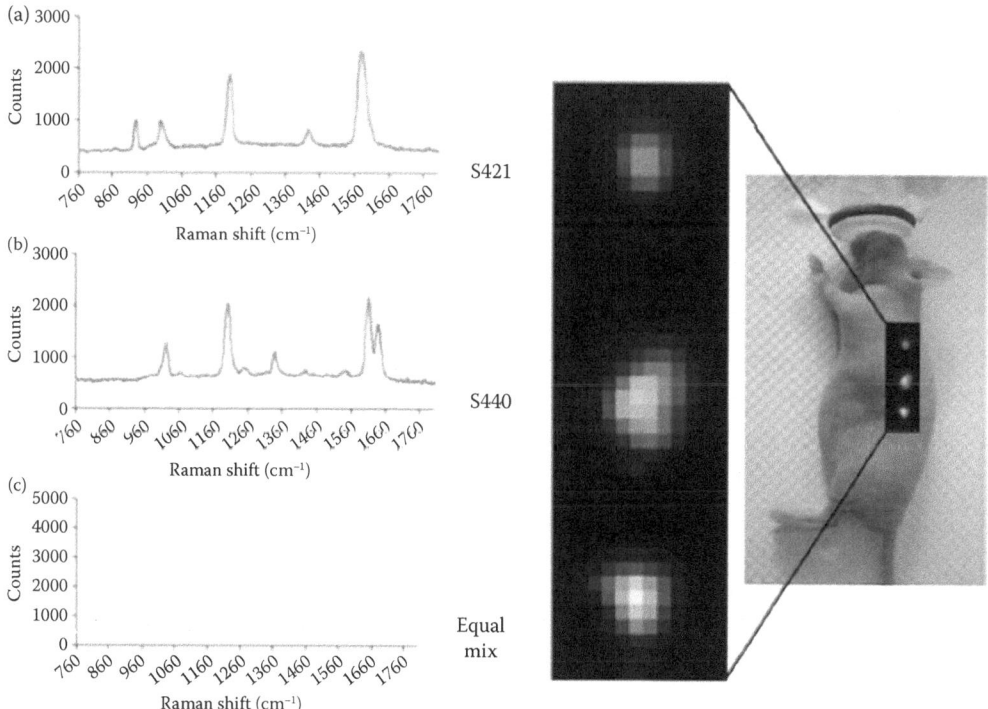

FIGURE 8.2 Raman spectroscopy study of Au NPs coated with a layer of Raman-active material (S421 and S440, Nanoplex Biotags, Oxonical). (a) Raman spectroscopy acquired from first s.c. injection of S421 NPs, (b) Raman spectroscopy acquired from second s.c. injection of S440, and (c) Raman spectroscopy acquired from third s.c. injection of an equal mix of S421 and S440. In the right panel, the image designated "Equal mix" was calculated by the analysis software to represent an equal mix of the S421 and S440 NPs in the map. (Reprinted with permission from Keren, S. et al. *Proc. Natl. Acad. Sci. USA* 105: 5844–5849.)

after the NM suspension is centrifuged. However, it deserves to be pointed out that the elemental concentration obtained by ICP-based techniques can be affected by free ions and by the ions that result from the decomposition of matter at the very high temperatures and/or acid environments present for an ICP-based measurement. That is, after the sample is injected for ICP-based measurement, the plasma's extreme temperature or strong acid causes the sample to separate into individual atoms (atomization); these atoms are then ionized (M $\rightarrow$ M$^+$ + e$^-$) so that they can be detected by the mass spectrometer. As a result, the digestion procedures involved in ICP-based techniques make it almost impossible to differentiate the ions formed as a result of the dissolution of materials from the NMs per se. We found by TEM that the removal of NMs, such as ZnO NPs, by centrifugation, the procedure currently used for the estimation of the released ion concentration, was incomplete even at a relative centrifugal force of 150,000g [83]. Thus, the Zn concentration in supernatant as measured by ICP-based methods cannot be regarded as the concentration of free Zn^{2+} ions, which were released from ZnO NPs in cell culture medium. We also found that the toxic contribution of Zn^{2+} ions released from ZnO NPs to the A549 cell lines was estimated to be only about 10%.

There are other techniques used to study NMs in a biological matrix. Synchrotron microfocused x-ray fluorescence (μ-XRF) and micro-x-ray absorption near-edge structure (μ-XANES) were used to study the chemical form and localization of titanium in cucumber plants treated with TiO_2 NPs [84], the form of CeO_2 in the roots of corn seedlings [85], and the forms of Zn and Ce within soybean tissues that were obtained from the soybeans grown in organic farm soil spiked with ZnO or CeO_2 NPs [86]. A quantitative photoacoustic technique is being developed to image NMs in cells and tissues [87]. Multifocal two-photon microscopy was used to track gold nanorods in cells [88]. However, the resolution of the μ-XRF and μ-XANES is generally at the micrometer level. The signals of the other two techniques are mostly dependent upon the characteristics of NMs, nonplasmonic NMs for photoacoustic imaging, and plasmonic NMs for two-photon microscopy.

8.3 CONCLUSION

The characterization of NMs in a biological matrix, while preserving the spatial information and chemical components, is very challenging. An inappropriate analytical method can strongly influence the interpretation of the experiment, and does not provide information on the direct relevance from a toxicological perspective. The further progress of nanotoxicology requires breakthroughs in analytical methods for the accurate tracking, detecting, and identifying of NMs as well as their transformation in biological matrices and interactions with biological systems.

In this chapter, we have proposed a nomenclature for NMs collected from the existing literature and based on our own experience for the accurate description and characterization of NMs. We have emphasized the advantages and shortcomings of the techniques commonly used for the characterizations of size, shape, chemical composition, and electronic/optical properties of NMs in a dry form. There is no direct method for the investigation of the dissolution of NMs in biological environments. At this moment, we believe that electron microscopy and Raman spectroscopy are among the most appropriate techniques for the characterization of NMs in biological matrices.

Hopefully, this chapter is a useful guide to the choice of analytical methods to characterize NMs, not only in a dry form, but also in a biological matrix, and mostly promoting the development of new techniques and methods for the *in situ* characterization of NMs in biological matrices.

ACKNOWLEDGMENTS

This work was partially supported by the Program of International S & T Cooperation (No. 2013DFG52800), by Zhejiang Provincial Natural Science Foundation (Youth Talent Program: R4110030), Science and Technology Department of Zhejiang Provincial (Qianjiang Talent Program: 2011R10077), by Program for New Century Excellent Talents in University (NCET-12-0494) of China, by the Interdisciplinary Laboratory for Nanoscale Science and Technology of National

Institute for Materials Science (NIMS), Japan, and by the MEXT Program for the "Development of Environmental Technology using Nanotechnology," Japan.

REFERENCES

1. Xu, M. S., Li, J., Iwai, H. et al. 2012. Formation of nano-bio-complex as nanomaterials dispersed in a biological solution for understanding nanobiological interactions. *Sci. Rep.* 2: 406.
2. Walkey, C. D. and Chan, W. C. W. 2012. Understanding and controlling the interaction of nanomaterials with proteins in a physiological environment. *Chem. Soc. Rev.* 41: 2780–2799.
3. Xu, M. S., Fujita, D., Kajiwara, S. et al 2010. Contribution of physicochemical characteristics of nano-oxides to cytotoxicity. *Biomaterials* 31: 8022–8031.
4. Hanagata N., Zhuang, F., Connolly, S., Li, J., Ogawa, N., and Xu, M. S. 2011. Molecular responses of human lung epithelial cells to the toxicity of copper oxide nanoparticles inferred from whole genome expression analysis. *ACS Nano* 5: 9326–9338.
5. Reddy, L. H., Arias, J. L., Nicolas, J., and Couvreur, P. 2012. Magnetic nanoparticles: Design and characterization, toxicity and biocompatibility, pharmaceutical and biomedical applications. *Chem. Rev.* 112: 5818–5878.
6. Sharifi, S., Behzadi, S., Laurent, S., Forrest, M. L., Stroeve, P., and Mahmoudi, M. 2012. Toxicity of nanomaterials. *Chem. Soc. Rev.* 41: 2323–2343.
7. Albanese, A., Tang, P. S., and Chan, W. C. W. 2012. The effect of nanoparticle size, shape, and surface chemistry on biological systems. *Annu. Rev. Biomed. Eng.* 14: 1–16.
8. Griffitt, R. J., Luo, J., Gao, J., Bonzongo, J. C., and Barber, D. S. 2008. Effects of particle composition and species on toxicity of metallic nanomaterials in aquatic organisms. *Environ. Toxicol. Chem.* 27: 1972–1978.
9. Vandebriel, R. J. and De Jong, W. H. 2012. A review of mammalian toxicity of ZnO nanoparticles. *Nanotech. Sci. Appl.* 5: 61–71.
10. Moos, P. J., Chung, K., Woessner, D., Honeggar, M., Shane Cutler, N., and Veranth, J. M. 2010. ZnO particulate matter requires cell contact for toxicity in human colon cancer cells. *Chem. Res. Toxicol.* 23: 733–739.
11. Bouwmeester, H., Lynch, I., Marvin, H. J. P. et al. 2011. Minimal analytical characterization of engineered nanomaterials needed for hazard assessment in biological matrices. *Nanotoxicology* 5: 1–11.
12. Zhang, H. Y., Ji, Z. X., Xia, T. et al. 2012. Use of metal oxide nanoparticle band gap to develop a predictive paradigm for oxidative stress and acute pulmonary inflammation. *ACS Nano* 6: 4349–4368.
13. Xu, M. S., Fujita, D., and Onishi, K. 2009. Reconstruction of atomic force microscopy image by using nanofabricated tip characterizer toward the actual sample surface topography. *Rev. Sci. Instrum.* 80: 043703.
14. Xu, M. S., Fujita, D., Sagisaka, S., Watanabe, E., and Hanagata, N. 2011. Production of extended single-layer graphene. *ACS Nano* 5: 1522–1528.
15. Ramachandran, S., Arce, F. T., and Lal, R. 2011. Potential role of atomic force microscopy in systems biology. *Wiley Interdisciplinary Reviews: Systems Biology and Medicine* 3: 702–716.
16. Dudkiewicz, A., Tiede, K., Loeschner, K. et al. 2011. Characterization of nanomaterials in food by electron microscopy. *Trend. Anal. Chem.* 30: 28–43.
17. Xu, M. S., Fujita, D., Gao, J. H., and Hanagata, N. 2010. Auger electron spectroscopy: A rational method for determining thickness of graphene films. *ACS Nano* 4: 2937–2945.
18. Lesniak, A., Fenaroli, F., Monopoli, M. P., Aberg, C., Dawson, K. A., and Salvati, A. 2012. Effects of the presence or absence of a protein corona on silica nanoparticle uptake and impact on cells. *ACS Nano* 6: 5845–5857.
19. Lovestam, G., Rauscher, H., Roebben, G. et al. 2010. *Considerations on a Definition of Nanomaterial for Regulatory Purposes*. Luxembourg: Publications Office of the European Union. DOI:10.2788/98686.
20. http://www.iso.org/iso/home.htm.
21. https://www.iso.org/obp/ui/.
22. Vecitis, C. D., Zodrow, K. R., Kang, S., and Elimelech, M. 2010. Electronic-structure-dependent bacterial cytotoxicity of single-walled carbon nanotubes. *ACS Nano* 4: 5471–5479.
23. Burello, E. and Worth, A. P. 2011. A theoretical framework for predicting the oxidative stress potential of oxide nanoparticles. *Nanotoxicology* 5: 228–235.
24. Nel, A., Xia, T., Madler, L., and Li, N. 2006. Toxic potential of materials at the nanolevel. *Science* 311: 622–627.

25. Moller, P., Jacobsen, N. R., Folkmann, J. K. et al. 2010. *Free Radical Res.* 44: 1–46.
26. Kohen, R. and Nyska, A. 2002. Oxidation of biological systems: Oxidative stress phenomena, antioxidants, redox reactions, and methods for their quantification. *Toxicol. Pathol.* 30:620–650.
27. Giessibl, F. J. 2003. Advances in atomic force microscopy. *Rev. Mod. Phys.* 75: 949–983.
28. Hinterdorfer, P. and Dufrene, Y. F. 2006. Detection and localization of single molecular recognition events using atomic force microscopy. *Nat. Meth.* 3: 347–355.
29. Xu, M. S., Pathak, Y., Fujita, D., Ringor, C., and Miyazawa, K. 2008. Covered conduction of individual C_{60} nanowhiskers. *Nanotechnology* 19: 075712.
30. Mohamed, H. D. A., Watson, S. M. D., Horrocks, B. R., and Houlton, A. 2012. Magnetic and conductive magnetite nanowires by DNA-templating. *Nanoscale* 4: 5936–5945.
31. Hasseloev, M. and Kaegi, R. 2009. Analysis and characterization of manufactured nanoparticles in aquatic environments. *Environmental and Human Health Impact of Nanotechnology*, Lead, J. R. Ed., Smith, E. Blackwell Publishing Ltd.
32. Domingos, R., Ballousha, M. A., Nam, Y. et al. 2009. Characterizing manufactured nanoparticles in the environment: Multimethod determination of particle sizes. *Environ. Sci. Technol.* 43: 7277–7284.
33. Brar, S. K. and Verma, M. 2011. Measurement of nanoparticles by light-scattering techniques. *Trend. Anal. Chem.* 30: 4–17.
34. Monopoli, M. P., Walczyk, D., Campbell, A. et al. 2011. Physical-chemical aspects of protein corona: Relevance to *in vitro* and *in vivo* biological impacts of nanoparticles. *J. Am. Chem. Soc.* 133: 2525–2534.
35. Filipe, V., Hawe, A., and Jiskoot, W. 2010. Critical evaluation of nanoparticle tracking analysis (NTA) by NanoSight for the measurement of nanoparticles and protein aggregates. *Pharm. Res.* 27: 796–810.
36. Walker, J. G. 2012. Improved nano-particle tracking analysis. *Meas. Sci. Technol.* 23: 065605.
37. Hassellov, M., Readman, J. M., Ranville, J. F., and Tiede, K. 2008. Nanoparticle analysis and characterization methodologies in environmental risk assessment of engineered nanoparticles. *Ecotoxicology* 17: 344–361.
38. Laidlaw, I. and Steinmetz, M. 2005. Introduction to differential sedimentation. *Analytical Ultracentrifugation: Techniques and Methods*, Scott D. J., Harding S. E., Rowe A.J., Eds. 1st ed. Cambridge: The Royal Society of Chemistry, pp 270–290.
39. Thomas, J. C., Middelberg, A. P., Hamel, J. F., and Snoswell, M. A. 1991. High-resolution particle size analysis in biotechnology process control. *Biotechnol. Prog.* 7: 377–379.
40. Zolls, S., Tantipolphan, R., Wiggenhorn, M., Winter, G., Jiskoot, W., Friess, W., and Hawe, A. 2012. Particles in therapeutic protein formulations, Part 1: Overview of analytical methods. *J. Pharm. Sci.* 101: 914–935.
41. von der Kammer, F., Ferguson, P. L., Holden, P. A. et al. 2012. Analysis of engineered nanomaterials in complex matrices (environment and biota): General considerations and conceptual case studies. *Environ. Toxicol. Chem.* 31: 32–49.
42. Reed, R. B., Higgins, C. P., Westerhoff, P., Tadjiki, S., and Ranville, J. F. 2012. Overcoming challenges in analysis of polydisperse metal-containing nanoparticles by single particle inductively coupled plasma mass spectrometry. *J. Anal. At. Spectrom.* 27: 1093–1100.
43. Gray, E. P., Bruton, T. A., Higgins, C. P., Halden, R. U., Westerhoff, P., and Ranville, J. F. 2012. Analysis of gold nanoparticle mixtures: A comparison of hydrodynamic chromatography (HDC) and asymmetrical flow field-flow fractionation (AF4) coupled to ICP-MS. *J. Anal. At. Spectrom.* 27: 1532–1539.
44. SCCS/1448/12 (Scientific Committee on Consumer Safety), Guidance on safety assessment of nanomaterials in cosmetics, June 26–27, 2012.
45. Walczyk, D., Bombelli, F. B., Monopoli, M. P., Lynch, I., and Dawson, K. A. 2010. What the cell "see" in bionanoscience. *J. Am. Chem. Soc.* 132: 5761–5768.
46. Azizeh-Mitra Yousefi, A. M., Zhou, Y. L., Querejeta-Fernández, A., Sun, K., and Kotov, N. A. 2012. Streptavidin inhibits self-assembly of CdTe nanoparticles. *J. Phys. Chem. Lett.* 3: 3249–3256.
47. Gratton, S. E. A., Ropp, P. A., Pohlhaus, P. D. et al. 2008. The effect of particle design on cellular internalization pathways. *Proc. Natl. Acad. Sci. USA* 105: 11613–11618.
48. Belade, E., Armand, L., Matinon, L. et al. 2012. A comparative transmission electron microscopy study of titanium dioxide and carbon black nanoparticles uptake in human lung epithelial and fibroblast cell lines. *Toxicol. In Vitro.* 26: 57–66.
49. Javier, A. M., Kreft, O., Semmling, M. et al. 2008. Uptake of colloidal polyelectrolyte-coated particles and polyelectrolyte multilayer capsules by living cells. *Adv. Mater.* 20: 4281–4287.
50. Kuntsche, J., Horst, J. C., and Bunjes, H. 2011. Cryogenic transmission electron microscopy (cryo-TEM) for studying the morphology of colloidal drug delivery systems. *Int. J. Pharm.* 417: 120–137.

51. Hondow, N., Brydson, R., Wang, P. Y. et al. 2012. Quantitative characterization of nanoparticle agglomeration within biological media. *J. Nanopart. Res.* 14: 977.
52. Win, K. Y. and Feng, S. S. 2005. Effects of particle size and surface coating on cellular uptake of polymeric nanoparticles for oral delivery of anticancer drugs. *Biomaterials* 26: 2713–2722.
53. Drescher, D. and Kneipp, J. 2012. Nanomaterials in complex biological systems: Insights from Raman spectroscopy. *Chem. Soc. Rev.* 41: 5780–5799.
54. Downes, A. and Elfick, A. 2010. Raman spectroscopy and related techniques in biomedicine. *Sensors* 10: 1871–1889.
55. Braun, G., Lee, S. J., Dante, M. et al. 2007. Surface enhanced Raman spectroscopy for DNA detection by nanoparticle assembly onto smooth metal films. *J. Am. Chem. Soc.* 129: 6378–6379.
56. Bell, S. E. J. and Sirimuthu, N. M. S. 2006. Surface-enhanced Raman spectroscopy (SERS) for submicromolar detection of DNA/RNA mononucleotides. *J. Am. Chem. Soc.* 128: 15580–15581.
57. Bizzarri, A. R. and Cannistraro, S. 2007. SERS detection of thrombin by protein recognition using functionalized gold nanoparticles. *Nanomed. Nanotechnol. Biol. Med.* 3: 306–310.
58. Kim, J. H., Kim, J. S., Choi, H. et al. 2006. Nanoparticle probes with surface enhanced Raman spectroscopic tags for cellular cancer targeting. *Anal. Chem.* 78: 6967–6973.
59. Sathuluri, R. R., Yoshikawa, H., Shimizu, E., Saito, M., and Tamiya, E. 2011 Gold nanoparticle-based surface-enhanced Raman scattering for noninvasive molecular probing of embryonic stem cell differentiation. *PLoS ONE* 6: e22802.
60. Sun L., Sung K. B., Dentinger C., Lutz B., Nguyen L. et al. 2007. Composite organic-inorganic nanoparticles as Raman labels for tissue analysis. *Nano Lett.* 7: 351–356.
61. Yu K. N., Lee S. M., Han J. Y. et al. 2007. Multiplex targeting, tracking, and imaging of apoptosis by fluorescent surface enhanced Raman spectroscopic dots. *Bioconjug. Chem.* 18: 1155–1162.
62. Keren, S., Zavaleta, C., Cheng, Z., dela Zerda, A., Gheysens, O., and Gambhir, S. S. 2008. Noninvasive molecular imaging of small living subjects using Raman spectroscopy. *Proc. Natl. Acad. Sci. USA* 105: 5844–5849.
63. Kang, J. W., Nguyen, F. T., Lue, N., Dasari, R., and Heller, D. A. 2012. Measuring uptake dynamics of multiple identifiable carbon nanotube species via high-speed confocal Raman imaging of live cells. *Nano Lett.* DOI:10.1021/nl302991y.
64. Liu, Z., Davis, C., Cai, W. B., He, L., Chen, X. Y., and Dai, H. J. 2008. Circulation and long-term fate of functionalized, biocompatible single-walled carbon nanotubes in mice probed by Raman spectroscopy. *Proc. Natl. Acad. Sci. USA* 105: 1410–1415.
65. Dorney, J. Bonnier, F., Garcia, A., Casey, A., Chambers, G., and Byrne, H. J. 2012. Identifying and localizing intracellular nanoparticles using Raman spectroscopy. *Analyst* 137: 1111–1119.
66. Zhang, Y. B., Xu, Y., Li, Z. G. et al. 2011. Mechanistic toxicity evaluation of uncoated and PEGylated single-walled carbon nanotubes in neuronal PC12 cells. *ACS Nano* 5: 7020–7033.
67. Andersson, P. A., Lejon, C., Ekstrand-Hammarstrom, B. et al. 2011. Polymorph- and size-dependent uptake and toxicity of TiO_2 nanoparticles in living lung epithelial cells. *Small* 7: 514–523.
68. Zavaleta, C., dela Zerda, A., Liu, Z. et al. 2008. Noninvasive Raman spectroscopy in living mice for evaluation of tumor targeting the carbon nanotubes. *Nano Lett.* 8: 2800–2805.
69. Luyts, K., Napierska, D., Nemery, B., and Hoet, P. H. M. 2013. How physico-chemical characteristics of nanoparticles cause their toxicity: Complex and unresolved interrelations. *Environ. Sci. Processes Impacts* 15: 23–38.
70. Xia, T., Kovochich, M., Liong, M. et al. 2008. Comparison of the mechanism of toxicity of zinc oxide and cerium oxide nanoparticles based on dissolution and oxidative stress properties. *ACS Nano* 2: 2121–2134.
71. Brunner, T. J., Wick, P., Manser, P. et al. 2006. *In vitro* cytotoxicity of oxide nanoparticles: Comparison to asbestos, silica, and the effect of particle solubility. *Environ. Sci. Technol.* 40: 4374–4381.
72. Franklin, N. N., Rogers, N. J., Apte, S. C., Batley, G. E., Gadd, G. E., and Casey, P. S. 2007. Comparative toxicity of nanoparticulate ZnO, bulk ZnO, and $ZnCl_2$ to a freshwater microalga (*Pseudokirchneriella subcapitata*): The importance of particle solubility. *Environ. Sci. Technol.* 41: 8484–8490.
73. Gilbert, B., Fakra, S. C., Xia, T., Pokhrel, S., Madler, L., and Nel, A. E. 2012. The fate of ZnO nanoparticles administered to human bronchial epithelial cells. *ACS Nano* 6: 4921–4930.
74. Magdolenova, Z., Collins, A. R., Kumar, A., Dhawam, A., Stone, V., and Dusinska, M. 2014. Mechanisms of genotoxicity: Review of recent *in vitro* and *in vivo* studies with engineered nanoparticles. *Nanotoxicology.* 8:233–237.
75. Studer, A. M., Limbach, L. K., Van Duc, L. et al. 2010. Comparison of stabilized copper metal and degradable copper oxide nanoparticles. *Toxicol. Lett.* 197: 169–174.

76. Gunawan, C., Teoh, W. Y., Marquis, C. P., and Amal, R. 2011. Cytotoxic origin of copper(II) oxide nanoparticles: Comparative studies with micron-sized particles, leachate, and metal Salts. *ACS Nano* 5: 7214–7225.
77. Volker, C., Oetken, M., and Oehlmann, J. 2013. The biological effects and possible modes of action of nanosilver. In *Rev. Environ. Contam. Toxi.* Vol. 223. D. M. Whitacre, Ed. Springer, New York. DOI:10.1007/978-1-4614-5577-6_4. P.81-106.
78. Mortimer, M., Kasemets, K., and Kahru, A. 2010. Toxicity of ZnO and CuO nanoparticles to ciliated protozoa *Tetrahymena thermophila*. *Toxicology* 269: 182–189.
79. Midander, K., Cronholm, P., Karlsson, H. L. et al. 2009. Surface characteristics, copper release, and toxicity of nano- and micrometer-sized copper and copper(II) oxide particles: A cross-disciplinary study. *Small* 5: 389–399.
80. Wang, Z. Y., Li, N., Zhao, J., White, J. C., Qu, P., and Xing, B. S. 2012. CuO nanoparticle interaction with human epithelial cells: Cellular uptake, location, export, and genotoxicity. *Chem. Res. Toxicol.* 25: 1512–1521.
81. Ammann, A. A. 2007. Inductively coupled plasma mass spectrometry (ICP MS): A versatile tool. *J. Mass Spectrum.* 42: 419–427.
82. Manning, T. J. and Grow, W. R. 1997. Inductively coupled plasma—atomic emission spectrometry. *Chemical Educator* 2: 1–19.
83. Xu, M. S., Li, J., Hanagata, N., Chen, H. Z., and Fujita, D. 2013. Challenge to assess the toxic contribution of metal cation released from nanomaterials for nanotoxicology—The case of ZnO nanoparticles. *Nanoscale* 5: 4763–4769.
84. Servin, A. D., Castillo-Michel, H., Hernandez-Viezcas, J. A., Corral Diaz, B., Peralta-Videa, J. R., and Gardea-Torresdey, J. L. 2012. Synchrotron micro-XRF and micro-XANES confirmation of the uptake translocation of TiO_2 nanoparticles in cucumber (*Cucumis sativus*) plants. *Environ. Sci. Technol.* 46: 7637–7643.
85. Zhao, L., Peralta-Videa, J. R., Varela-Ramirez, A. et al. 2012. Effect of surface coating and organic matter on the uptake of CeO_2 NPs by corn plants grown in soil: Insight into the uptake mechanism. *J. Hazard. Mater.* 225–226: 131–138.
86. Hernandez-Viezcas, J. A., Castillo-Michel, H., Andrews, J. C. et al. 2013. *In situ* synchrotron X-ray fluorescence mapping and speciation of CeO_2 and ZnO nanoparticles in soil cultivated soybean (glycine mas). *ACS Nano* 7: 1415–1423.
87. Cook, J. R., Frey, W., and Emelianov, S. 2013. Quantitative photoacoustic imaging of nanoparticles in cells and tissues. *ACS Nano* 7: 1277–1280.
88. van den Broek, B., Ashcroft, B., Oosterkamp, T. H., and van Noort, J. 2013. Parallel nanometric 3D tracking of intracellular gold nanorods using multifocal two-photon microscopy. *Nano Lett.* 13: 980–986.

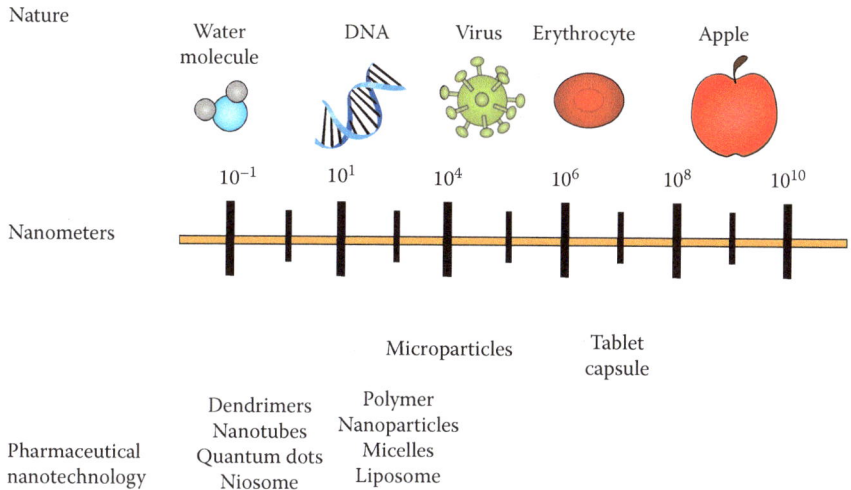

FIGURE 1.1 Dimensions scale of nanotechnology.

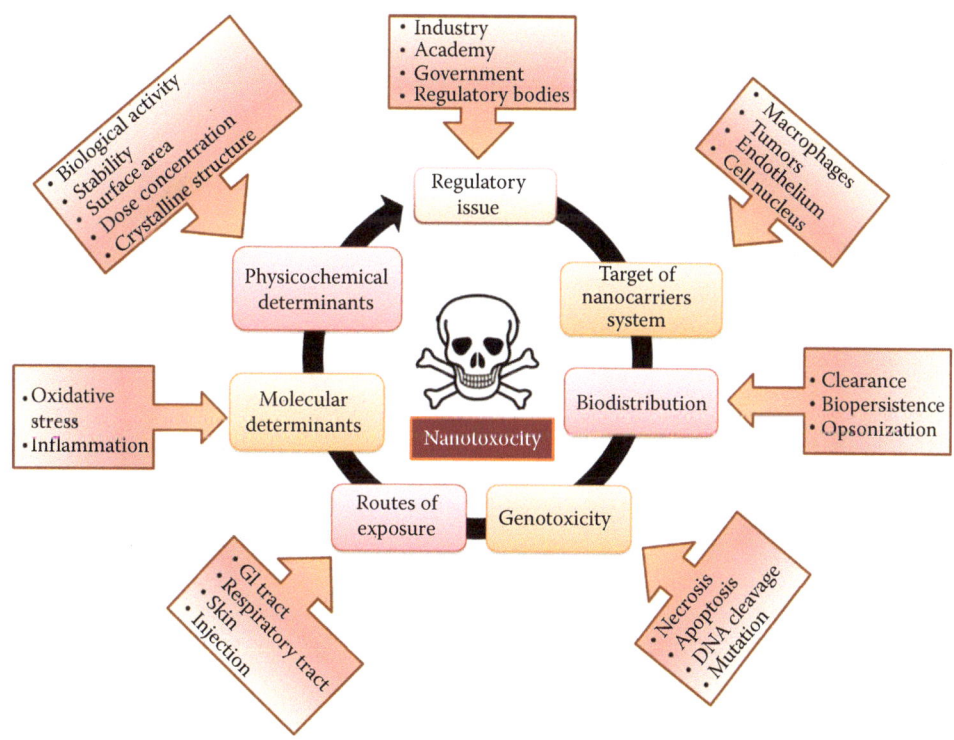

FIGURE 3.4 Multidimensional issues affecting nanotoxicity.

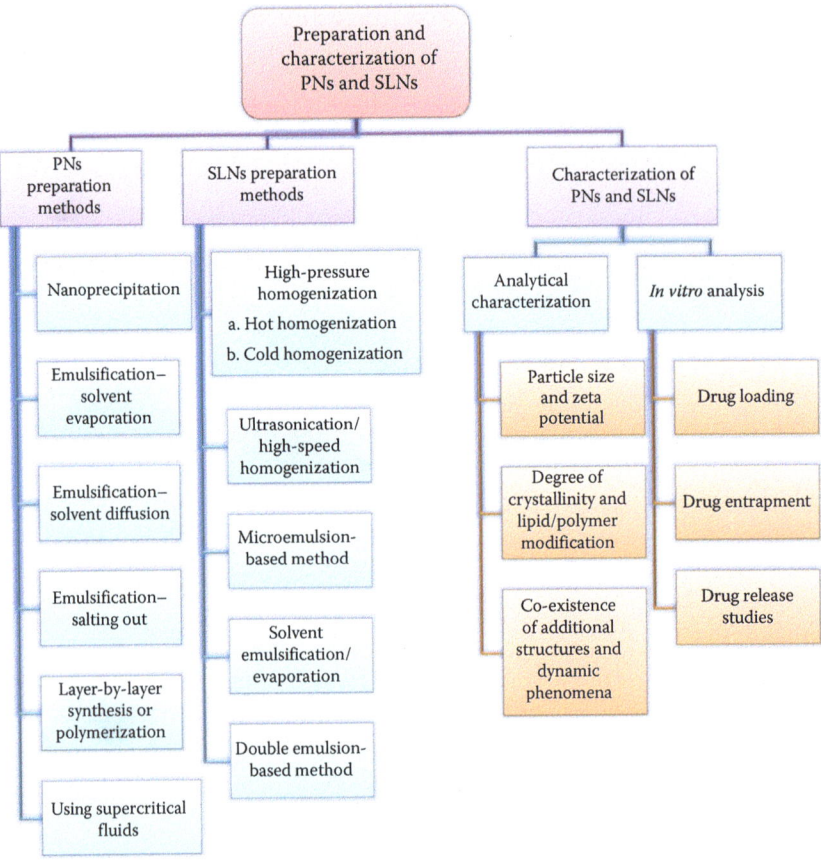

FIGURE 7.1 Preparation and characterization techniques of PNs and SLNs.

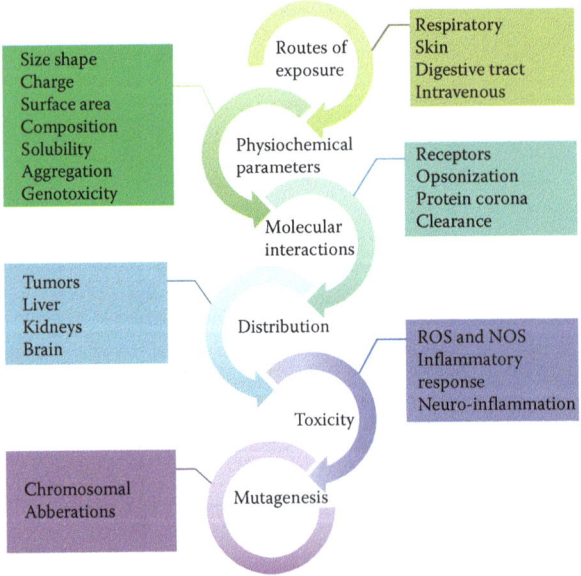

FIGURE 9.2 Concepts in nanotoxicology. The toxicity of nanomaterials is most commonly determined by the route of exposure and physiochemical parameters of the nanomaterials. Further, the type of molecular interactions, based on the properties of the nanoparticle surface, define the distribution in the body, and the extent and location of toxicity.

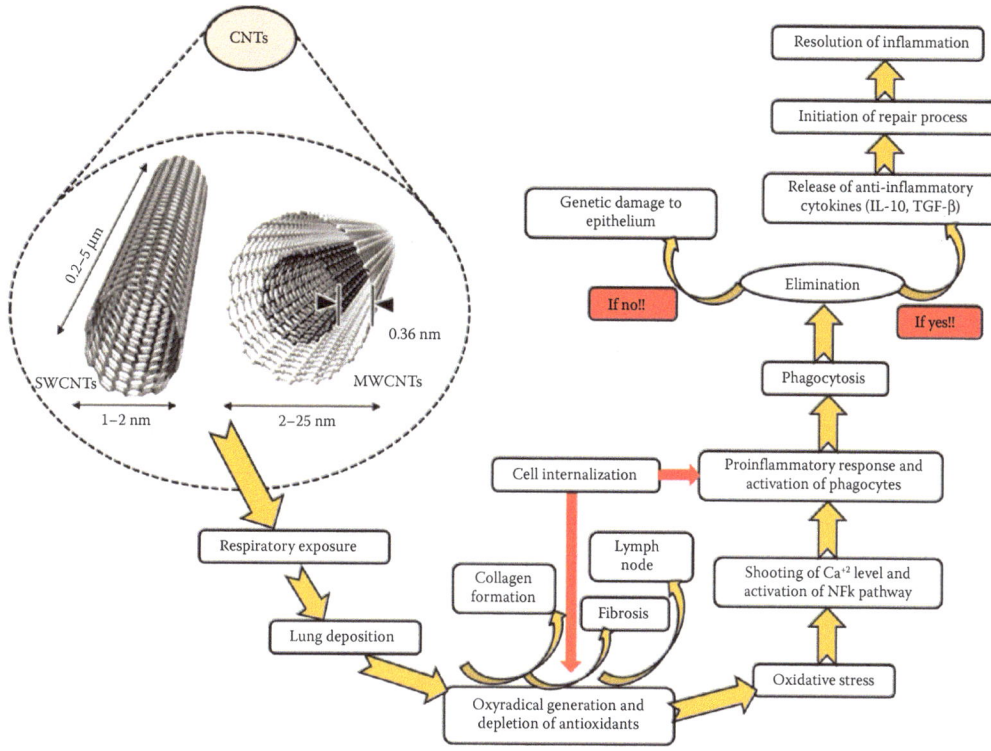

FIGURE 11.4 Possible pathways following the inhalation of CNTs.

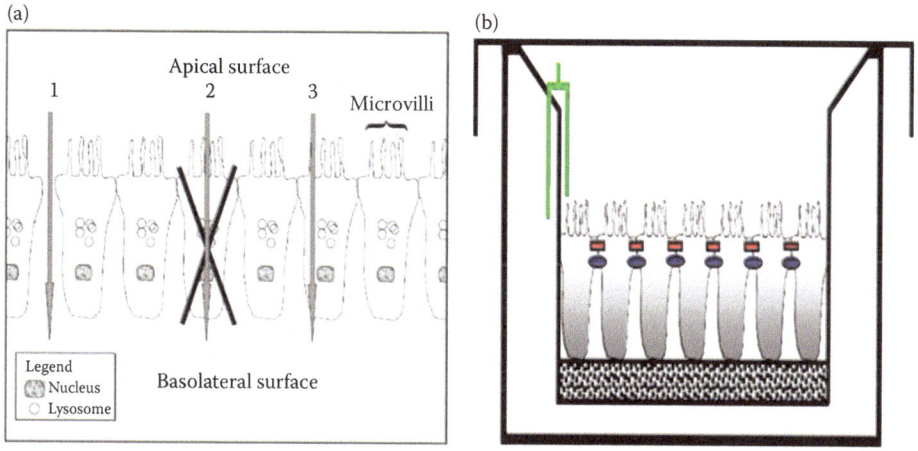

FIGURE 13.4 (a) Diagram illustrating substantial routes of epithelial transport. An epithelium is polarized into an apical and basolateral surface with the apical surface covered with microvilli to increase the surface area for absorption. Nuclei and other organelles inside the cell are also polarized with the nuclei polarized closer to the basolateral surface. In between the cells, tight junctions and adherent junctions are present at the apical surface that inhibit the free flow of materials between the apical and basolateral spaces and provide for epithelial integrity. The nanomaterials can flow between the apical and basolateral spaces if (1) epithelial integrity is disrupted (pathway 1), (2) cells within the epithelium are killed, providing holes for the flow of particles (pathway 2), and (3) nanomaterials are moved by transcytosis—a cellular process where materials are picked up at the apical surface and transported to the basolateral surface without being metabolized by the cell (pathway 3). (b) Diagram of cells in the Transwell chamber. Cells are shown on a membrane support, which permits the partitioning of the apical and basolateral chambers for measurements of electrical resistance. Red squares represent tight junctions and blue ellipses represent adherent junctions.

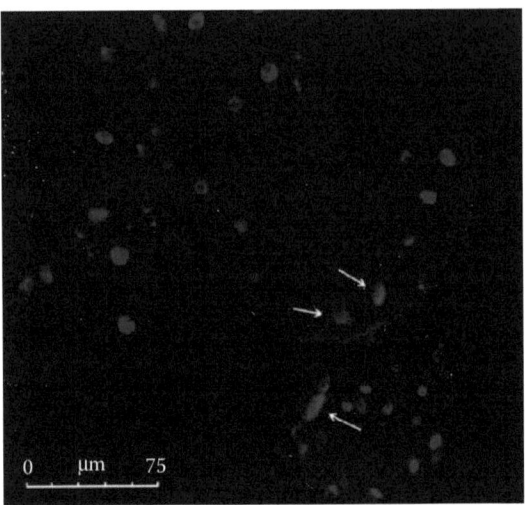

FIGURE 13.7 Representative confocal microscopy image of *Drosophila* midgut in flies treated with 15 nm Au NPs (100 pmol/L). Nuclei are stained with Hoechst 33 342 (blue) while cells containing DNA strand nicks are detected by TUNEL assay and fluorescent red (highlighted by the white arrows).

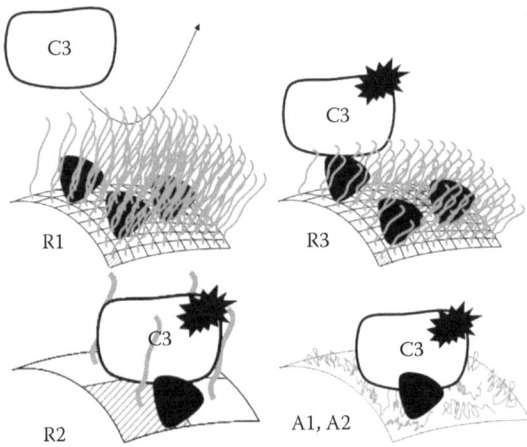

FIGURE 18.5 A scheme illustrating the interactions of proteins with a series of nanoparticles having corona of dextran with different characteristics. The dark spot included in C3 indicates that the component C3 of the complement system is activated. Albumin adsorbed on the surface of the nanoparticle core appears as a dark triangle.

9 Nanoparticles and Human Health
A Review of Epidemiological Studies

Vijaykumar B. Sutariya, Ana Groshev, Vivek Dave, Hardeep Saluja, Deepak Bhatia, Prabodh Sadana, and Yashwant Pathak

CONTENTS

9.1 Are Nanoparticles Safe for Human Health? 175
    9.1.1 Predicting and Identifying Health Risks 176
9.2 Toxicokinetics in the Body 180
    9.2.1 Role of the Route of Exposure in Toxicity 180
    9.2.2 Role of Specific Nanoparticle Characteristics in Determining Health Effects 182
    9.2.3 Nanoparticle Biointeractions and Mechanisms of Toxicity 184
    9.2.4 Design Biosafety 189
9.3 Health Effects by Category 190
    9.3.1 Health Effects of Fullerenes 190
    9.3.2 Health Effects of Carbon Nanotubes 191
    9.3.3 Health Effects of Inorganic Nanoparticles 192
    9.3.4 Health Effects of Organic Nanoparticles 193
    9.3.5 Health Effects of Quantum Dots 194
9.4 Conclusion 194
9.5 Summary 195
References 195

9.1 ARE NANOPARTICLES SAFE FOR HUMAN HEALTH?

In the last few decades, nanotechnology has become a very rapidly growing field with ever-expanding areas of application; thus, understanding the implications of nanomaterials for human health is important, more so now than ever. With tremendous economic impacts, which are projected to increase, nanomaterials are utilized in many commonly used materials such as sunscreens, cosmetics, and tennis balls. To continue reaping the benefits of nanotechnology, it is necessary to understand the risks and returns of its application. Nanotechnology is a unique field because, when operating on a nanoscale size, many materials begin to exhibit different properties due to higher involvement of van der Waal's forces, hydrophobic interactions, and induced dipole forces that are not as important on a physical scale. This has direct implications in the human body, because, as familiar as we are with gross anatomy and physiology, nanoparticles experience a multitude of specific biointeractions with cell membranes, proteins, and other biological fluid compounds, which may alter their behavior and properties based on the composition of the coating acquired in the biological environment. Furthermore, localization of the nanoparticles is not always straightforward due to their high mobility and ability to penetrate defense systems and translocate from the original port of entry. This phenomenon can be beneficial in the delivery of medicines, but may also be potentially very detrimental to health when exposure to nanomaterials is not intentional or when the dosage and parameters of the nanoparticles, intended for medical use, are not appropriately designed.

Concerns about the impact of nanotechnology on human health have been voiced repeatedly. In 2004, Donaldson was the first to emphasize the need for the development of a specific field of study that would assess nanotoxicology, and develop and improve protocols for assessment of toxicology profiles [1]. As the field of nanotechnology is very broad, enveloping multiple areas of study and application ranging from computer technology to medical applications, there is expected to be an increase in exposure to many different types of nanomaterials with an increase in their use. Thus, understanding nanotoxicology becomes more of a priority. The characterization of specific toxicity parameters of nanomaterials requires the collaboration of multiple disciplines, as the physicochemical properties change from physical scale to quantum scale properties. Another layer of complexity is added when the entrance of these nanoparticles into the body is considered. First, the characterization of these nanoparticles is primary in the understanding of their behavior. The chemical description, including size, charge, surface area, surface properties, and the presence of receptor ligands, is helpful in predicting the biointeractions of nanoparticles in the body. More specifically, the transport of the nanoparticle throughout the body lends understanding of the potential accumulation of nanoparticles that may impose a risk of toxicity. The projection of health effects is directly linked to the understanding of the biological systems and its metabolic pathways. Furthermore, normal metabolism may be modified by the presence of nanoparticles; thus, it is important to understand the impact of the interaction of the nanoparticle with the biological system in order to determine toxicological effects of nanoparticles on overall health, which may be done by either adjusting the design of the nanoparticle or the design of treatment alleviating the toxicological effect.

Although consumer reports and media outlets have been on the skeptical side, currently nurturing a phobic atmosphere concerning any application of nanotechnology, this field offers a plethora of exciting possibilities of application of this innovative technology [2]. However, the success of nanotechnology goes hand in hand with the development of the nanotoxicology field, as it is necessary for establishing the border between the safe and dangerous. The parameters that are important in understanding the toxicokinetics of nanoparticles are discussed in the sections to follow.

9.1.1 Predicting and Identifying Health Risks

After having considered the scope of the field of nanotechnology, the challenges of predicting health risks are very obvious. Nanoparticles are made for a variety of applications, ranging from preservation of freshness in the food industry to delivering active pharmacological ingredients (APIs) in cancer therapy. For instance, metal nanoparticles, especially silver nanoparticles, are widely used for their antibacterial properties. Such applications include wound dressings, food packaging, cosmetic creams, textiles, and nasal sprays [3–7]. Manufactured nanoparticles inevitably escape into the waste system and environment, where they can accumulate and enter the food chain. As companies begin to embrace the use of silver nanoparticles, it becomes increasingly important to ensure the scope of effective and safe dosages, and establish a known toxicological profile. In response to the concerns about health effects, a large number of studies have been published assessing the toxicity of silver and the extent of their antibacterial properties. Studies have been published determining the antibacterial properties in various conditions to range 0.1–20 mg/L; on the other hand, the World Health Organization has identified no observed adverse effect level at 10 g (total cumulative dose over a lifetime) [8]. However, excessively high dosages of silver lead to blue or bluish gray skin

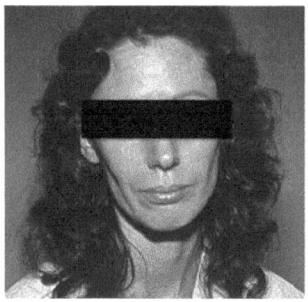 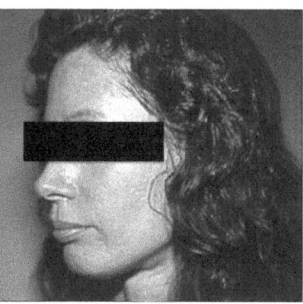

FIGURE 9.1 A sufferer from argyria developed from improper dosage of nasal drops containing silver in childhood. (Reproduced from Jacobs R. Argyria: My life story. *Clinics in Dermatology*. 2006;24(1):66–9; discussion 9. Epub 2006/01/24.)

pigmentation, a characteristic of argyria (Figure 9.1) [9–11]. For example, a case of argyria has been reported due to the chewing of photographic film [12].

Another source of exposure is intentional application, such as the application of sunblock to the skin. More companies have been using nanoparticles in sunblock and topical creams [13]. Because of repeated application, the penetration and toxicity of nanoparticles through the skin becomes a concern. The goal of any topical treatment applications is to ensure the safe and effective delivery of a therapeutic effect without compromising the integrity of the barrier and the health of deeper layers of tissues and organs; this also applies to nanoparticles designed for skin application. There are numerous reports that have been published that assess the impact of nanoparticle parameters on the ability to penetrate the stratum corneum [14–17]. At first glance, published data may seem contradicting, perhaps due to variations in the models used and the nature of nanoparticles. Also, skin nanoparticle absorption capacity may correlate with its physiochemical properties. The ability of the skin to absorb liposomes has already been established [18–20]. The nanoparticles most commonly utilized in the UV-exposure protective creams are titanium oxide (TiO_2) and zinc oxide (ZnO) nanoparticles, which render their application as photon blockers through the ability to reflect, scatter, and absorb UV radiation. Studies on porcine skin demonstrated that metal nanoparticles on average of 50 nm do not permeate intact stratum corneum; however, cracked and razor-treated skin may be susceptible to nanomaterial penetration [13].

On the other hand, it is has been repeatedly shown that, depending on the properties of the nanomaterials, the nanoparticles may penetrate the skin via hair follicles, which may act as a shunt for the transport and storage of pharmacological substances [21]. This ability renders an opportunity for the design of topical medications with the capacity for sustained release of API with the advantages of improved patient compliance, due to a decrease in the frequency of applications. In addition, topical administration reduces the risk of adverse effects and overdosing as compared with other routes of exposure, such as intravenous injection and inhalation. The potential for using topical administration in drug delivery methods is greatly due to the utilization ability of hair shafts as a shunt for the transport of nanomaterials, as well as the structural capacity of the hair shaft, specifically the follicular infandibulum, to act as a reservoir for nanoparticles [21]. Furthermore, recent developments in nanotechnology allow for the design of nanoparticles with parameters specifically tailored to be taken up by skin cells. Owing to the high versatility of the materials utilized in nanoparticle synthesis and their properties, nanotechnology allows for a multitude of therapeutic delivery designs, enhancing the barrier integrity of the stratum corneum in dermatitis, improved moisturizers for eczema, sustained release antiseptics in wound care, improved efficacy hair treatments for apolecia, and sebaceous gland targeting for rosacea, among many other applications. Numerous topical application products are already available on the market—such as ActisorbSilver 220® wound dressing, Nanoclorex®, Dermaviduals® (USA) eczema cream, and Regenerationscheme Intensiv® regenerative cream—the list is ever expanding. More studies are required to understand the benefits and potential risks, and for establishing specific parameters for facilitating skin absorption and the establishment of safe exposure ranges, testing the possibility of nanoparticle accumulation in crevices, hair follicles, and/or epidermis and underlying layers of connective tissues.

On the other hand, exposure to nanomaterials can also be unintentional. As the production and usage of nanotechnologies increase, more nanoparticulates are expected to be released into the environment. For example, because silver nanoparticles are increasingly being used as an antibacterial in textiles, some of the nanoparticles end up in wastewater due to leaching during the washing process [22]. In addition, it has already been demonstrated that nanoparticles are already present in air pollution as a by-product of burning fuel and some manufacturing processes [23,24]. Repeated concerns have been raised and investigated about the health implications of exposure to nanoparticles in the environment and workplace. Specifically, air pollution presents a substantial source of an assortment of particulates, including sulfate-, nitrate-, and carbon-containing particulates [25]. In the case of airborne particulates, size determines half-life and tissue penetration. Numerous scientific studies have demonstrated that the size of the nanoparticle determines the location of deposition in the lungs, such that the smaller nanoparticles (1–10 nm) travel all the way to the alveoli where they may be absorbed, whereas larger particles become trapped in the nasal cavity and larynx [26].

However, one of the largest challenges in the process of assessing the toxicological profile for specific types of the nanoparticles is the development of accurate models; toxicology findings are often inconsistent between *in vitro* and *in vivo*. The differences may be due to the high reactivity of the nanomaterials, resulting in significantly more complex biointeractions *in vivo* as compared with cell culture systems. Specifically, the complex microsystems of different cell types are difficult to mimic *in vitro*, as cell/cell and cell/matrix interactions, diversity of cell types, and complex hormonal signaling are missing *in vitro* [27,28]. In addition, the interactions with nanoparticles may differ *in vitro*, as *in vitro* tests are often quite static whereas nanoparticles are very mobile, often translocating in the body and accumulating in specific locations such as tumors. However, *in vitro* assays offer a quick and cost-effective method for testing toxicological endpoints and allowing for the deduction and testing of specific primary mechanisms involved without the ambiguity introduced through physiological compensatory mechanisms [29]. Furthermore, *in vitro* testing allows for the determination of primary effects without the secondary effects due to inflammation. Other well-known benefits of the use of *in vitro* assays for toxicology studies include reproducibility, cost-effectiveness, and the reduced amount of test materials required. Most importantly, *in vitro* toxicology studies minimize the need for animal testing, and assist with identifying the areas and scopes for improvements in subsequent *in vivo* studies, which are usually more expensive and with results that may be more difficult to interpret [30]. On the other hand, the shortcomings of *in vitro* tests for toxicological studies are reportedly nanomaterial interactions with the classical dye-based [31] and fluorescence-based assays [32], especially in the case of using metal and magnetic nanoparticles at high doses. The nanoparticle may interact with dyes such as 3-(4,5-dimethylthiazole-2-yl)-2,5-biphenyl tetrazolium bromide (MTT), neutral red, and Coomassie blue among others, stabilizing the structures by preventing solubilization, inducing chemical reactions and false-positives, and/or absorbing the dye, resulting in erroneous reports [28]. If such interactions are suspected, alternative methods need to be utilized. Other available methods include commonly used assessments of DNA damage such as the comet assay, Ames test, and chromosome aberration assays, among others [33,34]. Other aspects contributing to toxicology are the induction of oxidative stress and inflammation, which can also be assessed by biochemical assays. However, the interactions of assays with the nanomaterial cannot be excluded in the case of DNA damage tests, oxidative stress, and inflammation. Therefore, the inclusion of appropriate controls, both negative and positive, becomes more important in ascertaining the absence of interference. Further, the usage of a battery of toxicity assays testing different endpoints may improve the understanding of toxicity-inducing parameters of the specific nanomaterials in question.

Although *in vitro* methods are faster, cheaper, more convenient, and do not involve ethical issues, *in vitro* tests do not always provide satisfactory methods of toxicology assessment. Modeling complex biological interaction systems is challenging and long-term testing is not possible *in vitro* [28]. On the other hand, *in vivo* tests allow for the accounting of multiple factors that may affect the toxicity of the nanomaterials in a biological system. For instance, it has been reported that in serum, nanoparticles tend to have a "protein corona" formation around their surface, which may directly impact the properties exhibited by the nanoparticle and the nature of the interactions with the cell membrane [35]. On the other hand, the interactions of the nanoparticle with the biological proteins, receptors, and membranes may further complicate the interpretation of *in vivo* results; thus, thorough characterization is necessary in order to adequately understand the biointeractions of the nanomaterials. Identifying and understanding the processes and pathways leading to toxicity in an organism may be challenging in the *in vivo* setting.

The design of the delivery vehicle is another noteworthy component of *in vivo* experiments. Because of their increased specific surface area and high reactivity, the nanoparticles tend to agglomerate. Thus, the vehicle is very important in maintaining stable homogeneous solution. Physiological solutions, the presence of proteins, and pH changes may result in agglomeration, which may alter the behavior of nanomaterials in a biological system [36]. Specifically for *in vivo* tests, the delivery phase needs to be isotonic and nontoxic with well-dispersed nanoparticles. For biocompatibility reasons, phosphate-buffered saline solution has been used most commonly for *in vitro* treatments although it may not prevent agglomeration [37]. Other delivery vehicles containing lipids and proteins have been designed and tested; however, the composition of the delivery vehicle depends more on the specific route of nanomaterial exposure: pulmonary, ingestion, topical application, and intravenous administration [38]. Another challenge for *in vivo* studying is determining the treatment dose. It is difficult to determine the exposure concentration due to many external sources, such as food, air, and the environment; a high treatment dose may not have relevant physiological consequences. Furthermore, the differences between species and in animal models require different dosing, and the physiological distinctions are sufficient to yield results that are different from human subjects. Overall, toxicity from nanomaterials can be most optimally studied by combining experiments that address multiple components of nanoparticle properties. The thorough characterization of chemical and physical parameters is necessary for fully understanding the behavior of nanomaterials in biological systems. *In vitro* studies allow for the identification of specific pathways involved in metabolism and the extrapolation of pharmacokinetics through the use of cell proliferation assessments, reactive oxygen and nitrogen species generation, apoptosis and necrosis assays, microscopic imaging in tissues, gene profiling, and genotoxicity evaluations [39]. *In vivo* studies are often deemed necessary, as *in vitro* and *in vivo* studies do not always correlate and published literature findings are often contradicting. This may be due to the need of further developing sophisticated *in vitro* cell culture systems for the adequate presentation of biological systems, and to address the unique nature of nanoparticles.

The study of toxicokinetics certainly presents its own challenges of multivariable relationships in nanomaterial parameters and biological systems. A plethora of studies have been accomplished to address the question of nanomaterial toxicity; however, owing to the vast number of combinations and variations, each nanoformulation has to be tested for each biological system individually, as it is the description of specific parameters and their impact on health that lends the most assistance in the design of safe nanomaterials. The field of nanotoxicology is still lacking predictive capabilities. More recently, more studies on the applications of mathematical models using multivariate linear regression have been published as potentially beneficial methods in planning and directing toxicological studies [40]. Toxicological assessments using theoretical models before the development of *in vitro* or *in vivo* experiments would allow for the identification of relevant pathways, prediction of toxicity, and determination of toxic dosages. To most effectively harness the ability to test toxicological properties of nanoparticles in cell lines, mathematical models along with mechanism-centered high-throughput testing may significantly benefit the further development of the field of toxicogenomics [41].

9.2 TOXICOKINETICS IN THE BODY

Nanotechnology is a particularly exciting area of study due to the versatility of variable and adjustable parameters of nanomaterials, providing new approaches to the current challenges in medicine. However, in order to fully materialize the potential benefits of nanotechnology, the risks have to be appropriately identified and addressed. The flowthrough of nanotoxicity is understood to begin with the route of exposure, where the physiochemical parameters of the nanomaterials determine the types of molecular interactions and their distribution in the body, inducing toxicity and perhaps even mutagenesis (Figure 9.2). At each of these components of nanotoxicity, it is imperative to understand the factors inducing toxicity and regulate them appropriately, where the benefits are maximized and the risks are minimized. The route of entry plays an important role in determining the distribution, metabolism, and interactions of the nanomaterials. Therefore, the limitation of undesired effects of the nanoparticles can only be accomplished by the careful design of nanomaterial characteristics, with consideration to the route of exposure that could help target the desired effect.

9.2.1 Role of the Route of Exposure in Toxicity

The exposure of nanoparticles to humans occurs usually through the dermal or inhalation routes. Individuals who are most prone to exposure are most likely to be in a manufacturing or research setting [42]. The literature provides strong evidence of the injurious effects of inhaled nanoparticles. For example, a consistent correlation was seen between urban air pollution and a decrease in expiratory peak flows in asthmatic patients along with impaired lung function in children [43]. Concerns have been raised over occupational safety for individuals exposed to asbestos [44], quartz

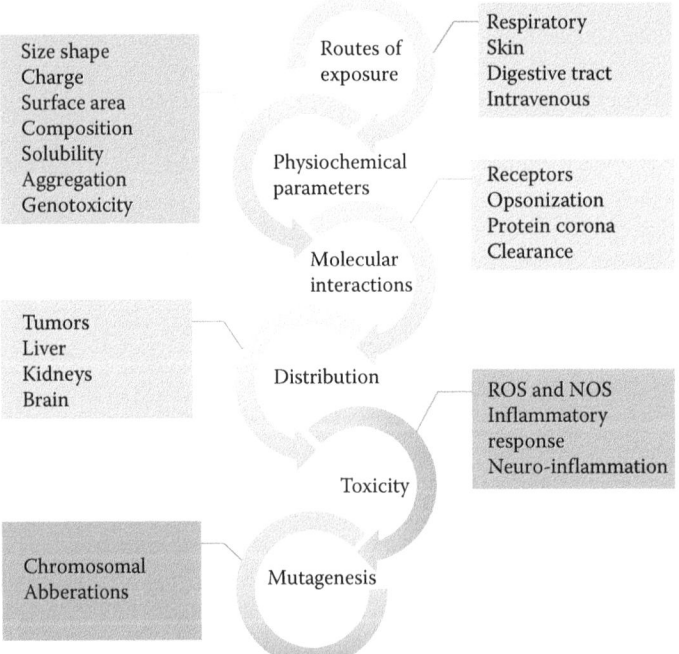

FIGURE 9.2 (See color insert.) Concepts in nanotoxicology. The toxicity of nanomaterials is most commonly determined by the route of exposure and physiochemical parameters of the nanomaterials. Further, the type of molecular interactions, based on the properties of the nanoparticle surface, define the distribution in the body, and the extent and location of toxicity.

[45], carbon particles, and nanotubes [46,47]. It is well known that asbestos exposure can cause mesothelioma (cancer of the pleura and peritoneum). Both asbestos fibers and long, multi-walled carbon nanotubes (CNTs) demonstrates comparable carcinogenic potential. Their similar geometry, biopersistence (stability in physiological environments), and potential to generate free radicals is responsible for causing serious tissue damage. The two main physical properties of nanoparticles, fiber length and biopersistence, were identified as key contributors to toxicity and carcinogenicity [48]. Kane et al. reported that fibrous materials longer than 10–20 μm undergo incomplete engulfment by macrophages, which "impairs macrophage mediated clearance," stimulating the release of free radicals, inflammatory mediators, and growth factors. Fibers of asbestos can also lead to the aforementioned response, and long-term exposure can cause a constant release of inflammatory mediators. These free radicals may also cause DNA damage and mutations, favoring tumor development and progression [44].

Human exposure to dust containing silicon dioxide, like quartz, can cause pneumoconiosis in the form of acute or chronic silicosis. It may further lead to the induction of malignant diseases that may result in obstructive and restrictive lung diseases [47]. One of the most important factors that determine the lung toxicity of these inhaled nanoparticles is size. Depending on the size, these particles may deposit at different locations inside the respiratory tract. In general, small particles (<100 nm) can more efficiently reach distal airways compared with larger particles. Also, small particles are retained longer than large particles [49]. Apart from size, shape and structure can also influence inhalation toxicity. Studies have shown that the diameter, length, and surface characteristics of asbestos fibers dictate their harmful effects [50,51]. For example, a reactive fiber surface with a small diameter can penetrate deep into the lungs, with the length of the fiber affecting its elimination [48]. This will determine the development of inflammation, fibrosis, and carcinogenesis to the affected area in the body [42]. CNTs also exhibit similar pulmonary toxicity because of the similarity of the aforementioned properties [44].

Inhaled nanoparticles can also cross cell membranes and reach systemic circulation. These particles can then penetrate and deposit into other organs, such as the liver, spleen, kidneys, heart, and brain, and can produce toxic effects to these vital organs. These particles can induce inflammatory and prothrombotic responses, and can cause atherosclerosis and thrombogenesis. Inside the brain, these deposited particles can cause neurotoxicity and neurodegenerative diseases. Studies provide evidence that these particles translocate to various vital organs, producing serious, unwanted, toxic effects [52,53].

Nanoparticles, once cleared from the respiratory tract, can be ingested into the gastrointestinal (GI) tract [54]. They can also be ingested directly into the GI tract with food, water, drugs, or drug delivery devices [55]. The effect of nanoparticles on the GI system has not been extensively studied. Only a few studies have examined the effects of ingested nanoparticles [42]. Zhen et al. studied the toxicity of copper nanoparticles (23.5 nm) and micro copper particles (17 μm) *in vivo*. The copper particles were administered in mice via oral gavage and LD50 was found to be 41 mg/kg for copper nanoparticles compared with >5000 mg/kg for micro copper particles. The copper nanoparticles produced serious injuries to the liver, kidneys, and spleen [56]. Yamago et al. [57] showed that water-soluble fullerenes, upon oral administration in rats, were not efficiently absorbed, and were primarily excreted in the feces.

The skin, as the largest defense barrier consisting of three layers—the epidermis, dermis, and the subcutaneous layer—plays a very important role in determining the cytotoxic effects of nanomaterials by impeding their permeation. The keratinized *stratum corneum* of the epidermis layer is one of the most important components of the tight mechanical barrier, as it significantly restricts permeation of foreign materials. However, skin structures such as hair follicles, sebaceous glands, and sweat glands may act as reservoirs and potentials routes of penetration, with the potential for therapeutic and drug delivery applications or toxicological effects [58]. Once the nanomaterials have penetrated the top keratinous layers of the skin, possible phagocytosis by macrophages, Langerhans cells, and keratinocytes may occur, potentially triggering the immune system [59,60]. Although the distribution of nanomaterials is more than likely to remain localized, a possible distribution in the rest of the body may occur if the nanomaterial reaches deeper layers of skin in sufficient concentrations.

Exposure to nanomaterials via the dermal route most commonly arises from topical applications as, more recently, nanotechnology is consistently being incorporated into wound dressings, textiles, cosmetic products, and sunblock, among others [42,54,55]. Occupational exposure to nanomaterials also presents a concern, as repeated exposure may alter the integrity of the integumentary barrier and increase the permeation of nanomaterials. For example, a study by Rouse et al. [61] demonstrated skin penetration of fullerene-based peptides *in vitro* using a porcine skin model subjected to repeated mechanic stress. On the other hand, an exhaustive study by Baroli et al. [62] demonstrated that metallic nanoparticles passively penetrate the hair follicle and *stratum corneum*; however, the nanoparticles were unable to penetrate the full thickness of the skin and remained in the dermis. The majority of the current literature reports limited to none skin penetration by metallic nanoparticles often used in topical crèmes [63,64]. Smaller-sized nanoparticles (less than 10 nm) are favored in permeation to the dermis layer, but nanoparticles as large as 200 nm have been shown in the dermis layer via the penetration of hair follicles [64]. However, limited studies are available on the role of specific nanoparticle parameters on skin permeation, and the further building of the basis of understanding the mechanisms of permeation is necessary. Overall, dermal delivery presents an exciting area of drug delivery research, where potential therapeutic applications of local drug delivery would have very limited toxic effects; however, further studies are warranted to understand the mechanisms of nanoparticle translocation and metabolism in all of the components of the skin.

9.2.2 Role of Specific Nanoparticle Characteristics in Determining Health Effects

The size of the nanoparticle yields a whole assortment of unique characteristics. With a decrease in the average size of the particles, the surface-area-to-mass ratio drastically increases in particles sized 100 nm and below in diameter (Figure 9.3). Because of the large specific surface area, nanoparticles tend to be highly reactive due to a high number of interaction points that could participate in biochemical interactions. Thus, when considering the concentration of nanoparticles, the number of particles and their size (i.e., surface area) provides more insight into the expected scope of interactions.

The makeup and morphology of nanoparticles is another aspect that determines its physiochemical properties and the nature of its interaction with the environment. For instance, the shape of nanoparticles plays a role in its uptake, such that rods experience the highest uptake by the cells, followed by spheres, cylinders, and then cubes in nanoparticles sized larger than 100 nm [65]. This phenomenon can be due to the number of interactions with the membrane, as rods have more interactions with the membrane and are thus more likely to be taken up by the cell. However, at the same time, nanoparticle spheres less than 100 nm in size are taken up better than rods due to the fact that nanorods have two different orientations of interaction [66–68]. More interactions with the membrane would facilitate phagocytosis; however, a vertical axis of interaction may be highly unfavorable for uptake. Thus, nanorods require another level of specification: the placement and orientation of their ligands [69,70].

Similar to the shape of the nanoparticle, the size also plays a role in determining the interactions with the cell membrane. The smaller the nanoparticle, the larger the ratio of specific surface area to size, making uptake more likely. However, normally a cell has a limited number of specific receptors; if the nanoparticle size is too small and the concentration too high, localized overloading may occur, such that the uptake may not be as efficient. Mathematical models dictate that the optimal size occurs when there is no shortage of the ligand on the nanoparticle, with no localized shortage of the receptor on the cell membrane [71]. The best size for cell uptake of spherical gold, silica, single-walled carbon nanoparticles, and quantum dots has been experimentally determined to be 50 nm

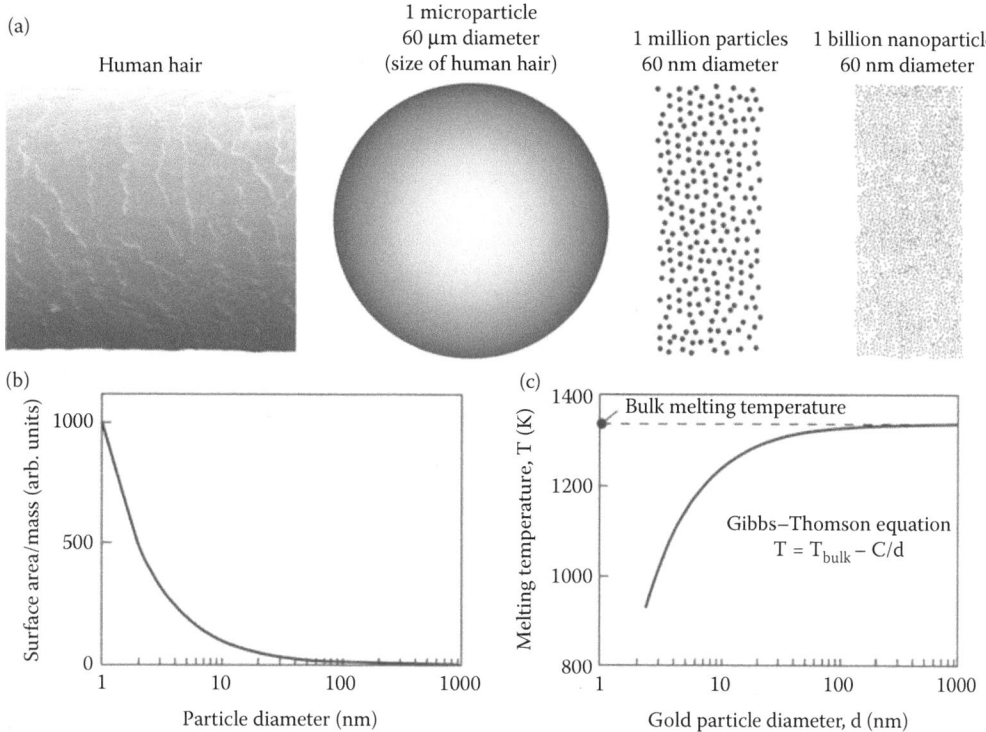

FIGURE 9.3 Correlation of size and specific surface area. (a) An illustration of the increase in surface area with the decrease of surface area. The image of human hair is shown to scale, with a 60 μm diameter representing the approximate size of human hair. In comparison, the same amount of particles, by mass, but of a smaller diameter (600 and 60 nm) has a much larger surface area. (b) Graph representing correlation of particle diameter and surface-area-to-mass ratio. (c) Impact of diameter on melting temperature provides evidence that physiochemical properties become altered with the change in particle diameter. (Reprinted with permission from Buzea C, Pacheco, II, Robbie K. Nanomaterials and nanoparticles: Sources and toxicity. *Biointerphases* 2007;2(4):MR17–71.)

in diameter [69,72,73]. However, this may vary based on the targeted cell type and other factors. Specifically, the ligand is a more important factor in cell uptake, as the rate of uptake depends on the targeted ligand. Studies have shown that same nanoparticle coated in different protein but targeting the same receptor has different uptake rates [74].

Depending on the material and the process of synthesis, the nanoparticles may carry a charge. As the intracellular side of the membrane carries a slightly negative charge, positively charged nanoparticles uptake more readily, perhaps due to the attraction forces or the types of components present in the protein corona in the biological system [73–76]. The interactions of the nanoparticles with the cell membrane on a molecular level are especially important to consider in the assessment of their health effects. For instance, even very small (2 nm) nanoparticles bearing a positive charge can perturb the cell membrane potential and may cause the influx of calcium ions (Ca^{2+}) into the cell [77]. As a result, cell proliferation may be inhibited. On the other hand, larger, charged nanoparticles (4–20 nm) may even induce reconstruction of lipid bilayers [78]. Furthermore, interactions with the cell membrane also differ based on the charge of the nanoparticle, as the binding of negatively charged nanoparticles induce local gelation, while the binding of positively charge nanoparticles induces fluidity. However, the surface charge and properties of the nanoparticle may be altered through interactions with the environment, which include interactions with proteins in serum and the formation of a corona made up of multiple proteins (Figures 9.4 and 9.5) [68,69,79–81]. In fact,

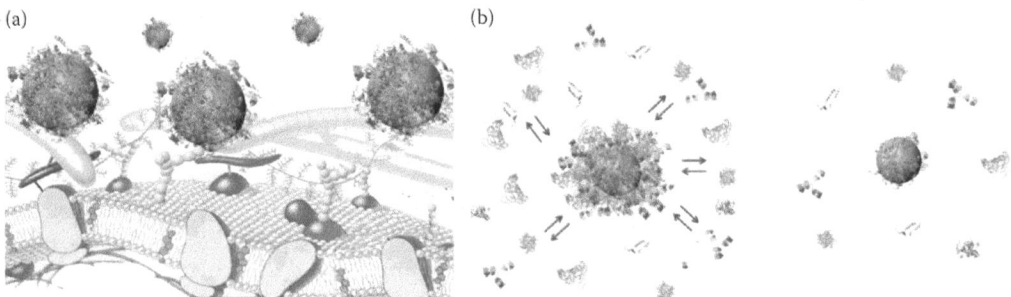

FIGURE 9.4 (a) Nanoparticle interactions with their environment, (b) the formation of a protein corona around the nanoparticle. (Reproduced with permission from Walczyk D et al. What the cell "sees" in bionanoscience. *Journal of American Chemical Society* 2010;132(16):5761–8.)

the interactions of nanoparticles with the lipids and protein receptors on the cell membrane are highly dependent on the protein corona and the ability of the nanoparticle to adsorb the biological molecules [82,83]. Thus, it is important to not only understand the structure and characterization of the nanoparticles, but also its environment and its interactions with the environment [82,84].

Furthermore, the charge of the nanoparticle has a direct impact on its stability in solution. Charged nanoparticles stay well suspended in a solution due to repulsive forces. Neutral nanoparticles, however, have the longest half-life in the blood, as charged nanoparticles get quickly cleared and may cause complications such as hemolysis and platelet coagulation [63]. However, the interactions with blood cells may be more dependent on the functional groups on the nanoparticle surface [85]. Thus, hemolysis is one of the aspects of nanotoxicology to be considered, although many nanoparticles do not exhibit interactions with erythrocytes or hemolysis [62]. Overall, the half-life property depends on interactions of the nanoparticles with immunoglobulin, lipoprotein, complement and coagulation factors, acute phase proteins, and metal-binding and sugar-binding proteins [80,82]. Thus, PEGylation helps extend the half-life by limiting the interactions and preventing detection by macrophages and the complement system [58,64,86]. The dimensions of nanoparticles also impact the clearance rate. Studies have shown that rod-shaped nanoparticles circulate 10 times longer than spherical nanoparticles, which may be related to the rate of cell uptake, as spherical shape makes uptake more likely [87]. Size also plays a role in the clearance rate from the blood. Kidneys, as one of the sites of major blood filtration in the body, quickly clear small nanoparticles (less than 6 nm) [88]. The most important factor in the determination of biointeractions is the surface of the nanoparticle. By specifying the materials from which the nanoparticle is made, the biointeractions can be more controlled (Table 9.1).

9.2.3 Nanoparticle Biointeractions and Mechanisms of Toxicity

As the nanoparticle enters a biological system, it is bound to exert some effect on the system [89]. The beauty of nanotechnology is that the manipulations of very specific aspects of the system can occur on a very small scale to minimize nonspecific interactions and off-target actions. Nanomaterials offer the potential to deliver nucleic acids for the modification or regulation of faulty pathways and for the transcription of specific proteins. Small molecules can be shuttled to specific targets, such as tumors, and natural barriers, such as the blood–brain barrier, can be overcome. Owing to their surface area, nanoparticles are very reactive and may demonstrate toxicity at sufficient dosages and accumulation, even if the material itself is not toxic [90].

The design of the nanomaterial plays a crucial role in determining its level of toxicity. As mentioned earlier, particle size and surface properties are crucial in determining the stability, translocation, and types of interactions in a biological system (Figures 9.4 and 9.5). However, it is necessary to account for the interactions of the nanomaterials with their suspension medium, such as the

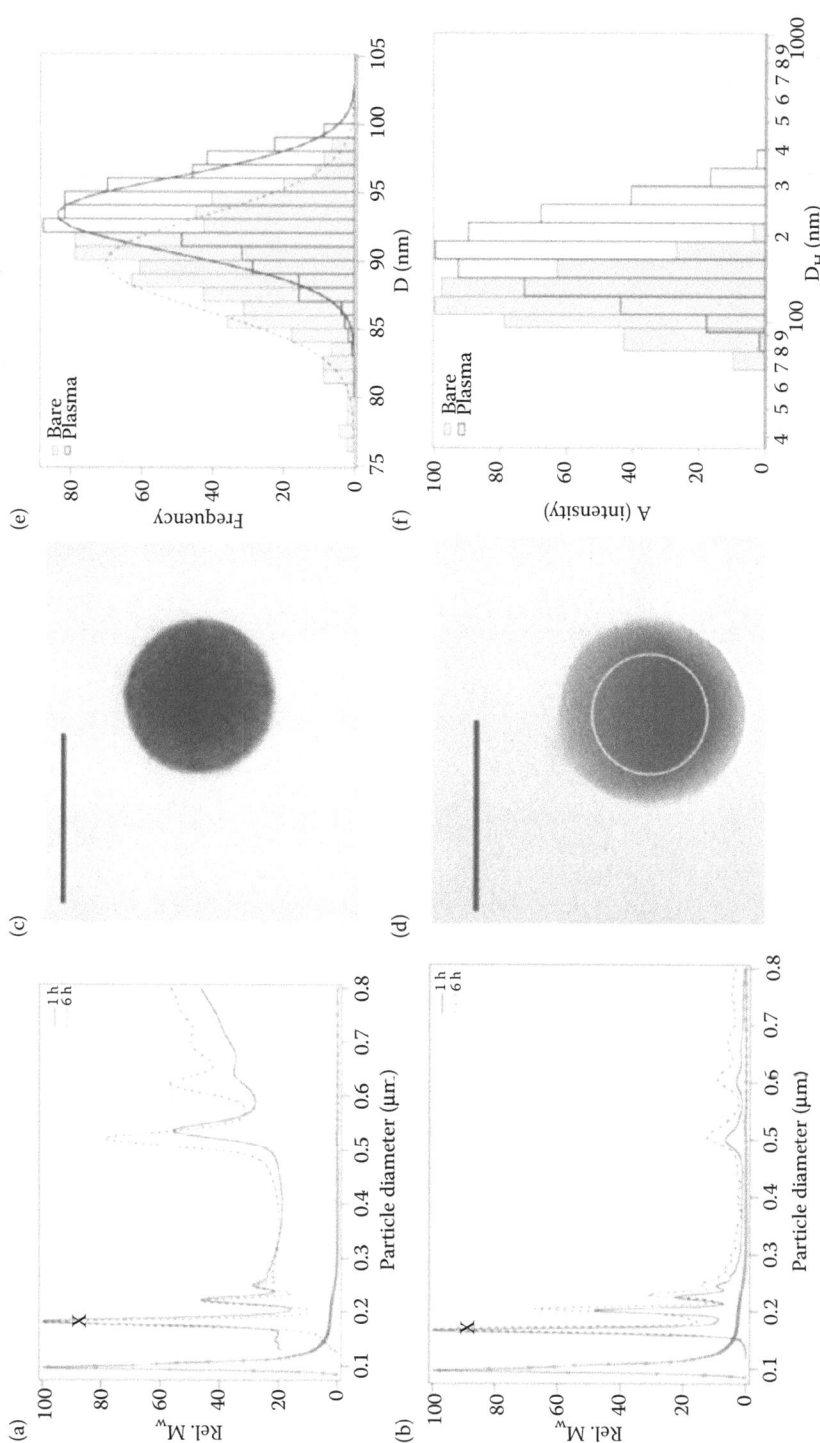

FIGURE 9.5 Formation of protein corona around a spherical nanoparticle. (a) Time-dependent measurements of differential scanning calorimeter for 100 nm nanoparticle–protein complexes, 1 h (full line) versus 6 h (dotted line) in full plasma; (b) time-dependent measurements of differential scanning calorimeter for 100 nm nanoparticle–protein complexes, 1 h (full line) versus 6 h (dotted line) without excess plasma; (c) TEM image of nanoparticle, scale bar is 100 nm; (d) TEM image of protein–nanoparticle complex, scale bar is 100 nm; (e) distribution curve representing the size distribution of different nanoparticle formations based on TEM analysis; (f) size distribution of 100 nm bare nanoparticles and nanoparticle–protein complexes without excess plasma. (Reproduced with permission from Walczyk D et al. What the cell "sees" in bionanoscience. *Journal of American Chemical Society* 2010;132(16):5761–8.)

TABLE 9.1
Toxicological Effects of Some Widely Used Nanoparticles

	Health Effects	
Type of NP	*In Vitro*	*In Vivo*
Quantum dots	Inflammatory response Oxidative stress [197–203] Genotoxicity [204]	Transfer across the placental barrier Negative impact on mouse oocyte development Skin penetration post-UV irradiation [205–208]
Gold NPs	Mitochondrial damage Oxidative stress Autophagy [209–212] Change of secondary structures of the proteins [213] Strong photothermal effect upon x-ray irradiation facilitating cell death [214]	Adverse effect on human sperm motility; penetration into sperm head and tail effecting motility [215–217] Increase in white blood cells count [218]
Silver NPs	Generation of ROS, genotoxicity, and apoptosis Intracellular calcium transients JNK activation [219–225] Strong antibacterial properties against Gram-negative and Gram-positive bacteria [226]	Generation of ROS, genotoxicity, and apoptosis Blood–brain barrier disruption Neuronal degeneration and brain edema [227–229] Adverse effects on embryonic development Lysosomal destabilization [230–235] Anti-inflammatory effect in mouse macrophages infected with *Chlamydia trachomatis* [236] Antimicrobial effect [237]
SWCNT	Impair human macrophage engulfment of apoptotic cell corpses Fibrogenic effects in lung cells Suppress inflammatory mediator responses in human lung epithelium Disrupt actin filament integrity and VE-cadherin distribution in human aortic endothelial cells Activate MAPKs, AP-1, NFκB, and Akt in normal and malignant human mesothelial cells Oxidative stress Inflammation Apoptosis Induction of micronuclei and double-strand breaks of DNA [238–245] Inactivation of microflora of river and waste waters [246]	Lung inflammation and genotoxicity Increased levels of 8-oxo-dG in liver and lung Activate platelets and accelerate thrombus formation in the microcirculation Allergic response [247–250] Improvement of viability of human stem cells *ex vivo* [251]
Multiwalled carbon nanotubes (MWCNTs)	Generation of ROS, genotoxicity, and apoptosis Disruption of actin filaments and VE-cadherin distribution in human aortic endothelial cells [238,252–255] Antibacterial effect Biocompatibility and support of neural growth and long-term survival of hippocampal neurons [256] Suppression of IL-27 in Th17 cells [257]	Genotoxicity Pulmonary toxicity Asbestos-like, length-dependent, toxicity [258] Inflammation, granuloma, and fibrosis in lungs Apoptosis of alveolar macrophages Allergic response in mice Suppression of systemic immune function Hepatocyte mitochondrial swelling [239,247,259–268]
Metal oxide NPs	Inflammatory response Generation of ROS, genotoxicity, and apoptosis Membrane disruption Inhibition of osteogenic differentiation Disruption of hippocampal neurons Antibacterial effect [269–285]	

TABLE 9.1 (continued)
Toxicological Effects of Some Widely Used Nanoparticles

Type of NP	Health Effects	
	In Vitro	*In Vivo*
Fullerenes	Oxidative stress and genotoxicity [152,286,287] Antibacterial activity through reactive oxygen species production [288,289]	Increase in proinflammatory cytokines and Th1 cytokines in BAL fluid Increased T-cell distribution in lungs Elevated levels of 8-oxo-dG in the liver and lung [249,265,290–294]

Source: Adapted from Dhawan A, Sharma V. *Analytical and Bioanalytical Chemistry.* 2010;398(2):589–605. Epub 2010/07/24.

formation of a protein corona, when considering possible biointeractions [91]. As the nanoparticle enters a biological solution, such as blood, it becomes coated with proteins from the serum. The proteins may change their folding and confirmations as they interact with the nanoparticle, leading to the exposure of new epitopes and, perhaps, altered function [92,93]. The composition of the proteins would depend on the nature of the biological fluid, such that the blood, a high-protein environment, would result in a different protein corona composition as compared with ocular fluid, a low-protein environment [89]. The properties of the nanoparticle surface, as well as the properties of the proteins, also play a role in determining the affinity and composition of protein binding in the corona. The nanoparticle may already possess some of the ligands intentionally bound in the process of synthesis or may have residues from the manufacturing process. Coupled with other properties of nanoparticle surface, such as the degree of hydrophobicity, the nanoparticle's surface determines the composition of its corona.

Once the nanoparticle comes into contact with a cell membrane, a whole new layer of complexity arises, based on the properties of the membrane, such as its fluidity and thermodynamics, heterogeneous composition, and its high activity [94]. The specific mechanism of interaction depends on the type and number of ligands present on the nanoparticle and the availability of the receptors on the cell membrane. If endocytosis occurs, its mechanics, including the "wrapping time," can be described as a mathematical function of numerous factors pertaining to the components of the process—the particle properties, such as particle size and shape; the environment properties, such as the energy of the system; and the cell membrane properties, such as the rate of receptor diffusion and elasticity of the membrane [68,95,96]. Penetration of the membrane does not solely occur based on a receptor-mediated basis; charge-dependent reconstruction of the phospholipid bilayer has been observed where negatively charged particles bind to liquid membrane and induce gelation, whereas positively charged nanoparticles caused gelled areas to turn into a liquid state [97]. This enables some cationic and gold nanoparticles with amphipathic properties to generate transient holes in the membrane without disrupting the bilayer [66,98]. However, if the cationic properties are not tightly controlled, the nanoparticles may cause disturbances of the membrane by altering its structure, as a direct relationship of defects in the cell membrane and the concentration of nanoparticles has been demonstrated in scientific reports [91,99]. Charged nanoparticles may result in the disruption of membrane activities, such as depolarization [100] and transport processes [101].

A limited amount of information is available on the processes that occur after the nanoparticle enters the cell (Figure 9.6). Reports have shown that nanoparticles retain high activity resulting in effects on cytoplasmic proteins, mitochondria, and nucleus especially with genetic material. An understanding in the area of distribution of nanoparticles in the cell is especially important for the design of effective drug delivery methods. Based on uptake mechanisms, the nanoparticles may become localized in specific cell compartments. For instance, a clathrin-dependent mechanism of

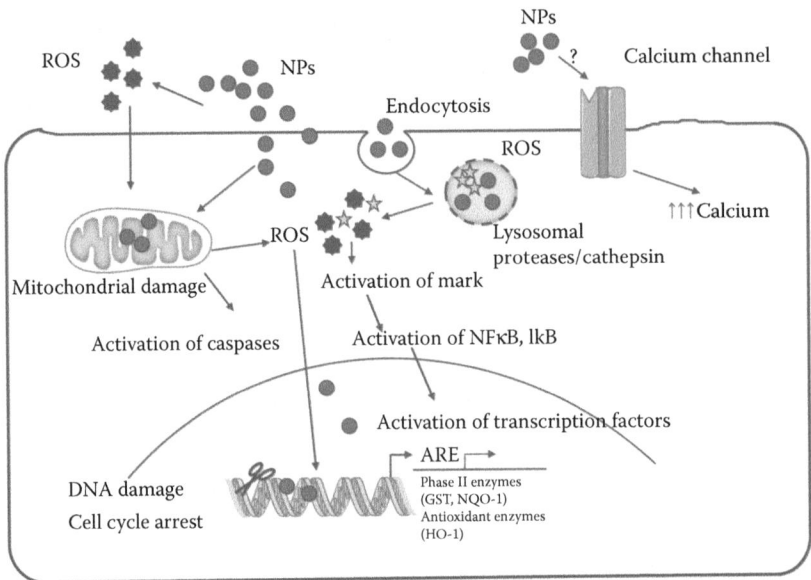

FIGURE 9.6 Schematic of biointeractions of nanoparticles with the cell and cell membrane. (Reproduced with permission from Marano F et al. Nanoparticles: Molecular targets and cell signalling. *Archives of Toxicology*. 2011;85(7):733–41.)

nanoparticle uptake, results in lysosomal localization, such as demonstrated in the case of silver and SiO_2 nanoparticle uptake [102,103]. The size of the nanoparticle also plays role in determining its location in the cell; it has been shown that gold nanoparticle translocation is limited by size as it diffuses throughout the cell [104]. Further, nanoparticles also have an ability to accumulate in the nucleus [105–107]. The mechanism of translocation to the nucleus may involve diffusion, but it has also been shown that gold nanoparticles close to 39 nm in size can be transported into the nucleus by nuclear pore complexes [108].

In the cell, as a result of protein–nanoparticle interactions, protein confirmations may occur and more significant alterations can be expected with a larger-sized nanoparticle [92,109]. These protein conformation changes may be irreversible and can result in the loss of function and induce, therefore, cytotoxicity [110–112]. Furthermore, the nanoparticles have the ability to penetrate into the nucleus and may alter the structure and function of nuclear proteins. For instance, SiO_2 nanoparticles have been shown to cause the aggregation and fibrillation of nucleoplasmic proteins, impacting nuclear function [113]. However, the nanoparticles may also result in increased enzymatic activity. The effect of the interactions depends on the location and nature of the binding. If the interaction is close to the active site, the protein will be inactivated, but if the protein is encapsulated or otherwise stabilized, the activity of the enzyme may be increased [114–116]. The ability of the nanoparticles to stabilize proteins and trigger refolding can also facilitate therapeutic intervention for diseases caused by misfolded proteins or their aggregates, such as Alzheimer's [117].

Cell signaling may be effected as a result of nanoparticle and cellular protein interactions [118]. As almost any foreign object entering a body, the induction of an inflammation response is one of the most common adverse effects of nanoparticles considered. Following the uptake by cells such as epithelial cells or macrophages, nanoparticles directly or indirectly produce biological responses that involve the generation of intra- or extracellular reactive oxygen and nitrogen species (ROS and NOS) (Figure 9.6) [119,120]. Some of the nanoparticles, namely, fullerenes and CNTs, are shown to favor localization in mitochondria, where they may facilitate the generation of ROS in connection with the leakage of electrons from the electron transport chain [121,122]. Through the generation of ROS, an inflammation signaling pathway (NF-kB, AB-1) may be activated, resulting in free radical formation, further increasing the inflammatory response [119,120,123]. The generation of ROS and inflammation may further result in mitochondrial damage and apoptosis [124]. In the case that oxidative stress overwhelms the cell's capacity to restore the redox balance, it may lead to cell death by necrosis, apoptosis, and adaptive proinflammatory responses. Repair processes and antioxidant enzymes may be activated. These cellular responses are usually targeted in the design of cancer treatments, making nanomaterials attractive in drug development and delivery. However, the scope of inflammatory action depends on the type of the nanomaterial and its physiochemical properties.

The interactions of nanoparticles, activation of inflammatory responses, and modulation of pathways may facilitate a pathological process but could also be utilized to achieve therapeutic goals. Some nanoparticles, such as fullerenes investigated by Lao et al. in fact have the capacity to attenuate the effects of oxidative stress and reduce mitochondrial swelling, characteristic of apoptosis [125]. Thus, specific biointeractions are more dependent on the interplay of multiple factors relating to nanoparticle makeup and surface properties, the characteristics of the suspending media, the properties of nanoparticle–media interface, and the nano–bio interface of the nanoparticle and its shell with the cell membrane and its proteins and receptors [89].

9.2.4 Design Biosafety

Nanotoxicology is a very broad discipline. With a countless number of combinations and variables, it may seem like a daunting task to account for all of the factors at play. Additional factors play a role, such as the difficulty in determining the effective dose for a toxicology treatment, as a very high dose might not have a physiological significance. Thus, it is very important that the optimal dose is determined in a gradient, and not simply the highest dose to get the desired results [126–129]. Furthermore, *in vivo* experimental data might be difficult to extrapolate into human applications, as the animal model might not necessarily provide an adequate analog. How is it best to go about the design and toxicology testing of nanomaterials?

Designing nanoparticle delivery systems is somewhat similar to a Rubik's cube puzzle. Although there are multiple parameters that can be specified, victory lies in the selection of the appropriate characteristics for targeting specific actions and localizations in the body. As the large surface area of the nanoparticle determines most of its properties, understanding the biointeractions at the surface becomes crucial. The selection of material plays a crucial role, yet, even nontoxic substances may possess toxicity in the form of a nanoparticle. Thus, through the manipulation of the nanoparticle surface, biocompatibility can be secured. Coating the nanoparticle with biocompatible material could help manage cytotoxicity and regulate cell uptake [89]. Furthermore, coating of the nanoparticle would offer control over the composition of proteins in the corona by steric hindrance. So far, the focus has been on the cytotoxicity of the nanoparticle; however, the nanoparticle may carry toxic compounds either by design, such as API, or through the manufacturing process—DMSO, and so on [130,131]. The stability of the coatings would be crucial in ensuring biocompatibility and the containment of toxic cargo. Appropriate shells could be either biocompatible organic or inorganic substances, such as PEG, PEG–SiO_2, gold, and biocompatible polymers [132].

Decreasing hydrophobicity and modulating the surface charge is another aspect that would allow for the attenuation of cytotoxicity by decreasing the adverse interactions with proteins, such as a strong, irreversible binding, a change of confirmation, a loss of activity, and fibrillation [92,133]. Agglomeration is another aspect that could help lower protein binding and the number of biointeractions. Further, adjusting the surface charge can help regulate cell uptake. For example, a layer-by-layer polyelectrolyte coating allows control over the rate of cell uptake, surface charge, and surface functionality [93]. Furthermore, functionalizations with hydrophilic groups can be used to increase urinary excretion rates as an additional safety measure to prevent the accumulation of potentially toxic materials [134,135]. It has also been postulated that an increase in nanomaterial solubility might lower the production of ROS and decrease their toxicity.

9.3 HEALTH EFFECTS BY CATEGORY

9.3.1 Health Effects of Fullerenes

Fullerenes are spherical nanomaterials having a diameter of about 1 nm. They are also termed C_{60}, *buckminsterfullerene,* or *buckyballs.* Structurally, the simplest fullerenes are composed of 60 linked carbon atoms in a stable icosahedron containing 60 vertices and 32 faces, and hence the name C_{60}. The faces are hexagonal (21) and pentagonal (12) in shape. These nanoparticles were first described by Kroto et al. in 1985 [136].

As reported by Powell and Karanek [137], fullerene nanoparticles are naturally produced during combustion processes, such as the burning of fuel or fires. Fullerenes can also be produced artificially and have recently found applications in pharmaceutical drug delivery and cosmetic products [138]. Considering these applications and direct human exposure, it is important to identify and assess the biocompatibility and magnitude of the deleterious effects of fullerenes. Also important is the fact that although fullerenes are described as nanoparticles consisting of 60 or more carbon atoms, typically they are found as larger agglomerates or clusters frequently termed as colloidal fullerenes. In addition, fullerenes are available in a variety of physical forms, such as fullerene derivatives, based on the number of carbon atoms, additional molecules attached to the surface, and the process of fullerene preparation. Thus, identifying the properties of fullerenes that contribute to their toxicity as well as their mechanism of such toxicity is highly significant. This requires a detailed analysis of the physicochemical properties of a given fullerene along with the toxicological evaluation [139].

Studies suggest several mechanistic pathways that may be involved in the manifestation of toxic effects of fullerenes. The commonly identified potential mechanisms for fullerene toxicities include the induction of an inflammatory response, an increase in oxidative stress, as well as toxicities due to genetic modifications. Several researchers have attempted to determine possible mechanisms of fullerene toxicity. Data from these studies are frequently speculative, often inconclusive, and occasionally even contradictory.

Nanoparticles as a general class have been shown to produce inflammatory responses, and this is thought to be a primary mechanism common to the members of this class, including fullerenes [140–142]. Rouse et al. [143] studied the potential proinflammatory responses of fullerene–amino acid complexes *in vitro* using human epidermal keratinocyte cells. The authors demonstrated an increase in the production of proinflammatory mediators like interleukin-8 and tumor necrosis factor α (TNF-α) upon exposure to fullerene complexes. Harhaji et al. [144] investigated the influence of fullerenes (C_{60}/C_{70}) and polyhydroxylated preparations on TNF-α-induced toxicity in the mouse, L929 fibroblast cell line. The authors demonstrated a synergistic interaction between C_{60}/C_{70} and TNF-α in producing the cytotoxic effects in the cells. However, functionalized fullerenes, that is polyhydroxylated fullerenes appeared to have a protective activity on the TNF-α-mediated cell death. The main mechanism involved in the toxicity was speculated to be the modulation of TNF-α-mediated reactive oxygen species (ROS) production. The study highlighted the fact that the magnitude of observed toxicities due to fullerenes depends on the type of preparation and it is difficult to generalize the predictability of toxic effects of fullerenes.

In contrast, Roursgaard et al. [145] showed that at lower concentrations in the lungs, fullerols had anti-inflammatory effects, whereas they showed proinflammatory effects at higher concentrations [145]. Similarly, Huang et al. [146] demonstrated the anti-inflammatory effects of fullerene derivatives *in vitro.* The authors demonstrated the free radical scavenging effects of fulleropyrrolidine–xanthine complex in lipopolysaccharide (LPS)-stimulated J774 macrophage-like cells. Fullerenes were able to reduce LPS-induced nitric oxide and TNF-α production in these cells. Tsao et al. [147] studied the anti-inflammatory effects of carboxyfullerene (up to 40 mg/kg) in protection against *Escherichia coli*-mediated meningitis in mice after intraperitoneal injection. The authors suggested that carboxyfullerene was able to cross the blood–brain barrier and was found to be more effective in attenuating *E. coli*-induced meningitis compared with dexamethasone.

Nanoparticles in general are known to increase in the oxidative stress by the virtue of enhancing the production of ROS within cells [148]. Since fullerenes, by the virtue of their size, belong to a subcategory of nanoparticles, it is important to understand their effects in this context. Sayes et al. [149], using a variety of cell lines, such as dermal fibroblasts, hepatocytes, and astrocytes, demonstrated cytotoxicity of $nanoC_{60}$ mediated via increased ROS production. The ROS-mediated mechanism of toxicity was ascertained by the protective effects observed upon the administration of an antioxidant, that is, ascorbic acid, against fullerene-induced cytotoxicity. Similarly, Kamat et al. [150] observed the oxidative toxicities of fullerenes (C_{60}) and polyhydroxylated fullerenes $(C_{60})OH_{22-26}$, within isolated rat liver microsomes. The production of oxidized metabolites, along with blockage of toxic effects by antioxidants, confirmed the oxidative mechanism of toxicity. In a contradictory study, Xia et al. [151] demonstrated that polyhydroxy derivatives of fullerenes failed to produce an oxidative response as measured by ROS production, glutathione depletion, or heme oxygenase 1 stimulation in macrophages, although mitochondrial damage was observed. In the same study, the authors evaluated a variety of nanoparticles and found that there was a significant difference in the abilities of these particles to increase oxidative stress and the magnitude of induced toxicity.

Several researchers have utilized standard genotoxicity tests and assays to evaluate the potential mutagenic effects of fullerenes on DNA. The commonly used tests for genotoxicity include the Ames test, comet assay, and the influence on tumor formation in biological systems.

Using a comet assay, Dhawan et al. [152] evaluated the DNA damaging potential of C_{60} within human lymphocytes. The authors demonstrated that fullerene particles prepared using different dispersion solvents, that is water or ethanol, showed a genotoxic response within cells, with a higher degree of toxicity elicited by fullerenes dispersed in water. In a different study, Sera et al. [153] investigated the influence of light and dark conditions on the mutagenic effects of fullerenes using the Ames test. The authors observed that fullerenes produced mutagenic effects on *Salmonella typhimurium* DNA in the presence of light via the production of ROS. DNA damage due to oxidative processes was demonstrated by increases in lipid peroxidation and the formation of 8-hydroxydeoxyguanosine.

9.3.2 Health Effects of Carbon Nanotubes

The similarities in physical properties between CNTs and asbestos have led to questions regarding the biological safety of CNT exposure [154]. Indeed, studies of single-walled CNT (SWCNT) and multiwalled CNT (MWCNT) in biological systems—both *in vitro* and *in vivo*—have resulted in DNA strand breakages and oxidative damages, chromosomal damages and mutations, and tissue inflammatory responses among other biological responses [154,155]. Besides being able to induce DNA breakages in a variety of *in vitro* models, SWCNTs and MWCNTs also showed similar effects in *in vivo* models [154]. CNTs induced the most breakage in DNA strands when investigating the effect of nano-sized biomaterials in fibroblasts [156]. This was validated by another study that showed purified functionalized MWCNTs induced greater DNA strand damage than nonfunctionalized MWCNTs, demonstrating stronger genotoxic potential [157]. Furthermore, pulmonary and hepatic production of 8-oxo-2′-deoxyguanosine (8-OHdG), an indicator of DNA oxidation, was detected in rats given SWCNTs (0.064 and 0.64 mg/kg) orally [158]. Significant DNA damage was also exhibited in lung and tracheal tissues, through inhalation exposure and injection of CNTs, respectively, among various other *in vivo* toxicology studies [154,159–161].

The genotoxic potential in CNTs may also contribute to chromosomal changes in biological systems, possibly as a result of structural defects in the CNTs [162]. CNTs were shown to disturb mitotic spindle formation, which could contribute to chromosomal changes and aneuploidy, along with multinucleation, in *in vitro* and *in vivo* biological systems exposed to CNTs [154,163], possibly a result of the CNTs' physical similarity to cellular microtubules [164]. Indeed, Sargent et al. [165] observed the presence and interaction of SWCNTs with centrosomes and mitotic spindles inside the nuclei of SWCNT-treated human epithelial cells. Aside from inducing micronucleation and enhanced sister chromatidic exchanges in A549 cells and CHO AA8 *in vitro* biological systems, respectively [160], CNTs showed similar altering effects in the lung and peritoneal *in vivo* biological systems, among others [154]. The genotoxic potential of the CNTs induced multinucleation and aneuploidy in Chinese hamster lung cells [166]. Three days following intratracheal instillation of 0.5–2 mg MWCNTs, rat lung epithelial cells displayed a significant dose-dependent increase of micronucleus formation [167]. Changes to chromosomal structure and formation of micronuclei in Swiss-Webster mice leukocytes were induced by intraperitoneal CNT injection, the effects of which were enhanced by functionalized MWCNTs [157]. Such chromosomal changes may be able to induce molecular, oncogenic changes in tissues. C57BL/6 mice exposed to 5 mg/m^3 SWCNTs through inhalation displayed enhanced pulmonary mutation of the K-ras gene, a known proto-oncogene [168].

Apart from causing systemic immune function and spleen damage, CNTs also have the potential to cause genotoxicity through pulmonary inflammation [155]. Damage to the upper respiratory tract and lymph nodes could be caused by high-dose MWCNT instillations through the oral and intratracheal routes [169]. Dermal contact with CNTs could cause an inflammatory response, leading to increases in dermal cell number, localized alopecia, and thickening of the skin [170]. MWCNTs and SWCNTs caused histiocytic and neutrophilic inflammation responses in mice lung tissues, along with injuring lung tissue and inducing early tumor formations, including granulomas [171–174].

In light of the aforementioned genotoxic potential of MWCNTs and SWCNTs, it is important to note that these hazards are highly variable and depend on the physiochemical properties of the tissue, CNTs, and experimental procedures under which the toxicological induction was observed [154]. However, this does not discount the hazards posed by CNTs; it behooves scientists to further investigate CNTs in efforts to avoid similar detrimental health defects observed by then-harmless asbestos use.

9.3.3 Health Effects of Inorganic Nanoparticles

The health risks associated with inorganic nanoparticle exposure are similar to those associated with the exposure to CNTs. Experiments have shown that inhaled nanoparticles enter the alveolar epithelium after phagocytization by alveolar macrophages, thereby making it harder for the nanoparticles to be removed from the lung tissue [175]. Indeed, silicon dioxide (SiO_2) particles were shown to cause silicosis among other respiratory abnormalities in individuals exposed to these particles [176]. These foreign, residual particles may stimulate inflammatory reactions and other related diseases in lung tissue [175]. Lung tissues exposed to iron oxide nanoparticles displayed acute inflammatory reactions via protein effusion, follicular hyperplasia of the lymph node, pulmonary capillary hyperemia, and alveolar lipoproteinosis, suggesting that the inhalation and intratracheal instillation of these nanoparticles may increase microvascular permeability and cell lysis in lung epithelium by initiating acute lung injury responses [175]. Furthermore, the residual nanoparticles induced the thickening of alveolar walls and the formation of collagen in lung epithelium, both of which are early signs of lung fibrosis [175]. In addition, the further long-term deposition of nanoparticles in lung tissues led to initial signs of emphysema, which progressed with time and exposure to the nanoparticles [175].

The seemingly physiological sequestering of nanoparticles in lung tissue by alveolar macrophages may induce oxidative stress aside from the induction of inflammatory responses, which may be caused and exacerbated by the metallic nature of most inorganic nanoparticles [175]. Apart from being found inside cytosolic organelles, especially nuclei and mitochondria, inorganic nanoparticles showed signs of damaging DNA through its breakage and oxidation [177–179]. The cytotoxic potential of SiO_2 nanoparticles dealt with the production of ROS, which induced DNA damage [177]. Human lung cells exposed to titanium dioxide (TiO_2) nanoparticles displayed significant levels of 8-OHdG, an indicator of oxidative DNA damage [178]. In addition, TiO_2 induced DNA adduct formation by the production of elevated free radical levels in lung tissue [178]. However, TiO_2 nanoparticles were not able to induce DNA breakage, as displayed by ferric oxide nanoparticles [175,178]. DNA damage was shown to be dose dependent in silver nanoparticle activity [179]. TEM images indicating the presence of silver nanoparticles in cellular nuclei and mitochondria implicate these nanoparticles in the production of ROS, leading to mitochondrial toxicity and DNA damage [179]. Cells exposed to silver nanoparticles spent more time in the G_2/M cell cycle phase to repair the damage done to DNA [179].

9.3.4 HEALTH EFFECTS OF ORGANIC NANOPARTICLES

An advantage of the use of organic nanoparticles as opposed to inorganic nanoparticles is the possibility of reduced cytotoxicity due to oxidative stress seen in inorganic nanoparticles. A number of organic nanoparticles have been developed and used, and have been relatively successful in reducing cytotoxic effects in cells. However, the chemicals used in the preparation of organic nanoparticles have the ability to induce cytotoxic effects in *in vitro* models.

Solid lipid nanoparticles (SLNs) are made of biocompatible lipids and waxes with various sizes and alkyl polymer length, depending on the preparation method employed [180]. However, owing to the hydrophobic surface properties of these nanoparticles, they are prone to becoming complexed with plasma proteins and cytokines, facilitating their clearance from the blood stream by phagocytic cells [180]. However, the use of surfactants in the synthesis of these polymers in efforts to modify the surface properties of SLN could potentially affect phagocytic activity, thereby enhancing the particle's therapeutic efficiency in biological settings [180]. The use of surfactants previously shown to be relatively safe would reduce any cytotoxic potential in the nanoparticles. However, the size of the SLN and alkyl chain length may also contribute to its cytotoxicity: longer chain lengths and smaller particles tend to be more toxic [180]. Moreover, a similar surfactant-induced cytotoxicity effect is displayed by certain surfactants used in the production of edible lipid nanoparticles—used in the food industry, and PLGA nanoparticles (polymers of polylactic acid and polyglycolic acid) [181,182]. In the case of the PLGA nanoparticles, the modified surface charges of the nanoparticles were also able to induce inflammatory responses in addition to the responses induced by the surfactants, as measured by the release of cytokines and interleukins [181]. Related to the cytotoxic effects of functional groups in SLNs, a similar phenomenon was displayed in acylated starch nanoparticles, which induced higher cytotoxicity via aromatic group substitution as compared with aliphatic group substitution [183].

Chitosan, a biopolymer produced by the deacetylation of chitin obtained from crustacean shells, is widely used in the production of nanoparticles [184,185]. Owing to its poor solubility at physiological pH, the modification of chitosan with oligosaccharides enables its solubility in water and several organic solvents, enabling effective drug delivery in low cytotoxic levels [184,186]. Unmodified chitosan displays an almost-zero zeta potential, imparting bio-inert qualities that would reduce irritation to surrounding tissue [185]. It shows great biocompatibility, very low cytotoxicity, and biodegradability [185]. In addition, its cationic surface charge allows it to complex with DNA, imparting stability and functioning as a possible transfection agent following surface modification [185]. However, because chitosan is a carbohydrate-based compound derived from crustacean shells, it has the possibility of inducing an immune response in the presence of antigens to surface modifications [187]. However, these would vary by individual and targeted tissue.

Nanomicelles, self-assembling nanoparticles with a hydrophobic core and hydrophilic shell, are an emerging therapeutic vector [188]. Many hydrophobic drugs are blocked from uptake by tissue because of the hydrophilic nature of the tissue; nanomicelles entrap the hydrophobic drug in its core and the hydrophilic surface enables the particle to penetrate hydrophilic tissues [188]. Relatively high levels of entrapment can be achieved by this method and the particles are stable in tissue for an extended period of time, enhancing therapeutic effects [188]. Nanomicelles' small size and very low toxicity provide for an emerging therapeutic model that should be considered in future *in vitro* and *in vivo* testing to allow its large-scale therapeutic development [188].

9.3.5 Health Effects of Quantum Dots

The biomedical engineering potential of quantum dots begs investigation into its toxicity in biological systems. It seems that the variability in composition of quantum dots also provides variability in its potential toxicities.

Near-infrared-emitting quantum dots (NIR QDs) are models of imaging probes made in nanoscale sizes, which have applications in imaging body tissues for potential tumor detection and therapies [189]. *In vivo* studies revealed that there was no long-term overt toxicity in mice over a 94-day exposure period using histological and biochemical analyses along with body weight measurements [189]. NIR QDs were found to have accumulated in the liver, spleen, and lung up to 4 h post-injection, but were increasingly absorbed in the blood, liver, and kidneys up to 94 days post-injection in a time-dependent manner [189]. A similar biocompatibility was exhibited by graphene quantum dots (GQD) in multiple mouse organs [190]. Postintravenous injection of the citric acid-complexed, carboxylated GQD revealed accumulation of GQD in the liver, spleen, lung, and kidney without any mentionable toxicity, biological interactions, morphological changes, or inflammatory responses through histological analyses up to 21 days post-injection [190]. In addition, the complex formed with citric acid allowed the GQD to be excreted through the kidneys [190].

Contrarily, there was shown to be some toxicity and induction of an inflammatory response by treatment of cadmium-complexed quantum dots [191–194]. In some cases, cadmium, which is a highly reactive and potentially toxic metal, can induce cytotoxicity by freeing itself from the quantum dot complex [193,194]. Free or complexed cadmium can increase levels of ROS, cause oxidative stress, interfere with antioxidant defenses, and activate intrinsic and extrinsic pathways causing apoptosis in effected cells [191]. In addition, rats displayed dose-dependent lung injury and inflammation after intratracheal instillation of cadmium-complexed quantum dots, with the cadmium remaining in biological systems for up to 28 days post-instillation [193]. The reactivity of macrophages and monocytes in the inflammation pathway in the presence of cadmium quantum dots implicates the sensitivity of adaptive immunity in such cases [192]. Improved methods for complexing cadmium to quantum dots may prove effective in reducing instances of cytotoxicity in biological systems treated with cadmium quantum dots.

9.4 CONCLUSION

As the applications of nanoparticles grow, it is important that the field of nanotoxicology receives more attention. The wide use of different types of nanoparticles in a variety of applications—cosmetic, medicinal, industrial, and so on—increases the public's exposure to nanoparticles. Because of the relatively recent introduction of these materials into common use, we are just beginning to understand the health effects these particles may pose.

The variety of applications that these nanoparticles have also increases the frequency that their exposure may lead to detrimental health effects. The use of silver nanoparticles as antibiotics on textiles may lead to their introduction into wastewaters and food chains, possibly disrupting ecological balances and unknowingly being introduced to our bodies. Nanoparticles in creams and cosmetics can escape into deeper layers of the skin through hair follicles, and even enter systemic circulation in some circumstances, thereby opening the door to a variety of effects of which we may not be aware. Nanoparticles, of which we risk unknown exposure, have a variety of physical properties and applications. These qualities may influence the nanoparticle's interactions with our cells in a number of ways—some clandestinely beneficial, others horribly detrimental.

Scientists can carry out *in vitro* and *in vivo* tests to investigate the possible effects that nanoparticles may have on biological systems. However, there are drawbacks to these tests. Cells that are exposed to nanoparticles *in vitro* may not react in the same way that they would in a normal, biological setting. Furthermore, when attempting to obtain results after exposing cells to nanomaterials, it is very possible that the nanomaterials may react with the substrates and reagents essential for the assays being carried out. *In vivo* tests may be carried out to avoid the obstacles posed by *in vitro* testing, but they may be very expensive and may only be approved by institutions to be carried out in a limited number of organisms. Moreover, the results obtained from *in vivo* tests may not be applicable to humans, as, although the organ systems may be comparable between humans and other organisms, the way tissues react may not be exactly the same. However, this does not take away from the contributions that numerous *in vivo* and *in vitro* tests have made to the field of nanotoxicology.

Indeed, through various *in vitro* and *in vivo* tests, scientists have found that the physical properties of nanoparticles impart characteristic, cytotoxic potential. The nanoparticle surface charge may indicate its length of exposure and excretion from body systems, along with the nature of its interaction with cells and cell membranes. The nanoparticle may be able to pass through the cell membrane into the cell, or may change the physical properties of the cell membrane altogether, inducing apoptosis in some cases. In addition, the surface charge of the nanoparticles may impart the ability of the nanoparticle to attract and complex with a number of proteins native to the biological system in contact with the particle. This type of interaction may change the conformation of the protein, changing its effect on the host tissue, and introducing a novel process by which cytotoxicity may be induced. The corona of proteins that the nanoparticle attracts to its surface may also dictate the type of interactions that the nanoparticle has with neighboring cells and its excretion from the body. The physical properties of the nanoparticles also dictate its location in the body and cell. Different organs and organelles act as specific reservoirs for nanoparticles, based on the physical properties of the nanoparticles. This contributes to the production of ROS, induction of DNA breakage and oxidation, and stimulation of an inflammatory response in the effected cells and tissues. All of these contribute to the nanoparticle's cytotoxic potential.

The effects that nanoparticles may have on an individual depend on a number of factors. *In vitro* and *in vivo* biological models introduce scientists to the properties that may become cytotoxic in humans, but the exact reaction that the body has may not be as easy to predict, especially with the variations seen in extrapolating *in vitro* results to *in vivo* systems. However, such models have helped scientists obtain invaluable information in the field of nanotoxicity. These tests have enabled scientists to expand the field and obtain data regarding basic properties in nanoparticles that may induce cytotoxic effects. This indispensable first line of testing has, and continues to, further the field of nanotoxicology and ensures that nanoparticles will become less toxic and more efficient. The field of nanotoxicity is still one that is growing. It behooves scientists to help further the field, as the applications of nanomaterials are numerous and may turn out to have applications that can have a profound effect on human health and well-being.

9.5 SUMMARY

In this chapter, the health implications of nanomaterials are considered with regard to their toxicological, and potentially beneficial, properties. The field of nanotoxicology, a relatively new field, is essential to the further development of the exciting applications of nanotechnology. Careful study and investigation of the parameters of nanomaterials and their interactions with biomembranes and biological fluids is required for the adequate design of all the components of nanomaterials to minimize and circumvent potential toxicological effects, and to ensure optimal execution of the beneficial effect. In addition, the routes of exposure and their absorption, digestion, metabolism, and elimination (ADME) parameters also influence the physiological effects of the nanomaterials. Nanomaterial properties, route of exposure, and their ADME parameters ultimately determine their biocompatibility.

REFERENCES

1. Donaldson K, Stone V, Tran CL, Kreyling W, Borm PJ. Nanotoxicology. *Occupational and Environmental Medicine*. 2004;61(9):727–8. Epub 2004/08/20.
2. Handy RD, Shaw BJ. Toxic effects of nanoparticles and nanomaterials: Implications for public health, risk assessment and the public perception of nanotechnology. *Health Risk & Society*. 2007;9(2): 125–44.

3. Merrifield DL, Shaw BJ, Harper GM, Saoud IP, Davies SJ, Handy RD et al. Ingestion of metal-nanoparticle contaminated food disrupts endogenous microbiota in zebrafish (*Danio rerio*). *Environmental Pollution.* 2013;174:157–63. Epub 2012/12/25.
4. Chernousova S, Epple M. Silver as antibacterial agent: Ion, nanoparticle, and metal. *Angewandte Chemie.* 2012. Epub 2012/12/21.
5. Miller CN, Newall N, Kapp SE, Lewin G, Karimi L, Carville K et al. A randomized-controlled trial comparing cadexomer iodine and nanocrystalline silver on the healing of leg ulcers. *Wound Repair and Regeneration.* 2010;18(4):359–67. Epub 2010/07/20.
6. Chen MY, Millwood IY, Wand H, Poynten M, Law M, Kaldor JM et al. A randomized controlled trial of the safety of candidate microbicide SPL7013 gel when applied to the penis. *Journal of Acquired Immune Deficiency Syndrome.* 2009;50(4):375–80. Epub 2009/02/14.
7. Rigo C, Ferroni L, Tocco I, Roman M, Munivrana I, Gardin C et al. Active silver nanoparticles for wound healing. *International Journal of Molecular Sciences.* 2013;14(3):4817–40. Epub 2013/03/05.
8. World Health Organization. *Health Criteria and Other Supporting Information.* 2nd ed. Geneva: World Health Organization; 1996.
9. Alexander JW. History of the medical use of silver. *Surgical Infections.* 2009;10(3):289–92. Epub 2009/07/02.
10. Alexander JW, Rahn R, Goodman HR. Prevention of surgical site infections by an infusion of topical antibiotics in morbidly obese patients. *Surgical Infections.* 2009;10(1):53–7. Epub 2009/02/28.
11. Fung MC, Bowen DL. Silver products for medical indications: Risk-benefit assessment. *Journal of Toxicology Clinical Toxicology.* 1996;34(1):119–26. Epub 1996/01/01.
12. Plack W, Bellizzi R. Generalized argyria secondary to chewing photographic film. Report of a case. *Oral Surgery, Oral Medicine, and Oral Pathology.* 1980;49(6):504–6. Epub 1980/06/01.
13. Kimura E, Kawano Y, Todo H, Ikarashi Y, Sugibayashi K. Measurement of skin permeation/penetration of nanoparticles for their safety evaluation. *Biological & Pharmaceutical Bulletin.* 2012;35(9):1476–86. Epub 2012/09/15.
14. Samberg ME, Oldenburg SJ, Monteiro-Riviere NA. Evaluation of silver nanoparticle toxicity in skin *in vivo* and keratinocytes *in vitro*. *Environmental Health Perspectives.* 2010;118(3):407–13. Epub 2010/01/13.
15. Brabez N, Lynch RM, Xu L, Gillies RJ, Chassaing G, Lavielle S et al. Design, synthesis, and biological studies of efficient multivalent melanotropin ligands: Tools toward melanoma diagnosis and treatment. *Journal of Medicinal Chemistry.* 2011;54(20):7375–84. Epub 2011/09/21.
16. Nguyen TH, Kim YH, Song HY, Lee BT. Nano Ag loaded PVA nano-fibrous mats for skin applications. *Journal of Biomedical Materials Research B Application Biomaterials.* 2011;96(2):225–33. Epub 2011/01/07.
17. Kokura S, Handa O, Takagi T, Ishikawa T, Naito Y, Yoshikawa T. Silver nanoparticles as a safe preservative for use in cosmetics. *Nanomedicine.* 2010;6(4):570–4. Epub 2010/01/12.
18. Monteiro MS, Ozzetti RA, Vergnanini AL, de Brito-Gitirana L, Volpato NM, de Freitas ZM et al. Evaluation of octyl p-methoxycinnamate included in liposomes and cyclodextrins in anti-solar preparations: Preparations, characterizations and *in vitro* penetration studies. *International Journal of Nanomedicine.* 2012;7:3045–58. Epub 2012/07/13.
19. Dragicevic-Curic N, Fahr A. Liposomes in topical photodynamic therapy. *Expert Opinion on Drug Delivery.* 2012;9(8):1015–32. Epub 2012/06/27.
20. Golmohammadzadeh S, Jaafarixx MR, Khalili N. Evaluation of liposomal and conventional formulations of octyl methoxycinnamate on human percutaneous absorption using the stripping method. *Journal of Cosmetic Science.* 2008;59(5):385–98. Epub 2008/10/09.
21. Papakostas D, Rancan F, Sterry W, Blume-Peytavi U, Vogt A. Nanoparticles in dermatology. *Archives of Dermatological Research.* 2011;303(8):533–50. Epub 2011/08/13.
22. Pasricha A, Jangra SL, Singh N, Dilbaghi N, Sood KN, Arora K et al. Comparative study of leaching of silver nanoparticles from fabric and effective effluent treatment. *Journal of Environmental Sciences.* 2012;24(5):852–9. Epub 2012/08/17.
23. Whitlow TH, Hall A, Zhang KM, Anguita J. Impact of local traffic exclusion on near-road air quality: Findings from the New York City "Summer Streets" campaign. *Environmental Pollution.* 2011;159(8–9):2016–27. Epub 2011/03/25.
24. Li R, Ning Z, Cui J, Khalsa B, Ai L, Takabe W et al. Ultrafine particles from diesel engines induce vascular oxidative stress via JNK activation. *Free Radical Biology & Medicine.* 2009;46(6):775–82. Epub 2009/01/22.
25. Harrison RM, Yin J. Particulate matter in the atmosphere: Which particle properties are important for its effects on health? *The Science of the Total Environment.* 2000;249(1–3):85–101. Epub 2000/05/17.

26. Kreyling W, Semmler-Behnke M, M√∂ller W. Health implications of nanoparticles. *Journal of Nanoparticle Research.* 2006;8(5):543–62.
27. Warheit DB, Sayes CM, Reed KL. Nanoscale and fine zinc oxide particles: Can *in vitro* assays accurately forecast lung hazards following inhalation exposures? *Environmental Science & Technology.* 2009;43(20):7939–45. Epub 2009/11/20.
28. Dhawan A, Sharma V. Toxicity assessment of nanomaterials: Methods and challenges. *Analytical and Bioanalytical Chemistry.* 2010;398(2):589–605. Epub 2010/07/24.
29. Arora S, Rajwade JM, Paknikar KM. Nanotoxicology and *in vitro* studies: The need of the hour. *Toxicology and Applied Pharmacology.* 2012;258(2):151–65. Epub 2011/12/20.
30. National Research Council (US). *Toxicity Testing in the 21st Century: A Vision and a Strategy.* The National Academies Press; 2007.
31. Monteiro-Riviere NA, Inman AO, Zhang LW. Limitations and relative utility of screening assays to assess engineered nanoparticle toxicity in a human cell line. *Toxicology and Applied Pharmacology.* 2009;234(2):222–35. Epub 2008/11/06.
32. Griffiths SM, Singh N, Jenkins GJ, Williams PM, Orbaek AW, Barron AR et al. Dextran coated ultrafine superparamagnetic iron oxide nanoparticles: Compatibility with common fluorometric and colorimetric dyes. *Analytical Chemistry.* 2011;83(10):3778–85. Epub 2011/04/08.
33. Karlsson HL. The comet assay in nanotoxicology research. *Analytical and Bioanalytical Chemistry.* 2010;398(2):651–66. Epub 2010/07/20.
34. Singh N, Manshian B, Jenkins GJ, Griffiths SM, Williams PM, Maffeis TG et al. NanoGenotoxicology: The DNA damaging potential of engineered nanomaterials. *Biomaterials.* 2009;30(23–24):3891–914. Epub 2009/05/12.
35. Lai ZW, Yan Y, Caruso F, Nice EC. Emerging techniques in proteomics for probing nano-bio interactions. *ACS Nano.* 2012;6(12):10438–48. Epub 2012/12/12.
36. Buford MC, Hamilton RF, Jr., Holian A. A comparison of dispersing media for various engineered carbon nanoparticles. *Particle and Fibre Toxicology.* 2007;4:6. Epub 2007/07/28.
37. Sayes CM, Reed KL, Warheit DB. Assessing toxicity of fine and nanoparticles: Comparing *in vitro* measurements to *in vivo* pulmonary toxicity profiles. *Toxicological Sciences: An Official Journal of the Society of Toxicology.* 2007;97(1):163–80. Epub 2007/02/16.
38. Chen C, Xing G, Wang J, Zhao Y, Li B, Tang J et al. Multihydroxylated [Gd Antineoplastic activity of high efficiency and low toxicity. *Nano Letters* 2005;5(10):2050–7. Epub 2005/10/13.
39. Gonzalez L, Sanderson BJ, Kirsch-Volders M. Adaptations of the *in vitro* MN assay for the genotoxicity assessment of nanomaterials. *Mutagenesis.* 2011;26(1):185–91. Epub 2010/12/18.
40. Sayes C, Ivanov I. Comparative study of predictive computational models for nanoparticle-induced cytotoxicity. *Risk Analysis: An official publication of the Society for Risk Analysis.* 2010;30(11):1723–34. Epub 2010/06/22.
41. North M, Vulpe CD. Functional toxicogenomics: Mechanism-centered toxicology. *International Journal of Molecular Sciences.* 2010;11(12):4796–813. Epub 2010/01/01.
42. Curtis J, Greenberg M, Kester J, Phillips S, Krieger G. Nanotechnology and nanotoxicology: A primer for clinicians. *Toxicological Reviews.* 2006;25(4):245–60. Epub 2007/02/10.
43. Peters A, Wichmann HE, Tuch T, Heinrich J, Heyder J. Respiratory effects are associated with the number of ultrafine particles. *American Journal of Respiratory and Critical Care Medicine.* 1997;155(4):1376–83.
44. Kane AB, Hurt RH. Nanotoxicology: The asbestos analogy revisited. *Nature Nanotechnology.* 2008;3(7):378–9. Epub 2008/07/26.
45. Donaldson K, Stone V, Duffin R, Clouter A, Schins R, Borm P. The quartz hazard: Effects of surface and matrix on inflammogenic activity. *Journal of Environmental Pathology, Toxicology and Oncology: Official Organ of the International Society for Environmental Toxicology and Cancer.* 2001;20(Suppl 1):109–18. Epub 2001/09/26.
46. Alfaro-Moreno E, Nawrot TS, Nemmar A, Nemery B. Particulate matter in the environment: Pulmonary and cardiovascular effects. *Current Opinion in Pulmonary Medicine.* 2007;13(2):98–106. Epub 2007/01/27.
47. Gillissen A, Gessner C, Hammerschmidt S, Hoheisel G, Wirtz H. [Health significance of inhaled particles]. *Deutsche medizinische Wochenschrift.* 2006;131(12):639–44. Epub 2006/03/18. Gesundheitliche Bedeutung inhalierter Staube.
48. Warheit DB, Reed KL, Delorme MP. Embracing a weight-of-evidence approach for establishing NOAELs for nanoparticle inhalation toxicity studies. *Toxicologic Pathology.* 2012. Epub 2012/12/18.
49. Beckett WS, Chalupa DF, Pauly-Brown A, Speers DM, Stewart JC, Frampton MW et al. Comparing inhaled ultrafine versus fine zinc oxide particles in healthy adults: A human inhalation study. *American Journal of Respiratory and Critical Care Medicine.* 2005;171(10):1129–35.

50. Mossman BT, Bignon J, Corn M, Seaton A, Gee JB. Asbestos: Scientific developments and implications for public policy. *Science (New York, NY)*. 1990;247(4940):294–301.
51. Brown RC, Hoskins JA, Miller K, Mossman BT. Pathogenetic mechanisms of asbestos and other mineral fibres. *Molecular Aspects of Medicine*. 1990;11(5):325–49.
52. Peters A, Veronesi B, Calderón-Garcidueñas L, Gehr P, Chen LC, Geiser M et al. Translocation and potential neurological effects of fine and ultrafine particles a critical update. *Particle and Fibre Toxicology*. 2006;3:13.
53. Geiser M, Rothen-Rutishauser B, Kapp N, Schürch S, Kreyling W, Schulz H et al. Ultrafine particles cross cellular membranes by nonphagocytic mechanisms in lungs and in cultured cells. *Environmental Health Perspectives*. 2005;113(11):1555–60.
54. Hagens WI, Oomen AG, de Jong WH, Cassee FR, Sips AJ. What do we (need to) know about the kinetic properties of nanoparticles in the body? *Regulatory Toxicology and Pharmacology: RTP*. 2007;49(3):217–29. Epub 2007/09/18.
55. Oberdorster G, Oberdorster E, Oberdorster J. Nanotoxicology: An emerging discipline evolving from studies of ultrafine particles. *Environmental Health Perspectives*. 2005;113(7):823–39. Epub 2005/07/09.
56. Chen Z, Meng H, Xing G, Chen C, Zhao Y, Jia G et al. Acute toxicological effects of copper nanoparticles *in vivo*. *Toxicology Letters*. 2006;163(2):109–20.
57. Yamago S, Tokuyama H, Nakamura E, Kikuchi K, Kananishi S, Sueki K et al. *In vivo* biological behavior of a water-miscible fullerene: 14C labeling, absorption, distribution, excretion and acute toxicity. *Chemistry & Biology*. 1995;2(6):385–9.
58. Alvarez-Román R, Naik A, Kalia YN, Guy RH, Fessi H. Skin penetration and distribution of polymeric nanoparticles. *Journal of Controlled Release*. 2004;99(1):53–62.
59. Tinkle SS, Antonini JM, Rich BA, Roberts JR, Salmen R, DePree K et al. Skin as a route of exposure and sensitization in chronic beryllium disease. *Environmental Health Perspectives*. 2003;111(9):1202–8. Epub 2003/07/05.
60. Monteiro-Riviere NA, Nemanich RJ, Inman AO, Wang YY, Riviere JE. Multi-walled carbon nanotube interactions with human epidermal keratinocytes. *Toxicology Letters*. 2005;155(3):377–84. Epub 2005/01/15.
61. Rouse JG, Yang J, Ryman-Rasmussen JP, Barron AR, Monteiro-Riviere NA. Effects of mechanical flexion on the penetration of fullerene amino acid-derivatized peptide nanoparticles through skin. *Nano Letters*. 2007;7(1):155–60. Epub 2007/01/11.
62. Baroli B, Ennas MG, Loffredo F, Isola M, Pinna R, Lopez-Quintela MA. Penetration of metallic nanoparticles in human full-thickness skin. *The Journal of Investigative Dermatology*. 2007;127(7):1701–12. Epub 2007/03/24.
63. Larese FF, D'Agostin F, Crosera M, Adami G, Renzi N, Bovenzi M et al. Human skin penetration of silver nanoparticles through intact and damaged skin. *Toxicology*. 2009;255(1Äì2):33–7.
64. Mostafalou S, Mohammadi H, Ramazani A, Abdollahi M. Different biokinetics of nanomedicines linking to their toxicity; An overview. *Daru: Journal of Faculty of Pharmacy, Tehran University of Medical Sciences*. 2013;21(1):14. Epub 2013/02/26.
65. Albanese A, Tang PS, Chan WC. The effect of nanoparticle size, shape, and surface chemistry on biological systems. *Annual Review of Biomedical Engineering*. 2012;14:1–16. Epub 2012/04/25.
66. Gratton SE, Ropp PA, Pohlhaus PD, Luft JC, Madden VJ, Napier ME et al. The effect of particle design on cellular internalization pathways. *Proceedings of the National Academy of Sciences of the United States of America*. 2008;105(33):11613–8. Epub 2008/08/14.
67. Qiu Y, Liu Y, Wang L, Xu L, Bai R, Ji Y et al. Surface chemistry and aspect ratio mediated cellular uptake of Au nanorods. *Biomaterials*. 2010;31(30):7606–19. Epub 2010/07/27.
68. Chithrani BD, Ghazani AA, Chan WC. Determining the size and shape dependence of gold nanoparticle uptake into mammalian cells. *Nano Letters*. 2006;6(4):662–8. Epub 2006/04/13.
69. Chithrani BD, Chan WC. Elucidating the mechanism of cellular uptake and removal of protein-coated gold nanoparticles of different sizes and shapes. *Nano Letters*. 2007;7(6):1542–50. Epub 2007/05/01.
70. Hutter E, Boridy S, Labrecque S, Lalancette-Hebert M, Kriz J, Winnik FM et al. Microglial response to gold nanoparticles. *ACS Nano*. 2010;4(5):2595–606. Epub 2010/03/25.
71. Yuan H, Li J, Bao G, Zhang S. Variable nanoparticle-cell adhesion strength regulates cellular uptake. *Physical Review Letters*. 2010;105(13):138101. Epub 2011/01/15.
72. Lu F, Wu SH, Hung Y, Mou CY. Size effect on cell uptake in well-suspended, uniform mesoporous silica nanoparticles. *Small*. 2009;5(12):1408–13. Epub 2009/03/20.
73. Jin H, Heller DA, Sharma R, Strano MS. Size-dependent cellular uptake and expulsion of single-walled carbon nanotubes: Single particle tracking and a generic uptake model for nanoparticles. *ACS Nano*. 2009;3(1):149–58. Epub 2009/02/12.

74. Wang J, Tian S, Petros RA, Napier ME, Desimone JM. The complex role of multivalency in nanoparticles targeting the transferrin receptor for cancer therapies. *Journal of the American Chemical Society.* 2010;132(32):11306–13. Epub 2010/08/12.
75. Thorek DL, Tsourkas A. Size, charge and concentration dependent uptake of iron oxide particles by non-phagocytic cells. *Biomaterials.* 2008;29(26):3583–90. Epub 2008/06/06.
76. Slowing I, Trewyn BG, Lin VS. Effect of surface functionalization of MCM-41-type mesoporous silica nanoparticles on the endocytosis by human cancer cells. *Journal of the American Chemical Society.* 2006;128(46):14792–3. Epub 2006/11/16.
77. Arvizo RR, Miranda OR, Thompson MA, Pabelick CM, Bhattacharya R, Robertson JD et al. Effect of nanoparticle surface charge at the plasma membrane and beyond. *Nano Letters.* 2010;10(7):2543–8. Epub 2010/06/11.
78. Wang B, Zhang L, Bae SC, Granick S. Nanoparticle-induced surface reconstruction of phospholipid membranes. *Proceedings of the National Academy of Sciences of the United States of America.* 2008;105(47):18171–5. Epub 2008/11/18.
79. Giljohann DA, Seferos DS, Patel PC, Millstone JE, Rosi NL, Mirkin CA. Oligonucleotide loading determines cellular uptake of DNA-modified gold nanoparticles. *Nano Letters.* 2007;7(12):3818–21. Epub 2007/11/14.
80. Cedervall T, Lynch I, Lindman S, Berggard T, Thulin E, Nilsson H et al. Understanding the nanoparticle-protein corona using methods to quantify exchange rates and affinities of proteins for nanoparticles. *Proceedings of the National Academy of Sciences of the United States of America.* 2007;104(7):2050–5. Epub 2007/02/03.
81. Lynch I, Cedervall T, Lundqvist M, Cabaleiro-Lago C, Linse S, Dawson KA. The nanoparticle-protein complex as a biological entity; a complex fluids and surface science challenge for the 21st century. *Advances in Colloid and Interface Science.* 2007;134–135:167–74. Epub 2007/06/19.
82. Lynch I, Salvati A, Dawson KA. Protein-nanoparticle interactions: What does the cell see? *Nature Nanotechnology.* 2009;4(9):546–7. Epub 2009/09/08.
83. Hellstrand E, Lynch I, Andersson A, Drakenberg T, Dahlback B, Dawson KA et al. Complete high-density lipoproteins in nanoparticle corona. *The FEBS Journal.* 2009;276(12):3372–81. Epub 2009/05/15.
84. Elsaesser A, Howard CV. Toxicology of nanoparticles. *Advanced Drug Delivery Reviews.* 2012;64(2):129–37. Epub 2011/09/20.
85. Cross SE, Innes B, Roberts MS, Tsuzuki T, Robertson TA, McCormick P. Human skin penetration of sunscreen nanoparticles: *In-vitro* assessment of a novel micronized zinc oxide formulation. *Skin Pharmacology and Physiology.* 2007;20(3):148–54. Epub 2007/01/19.
86. Madan J, Pandey RS, Jain V, Katare OP, Chandra R, Katyal A. Poly (ethylene)-glycol conjugated solid lipid nanoparticles of noscapine improve biological half-life, brain delivery and efficacy in glioblastoma cells. *Nanomedicine (London).* 2013;9(4):492–503. Epub 2012/11/03.
87. Geng Y, Dalhaimer P, Cai S, Tsai R, Tewari M, Minko T et al. Shape effects of filaments versus spherical particles in flow and drug delivery. *Nature Nanotechnology.* 2007;2(4):249–55. Epub 2008/07/26.
88. Choi HS, Liu W, Misra P, Tanaka E, Zimmer JP, Itty Ipe B et al. Renal clearance of quantum dots. *Nature Biotechnology.* 2007;25(10):1165–70. Epub 2007/09/25.
89. Nel AE, Madler L, Velegol D, Xia T, Hoek EM, Somasundaran P et al. Understanding biophysicochemical interactions at the nano-bio interface. *Nature Materials.* 2009;8(7):543–57. Epub 2009/06/16.
90. Donaldson K, Tran L, Jimenez LA, Duffin R, Newby DE, Mills N et al. Combustion-derived nanoparticles: A review of their toxicology following inhalation exposure. *Particle and Fibre Toxicology.* 2005;2:10. Epub 2005/10/26.
91. Walczyk D, Bombelli FB, Monopoli MP, Lynch I, Dawson KA. What the cell "sees" in bionanoscience. *Journal of the American Chemical Society.* 2010;132(16):5761–8. Epub 2010/04/02.
92. Yang ST, Liu Y, Wang YW, Cao A. Biosafety and bioapplication of nanomaterials by designing protein-nanoparticle interactions. *Small.* 2013. Epub 2013/01/24.
93. Ryman-Rasmussen JP, Riviere JE, Monteiro-Riviere NA. Surface coatings determine cytotoxicity and irritation potential of quantum dot nanoparticles in epidermal keratinocytes. *The Journal of Investigative Dermatology.* 2007;127(1):143–53. Epub 2006/08/12.
94. Su Y, He Y, Lu H, Sai L, Li Q, Li W et al. The cytotoxicity of cadmium based, aqueous phase—synthesized, quantum dots and its modulation by surface coating. *Biomaterials.* 2009;30(1):19–25. Epub 2008/10/14.
95. Tang M, Xing T, Zeng J, Wang H, Li C, Yin S et al. Unmodified CdSe quantum dots induce elevation of cytoplasmic calcium levels and impairment of functional properties of sodium channels in rat primary cultured hippocampal neurons. *Environmental Health Perspectives.* 2008;116(7):915–22. Epub 2008/07/17.

96. Wang L, Nagesha DK, Selvarasah S, Dokmeci MR, Carrier RL. Toxicity of CdSe Nanoparticles in Caco-2 Cell Cultures. *Journal of Nanobiotechnology*. 2008;6:11. Epub 2008/10/25.
97. Zhang Y, Chen W, Zhang J, Liu J, Chen G, Pope C. In vitro and in vivo toxicity of CdTe nanoparticles. *Journal of Nanoscience and Nanotechnology*. 2007;7(2):497–503. Epub 2007/04/25.
98. Jones P, Sugino S, Yamamura S, Lacy F, Biju V. Impairments of cells and genomic DNA by environmentally transformed engineered nanomaterials. *Nanoscale*. 2013. Epub 2013/07/23.
99. Chu M, Wu Q, Yang H, Yuan R, Hou S, Yang Y et al. Transfer of quantum dots from pregnant mice to pups across the placental barrier. *Small*. 2010;6(5):670–8. Epub 2010/02/10.
100. Farkas J, Christian P, Urrea JA, Roos N, Hassellov M, Tollefsen KE et al. Effects of silver and gold nanoparticles on rainbow trout (*Oncorhynchus mykiss*) hepatocytes. *Aquatic Toxicology*. 2010;96(1):44–52. Epub 2009/10/27.
101. Li JJ, Hartono D, Ong CN, Bay BH, Yung LY. Autophagy and oxidative stress associated with gold nanoparticles. *Biomaterials*. 2010;31(23):5996–6003. Epub 2010/05/15.
102. Pan Y, Leifert A, Ruau D, Neuss S, Bornemann J, Schmid G et al. Gold nanoparticles of diameter 1.4 nm trigger necrosis by oxidative stress and mitochondrial damage. *Small*. 2009;5(18):2067–76. Epub 2009/07/31.
103. Tarantola M, Schneider D, Sunnick E, Adam H, Pierrat S, Rosman C et al. Cytotoxicity of metal and semiconductor nanoparticles indicated by cellular micromotility. *ACS Nano*. 2009;3(1):213–22. Epub 2009/02/12.
104. Deng J, Sun M, Zhu J, Gao C. Molecular interactions of different size AuNP-COOH nanoparticles with human fibrinogen. *Nanoscale*. 2013. Epub 2013/07/26.
105. Liu CP, Lin FS, Chien CT, Tseng SY, Luo CW, Chen CH et al. In-situ formation and assembly of gold nanoparticles by gum arabic as efficient photothermal agent for killing cancer cells. *Macromolecule Bioscience*. 2013. Epub 2013/07/19.
106. Cho WS, Cho M, Jeong J, Choi M, Cho HY, Han BS et al. Acute toxicity and pharmacokinetics of 13 nm-sized PEG-coated gold nanoparticles. *Toxicology and Applied Pharmacology*. 2009;236(1):16–24. Epub 2009/01/24.
107. Hoshino A, Fujioka K, Oku T, Nakamura S, Suga M, Yamaguchi Y et al. Quantum dots targeted to the assigned organelle in living cells. *Microbiology and Immunology*. 2004;48(12):985–94. Epub 2004/12/22.
108. Pante N, Kann M. Nuclear pore complex is able to transport macromolecules with diameters of about 39 nm. *Molecular Biology of the Cell*. 2002;13(2):425–34. Epub 2002/02/21.
109. Dobson CM. Protein folding and misfolding. *Nature*. 2003;426(6968):884–90. Epub 2003/12/20.
110. Lasagna-Reeves C, Gonzalez-Romero D, Barria MA, Olmedo I, Clos A, Sadagopa Ramanujam VM et al. Bioaccumulation and toxicity of gold nanoparticles after repeated administration in mice. *Biochemical and Biophysical Research Communications*. 2010;393(4):649–55. Epub 2010/02/16.
111. Wiwanitkit V, Sereemaspun A, Rojanathanes R. Effect of gold nanoparticles on spermatozoa: The first world report. *Fertility and Sterility*. 2009;91(1):e7–8. Epub 2007/12/07.
112. Sengupta J, Datta P, Patra HK, Dasgupta AK, Gomes A. In vivo interaction of gold nanoparticles after acute and chronic exposures in experimental animal models. *Journal of Nanoscience and Nanotechnology*. 2013;13(3):1660–70. Epub 2013/06/13.
113. Asharani PV, Hande MP, Valiyaveettil S. Anti-proliferative activity of silver nanoparticles. *BMC Cell Biology*. 2009;10:65. Epub 2009/09/19.
114. AshaRani PV, Low Kah Mun G, Hande MP, Valiyaveettil S. Cytotoxicity and genotoxicity of silver nanoparticles in human cells. *ACS Nano*. 2009;3(2):279–90. Epub 2009/02/25.
115. Foldbjerg R, Dang DA, Autrup H. Cytotoxicity and genotoxicity of silver nanoparticles in the human lung cancer cell line, A549. *Archives of Toxicology*. 2011;85(7):743–50. Epub 2010/04/30.
116. Hsin YH, Chen CF, Huang S, Shih TS, Lai PS, Chueh PJ. The apoptotic effect of nanosilver is mediated by a ROS- and JNK-dependent mechanism involving the mitochondrial pathway in NIH3T3 cells. *Toxicology Letters*. 2008;179(3):130–9. Epub 2008/06/13.
117. Kawata K, Osawa M, Okabe S. In vitro toxicity of silver nanoparticles at noncytotoxic doses to HepG2 human hepatoma cells. *Environmental Science & Technology*. 2009;43(15):6046–51. Epub 2009/09/08.
118. Miura N, Shinohara Y. Cytotoxic effect and apoptosis induction by silver nanoparticles in HeLa cells. *Biochemical and Biophysical Research Communications*. 2009;390(3):733–7. Epub 2009/10/20.
119. Yang W, Shen C, Ji Q, An H, Wang J, Liu Q et al. Food storage material silver nanoparticles interfere with DNA replication fidelity and bind with DNA. *Nanotechnology*. 2009;20(8):085102. Epub 2009/05/07.
120. Noh HJ, Kim HS, Jun SH, Kang YH, Cho S, Park Y. Biogenic silver nanoparticles with chlorogenic acid as a bioreducing agent. *Journal of Nanoscience and Nanotechnology*. 2013;13(8):5787–93. Epub 2013/07/26.

121. Tang J, Xiong L, Wang S, Wang J, Liu L, Li J et al. Distribution, translocation and accumulation of silver nanoparticles in rats. *Journal of Nanoscience and Nanotechnology*. 2009;9(8):4924–32. Epub 2009/11/26.
122. Rahman MF, Wang J, Patterson TA, Saini UT, Robinson BL, Newport GD et al. Expression of genes related to oxidative stress in the mouse brain after exposure to silver-25 nanoparticles. *Toxicology Letters*. 2009;187(1):15–21. Epub 2009/05/12.
123. Ringwood AH, McCarthy M, Bates TC, Carroll DL. The effects of silver nanoparticles on oyster embryos. *Marine Environmental Research*. 2010;69(Suppl):S49–51. Epub 2009/11/17.
124. Scown TM, Santos EM, Johnston BD, Gaiser B, Baalousha M, Mitov S et al. Effects of aqueous exposure to silver nanoparticles of different sizes in rainbow trout. *Toxicological Sciences: An Official Journal of the Society of Toxicology*. 2010;115(2):521–34. Epub 2010/03/12.
125. Wise JP, Sr., Goodale BC, Wise SS, Craig GA, Pongan AF, Walter RB et al. Silver nanospheres are cytotoxic and genotoxic to fish cells. *Aquatic Toxicology*. 2010;97(1):34–41. Epub 2010/01/12.
126. Ahamed M, Posgai R, Gorey TJ, Nielsen M, Hussain SM, Rowe JJ. Silver nanoparticles induced heat shock protein 70, oxidative stress and apoptosis in Drosophila melanogaster. *Toxicology and Applied Pharmacology*. 2010;242(3):263–9. Epub 2009/10/31.
127. Roh JY, Sim SJ, Yi J, Park K, Chung KH, Ryu DY et al. Ecotoxicity of silver nanoparticles on the soil nematode *Caenorhabditis elegans* using functional ecotoxicogenomics. *Environmental Science & Technology*. 2009;43(10):3933–40. Epub 2009/06/24.
128. Yilma AN, Singh SR, Dixit S, Dennis VA. Anti-inflammatory effects of silver-polyvinyl pyrrolidone (Ag-PVP) nanoparticles in mouse macrophages infected with live Chlamydia trachomatis. *International Journal of Nanomedicine*. 2013;8:2421–32. Epub 2013/07/25.
129. Gnanadhas DP, Thomas MB, Thomas R, Raichur AM, Chakravortty D. Interaction of silver nanoparticles with serum proteins affects their antimicrobial activity *in vivo*. *Antimicrobial Agents and Chemotherapy*. 2013. Epub 2013/07/24.
130. Cveticanin J, Joksic G, Leskovac A, Petrovic S, Sobot AV, Neskovic O. Using carbon nanotubes to induce micronuclei and double strand breaks of the DNA in human cells. *Nanotechnology*. 2010;21(1):015102. Epub 2009/12/01.
131. Kirchner C, Liedl T, Kudera S, Pellegrino T, Munoz Javier A, Gaub HE et al. Cytotoxicity of colloidal CdSe and CdSe/ZnS nanoparticles. *Nano Letters*. 2005;5(2):331–8. Epub 2005/03/30.
132. Jain TK, Morales MA, Sahoo SK, Leslie-Pelecky DL, Labhasetwar V. Iron oxide nanoparticles for sustained delivery of anticancer agents. *Molecular Pharmaceutics*. 2005;2(3):194–205. Epub 2005/06/07.
133. Walker VG, Li Z, Hulderman T, Schwegler-Berry D, Kashon ML, Simeonova PP. Potential *in vitro* effects of carbon nanotubes on human aortic endothelial cells. *Toxicology and Applied Pharmacology*. 2009;236(3):319–28. Epub 2009/03/10.
134. Lindberg HK, Falck GC, Suhonen S, Vippola M, Vanhala E, Catalan J et al. Genotoxicity of nanomaterials: DNA damage and micronuclei induced by carbon nanotubes and graphite nanofibres in human bronchial epithelial cells *in vitro*. *Toxicology Letters*. 2009;186(3):166–73. Epub 2008/12/31.
135. Herzog E, Byrne HJ, Casey A, Davoren M, Lenz AG, Maier KL et al. SWCNT suppress inflammatory mediator responses in human lung epithelium *in vitro*. *Toxicology and Applied Pharmacology*. 2009;234(3):378–90. Epub 2008/12/02.
136. Migliore L, Saracino D, Bonelli A, Colognato R, D'Errico MR, Magrini A et al. Carbon nanotubes induce oxidative DNA damage in RAW 264.7 cells. *Environmental and Molecular Mutagenesis*. 2010;51(4):294–303. Epub 2010/01/22.
137. Murray AR, Kisin E, Leonard SS, Young SH, Kommineni C, Kagan VE et al. Oxidative stress and inflammatory response in dermal toxicity of single-walled carbon nanotubes. *Toxicology*. 2009;257(3):161–71. Epub 2009/01/20.
138. Pacurari M, Yin XJ, Zhao J, Ding M, Leonard SS, Schwegler-Berry D et al. Raw single-wall carbon nanotubes induce oxidative stress and activate MAPKs, AP-1, NF-kappaB, and Akt in normal and malignant human mesothelial cells. *Environmental Health Perspectives*. 2008;116(9):1211–7. Epub 2008/09/17.
139. Wang L, Mercer RR, Rojanasakul Y, Qiu A, Lu Y, Scabilloni JF et al. Direct fibrogenic effects of dispersed single-walled carbon nanotubes on human lung fibroblasts. *Journal of Toxicology and Environmental Health Part A*. 2010;73(5):410–22. Epub 2010/02/16.
140. Kang S, Mauter MS, Elimelech M. Microbial cytotoxicity of carbon-based nanomaterials: Implications for river water and wastewater effluent. *Environmental Science & Technology*. 2009;43(7):2648–53. Epub 2009/05/21.
141. Nygaard UC, Hansen JS, Samuelsen M, Alberg T, Marioara CD, Lovik M. Single-walled and multi-walled carbon nanotubes promote allergic immune responses in mice. *Toxicological Sciences: An Official Journal of the Society of Toxicology*. 2009;109(1):113–23. Epub 2009/03/19.

142. Bihari P, Holzer M, Praetner M, Fent J, Lerchenberger M, Reichel CA et al. Single-walled carbon nanotubes activate platelets and accelerate thrombus formation in the microcirculation. *Toxicology*. 2010;269(2–3):148–54. Epub 2009/08/25.
143. Bari S, Chu PP, Lim A, Fan X, Gay FP, Bunte RM et al. Protective role of functionalized single walled carbon nanotubes enhance *ex vivo* expansion of hematopoietic stem and progenitor cells in human umbilical cord blood. *Nanomedicine (London)*. 2013. Epub 2013/06/05.
144. Cheng C, Muller KH, Koziol KK, Skepper JN, Midgley PA, Welland ME et al. Toxicity and imaging of multi-walled carbon nanotubes in human macrophage cells. *Biomaterials*. 2009;30(25):4152–60. Epub 2009/05/29.
145. Patlolla A, Patlolla B, Tchounwou P. Evaluation of cell viability, DNA damage, and cell death in normal human dermal fibroblast cells induced by functionalized multiwalled carbon nanotube. *Molecular and Cellular Biochemistry*. 2010;338(1–2):225–32. Epub 2009/12/18.
146. Ravichandran P, Periyakaruppan A, Sadanandan B, Ramesh V, Hall JC, Jejelowo O et al. Induction of apoptosis in rat lung epithelial cells by multiwalled carbon nanotubes. *Journal of Biochemical and Molecular Toxicology*. 2009;23(5):333–44. Epub 2009/10/15.
147. Reddy AR, Reddy YN, Krishna DR, Himabindu V. Multi wall carbon nanotubes induce oxidative stress and cytotoxicity in human embryonic kidney (HEK293) cells. *Toxicology*. 2010;272(1–3):11–6. Epub 2010/04/08.
148. Mattson MP, Haddon RC, Rao AM. Molecular functionalization of carbon nanotubes and use as substrates for neuronal growth. *Journal of Molecular Neuroscience*. 2000;14(3):175–82.
149. Moraes AS, Paula RF, Pradella F, Santos MP, Oliveira EC, von Glehn F et al. The suppressive effect of IL-27 on encephalitogenic Th17 cells induced by multiwalled carbon nanotubes reduces the severity of experimental autoimmune encephalomyelitis. *CNS Neuroscience & Therapeutics*. 2013. Epub 2013/06/05.
150. Toyokuni S. Genotoxicity and carcinogenicity risk of carbon nanotubes. *Advanced Drug Delivery Reviews*. 2013. Epub 2013/06/12.
151. Xia T, Kovochich M, Brant J, Hotze M, Sempf J, Oberley T et al. Comparison of the abilities of ambient and manufactured nanoparticles to induce cellular toxicity according to an oxidative stress paradigm. *Nano Letters*. 2006 Aug;6(8):1794–807.
152. Dhawan A, Taurozzi JS, Pandey AK, Shan W, Miller SM, Hashsham SA et al. Stable colloidal dispersions of C_{60} fullerenes in water: Evidence for genotoxicity. *Environmental Science and Technology*. 2006;40(23):7394–401. Epub 2006/12/22.
153. Sera N, Tokiwa H, Miyata N. Mutagenicity of the fullerene C_{60}-generated singlet oxygen dependent formation of lipid peroxides. *Carcinogenesis*. 1996 Oct;17(10):2163–9.
154. van Berlo D, Clift M, Albrecht C, Schins R. Carbon nanotubes: An insight into the mechanisms of their potential genotoxicity. *Swiss Medical Weekly*. 2012;142:w13698.
155. Du J, Wang S, You H, Zhao X. Understanding the toxicity of carbon nanotubes in the environment is crucial to the control of nanomaterials in producing and processing and the assessment of health risk for human: A review. *Environmental Toxicology and Pharmacology*. 2013;36:451–462.
156. Yang H, Liu C, Yang D, Zhang H, Xi Z. Comparative study of cytotoxicity, oxidative stress and genotoxicity induced by four typical nanomaterials: The role of particle size, shape and composition. *Journal of Applied Toxicology*. 2009;29(1):69–78.
157. Patlolla AK, Hussain SM, Schlager JJ, Patlolla S, Tchounwou PB. Comparative study of the clastogenicity of functionalized and nonfunctionalized multiwalled carbon nanotubes in bone marrow cells of Swiss-Webster mice. *Environmental Toxicology*. 2010;25(6):608–21.
158. Folkmann JK, Risom L, Jacobsen NR, Wallin H, Loft S, Møller P. Oxidatively damaged DNA in rats exposed by oral gavage to C_{60} fullerenes and single-walled carbon nanotubes. *Environmental Health Perspectives*. 2009;117(5):703.
159. Jacobsen NR, Moller P, Jensen KA, Vogel U, Ladefoged O, Loft S et al. Lung inflammation and genotoxicity following pulmonary exposure to nanoparticles in ApoE-/-mice. *Part Fibre Toxicology*. 2009;6(2):2.
160. Kato T, Totsuka Y, Ishino K, Matsumoto Y, Tada Y, Nakae D et al. Genotoxicity of multi-walled carbon nanotubes in both *in vitro* and *in vivo* assay systems. *Nanotoxicology*. 2013;7(4):452–61.
161. Kim JS, Sung JH, Song KS, Lee JH, Kim SM, Lee GH et al. Persistent DNA damage measured by Comet Assay of Sprague Dawley rat lung cells after five days of inhalation exposure and 1 month post-exposure to dispersed multi-wall carbon nanotubes (MWCNTs) generated by new MWCNT aerosol generation system. *Toxicological Sciences*. 2012;128(2):439–48.
162. Muller J, Huaux Fo, Fonseca A, Nagy JB, Moreau N, Delos M et al. Structural defects play a major role in the acute lung toxicity of multiwall carbon nanotubes: Toxicological aspects. *Chemical Research in Toxicology*. 2008;21(9):1698–705.

163. Gonzalez L, Lison D, Kirsch-Volders M. Genotoxicity of engineered nanomaterials: A critical review. *Nanotoxicology*. 2008;2(4):252–73.
164. Sargent LM, Reynolds SH, Castranova V. Potential pulmonary effects of engineered carbon nanotubes: *In vitro* genotoxic effects. *Nanotoxicology*. 2010;4(4):396–408.
165. Sargent L, Shvedova A, Hubbs A, Salisbury J, Benkovic S, Kashon M et al. Induction of aneuploidy by single-walled carbon nanotubes. *Environmental and Molecular Mutagenesis*. 2009;50(8):708–17.
166. Asakura M, Sasaki T, Sugiyama T, Takaya M, Koda S, Nagano K et al. Genotoxicity and cytotoxicity of multi-wall carbon nanotubes in cultured Chinese hamster lung cells in comparison with chrysotile A fibers. *Journal of Occupational Health*. 2010(0):1003290122.
167. Muller J, Decordier I, Hoet PH, Lombaert N, Thomassen L, Huaux F et al. Clastogenic and aneugenic effects of multi-wall carbon nanotubes in epithelial cells. *Carcinogenesis*. 2008;29(2):427–33.
168. Shvedova AA, Kisin E, Murray AR, Johnson VJ, Gorelik O, Arepalli S et al. Inhalation vs. aspiration of single-walled carbon nanotubes in C57BL/6 mice: Inflammation, fibrosis, oxidative stress, and mutagenesis. *American Journal of Physiology—Lung Cellular and Molecular Physiology*. 2008;295(4):L552–L65.
169. Pauluhn J. Subchronic 13-week inhalation exposure of rats to multiwalled carbon nanotubes: Toxic effects are determined by density of agglomerate structures, not fibrillar structures. *Toxicological Sciences*. 2010;113(1):226–42.
170. Koyama S, Kim YA, Hayashi T, Takeuchi K, Fujii C, Kuroiwa N et al. *In vivo* immunological toxicity in mice of carbon nanotubes with impurities. *Carbon*. 2009;47(5):1365–72.
171. Lam C-W, James JT, McCluskey R, Hunter RL. Pulmonary toxicity of single-wall carbon nanotubes in mice 7 and 90 days after intratracheal instillation. *Toxicological Sciences*. 2004;77(1):126–34.
172. Muller J, Huaux F, Lison D. Respiratory toxicity of carbon nanotubes: How worried should we be? *Carbon*. 2006;44(6):1048–56.
173. Shvedova AA, Kisin ER, Mercer R, Murray AR, Johnson VJ, Potapovich AI et al. Unusual inflammatory and fibrogenic pulmonary responses to single-walled carbon nanotubes in mice. *American Journal of Physiology—Lung Cellular and Molecular Physiology*. 2005;289(5):L698–L708.
174. Carrero-Sanchez J, Elias A, Mancilla R, Arrellin G, Terrones H, Laclette J et al. Biocompatibility and toxicological studies of carbon nanotubes doped with nitrogen. *Nano Letters*. 2006;6(8):1609–16.
175. Zhu M-T, Feng W-Y, Wang B, Wang T-C, Gu Y-Q, Wang M et al. Comparative study of pulmonary responses to nano-and submicron-sized ferric oxide in rats. *Toxicology*. 2008;247(2):102–11.
176. Brown T. Silica exposure, smoking, silicosis and lung cancer—complex interactions. *Occupational Medicine*. 2009;59(2):89–95.
177. De Jong WH, Borm PJ. Drug delivery and nanoparticles: Applications and hazards. *International Journal of Nanomedicine*. 2008;3(2):133.
178. Bhattacharya K, Davoren M, Boertz J, Schins R, Hoffmann E, Dopp E. Titanium dioxide nanoparticles induce oxidative stress and DNA-adduct formation but not DNA-breakage in human lung cells. *Part Fibre Toxicology*. 2009;6:17.
179. AshaRani P, Low Kah Mun G, Hande MP, Valiyaveettil S. Cytotoxicity and genotoxicity of silver nanoparticles in human cells. *ACS Nano*. 2009;3(2):279–90. Epub 2009/02/25.
180. Blasi P, Giovagnoli S, Schoubben A, Ricci M, Rossi C. Solid lipid nanoparticles for targeted brain drug delivery. *Advanced Drug Delivery Reviews*. 2007;59(6):454–77.
181. Grabowski N, Hillaireau H, Vergnaud J, Aragao LS, Kerdine-Romer S, Pallardy M et al. Toxicity of surface-modified plga nanoparticles towards lung alveolar epithelial cells. *International Journal of Pharmaceutics*. 2013;454:686–694.
182. McClements DJ. Edible lipid nanoparticles: Digestion, absorption, and potential toxicity. *Progress in Lipid Research*. 2013;52:409–423.
183. Thakore S, Valodkar M, Soni JY, Vyas K, Jadeja RN, Devkar RV et al. Synthesis and cytotoxicity evaluation of novel acylated starch nanoparticles. *Bioorganic Chemistry*. 2013;46:26–30.
184. Termsarasab U, Cho H-J, Kim DH, Chong S, Chung S-J, Shim C-K et al. Chitosan oligosaccharide–arachidic acid–based nanoparticles for anti-cancer drug delivery. *International Journal of Pharmaceutics*. 2012;441:373–380.
185. Xu Q, Wang C-H, Pack DW. Polymeric carriers for gene delivery: Chitosan and poly (amidoamine) dendrimers. *Current Pharmaceutical Design*. 2010;16(21):2350.
186. Vishu Kumar A, Varadaraj MC, Lalitha RG, Tharanathan R. Low molecular weight chitosans: Preparation with the aid of papain and characterization. *Biochimica et Biophysica Acta (BBA)—General Subjects*. 2004;1670(2):137–46.
187. Prow TW. Toxicity of nanomaterials to the eye. *Wiley Interdisciplinary Reviews: Nanomedicine and Nanobiotechnology*. 2010;2(4):317–33.

188. Vadlapudi AD, Mitra AK. Nanomicelles: An emerging platform for drug delivery to the eye. *Therapeutic Delivery*. 2013;4(1):1–3.
189. Lu Y, Zhou, Y., Wang, J., Peng, F., Zhong, Y., Huang, Q., Fan, C., He, Y. *In vivo* behavior of near infrared-emitting quantum dots. *Biomaterials*. 2013;34(17). Epub 2013 Mar 13.
190. Nurunnabi M, Khatun Z, Huh KM, Park SY, Lee DY, Cho KJ et al. *In vivo* biodistribution and toxicology of carboxylated graphene quantum dots. *ACS Nano*. 2013;7:6858–6867.
191. Nguyen KC, Willmore WG, Tayabali AF. Cadmium telluride quantum dots cause oxidative stress leading to extrinsic and intrinsic apoptosis in hepatocellular carcinoma HepG2 cells. *Toxicology*. 2013;306:114–123.
192. Bruneau A, Fortier M, Gagne F, Gagnon C, Turcotte P, Tayabali A et al. *In vitro* immunotoxicology of quantum dots and comparison with dissolved cadmium and tellurium. *Environmental Toxicology*. 2013. DOI: 10.1002/tox.21890.
193. Roberts JR, Antonini JM, Porter DW, Chapman RS, Scabilloni JF, Young S-H et al. Lung toxicity and biodistribution of Cd/Se-ZnS quantum dots with different surface functional groups after pulmonary exposure in rats. *Particle and Fibre Toxicology*. 2013;10(1).
194. Wiecinski PN, Metz KM, King Heiden TC, Louis KM, Mangham AN, Hamers RJ, Heideman W, Peterson RE, Pedersen JA. Toxicity of oxidatively degraded quantum dots to developing zebrafish (Danio rerio). *Environmental Science & Technology*. 2013;47(16): 9132–9139.
195. Jacobs R. Argyria: My life story. *Clinics in Dermatology*. 2006;24(1):66–9; discussion 9. Epub 2006/01/24.
196. Buzea C, Pacheco, II, Robbie K. Nanomaterials and nanoparticles: Sources and toxicity. *Biointerphases*. 2007;2(4):MR17–71. Epub 2007/12/01.
197. Cho SJ, Maysinger D, Jain M, Roder B, Hackbarth S, Winnik FM. Long-term exposure to CdTe quantum dots causes functional impairments in live cells. *Langmuir: The ACS Journal of Surfaces and Colloids*. 2007;23(4):1974–80. Epub 2007/02/07.
198. Li KG, Chen JT, Bai SS, Wen X, Song SY, Yu Q et al. Intracellular oxidative stress and cadmium ions release induce cytotoxicity of unmodified cadmium sulfide quantum dots. *Toxicology In Vitro: An International Journal Published in Association with BIBRA*. 2009;23(6):1007–13. Epub 2009/06/23.
199. Elgrabli D, Abella-Gallart S, Robidel F, Rogerieux F, Boczkowski J, Lacroix G. Induction of apoptosis and absence of inflammation in rat lung after intratracheal instillation of multiwalled carbon nanotubes. *Toxicology*. 2008;253(1–3):131–6. Epub 2008/10/07.
200. Han SG, Andrews R, Gairola CG. Acute pulmonary response of mice to multi-wall carbon nanotubes. *Inhalation Toxicology*. 2010;22(4):340–7. Epub 2010/01/13.
201. Ma-Hock L, Treumann S, Strauss V, Brill S, Luizi F, Mertler M et al. Inhalation toxicity of multiwall carbon nanotubes in rats exposed for 3 months. *Toxicological Sciences: An Official Journal of the Society of Toxicology*. 2009;112(2):468–81. Epub 2009/07/09.
202. Porter DW, Hubbs AF, Mercer RR, Wu N, Wolfarth MG, Sriram K et al. Mouse pulmonary dose- and time course-responses induced by exposure to multi-walled carbon nanotubes. *Toxicology*. 2010;269(2–3):136–47. Epub 2009/10/28.
203. Poland CA, Duffin R, Kinloch I, Maynard A, Wallace WA, Seaton A et al. Carbon nanotubes introduced into the abdominal cavity of mice show asbestos-like pathogenicity in a pilot study. *Nature Nanotechnology*. 2008;3(7):423–8. Epub 2008/07/26.
204. Ji Z, Zhang D, Li L, Shen X, Deng X, Dong L et al. The hepatotoxicity of multi-walled carbon nanotubes in mice. *Nanotechnology*. 2009;20(44):445101. Epub 2009/10/06.
205. Park EJ, Cho WS, Jeong J, Yi J, Choi K, Park K. Pro-inflammatory and potential allergic responses resulting from B cell activation in mice treated with multi-walled carbon nanotubes by intratracheal instillation. *Toxicology*. 2009;259(3):113–21. Epub 2009/05/12.
206. Hsieh MS, Shiao NH, Chan WH. Cytotoxic effects of CdSe quantum dots on maturation of mouse oocytes, fertilization, and fetal development. *International Journal of Molecular Sciences*. 2009;10(5):2122–35. Epub 2009/07/01.
207. Mortensen LJ, Oberdorster G, Pentland AP, Delouise LA. *In vivo* skin penetration of quantum dot nanoparticles in the murine model: The effect of UVR. *Nano Letters*. 2008;8(9):2779–87. Epub 2008/08/09.
208. Kim J, Park Y, Yoon TH, Yoon CS, Choi K. Phototoxicity of CdSe/ZnSe quantum dots with surface coatings of 3-mercaptopropionic acid or tri-n-octylphosphine oxide/gum Arabic in *Daphnia magna* under environmentally relevant UV-B light. *Aquatic Toxicology*. 2010;97(2):116–24. Epub 2010/01/20.
209. Patlolla AK, Hussain SM, Schlager JJ, Patlolla S, Tchounwou PB. Comparative study of the clastogenicity of functionalized and nonfunctionalized multiwalled carbon nanotubes in bone marrow cells of Swiss-Webster mice. *Environmental Toxicology*. 2010;25(6):608–21. Epub 2010/06/16.

210. Asharani PV, Serina NG, Nurmawati MH, Wu YL, Gong Z, Valiyaveettil S. Impact of multi-walled carbon nanotubes on aquatic species. *Journal of Nanoscience and Nanotechnology*. 2008;8(7):3603–9. Epub 2008/12/05.
211. Kang S, Mauter MS, Elimelech M. Physicochemical determinants of multiwalled carbon nanotube bacterial cytotoxicity. *Environmental Science & Technology*. 2008;42(19):7528–34. Epub 2008/10/23.
212. Ma H, Bertsch PM, Glenn TC, Kabengi NJ, Williams PL. Toxicity of manufactured zinc oxide nanoparticles in the nematode *Caenorhabditis elegans*. *Environmental Toxicology and Chemistry/SETAC*. 2009;28(6):1324–30. Epub 2009/02/06.
213. Miller RJ, Lenihan HS, Muller EB, Tseng N, Hanna SK, Keller AA. Impacts of metal oxide nanoparticles on marine phytoplankton. *Environmental Science & Technology*. 2010;44(19):7329–34. Epub 2010/05/18.
214. Zhu X, Wang J, Zhang X, Chang Y, Chen Y. The impact of ZnO nanoparticle aggregates on the embryonic development of zebrafish (*Danio rerio*). *Nanotechnology*. 2009;20(19):195103. Epub 2009/05/08.
215. Chen YC, Hsiao JK, Liu HM, Lai IY, Yao M, Hsu SC et al. The inhibitory effect of superparamagnetic iron oxide nanoparticle (Ferucarbotran) on osteogenic differentiation and its signaling mechanism in human mesenchymal stem cells. *Toxicology and Applied Pharmacology*. 2010;245(2):272–9. Epub 2010/03/27.
216. Choi SJ, Oh JM, Choy JH. Toxicological effects of inorganic nanoparticles on human lung cancer A549 cells. *Journal of Inorganic Biochemistry*. 2009;103(3):463–71. Epub 2009/02/03.
217. Eom HJ, Choi J. Oxidative stress of CeO_2 nanoparticles via p38-Nrf-2 signaling pathway in human bronchial epithelial cell, Beas-2B. *Toxicology Letters*. 2009;187(2):77–83. Epub 2009/05/12.
218. Fahmy B, Cormier SA. Copper oxide nanoparticles induce oxidative stress and cytotoxicity in airway epithelial cells. *Toxicology In Vitro: An International Journal Published in Association with BIBRA*. 2009;23(7):1365–71. Epub 2009/08/25.
219. Falck GC, Lindberg HK, Suhonen S, Vippola M, Vanhala E, Catalan J et al. Genotoxic effects of nano-sized and fine TiO_2. *Human & Experimental Toxicology*. 2009;28(6–7):339–52. Epub 2009/09/17.
220. Huang CC, Aronstam RS, Chen DR, Huang YW. Oxidative stress, calcium homeostasis, and altered gene expression in human lung epithelial cells exposed to ZnO nanoparticles. *Toxicology In Vitro: An International Journal Published in Association with BIBRA*. 2010;24(1):45–55. Epub 2009/09/17.
221. Hussain S, Thomassen LC, Ferecatu I, Borot MC, Andreau K, Martens JA et al. Carbon black and titanium dioxide nanoparticles elicit distinct apoptotic pathways in bronchial epithelial cells. *Particle and Fibre Toxicology*. 2010;7:10. Epub 2010/04/20.
222. Karlsson HL, Gustafsson J, Cronholm P, Moller L. Size-dependent toxicity of metal oxide particles—A comparison between nano- and micrometer size. *Toxicology Letters*. 2009;188(2):112–8. Epub 2009/05/19.
223. Midander K, Cronholm P, Karlsson HL, Elihn K, Moller L, Leygraf C et al. Surface characteristics, copper release, and toxicity of nano- and micrometer-sized copper and copper(II) oxide particles: A cross-disciplinary study. *Small*. 2009;5(3):389–99. Epub 2009/01/17.
224. Ogami A, Morimoto Y, Myojo T, Oyabu T, Murakami M, Todoroki M et al. Pathological features of different sizes of nickel oxide following intratracheal instillation in rats. *Inhalation Toxicology*. 2009;21(10):812–8. Epub 2009/02/20.
225. Tsaousi A, Jones E, Case CP. The *in vitro* genotoxicity of orthopaedic ceramic (Al2O3) and metal (CoCr alloy) particles. *Mutation Research*. 2010;697(1–2):1–9. Epub 2010/02/09.
226. Hauck TS, Ghazani AA, Chan WC. Assessing the effect of surface chemistry on gold nanorod uptake, toxicity, and gene expression in mammalian cells. *Small*. 2008;4(1):153–9. Epub 2007/12/18.
227. Feick JD, Chukwumah N, Noel AE, Velegol D. Altering surface charge nonuniformity on individual colloidal particles. *Langmuir: The ACS Journal of Surfaces and Colloids*. 2004;20(8):3090–5. Epub 2005/05/07.
228. Decuzzi P, Ferrari M. The role of specific and non-specific interactions in receptor-mediated endocytosis of nanoparticles. *Biomaterials*. 2007;28(18):2915–22. Epub 2007/03/17.
229. Sharma HS, Hussain S, Schlager J, Ali SF, Sharma A. Influence of nanoparticles on blood-brain barrier permeability and brain edema formation in rats. *Acta Neurochirurgica Supplement*. 2010;106:359–64. Epub 2009/12/05.
230. Choi JE, Kim S, Ahn JH, Youn P, Kang JS, Park K et al. Induction of oxidative stress and apoptosis by silver nanoparticles in the liver of adult zebrafish. *Aquatic Toxicology*. 2010;100(2):151–9. Epub 2010/01/12.
231. Gao H, Shi W, Freund LB. Mechanics of receptor-mediated endocytosis. *Proceedings of the National Academy of Sciences of the United States of America*. 2005;102(27):9469–74. Epub 2005/06/24.
232. Dawson KA, Salvati A, Lynch I. Nanotoxicology: Nanoparticles reconstruct lipids. *Nature Nanotechnology*. 2009;4(2):84–5. Epub 2009/02/07.

233. Verma A, Uzun O, Hu Y, Han HS, Watson N, Chen S et al. Surface-structure-regulated cell-membrane penetration by monolayer-protected nanoparticles. *Nature Materials.* 2008;7(7):588–95. Epub 2008/05/27.
234. Leroueil PR, Berry SA, Duthie K, Han G, Rotello VM, McNerny DQ et al. Wide varieties of cationic nanoparticles induce defects in supported lipid bilayers. *Nano Letters.* 2008;8(2):420–4. Epub 2008/01/26.
235. Navarro E, Baun A, Behra R, Hartmann NB, Filser J, Miao AJ et al. Environmental behavior and ecotoxicity of engineered nanoparticles to algae, plants, and fungi. *Ecotoxicology.* 2008;17(5):372–86. Epub 2008/05/08.
236. Ovrevik J, Lag M, Schwarze P, Refsnes M. p38 and Src-ERK1/2 pathways regulate crystalline silica-induced chemokine release in pulmonary epithelial cells. *Toxicological Sciences: An Official Journal of the Society of Toxicology.* 2004;81(2):480–90. Epub 2004/07/09.
237. Greulich C, Diendorf J, Simon T, Eggeler G, Epple M, Koller M. Uptake and intracellular distribution of silver nanoparticles in human mesenchymal stem cells. *Acta Biomaterialia.* 2011;7(1):347–54. Epub 2010/08/17.
238. Al-Rawi M, Diabate S, Weiss C. Uptake and intracellular localization of submicron and nano-sized SiO(2) particles in HeLa cells. *Archives of Toxicology.* 2011;85(7):813–26. Epub 2011/01/18.
239. Parfenov AS, Salnikov V, Lederer WJ, Lukyanenko V. Aqueous diffusion pathways as a part of the ventricular cell ultrastructure. *Biophysical Journal.* 2006;90(3):1107–19. Epub 2005/11/15.
240. Williams Y, Sukhanova A, Nowostawska M, Davies AM, Mitchell S, Oleinikov V et al. Probing cell-type-specific intracellular nanoscale barriers using size-tuned quantum dots. *Small.* 2009;5(22):2581–8. Epub 2009/08/18.
241. Nabiev I, Mitchell S, Davies A, Williams Y, Kelleher D, Moore R et al. Nonfunctionalized nanocrystals can exploit a cell's active transport machinery delivering them to specific nuclear and cytoplasmic compartments. *Nano Letters.* 2007;7(11):3452–61. Epub 2007/10/24.
242. Mahmoudi M, Shokrgozar MA, Sardari S, Moghadam MK, Vali H, Laurent S et al. Irreversible changes in protein conformation due to interaction with superparamagnetic iron oxide nanoparticles. *Nanoscale.* 2011;3(3):1127–38. Epub 2011/01/07.
243. Shang W, Nuffer JH, Dordick JS, Siegel RW. Unfolding of ribonuclease A on silica nanoparticle surfaces. *Nano Letters.* 2007;7(7):1991–5. Epub 2007/06/15.
244. Peng ZG, Hidajat K, Uddin MS. Adsorption of bovine serum albumin on nanosized magnetic particles. *Journal of Colloid and Interface Science.* 2004;271(2):277–83. Epub 2004/02/20.
245. Chen M, von Mikecz A. Formation of nucleoplasmic protein aggregates impairs nuclear function in response to SiO_2 nanoparticles. *Experimental Cell Research.* 2005;305(1):51–62. Epub 2005/03/22.
246. Yang ST, Wang H, Guo L, Gao Y, Liu Y, Cao A. Interaction of fullerenol with lysozyme investigated by experimental and computational approaches. *Nanotechnology.* 2008;19(39):395101. Epub 2008/10/01.
247. Cao A, Ye Z, Cai Z, Dong E, Yang X, Liu G et al. A facile method to encapsulate proteins in silica nanoparticles: Encapsulated green fluorescent protein as a robust fluorescence probe. *Angewandte Chemie.* 2010;49(17):3022–5. Epub 2010/03/24.
248. Cai Z, Ye Z, Yang X, Chang Y, Wang H, Liu Y et al. Encapsulated enhanced green fluorescence protein in silica nanoparticle for cellular imaging. *Nanoscale.* 2011;3(5):1974–6. Epub 2011/03/04.
249. Folkmann JK, Risom L, Jacobsen NR, Wallin H, Loft S, Moller P. Oxidatively damaged DNA in rats exposed by oral gavage to C_{60} fullerenes and single-walled carbon nanotubes. *Environmental Health Perspectives.* 2009;117(5):703–8. Epub 2009/05/30.
250. Yang ST, Wang X, Jia G, Gu Y, Wang T, Nie H et al. Long-term accumulation and low toxicity of single-walled carbon nanotubes in intravenously exposed mice. *Toxicology Letters.* 2008;181(3):182–9. Epub 2008/09/02.
251. De M, Rotello VM. Synthetic 'chaperones': Nanoparticle-mediated refolding of thermally denatured proteins. *Chemical Communications.* 2008;(30):3504–6. Epub 2008/07/26.
252. Marano F, Hussain S, Rodrigues-Lima F, Baeza-Squiban A, Boland S. Nanoparticles: Molecular targets and cell signalling. *Archives of Toxicology.* 2011;85(7):733–41. Epub 2010/05/27.
253. Li N, Xia T, Nel AE. The role of oxidative stress in ambient particulate matter-induced lung diseases and its implications in the toxicity of engineered nanoparticles. *Free Radical Biology & Medicine.* 2008;44(9):1689–99. Epub 2008/03/04.
254. Li N, Sioutas C, Cho A, Schmitz D, Misra C, Sempf J et al. Ultrafine particulate pollutants induce oxidative stress and mitochondrial damage. *Environmental Health Perspectives.* 2003;111(4):455–60. Epub 2003/04/05.

255. Foley S, Crowley C, Smaihi M, Bonfils C, Erlanger BF, Seta P et al. Cellular localisation of a water-soluble fullerene derivative. *Biochemical and Biophysical Research Communications.* 2002;294(1):116–9. Epub 2002/06/11.
256. Zhu Y, Zhao Q, Li Y, Cai X, Li W. The interaction and toxicity of multi-walled carbon nanotubes with *Stylonychia mytilus. Journal of Nanoscience and Nanotechnology.* 2006;6(5):1357–64. Epub 2006/06/24.
257. Uchino T, Tokunaga H, Ando M, Utsumi H. Quantitative determination of OH radical generation and its cytotoxicity induced by TiO(2)-UVA treatment. *Toxicology In Vitro: An International Journal Published in Association with BIBRA.* 2002;16(5):629–35. Epub 2002/09/11.
258. Nel A, Xia T, Madler L, Li N. Toxic potential of materials at the nanolevel. *Science.* 2006;311(5761):622–7. Epub 2006/02/04.
259. Lao F, Chen L, Li W, Ge C, Qu Y, Sun Q et al. Fullerene nanoparticles selectively enter oxidation-damaged cerebral microvessel endothelial cells and inhibit JNK-related apoptosis. *ACS Nano.* 2009;3(11):3358–68. Epub 2009/10/21.
260. Oberdorster G. Safety assessment for nanotechnology and nanomedicine: Concepts of nanotoxicology. *Journal of Internal Medicine.* 2010;267(1):89–105. Epub 2010/01/12.
261. Wittmaack K. In search of the most relevant parameter for quantifying lung inflammatory response to nanoparticle exposure: Particle number, surface area, or what? *Environmental Health Perspectives.* 2007;115(2):187–94. Epub 2007/03/27.
262. Elsaesser A, Barnes CA, McKerr G, Salvati A, Lynch I, Dawson KA et al. Quantification of nanoparticle uptake by cells using an unbiased sampling method and electron microscopy. *Nanomedicine.* 2011;6(7):1189–98. Epub 2011/09/21.
263. Elsaesser A, Taylor A, de Yanes GS, McKerr G, Kim EM, O'Hare E et al. Quantification of nanoparticle uptake by cells using microscopical and analytical techniques. *Nanomedicine.* 2010;5(9):1447–57. Epub 2010/12/07.
264. Xia T, Kovochich M, Liong M, Madler L, Gilbert B, Shi H et al. Comparison of the mechanism of toxicity of zinc oxide and cerium oxide nanoparticles based on dissolution and oxidative stress properties. *ACS Nano.* 2008;2(10):2121–34. Epub 2009/02/12.
265. Khan JA, Pillai B, Das TK, Singh Y, Maiti S. Molecular effects of uptake of gold nanoparticles in HeLa cells. *Chembiochem: A European Journal of Chemical Biology.* 2007;8(11):1237–40. Epub 2007/06/15.
266. Kostarelos K. The long and short of carbon nanotube toxicity. *Nature Biotechnology.* 2008;26(7):774–6. Epub 2008/07/10.
267. Lacerda L, Herrero MA, Venner K, Bianco A, Prato M, Kostarelos K. Carbon-nanotube shape and individualization critical for renal excretion. *Small.* 2008;4(8):1130–2. Epub 2008/07/31.
268. Kroto HW, Heath JR, O'Brien SC, Curl RF, Smalley RE. C_{60}: Buckminsterfullerene. *Nature.* 1985;318(6042):162–3.
269. Powell MC, Kanarek MS. Nanomaterial health effects—Part 1: Background and current knowledge. *WMJ* 2006 Mar;105(2):16–20.
270. Gaiser BK, Fernandes TF, Jepson M, Lead JR, Tyler CR, Stone V. Assessing exposure, uptake and toxicity of silver and cerium dioxide nanoparticles from contaminated environments. *Environmental Health: A Global Access Science Source.* 2009;8(Suppl 1):S2. Epub 2010/02/05.
271. Kasemets K, Ivask A, Dubourguier HC, Kahru A. Toxicity of nanoparticles of ZnO, CuO and TiO_2 to yeast Saccharomyces cerevisiae. *Toxicology In Vitro: An International Journal Published in Association with BIBRA.* 2009;23(6):1116–22. Epub 2009/06/03.
272. Halford B. Fullerene for the face. *Chemical and Engineering News.* 2006;84:47.
273. Warheit DB. How Meaningful are the Results of Nanotoxicity Studies in the Absence of Adequate Material Characterization? *Toxicological Sciences.* 2008;101(2):183–5.
274. Donaldson K, Stone V. Current hypotheses on the mechanisms of toxicity of ultrafine particles. *Annali dell'Istituto superiore di sanita.* 2003;39(3):405–10.
275. Donaldson K, Tran L, Jimenez LA, Duffin R, Newby DE, Mills N et al. Combustion-derived nanoparticles: A review of their toxicology following inhalation exposure. *Part Fibre Toxicology.* 2005 Oct 21;2:10.
276. Kagan VE, Bayir H, Shvedova AA. Nanomedicine and nanotoxicology: Two sides of the same coin. *Nanomedicine* 2005 Dec;1(4):313–6.
277. Rouse JG, Yang J, Barron AR, Monteiro-Riviere NA. Fullerene-based amino acid nanoparticle interactions with human epidermal keratinocytes. *Toxicology In Vitro.* 2006 Dec;20(8):1313–20 Epub 2006 May 3.
278. Harhaji L, Isakovic A, Vucicevic L, Janjetovic K, Misirkic M, Markovic Z et al. Modulation of tumor necrosis factor-mediated cell death by fullerenes. *Pharmaceutical Research* 2008 Jun;25(6):1365–76.

279. Roursgaard M, Poulsen SS, Kepley CL, Hammer M, Nielsen GD, Larsen ST. Polyhydroxylated C_{60} fullerene (fullerenol) attenuates neutrophilic lung inflammation in mice. *Basic Clinical Pharmacology and Toxicology.* 2008 Oct;103(4):386–8 doi: 101111/j1742–7843200800315x.
280. Huang ST, Liao JS, Fang HW, Lin CM. Synthesis and anti-inflammation evaluation of new C_{60} fulleropyrrolidines bearing biologically active xanthine. *Bioorganic and Medicinal Chemistry Letters.* 2008 Jan 1;18(1):99–103 Epub 2007 Nov 5.
281. Tsao N, Kanakamma PP, Luh TY, Chou CK, Lei HY. Inhibition of *Escherichia coli*-induced meningitis by carboxyfullerence. *Antimicrobial Agents and Chemotherapy.* 1999 Sep;43(9):2273–7.
282. Stone V, Shaw J, Brown DM, Macnee W, Faux SP, Donaldson K. The role of oxidative stress in the prolonged inhibitory effect of ultrafine carbon black on epithelial cell function. *Toxicology In Vitro.* 1998 Dec;12(6):649–59.
283. Sayes CM, Gobin AM, Ausman KD, Mendez J, West JL, Colvin VL. Nano-C_{60} cytotoxicity is due to lipid peroxidation. *Biomaterials.* 2005 Dec;26(36):7587–95.
284. Kamat JP, Devasagayam TP, Priyadarsini KI, Mohan H. Reactive oxygen species mediated membrane damage induced by fullerene derivatives and its possible biological implications. *Toxicology.* 2000 Nov 30;155(1–3):55–61.
285. Zhao J, Xu L, Zhang T, Ren G, Yang Z. Influences of nanoparticle zinc oxide on acutely isolated rat hippocampal CA3 pyramidal neurons. *Neurotoxicology.* 2009;30(2):220–30. Epub 2009/01/17.
286. Zhang LW, Yang J, Barron AR, Monteiro-Riviere NA. Endocytic mechanisms and toxicity of a functionalized fullerene in human cells. *Toxicology Letters.* 2009;191(2–3):149–57. Epub 2009/09/03.
287. Jacobsen NR, Pojana G, White P, Moller P, Cohn CA, Korsholm KS et al. Genotoxicity, cytotoxicity, and reactive oxygen species induced by single-walled carbon nanotubes and C(60) fullerenes in the FE1-Mutatrade markMouse lung epithelial cells. *Environmental and Molecular Mutagenesis.* 2008;49(6):476–87. Epub 2008/07/12.
288. Brunet L, Lyon DY, Hotze EM, Alvarez PJ, Wiesner MR. Comparative photoactivity and antibacterial properties of C60 fullerenes and titanium dioxide nanoparticles. *Environmental Science & Technology.* 2009;43(12):4355–60. Epub 2009/07/17.
289. Cho M, Fortner JD, Hughes JB, Kim JH. *Escherichia coli* inactivation by water-soluble, ozonated C_{60} derivative: Kinetics and mechanisms. *Environmental Science & Technology.* 2009;43(19):7410–5. Epub 2009/10/24.
290. Canesi L, Ciacci C, Vallotto D, Gallo G, Marcomini A, Pojana G. In vitro effects of suspensions of selected nanoparticles (C_{60} fullerene, TiO_2, SiO_2) on *Mytilus* hemocytes. *Aquatic Toxicology.* 2010;96(2):151–8. Epub 2009/11/11.
291. Ringwood AH, Levi-Polyachenko N, Carroll DL. Fullerene exposures with oysters: Embryonic, adult, and cellular responses. *Environmental Science & Technology.* 2009;43(18):7136–41. Epub 2009/10/08.
292. Tao X, Fortner JD, Zhang B, He Y, Chen Y, Hughes JB. Effects of aqueous stable fullerene nanocrystals (nC_{60}) on Daphnia magna: Evaluation of sub-lethal reproductive responses and accumulation. *Chemosphere.* 2009;77(11):1482–7. Epub 2009/11/10.
293. Yang XY, Edelmann RE, Oris JT. Suspended C_{60} nanoparticles protect against short-term UV and fluoranthene photo-induced toxicity, but cause long-term cellular damage in Daphnia magna. *Aquatic Toxicology.* 2010;100(2):202–10. Epub 2009/10/27.
294. Zhu X, Zhu L, Lang Y, Chen Y. Oxidative stress and growth inhibition in the freshwater fish *Carassius auratus* induced by chronic exposure to sublethal fullerene aggregates. *Environmental Toxicology and Chemistry/SETAC.* 2008;27(9):1979–85. Epub 2008/12/17.

10 Toxicogenomic Approaches to Understanding the Toxicity of Nanoparticles

Qiwen Shi, Mahavir B. Chougule, Vijaykumar B. Sutariya, and Deepak Bhatia

CONTENTS

10.1 Introduction 209
10.2 Nanotoxicology 210
10.3 *In Vitro* Evaluation of Nanoparticle Toxicity 211
    10.3.1 *In Vitro* Cytotoxicity Screening 211
    10.3.2 Assays of ROS 211
    10.3.3 Microscopic Evaluation of Intracellular Localization 212
    10.3.4 Gene Expression Analysis 212
    10.3.5 *In Vitro* Hemolysis Test 214
    10.3.6 Genotoxicity Testing 214
    10.3.7 Inflammation Assay 215
    10.3.8 Advantage and Disadvantage of *In Vitro* Assay 215
10.4 *In Vivo* Evaluation of Nanoparticle Toxicity 216
    10.4.1 ADME of Nanoparticles 216
    10.4.2 Animal Models 216
    10.4.3 Carcinogenicity Studies of Nanoparticles 216
    10.4.4 Predication of *In Vivo* Toxicity by *In Vitro* Data 217
10.5 *In Silico* Evaluation of Nanoparticle Toxicity 217
    10.5.1 High-Throughput Screening 218
    10.5.2 Quantitative Structure Activity Relationship 218
    10.5.3 Global Models versus Local Models 219
    10.5.4 Artificial Intelligence 221
    10.5.5 Nanoinformatics 221
10.6 Conclusion 221
References 222

10.1 INTRODUCTION

Since nanoscience is sustainably turning into nanotechnology, the potential widespread use of man-made nanoparticles has been discussed. The main applications of engineered nanoparticles involve two areas, industry (chemical sensing, fuel catalysis, etc.) and biomedicine (drug delivery, medical imaging, etc.). Recently, some anticipated applications are beginning to be realized, such as sensors, biolabels, tips for scanning probe microscopy, electrochemical actuators, batteries, and so on (Barnard, 2009). During the transition, problems related to nanoparticle toxicity have arisen. To manufacture efficient, safe, and environmentally friendly nanoproducts, a predictable and reliable way to assess nanotoxicity should be established before the omnipresence of nanoparticles. Not only the interactions of nanoparticles with biological systems and natural ecosystems, but also the interactions between nanoparticles and other common compounds in the human body should be estimated during toxic studies.

Toxicogenomics is a combination of toxicology and genomics or other high-throughput molecular profiling technologies to integrate changes in gene, protein, and metabolite expressions related to chemical toxicity. It investigates the adverse effects of environmental and pharmaceutical chemicals on human health and the environment. The components of toxicogenomics include microarray, proteomics, metabolomics, and cytomics (Waters et al., 2003). Compared to traditional toxicity evaluation systems, the information obtained and analyzed by toxicogenomics can be more discriminating, predictive, and sensitive. Thus, toxicogenomics has a huge potential in improving risk assessments and hazard screenings.

The focus of this chapter is on current nanotoxicological aspects on *in silico*, *in vitro*, and *in vivo* studies via toxicogenomic approaches.

10.2 NANOTOXICOLOGY

Nanotoxicology is a new branch of toxicology focusing on the adverse health effects specifically caused by nanomaterials. The study of nanotoxicology aims at learning the physicochemical properties of nanomaterials, its routes of exposure, absorption, distribution, metabolism, excretion (ADME), molecular determinants, genotoxic and immunogenic potential, and safety regulation (Figure 10.1).

The physicochemical properties of nanoparticles include their size, chemical composition, surface structure, solubility, shape, and aggregation. A tiny change in any of these can lead to unique biological effects. People can be exposed to engineered nanomaterials by direct substance exchanges with the environment, such as the skin, respiratory tract, and gastrointestinal tract, or by administration as drug vectors. Nanoparticles, due to their unique physicochemical characteristics,

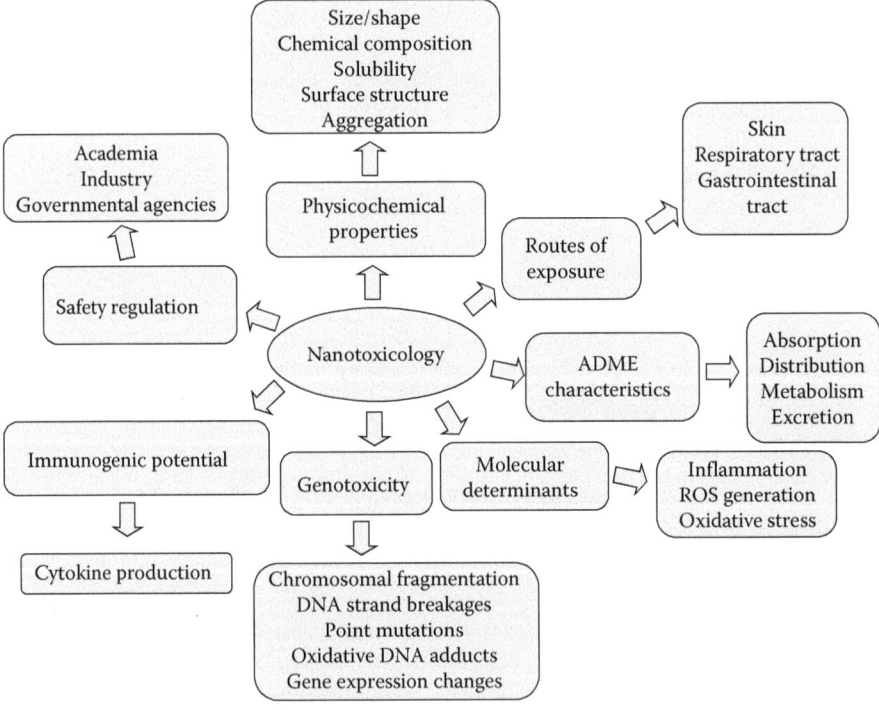

FIGURE 10.1 The area of nanotoxicology research.

have different biodistribution profiles, biological accumulation and interactions, and physical and chemical clearance processes. Understanding the relationship between the physicochemical properties and their ADME characteristics is critical in achieving desired biological effects and also minimizing potential toxicities. On the other hand, a number of nanoparticles promote the formation of pro-oxidants, and, subsequently, impair the balance between the biological system's ability to produce and detoxify reactive oxygen species (ROS). The increased levels of ROS can result in cell apoptosis or necrosis by reacting with DNA, proteins, carbohydrates, and lipids in a destructive manner. Primary genotoxicity can be a result of the direct exposure to nanoparticles, while secondary genotoxicity can be related to the interactions of nanoparticles with cells or tissues, and the adverse effects, like inflammation and oxidative stress, caused by factors released via these interactions.

Considering the specific biological characteristics of nanoparticles, multidisciplinary studies are encouraged in establishing nanoparticle classifications and testing procedures emphasizing safety regulations.

10.3 *IN VITRO* EVALUATION OF NANOPARTICLE TOXICITY

In vitro studies are conducted using established cell lines and primary cells derived from target tissues. As there is an absence of clear guidelines about how to test and evaluate nanoparticles, *in vitro* studies are extremely important to provide information on the toxicity of nanoparticles. Generally, for nanotoxicological studies, the experimental techniques are the same as the ones employed for biological studies and toxicological studies. There are seven main categories of traditional *in vitro* studies, including both genomic methods and conventional methods.

10.3.1 *In Vitro* Cytotoxicity Screening

A variety of approaches could be applied to determine the damage from nanoparticles on cell viability. It is important to choose an appropriate cytotoxicity assay when testing, since nanoparticles can influence results by adsorbing dyes or play roles in redox reactions. For example, in terms of the MTT (3-(4,5-dimethylthiazol-2-yl)-2,5-diphenyltetrazolium bromide) assay, the over- or underestimation of cell viability may happen due to the characteristics of nanoparticles. Overestimation is probably caused by the particles, which generate an absorbance at the same wavelength as that used to quantify the colored product, and underestimation may be because of the large surface area, or other surface properties, causing a high adsorptive ability that allows the nanoparticles to effectively extract the colored product from the cell extract (Worle-Knirsch et al., 2006). Considering the above possibilities, a positive- and negative-control particle should be included as a benchmark to the tested nanoparticle. Moreover, a suspension containing cell debris, the dissolved formazan, and particulates would be generated when testing nanoparticles. Thus, in order to read the absorbance of the supernatant without particles, cell debris, and other background interference, the centrifugation and transfer of the sample to a fresh, 96-well plate is recommended. Table 10.1 summarizes several of the most commonly used cytotoxicity assays.

10.3.2 Assays of ROS

ROS are electrophilic molecules or free radicals containing an oxygen atom. They can occur naturally in the body as intermediates in metabolic reactions, or as a result of toxic insults.

One of the most common ways to assess the roles of nanoparticles in oxidative stress is to use the fluorescent marker H2DCF-DA (2′, 7′-dichlorodihydrofluorescein diacetate). Fluorescence increases when H2DCF-DA is oxidized to DCF by ROS (2′, 7′-dichlorofluorescein). This assay is carried out to test the toxicity of various nanoparticles such as Ag-NPs (Singh and Ramarao, 2012), TiO_2, Fe_2O_3, Co_3O_4, Mn_3O_4 (Limbach et al., 2007), and so on.

TABLE 10.1
Summary of Selected Cytotoxicity Assays for Different Nanoparticles

Assay	Principle	Application	References
Tetrazolium salts (MTT, MTS, XT, WST)	Mitochondrial function	Ag-NP TiO_2 CoCr ZnO, Fe_2O_3, MCM-41	Ahamed et al. (2008) Liu et al. (2010) Papageorgiou et al. (2007) Dua et al. (2011)
Lactate dehydrogenase	Membrane integrity	CoCr CI, CS, AS,[a] ZnO	Papageorgiou et al. (2007) Sayes et al. (2007)
Trypan blue	Membrane integrity	Ag-NP	Hackenberg et al. (2011)
ATP assay	Actively growing cells	Ag-NP	AshaRani et al. (2008)

[a] CI, carbonyl iron; CS, crystalline silica; AS, amorphous silica.

Glutathione is a naturally occurring antioxidant in cells and biological fluids throughout the body. GSH (reduced glutathione) can react directly with ROS to neutralize them, and then be rapidly converted back into GSH. The oxidative stress induced by Zn, Fe, and Si nanoparticles has been assessed by measuring the glutathione production, demonstrating that the presence of these nanoparticles barely increased the level of ROS (Cha and Myung, 2007). The concentration of GSH, rather than the GSH: GSSG (oxidized) ratio, is selected because cells often actively export GSSG as a protective mechanism, making GSSG concentrations very low and lowering the ability to measure GSSG in cells.

Electron paramagnetic resonance (EPR) can quantify and specifically identify the free radical species generated via the use of specific spin traps or probes in combination with specific reagents, whereas DCFH assay does not possibly detect such levels of specificity. However, the measurement may be confounded by chemical or physical interferences with spin-trapping agents.

The other methods to test the levels of ROS are through the measurements of lipid peroxidation, the plasmid assay, the Trolox equivalent antioxidant capacity assay (TEAC), and the measurement of mRNA expression changes in oxidative stress-dependent genes (Stone et al., 2009).

10.3.3 MICROSCOPIC EVALUATION OF INTRACELLULAR LOCALIZATION

The microscopic evaluation techniques employed in nanotoxicology studies include scanning electron microscope (SEM), transmission electron microscope (TEM), atomic force microscopy (AFM), fluorescence spectroscopy, video-enhanced differential interference contrast microscopy (VEDICM), and so on. Figure 10.2 presents the images taken by TEM, showing that Ag-NP agglomerates (100–300 nm in diameter) are within the cytoplasm (within compartments resembling endosomes and lysosomes) and nuclei of HepG2 cells (Kim et al., 2009). Figure 10.3 is the SEM of live, adherent fibroblasts, demonstrating that nanoparticles of cobalt-chromium (CoCr) are dispersed more randomly within the cytoplasmic space (a) as compared to the micron-sized particles (b) (Papageorgiou et al., 2007). Figure 10.4 is the AFM images of the Ag NP-treated BHK21 and HT29 cells showing higher degrees of surface roughness with pit-like structures spread over the cell surface. These changes could be attributed to the aggregation of membrane proteins and the accompanying randomization of membrane lipids, proving the toxic effects of Ag-NPs on the cell membrane (Gopinath et al., 2010).

10.3.4 GENE EXPRESSION ANALYSIS

DNA microarrays can be used to measure changes in gene expression levels. Practically, the expression of ~40,000 gene spots and replicates can be simultaneously analyzed on a couple of glass arrays

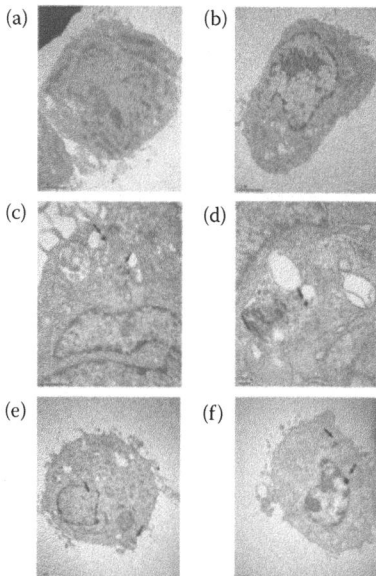

FIGURE 10.2 TEM images of 80 nm sections of HepG2 cells treated with AgNPs. AgNPs are visible in the cells as black and electron-dense spots indicated by arrows. Panels (a) and (b) are representative images of untreated controls and panels (c), (d), (e), and (f) are representative images of cells treated with 1 μg/mL AgNPs for 24 h. Scale bars represent 1 μm in panel (c), 200 nm in panel (d), and 2 μm in panels (a), (b), (e), and (f). (Reprinted with permission from Kim, S. et al. 2009. *Toxicol. In Vitro.* 23, 1076–1084.)

in a single experiment. The data from microarray technology help to develop a more complete understanding of gene expression, which can be used for transcriptional regulation and interactions as well as functional genomics. However, the data should be verified by substantial complementary investigations. Dua et al. (2011) performed DNA microarray analysis of HEK293 cells exposed to IC20 concentrations of Fe_2O_3, MCM-41, and ZnO NPs for 24 h, finding that both Fe_2O_3 and MCM-41 induced the expression of many genes encoding for 40S and 60S ribosomal protein homologs as well as histones involved in chromatin remodeling, while ZnO resulted in the altered expression of many genes involved in cell death and apoptosis. In addition, following real-time PCR, the changes in the expression of eight selected genes were validated. Although microarray technology is very successful *in vitro* studies, its exploitation is questioned in *in vivo* investigations, since it can be confounded by a number of variables, such as the type of target organ, the effects of pharmacokinetics, and/or pharmacodynamics parameters.

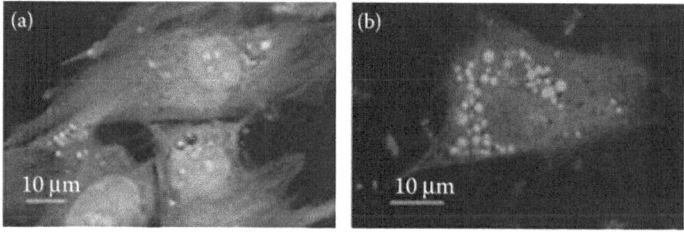

FIGURE 10.3 Scanning electron microscopy of live cells exposed to nanoparticles (a) and micron-sized particles (b) of CoCr for 24 h and viewed in quantomix capsules. Scale bars represent 10 μm. (Reprinted with permission from Papageorgiou, I., Brown, C., Schins, R. et al. 2007. *Biomaterials.* 28, 1946–1958.)

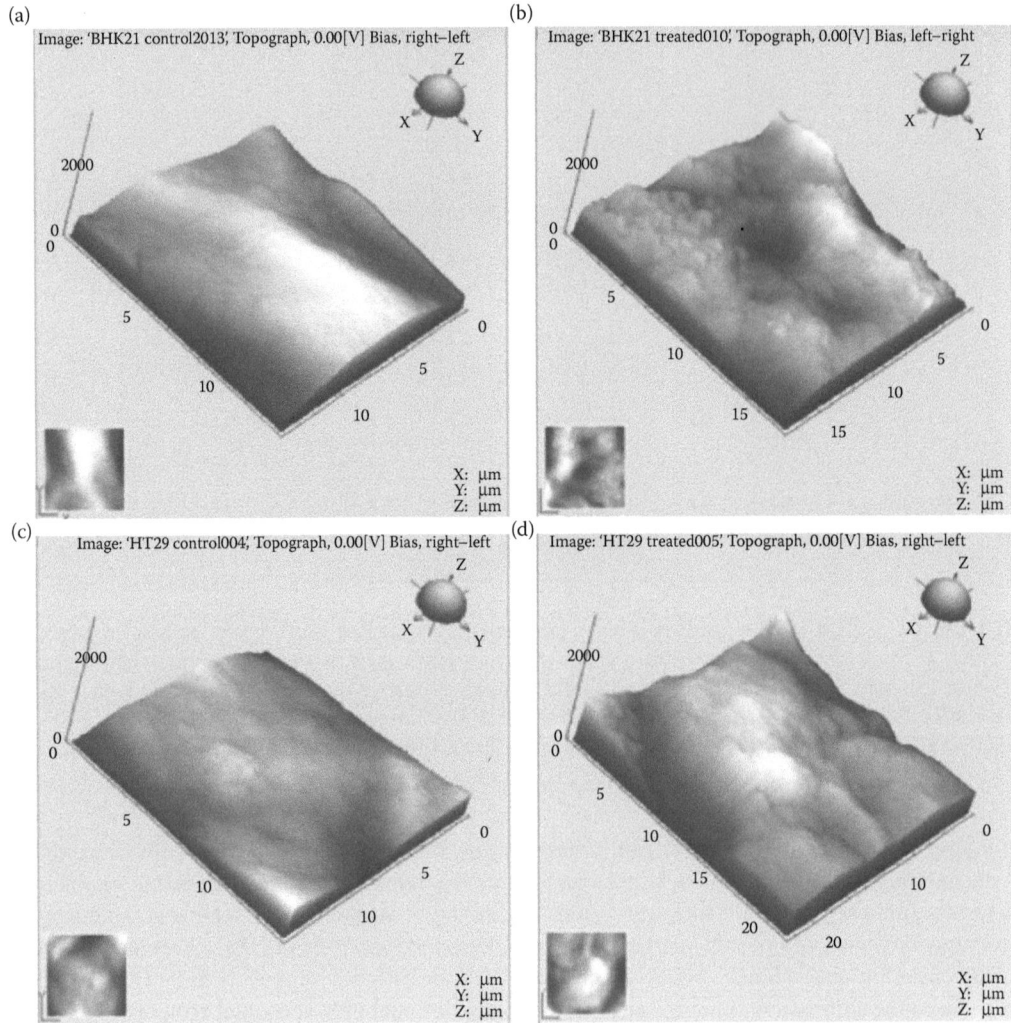

FIGURE 10.4 AFM analysis of AgNPs treated BHK21 and HT29 cells. Untreated cells: (a) BHK21 (c) HT29. AgNPs-treated cells: (b) BHK21 (d) HT29. (Reprinted with permission from Gopinath, P. et al. 2010. *Colloids. Surf. B. Biointerfaces.* 77, 240–245.)

10.3.5 IN VITRO HEMOLYSIS TEST

The aim of this test is to test the biocompatibility of nanoparticles. The results reveal the impact of nanoparticle physicochemical characteristics on human red blood cells. Sayes et al. (2007) have found that the incubation of amorphous- or crystalline-silica nanoparticles produce the hemolysis of erythrocytes.

10.3.6 GENOTOXICITY TESTING

Genotoxicity tests are conducted to identify potential genotoxic carcinogens and germ cell mutagens. "Safe" nanotechnology can be established only when it is proven to have a nongenotoxic nature.

The single-cell gel electrophoresis assay (Comet assay) is a fundamental and sensitive technique for the detection of DNA damage and repair at the level of the individual cell. It is widely applied in the genotoxicity testing of novel chemicals and pharmaceuticals, as well as nanoparticles. It is called "Comet" because the microscopic detection of damaged DNA fragments in individual cells appears as comets upon cell lysis and subsequent DNA denaturation and electrophoresis. Kim et al. (2011) have found that BEAS-2B cells treated with Ag NPs exhibited a dose-dependent increase of DNA breakage by the Comet assay. In addition, there is a review that concludes at least 46 cellular, *in vitro* studies using the Comet assay (Karlsson, 2010). The major disadvantage of this assay is that it does not measure fixed mutations.

The micronucleus (MN) assay is one of the most successful and reliable tests for genotoxic carcinogens, and is quite appropriate for the toxicological screening of potential genotoxic compounds. The basis of this assay is the microscopic detection of a cell's chromosome or chromosome fragment that has failed to integrate into the nucleus of its daughter cell after division. Chromosomal breakages in Ag NP-treated cells are found by the MN assay (AshaRani et al., 2008). And there are several other types of nanomaterials proven to induce a significant increase of MN frequencies. The MN assay can detect both chromosomal and genomic mutations, while it can only be applied to dividing cells.

The Ames test (or bacterial reversion mutation test) is usually used as an adjunct technique. The test is performed in prokaryotic systems, so it is complicated to translate to a eukaryotic system. Moreover, some nanoparticles might not be able to cross the bacterial cell wall or they might only be harmful to bacteria.

10.3.7 INFLAMMATION ASSAY

Inflammation involves the complex interactions of multiple cell types, so it is only possible to measure the markers of proinflammatory signaling and gene expression *in vitro*. The most common technique is to measure the cytokines and/or chemokines produced by cells via enzyme-linked immunosorbent assay (ELISA). The release of inflammation markers (IL1α, IL1β, IL6, IL10, MIP1α, MIP1β, MIP2, GCSF, and TNFα) were measured in RAW 246.7 cells exposed to various sized Ag-NPs. All markers were induced the most by 20 nm Ag-NPs (Park et al., 2011).

10.3.8 ADVANTAGE AND DISADVANTAGE OF *IN VITRO* ASSAY

Cell-culture studies are practical approaches in understanding how an agent will react in the body. They can reveal the primary effects of target cells in the absence of secondary effects caused by inflammation, and the identification of the primary mechanisms of toxicity in the absence of the physiological and compensatory factors that confound the interpretation of whole *in vivo* studies (Arora et al., 2012).

Compared to animal studies, they are easier to control, perform, interpret, and reproduce. Controlling the experimental condition is crucial since all the other factors, except nanoparticles, causing cell death should be minimized or eliminated. Besides, *in vivo* toxicity studies have relatively lower costs and faster speeds and, more importantly, they arise less ethical issues. *In vitro* systems are simple when only one cell line is tested, and the complexity of these *in vitro* systems increases when multiple cell types are included with the purpose to better mimic the *in vivo* environment. The co-culture of different types of cells, for example dendritic cells and epithelial cells, are cultured together to mimic the lung surface (Rothen-Rutishauser et al., 2005), possibly generates data with more meaning but is different to interpret.

Inevitably, *in vitro* systems have a bunch of disadvantages. The main shortcoming is that they cannot fully replicate the complex interactions that occur between multiple cell types *in vivo*, both within an organ and between organs, and cannot be used for true pharmacokinetic and toxicokinetic studies even though they can examine the potential for particles to cross cell boundaries (Stone et al., 2009). Moreover, cells in culture do not experience the range of pathogenic effects observed in the human body. Therefore, unfortunately, the data from cell culture systems could be misleading and invalid.

10.4 IN VIVO EVALUATION OF NANOPARTICLE TOXICITY

In vivo systems are extremely complex, and the interactions between nanomaterials and biological components can cause unique biodistributions, clearances, immune responses, and metabolisms. The major challenges for *in vivo* assays include dosimetry, the optimization of dispersion, the evaluation of interactions and biodistribution, and so on.

10.4.1 ADME OF NANOPARTICLES

The assessment of pharmacokinetics (PK) provides information regarding nanotoxicity and guides future investigations. The time and concentration of the exposure is determined by PK studies. The residence time and accumulation locations of nanoparticles can mean the difference between avoiding and experiencing toxic responses.

The small size of nanoparticles allows them to enter tissues easily but, unfortunately, they still cannot go freely into all biological systems. Nanoparticles barely enter the brain because of the protection of the blood–brain barrier, unless aided by tailored surface functionalizations. Whenever nanoparticles enters into the body via six major ways (oral, intravenous, dermal, subcutaneous, inhalation, and intraperitoneal), they interact with biological components (proteins and cells) first, and then distribute to various organs or tissues with the same or modified/metabolized structure. Afterwards, they reside in the cells of the organ for an unknown amount of time and dose before leaving. Excretion can happen at any time after absorption via the kidneys and liver/bile duct.

Although the number of current studies is not sufficient to draw any conclusion about nanopharmacokinetics, new information appears from time to time. For example, a study regarding organ coefficients of TiO_2 in mice indicated that the coefficients of the liver, kidney, and spleen increased while the coefficients of the lung and brain dropped with an increasing dose of nano-anatase TiO_2 (Liu et al., 2009). In addition, some *in vivo* PK data found a correlation of nanostructure–protein interactions. These data allow setting up a system to examine and predict structure–activity relationships (Fischer and Chan, 2007).

10.4.2 ANIMAL MODELS

Mice, rats, zebra fish, *Caenorhabditis elegans*, and rainbow trout are often-seen models. These models are employed to investigate the biodistribution and clearance of nanoparticles, environmental hazards, carcinogenicity, and acute toxicity. A DNA microarray of 207 stress-related genes performed in rainbow trout found that 13% of tested genes responded to both dissolved and nano-Ag, while about 12% of genes changed upon nano-Ag treatment specifically; the levels of vitellogenin-like proteins and DNA strand breaks were significantly reduced by both forms of Ag (Gagne et al., 2012). An *in vivo* measurement of nano-Zn, Fe, and Si acute toxicity in mice compared to microsized particles demonstrated that the low-level toxicity of nanoparticles was due to the presence of the inorganic particles themselves, not to the nanometer size (Cha and Myung, 2007). An acute toxicity of Cu-NP was reported in zebra fish, indicating that the gill was the primary target and the transcriptional response induced by Cu-NP was highly divergent (Griffitt et al., 2007). *In vivo* studies with adult Sprague–Dawley rats showed that Ag-NP did not induce genetic toxicity in either male or female rat bone marrow, but it did damage the liver (Kim et al., 2008). In contrast, tests of Ag-NPs on *Caenorhabditis elegans* highlighted that the toxicity of Ag-NP was not greater than silver ions when investigating the mechanisms of toxicity via pharmacological, genetic, and physicochemical ways (Yang et al., 2011).

10.4.3 CARCINOGENICITY STUDIES OF NANOPARTICLES

Carcinogenicity can be induced through genotoxic and nongenotoxic ways. Some scientists assume that, due to their unique properties, all nanoparticles may be carcinogenic no matter what their

TABLE 10.2
Summary of Selected *In Vivo* Carcinogenic Studies for Different Nanoparticles

NPs	Animals	Route	Result	References
TiO_2	Female rats	Inhalation	Lung tumors, benign and malignant squamous, and alveolar cell tumors combined	Heinrich et al. (1995)
Carbon black	Female rats	Inhalation	Lung tumors, benign and malignant squamous cell tumors, and bronchio-alveolar cell tumors combined	Heinrich et al. (1995)
MWCNT	Male rats	A single intrascrotal injection	Disseminated mesothelioma in the peritoneal cavity	Sakamoto et al. (2009)
SWCNT	Male rats	Intratracheal instillation	No tumors	Warheit et al. (2004)

components. The basis of this assumption is mechanistic considerations and a number of animal studies using relatively high doses. Although some experimental evidences do prove that some nanoparticles have a carcinogenic potential or higher carcinogenic potential compared to larger particles of the same material, it is still hard to say whether nanoparticles are naturally carcinogenic. In fact, current *in vivo* experiments have significant variations regarding the sample preparations of nanoparticles, the way of administration, dosages, animal species, and experimental design, and only a handful of the studies on select nanoparticles meet the standardization and quality criteria necessary in order to consider them for regulatory assessments. On the other hand, it is inadequate to consider the particle size alone when investigating the carcinogenic effect of nanoparticles. Other factors, such as the interactions of nanoparticles with cellular structures and biomolecules, can also result in toxic effects. Table 10.2 concludes a number of reports available to present a partial picture of the potential carcinogens entering the market place. However, the carcinogenic risks of other nanoparticles are still masked, as there is a lack of sufficient *in vivo* data.

10.4.4 Predication of *In Vivo* Toxicity by *In Vitro* Data

In vitro studies do not necessarily reflect the result of *in vitro* studies, but if simple high-throughput assays have been developed and validated, ethical and economic problems in *in vivo* studies could be avoided. A report from Rushton et al. (2010) evaluated two cell-free and two-cell-based assays for their reliability in *in vivo* toxicity predications. The cell-free systems used fluorescence- and electron spin resonance (ESR)-based assays of oxidant activity, while the cell-based systems used ESR and luciferase assays. *In vivo* validation was conducted by acute pulmonary inflammatory responses in rats. This group of scientists suggested that the A549 Luc1 cell line might be well suited for use in a high-throughput assay.

10.5 *IN SILICO* EVALUATION OF NANOPARTICLE TOXICITY

The use of computational models to predict the behavior of nanomaterials in biological systems is promising, because they can avoid time-consuming and resource-intensive toxicological tests. The integration of mathematical, statistical, modeling, and computer science tools provide a better understanding of the toxic mechanisms. Computational toxicology takes advantage of three modern techniques: High-information-content data streams, such as microarray; novel biostatistical methods; and computational data analysis (Rusyn and Daston, 2010). Computational approaches to toxicity predications can be applied in various areas. The pharmaceutical industry can apply these models to screen new compounds and eliminate the toxic effects of new chemical entities (Cronin, 2002).

Regulatory agencies can take advantage of these models in the regulation of new and existing chemicals. To date, a number of *in silico* techniques have been applied to detect and predict the toxicity and fate of chemicals.

10.5.1 High-Throughput Screening

High-throughput screening (HTS) enables the rapid generation of large data sets for the assessment of nanoparticles toxicity via using single-cell lines and simple organisms, laboratory automation, and robotic equipment. In HTS, simple assays, for example, the receptor binding assay, are performed in multiwell-plate formats to test thousands of chemicals and their effects on a single biological response at once (Houck and Kavlock, 2008). Unfortunately, the ultimate results are the lack of activity for key toxicity targets, which means that it is inappropriate to use HTS as an indicator of hazard or to define the mode of action. Recently, the emergence of multiparametric assays, such as multiple concentrations, exposure times, target cell lines, and toxicity end points, facilitates nanotoxicity screenings by HTS. However, this technique is vulnerable due to both systematic and random errors. Random errors can be minimized by replicate measurements and procedural quality controls, while systematic errors are more difficult to reduce. The latter ones are usually induced by across-plate and within-plate row and column biases, and require the use of control wells to assess plate-to-plate variabilities in multiplate assays and establish proper assay background levels. The normalization of HTS raw data can remove systematic plate-to-plate variability and requires the identification and elimination of data outliers. Moreover, it should be accomplished with the clear interpretation of statistical parameters.

The exploratory analysis of HTS data can extract important information about possible toxicity mechanisms and relationships among different cell responses. Heat maps combined with hierarchical clustering are basic level analyses, providing ordered representations of data. Topology-preserving mapping techniques are more comprehensive. For example, the self-organizing map (SOM) provides an ordered, 2D projection of data vectors containing the cell response information. It is both used to identify similar nanoparticles and analyze nanoparticle biological activity profiles, including multiple assay conditions, cell types, and cell responses (Cohen et al., 2013).

Although HTS data is far away from clearly identifying nanoparticles' effects on biological systems, they are indispensable for the development of other *in silico* toxicity models, such as quantitative structure–activity relationships (QSAR).

10.5.2 Quantitative Structure Activity Relationship

QSAR models are regression or classification models that relate a set of structural parameters to the potency of biological activities. Figure 10.5 represents the workflow of a QSAR paradigm (Burello and Worth, 2011). The application of QSARs encompasses both the human health effects and the environmental impact of chemicals.

The first QSARs were based on the premise that toxicity could be related to certain molecular properties of chemicals by mathematical methods. Early attempts were not successful, partly due to the limited number of parameters that could be modeled, especially investigating complex toxicities encompassing many different mechanisms of action. Although there have been progresses in modeling chemical–biological interactions which could improve QSAR models, it is still an optimistic idea that QSAR models will be able to play a major role in prediction of chemical toxicity. To date, only a few nano-QSARs have been published, which are limited to small data sets and mainly focus on metal- and metal oxide-nanoparticles (Cohen et al. 2013).

Puzyn et al. (2011) developed a model to describe the cytotoxicity of 17 different types of metal oxide nanoparticles to *Escherichia coli*, which is expressed as the EC50. This report suggests that the properties of nanoparticles are responsible for their activity and adverse effects on living

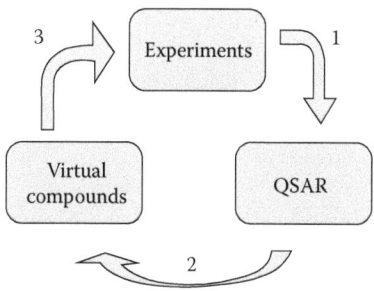

FIGURE 10.5 Workflow of a QSAR paradigm. (1) Data collected from experiments and descriptors calculated from the structure of nanomaterials are combined to generate a QSAR model. (2) The model is then used to predict the activity/toxicity of untested compounds that lie within the applicability domain of the model. (3) Computed toxicities are used to streamline and prioritize new experiments. (Reprinted with permission from Burello, E. and Worth, A. 2011. *Nat. Nanotechnol.* 6, 138–139.)

organisms, since the only descriptor in this prediction model is the enthalpy of formation of a gaseous cation with the same oxidation state as that in the tested metal oxide structure.

A report from Sayes and Ivanov (2010) claimed that a model has been set up using six different physicochemical properties/features (engineered size, size in water, size in phosphate-buffered saline (PBS) solution, size in cell culture medium, particle concentration, and zeta potential) for both TiO_2 and ZnO nanoparticles to predict cellular membrane damage (lactate dehydrogenase release). Two commonly used pattern recognition mathematical models, multivariate linear regression, and linear discriminant analyses (LDA) classification, were selected in designing predictive models. Taking TiO_2 data from this report as an example, if the data are limited to the five TiO_2 features listed in Table 10.3, linear regression cannot provide the correct framework to model cell membrane damage. Therefore, the LDA classification was performed and its results showed that there was a significant discrimination between a dense cellular membrane and a leaky cellular membrane. Table 10.3 is the measurement of the resubstitution error based on the LDA classification. Figure 10.6 shows that the combination of the nanomaterial's zeta potential with its engineered size and concentration in ultrapure water produced a classifier with 0 resubstitution error. The multiple linear regression model $y = \beta_0 + \beta_1 x_0 + \beta_2 x_4 + \beta_3 x_5$ was tested to two different data sets. The coefficient of determination for the first set, which is composed of the entire available data, was $r^2 = 0.70$, while the same model produces $r^2 = 0.77$ when fit to the second set, which includes only dense- and disrupted-cell membranes (Table 10.4). Although additional data are required for a thorough testing of existing hypotheses, the possibility of using different QSARs models for different categories of biological responses cannot be ignored. The authors also mentioned that the evaluation using additional features, such as water solubility, bioavailable metals, and time-course evaluations, should be planned in the future.

10.5.3 Global Models versus Local Models

Defined by Cronin et al. (2009), global models are developed from data for large numbers of compounds, crossing broad structural classes, and often mechanisms and modes of action, while local models are more likely to be built on smaller numbers of compounds, often with some element of structural and/or mechanistic similarity to them. Global models, definitely, cover broad chemical space, but may include a number of descriptors, which do not have a direct, physicochemical significance and may be formed with nonlinear techniques. There are many global models for toxicity, such as the TOPKAT, M-CASE, and CAESAR models. In contrast, local models restrict the domain of the model, but their advantages are that they may simply involve read-across or a simple linear technique and are more accurate. There are various types of read-across and QSARs can be regarded as sources of local models.

TABLE 10.3
Physical Features of TiO$_2$ Nanoparticles and Membrane State Values after TiO$_2$ Cellular Exposures Used in the Multivariate Linear Regression Analyses and LDA Classification

x_0 Eng. Size (nm)	x_1 Size in Water (nm)	x_2 Size in PBS (nm)	x_4 Conc. (mg/L)	x_5 Zeta Potential (mV)	y Membrane Damage (units/L)
30	125	1250	25	−10	0.90
30	102	987	25	−12	1.00
30	281	1543	50	−15	0.75
30	101	1045	50	−9	0.70
30	299	1754	100	−11	1.04
30	134	961	100	−11	1.09
30	600	1876	200	−12	1.15
30	298	1165	200	−12	1.20
45	129	2567	25	−9	0.90
45	129	2309	25	−10	0.85
45	201	2431	50	−9	0.75
45	201	2987	50	−11	0.78
45	451	2941	100	−11	1.40
45	451	1934	100	−9	1.50
45	876	1965	200	−11	1.35
45	876	2109	200	−10	1.40
125	136	3215	25	−11	1.25
125	136	2667	25	−10	1.17
125	149	3782	50	−10	1.00
125	149	2144	50	−15	1.10
125	343	3871	100	−12	1.50
125	343	2890	100	−9	1.42
125	967	3813	200	−9	1.60
125	967	2671	200	−8	1.65

Source: Reprinted with permission from Sayes, C. and Ivanov, I. 2010. *Risk. Anal.* 30, 1723–1734. With permission.

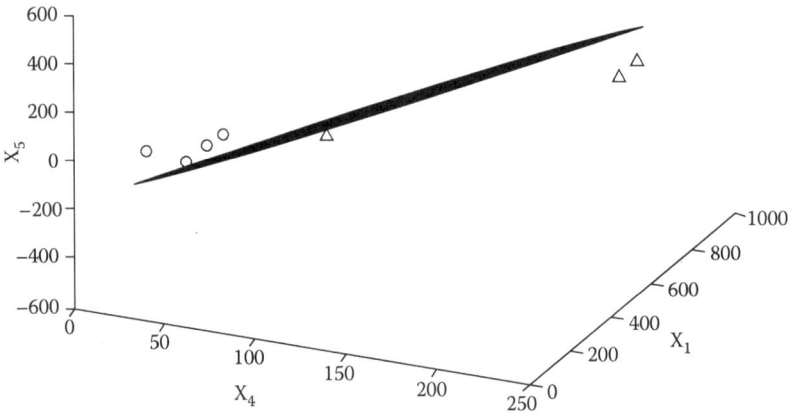

FIGURE 10.6 Classification of dense cell membrane samples versus disrupted cell membrane samples after exposure to TiO$_2$ using the combination of three features: X_1 = size in water, X_4 = concentration, and X_5 = zeta potential. Circles represent no membrane damage and triangles represent damaged membrane. (Reprinted with permission from Sayes, C. and Ivanov, I. 2010. *Risk. Anal.* 30, 1723–1734.)

TABLE 10.4
Classification of Dense Cell Membrane Samples Versus Disrupted Cell Membranes of TiO_2

Features			ε_{resub}
x_1	x_4	x_5	0
x_0	x_1	x_2	0.067
x_1	x_2	x_4	0.067
x_1	x_2	x_5	0.067
x_2	x_4	x_5	0.067

Source: Reprinted with permission from Sayes, C. and Ivanov, I. 2010. *Risk. Anal.* 30, 1723–1734.

Note: Triplet-wise LDA classifiers are shown. ε_{resub} denotes the resubstitution error for the respective classifier. For the legend of the labeling of the features, see Table 10.1. With permission.

10.5.4 ARTIFICIAL INTELLIGENCE

A new avenue has opened to QSAR models recently through the use of artificial intelligence. One group of scientists applied neural and fuzzy-neural networks with the QSAR approach (Mazzatorta et al., 2003). This study was conducted on 562 organic compounds in order to establish models for predicting the acute toxicities in fish. In the discussion section, they stressed the problems in neural network training, such as over fitting, and illustrated the difficulties of modeling the toxicity of chemicals. Although introducing artificial intelligence into QSAR approaches has been criticized, when the problems caused by parameters are settled, it would eventually be a powerful method.

10.5.5 NANOINFORMATICS

Nanoinformatics is a relatively new field that has emerged over the last few years. Its purpose is to integrate information relevant to the nanoscale science and engineering communities to develop and implement effective mechanisms for collecting, validating, storing, sharing, analyzing, modeling, and applying that information. Nanoinformatics can be applied in five major areas: Delivery systems, implantable devices, diagnosis and prevention, therapeutics, and materials (Maojo et al., 2012). All of the *in silico* methods are the basis of nanoinformatics. However, it still faces a number of challenges, mainly due to the rapid emergence of new and revised nanoparticles, data from uncertain and inaccurate methods and protocols, and various types of risks.

10.6 CONCLUSION

Current research on nanotoxicity is very dispensing. Various cell lines and animal models are being used, and different mechanisms of action are being tested. Moreover, the inconsistencies in nanoparticle preparation and dosimetry make it even harder to draw any conclusions from present data. Thus, there is an urgent need to standardize nanotoxicity studies in order to fairly assess the potential risk of new and existing nanoparticles.

Toxicogenomic approaches with proteomics and metabolomics provide a comprehensive assessment of gene expression, protein expression, or metabolite generation in a particular tissue, organ, or organism, in response to a perturbation (Rusyn and Daston, 2010). In addition, the information from toxicogenomics can also be used to identify pathways, which are responsible for the toxic properties. Combined with the conventional methods of toxicity studies, toxicogenomics has the potential to reveal the underlying mechanisms of nanotoxicity and establish a system to predict nanotoxicity for both government and industry. To date, there are many predictive models for conventional chemicals, which are convenient, fast, and relatively reliable, but not specific for nanomaterials. These models, with certain adjustments, may be evolved to be applicable for nanoparticles.

REFERENCES

Ahamed, M., Karns, M., Goodson, M. et al. 2008. DNA damage response to different surface chemistry of silver nanoparticles in mammalian cells. *Toxicol. Appl. Pharmacol.* 233, 404–410.

Arora, S., Rajwade, J. M., Paknikar, K. M. 2012. Nanotoxicology and *in vitro* studies: The need for the hour. *Toxicol. Appl. Pharmacol.* 258, 151–165.

AshaRani, P. V., Mun, G. L. K., Hande, M. P. and Valiyaveettil, S. 2008. Cytotoxicity and genotoxicity of silver nanoparticles in human cells. *ACS. Nano.* 24, 279–290.

Barnard, A. S. 2009. Computational strategies for predicting the potential risks associated with nanotechnology. *Nanoscale.* 1, 89–95.

Burello, E. and Worth, A. 2011. Predicting toxicity of nanoparticles. *Nat. Nanotechnol.* 6, 138–139.

Cohen, Y., Rallo, R., Liu, R. and Liu, H. H. 2013. In silico analysis of nanomaterials hazard and risk. *Acc. Chem. Res.* 46(3), 802–812.

Cornin, M. T. D., Enoch, S. J., Hewitt, M., and Madden, J. C. 2011. Formation of mechanistic categories and local models to facilitate the predication of toxicity. *ALTEX.* 28, 45–49.

Cronin, M. T. D., Enoch, S. J., Hewitt, M., and Madden, J. C. 2009. Formation of Mechanistic Categories and Local Models to Facilitate the Prediction of Toxicity. Highlights of WC7—7th Worldcongress in Rome 2009. *ALTEX.* 28, 45–49.

Cronin, M. T. D. 2002. The current status and future applicability of quantitative structure–activity relationships (QSARs) in predicting toxicity. *Altern. Lab. Anim.* 30, 81–84.

Dua, P., Chaudhari, K. N., Lee, C. H. et al. 2011. Evaluation of toxicity and gene expression changes triggered by oxide nanoparticles. *Bull. Korean. Chem. Soc.* 32, 2051–2057.

Fischer, H. C. and Chan, W. C. W. 2007. Nanotoxicity: The growing need for *in vivo* study. *Curr. Opin. Biotechnol.* 18, 565–571.

Gagne, F., Andre, C., Skirrow, R. et al. 2012. Toxicity of silver nanoparticles to rainbow trout: A toxicogenomic approach. *Chemosphere.* 89, 615–622.

Gopinath, P., Gogoi, S. K., Sanpui, P. et al. 2010. Signaling gene cascade in silver nanoparticle induced apoptosis. *Colloids Surf. B. Biointerfaces.* 77, 240–245.

Griffitt, R. J., Weild, R., Hyndman, K. A. et al. 2007. Exposure to copper nanoparticles causes gill injury and acute lethality in zebrafish. *Environ. Sci. Technol.* 41(23), 8178–8186.

Hackenberg, S., Scherzed, A., Kessler, M. et al. 2011. Silver nanoparticles: Evaluation of DNA damage, toxicity and functional impairment in human mesenchymal stem cells. *Toxicol. Lett.* 201, 27–33.

Heinrich, U., Fuhst, R., Rittinghausen, S. et al. 1995. Chronic inhalation exposure of Wistar rats and two different strains of mice to diesel engine exhaust, carbon black, and titanium dioxide. *Inhal. Toxicol.* 7, 533–556.

Houck, K. A. and Kavlock, R. J. 2008. Understanding mechanisms of toxicity: Insights from drug discover research. *Toxicol. Appl. Pharmacol.* 227(2), 163–178.

Karlsson, H. L. 2010. The comet assay in nanotoxicology research. *Anal. Bioanal. Chem.* 398, 651–666.

Kim, H. R., Kim, M. J., Lee, S. Y. et al. 2011. Genotoxic effects of sliver nanoparticles stimulated by oxidative stress in human normal bronchial epithelial (BEAS-2B) cells. *Mut. Res.* 736, 129–135.

Kim, S., Choi, J. E., Choi, J. et al. 2009. Oxidative stress-dependent toxicity of silver nanoparticles in human hepatoma cells. *Toxicol. In Vitro.* 23, 1076–1084.

Kim, Y. S., Kim, J. S., Cho, H. S. et al. 2008. Twenty-eight-day oral toxicity, genotoxicity, and gender-related tissue distribution of silver nanoparticles in Sprague-Dawley rats. *Inhal. Toxicol.* 20(6), 575–583.

Kyung Eun, C. and Myung, H. 2007. Cytotoxic effects of nanoparticles assessed *in vitro* and in vivo. *J. Microbiol. Biotechnol.* 17, 1573–1578

Limbach, L. K., Wick, P., Manser, P. et al. 2007. Exposure of engineered nanoparticles to human lung epithelial cells: Influence of chemical composition and catalytic activity on oxidative stress. *Environ. Sci. Technol.* 41, 4158–4163.

Liu, H., Ma, L., Zhao J. et al. 2009. Biochemical toxicity of nano-nantase TiO2 particles in mice. *Biol. Trace. Elem. Res.* 129, 170–180.

Liu, S., Xu, L., Zhang, T. et al. 2010. Oxidative stress and apoptosis induced by nanosized titanium dioxide in PC12 cells. *Toxicology.* 267, 172–177.

Maojo, V., Fritts, M., de la Iglesia, D. et al. 2012. Nanoinformatics: A new area of research in nanomedicine. *Int. J. Nanomedicine.* 7, 3867–3890.

Mazzatorta, P., Benfenati, E., Neagu, C. D., and Gini, G. 2003. Tunning neural and fuzzy-neural networks for toxicity modeling. *J. Chen. Inf. Comput. Sci.* 43, 513–518.

Papageorgiou, I., Brown, C., Schins, R. et al. 2007. The effect of nano- and micro-sized particles of cobalt-chromium alloy on human fibroblasts in vitro. *Biomaterials.* 28, 1946–1958.

Park, M. V., Neigh, A. M., Vermeulen, J. P. et al. 2011. The effect of particle size on cytotoxicity, inflammation, developmental toxicity and genotoxicity of silver nanoparticles. *Biomaterials.* 32(36), 9810–9817.

Puzyn, T., Rasulev, B., Gajewicz, A. et al. 2011. Using nano-QSAR to predict the cytotoxicity of metal oxide nanoparticles. *Nat. Nanotechnol.* 6, 175–178.

Rothen-Rutishauser. B. M., Kiama, S. G., and Gehr, P. 2005. A three-dimensional cellular model of the human respiratory tract to study the interaction with particles. *Am. J. Respir. Cell. Mol. Biol.* 32, 281–289.

Rushton, E. K., Jiang, J., Leonard, S. S. et al. 2010. Concept of assessing nanoparticle hazards considering nanoparticle dosimetric and chemical/biological response metrics. *J. Toxicol. Environ. Health.* 73, 445–461.

Rusyn, I. and Daston, G. P. 2010. Computational toxicology: Realizing the promise of the toxicity testing in the 21st century. *Environ. Health Perspect.* 118, 1047–1050.

Sakamoto, Y., Nakae, D., Fukumori, N. et al. 2009. Induction of mesothelioma by a single intrascrotal administration of multi-wall carbon nanotube in intact male Fischer 344 rats. *J. Toxicol. Sci.* 34, 65–76.

Sayes, C. and Ivanov, I. 2010. Comparative study of predictive computational models for nanoparticle-induced cytotoxicity. *Risk. Anal.* 30, 1723–1734.

Sayes, C. M., Reed, K. L., and Warheit, D. B. 2007. Assessing toxicity of fine and nanoparticles: Comparing *in vitro* measurements to *in vivo* pulmonary toxicity profiles. *Toxicol. Sci.* 97, 163–180.

Singh, R. P. and Ramarao, P. 2012. Cellular uptake, intracellular trafficking and cytotoxicity of silver nanoparticles. *Toxicol. Lett.* 213, 249–259.

Stone, V., Johnston, H. and Schins, R. P. F. 2009. Development of *in vitro* systems for nanotoxicology: Methodological considerations. *Crit. Rev. Toxicol.* 39, 613–626.

Warheit, D. B., Laurence, B. R., Reed, K. L. et al. 2004. Comparative pulmonary toxicity assessment of single-wall carbon nanotubes in rats. *Toxicol. Sci.* 77, 117–125.

Waters, M. D., Olden, K., and Tennant, R. W. 2003. Toxicogenomic approach for assessing toxicant-related disease. *Mut. Res.* 544, 415–424.

Worle-Knirsch, J. H., Pulskamp, K., and Krug, H. F. 2006. Oops they did it again! Carbon nanotubes hoax scientists in viability assays. *Nano. Lett.* 6, 1261–1268.

Yang, X., Gondikas, A. P., Marinakos, S. M. et al. 2011. Mechanism of silver nanoparticle toxicity is dependent on dissolved silver and surface coating in *Caenorbabditis elegans*. *Environ. Sci. Technol.* 46, 1119–1127.

11 Nanomaterial-Based Gene and Drug Delivery
Pulmonary Toxicity Considerations

Mahavir B. Chougule, Rakesh K. Tekade, Peter R. Hoffmann, Deepak Bhatia, Vijaykumar B. Sutariya, and Yashwant Pathak

CONTENTS

11.1 Introduction 225
    11.1.1 Pulmonary Drug Delivery: Historical Prospective 226
    11.1.2 Why Pulmonary Drug Delivery? 226
11.2 Anatomy and Physiology of the Pulmonary System 227
11.3 Nanotechnology in the Arena of Pulmonary Therapy 230
11.4 Various Nanomaterials Used in Pulmonary Delivery: Toxicity Concerns 230
    11.4.1 Carbon Nanotubes 231
        11.4.1.1 SWCNTs and Pulmonary Toxicity Issues 233
        11.4.1.2 MWCNTs and Pulmonary Toxicity Issues 233
    11.4.2 Buckminsterfullerene (C_{60} Fullerene) 235
    11.4.3 Titanium Dioxide 236
    11.4.4 Albumin 237
    11.4.5 Silica 239
11.5 Conclusion 240
Acknowledgments 241
References 241

11.1 INTRODUCTION

There is a rising rate of pulmonary disorders with high rates of death and morbidity. In this line, pulmonary drug delivery is emerging as a smart, noninvasive move for the treatment of an assortment of pathogenic disarrays. New classes of pharmaceuticals and biologic agents (DNA, proteins, peptides, etc.) are fueling the rapid evolution of drug delivery technologies for pulmonary disorders. Typically, these new therapeutic candidates cannot be delivered effectively by conventional methods (Figure 11.1). Furthermore, for many conventional pharmaceutical therapies, it has been determined that the efficacy may be improved and side effects reduced if the therapy is administered via a sustained and controlled fashion rather than through conventional burst release techniques (oral, intravenous, intraperitoneal, etc.).

The pulmonary route of administration is increasingly explored as a means for the systemic administration of therapeutic agents. Directing drugs into the lungs is the most fitting route for the treatment of pulmonary disorders such as asthma, lung cancer, tuberculosis, and chronic obstructive pulmonary disease. Numerous studies have focused on the local application of low- and high-molecular-weight drugs for the treatment of these ailments, which conclusively infers how pulmonary delivery may offer great potentials for small and large molecules, such as proteins and peptides, for

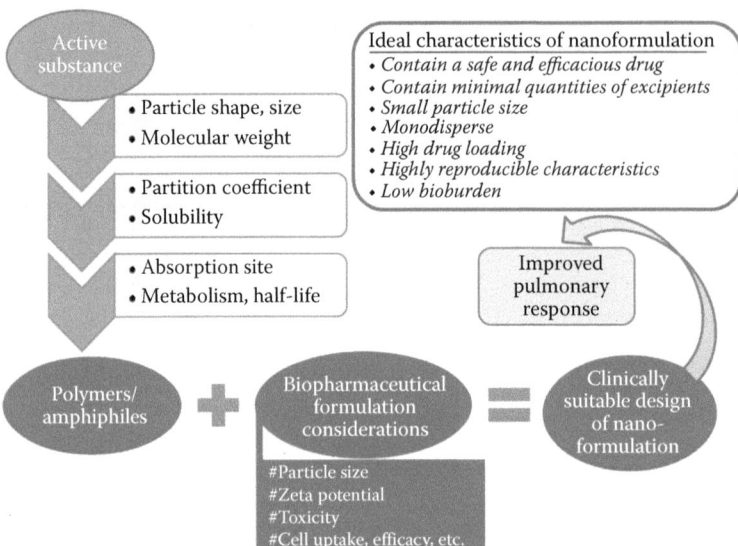

FIGURE 11.1 Design of pulmonary drug delivery systems.

both local targeting as well as systemic effect [1,2]. Gene delivery to the lungs is mainly focused on localized deliveries to the site of action against the genetic disorders affecting the airways such as antitrypsin deficiency, asthma, and cystic fibrosis, as well as obstructive lung diseases.

11.1.1 Pulmonary Drug Delivery: Historical Prospective

The history of engaging the respiratory route for drug delivery dates back to the ancient Egyptian civilization, when various substances were inhaled either ritually or for healing purposes. Inhalation therapies engrossed the exploitation of leaves from aromatic plants. With the discovery of liquid nebulizers around the turn of the twentieth century, this strategy developed into a valid pharmaceutical therapy [3,4]. The adrenaline nebulizer solution (1920s), porcine insulin nebulizer (1925s), pulmonary penicillin (1945s), and steroidal nebulizer (1950s) were among some of the primary examples in this line. In 1956, the pressurized metered dose inhaler was introduced; along with advances in molecule design and drug discovery, this route became the mainstay for the treatment of various disorders [5].

Administration through the pulmonary route has many advantages over other routes for the treatment of particular disease states. Specifically, lung-associated, large protein molecules administered through an oral route may degrade within the gastrointestinal tract or may be diverted through the first-pass metabolism into the liver, which can be avoided through the pulmonary route if deposited in the respiratory passage of the lungs. It must be noted that certain drugs taken by the pulmonary route are readily absorbed by the alveolar region and then directed into blood circulation. The pulmonary route has also been investigated as a possible mode of administration for drugs that act systemically, rather than locally, in the lungs. The driving force behind this was the observation that peptides and proteins could be absorbed systemically when delivered via the lungs (like morphine, sildenafil, triptans, etc.) [6–8].

11.1.2 Why Pulmonary Drug Delivery?

There is an increasing interest in pulmonary drug delivery owing to its high propensity to enhance the bioavailability of molecules of vivid origins, nature, and molecular weights, in contrast with other routes of administration (oral, intravenous, buccal, transdermal, rectal, etc.). For the local treatment of various lung-related disorders (viz. asthma, cystic fibrosis, bronchitis, cancer), pulmonary drug delivery allows the administration of high concentrations of drugs directly to the intended site of action. In this way, it provides a rapid onset of action with minimal side effects that are usually elicited by the systemic administration of drugs [9]. Furthermore, it also improves the therapeutic efficacy by by-passing the hepatic first-pass metabolism of the liver as well as the poor absorption of the intestines that is associated with oral drug delivery, thus allowing for similar therapeutic effects from a smaller dose.

For systemic administration, pulmonary drug delivery offers a noninvasive mode of delivery with low enzymatic activities or hepatic first-pass metabolism of drugs (Byron and Patton, 1996). It should also be considered that the macromolecules delivered by other routes are seldom absorbed into systemic circulation, while the delivery by the lungs allows for more effective absorption [10,11]. Since the pulmonary route avoids the introduction of bioactives to the gastrointestinal tract, reproducible absorption kinetics usually are achieved due to the lack of interference from the variations in an individual's diet and metabolic system [12,13].

Among the various drug delivery systems considered for pulmonary applications, nanocarriers demonstrate several advantages for the treatment of respiratory diseases. For example, nanocarriers may provide prolonged drug release and cell-specific, targeted, drug delivery. The development of an innovative nanocarrier, able to deliver the drug to the desired site of action, is highly dependent on the nature of the active substance and on its desired site and mode of action (Figure 11.1). The pulmonary system is in direct contact with the environment and represents a promising gate of entry into the body for therapeutic compounds.

11.2 ANATOMY AND PHYSIOLOGY OF THE PULMONARY SYSTEM

By removing metabolic wastes (CO_2) and maintaining the pH of the body, the pulmonary system is one of the most crucial organ systems of the body. The respiratory zone is chiefly divided into the upper and lower airways with the larynx and trachea as the line of junction [14,15]. Together with the nose, mouth, pharynx, and larynx, the upper airways frame the air transportation system, while the lower respiratory tract consists of trachea-bronchial, gas-conducting airways, and gas-exchanging acini. The lower airway is further divided into the conducting, transitional, and respiratory zones. The conducting zone is responsible for the bulk movement of air and blood. The conducting airways exhibit 16 bifurcations, followed by another 6 bifurcations of the respiratory bronchioles, facilitating the passage of air to the respiratory zone where the alveolar ducts—with alveolar sacs—finally branch off.

The respiratory zone is mainly composed of respiratory bronchioles and alveoli where gas exchange takes place [16]. The bronchial tree begins with the trachea, which bifurcates to form the main, left, and right primary bronchi (Figure 11.2). Each primary bronchus divides into still smaller secondary bronchi, or lobar bronchi—one for each lobe of the lung. The secondary bronchi branch into many tertiary bronchi that further branch several times, ultimately giving rise to tiny bronchioles that subdivide many times, finally forming terminal bronchioles and respiratory bronchioles. Each respiratory bronchiole subdivides into several alveolar ducts that end in clusters of small, thin-walled air sacs called alveoli, which open into a chamber called the alveolar sac [17,18].

Alveoli are the terminal air spaces of the respiratory system and are the actual sites of gas exchange between the air and the blood. Approximately 100 million alveoli are found in each lung. Each alveolus is a thin-walled, polyhedral chamber of approximately 0.2 mm in diameter (Figure 11.2). Each alveolus is confluent with a respiratory bronchiole at some point by means of an alveolar duct and an alveolar sac (Figure 11.3). Airway epithelium is composed of a variety of cell types, the distribution of which confers different functions according to the airway region [18,19]. The human lung consists of 5 lobules and 10 bronchopulmonary segments, and, adjacent to each segment, lung lobules are present that are composed of 3–5 terminal bronchioles. Each bronchiole is composed of the smallest structural unit of the lung, the acinus, which consists of alveolar ducts, alveolar sacs, and alveoli.

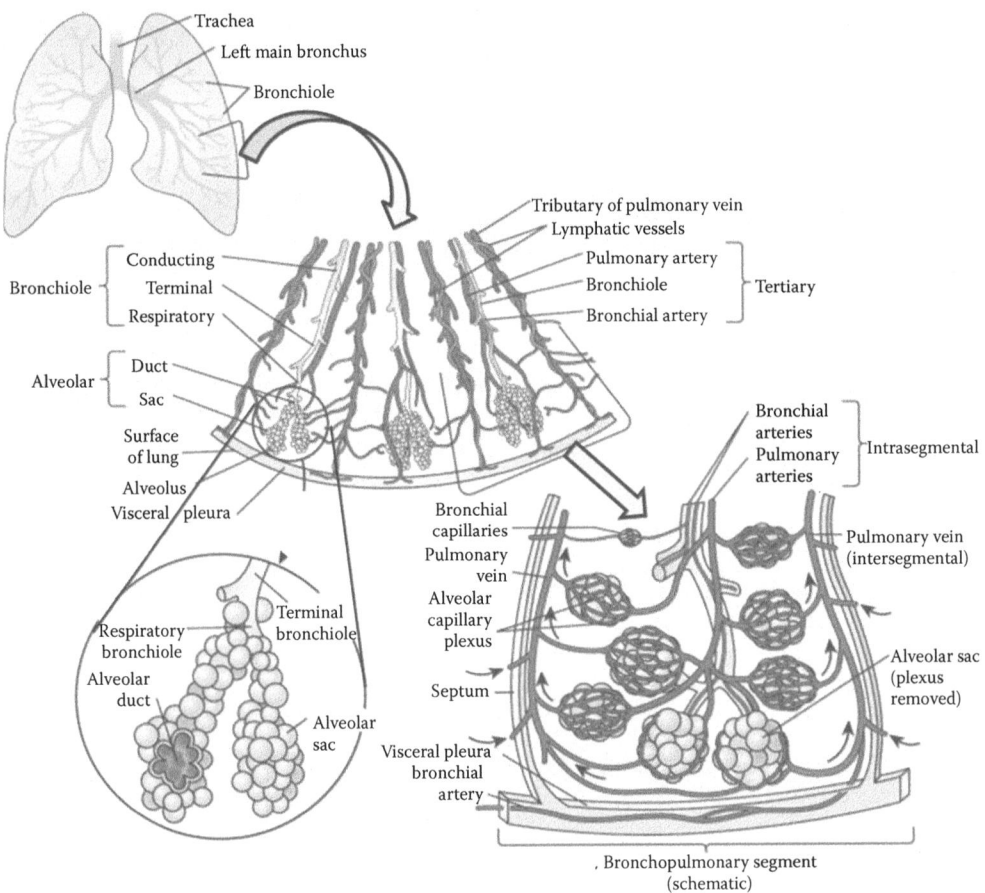

FIGURE 11.2 Structural organization of lungs. (Reproduced from *Pediatric Critical Care*. 14 ed. D'Angelis CA et al., Structure of the respiratory system. Lower respiratory tract. Copyright 2011, with permission from Elsevier Saunders.)

The walls of the conducting airways are coated by an adhesive, viscoelastic mucus layer (thickness: 5–55 µm), secreted by mucus cells. The major components of mucus are glycoproteins and water. This mucus fulfills important functions, including the protection of respiratory epithelium from dehydration, the saturation of inhaled air with water in the mucus, the exertion of antibacterial effects via mucus proteins and peptides such as defensins and lysozyme, and the protection of the airways from inhaled xenobiotics or chemicals [20–22]. The "mucociliary escalator" serves as an important protective mechanism for removing small, inhaled particles from the lungs. The composition, thickness, viscosity, and clearance of mucus are often altered in patients suffering from airway diseases. The majority of insoluble particles deposited in the upper airways are eliminated by mucociliary clearance [23–25]. The most prominent defense mechanism of the respiratory region is macrophage clearance. The particles deposited in the deeper lung will be taken up by alveolar macrophages, which slowly migrate out of the lung, either following the broncho-tracheal escalator or by migrating through the lymphatic system. Particle clearance by macrophages appears most efficient for particles having a geometric size between 1 and 3 µm [26].

The blood supply to the lung is provided by both pulmonary and systemic circulations. Both circulations are anatomizing at the level of the junction between the conducting and respiratory

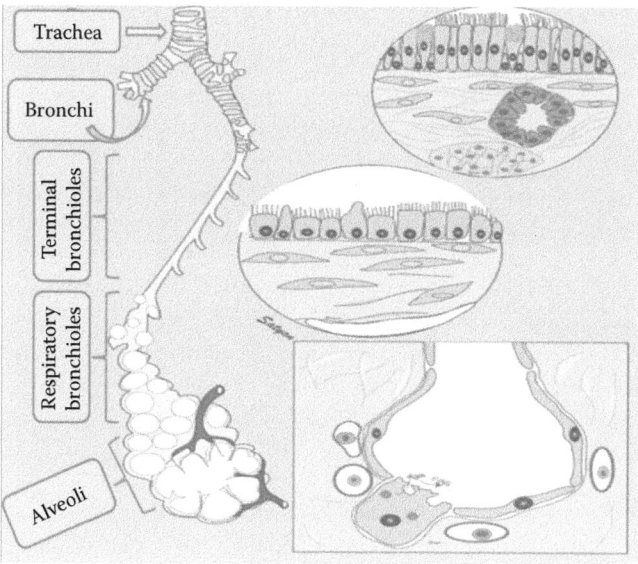

FIGURE 11.3 Illustration of the alveoli. (Reproduced from *Pediatric Critical Care*. 14 ed. D'Angelis CA et al., Structure of the respiratory system. Lower respiratory tract. Copyright 2011, with permission from Elsevier Saunders.)

passages. Hence, drugs delivered to the lower airways can enter the systemic circulation through the absorption into the systemic circulation or into the alveolar capillaries of the pulmonary vascular bed [20,27].

Traditionally, drug delivery has been carried out in the form of injections, infusions, ingestions, and inhalations, with additional variations in each category. The challenge for both drug and drug delivery companies is to deliver both existing and emerging drug technologies in a manner that improves the benefits to patients and the healthcare system. An ideal drug therapy must bear the following attributes:

- Improved efficacy
- Reduced side effects
- Sustained dosing (continuous release)
- Increased ease of use
- Reduced pain from administration
- Increased patient compliance
- Improved mobility
- Decreased involvement of healthcare workers
- Improved safety
- Safe elimination

To provide these benefits, a number of approaches are being explored to determine which of the approaches best meets the needs; this is a complex problem. Although ingestion is probably the most widely accepted form of delivery, it presents difficulties for a number of important classes of drugs. In the case of proteins and peptides, oral formulations can only deliver bioactives with a fewer bioavailability. Even though oral delivery meets the need for self-administered drugs, targeted, sustained releases and increased bioavailabilities represent the areas of difficulty in meeting the emerging value proposition. To address this difficulty, nano-fabricated drug delivery systems are being explored.

11.3 NANOTECHNOLOGY IN THE ARENA OF PULMONARY THERAPY

Nanotechnology holds an enormous promise for improving the health as well as eminence of life in people throughout the world. The beneficial effects of nanotechnology in numerous medical applications seem extremely promising. In recent years, there is noteworthy attention in developing biodegradable nanoparticles (NPs) as drug-and-gene delivery systems [28,29]. Basically, the term "nanotechnology" refers to the exploitation of materials at the nanometric scale, that is, to a billionth part of a meter. It deals specifically with the design, characterization, and production of novel nanostructures, nanodevices, and nanosystems, with controlled shapes, sizes, forms, and properties of matter at aforementioned stated scales for varying purposes. These colloidal particles are usually formulated using biodegradable polymers in which a therapeutic agent can be entrapped, adsorbed, or chemically coupled. There are considerable advantages with using NPs as drug delivery carriers, which chiefly include [30]

- A small size [31]
- The propensity to penetrate through smaller capillaries/pores
- Efficiencies in the resulting drug accumulation at the site of action [32]
- Allowing for a sustained drug release over a period of days or even weeks [33,34]
- Modifiable surfaces: surfaces can be modified to be conjugated to drugs, proteins, or peptides
- They can be modified by ligands, probes, or proteins to alter the biodistribution of drugs and to achieve target-specific drug delivery [35,36] U.S.F.D.A.-approved polymers are available for human use [37]

In 2004, the nanotechnology market was valued at approximately $13 billion, which increased to nearly $50 billion by 2006. Estimates for product sales in 2008 fix that value between $100 and $150 billon, with market expectations for 2010 set at $500 billion, and between $1 and $2 trillion by 2015. This "nano boom" clearly reflects the expected technological construction of a defined architecture that bears ample associated applications [38]. Engineered nanomaterials such as the quantum dots, dendrimers, carbon nanotubes (CNTs), and fullerenes exemplify nanomaterials that have diameters less than 100 nm and can be compared with the sizes of several living structures (Figure 11.1). Nanotech industries are reaching new horizons with the progression of these innovative classes of nanomaterials, extending applications to advance the excellence of healthcare commodities with superior patient compliance (Figure 11.2).

11.4 VARIOUS NANOMATERIALS USED IN PULMONARY DELIVERY: TOXICITY CONCERNS

Nanoparticulate entities can enter living organisms through ingestion (oral route), dermal absorption (via skin), injection (blood circulation), and inhalation (respiratory route). Currently, biopharmaceuticals and conventional drugs are frequently incorporated into nanocarriers to direct their fate in preferential pathways. Nanotechnology is believed to bring an elementary modification in manufacturing over the next few years and will have a dramatic impact on life sciences [39,40]. However, it must be noted that this promising utilization of nanotechnology for advances in healthcare is offset by the associated toxicity issues and potential adverse effects on human health. As a result of their small size and unique physicochemical properties, the toxicological profiles of NPs may differ considerably from those of larger particles composed of the same materials [41,42]. Furthermore, NPs of different materials (e.g., gold, silica, titanium, CNTs, dendrimers, and quantum dots) are not expected to interact with, or adversely affect, biological systems. Hence, the toxic potential of NPs cannot be predicted by any single mechanism. Once entered into the body, nanotherapeutics can interact with the local tissues and may provoke dysfunctions in the organs of contact [43]. These nanomaterials may sometimes enter the bloodstream and translocate to other organs. It remains unclear, due to limited data, how extensive these potential adverse effects may be, and the literature suggests that adequate precautions must be exercised when it comes to the application of nanomaterials. Therefore, the role of engineered nanomaterials in therapeutic settings needs to be completely understood before their full potential can be realized.

The respiratory system symbolizes a distinctive target for the possible toxicity of NPs due to the fact that, in addition to being the port of entry for inhaled particles, it also receives the entire cardiac output. As such, there is a potential for exposure to the lungs for NPs that are introduced to the body via vivid exposure routes (like dermal, gastrointestinal, injection, etc.) that may result in its systemic distribution [44,45]. The principle machinery responsible for the deposition of inhaled nanosized particles in the respiratory tract is diffusion. Additionally, numerous defense mechanisms exist throughout the respiratory tract aimed at keeping the mucosal surfaces free of cell debris and particles deposited by inhalation. Once deposited, nanosized particles translocate readily to extrapulmonary locations and reach other target organs by various transfer routes and mechanisms. Once the particles arrive at pulmonary interstitial sites, their uptake into blood circulation (in addition to lymphatic pathways) can also occur.

Nanomaterials can gain access to the bloodstream following inhalation and ingestion, whereupon they can be taken up by organs and tissues, including the brain, heart, liver, kidneys, spleen, bone marrow, and nervous system [46,47]. Hence, in addition to lung effects, within the last few decades, considerations of particle toxicology have also encompassed possible systemic effects [48,49]. An interest in the respiratory system as a target for the potential effects, both beneficial and adverse, of NPs is reflected by the steady increase in the number of scientific publications on these subjects during the past few decades. The studies have shown that, in the case of lung diseases, the buildup of even harmless matter may harm its function and enhance its damage. CNTs were found to be more toxic, and recent studies have shown it to cause the death of kidney cells and to inhibit cell growth by decreasing cellular adhesion [50,51]. Recently, researchers throughout the globe are exploring toxicity issues associated with various nanomaterials that are proficient in carrying, as well as delivering, loaded bioactives (drug, gene, diagnostic agents, etc.) to sites of action.

The purpose of this chapter is to complement and expand on the understanding of the pulmonary effects of various nanomaterials by providing an overview of potential applications as well as associated toxicity concerns, with a special emphasis on drug and gene delivery.

11.4.1 Carbon Nanotubes

CNTs are essentially cylindrical molecules composed exclusively of carbon atoms [52]. They basically exist in two distinct classes, namely, single-walled carbon nanotubes (SWCNTs) and multiwalled carbon nanotubes (MWCNTs). SWCNTs are long, wrapped, graphene sheets that are mostly considered one-dimensional [53], while MWCNTs are larger structures that are composed of several SWCNTs arranged one inside the other (Figure 11.4). CNTs could be one of the most promising nanovectors for the proficient delivery of drugs and biomolecules due to their huge surface area and exceptional properties. They can be conjugated noncovalently or covalently with drugs or biomolecules toward the development of a new-generation delivery system. CNTs bear the capacity to interact with mammalian cells and enter cells via cytoplasmic translocation; hence, they can deliver a range of therapeutic reagents into the cell. For example, plasmid DNA contained within the CNT may be internalized by the cell, and the expression of the plasmid-carried marker genes has been shown to be enhanced [54–56]. The unique and diverse properties of CNTs, including their capability to undergo chemical modification for a variety of applications, make them an excellent and widely explored nanomaterial. They are of a nanometeric size and bear the immense capacity to interact with macromolecules such as drugs, proteins, peptides, DNA, RNA, and siRNA [57–59].

In addition to the wide range of literature describing the biomedical application of CNTs, reports are also available on toxicity issues associated with this application. The toxicity of CNTs is closely

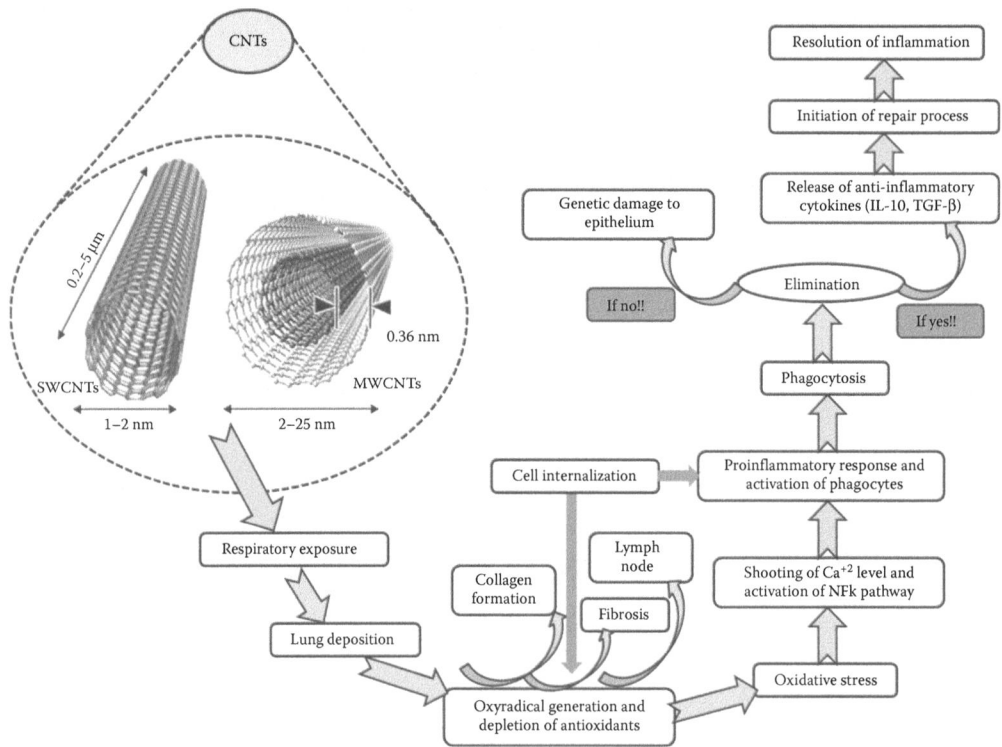

FIGURE 11.4 (**See color insert.**) Possible pathways following the inhalation of CNTs.

related to the properties of the CNT material, such as their structure, length, surface area, degree of aggregation, extent of oxidation, bound functional groups, and method of manufacturing (that can leave the catalytic residues and produced impurities), as well as their dose. Several *in vitro* and *in vivo* studies have inferred that CNTs and/or associated contaminants or catalytic materials that arise during the production process may induce oxidative stress and prominent pulmonary inflammation. Some studies even suggest comparable similarities between the pathogenic properties of MWCNTs and those of asbestos fibers [60]. Explorations focused on the elucidation of CNTs toxicity have mainly focused on the pulmonary effects of CNTs administered through the pharyngeal or intratracheal routes. Several studies demonstrated that both SWCNTs and MWCNTs might induce cytotoxic effects and apoptosis in different cell types [61–63]. The main reasons behind the toxicity of CNTs include (i) their nanoparticulate nature; (ii) fiber-shaped architecture and, hence, behavior like asbestos and other pathogenic fibers, which have toxicities associated with their needle-like shape; and (iii) they are essentially graphitic and, hence, believed to be biopersistent [64,65].

Muller et al. reported that the agglomerates of intact CNTs remained entrapped in the largest airway, whereas ground nanotubes were much better dispersed in the lung tissue. CNTs can cause pulmonary inflammation, fibrosis, stimulate the accumulation of neutrophils and eosinophils, and increase an assortment of cell-lysing markers in the lungs [66]. These include a significant increase in total bronchoalveolar lavage cells and polymorphonuclear leukocytes, as well as proteins, lactate dehydrogenase (LDH), tumor necrosis factor-α, interleukin-1, and mucin levels [67]. In another breakthrough study, Jacobsen et al. compared the effects of the instillation of three carbonaceous particles: carbon black, fullerenes C_{60} (C_{60}), and SWCNTs, as well as gold particles and quantum dots in mice. Characterizations of the instillation media revealed that all particles were delivered as aggregates as well as agglomerates. Significant increases in messenger RNA (Mip-2 and Mcp-1) were detected in lung tissue, 3 and 24 h following the instillation of SWCNTs. The administration of CNTs induced a dose-dependent increase in LDH release, which is a marker of cell toxicity during lung inflammation. CNTs induced a significant increase in lung type I collagen levels in comparison with control mice. The authors also reported that a significant fraction of the administered dose of CNTs (~40%) remained in the lung after 60 days [66,68]. Such reports strongly demand careful attention to be paid for the application of SWCNTS and MWCNTs.

11.4.1.1 SWCNTs and Pulmonary Toxicity Issues

Alveolar macrophages constitute the first line of immunological defense against invading particles in the lung. Researchers have conducted a cytotoxicity study of CNTs with macrophages and found that SWCNTs can induce pulmonary injury in mice. Intratracheal instillations of SWCNTs in the lungs of rats resulted in the formation of lung granulomas and produced mortality in almost 15% of rats within 24 h postinstillation due to the enhanced blockage of the large airways. The intratracheal instillation of 0.5 mg of SWCNTs into male imprinting control region (ICR) mice induced the activation of alveolar macrophages, various chronic inflammatory responses, and severe pulmonary granuloma formations [69,70]. The SWCNTs have a strong tendency to agglomerate following intratracheal exposures with the propensity to induce interstitial granulomas and pulmonary injuries in a dose-dependent manner [71]. The macrophage uptake of SWCNTs was observed after the intratracheal instillation of 0.5 mg of SWCNTs, which activated various transcription factors, such as activator protein 1 (AP-1) and nuclear factor κB (NF-κB). This leads to the induction of protective and antiapoptotic gene expression, oxidative stress, activation of T cells, recruitment of leukocytes, and release of proinflammatory cytokines [70].

Investigators also assessed the health risks of CNTs on the human respiratory system by using cocultures of normal bronchial epithelial cells and normal human fibroblasts. An aqueous solution of SWCNTs was incubated with the above coculture. The results indicated the increased production of nitrous oxide and decreased cell viability after exposure to different concentrations of SWCNTs. The above observations were associated with the inflammatory and cytotoxic effects of SWCNTs, respectively. The dose administered by inhalation produced greater respiratory toxicities than the same dose administered by aspiration. The SWCNTs in both a single-dose aspiration study and a 4-day inhalation study caused an early inflammatory response followed by granulomas, fibrosis, and decreased rates of respiration, as well as the activation of a gene that produces lung cancer [72]. By using adult female C57BL mice, investigators detected two morphologically distinct responses in the lung as early as 7 days postexposure of SWCNTs delivered by pharyngeal aspiration. CNTs were clearly associated with an elicited granulomatous inflammatory response, and interfacing bundles of fibrous connective tissue were observed within discrete granulomas.

Reports are also available that confirm the detection of signs of gill irritation and mucus secretion after the exposure of fish to SWCNTs. Visual inspection of the gills and mucus smears on day 4 showed a thin layer of secreted mucus on the surface of gills from fish treated with SWCNTs, but not the controls. It was found that the exposure to SWCNTs caused respiratory toxicity in trout, increased mucus secretion, and enlarged mucocytes on the gills; elevated ventilation rates compared with the controls were observed, but no mortalities and gill injuries (swollen tips of lamellae, enlarged mucocytes, edema) had occurred by the fourth day [73,74]. In the same fashion, serious toxicity setbacks are also associated with MWCNTs.

11.4.1.2 MWCNTs and Pulmonary Toxicity Issues

With a large size and high phase ratio, MWCNTs caused significant protein exudates and granulomas on the peritoneal side of the diaphragm due to inflammation. The MWCNTs were found to be present in the lung after 60 days, and both induced inflammatory and fibrotic reactions. After 2 months, collagen-rich granulomas were observed protruding in the bronchial lumen, which was also associated with alveolitis in the surrounding tissues of the pulmonary tract. This class of CNTs also stimulated the production of TNF-α in the lungs of treated animals. *In vitro*, ground CNTs initiated the overproduction of TNF-α by macrophages, and results suggested that CNTs may be potentially toxic to humans. Experimental studies signified that CNTs have the potential to induce adverse pulmonary effects, including alveolitis, fibrosis, and genotoxicity in epithelial cells. *In vitro* experiments on rat lung epithelial cells showed that the acute pulmonary toxicity and the genotoxicity of CNTs were reduced upon heating but restored upon grinding, suggesting that the intrinsic toxicity of CNTs is mainly mediated by the presence of defective sites in their carbon framework [75,76]. After 30- and 60-day inhalation exposures, the pulmonary toxicity of MWCNTs was assessed using biochemical indices in bronchoalveolar lavage fluid and pathological examinations [77]. It was found that the aerosolized MWCNTs did not induce obvious pulmonary toxicities in the 30-day exposure group, but induced severe pulmonary toxicity in the 60-day exposure group.

The drastic, dose-dependent cytotoxicity of SWCNTs was noticed in alveolar macrophages separated from guinea pigs *in vitro* for 6 h. The macrophages exposed to SWCNTs or MWCNTs showed characteristic features of apoptosis at different dosages, toxic responses being more with SWCNTs than with the MWCNTs, quartz, or fullerene used in this study [48,78]. There are reports showing contradictory results of CNT cytotoxicity to macrophages. The intratracheal or pharyngeal instillations of a SWCNT suspension in mice caused a persistent accumulation of CNT aggregates in the lung, followed by the rapid formation of pulmonary granulomatous and fibrous tissues at the site; it also produced cardiovascular toxicity. A brief account of toxicity issues associated with CNTs are briefed in Table 11.1.

TABLE 11.1
Pulmonary Toxicities of SWCNTs and MWCNTs

Type of CNT	Studies Performed on Animal Species/Cells	Toxicity Noticed	Reference
SWCNTs	B6C3F1 mice	Generation of interstitial granulomas and associated lung injuries	[71]
	C57BL mice	Production of granulomatous fibrosis with thickening of alveolar walls, increment in count of alveolar type II cells and damage of pulmonary cells	[74]
	Albino Wistar rats	Formation of lung granuloma with approximately 15% death of instilled rats within one day	[109]
	C57BL mice	Production of pulmonary fibrous tissues with cardiovascular toxicity	[48]
	C57BL mice	Raised count of inflammatory cells, raised LDH release, leukocytes count, and raised levels of proinflammatory cytokines (tumor necrosis factor-α (TNF-α) and interleukin-6, (IL-6)) with enhanced profibrotic response	[74]
	CR mice	Activation of alveolar macrophages, initiation of several chronic inflammatory responses with severe granuloma	[70]
	Trout fish	Severe respiratory toxicity including raised mucus secretion, bulged mucocytes on gills with significantly increased ventilation rates	[73]
MWCNTs	Sprague-Dawley rats	Authors observed significant protein exudation on peritoneal side of diaphragm with drastically increased inflammatory and fibrotic response. Stimulated production of TNF-α was also reported	[66]
	Albino Wistar rats	Alveolitis, fibrosis with genotoxicity in epithelial cells	[75]
	Wistar rats	Acute pulmonary toxicity	[66]
	Wistar rats/lung epithelial cells	Adverse pulmonary effects, including alveolitis, fibrosis, and genotoxicity in epithelial cells. *In vitro* experiments showed acute pulmonary toxicity and the genotoxicity of CNTs	[76]
	NR8383 and human A549 lung cell lines	Formation of reactive oxygen species in rat macrophages	[60,72]
	Epidermal keratinocytes (HaCaTs) cells	Loss of cell viability, including ultrastructural and morphological changes	[74]

11.4.2 Buckminsterfullerene (C_{60} Fullerene)

A fullerene is a molecule composed completely of carbon and in the shape of a hollow sphere, ellipsoid, or tube. At Rice University in 1985, spherical fullerenes, also called buckyballs or buckminsterfullerene (C_{60}; Figure 11.5), were the first fullerene molecules to be discovered [79]. In 1996, Sir Harry Kroto received the Nobel Prize in Chemistry for the discovery of buckyballs [80]. C_{60} is a truncated icosahedron containing 60 carbon atoms with C5–C5 single bonds forming pentagons and C5–C6 double bonds forming hexagons. The diameter of a C_{60} fullerene molecule is less than 1 nm (average bond length ≈1.5 Å) and, hence, is an important member of the nanomaterial family [81].

They have been the subject of intense research, both for their unique chemistry and for their technological applications, especially in materials science and nanotechnology.

They have captured the imagination of pharmaceutical researchers for their unique properties as drug candidates. Derivatized fullerene can cross the cell membrane and bind to the mitochondria as demonstrated by Foley et al. [82]. Moreover, DNA-functionalized fullerenes are able to enter COS-1 cells and show comparable, or even better, efficiency than that of commercially available, lipid-based vectors [83]. A lipophilic, slow-release, drug delivery system, which employs fullerene derivatives to enhance therapeutic efficacy in tissue culture, was designed by Zakharian et al. [84]. Modified fullerenes have the potential to provide such a lipophilic, slow-release system, and are composed of significant anticancer activities in cell culture as demonstrated with a C_{60}–paclitaxel conjugate.

Similarly, several biomedical uses have been explored using water-soluble C_{60}-derivatives. However, the employment of fullerenes for drug delivery is still at an early stage of development primarily because of the associated toxicity issues. One study was performed to determine the genotoxicity of fullerenes (a mixture of C_{60} and C_{70}) in a bacterial reverse mutation assay, including the chromosomal aberration test, in hamster lung cells, followed by the acute, oral, median lethal dose of fullerenes when applied to rats [85]. The results revealed that fullerenes did not have the ability to induce acute oral toxicity or *in vitro* genotoxicity. Although water-soluble fullerenes are not acutely toxic, they are retained in the organism for long periods, raising concerns about chronic toxic effects [86]. Underivatized fullerenes aggregate in water where they are supposed to cause oxidative damage to cellular membranes even at relatively low concentrations.

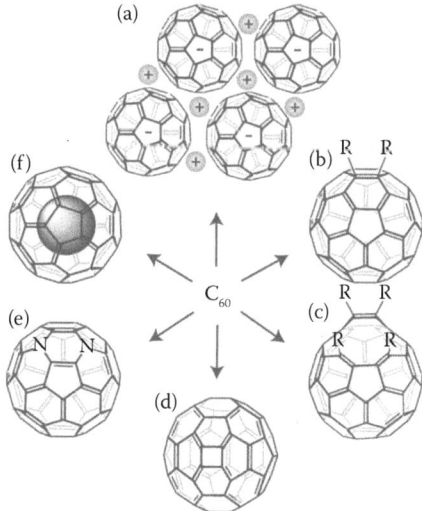

FIGURE 11.5 Structure of fullerenes: (a) fullerene salts; (b) exohedral adducts; (c) open-cage fullerenes; (d) quasi-fullerenes; (e) heterofullerenes; (f) endohedral fullerenes. (Adapted with permission from Hirsch A. The era of carbon allotropes. *Nature Materials*. 2010;9:868–71.)

Oberdorster performed the foremost investigations that inferred uncoated fullerenes to cause oxidative stress in the brain and the depletion of glutathione levels in juvenile largemouth bass species [87,88]. Nielsen et al. described fullerenes as a "sword" that elicits useful effects at low concentrations, but at high concentrations they may be able to induce inflammation and serious disorders like cancer [89]. They concluded that direct DNA-damaging effects are low, but the formation of reactive oxygen species (ROS) may cause inflammation and genetic damage. Several authors reviewed the literature involving fullerene toxicity beginning in the early 1990s to present and conclude that very little evidence gathered since the discovery of fullerenes indicate that C_{60} is toxic [90,91]. A study on fullerenes introduced into the abdominal cavity of mice concluded that there was evidence of asbestos-like pathogenicity [92].

CoSayes et al. found that *in vivo* inhalation of $C_{60}(OH)_{24}$ and nano-C_{60} in rats gave no effect, whereas, in comparison, quartz particles produced an inflammatory response under the same conditions. Utilizing these water-soluble, C_{60} fullerene suspensions, several studies have reported that exposure to C_{60} fullerenes causes toxicity in various organisms. Juvenile bass fish were reported to have increased lipid peroxidation in the brain and glutathione depletion in their gills after being placed in 0.5 ppm water-soluble C60 fullerenes for 48 h [93]. *In vitro* studies have reported cytotoxicity in human cells exposed to water-soluble C_{60} fullerenes due to the production of ROS and lipid peroxidation [94,95]. One *in vitro* study reports that less than 10% of the suspension is residual solvent and that controls for this residual solvent did not contribute to reactive oxygen production or cell death [94].

The exponential increase in patent filings and publications indicates a growing industrial interest in fullerenes that parallels an academic interest. The discovery of fullerenes has been compared to the discovery of benzene by many researchers. Fullerenes bring novel, three-dimensional carbon structures to medicine that can be made tissue selective as well as act as potential therapeutic agents. However, before reaching a conclusive, approved formulation, its toxicity issues need to be met or comprehended completely.

11.4.3 Titanium Dioxide

Titanium dioxide (TiO_2) has been extensively documented as a food additive or food colorant. With the latest advancement in nanotechnology, the use of nanosized TiO_2 has been accelerated in the field of cosmetics and pharmaceuticals and they are finding key roles in nanotech research due to their unique physiochemical properties. TiO_2 nanocomposites have advanced medical biotechnology in the same fashion as they are improving microarray and imaging technologies [96–98]. Paunesku et al. described the *in vivo* and *in vitro* behavior of TiO_2 nanocomposites combined with oligonucleotide DNA (size ~45 Å). These nanocomposites not only retained the intrinsic photocatalytic capacity of TiO_2 and the bioactivity of covalently attached oligonucleotide DNA, but also possessed the chemically and biologically unique novel property of a light-inducible nucleic acid endonuclease, which could become a new tool for gene therapy [99].

Literature infers the association of the alveolar uptake of TiO_2 NPs with inflammation, fibrosis, and pulmonary damage, both *in vivo* and *in vitro* [100]. These toxic effects are deemed mainly to be due to alveolar macrophages and polymorphonuclear leukocytes, which produce excessive amounts of mediators such as ROS, proteases, cytokines, and so on [101,102]. In addition, cytotoxicity mediated by TiO_2 NPs themselves has also been reported in several cultured cell lines, mainly resulting from the formation of ROS [103,104]. Studies are available that demonstrates that the exposure to TiO_2 NPs also results in elevated levels of lipid peroxidation; alveolar macrophage numbers; and increased activities of glutathione peroxidase, glutathione reductase 6-phosphate glucose dehydrogenase, and glutathione *S*-transferase in rats [100]. Grassian et al. demonstrated that exposing mice to TiO_2 NPs was essentially negative and showed reversible inflammation characterized by an increase in alveolar macrophages in lungs [105]. It is believed that exposure to TiO_2 NPs can initiate an inflammatory reaction and induce the inflammasome activation and release of inflammatory cytokines through a cathepsin-B-mediated mechanism in mouse lungs [106].

A very recent study by Sun et al. suggested that the intratracheal instillation of TiO_2 NPs for 90 consecutive days leads to its successive accumulation in mouse lungs. Thereafter, the accumulated TiO_2 NPs can significantly increase lung indices and induce histopathological changes in the lung (including emphysema, edema, inflammatory cell infiltration, congestion of blood vessels, and pulmonary bleeding [107]). Adult male ICR mice exposed to a single intratracheal dose of TiO_2 NPs showed emphysema, macrophage accumulation, extensive disruption of alveolar septa, type II pneumocyte hyperplasia, and epithelial cell apoptosis in the lung [108]. Warheit et al. demonstrated that exposure to intratracheal instillations of TiO_2 NPs produced pulmonary inflammatory responses in the rat lung [109,110]. Chen et al. observed alveolar septal thickening, neutrophil infiltration, and thrombosis in the pulmonary vascular system in mice after an intraperitoneal injection of TiO_2 NPs (3.6 nm) for 7 days [111].

Several studies have addressed the pulmonary effects induced by TiO_2 during inhalation experiments in rats [112,113]. A few studies have also investigated how agglomeration influences the relationship between the NP exposure dose and the induction of pulmonary toxicity [105]. However, the initial size of the agglomerates is a factor that determines their deposition in the lung, their ability to cross biological barriers, and their capacity to reach and enter cellular targets [45,114]. For instance, bigger NP agglomerates (>100 nm) are more likely to promote pulmonary clearance by alveolar macrophages than smaller agglomerates, thus reducing persistence time in the airway. Alternatively, small NP agglomerates (<100 nm) may escape the pulmonary defense systems and induce deleterious effects by interacting with lung cells [105].

After administration, NPs may translocate to blood circulation and become entrapped in other tissues/organs [115,116]. Berry et al. first confirmed the translocation of 30 nm particles across the alveolar epithelium into pulmonary capillaries by intratracheal instillation. There was a rapid translocation (~30%) of instilled 99mTc-labeled NPs in 5 min from the lung to the bloodstream in hamsters [117,118]. Both *in vivo* and *in vitro* studies have shown that some NPs (including TiO_2), can cause injurious effects on biological systems, whereas larger particles of the same substance are relatively less toxic [119,120]. Differences in toxicity have been attributed to the small size of the NP, their large surface area or high surface reactivity, their crystal phases, and their prolonged residence times in the lung [110]. The measurement of the surface area of NPs has been widely studied and has shown the potential for relating NP exposure dose and pulmonary responses [64]. Hence, it can be stated that the continued use of TiO_2 NPs as a biomaterial requires its biological impact and toxicity to be assessed accurately and alleviated where possible.

11.4.4 Albumin

The name albumin derives from the early German term "albumen," generally indicating proteins. Albumen, on the other hand, derives from the Latin word, albus (white), indicating the white part of the cooked egg surrounding the yolk. It is basically a polymer that consists of a single polypeptide chain and exists mainly in the α-helical form [121,122]. Albumin can be found in egg whites and in blood. The three-dimensional structure of human serum albumin (HAS) has been determined by x-ray crystallography to a resolution of 2.5 Å. It is composed of three homologous domains that gather to form a heart-shaped molecule [123,124]. Each domain is a product of two subdomains that bears a universal structural design. The primary binding regions for ligands to HSA are located in hydrophobic cavities in subdomains IIA and IIIA, which exhibit similar chemistry. Structurally, the serum albumins are similar, with each domain containing 5–6 internal disulfide bonds [124]. Albumin has numerous unique properties that make it an attractive drug vehicle in medicine. It is a versatile carrier for an array of hydrophobic drugs and biomolecules, including hormones, vaccines, DNA, siRNA, and vitamins [125–127].

NPs made of HSA offer numerous advantages, including ease of preparation, reproducibility, and propensity to undergo surface modifications (ligand binding), as well as its high tolerability and biodegradability in a given bioenvironment [128–130]. Impending drug molecules can be integrated into the particle matrix, covalently bound to surface functional groups, or adsorbed into the particle. Albumin was well reported to facilitate the endothelial transcytosis of bound constituents principally through binding to a cell surface glycoprotein receptor (albondin). This glycoprotein binds to caveolin-1 (an intracellular protein) with the successive formation of transcytotic vesicles (known as caveolae). Caveolin-1 is often present in some neoplasms (breast, lung, and prostate cancer), which could explain why albumin is known to accumulate in some tumors and thus facilitates intratumor accumulation of albumin-bound drugs [125,131]. In addition, the surface modification and introduction of targeting ligands can be attained with great ease to make the NP target specific (Figures 11.6 and 11.7) [129,130,145,146].

Albumin-bound paclitaxel (Abraxane®; Abraxis BioScience and AstraZeneca) is another example of an EPR-based nanovector application for breast cancer. It represents one of the strategies adopted to overcome the solvent-related problems of paclitaxel and it has been recently approved by the U.S. Food and Drug Administration for pretreated metastatic breast cancer patients. This novel, albumin-bound formulation of paclitaxel (size 130 nm) is used as a colloidal suspension derived from the lyophilized formulation of paclitaxel and HSA diluted in saline solution (0.9% NaCl) [132]. Preclinical studies, conducted in athymic mice with human breast cancer, demonstrated that this product has a higher penetration into tumor cells with an increased antitumor activity compared to an equal dose of standard paclitaxel [133,134]. Similarly, albumin microspheres also represent an inimitable class of biodegradable colloidal particles, which have been used for lung scanning and circulating studies in both animal and human subjects. Bovine serum albumin (BSA) had been successfully employed to prepare dry powder formulations for therapeutics [121]. BSA microspheres loaded with Ciprofloxacin were prepared for lung delivery by a spray drying technique [122]. In another study, BSA was used to deliver and retain tetrandrine in the lung so as to reduce any antisilicotic effects to other organs [135].

In spite of these delivery applications, several investigations related to the toxicity, distribution, physiological response, and metabolism of the aggregated albumin have been published. Albumin

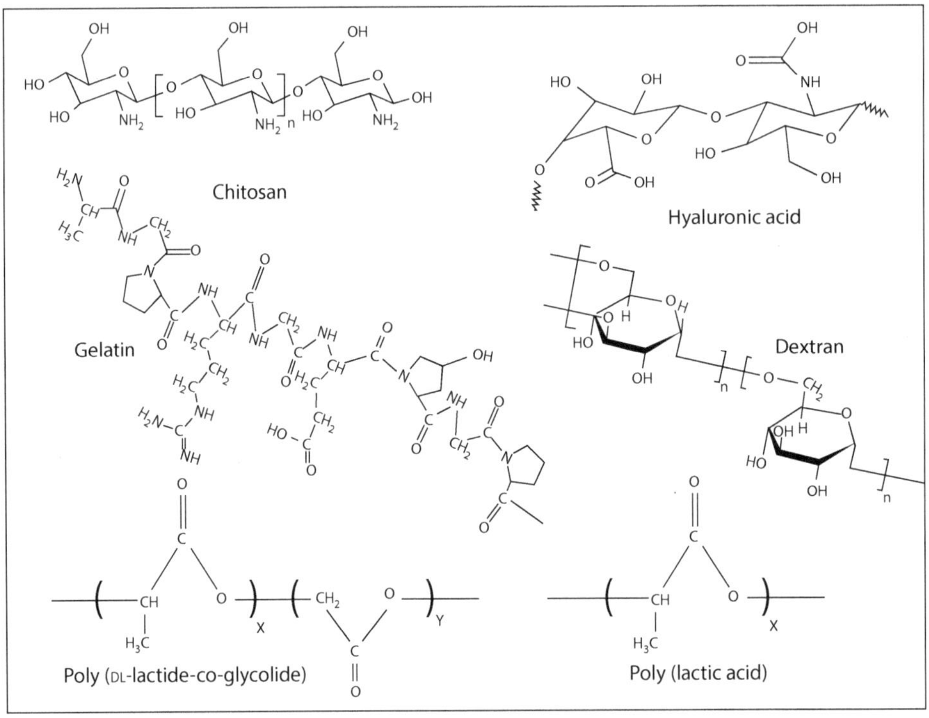

FIGURE 11.6 Various biomaterials used as gene and drug delivery agents.

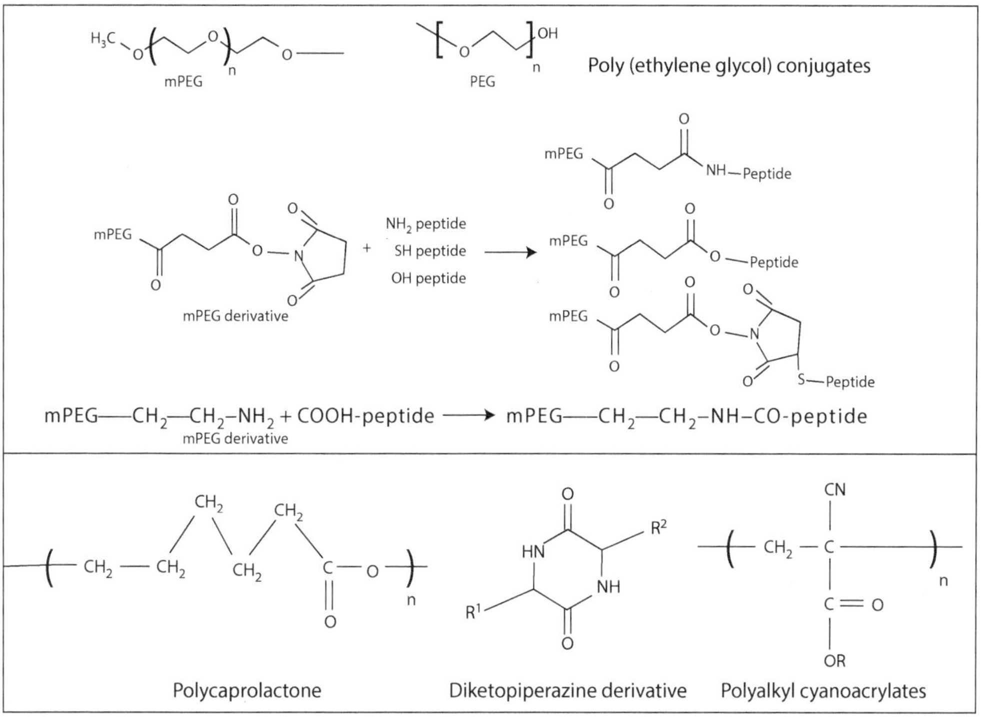

FIGURE 11.7 Various biomaterials used as gene and drug delivery agents.

can exhibit glutamate toxicity similar to casein. It is concluded that the minor effects of glutamate toxicity could be considerably aggravated by serum albumin, as well as in pathological situations *in vivo*. Delayed toxicity has been observed in cultures of prenatal hippocampal, which was elicited in serum containing media [136]. Work done in animals with albumin suspension particles demonstrated initial pulmonary retention, followed by its clearance from the lungs and transposition to the liver and spleen. The smaller mean size of the suspension led to a faster clearance from the lungs and subsequent appearance in the liver and spleen. Likewise, the incidence of capillary embolization in the lungs and other organs was greatly diminished, as the mean size and the proportion of larger particles are reduced. Hence, in spite of this material's widely reported applicability, clinical trials must be performed with animal investigation to ensure a wide margin of safety [137].

11.4.5 Silica

Another important nanomaterial is mesoporous silica NPs, which is composed of a nanometer-sized ridged assembly of silica channels (pores) capable of adsorbing small molecules [138]. The surface of mesoporous silica can be easily functionalized, and this material has tremendous potential toward the adsorption of drugs and other therapeutic bioactives. Investigators showed that mesoporous silica has the ability to absorb and release ibuprofen [139]. Mesoporous silica NPs were used to deliver a hydrophobic anticancer agent to human cancer cells [140,141]. However, reports have shown that silica architects cause the generation of ROS, which can produce direct damage to the lung epithelium, induce the expression of inflammatory genes, and, finally, lead to an enhanced lung inflammatory response. Ultimately, the sustained inflammatory and proliferative response can lead to the long-term damage described above for particles of low cytotoxicity [64,76,142].

Exposures to silica at submicron sizes is associated with the development of several autoimmune diseases, including systemic sclerosis, rheumatoid arthritis, lupus, and chronic renal disease, while certain crystalline silica polymorphs may cause silicosis and lung cancer [64,142]. Several reports have demonstrated that SiO_2 NPs caused anomalous proinflammatory stimulations of endothelial cells and fibrogenesis in Wistar rats [143,144]. In an *in vivo* study, Warheit et al. also found that nanosized SiO_2 NPs produced a greater pulmonary inflammatory response than quartz particles after the instillation of the particles into the lungs of rats at doses of up to 5 mg/kg [145]. Lin et al. also investigated the toxicity of SiO_2 NPs in cultured human bronchoalveolar carcinoma-derived cells along with crystalline silica as a positive control. Exposure to crystalline silica results in a significantly reduced cell viability [146]. Kaewamatawong et al. found that on an equal mass basis, silica NPs cause more severe bronchial cell necrosis and lung inflammation than similar particles of a larger diameter (230 nm) [147].

Future investigations on the uptake and response of silica NPs will allow a better understanding of the nanomaterial's sites of action. Such investigations must also elucidate strategies for enhancing the beneficial applications of this material in biological systems, whereas reducing potential adverse effects mediated through macrophage interactions.

11.5 CONCLUSION

Nanotechnology is a rapidly growing field that permits the development of materials with desired property attributes. The properties of nanomaterials are currently being extensively engaged in medicine in the form of vectors to transport drugs to targeted sites in the body. However, some of these NPs have exhibited highly toxic effects. The properties of nanomaterials, such as their highly reactive surfaces as well as their competence to cross through membranes, may result in significant dangers, particularly in relation to their potentially high levels of toxicity. Toxicity and the biodegradability of nanostructures and their deleterious effect on various organs (acute and chronic) is of pivotal interest along with their probable interventions with other vital body functions. In the case of nanomaterials, attention should not only be paid to its application but consideration is also given to associated toxicity issues. A majority of the literature admits and clearly specifies the need for the detailed toxicity profiling of the developed nanomaterials before claiming their clinical use.

Several studies have clearly shown that the toxic effects measured are directly related to the surface of the nanomaterials. These entities have a natural tendency to agglomerate, meaning that they group together to form much larger particles. Most pharmaceutical applications require unagglomerated NPs. In such a line, scientists are expediting vivid formulation strategies to stimulate deaggregation and attain individual, NP-based formulations. Preferably, the NP surface is modified by solubilizers, targeting ligands, diagnostic agents, and imaging agents, as well as for intended application. In addition to base surface properties, these surface modifications also elicit a major impact on NP toxicity or safety. Although some of these functionalized formulations may provide great improvements over existing medications and improve the quality of life for patients with severe diseases such as cancer, asthma, and so on. All new technologies may bear a high amount of unknown factors along with a lack of long-term exposure risks, but if the benefits offered are dramatic then they could overcome certain drawbacks. Ideally, during the formulation of NPs, investigators must consider these points that can influence the toxicity profile of the product. Details are available on the pharmacology of these biomaterials, but little information exists on their impact on toxicology.

Toxic effects have already been widely reported at the cellular as well as organ-system levels (pulmonary, reproductive, renal, cutaneous cardiac). It is important to realize that the NP deposition site in the lungs will be affected greatly by NP dimensions, which can change substantially throughout the production process. Because of their very small size, these particles offer a large contact surface per mass unit. It has been shown clearly that the degree of toxicity is linked to this surface and to the surface properties of these NPs, rather than their mass.

In line with pulmonary drug delivery, the facets of nanotechnology have grown enormously, and with this advancement in drug delivery approaches, investigators and healthcare professionals have a wide variety of choice and formulation strategies to target specific regions of the lung to be retained within the lung for longer periods. The first feature of NPs is their pulmonary deposition mode, whereby the particles will be deposited throughout the alveolar region. Because of their size, nanometric particles can enter the extrapulmonary organs. Thus, insoluble NPs pass through the barriers and are distributed to various organs throughout the body, including the brain. Despite the many promising proofs of concepts of various delivery technologies, there is still a long way ahead that must be covered. There are still many challenges that are being faced by the academic and industrial scientists to improve formulations and make a decisive impact.

There are currently no established guidelines for determining the potential toxicity of engineered nanomaterials in the lung or any other organ. Improvements in the diagnosis and treatment of respiratory diseases as a result of the application of nanotechnology are anticipated, and experimental evidence indicates that engineered NPs have unique properties that may render them beneficial in visualizing disease processes earlier and in delivering therapeutics to the lung, possibly even to specific areas within the lung. It is anticipated that the continued investigation into the mechanisms underlying the adverse *in vitro* and *in vivo* effects summarized in this chapter and their relevance to lung physiology will lead to a better understanding of the potential hazards associated with NP exposures and to the development of safe and effective respiratory medical applications and therapeutics based on nanotechnology. Further research efforts are needed to ensure the safety of long-term, *in vivo* applications and the scaled-up development from the laboratory to the industry in order to reach, within a few years, the safety and large-scale production at affordable costs of innovative lung delivery technologies. Also, complete knowledge of the possible biodistribution fate, as well as toxicity profiling of nanomedicines, is immensely important to regard nanotherapeutics as generally regarded as safe (GRAS).

ACKNOWLEDGMENTS

We would like to acknowledge the UH-Hilo College of Pharmacy for providing start-up financial support to our research group. We acknowledge the research support of Leahi Fund to Treat & Prevent Pulmonary Disease of the Hawai'i Community Foundation, Honolulu, HI, USA and seed grant from The Research Corporation of the University of Hawai'i at Hilo, HI, USA.

REFERENCES

1. Cefalu WT. Concept, strategies, and feasibility of noninvasive insulin delivery. *Diabetes Care*. 2004;27:239–46.
2. Illum L. Nasal drug delivery: New developments and strategies. *Drug Discovery Today*. 2002;7:1184–9.
3. Watts AB, Wang YB, Johnston KP, Williams RO, 3rd. Respirable low-density microparticles formed *in situ* from aerosolized brittle matrices. *Pharmaceutical Research*. 2013;30:813–25.
4. Son YJ, McConville JT. Advancements in dry powder delivery to the lung. *Drug Development and Industrial Pharmacy*. 2008;34:948–59.
5. Djupesland PG. Nasal drug delivery devices: Characteristics and performance in a clinical perspective—A review. *Drug Delivery and Translational Research*. 2013;3:42–62.
6. Watts AB, McConville JT, Williams RO, 3rd. Current therapies and technological advances in aqueous aerosol drug delivery. *Drug Development and Industrial Pharmacy*. 2008;34:913–22.
7. Dolovich M. New propellant-free technologies under investigation. *Journal of Aerosol Medicine: The Official Journal of the International Society for Aerosols in Medicine*. 1999;12 Suppl 1:S9–17.
8. Krajnik M, Podolec Z, Siekierka M, Sykutera M, Pufal E, Sobanski P et al. Morphine inhalation by cancer patients: A comparison of different nebulization techniques using pharmacokinetic, spirometric, and gasometric parameters. *Journal of Pain and Symptom Management*. 2009;38:747–57.
9. Scheuch G, Siekmeier R. Novel approaches to enhance pulmonary delivery of proteins and peptides. *Journal of Physiology and Pharmacology: An Official Journal of the Polish Physiological Society*. 2007;58 Suppl 5:615–25.
10. Beauchesne PR, Chung NS, Wasan KM. Cyclosporine A. A review of current oral and intravenous delivery systems. *Drug Development and Industrial Pharmacy*. 2007;33:211–20.

11. Degim IT, Celebi N. Controlled delivery of peptides and proteins. *Current Pharmaceutical Design*. 2007;13:99–117.
12. Labiris NR, Dolovich MB. Pulmonary drug delivery. Part I: Physiological factors affecting therapeutic effectiveness of aerosolized medications. *British Journal of Clinical Pharmacology*. 2003;56:588–99.
13. Wolff RK. Safety of inhaled proteins for therapeutic use. *Journal of Aerosol Medicine: The Official Journal of the International Society for Aerosols in Medicine*. 1998;11:197–219.
14. Mygind N. Structure and function of the respiratory tract: In *Aerosols in Medicine*. New York: Elsevier; 1985. p. 52.
15. Wiebel. E. *The Lung: Scientific Foundations*. New York: Raven Press; 1991.
16. Yu J, Chien YW. Pulmonary drug delivery: Physiologic and mechanistic aspects. *Critical Reviews in Therapeutic Drug Carrier Systems*. 1997;14:395–453.
17. Parks WR. Morphology of the respiratory tract. *Occupational Lung Disorders*. Oxford: Butterworth-Heinemann; 1994.
18. Thompson AJHaDC. *Physiology of the Airways*. Pharmaceutical Inhalation Technology. New York: Marcel Dekker; 2004. p. 1.
19. Stone KC, Mercer RR, Gehr P, Stockstill B, Crapo JD. Allometric relationships of cell numbers and size in the mammalian lung. *American Journal of Respiratory Cell and Molecular Biology*. 1992;6:235–43.
20. Lai SK, Wang YY, Wirtz D, Hanes J. Micro- and macrorheology of mucus. *Advanced Drug Delivery Reviews*. 2009;61:86–100.
21. Finkbeiner WE. Physiology and pathology of tracheobronchial glands. *Respiration Physiology*. 1999;118:77–83.
22. Schutte BC, McCray PB, Jr. [Beta]-defensins in lung host defense. *Annual Review of Physiology*. 2002;64:709–48.
23. Chow AH, Tong HH, Chattopadhyay P, Shekunov BY. Particle engineering for pulmonary drug delivery. *Pharmaceutical Research*. 2007;24:411–37.
24. Patton JS, Byron PR. Inhaling medicines: Delivering drugs to the body through the lungs. *Nature Reviews Drug Discovery*. 2007;6:67–74.
25. Van der Schans CP. Bronchial mucus transport. *Respiratory Care*. 2007;52:1150–6; discussion 6–8.
26. Geiser M. Update on macrophage clearance of inhaled micro- and nanoparticles. *Journal of Aerosol Medicine and Pulmonary Drug Delivery*. 2010;23:207–17.
27. Ross MH, Romrell LJ, Kaye GI. Respiratory system: In *Histology—A Text and Atlas*. Philadelphia: Wiliams & Wilkins; 1995.
28. Panyam J, Labhasetwar V. Biodegradable nanoparticles for drug and gene delivery to cells and tissue. *Advanced Drug Delivery Reviews*. 2003;55:329–47.
29. Kubik T, Bogunia-Kubik K, Sugisaka M. Nanotechnology on duty in medical applications. *Current Pharmaceutical Biotechnology*. 2005;6:17–33.
30. Tekade RK, Kumar PV, Jain NK. Dendrimers in oncology: An expanding horizon. *Chemical Reviews*. 2009;109:49–87.
31. Tekade RK, Dutta T, Gajbhiye V, Jain NK. Exploring dendrimer towards dual drug delivery: pH responsive simultaneous drug-release kinetics. *Journal of Microencapsulation*. 2009;26:287–96.
32. Tekade RK, Dutta T, Tyagi A, Bharti AC, Das BC, Jain NK. Surface-engineered dendrimers for dual drug delivery: A receptor up-regulation and enhanced cancer targeting strategy. *Journal of Drug Targeting*. 2008;16:758–72.
33. Panyam J, Labhasetwar V. Sustained cytoplasmic delivery of drugs with intracellular receptors using biodegradable nanoparticles. *Molecular Pharmaceutics*. 2004;1:77–84.
34. Prabha S, Labhasetwar V. Nanoparticle-mediated wild-type p53 gene delivery results in sustained antiproliferative activity in breast cancer cells. *Molecular Pharmaceutics*. 2004;1:211–9.
35. Sahoo SK, Ma W, Labhasetwar V. Efficacy of transferrin-conjugated paclitaxel-loaded nanoparticles in a murine model of prostate cancer. *International Journal of Cancer Journal International du Cancer*. 2004;112:335–40.
36. Moghimi SM, Hunter AC. Capture of stealth nanoparticles by the body's defences. *Critical Reviews in Therapeutic Drug Carrier Systems*. 2001;18:527–50.
37. Bala I, Hariharan S, Kumar MN. PLGA nanoparticles in drug delivery: The state of the art. *Critical Reviews in Therapeutic Drug Carrier Systems*. 2004;21:387–422.
38. Williams D. Nanotechnology: A new look. *Medical Device Technology*. 2004;15:9–10.
39. Yuan X, Naguib S, Wu Z. Recent advances of siRNA delivery by nanoparticles. *Expert Opinion on Drug Delivery*. 2011;8:521–36.
40. Farokhzad OC, Langer R. Impact of nanotechnology on drug delivery. *ACS Nano*. 2009;3:16–20.

41. Borm PJ, Kreyling W. Toxicological hazards of inhaled nanoparticles—Potential implications for drug delivery. *Journal of Nanoscience and Nanotechnology*. 2004;4:521–31.
42. Nel A, Xia T, Madler L, Li N. Toxic potential of materials at the nanolevel. *Science (New York, NY)*. 2006;311:622–7.
43. Makarucha AJ, Todorova N, Yarovsky I. Nanomaterials in biological environment: A review of computer modelling studies. *European Biophysics Journal: EBJ*. 2011;40:103–15.
44. Bergamaschi E, Bussolati O, Magrini A, Bottini M, Migliore L, Bellucci S et al. Nanomaterials and lung toxicity: Interactions with airways cells and relevance for occupational health risk assessment. *International Journal of Immunopathology and Pharmacology*. 2006;19:3–10.
45. Donaldson K, Aitken R, Tran L, Stone V, Duffin R, Forrest G et al. Carbon nanotubes: A review of their properties in relation to pulmonary toxicology and workplace safety. *Toxicological Sciences: An Official Journal of the Society of Toxicology*. 2006;92:5–22.
46. Baker SE, Sawvel AM, Fan J, Shi Q, Strandwitz N, Stucky GD. Blood clot initiation by mesocellular foams: Dependence on nanopore size and enzyme immobilization. *Langmuir: The ACS Journal of Surfaces and Colloids*. 2008;24:14254–14260.
47. Lewinski J, Suwala K, Kaczorowski T, Galezowski M, Gryko DT, Justyniak I et al. Oxygenation of alkylzinc complexes with pyrrolylketiminate ligand: Access to alkylperoxide versus oxo-encapsulated complexes. *Chemical communications* (Cambridge, England). 2009;215–217.
48. Li Z, Hulderman T, Salmen R, Chapman R, Leonard SS, Young SH et al. Cardiovascular effects of pulmonary exposure to single-wall carbon nanotubes. *Environmental Health Perspectives*. 2007;115:377–82.
49. Shvedova AA, Castranova V, Kisin ER, Schwegler-Berry D, Murray AR, Gandelsman VZ et al. Exposure to carbon nanotube material: Assessment of nanotube cytotoxicity using human keratinocyte cells. *Journal of Toxicology and Environmental Health Part A*. 2003;66:1909–26.
50. Cha KE, Myung H. Cytotoxic effects of nanoparticles assessed *in vitro* and *in vivo*. *Journal of Microbiology and Biotechnology*. 2007;17:1573–8.
51. Firme CP, 3rd, Bandaru PR. Toxicity issues in the application of carbon nanotubes to biological systems. *Nanomedicine: Nanotechnology, Biology, and Medicine*. 2010;6:245–56.
52. Kelly KF, Billups WE. Synthesis of soluble graphite and graphene. *Accounts of Chemical Research*. 2013;46:4–13.
53. Daenen M, de Fouw RD, Hamers B, Janssen PGA, Schouteden K, and Veld MAJ. The wondrous world of carbon nanotubes. *Chemistry and Applied Physics*, 2003;1–63.
54. Kostarelos K, Lacerda L, Pastorin G, Wu W, Wieckowski S, Luangsivilay J et al. Cellular uptake of functionalized carbon nanotubes is independent of functional group and cell type. *Nature Nanotechnology*. 2007;2:108–13.
55. Cai D, Mataraza JM, Qin ZH, Huang Z, Huang J, Chiles TC et al. Highly efficient molecular delivery into mammalian cells using carbon nanotube spearing. *Nature Methods*. 2005;2:449–54.
56. Pantarotto D, Singh R, McCarthy D, Erhardt M, Briand JP, Prato M et al. Functionalized carbon nanotubes for plasmid DNA gene delivery. *Angewandte Chemie (International ed in English)*. 2004;43:5242–6.
57. Meredith JR, Jin C, Narayan RJ, Aggarwal R. Biomedical applications of carbon-nanotube composites. *Frontiers in Bioscience (Elite edition)*. 2013;5:610–21.
58. Kayat J, Gajbhiye V, Tekade RK, Jain NK. Pulmonary toxicity of carbon nanotubes: A systematic report. *Nanomedicine: Nanotechnology, Biology, and Medicine*. 2011;7:40–9.
59. O'Connell MJ, Bachilo SM, Huffman CB, Moore VC, Strano MS, Haroz EH et al. Band gap fluorescence from individual single-walled carbon nanotubes. *Science (New York, NY)*. 2002;297:593–6.
60. Shvedova AA, Kisin ER, Porter D, Schulte P, Kagan VE, Fadeel B et al. Mechanisms of pulmonary toxicity and medical applications of carbon nanotubes: Two faces of Janus? *Pharmacology & Therapeutics*. 2009;121:192–204.
61. Bottini M, Bruckner S, Nika K, Bottini N, Bellucci S, Magrini A et al. Multi-walled carbon nanotubes induce T lymphocyte apoptosis. *Toxicology Letters*. 2006;160:121–6.
62. Maynard AD, Aitken RJ, Butz T, Colvin V, Donaldson K, Oberdorster G et al. Safe handling of nanotechnology. *Nature*. 2006;444:267–9.
63. Cui D, Tian F, Ozkan CS, Wang M, Gao H. Effect of single wall carbon nanotubes on human HEK293 cells. *Toxicology Letters*. 2005;155:73–85.
64. Monteiller C, Tran L, MacNee W, Faux S, Jones A, Miller B et al. The pro-inflammatory effects of low-toxicity low-solubility particles, nanoparticles and fine particles, on epithelial cells *in vitro*: The role of surface area. *Occupational and Environmental Medicine*. 2007;64:609–15.
65. Vardharajula S, Ali SZ, Tiwari PM, Eroglu E, Vig K, Dennis VA et al. Functionalized carbon nanotubes: Biomedical applications. *International Journal of Nanomedicine*. 2012;7:5361–74.

66. Muller J, Huaux F, Moreau N, Misson P, Heilier JF, Delos M et al. Respiratory toxicity of multi-wall carbon nanotubes. *Toxicology and Applied Pharmacology*. 2005;207:221–31.
67. Han SG, Andrews R, Gairola CG, Bhalla DK. Acute pulmonary effects of combined exposure to carbon nanotubes and ozone in mice. *Inhalation Toxicology*. 2008;20:391–8.
68. Jacobsen NR, Moller P, Jensen KA, Vogel U, Ladefoged O, Loft S et al. Lung inflammation and genotoxicity following pulmonary exposure to nanoparticles in ApoE-/- mice. *Particle and Fibre Toxicology*. 2009;6:2.
69. Warheit DB, Laurence BR, Reed KL, Roach DH, Reynolds GA, Webb TR. Comparative pulmonary toxicity assessment of single-wall carbon nanotubes in rats. *Toxicological Sciences: An Official Journal of the Society of Toxicology*. 2004;77:117–25.
70. Chou CC, Hsiao HY, Hong QS, Chen CH, Peng YW, Chen HW et al. Single-walled carbon nanotubes can induce pulmonary injury in mouse model. *Nano Letters*. 2008;8:437–45.
71. Lam CW, James JT, McCluskey R, Hunter RL. Pulmonary toxicity of single-wall carbon nanotubes in mice 7 and 90 days after intratracheal instillation. *Toxicological Sciences: An Official Journal of the Society of Toxicology*. 2004;77:126–34.
72. Shvedova AA, Kisin E, Murray AR, Johnson VJ, Gorelik O, Arepalli S et al. Inhalation vs. aspiration of single-walled carbon nanotubes in C57BL/6 mice: Inflammation, fibrosis, oxidative stress, and mutagenesis. *American Journal of Physiology Lung Cellular and Molecular Physiology*. 2008;295:L552–65.
73. Smith CJ, Shaw BJ, Handy RD. Toxicity of single walled carbon nanotubes to rainbow trout, (*Oncorhynchus mykiss*): Respiratory toxicity, organ pathologies, and other physiological effects. *Aquatic Toxicology (Amsterdam, Netherlands)*. 2007;82:94–109.
74. Shvedova AA, Kisin ER, Mercer R, Murray AR, Johnson VJ, Potapovich AI et al. Unusual inflammatory and fibrogenic pulmonary responses to single-walled carbon nanotubes in mice. *American Journal of Physiology Lung Cellular and Molecular Physiology*. 2005;289:L698–708.
75. Muller J, Huaux F, Fonseca A, Nagy JB, Moreau N, Delos M et al. Structural defects play a major role in the acute lung toxicity of multiwall carbon nanotubes: Toxicological aspects. *Chemical Research in Toxicology*. 2008;21:1698–705.
76. Fenoglio I, Greco G, Tomatis M, Muller J, Raymundo-Pinero E, Beguin F et al. Structural defects play a major role in the acute lung toxicity of multiwall carbon nanotubes: Physicochemical aspects. *Chemical Research in Toxicology*. 2008;21:1690–7.
77. Li JG, Li QN, Xu JY, Cai XQ, Liu RL, Li YJ et al. The pulmonary toxicity of multi-wall carbon nanotubes in mice 30 and 60 days after inhalation exposure. *Journal of Nanoscience and Nanotechnology*. 2009;9:1384–7.
78. Jia G, Wang H, Yan L, Wang X, Pei R, Yan T et al. Cytotoxicity of carbon nanomaterials: Single-wall nanotube, multi-wall nanotube, and fullerene. *Environmental Science & Technology*. 2005;39:1378–83.
79. Klumpp C, Kostarelos K, Prato M, Bianco A. Functionalized carbon nanotubes as emerging nanovectors for the delivery of therapeutics. *Biochimica et Biophysica Acta*. 2006;1758:404–12.
80. Burghardt S, Hirsch A, Schade B, Ludwig K, Bottcher C. Switchable supramolecular organization of structurally defined micelles based on an amphiphilic fullerene. *Angewandte Chemie (International ed in English)*. 2005;44:2976–9.
81. Schoenhagen P, Conyers JL. Nanotechnology and atherosclerosis imaging: Emerging diagnostic and therapeutic applications. *Recent Patents on Cardiovascular Drug Discovery*. 2008;3:98–104.
82. Foley S, Crowley C, Smaihi M, Bonfils C, Erlanger BF, Seta P et al. Cellular localisation of a water-soluble fullerene derivative. *Biochemical and Biophysical Research Communications*. 2002;294:116–9.
83. Nakamura E, Isobe H. Functionalized fullerenes in water. The first 10 years of their chemistry, biology, and nanoscience. *Accounts of Chemical Research*. 2003;36:807–15.
84. Zakharian TY, Seryshev A, Sitharaman B, Gilbert BE, Knight V, Wilson LJ. A fullerene-paclitaxel chemotherapeutic: Synthesis, characterization, and study of biological activity in tissue culture. *Journal of the American Chemical Society*. 2005;127:12508–9.
85. Mori T, Takada H, Ito S, Matsubayashi K, Miwa N, Sawaguchi T. Preclinical studies on safety of fullerene upon acute oral administration and evaluation for no mutagenesis. *Toxicology*. 2006;225:48–54.
86. Yamago S, Tokuyama H, Nakamura E, Kikuchi K, Kananishi S, Sueki K et al. In vivo biological behavior of a water-miscible fullerene: 14C labeling, absorption, distribution, excretion and acute toxicity. *Chemistry & Biology*. 1995;2:385–9.
87. Zhu S, Oberdorster E, Haasch ML. Toxicity of an engineered nanoparticle (fullerene, C60) in two aquatic species, Daphnia and fathead minnow. *Marine Environmental Research*. 2006;62 Suppl:S5–9.
88. Blickley TM, McClellan-Green P. Toxicity of aqueous fullerene in adult and larval *Fundulus heteroclitus*. *Environmental Toxicology and Chemistry/SETAC*. 2008;27:1964–71.

89. Markovic Z, Trajkovic V. Biomedical potential of the reactive oxygen species generation and quenching by fullerenes (C60). *Biomaterials*. 2008;29:3561–73.
90. Kolosnjaj J, Szwarc H, Moussa F. Toxicity studies of fullerenes and derivatives. *Advances in Experimental Medicine and Biology*. 2007;620:168–80.
91. Lens M, Medenica L, Citernesi U. Antioxidative capacity of C(60) (buckminster fullerene) and newly synthesized fulleropyrrolidine derivatives encapsulated in liposomes. *Biotechnology and Applied Biochemistry*. 2008;51:135–40.
92. Poland CA, Duffin R, Kinloch I, Maynard A, Wallace WA, Seaton A et al. Carbon nanotubes introduced into the abdominal cavity of mice show asbestos-like pathogenicity in a pilot study. *Nature Nanotechnology*. 2008;3:423–8.
93. Oberdorster E. Manufactured nanomaterials (fullerenes, C60) induce oxidative stress in the brain of juvenile largemouth bass. *Environmental Health Perspectives*. 2004;112:1058–62.
94. Sayes CM, Gobin AM, Ausman KD, Mendez J, West JL, Colvin VL. Nano-C60 cytotoxicity is due to lipid peroxidation. *Biomaterials*. 2005;26:7587–95.
95. Gharbi N, Pressac M, Hadchouel M, Szwarc H, Wilson SR, Moussa F. [60]fullerene is a powerful antioxidant *in vivo* with no acute or subacute toxicity. *Nano Letters*. 2005;5:2578–85.
96. Yu W, Zhang Y, Xu L, Sun S, Jiang X, Zhang F. Microarray-based bioinformatics analysis of osteoblasts on TiO_2 nanotube layers. *Colloids and Surfaces B, Biointerfaces*. 2012;93:135–42.
97. De Marni ML, Monegal A, Venturini S, Vinati S, Carbone R, de Marco A. Antibody purification-independent microarrays (PIM) by direct bacteria spotting on TiO_2-treated slides. *Methods (San Diego, California)*. 2012;56:317–25.
98. Doong RA, Shih HM. Array-based titanium dioxide biosensors for ratiometric determination of glucose, glutamate and urea. *Biosensors & Bioelectronics*. 2010;25:1439–46.
99. Paunesku T, Rajh T, Wiederrecht G, Maser J, Vogt S, Stojicevic N et al. Biology of TiO_2-oligonucleotide nanocomposites. *Nature Materials*. 2003;2:343–6.
100. Afaq F, Abidi P, Matin R, Rahman Q. Cytotoxicity, pro-oxidant effects and antioxidant depletion in rat lung alveolar macrophages exposed to ultrafine titanium dioxide. *Journal of Applied Toxicology: JAT*. 1998;18:307–12.
101. Churg A, Gilks B, Dai J. Induction of fibrogenic mediators by fine and ultrafine titanium dioxide in rat tracheal explants. *The American Journal of Physiology*. 1999;277:L975–82.
102. Oberdorster G, Ferin J, Gelein R, Soderholm SC, Finkelstein J. Role of the alveolar macrophage in lung injury: Studies with ultrafine particles. *Environmental Health Perspectives*. 1992;97:193–9.
103. Jin CY, Zhu BS, Wang XF, Lu QH. Cytotoxicity of titanium dioxide nanoparticles in mouse fibroblast cells. *Chemical Research in Toxicology*. 2008;21:1871–7.
104. Liu S, Xu L, Zhang T, Ren G, Yang Z. Oxidative stress and apoptosis induced by nanosized titanium dioxide in PC12 cells. *Toxicology*. 2010;267:172–7.
105. Grassian VH, O'Shaughnessy PT, Adamcakova-Dodd A, Pettibone JM, Thorne PS. Inhalation exposure study of titanium dioxide nanoparticles with a primary particle size of 2 to 5 nm. *Environmental Health Perspectives*. 2007;115:397–402.
106. Hamilton RF, Wu N, Porter D, Buford M, Wolfarth M, Holian A. Particle length-dependent titanium dioxide nanomaterials toxicity and bioactivity. *Particle and Fibre Toxicology*. 2009;6:35.
107. Sun Q, Tan D, Ze Y, Sang X, Liu X, Gui S et al. Pulmotoxicological effects caused by long-term titanium dioxide nanoparticles exposure in mice. *Journal of Hazardous Materials*. 2012;235–236:47–53.
108. Chen HW, Su SF, Chien CT, Lin WH, Yu SL, Chou CC et al. Titanium dioxide nanoparticles induce emphysema-like lung injury in mice. *FASEB Journal: Official Publication of the Federation of American Societies for Experimental Biology*. 2006;20:2393–5.
109. Warheit DB, Webb TR, Sayes CM, Colvin VL, Reed KL. Pulmonary instillation studies with nanoscale TiO_2 rods and dots in rats: Toxicity is not dependent upon particle size and surface area. *Toxicological Sciences: An Official Journal of the Society of Toxicology*. 2006;91:227–36.
110. Warheit DB, Webb TR, Reed KL, Frerichs S, Sayes CM. Pulmonary toxicity study in rats with three forms of ultrafine-TiO_2 particles: Differential responses related to surface properties. *Toxicology*. 2007;230:90–104.
111. Chen J, Dong X, Zhao J, Tang G. *In vivo* acute toxicity of titanium dioxide nanoparticles to mice after intraperitioneal injection. *Journal of Applied Toxicology: JAT*. 2009;29:330–7.
112. Rossi EM, Pylkkanen L, Koivisto AJ, Vippola M, Jensen KA, Miettinen M et al. Airway exposure to silica-coated TiO_2 nanoparticles induces pulmonary neutrophilia in mice. *Toxicological Sciences: An Official Journal of the Society of Toxicology*. 2010;113:422–33.

113. Leppanen M, Korpi A, Miettinen M, Leskinen J, Torvela T, Rossi EM et al. Nanosized TiO(2) caused minor airflow limitation in the murine airways. *Archives of Toxicology*. 2011;85:827–39.
114. Koike E, Kobayashi T. Chemical and biological oxidative effects of carbon black nanoparticles. *Chemosphere*. 2006;65:946–51.
115. Muhlfeld C, Geiser M, Kapp N, Gehr P, Rothen-Rutishauser B. Re-evaluation of pulmonary titanium dioxide nanoparticle distribution using the "relative deposition index": Evidence for clearance through microvasculature. *Particle and Fibre Toxicology*. 2007;4:7.
116. Olmedo D, Guglielmotti MB, Cabrini RL. An experimental study of the dissemination of titanium and zirconium in the body. *Journal of Materials Science Materials in Medicine*. 2002;13:793–6.
117. Berry JP, Arnoux B, Stanislas G, Galle P, Chretien J. A microanalytic study of particles transport across the alveoli: Role of blood platelets. *Biomedicine/[publiee pour l'AAICIG]*. 1977;27:354–7.
118. Nemmar A, Vanbilloen H, Hoylaerts MF, Hoet PH, Verbruggen A, Nemery B. Passage of intratracheally instilled ultrafine particles from the lung into the systemic circulation in hamster. *American Journal of Respiratory and Critical Care Medicine*. 2001;164:1665–8.
119. Shimada M, Wang WN, Okuyama K, Myojo T, Oyabu T, Morimoto Y et al. Development and evaluation of an aerosol generation and supplying system for inhalation experiments of manufactured nanoparticles. *Environmental Science & Technology*. 2009;43:5529–34.
120. Christensen FM, Johnston HJ, Stone V, Aitken RJ, Hankin S, Peters S et al. Nano-TiO(2)—Feasibility and challenges for human health risk assessment based on open literature. *Nanotoxicology*. 2011;5:110–24.
121. Moebus K, Siepmann J, Bodmeier R. Cubic phase-forming dry powders for controlled drug delivery on mucosal surfaces. *Journal of Controlled Release: Official Journal of the Controlled Release Society*. 2012;157:206–15.
122. Li FQ, Hu JH, Lu B, Yao H, Zhang WG. Ciprofloxacin-loaded bovine serum albumin microspheres: Preparation and drug-release *in vitro*. *Journal of Microencapsulation*. 2001;18:825–9.
123. Sugio S, Kashima A, Mochizuki S, Noda M, Kobayashi K. Crystal structure of human serum albumin at 2.5 A resolution. *Protein Engineering*. 1999;12:439–46.
124. He XM, Carter DC. Atomic structure and chemistry of human serum albumin. *Nature*. 1992;358:209–15.
125. Hawkins MJ, Soon-Shiong P, Desai N. Protein nanoparticles as drug carriers in clinical medicine. *Advanced Drug Delivery Reviews*. 2008;60:876–85.
126. Paal K, Muller J, Hegedus L. High affinity binding of paclitaxel to human serum albumin. *European Journal of Biochemistry/FEBS*. 2001;268:2187–91.
127. Purcell M, Neault JF, Tajmir-Riahi HA. Interaction of taxol with human serum albumin. *Biochimica et Biophysica Acta*. 2000;1478:61–8.
128. Wartlick H, Michaelis K, Balthasar S, Strebhardt K, Kreuter J, Langer K. Highly specific HER2-mediated cellular uptake of antibody-modified nanoparticles in tumour cells. *Journal of Drug Targeting*. 2004;12:461–71.
129. Weber C, Coester C, Kreuter J, Langer K. Desolvation process and surface characterisation of protein nanoparticles. *International Journal of Pharmaceutics*. 2000;194:91–102.
130. Langer K, Balthasar S, Vogel V, Dinauer N, von Briesen H, Schubert D. Optimization of the preparation process for human serum albumin (HSA) nanoparticles. *International Journal of Pharmaceutics*. 2003;257:169–80.
131. Minshall RD, Sessa WC, Stan RV, Anderson RG, Malik AB. Caveolin regulation of endothelial function. *American Journal of Physiology Lung Cellular and Molecular Physiology*. 2003;285:L1179–83.
132. Miele E, Spinelli GP, Miele E, Tomao F, Tomao S. Albumin-bound formulation of paclitaxel (Abraxane ABI-007) in the treatment of breast cancer. *International Journal of Nanomedicine*. 2009;4:99–105.
133. Desai N, Trieu V, Yao Z, Louie L, Ci S, Yang A et al. Increased antitumor activity, intratumor paclitaxel concentrations, and endothelial cell transport of cremophor-free, albumin-bound paclitaxel, ABI-007, compared with cremophor-based paclitaxel. *Clinical Cancer Research: An Official Journal of the American Association for Cancer Research*. 2006;12:1317–24.
134. Sparreboom A, Scripture CD, Trieu V, Williams PJ, De T, Yang A et al. Comparative preclinical and clinical pharmacokinetics of a cremophor-free, nanoparticle albumin-bound paclitaxel (ABI-007) and paclitaxel formulated in Cremophor (Taxol). *Clinical Cancer Research: An Official Journal of the American Association for Cancer Research*. 2005;11:4136–43.
135. Wang C, Wu QH, Li CR, Wang Z, Ma JJ, Zang XH et al. Interaction of tetrandrine with human serum albumin: A fluorescence quenching study. *Analytical Sciences: The International Journal of the Japan Society for Analytical Chemistry*. 2007;23:429–33.
136. Peterson C, Neal JH, Cotman CW. Development of N-methyl-D-aspartate excitotoxicity in cultured hippocampal neurons. *Brain Research Developmental Brain Research*. 1989;48:187–95.

137. Taplin GV, Johnson DE, Dore EK, Kaplan HS. Suspensions of radioalbumin aggregates for photoscanning the liver, spleen, lung and other organs. *Journal of Nuclear Medicine: Official Publication, Society of Nuclear Medicine*. 1964;5:259–75.
138. Vallet-Regi M. Ordered mesoporous materials in the context of drug delivery systems and bone tissue engineering. *Chemistry (Weinheim an der Bergstrasse, Germany)*. 2006;12:5934–43.
139. Charnay C, Begu S, Tourne-Peteilh C, Nicole L, Lerner DA, Devoisselle JM. Inclusion of ibuprofen in mesoporous templated silica: Drug loading and release property. *European Journal of Pharmaceutics and Biopharmaceutics: Official Journal of Arbeitsgemeinschaft fur Pharmazeutische Verfahrenstechnik eV*. 2004;57:533–40.
140. Lu J, Liong M, Zink JI, Tamanoi F. Mesoporous silica nanoparticles as a delivery system for hydrophobic anticancer drugs. *Small (Weinheim an der Bergstrasse, Germany)*. 2007;3:1341–6.
141. Slowing I, Trewyn BG, Lin VS. Effect of surface functionalization of MCM-41-type mesoporous silica nanoparticles on the endocytosis by human cancer cells. *Journal of the American Chemical Society*. 2006;128:14792–3.
142. Fubini B, Hubbard A. Reactive oxygen species (ROS) and reactive nitrogen species (RNS) generation by silica in inflammation and fibrosis. *Free Radical Biology & Medicine*. 2003;34:1507–16.
143. Peters K, Unger RE, Kirkpatrick CJ, Gatti AM, Monari E. Effects of nano-scaled particles on endothelial cell function *in vitro*: Studies on viability, proliferation and inflammation. *Journal of Materials Science Materials in Medicine*. 2004;15:321–5.
144. Chen M, von Mikecz A. Formation of nucleoplasmic protein aggregates impairs nuclear function in response to SiO_2 nanoparticles. *Experimental Cell Research*. 2005;305:51–62.
145. Warheit DB, Brock WJ, Lee KP, Webb TR, Reed KL. Comparative pulmonary toxicity inhalation and instillation studies with different TiO_2 particle formulations: Impact of surface treatments on particle toxicity. *Toxicological Sciences: An Official Journal of the Society of Toxicology*. 2005;88:514–24.
146. Lin W, Huang YW, Zhou XD, Ma Y. *In vitro* toxicity of silica nanoparticles in human lung cancer cells. *Toxicology and Applied Pharmacology*. 2006;217:252–9.
147. Kaewamatawong T, Kawamura N, Okajima M, Sawada M, Morita T, Shimada A. Acute pulmonary toxicity caused by exposure to colloidal silica: Particle size dependent pathological changes in mice. *Toxicologic Pathology*. 2005;33:743–9.
148. D'Angelis CA, Coalson JJ, Ryan RM, Fuhrman B, Zimmerman J. Structure of the respiratory system. Lower respiratory tract: In *Pediatric Critical Care*. 14 ed. Philadelphia: Elsevier Saunders; 2011.
149. Hirsch A. The era of carbon allotropes. *Nature Materials*. 2010;9:868–71.

12 Cardiovascular Toxicity of Nanomaterials

Saijie Zhu and Minghuang Hong

CONTENTS

12.1 Introduction 249
12.2 Cardiovascular Toxicity of Nanomaterials 249
    12.2.1 Carbonaceous Nanomaterials 249
        12.2.1.1 Carbon Nanotubes 250
        12.2.1.2 Graphene 251
    12.2.2 Silica Nanomaterials 251
    12.2.3 Metallic Nanomaterials 252
        12.2.3.1 Gold Nanomaterials 252
        12.2.3.2 Iron Oxide Nanomaterials 253
        12.2.3.3 Quantum Dots 254
        12.2.3.4 Other Metallic Nanomaterials 255
12.3 Conclusions and Perspectives 255
References 256

12.1 INTRODUCTION

Nanotechnology, one of the most important technologies in modern times, has been growing explosively throughout the world over the last few years, producing a diverse array of nanomaterials. Engineered nanomaterials are designed at the molecular level to take advantage of their high relative surface area and quantum effects not seen in their corresponding conventional forms. As a matter of fact, nanomaterials are already in commercial use, with a broad range of commercial products available in textiles, cosmetics, sunscreens, electronics, paints, and medicines. While benefits of nanotechnology and nanomaterials are widely publicized, concerns have been raised about the potential risks posed by the use of these engineered nanomaterials.[1,2]

The cardiovascular system consists of two components: (1) the heart, which pumps blood to the lung and tissue capillaries; and (2) the blood vessels through which the blood flows. The major functions of the cardiovascular system are transportation, protection, fluid balance, and thermoregulation, all of which are vital for the human body. Data on the toxicity of engineered nanomaterials against the cardiovascular system have been accumulating in the past decade, while the mechanisms have yet to be clarified. The current chapter summarizes recent publications regarding the toxicity of engineered nanomaterials on the cardiovascular system.

12.2 CARDIOVASCULAR TOXICITY OF NANOMATERIALS

12.2.1 CARBONACEOUS NANOMATERIALS

Since the discovery of C_{60} fullerene in 1985, a spherical molecule with carbon atoms arranged in a pattern like that of a geodesic dome, fullerenes of a larger size—as well as their derivatives—have been synthesized and studied.[3,4] Nowadays, the family of carbonaceous nanomaterials has expanded from fullerenes to nanotubes and graphene.

12.2.1.1 Carbon Nanotubes

Carbon nanotubes (CNTs) represent a novel class of nanomaterials and have drawn ever-increasing attention to the nanomedicine arena. CNTs exist in two forms, single-wall carbon nanotubes (SWCNTs) and multiwall carbon nanotubes (MWCNTs). To date, systemic endpoint effects, including effects on the cardiovascular systems following pulmonary CNTs exposure, have been observed in several studies. A number of systemic markers and indicators were recorded to evaluate the cardiovascular toxicity of CNTs. Furthermore, the possible mechanisms have also been investigated accordingly.

Early in 2005, the dose-dependent damage to mitochondrial DNA (deoxyribonucleic acid) in the aortas of mice was observed at 7, 28, and 60 days after pharyngeal aspiration CNT treatment.[5,6] It was concluded that these oxidative changes resulted from the altered expression of inflammatory genes, including intracellular cell adhesion molecule-1 (ICAM-1) and monocyte chemotactic protein-1 (MCP-1) in the heart. The authors further studied the effects of CNTs on the heart by pharyngeally instilling a total dose of 20 μg SWCNTs into hypercholesterolemic (ApoE$^{-/-}$) mice.[6,7] At the end of the exposure regiment, the percent of the aortic area covered by plaque was significantly increased in the CNT-treated mice compared to the vehicle-treated controls. A morphometric analysis revealed a significant increase in atherosclerotic lesions in the brachiocephalic arteries of the CNT-treated mice. The authors concluded that the effects in the heart might be caused by cytokines released from inflamed areas in the lungs and/or by CNTs that leave the lungs and enter systemic circulation.[5,7]

Erdely et al.[8] observed generalized stress responses in various extrapulmonary tissues, including acute sensitivity in the aorta within 4 h of CNT pharyngeal aspiration. Furthermore, MWCNTs produced a greater magnitude response than SWCNTs at an equal mass dose.[8] Therefore, they primarily focused on MWCNTs in a further study where elevated levels of serum interleukin-6 (IL-6) and the tissue inhibitor of matrix metalloproteinase-4 (TIMP-4) in the aortas were found in mice with pulmonary exposure to CNTs.[9] The former was reported to be an indicator of vascular dysfunction and prothrombotic potential,[10] and the latter was proposed to be a systemic marker for vascular inflammation.[11] A key mechanism proposed for the contribution of these downstream effects observed in CNT exposure is the release of soluble mediators from the lung into circulation.[8,12] It was also found that the lung's response was translocated to the periphery via the blood following pulmonary exposure to CNTs.[9]

A more recent study monitored the arterial baroreflex control of the sinus node as an index of cardiovascular autonomic control after pulmonary exposure to SWCNTs.[13] A significant decrease in the number of baroreflex sequences (from 502 at baseline to 194 after 4 weeks) was observed in SWCNT-instilled rats, whereas no significant change was detected in controls, suggesting that SWCNTs may affect autonomic cardiovascular control regulation.

Wingard and coworkers[14] demonstrated that MWCNT instillation in C57BL/6 mice resulted in the exacerbation of cardiac ischemia/reperfusion (IR) injuries in a dose-dependent manner at 1 day postexposure. Further mechanistic studies performed in the same laboratory revealed that MWCNTs activated mast cells through the IL-33/ST2 axis, altered pulmonary function, and exacerbated IR injury responses in the heart.[15]

The adverse health effects of CNTs are usually studied by intratracheal inhalation or instillation because pulmonary exposure is the major route for these lightweight fibrous materials to affect human health during their production and application. Recently, the field of CNT application is expanding to biomedical research where intravenous (IV) injections are a frequently used administration route. A number of studies have examined the effects of CNTs on the cardiovascular system upon IV injections. The accumulation of CNTs in the heart upon IV injection was relatively low; thus, it is often not discussed in the literatures.[16,17] The above fact also explains the low cardiovascular toxicity of CNTs. Tang et al.[18] demonstrated no significant and acute toxicities from CNTs to the heart when mice were intravenously injected with 100 µg of either short (s)-MWCNTs or polyethylene glycol (PEG)-modified ones (s-MWCNTs–PEG) as indicated by histological examination and blood levels of creatine kinase, an indicator of heart damage. Another study describing the use of functionalized MWCNTs as ultrasound contrast agents showed that IV injections of MWCNTs did not exert adverse effects on the heart by histological examination or the analysis of blood.[19] It should be noted that some blood parameters may be altered upon CNT treatments, which might be due to the effect of CNTs on other organs. For example, the level of aspartate transaminase (AST) in mouse blood was elevated when mice were intravenously injected with pristine (p) and acid-oxidized (f) MWCNTs with varying degrees of carboxylation. In addition, p-MWCNTs induced the highest level of AST because they showed the highest propensity of liver accumulation and caused maximum liver damage in mice.[20]

A few *in vitro* studies using cell models have been carried out to examine the potential effects of CNTs on the cardiovascular system. Cheng et al.[21] found that SWCNT exposure induced oxidative damage in rat aortic endothelial cells (RAECs), suggesting potential adverse effects on the cardiovascular system. It was additionally found that the expression levels of the stress protein, heme oxygenase-1 (HO-1), increased with the increasing concentration of SWCNTs. Contrarily, however, when Martinelli et al.[22] cultured neonatal rat ventricular myocytes (NRVMs) with MWCNT scaffolds, they observed the increased viability and proliferation of these NRVMs. Furthermore, alterations in the electrophysiological properties of cardiomyocytes suggest the ability of MWCNTs to promote cardiomyocyte maturation.[22]

12.2.1.2 Graphene

Since the first single layers of graphene were identified in 2004 by Geim and coworkers,[23] graphene family materials, including graphene, graphene oxide, and reduced graphene oxide, have attracted a considerable interest, both in the academic and industrial fields. So far, the toxicity studies of graphene family materials are still in their infancy, with even less literature regarding their cardiovascular toxicity.

By using a zebrafish model, Gollavelli and Ling[24] found that 0.1 ng/nL of graphene oxide and its functionalized counterpart (polyacrylic acid [PAA] and fluorescein *o*-methacrylate [FMA]-modified magnetic graphene oxide) induced 2% and 6% cardiac malfunction, respectively. In another study, Duch et al.[25] compared the effects of graphene and graphene oxide on mouse lungs 24 h after pulmonary instillation. Through histological examination, graphene oxide was found to induce severe lung inflammation, while graphene, either dispersed or aggregated, showed minimal lung inflammation. The authors pointed out the potentially increased risk of heart attacks and stroke following lung inflammation.

12.2.2 Silica Nanomaterials

Naturally occurring silica-based materials, such as asbestos and silica, have been investigated as air pollutants for many years. On the other hand, synthetic amorphous silica nanoparticles (SNPs), including the recently developed mesoporous silica nanoparticles (MSNs), are gaining an increasing popularity as the material of choice in the biomedical and biotechnological fields as drug carriers, for gene therapy, and for molecular imaging. SNPs are generally recognized as low toxic; thus, they were approved as a food additive by Food and Drug Administration (FDA),[26] and used as a low-toxic reference material in cytotoxicity experiments.[27] In this section, we will focus on the cardiovascular toxicity of synthetic silica nanomaterials.

The biomedical applications of SNPs as diagnostic, imaging, and drug-delivery agents were often achieved via IV administration; thus, most of the *in vivo* toxicity studies were performed through this administration route, including the studies for cardiovascular toxicity. The SNPs were relatively biocompatible with the cardiovascular system in terms of both acute and chronic toxicity based on the reports to date. Galagudza et al.[28] infused SNP suspensions (6–13 nm) into male Wistar rats through the femoral vein at a dose of 0.7 mg/100 g body weight, and recorded the hemodynamic parameters at the end of infusion. Mild changes in hemodynamic parameters, such as heart rate, mean arterial pressure (MAP), and pulse arterial pressure were observed, indicating the lack of acute toxicity from the tested nanoparticles. Nishimori et al.[29] found no histological injury in mice hearts after the chronic IV injection (twice a week for 4 weeks) of 70-nm silica particles. A more comprehensive study conducted by He et al.[30] revealed that intravenously injected MSNs of various particle sizes (80, 120, 200, and 360 nm) induced no pathological changes in major organs, including the heart, after 1 month, yielding good compatibility. Huang et al.[31] recorded hematology markers 24 h and 18 days after the IV injection of differently shaped MSNs. All hematology marker values were mostly within normal ranges and did not indicate a trend of toxicity associated with shape, indicating a good compatibility. The absence of cardiovascular toxicity from SNPs upon IV administration might be explained by their low accumulation in the heart.[30] Kumar et al.[32] prepared multimodal, organically modified SNPs with diameters of 20–25 nm. It was found that <5% of the total dose accumulated in the heart upon IV injection, was much less than in the liver and spleen.

Several *in vitro* studies have been performed to evaluate toxic effects on the cardiovascular system. The underlying, molecular mechanisms were revealed with the help of these *in vitro* models. Intravenously injected SNPs are directly exposed to the endothelial cells; thus, it is of interest to investigate the biological effects of SNPs on the endothelial cells. Liu et al.[33] found that SNPs (20 nm) markedly induced dysfunctions (ROS production, mitochondrial depolarization, and apoptosis) in human umbilical vein endothelial cells (HUVECs) 24 h after exposure, with the concentration ranging from 50 to 200 μg/mL; these effects were through oxidative stresses via the JNK, p53, and NF-κB pathways, suggesting a significant risk for the development of cardiovascular diseases such as atherosclerosis and thrombus. Corbalan et al.[34] found that SNPs (10, 50, 150, and 500 nm) induced platelet aggregation when incubated with isolated human platelets. Furthermore, the effects were inversely proportional to their sizes.[34] Mechanistic studies revealed that amorphous SNPs penetrated the platelet plasma membrane and stimulated a rapid and prolonged NO release, followed by the activation of the glycoprotein IIb/IIIa (GPIIb/IIIa), the expression of the selectin P (SELP) on the platelet surface membrane, and the aggregation of platelets.[34] An alveolar–capillary coculture model was used to evaluate the possible hazards of amorphous SNPs encountering the principal biological barrier in the lower respiratory tract.[35] The epithelial cell line, H441, and the endothelial cell line, ISO-HAS-1, were cocultured on the opposite sides of a filter membrane to mimic the alveolar–capillary barrier. It was found that the endothelial cell line (ISO-HAS-1) seemed to be more sensitive to exposure to SNPs than the epithelial cell line (H441) in terms of cell viability and the release of lactate dehydrogenase (LDH). Furthermore, cocultures showed a release of IL-6 and IL-8 even at concentrations that were 10–100-fold less than the toxic concentrations, indicating early inflammatory events in pulmonary alveoli after SNP inhalation.[35] It could be concluded that *in vitro* models were more sensitive than the *in vivo* models in detecting the adverse effects on the cardiovascular system.

12.2.3 Metallic Nanomaterials

12.2.3.1 Gold Nanomaterials

Recently, gold nanoparticles (GNPs) are being increasingly parenterally administered to animals and humans as carriers for the delivery of drugs, genetic materials, antigens, and diagnostic agents. Meanwhile, concerns were raised about the possible toxicity of these engineered GNPs to the organisms. Several review papers have been published regarding the *in vivo* and *in vitro* toxicity of GNPs.[36,37] However, special attention has not yet been paid to their cardiovascular toxicities. A summary of recent data about the adverse effects of GNPs on the cardiovascular systems are presented below.

Abdelhalim[38–41] conducted a series of *in vivo* experiments to evaluate the effects of GNPs on the cardiovascular system. 10 μg of GNPs of various particle sizes (10, 20, and 50 nm) were intraperitoneally administered to Wistar–Kyoto rats for 3 or 7 days. The disruption of the central vein intima of the hepatic tissues was observed in rats that received 10 and 20 nm GNPs, while less disruption was observed in rats that received 50 nm GNPs. It was also found that more damage was detected if the GNP exposure increased from 3 to 7 days. The above observations suggested the potential endothelial damage and vascular stress by GNPs exposure.[41] GNPs were also found to result in size- and time-dependent histological alterations of the heart tissues, with smaller size and longer exposures induced a higher degree of heart muscle disarray, foci of hemorrhage, scattered cytoplasmic vacuolization, and congested and dilated blood vessels.[38] Similar results were found in another study where GNPs of smaller sizes (10 and 20 nm) induced the congestion of the heart muscle in rats, but rats treated with 50 nm GNPs demonstrated benign, normal-looking heart muscle.[40] The author also found that GNPs of smaller sizes decreased the blood plasma viscosity to an even lower level than the larger size did, which may be attributed to decreases in hematocrit and hemoglobin concentrations in addition to erythrocyte deformabilities.[39] It could be concluded from Abdelhalim's work that GNPs caused size- and time-dependent cardiovascular toxicities when intraperitoneally administered to rats, with a smaller size and longer exposure that caused a more severe effect. A number of literatures have reported the biodistribution of variously sized GNPs. Smaller particles were found to be present in more organ systems and in higher levels than their larger counterparts were,[42,43] which might partly explain Abdelhalim's observations.

On the other hand, the absence of GNP-induced cardiovascular toxicity was observed in other *in vivo* studies using mice or zebrafish embryos as models. Thakor et al.[44] synthesized Raman-active PEGylated silica-coated GNPs. They examined the toxicity of the GNPs after IV administration in mice and found no changes in the electrocardiogram (ECG), blood pressure, or heart rate for 2 weeks. In addition, all plasma biochemical and hematological indices remained within their normal ranges. The differences in results from Abdelhalim's study might be due to the different administration routes and the different surface coatings of the nanoparticles. GNPs showed good biocompatibility in the zebrafish embryo model. Asharani et al.[45] examined the cardiovascular effects of GNPs (15–35 nm) at six different concentrations (10, 25, 50, 75, and 100 μg/mL) in zebrafish embryos, and found that GNPs did not result in pericardial edema or heart rate changes even after 72 h of incubation at the highest tested concentration. Wang et al.[46] evaluated the biocompatibility and toxicity of the surface-enhanced Raman scattering (SERS) GNPs by injecting around 3×10^6 particles into the embryonic cell at the one-cell stage. The SERS GNPs injected 7-day-old zebrafish embryos showed proper form and function of the heart and vasculature, indicating no toxic effects on the cardiovascular systems.

In an *in vitro* study, Freese et al.[47] found that 15 sequentially modified GNPs based on three different core sizes (18, 35, and 65 nm) and five polymeric coatings remained nontoxic to primary human dermal microvascular endothelial cells (HDMECs) up to 48 h at 250 μg/mL, the highest concentration tested. However, there was no group treated with the unmodified GNPs; thus, the results need to be interpreted with caution.

12.2.3.2 Iron Oxide Nanomaterials

Iron oxide nanomaterials, mainly existing as magnetite (Fe_3O_4) and hematite ($\alpha\text{-}Fe_2O_3$ and $\gamma\text{-}Fe_2O_3$) nanoparticles, have attracted an extensive interest in their application in terabit magnetic storage devices, pigments, catalysts, sensors, high-sensitivity biomolecular magnetic resonance imaging, drug and gene delivery, and labeling macromolecules and cells.[48] The increased production and use of iron oxide nanoparticles inevitably results in a greater exposure risk for both people and the environment. Studies regarding the cardiovascular effects of iron oxide nanomaterials were mostly focused on *in vitro* studies, using various types of cell models including endothelial cells, embryonic stem cells (ESCs), and cardiac myocytes.

Gojova et al.[49] found that the incubation of human aortic endothelial cells (HAECs) with different concentrations (0.001–50 μg/mL) of Fe_2O_3 nanoparticles for 1–8 h did not change the expression levels of mRNA and proteins of the three inflammatory markers (ICAM-1, IL-8, and MCP-1) as compared to the untreated control, indicating the absence of an inflammatory response. Sun et al.[50] found that Fe_2O_3 and Fe_3O_4 nanoparticles did not have significant effects on the cytotoxicity, permeability, and inflammation response in human cardiac microvascular endothelial cells (HCMECs) within the concentration range of 0.001–100 μg/mL after 12–24 h of exposure. Au et al.[51] demonstrated that ESCs labeled with 50 μg/mL superparamagnetic iron oxide nanoparticles (SPIONPs) for 24 h showed cardiogenic capacities similar to their unlabeled counterparts, and SPIONPs labeling did not affect the calcium-handling property of ESC-derived cardiomyocytes. In another study, Mahmoudi et al.[52] found that human cardiac myocytes (HCMs) were tolerable to various SPIONPs including negatively charged, positively charged, and bare SPIONPs, showing IC_{50} values between 20–30 mM.

On the other hand, magnemite γ-Fe_2O_3 nanoparticles of around 10 nm caused the death of HUVECs within 24 h of exposure, and an even more severe drop-off of cell viability was revealed after 48 h and persisted until 72 h.[53] It was interesting to note that this effect was not dose related, because a similar cell viability was observed at concentrations ranging from 1 to 100 μg/mL. When administered to Wistar rats via a single IV injection at the dose of 0.8 mg/kg, the γ-Fe_2O_3 nanoparticles did not induce any toxic effects after 2 weeks as examined by staining tissues with hematoxylin and eosin. As to the blood parameters at the end of the experiments, white blood cells showed an increased number, suggesting an inflammatory process, while the red blood cell number remained unchanged, indicating the low toxicity of these particles toward erythrocytes.[53] Iversen et al.[54] intravenously injected PAA-coated γ-Fe_2O_3 nanoparticles (10 mg/kg) into healthy BALB/cJ mice, a commonly use inbred which is susceptible to developing the demyelinating disease upon infection, and monitored the cardiovascular effects. It was found that the arterial acid–base status did not change over the next 12 h after injection, neither did the arterial PCO_2 or (HCO_3^-). However, the MAP decreased 12–24 h after injection (111.1 ± 11.5 vs. 123.0 ± 6.1/min), associated with a decreased contractility of small mesenteric arteries as revealed by myography to characterize endothelial function.

12.2.3.3 Quantum Dots

Semiconductor quantum dots (QDs) have drawn great interests because of their superior optical properties and wide utilization in biological and biomedical studies. Most QDs are composed of heavy metals such as cadmium (Cd), selenide (Se), and/or tellurium (Te), which have hepatic, renal, neurologic, and/or genetic toxicity as ions in solution.[55] Recently, there have been intense concerns on the toxicity assessments of QDs,[56] while their cardiovascular effects are still less frequently reported.

The IV application of multimodal, silica-shelled QDs in rats at the dose of 80 pmol did not show significant acute cardiovascular toxicity.[57] 10 min after the injection, the heart rate increased slightly with a subsequent normalization to baseline levels within 40 min. The blood pressure was stable within 20 min after the injection and decreased about 10% after 80 min. In addition, the QDs did not affect the diameter and structure of the blood vessels.[57] In another study, biocompatible, lysine cross-linked mercaptoundecanoic acid (MUA) $CdSe_{0.25}Te_{0.75}$/CdS QDs were used as a near-infrared (NIR) probe, and the long-term toxicity was investigated.[58] Results from histological analyses showed that no toxic effects were found 100 days after IV injection in the hearts of mice with 10.5 mg/kg of NIR QDs.

Haque et al.[59] intraperitoneally injected mercaptopropionic acid (MPA)-conjugated CdSe/CdS QDs into BALB/c mice every 3 days for a period of 15 days with various doses of QDs (0, 5, 10, and 25 mg/kg). The QDs were not observed in the heart, and the histopathological examination did not show any toxicity against the heart. Zhang et al.[60] demonstrated that the incubation of thioglycolic acid (TGA)–CdTe QDs affected the heartbeats of zebrafish embryos, indicating cardiovascular toxicity. Treatments with 25 and 50 nM QDs resulted in a significantly lowered heartbeat as compared to the control and 1 nM treatment; notably, even lower heartbeats were observed with 100, 200, and 300 nM treatments.

12.2.3.4 Other Metallic Nanomaterials

Besides the most commonly used gold nanomaterials, iron oxide nanomaterials, and QDs, a series of other metallic nanomaterials have fascinated scientists due to their potential utilization in engineering and biomedical sciences.

It is believed that platinum nanoparticles can release platinum ions via surface oxidation, which can contribute to their anticancerous properties.[61] Silver nanoparticles are well known for their antimicrobial properties.[62] However, Asharani et al.[45] found that both silver (5–35 nm) and platinum nanoparticles (3–10 nm) induced a concentration-dependent drop in heart rate when incubated with zebrafish embryos. Silver nanoparticles also induced other significant phenotypic changes, including pericardial effusion, abnormal cardiac morphology, and circulatory defects.[45] In another study, zebrafish was found to develop cardiac malformations when treated with 0.07–0.71 nM silver nanoparticles, with the highest occurrences at 0.66 nM.[63] Among all types of observed deformities, cardiac malfunctions occurred at the third highest rate.

Yttrium oxide (Y_2O_3), and zinc oxide (ZnO) nanoparticles were found to induce a pronounced inflammatory response in HAECs when the concentration exceeded a threshold of 10 μg/mL, as indicated by the elevated mRNA and protein levels of ICAM-1, IL-8, and MCP-1.[49] In addition, ZnO nanoparticles were found to be cytotoxic and led to considerable cell death at 50 μg/mL. Similarly, Sun et al.[50] found that several metallic nanoparticles, including ZnO nanoparticles, produced cytotoxicity in a concentration- and time-dependent manner, and elicited a permeability and inflammation response in HCMECs. The threshold concentration to initiate an inflammatory response was 5 μg/mL for ZnO and CuO nanoparticles and 100 μg/mL for MgO nanoparticles. On the other hand, the authors also found that aluminum oxide (Al_2O_3) nanoparticles did not show significant effects on cytotoxic, permeability, and inflammation responses in HCMECs at any of the concentrations tested (12–24 h and 0.001–100 μg/mL).[50]

Cerium dioxide (CeO_2) nanoparticles were found to be relatively less toxic to cardiovascular systems in several *in vitro* and *in vivo* studies. Kennedy et al.[64] found that CeO_2 particles elicited no inflammation responses at low concentrations and a weak response from 10 to 50 μg/mL after 4 h of incubation with HAECs. Niu et al.[65] assessed the effects of CeO_2 nanoparticles on cardiac function and remodeling as well as endoplasmic reticulum (ER) stress responses in this murine model of cardiomyopathy. It was found that CeO_2 nanoparticles protected against the progression of cardiac dysfunction and the remodeling by attenuation of myocardial oxidative stress, ER stress, and inflammatory processes, probably through their autoregenerative, antioxidant properties.

12.3 CONCLUSIONS AND PERSPECTIVES

The development of various nanomaterials may be expected to lead to revolutionary therapeutic approaches. However, the pharmacological and toxicological profiles of nanomaterials within the cardiovascular system are still limited, and the underlying mechanisms are yet to be clarified. A few points may need to be paid attention to in future studies. The models used to evaluate cardiovascular toxicity need to be standardized. To date, various models have been used in different experiments without criteria for general selection, which may affect the comparison of the results from different researchers and the generation of conclusions. In *in vivo* studies, different species of animals were used, including mouse, rat, and rabbit, while cells of various origins were tested in the *in vitro* studies, such as HAECs, HUVECs, HDMECs, HCMECs, and HCMs. The interpretation of the data from a composite system needs to be performed with caution. Nanomaterials are often surface modified to improve their surface properties and consequently improve their biocompatibility with the human body and dispersity in the aqueous solution. The toxicity data from such systems need to be carefully studied to reveal which material is indeed responsible for adverse effects and provide a chance for improvement.

REFERENCES

1. Oberdorster, G. 2010, Safety assessment for nanotechnology and nanomedicine: Concepts of nanotoxicology. *J Intern Med* 267:89–105.
2. Oberdorster, G., Oberdorster, E., Oberdorster, J. 2005, Nanotoxicology: An emerging discipline evolving from studies of ultrafine particles. *Environ Health Perspect* 113:823–39.
3. Kroto, H. W., Heath, J. R., O'Brien, S. C., Curl, R. F., Smalley, R. E. 1985, C_{60}: Buckminsterfullerene. *Nature* 318:162.
4. Lam, C. W., James, J. T., McCluskey, R., Arepalli, S., Hunter, R. L. 2006, A review of carbon nanotube toxicity and assessment of potential occupational and environmental health risks. *Crit Rev Toxicol* 36:189–217.
5. Li, Z., Salmen, R., Huldermen, T., Kisin, E., Shvedova, A., Luster, M. I., Simeonova, P. P. 2005, Pulmonary exposure to carbon nanotubes induces vascular toxicity. *Toxicol CD—Off J Soc Toxicol* 84:213 (abstr. 1045).
6. Li, Z., Hulderman, T., Salmen, R., Chapman, R., Leonard, S. S., Young, S.-H., Shvedova, A., Luster, M. I., Simeonova, P. P. 2007, Cardiovascular effects of pulmonary exposure to single-wall carbon nanotubes. *Environ Health Perspect* 115:377.
7. Li, Z. J., Chapman, R., Hulderman, T., Salmen, R., Shvedova, A., Luster, M. I., Simeonova, P. P. 2006, Relationship between pulmonary exposure to multiple doses of single wall carbon nanotubes and atherosclerosis in ApoE -/- mouse model. *Toxicologist–Suppl Toxicol Sci* 90:213 (abstract # 1555).
8. Erdely, A., Hulderman, T., Salmen, R., Liston, A., Zeidler-Erdely, P. C., Schwegler-Berry, D., Castranova, V., Koyama, S., Kim, Y.-A., Endo, M. 2008, Cross-talk between lung and systemic circulation during carbon nanotube respiratory exposure. Potential biomarkers. *Nano Letts* 9:36–43.
9. Erdely, A., Liston, A., Salmen-Muniz, R., Hulderman, T., Young, S. H., Zeidler-Erdely, P. C., Castranova, V., Simeonova, P. P. 2011, Identification of systemic markers from a pulmonary carbon nanotube exposure. *J Occup Environ Med* 53:S80–6.
10. Mutlu, G. M., Green, D., Bellmeyer, A., Baker, C. M., Burgess, Z., Rajamannan, N., Christman, J. W. et al. 2007, Ambient particulate matter accelerates coagulation via an IL-6-dependent pathway. *J Clin Invest* 117:2952–61.
11. Koskivirta, I., Rahkonen, O., Mäyränpää, M., Pakkanen, S., Husheem, M., Sainio, A., Hakovirta, H. et al. 2006, Tissue inhibitor of metalloproteinases 4 (TIMP4) is involved in inflammatory processes of human cardiovascular pathology. *Histochem Cell Biol* 126:335–42.
12. Mitchell, L. A., Lauer, F. T., Burchiel, S. W., Mac Donald, J. D. 2009, Mechanisms for how inhaled multiwalled carbon nanotubes suppress systemic immune function in mice. *Nat Nanotechnol* 4:451–6.
13. Valentini, F., Carbone, M., Palleschi, G. 2013, Carbon nanostructured materials for applications in nanomedicine, cultural heritage, and electrochemical biosensors. *Anal Bioanal Chem* 405:451–65.
14. Urankar, R. N., Lust, R. M., Mann, E., Katwa, P., Wang, X., Podila, R., Hilderbrand, S. C. et al. 2012, Expansion of cardiac ischemia/reperfusion injury after instillation of three forms of multi-walled carbon nanotubes. *Part Fibre Toxicol* 9:38.
15. Katwa, P., Wang, X., Urankar, R. N., Podila, R., Hilderbrand, S. C., Fick, R. B., Rao, A. M., Ke, P. C., Wingard, C. J., Brown, J. M. 2012, A carbon nanotube toxicity paradigm driven by mast cells and the IL-(3)(3)/ST(2) axis. *Small* 8:2904–12.
16. Jain, S., Thakare, V. S., Das, M., Godugu, C., Jain, A. K., Mathur, R., Chuttani, K., Mishra, A. K. 2011, Toxicity of multiwalled carbon nanotubes with end defects critically depends on their functionalization density. *Chem Res Toxicol* 24:2028–39.
17. Lacerda, L., Ali-Boucetta, H., Herrero, M. A., Pastorin, G., Bianco, A., Prato, M., Kostarelos, K. 2008, Tissue histology and physiology following intravenous administration of different types of functionalized multiwalled carbon nanotubes. *Nanomedicine (London)* 3:149–61.
18. Tang, S., Tang, Y., Zhong, L., Murat, K., Asan, G., Yu, J., Jian, R., Wang, C., Zhou, P. 2012, Short- and long-term toxicities of multi-walled carbon nanotubes *in vivo* and *in vitro*. *J Appl Toxicol* 32:900–12.
19. Delogu, L. G., Vidili, G., Venturelli, E., Ménard-Moyon, C., Zoroddu, M. A., Pilo, G., Nicolussi, P. et al. 2012, Functionalized multiwalled carbon nanotubes as ultrasound contrast agents. *Proc Natl Acad Sci USA* 109:16612–7.
20. Jain, S., Thakare, V. S., Das, M., Godugu, C., Jain, A. K., Mathur, R., Chuttani, K., Mishra, A. K. 2011, Toxicity of multiwalled carbon nanotubes with end defects critically depends on their functionalization density. *Chem Res Toxicol* 24:2028–39.
21. Cheng, W. W., Lin, Z. Q., Ceng, Q., Wei, B. F., Fan, X. J., Zhang, H. S., Zhang, W. et al. 2012, Single-wall carbon nanotubes induce oxidative stress in rat aortic endothelial cells. *Toxicol Mech Methods* 22:268–76.

22. Martinelli, V., Cellot, G., Toma, F. M., Long, C. S., Caldwell, J. H., Zentilin, L., Giacca, M. et al. 2012, Carbon nanotubes promote growth and spontaneous electrical activity in cultured cardiac myocytes. *Nano Lett* 12:1831–8.
23. Novoselov, K. S., Geim, A. K., Morozov, S. V., Jiang, D., Zhang, Y., Dubonos, S. V., Grigorieva, I. V., Firsov, A. A. 2004, Electric field effect in atomically thin carbon films. *Science* 306:666–9.
24. Gollavelli G., Ling, Y. C. 2012, Multi-functional graphene as an *in vitro* and *in vivo* imaging probe. *Biomaterials* 33:2532–45.
25. Duch, M. C., Budinger, G. R., Liang, Y. T., Soberanes, S., Urich, D., Chiarella, S. E., Campochiaro, L. A. et al. 2011, Minimizing oxidation and stable nanoscale dispersion improves the biocompatibility of graphene in the lung. *Nano Lett* 11:5201–7.
26. Barnes, C. A., Elsaesser, A., Arkusz, J., Smok, A., Palus, J., Leśniak, A., Salvati, A., Hanrahan, J. P., Jong, W. H. D., Dziubałtowska, E. B. 2008, Reproducible comet assay of amorphous silica nanoparticles detects no genotoxicity. *Nano Lett* 8:3069–74.
27. Brunner, T. J., Wick, P., Manser, P., Spohn, P., Grass, R. N., Limbach, L. K., Bruinink, A., Stark, W. J. 2006, In vitro cytotoxicity of oxide nanoparticles: Comparison to asbestos, silica, and the effect of particle solubility. *Environ Sci Technol* 40:4374–81.
28. Galagudza, M. M., Korolev, D. V., Sonin, D. L., Postnov, V. N., Papayan, G. V., Uskov, I. S., Belozertseva, A. V., Shlyakhto, E. V. 2010, Targeted drug delivery into reversibly injured myocardium with silica nanoparticles: Surface functionalization, natural biodistribution, and acute toxicity. *Int J Nanomed* 5:231–7.
29. Nishimori, H., Kondoh, M., Isoda, K., Tsunoda, S.-I., Tsutsumi, Y., Yagi, K. 2009, Histological analysis of 70-nm silica particles-induced chronic toxicity in mice. *Eur J Pharm Biopharm* 72:626–9.
30. He, Q., Zhang, Z., Gao, F., Li, Y., Shi, J. 2011, In vivo biodistribution and urinary excretion of mesoporous silica nanoparticles: Effects of particle size and PEGylation. *Small* 7:271–80.
31. Huang, X., Li, L., Liu, T., Hao, N., Liu, H., Chen, D., Tang, F. 2011, The shape effect of mesoporous silica nanoparticles on biodistribution, clearance, and biocompatibility *in vivo*. *ACS Nano* 5:5390–9.
32. Kumar, R., Roy, I., Ohulchanskky, T. Y., Vathy, L. A., Bergey, E. J., Sajjad, M., Prasad, P. N. 2010, In vivo biodistribution and clearance studies using multimodal organically modified silica nanoparticles. *ACS Nano* 4:699–708.
33. Liu, X., Sun, J. 2010, Endothelial cells dysfunction induced by silica nanoparticles through oxidative stress via JNK/P53 and NF-κB pathways. *Biomaterials* 31:8198–209.
34. Corbalan, J. J., Medina, C., Jacoby, A., Malinski, T., Radomski, M. W. 2012, Amorphous silica nanoparticles aggregate human platelets: Potential implications for vascular homeostasis. *Int J Nanomed* 7:631–9.
35. Kasper, J., Hermanns, M. I., Bantz, C., Maskos, M., Stauber, R., Pohl, C., Unger, R. E., Kirkpatrick, J. C. 2011, Inflammatory and cytotoxic responses of an alveolar–capillary coculture model to silica nanoparticles: Comparison with conventional monocultures. *Part Fibre Toxicol* 8:6.
36. Khlebtsov, N., Dykman, L. 2011, Biodistribution and toxicity of engineered gold nanoparticles: A review of *in vitro* and *in vivo* studies. *Chem Soc Rev* 40:1647–71.
37. Johnston, H. J., Hutchison, G., Christensen, F. M., Peters, S., Hankin, S., Stone, V. 2010, A review of the *in vivo* and *in vitro* toxicity of silver and gold particulates: Particle attributes and biological mechanisms responsible for the observed toxicity. *Crit Rev Toxicol* 40:328–46.
38. Abdelhalim, M. A. 2011, Gold nanoparticles administration induces disarray of heart muscle, hemorrhagic, chronic inflammatory cells infiltrated by small lymphocytes, cytoplasmic vacuolization and congested and dilated blood vessels. *Lipids Health Dis* 10:233.
39. Abdelhalim, M. A. 2011, The effects of size and period of administration of gold nanoparticles on rheological parameters of blood plasma of rats over a wide range of shear rates: In vivo. *Lipids Health Dis* 10:191.
40. Abdelhalim, M. A. 2011, Exposure to gold nanoparticles produces cardiac tissue damage that depends on the size and duration of exposure. *Lipids Health Dis* 10:205.
41. Abdelhalim, M. A., Jarrar, B. M. 2011, Gold nanoparticles administration induced prominent inflammatory, central vein intima disruption, fatty change and Kupffer cells hyperplasia. *Lipids Health Dis* 10:133.
42. Sonavane, G., Tomoda, K., Makino, K. 2008, Biodistribution of colloidal gold nanoparticles after intravenous administration: Effect of particle size. *Colloids Surf B Biointerfaces* 66:274–80.
43. De Jong, W. H., Hagens, W. I., Krystek, P., Burger, M. C., Sips, A. J., Geertsma, R. E. 2008, Particle size-dependent organ distribution of gold nanoparticles after intravenous administration. *Biomaterials* 29:1912–9.
44. Thakor, A. S., Luong, R., Paulmurugan, R., Lin, F. I., Kempen, P., Zavaleta, C., Chu, P., Massoud, T. F., Sinclair, R., Gambhir, S. S. 2011, The fate and toxicity of Raman-active silica–gold nanoparticles in mice. *Sci Transl Med* 3:79ra33.

45. Asharani, P. V., Lianwu, Y., Gong, Z., Valiyaveettil, S. 2011, Comparison of the toxicity of silver, gold and platinum nanoparticles in developing zebrafish embryos. *Nanotoxicology* 5:43–54.
46. Wang, Y., Seebald, J. L., Szeto, D. P., Irudayaraj, J. 2010, Biocompatibility and biodistribution of surface-enhanced Raman scattering nanoprobes in zebrafish embryos: *In vivo* and multiplex imaging. *ACS Nano* 4:4039–53.
47. Freese, C., Gibson, M. I., Klok, H.-A., Unger, R. E., Kirkpatrick, C. J. 2012, Size- and coating-dependent uptake of polymer-coated gold nanoparticles in primary human dermal microvascular endothelial cells. *Biomacromolecules* 13:1533–43.
48. Liu, G., Gao, J., Ai, H., Chen, X. 2013, Applications and potential toxicity of magnetic iron oxide nanoparticles. *Small* 9:1533–45.
49. Gojova, A., Guo, B., Kota, R. S., Rutledge, J. C., Kennedy, I. M., Barakat, A. I. 2007, Induction of inflammation in vascular endothelial cells by metal oxide nanoparticles: Effect of particle composition. *Environ Health Perspect* 115:403.
50. Sun, J., Wang, S., Zhao, D., Hun, F. H., Weng, L., Liu, H. 2011, Cytotoxicity, permeability, and inflammation of metal oxide nanoparticles in human cardiac microvascular endothelial cells. *Cell Biol Toxicol* 27:333–42.
51. Au, K.-W., Liao, S.-Y., Lee, Y.-K., Lai, W.-H., Ng, K.-M., Chan, Y.-C., Yip, M.-C., Ho, C.-Y., Wu, E. X., Li, R. A. 2009, Effects of iron oxide nanoparticles on cardiac differentiation of embryonic stem cells. *Biochem Biophys Res Commun* 379:898–903.
52. Mahmoudi, M., Laurent, S., Shokrgozar, M. A., Hosseinkhani, M. 2011, Toxicity evaluations of superparamagnetic iron oxide nanoparticles: Cell vision versus physicochemical properties of nanoparticles. *ACS Nano* 5:7263–76.
53. Hanini, A., Schmitt, A., Kacem, K., Chau, F., Ammar, S., Gavard, J. 2011, Evaluation of iron oxide nanoparticle biocompatibility. *Int J Nanomed* 6:787–94.
54. Iversen, N. K., Frische, S., Thomsen, K., Laustsen, C., Pedersen, M., Hansen, P. B., Bie, P., Fresnais, J., Berret, J.-F., Baatrup, E. 2012, Superparamagnetic iron oxide polyacrylic acid coated γ-Fe_2O_3 nanoparticles does not affect kidney function but causes acute effect on the cardiovascular function in healthy mice. *Toxicol Appl Pharmacol* 266:276–88.
55. Pelley, J. L., Daar, A. S., Saner, M. A. 2009, State of academic knowledge on toxicity and biological fate of quantum dots. *Toxicol Sci* 112:276–96.
56. Chen, N., He, Y., Su, Y., Li, X., Huang, Q., Wang, H., Zhang, X., Tai, R., Fan, C. 2012, The cytotoxicity of cadmium-based quantum dots. *Biomaterials* 33:1238–44.
57. Bakalova, R., Zhelev, Z., Aoki, I., Masamoto, K., Mileva, M., Obata, T., Higuchi, M., Gadjeva, V., Kanno, I. 2008, Multimodal silica-shelled quantum dots: Direct intracellular delivery, photosensitization, toxic, and microcirculation effects. *Bioconjug Chem* 19:1135–42.
58. Yong, K. T., Roy, I., Ding, H., Bergey, E. J., Prasad, P. N. 2009, Biocompatible near-infrared quantum dots as ultrasensitive probes for long-term *in vivo* imaging applications. *Small* 5:1997–2004.
59. Haque, M., Im, H. Y., Seo, J. E., Hasan, M., Woo, K., Kwon, O. S. 2013, Acute toxicity and tissue distribution of CdSe/CdS-MPA quantum dots after repeated intraperitoneal injection to mice. *J Appl Toxicol* 33:940–50.
60. Zhang, W., Lin, K., Miao, Y., Dong, Q., Huang, C., Wang, H., Guo, M., Cui, X. 2012, Toxicity assessment of zebrafish following exposure to CdTe QDs. *J Hazard Mater* 213–214:413–20.
61. Gao, J., Liang, G., Zhang, B., Kuang, Y., Zhang, X., Xu, B. 2007, FePt@CoS_2 yolk-shell nanocrystals as a potent agent to kill HeLa cells. *J Am Chem Soc* 129:1428–33.
62. Kim, J. S., Kuk, E., Yu, K. N., Kim, J.-H., Park, S. J., Lee, H. J., Kim, S. H., Park, Y. K., Park, Y. H., Hwang, C.-Y. 2007, Antimicrobial effects of silver nanoparticles. *Nanomedicine: NBM* 3:95–101.
63. Lee, K. J., Nallathamby, P. D., Browning, L. M., Osgood, C. J., Xu, X.-H. N. 2007, *In vivo* imaging of transport and biocompatibility of single silver nanoparticles in early development of zebrafish embryos. *ACS Nano* 1:133–43.
64. Kennedy, I. M., Wilson, D., Barakat, A. I. 2009, Uptake and inflammatory effects of nanoparticles in a human vascular endothelial cell line. *Res Rep Health Eff Inst* 136:3–32.
65. Niu, J., Azfer, A., Rogers, L. M., Wang, X., Kolattukudy, P. E. 2007, Cardioprotective effects of cerium oxide nanoparticles in a transgenic murine model of cardiomyopathy. *Cardiovasc Res* 73:549–59.

13 Toxicity of Nanomaterials on the Gastrointestinal Tract

Jayvadan Patel and Vibha Champavat

CONTENTS

- 13.1 Introduction ... 260
  - 13.1.1 Background ... 260
  - 13.1.2 Nanomaterial Toxicity: The Underlying Factors ... 261
    - 13.1.2.1 Size ... 261
    - 13.1.2.2 Surface ... 261
    - 13.1.2.3 Shape ... 261
  - 13.1.3 Exposure and Dose Metrics ... 261
- 13.2 Gastrointestinal Tract Uptake and Clearance of NPs ... 262
  - 13.2.1 Exposure Sources of NPs ... 263
  - 13.2.2 Size- and Charge-Dependent Uptake of NPs ... 263
  - 13.2.3 Translocation ... 264
  - 13.2.4 Adverse Health Effects of GI Tract Uptake ... 264
    - 13.2.4.1 Reaction-Reduced Toxicity ... 264
    - 13.2.4.2 Crohn's Disease, Ulcerative Colitis, and Cancer ... 264
  - 13.2.5 Treatment ... 265
- 13.3 Uptake of NPs through GI Barrier ... 265
  - 13.3.1 Acellular Layers of the Orogastrointestinal Tract ... 265
  - 13.3.2 Interaction of Nanomaterials with the Mucus Layer ... 266
  - 13.3.3 Epithelial Layers of the Orogastrointestinal Tract ... 267
  - 13.3.4 Permeation through Orogastrointestinal Barriers *In Vitro* ... 267
- 13.4 Behavior and Fate of Nanomaterials in the GI Tract ... 270
  - 13.4.1 GI Absorption of NPs ... 271
  - 13.4.2 Distribution ... 272
  - 13.4.3 Excretion/Elimination ... 272
- 13.5 Toxicity of Nanomaterials on GI Tract ... 272
  - 13.5.1 Toxicity of TiO_2 NPs ... 272
  - 13.5.2 Toxicity of Nanoscale Zinc Powder ... 273
  - 13.5.3 Toxicity of Copper Nanoparticles (Copper NPs) ... 274
  - 13.5.4 Toxicity of Single-Walled Carbon Nanotubes ... 274
  - 13.5.5 Toxicity of Metal NPs ... 275
  - 13.5.6 Toxicity of Cadmium–Selenium Quantum Dot ... 276
  - 13.5.7 Toxicity of Si and SiO Particles ... 276
  - 13.5.8 Toxicity of Chitosan NPs ... 278
  - 13.5.9 Toxicity of Gold NPs ... 278
- References ... 279

13.1 INTRODUCTION

13.1.1 Background

Nanotechnology can be considered to be the application of science that "steps across the limit" of miniaturization, where "new rules" become valid (Schmidt et al. 2003). More specifically, when the dimensions of a solid material become incredibly small, its physical and chemical properties can become very different from the larger, bulk form of the same material. This is one of the hallmarks of nanotechnology, it can be described as a research area in which this limit is reached and strategies are developed to exploit the regime of size-controlled properties (Paul et al. 2006). In the last couple of years, the term "nanotechnology" has been inflated and has almost become synonymous for things that are innovative and exceedingly promising. Alternatively, it is also the subject of considerable debate regarding the open question on the toxicological and environmental impact of nanoparticles (NPs) and nanotubes (Donaldson et al. 2004; Oberdörster et al. 2005).

The term "nanomaterial" is used to describe materials with one or more components that have at least one dimension in the range of 1–100 nm, including NPs, nanotubes and nanofibers, nanostructured surfaces, and composite materials. These include NPs as a subset of nanomaterials, currently defined by consensus as single particles with a diameter <100 nm. Agglomerates of NPs can be larger than 100 nm in diameter, but will be included in the discussion since they may break down in solvents or from weak mechanical forces. Nanofibers are a subclass of NPs (including nanotubes) that have two dimensions <100 nm, but the third (axial) dimension can be much larger (Paul et al. 2006).

The unique size-dependent properties of nanomaterials mean that they behave like new chemical substances in some ways. For instance, NPs can scatter and absorb short-wavelength UV (ultraviolet) radiation but leave longer, wavelength-visible light almost unaffected, a property that is exploited in transparent sunscreens. Fluorescent NPs absorb UV radiation while they emit visible light, and the color of the emitted light is different for NPs of different diameters. This result is exploited when NPs are designed as color-coded, fluorescent labels that can be used as diagnostic markers or attached to target molecules. The changes in optical and transport properties become very pronounced for NPs smaller than about 30 nm. Particles smaller than 30 nm are often called "quantum dots" because the size controls the separation (or quantization) of energy levels inside the particle (Paul et al. 2006).

Some nanomaterials have been produced in large volumes for a very long time. NP-sized carbon blacks have been in production for more than a century and are used for the manufacture of pigments and rubber products. Oxides such as titanium, alumina, and zirconium, and fumed silica have been produced as nanomaterials for over half a century and used as thixotropic agents in cosmetics and pigments. In recent times, they have been used as the basis for fine polishing powders in the microelectronics industry. To a great extent, this high-volume production is based on vapor-phase flame or plasma reactions carried out under highly controlled conditions (Paul et al. 2006).

For new nanomaterials, surfaces and interfaces are very important. The proportion of atoms found at the surface increases relative to the volume as soon as particles become smaller. This means that NPs can be more reactive, such as creating more efficient filler materials or more effective catalysts that allow for the reduction in the weight of composite materials. The higher surface energy can also make NPs stick together and interact strongly. If the nanomaterial building blocks are synthesized in such a way that some parts of the surface are sticky but other parts are nonsticky and passive, random Brownian motion in a fluid can cause the blocks to stick together in defined ways to make larger structures (Paul et al. 2006).

However, as the applications of nanomaterials increase, the risk of exposure to the general public will grow. It will be necessary to monitor products that incorporate NPs and nanofibers from manufacture to disposal to estimate the probability of environmental emissions, particularly from waste-management processes and disposal. A few products may involve the direct delivery of NPs to humans, such as the injection of smart drug-delivery systems and the application of cosmetics to the skin and diagnostic markers. In a few cases, there could be an unintended uptake, an example of which could be the ingestion of NPs used in food-packaging technology (Paul et al. 2006).

13.1.2 Nanomaterial Toxicity: The Underlying Factors

Nanomaterials have unique properties and characteristics relative to bulk materials (e.g., high surface-area-to-volume ratio) that may endow them with unique mechanisms of toxicity from xenobiotics. Particularly, toxicity has been thought to originate from nanomaterial size and surface area, shape, and composition as reviewed by Lanone et al. (2006). The three features—size, surface, and shape—discussed below, either independently or in combination, may ultimately be shown in the future to predict the toxicity of NPs.

13.1.2.1 Size

Owing to their small size, NPs can cross cell membranes and penetrate blood vessel walls and the blood–brain barrier via passive and active diffusion, eventually interfering with cellular functions (Geiser et al. 2005).

13.1.2.2 Surface

For the same mass of any particular material, the combined surface area of a particle is inversely proportional to particle size. If the toxic properties of particles are determined by interactions occurring at the interface between particles and biological systems, toxic responses should correlate with the total surface area of particles. In fact, it was observed in animal studies that the inflammatory response to inhaled TiO_2 particulates of different sizes, including the nanoscale range, varied as a function of the surface area (Oberdorster 2000).

13.1.2.3 Shape

One of the benefits of nanotechnology is the ability to control the material structure with atomic precision. This control of materials on a nanoscale results in our ability to generate an immense number of engineered NPs with different shapes. Examples of the simplest engineered NPs are spheres, tubes, wires, rods, belts, and flakes. Examples of more complex engineered NPs are tripods, flowers, and brushes. Finally, the most complex NPs are three-dimensional structures such as multifunctional nanoscale particles, for example, functionalized liposomes, virosomes, and dendrimers (Oberdorster 2000).

13.1.3 Exposure and Dose Metrics

Historically, a mass-based paradigm has been employed by industrial hygienists to assess worker exposures to airborne particulates. The exposure and dose metrics for engineered NPs in the workplace are now actively under investigation because NPs have such little mass to measure. Since NPs have little mass, a new exposure and dose metrics may be needed. Currently, particle number and particle surface area are being studied as an exposure and dose metric. An exposure and dose metric for engineered nanoscale materials, which have a range of structures, chemical compositions, or both, will depend on the mechanism of their toxicological and pharmacokinetic behavior (Nel et al. 2006). For example, poorly soluble, low-toxicity particles, which interact with biological systems at the particle surface, can have their exposure and dose expressed as a combined surface area. Consequently, experimental studies in rodents and cell cultures have shown that the toxicity of nanoscale particles is greater than that of the same mass of larger particles of a similar surface area; chemical composition correlates best with the observed toxicological responses (Oberdörster et al. 1994; Tran et al. 2000).

There have been numerous seminal reports and review articles detailing the scientific evidence that nanomaterials have unique toxicological properties and are more toxic than their bulk materials (Borm et al. 2006; Oberdörster et al. 2005). When considering nanomaterial toxicity, it is extremely important to recognize that there are many different types of nanomaterials, which negate a generic approach. Indeed, studies in both tissue cultures and laboratory animals have shown that seemingly slight changes to the surface chemistry of nanomaterials can result in significant changes in their toxicity.

For example, fullerenes and carbon nanotubes (CNTs) that were chemically treated in different ways showed remarkably different toxicities compared to their untreated counterparts (Carrero-Sanchez et al. 2006; Sayes et al. 2004). As a result, the detailed characterization of nanomaterial properties is now recognized as a critical component of quality nanotoxicology studies. Additionally, it seems unlikely that future studies will provide generalizations that can describe the toxicity of all nanomaterials (Borm et al. 2006).

The number of nanotoxicology studies being conducted is gradually increasing, but there are many knowledge gaps that need to be filled before appropriate risk assessments and workplace exposure standards can be established. Data concerning the effects of engineered nanomaterials in humans are limited and, to date, the majority of studies have been conducted in rodent models and tissue cultures using tumor cell lines. Nevertheless, the studies conducted so far have identified a number of fundamental issues concerning the toxicity of nanomaterials. The most common finding is that the surface chemistry, particle size, and surface area are all key determinants in the adverse effects caused by particulate matter.

The exact cause of nanomaterial-induced inflammation has not yet been clarified, although a number of hypotheses have emerged. The most prominent hypothesis is that cellular damage may be due to the ability of some nanomaterials to produce reactive oxygen species (ROS), which can damage cell membranes and proteins (Xia et al. 2006). Nanomaterials have the ability to adsorb many different environmental contaminants to their surfaces due to their large surface areas and chemical natures. Nanomaterial-induced toxicity and inflammation could be due to chemical "contaminants" that are adsorbed to the surface of nanomaterials (e.g., bacteria-derived molecules, catalyst metals, or combustion waste products from manufacturing processes). Furthermore, chemicals attached to nanomaterials may be presented to receptors on the surface of immune cells, which could result in inflammation (Becker et al. 2005; Vallhov et al. 2006). On the other hand, toxic chemicals adsorbed to particles, such as metals and combustion waste products, may also be delivered into cells resulting in toxic effects (Penn et al. 2005). Some environmental research groups are using the theoretical term "nanovectors," which reflects this characteristic of nanomaterials as a portal of entry for the cellular uptake of toxic moieties.

In general, these findings suggest that the mechanism of toxic action from nanomaterials may depend on the processes used to produce engineered nanomaterials and the materials' ability to adsorb chemicals to their surface. Therefore, it is possible that some processes used in the quality fabrication of nanomaterials, such as using dust-free clean rooms and sterile cabinets, could also reduce the toxicity of nanomaterials; however, this is yet to be proven. Currently, there is evidence for all hypotheses but future research will demonstrate whether the mechanisms are operating independently, synergistically, or if one mechanism is more damaging than the other.

13.2 GASTROINTESTINAL TRACT UPTAKE AND CLEARANCE OF NPS

As different tissues and organs have different compositions, structures, and functions, toxic responses are mostly different once NPs enter different organs. The gastrointestinal tract (GI tract), also known as the digestive tract, can uptake, transport, digest, and adsorb various substances such as nutrients, water, and vitamins from food. On the other hand, the GI tract is also designed as a barrier to restrain the entry of pathogens, undigested macromolecules, and toxins. As such, these potential exposure routes are likely to be the first portal of entry for NPs invading the human body.

The GI tract is one of the largest immunological organs of the body, containing more lymphocytes and plasma cells than the spleen, bone marrow, and lymph node. Ingestion is considered as an exogenous source of exposure, primarily through the mouth in the workplace. Alternatively, the NPs can be directly ingested via water, ingestion, food, drugs, or drug-delivery systems. In addition, NPs cleared from the respiratory tract via the mucociliary escalator can be subsequently ingested into the GI tract. Thus, GI tract is considered as an important target for NPs exposure (Zhao et al. 2010).

13.2.1 Exposure Sources of NPs

Endogenous sources of NPs in the GI tract are derived from intestinal calcium and phosphate secretion (Lomer et al. 2004). Exogenous sources are particles from food (such as colorants, titanium oxide), pharmaceuticals, water or cosmetics (toothpaste, lipstick) (Lomer et al. 2004), dental prosthesis debris (Ballestri et al. 2001), and inhaled particles (Takenaka et al. 2001). The dietary consumption of NPs in developed countries is estimated around 1012 particles/person/day (Oberdörster 2004). They mainly consist of TiO_2 and mixed silicates. The use of specific products, such as salad dressing, containing an NP, such as a TiO_2 whitening agent, can lead to an increase of the daily average intake by more than 40-fold (Oberdörster 2004). These NPs do not degrade in time and accumulate in macrophages. A database of foods and drugs containing NPs can be found in Lomer et al. (2004). A portion of the particles cleared by the mucociliary escalator can be subsequently ingested into the GI tract. Also, a small fraction of inhaled NPs was found to pass into the GI tract (Takenaka et al. 2001).

13.2.2 Size- and Charge-Dependent Uptake of NPs

The GI tract is a complex barrier-exchange system and is the most important route for macromolecules to enter the body. The epithelium of the small and large intestines is in close contact with ingested material, which is absorbed by the villi (Figure 13.1). The uptake of NPs and microparticles has been the focus of many investigations, the earliest dating from the mid-seventeenth century, while the entire issues of scientific journals have been devoted to the subject more recently (Hussain et al. 2001). The extent of particle absorption in the GI tract is affected by size, surface chemistry and charge, length of administration, and dose (Hoet et al. 2004).

The absorption of particles in the GI tract depends on their size, and the uptake diminishing for larger particles (Jani et al. 1990). A study of polystyrene particles with sizes between 50 nm and 3 μm indicated that the uptake decreases with increasing particle sizes from 6.6% for 50 nm, 5.8% for 100 nm NPs, 0.8% for 1 μm, to 0% for 3 μm particles. The time required for NPs to cross the colonic mucus layer depends on the particle size, with smaller particles crossing faster than larger

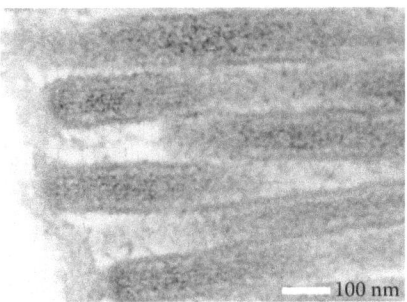

FIGURE 13.1 TEM image of a thin section cut through a segment of human small intestine epithelial cell. One notices densely packed microvilli, each microvillus being approximately 1 μm long and 100 nm in diameter. (The image is courtesy of Chuck Daghlian, Louisa Howard, Katherine Connollly.)

particles: 14-nm-diameter latex NPs cross within 2 min, 415 nm particles cross within 30 min, and 1000 nm particles do not pass this barrier (Hoet et al. 2004). Particles that penetrate the mucus layer reach the enterocytes and are able to translocate further (Hoet et al. 2004). Enterocytes are a type of epithelial cell of the superficial layer of the small and large intestine tissue that aid in the absorption of nutrients. When in contact with the submucosal tissue, NPs can enter the lymphatic system and capillaries, and are then able to reach various organs (Hoet et al. 2004).

Diseases such as diabetes may lead to a higher absorption of particles in the GI tract (Hoet et al. 2004). For example, rats with experimentally induced diabetes had a 100-fold increase in the absorption of 2 μm polystyrene particles (Hoet et al. 2004) relative to nondiabetic rats. Also, inflammation may lead to the uptake and translocation of larger particles of up to 20 μm (Ballestri et al. 2001).

The kinetics of particles in the GI tract strongly depends on the charge of the particles, as positively charged latex particles are trapped in the negatively charged mucus while negatively charged latex NPs diffused across the mucus layer and became available for interactions with epithelial cells (Hoet et al. 2004).

13.2.3 Translocation

Varying the characteristics of NPs, such as size, surface charge, attachment of ligands, or surfactant coatings, offers the possibility for the site-specific targeting of different regions of the GI tract. The fast transit of materials through the intestinal tract (on the order of hours), together with the continuous renewal of the epithelium, led to the hypothesis that nanomaterials will not remain there for indefinite periods (Hoet et al. 2004). Most of the studies of ingested NPs have shown that they are rapidly eliminated (within 48 h) with 98% in the feces and most of the remainder is eliminated via urine (Oberdörster et al. 2005). However, other studies indicate that certain NPs can translocate to blood, spleen, liver, bone marrow (Jani et al. 1990), lymph nodes, kidneys, lungs, and brain, and can also be found in the stomach and small intestine (Rae et al. 2005). The oral uptake of polystyrene spheres of various sizes (50 nm–3 μm) by rats resulted in a systemic distribution to the liver, spleen, blood, and bone marrow (Jani et al. 1990).

Particles larger than 100 nm did not reach the bone marrow, while those larger than 300 nm were absent from the blood (Jani et al. 1990). During this study, no particles were detected in the heart or lung tissue. Studies using iridium did not show significant uptake, while TiO_2 NPs were found in the blood and liver (Oberdörster et al. 2005). For several days following the oral inoculation of mice with a relatively biologically inert nanometer-sized plant virus (cowpea mosaic virus), the virus was found in a wide variety of tissues throughout the body, including the stomach, small intestine, lymph nodes, spleen, liver, lung, kidney, brain, and bone marrow (Rae et al. 2005).

The exact order of translocation from the GI tract to organs and blood is not known. However, a case study of dental prosthesis porcelain debris internalized by intestinal absorption suggests that the intestinal absorption of particles is followed by their clearance from the liver before they reach general circulation and the kidneys (Ballestri et al. 2001).

13.2.4 Adverse Health Effects of GI Tract Uptake

13.2.4.1 Reaction-Reduced Toxicity

There is a complex mix of compounds, enzymes, food, bacteria, and so on, that can interact with ingested particles and sometimes reduce their toxicity in the intestinal tract (Hoet et al. 2004). It was reported that particles are less cytotoxic *in vitro* and in a medium with high protein content.

13.2.4.2 Crohn's Disease, Ulcerative Colitis, and Cancer

NPs have been constantly found in the colon tissue of subjects affected by cancer, Crohn's disease, and ulcerative colitis, while NPs were absent in healthy subjects (Gatti 2004). The NPs present in diseased subjects had various chemical compositions and were not considered toxic in their bulk form. Microscopic and energy-dispersive spectroscopy analyses of colon mucosa indicated the presence of carbon, ceramic filosilicates, gypsum, sulfur, calcium, silicon, stainless steel, silver, and zirconium (Gatti 2004). The size of the debris varied from 50 nm to 100 μm; the smaller the particle, the further it is able to penetrate. The particles were found at the interface between healthy and cancerous tissues. On the basis of these findings, it was suggested that the GI barrier is not efficient for particles smaller than 20 μm (Ballestri et al. 2001).

Crohn's disease affects people primarily in developed countries, and occurs in both the native population and in immigrants from underdeveloped countries. It affects 1 in 1000 people (Lomer et al. 2002). Crohn's disease is believed to be caused by a genetic predisposition together with environmental factors (Lomer et al. 2002). It was recently suggested that there is an association between high levels of dietary NPs (100 nm–1 μm) and Crohn's disease (Lomer et al. 2002). Exogenous NPs were found in macrophages accumulated in the lymphoid tissue of the human gut, the lymphoid aggregates being the earliest sign of lesions in Crohn's disease (Lomer et al. 2002). Microscopy studies showed that macrophages located in lymphoid tissue uptake NPs of spherical anatase (TiO_2) with sizes ranging between 100 and 200 nm from food additives, 100 and 400 nm aluminosilicates typical of natural clay, and 100 and 700 nm environmental silicates with various morphologies (Powell et al. 1996). A diet low in exogenous particles seems to alleviate the symptoms of Crohn's disease (Lomer et al. 2002). This analysis is still controversial, with some proposing that an abnormal response to dietary NPs may be the cause of this disease and not an excess intake (Lomer et al. 2004). More precisely, some members of the population may have a genetic predisposition where they are more affected by the intake of NPs and, therefore, develop Crohn's disease (Oberdörster 2004).

Some evidence suggests that dietary NPs may exacerbate inflammation in Crohn's disease (Lomer et al. 2004). These studies measured the intake of dietary particles, but did not analyze the levels of indoor and outdoor NP pollution at the subjects' residences. As was described previously, significant quantities of NPs are cleared by the mucociliary escalator and subsequently swallowed, ultimately reaching the GI tract.

13.2.5 Treatment

The diseases associated with the GI uptake of NPs (such as Crohn's disease and ulcerative colitis) have no cure and often require surgical intervention. Treatments aim to keep the disease in remission and consist of anti-inflammatory drugs along with specially formulated liquid meals (Lomer et al. 2002). If dietary NPs are conclusively shown to cause these chronic diseases, their use in foods should be avoided or strictly regulated.

13.3 UPTAKE OF NPs THROUGH GI BARRIER

The GI barriers consist of cellular (epithelium) and acellular parts (dead cells, mucus). Permeation through the GI barrier has been shown for micro- and NPs. NP absorption is estimated to be about 15–250 times higher than that of microparticles (Desai et al. 1996).

13.3.1 Acellular Layers of the Orogastrointestinal Tract

Mucus represents an efficient acellular barrier for the entire GI tract, composed of the oral cavity, esophagus, stomach, and the intestine. The composition of mucus is primarily based on mucin proteins (highly glycosylated extracellular proteins with characteristic gel-forming properties), antiseptic proteins such as lysozyme, and other proteins, that is, lactoferrin, inorganic salts, and water. The major functions of mucus are the protection and lubrication of the underlying tissue.

Saliva, which is produced by the salivary glands, is mainly composed of water (up to 99.5%), proteins, inorganic salts, and mucins. The so-called mucus layer, constituting the acellular barrier of the oral cavity, is built by the binding of the high-molecular-weight mucin, MG1, to the surface of the epithelium (Bykov 1996, 1997). The thickness of this mucus layer is different before and after swallowing, and measures between 70 and 100 μm (Collins and Dawes 1987; Harris and Robinson 1992; Lagerlof and Dawes 1984). This mucus layer displays a thick gelatinous-like layer, structured as a three-dimensional network with a high water-holding capacity. Being highly viscoelastic, it displays a shear-thinning gel that acts as a lubricant. It defends the epithelial cell layers from pathogens, particles, and toxins, and enables the exchange of nutrients, gases, and water (Knowles and Boucher 2002). Once substances are swallowed, they pass through the esophagus. Esophageal glands located throughout the esophagus are involved in secreting mucus directly onto the surface (Squier and Kremer 2001). In addition, exocrine glands in the submucosa produce a secretion that is high in the concentration of bicarbonates. This is essential to neutralize refluxing stomach acid (Long and Orlando 1999). The mucus of the following organs, that is, stomach, small intestine, and large intestine, is chiefly produced by intraepithelial cells. Exocrine glands are also located in the submucosa of the first part of the small intestine (duodenum). The thickness of the mucus layer varies highly, depending on the location in the GI tract. The thickness of this layer increases from proximal to distal parts of the small and large intestine with the greatest thickness in the stomach (Atuma et al. 2001; Matsuo et al. 1997). The thickness of the mucus layer shows noticeable variations based on the method used for its determination. The fixation of the tissues is usually followed by shrinking, thus yielding lower values. Endoscopic ultrasound measurements indicate the thickness of the mucus layer in the stomach to be 897–1354 and 730–1136 μm in the rectum (Huh et al. 2003), but variations may be quite high because the thickness is dictated by the interplay between mucus secretions by goblet cells and mucus erosions by mechanical shear and bacterial digestion, particularly in the lower gut (Corfield et al. 1992; Hoskins and Boulding 1981). Additionally, the pH can vary. The pH of the mucus in the oral cavity is estimated to be around 6.6. Gastric mucus shows a broad pH range from 1 to 2 (luminal) to ~7 (epithelial surface) (Schreiber and Scheid 1997).

13.3.2 Interaction of Nanomaterials with the Mucus Layer

The characteristics facilitating the passage through the mucus are relatively well known: Electrostatic repulsions from negatively charged sugar moieties favor the penetration of positively charged, hydrophilic molecules; the passage of lipophilic compounds is slow (Avdeef and Testa 2002). It was thought that NPs are incapable of penetrating the mucus layer since recent studies demonstrated that particular viruses, for example, the Norwalk virus (38 nm) and human papilloma virus (55 nm), diffused into the mucus as rapidly as they do in water (Olmsted et al. 2001; Saltzman et al. 1994). The surfaces of the viruses able to permeate the mucus are thickly coated with positive and negative charges. Hence, this net neutral surface charge precludes mucoadhesion (Olmsted et al. 2001). Since the pore size is around 100 nm, it is proposed that small particles might also be capable of diffusing through the mucus. Olmsted et al. (2001) showed that small viruses diffused unhindered via the mucus, whereas polystyrene microspheres, ranging 59 nm in size and covalently modified with carboxyl groups, bound more tightly to mucins and clustered them into thick cables. The work by Dawson et al. reported that carboxyl- and amine-modified polystyrene particles (100, 200, and 500 nm) were embedded in cystic fibrosis sputum. Compared to negatively charged particles, positively charged particles penetrated more rapidly in the sputum. Moreover, smaller particles underwent a significantly faster transport (Dawson et al. 2003). Lai et al. investigated polystyrene particles ranging from 100 to 500 nm. The particle surface was covalently modified with a high density of low-molecular-weight polyethylene glycol (PEG) and the diffusion into the mucus was examined. The results demonstrated that the neutral surface charge increased the diffusion rate of all particles. Where larger particles (200 and 500 nm) demonstrated a sixfold and fourfold lower effective diffusion coefficient than that in water, the 100 nm particles were observed to be immobile in the mucus (Lai et al. 2007). It was demonstrated by a study that slightly negatively charged, 14 nm latex particles crossed the distal colon mucus gel layer within 2 min and 415 nm large particles crossed in 30 min, whereas 1 μm large particles did not cross (Szentkuri 1997). Nonbiodegradable latex particles can quickly permeate the mucus when they are coated with PEG. Surprisingly, 200 nm particles crossed the mucin layer faster than <100 nm nanomaterials (Wang et al. 2007a). These findings suggest that the surface charge plays a crucial role in the transport rates of NPs through a mucus layer.

The lifetime of the mucus is short and the fastest turnover (i.e., clearance time) is observed at surfaces with the thinnest mucus layers. Hence, NPs have to rapidly permeate through this barrier to reach the underlying epithelia (Cone 2009). Local effects after oral exposure to nanomaterials include abnormal mucus production, induced by TiO_2 NPs in cultured ChaGo-K1 cells (Chen et al. 2011) and by silver NPs *in vivo* (Jeong et al. 2010). In addition, pH changes induced by nanomaterials can change the pH-dependent aggregation of mucins (Bhaskar et al. 1991). Additionally, positively charged nanomaterials impede mucin swelling and thereby increase viscosity (Chen et al. 2010).

13.3.3 Epithelial Layers of the Orogastrointestinal Tract

The epithelium generally represents the highest resistance against the passage of chemical compounds and nanomaterials. Epithelial cells are polarized; they have an apical surface facing an internal or external surface and a basal site, where they confront the underlying tissue. The epithelia may consist of several layers and may vary in the height of the cells.

Penetration through a monostratified squamous epithelium, such as in the endothelia (Figure 13.2a), is easier than through the simple columnar epithelium in the stomach and intestine (Figure 13.2b) or the squamous epithelium layer of the oral cavity and esophagus (Figure 13.2c). The thickness of the nonkeratinized squamous epithelium in the oral cavity ranges between 550 and 800 µm (Collins and Dawes 1987; Harris and Robinson 1992; Lagerlof and Dawes 1984). The squamous epithelial layer of the esophagus shows a thickness of 300–500 µm (Takubo 2009). The epithelium of the esophagus has the same structure as that of the buccal mucosa, but is thinner and less variable (Diaz del Consuelo et al. 2005). The simple columnar epithelium found in the GI tract measures 20–25 µm (Atuma et al. 2001; Matsuo et al. 1997). Generally, only one cell type forms the structural basis of the barrier: keratinocytes for the oral cavity and the esophagus, enterocytes for the small and large intestine, and gastric epithelial cells for the stomach.

The epithelial cells are linked together by intercellular junctions, which provide mechanical strength to the epithelial layer and restrict the passage between cells. In the oral mucosa, immune cells (lymphocytes, Langerhans cells) are embedded in the keratinocytes layer (Figure 13.2d). The epithelium of the stomach contains mucus neck cells that produce mucus; gastric chief cells that produce pepsinogen and gastric acid; parietal cells that produce the intrinsic factor; and enteroendocrine cells that produce a variety of hormones such as gastrin, serotonin, somatostatin, and so on. (Figure 13.2e). Cells belonging to the immune system (M-cells), enteroendocrine cells, and goblet cells are embedded in a layer of enterocytes in the small intestine region (Figure 13.2f). M-cells are preferentially located in the epithelium overlying Peyer's patches, which is also known as the follicle-associated epithelium (FAE), and delivers foreign substances to the underlying tissues (mucosa lymphoid) to induce immune responses (Gerbert et al. 1996). However, M-cells are also a potential portal for NPs. The large intestine epithelium consists of goblet cells and enterocytes. As different cell types adjoin, the barrier property of the epithelium is altered because the location and structure of these junctions differ between cell types (Eom and Choi 2009). All epithelia rest on a basal membrane that separates them from the connective tissue lying underneath, which contains capillaries, peripheral nerves, lymph follicles, and lymph vessels. To reach systemic circulation via the capillaries, nanomaterials have to cross the basal membrane and connective tissue.

13.3.4 Permeation through Orogastrointestinal Barriers *In Vitro*

The epithelia can be permeated either by passages through cells (transcellular) or by passages between cells (paracellular). Physiological methods for evaluating interactions with biological

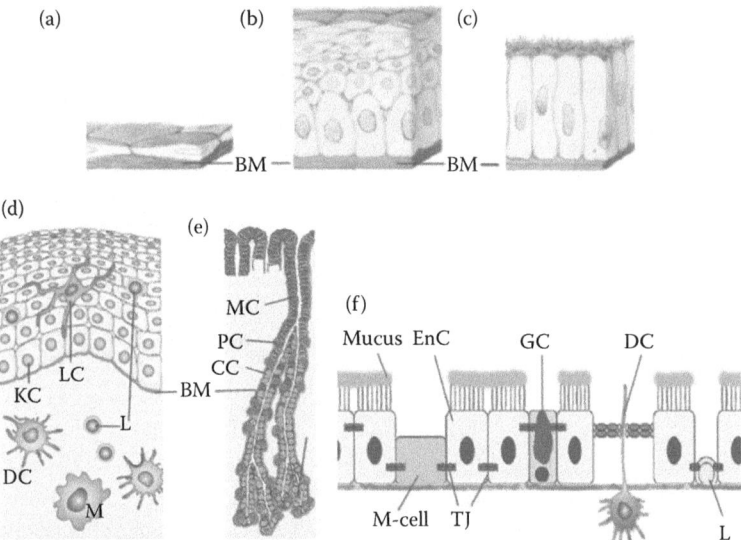

FIGURE 13.2 Different thickness and heterogeneity of orogastrointestinal epithelia (a) compared to the monostratified squamous epithelium of blood vessels and (b) simple columnar epithelium present in the stomach, small, and large intestine. (c) Epithelia of the orogastrointestinal tract are much thicker: stratified nonkeratinized squamous epithelium of the oral cavity and the esophagus. All epithelia reside on a basal lamina (BM). All epithelial layers are composed of different cell types. (d) In the epithelial layer of the oral cavity, the Langerhans cells (LC) and intraepithelial lymphocytes (L) in addition to keratinocytes (KC) are present. The connective tissue lying below the epithelial layer contains dendritic cells (DC), macrophages (M), and lymphocytes (L). (e) The mucosa of the stomach consists of mucus-producing cells (MC), gastric acid-releasing cells, that is, parietal cells (PC), pepsinogen-producing cells, that is, chief cells (CC), and endocrine hormone-producing enteroendocrine cells (EC). Cells of the immune systems are not indicated. (f) Enterocytes (EnC), microfold (M-cells), dendritic cells (DC), goblet cells (GC), and intraepithelial lymphocytes (L) are linked together by tight junctions (TJ) in the small intestine. These junctions show small differences in composition and location in the cell. (From Eom, H.J. and Choi, J. 2009. Oxidative stress of silica nanoparticles in human bronchial epithelial cell, Beas-2B. *Toxicol in Vitro* 23:1326–1332. With permission).

barriers and predicting the effect of NPs are highly needed. Studies addressing permeation usually use Transwell™ systems, where cells are cultivated on filters. Furthermore, diffusion cells can be used to evaluate the penetration/permeation of nanomaterials across excised tissues (Sudhakar et al. 2006). Studies on cell monolayers depict that polystyrene particles can readily permeate the alveolar epithelium (Yacobi et al. 2008). On the contrary, the rate of permeation of enterocyte (Caco-2 cell) monolayers by polystyrene particles without surface coatings appears low (Geiser et al. 2005; Pietzonka et al. 2002). Studies by Gaumet et al. demonstrated that small polystyrene particles were intracellularly observed in Caco-2 cells (Gaumet et al. 2009). In addition, TiO_2 NPs appear to cross Caco-2 monolayers without disrupting the junctional complexes and causing cytotoxicity (Koeneman et al. 2010).

As the plasma membrane of the cells forming the epithelial barrier is lipophilic in nature, lipophilic substances are passively taken up by the transcellular route whereas hydrophilic drug compounds use the paracellular route. The area of the paracellular route for penetration is extremely small compared to the transcellular route and restricted to polar substances ranging <1000 D. Paracellular transport is only passive in nature. NPs are not expected to be able to use the paracellular route, as they are considerably larger than 1000 D. To get an estimation about the relation of molecular weight and size: Serum albumin, 66 kD, has an almost sphere-like shape of $3 \times 8 \times 8$ nm (Takizawa et al. 1992).

Transcellular passages by passive diffusion appear to be rare: Although the passage of cells by 22 nm TiO$_2$ particles was suggested to occur passively (Geiser et al. 2005), other researchers demonstrated that Au-NPs in sizes of 5–8 nm could not enter cells by passive diffusion (Stoeger et al. 2006). The likely mode of cellular uptake for metal and metal oxide nanomaterials is believed to actively occur through endocytosis. Various endocytotic routes have been characterized, which are classified on the basis of the coating with clathrin and the involvement of dynamin in uptake. The chief mechanisms are termed as clathrin-mediated endocytosis, caveolae-dependent mechanism, and macropinocytosis. Various classifications are used for the clathrin-independent and caveolae-independent routes. The classification by Sahay et al. is mainly based on the GTPases involved (RhoA-dependent, Arf6-dependent, and Cdc42/Arf1-dependent endocytosis) and on the coat protein (flotillin dependent) (Sahay et al. 2010). Another nomenclature employs the term clathrin-dependent carriers/glycophosphatidylinositol (GPI)-anchored protein-enriched compartment (GEEC)-type endocytosis as a synonym for Cdc42/Arf1-dependent endocytosis and IL-2Rβ-dependent endocytosis for RhoA-dependent endocytosis (Doherty and McMahon 2009).

Independent of the route of entry, the cargo is mainly transported via endosomes to lysosomes (Figure 13.3). TiO$_2$, nonfunctionalized silver, and SiO$_2$ particles are mainly taken up by clathrin-mediated endocytosis (Chung et al. 2007; Greulich et al. 2011; He et al. 2009; Huang et al. 2005; Singh et al. 2007; Sun et al. 2008). NPs can leave the cells either by exocytosis or by transcytosis. Exocytosis of NPs is not well studied and conflicting results were obtained: Exocytosis of quantum dots was not consistently seen in the studies (Clift et al. 2008; Jiang et al. 2010). On the other hand, transcytosis can occur through receptor-mediated uptake or via adsorptive-mediated uptake. Receptors for BSA (bovine serum albumin), transferrin, and opioid peptide-functionalized nanomaterials are expressed on several cell types and BSA-coated NPs have been shown to transcytose

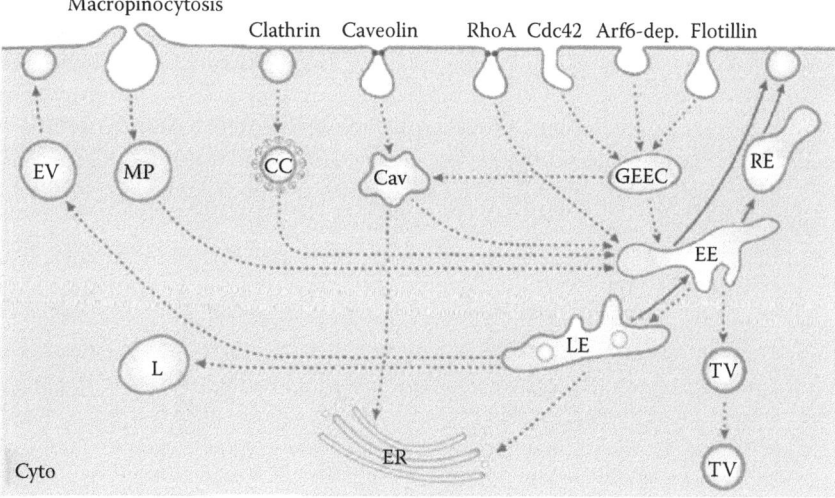

FIGURE 13.3 Active uptake mechanisms of nanomaterials into cells. Macropinocytosis, clathrin-mediated endocytosis, and caveolae-mediated and non-clathrin, non-caveolae-mediated uptake are the major routes identified. The latter are subclassified into RhoA- (or IL-2Rβ-) dependent endocytosis, Cdc42/Arf1 or clathrin-independent cargo/GPI-anchored protein-enriched compartment-dependent (GEEC) endocytosis, flotillin-dependent endocytosis, and Arf6-dependent endocytosis. Early endosomes (EE) and late endosomes (LE) transport the contents of macropinosomes (MP), clathrin-coated vesicles (CC), and GEEC to the lysosomes (L). Material taken up by caveolae-mediated endocytosis is transported via caveolosomes (Cav) either to the endoplasmic reticulum (ER) or to early endosomes. Nanomaterials may be removed from the cells by exocytotic vesicles (EV). Early endosomes may also fuse with the plasma membrane directly or through recycling endosomes (RE). In polarized cells, transcytosis occurs via transport vesicles (TV). Cyto: cytoplasm.

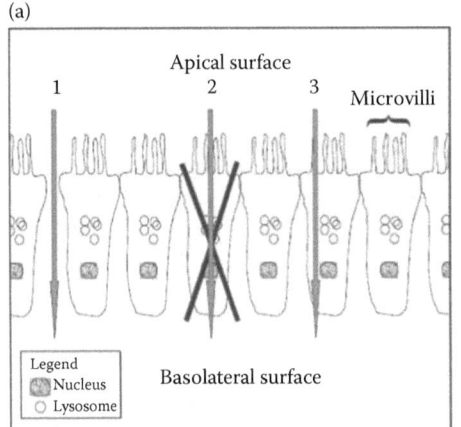 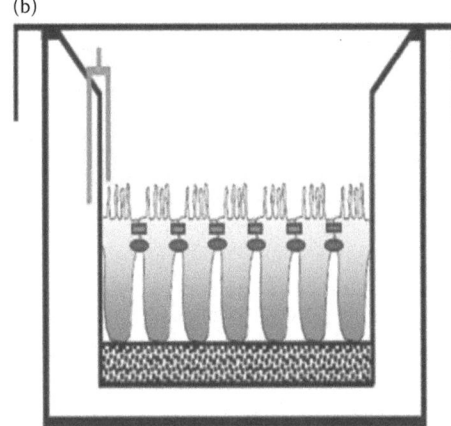

FIGURE 13.4 (**See color insert.**) (a) Diagram illustrating substantial routes of epithelial transport. An epithelium is polarized into an apical and basolateral surface with the apical surface covered with microvilli to increase the surface area for absorption. Nuclei and other organelles inside the cell are also polarized with the nuclei polarized closer to the basolateral surface. In between the cells, tight junctions and adherent junctions are present at the apical surface that inhibit the free flow of materials between the apical and basolateral spaces and provide for epithelial integrity. The nanomaterials can flow between the apical and basolateral spaces if (1) epithelial integrity is disrupted (pathway 1), (2) cells within the epithelium are killed, providing holes for the flow of particles (pathway 2), and (3) nanomaterials are moved by transcytosis—a cellular process where materials are picked up at the apical surface and transported to the basolateral surface without being metabolized by the cell (pathway 3). (b) Diagram of cells in the Transwell chamber. Cells are shown on a membrane support, which permits the partitioning of the apical and basolateral chambers for measurements of electrical resistance. Red squares represent tight junctions and blue ellipses represent adherent junctions.

through endothelial cells (Wang et al. 2009). For the GI tract, however, this type of uptake is not crucial. Absorption-dependent transcytosis is mediated by the interaction of positively charged substances with anionic sites of the plasma membrane: Cationic NPs had a greater potential than neutral or negatively charged NPs (Harush-Frenkel et al. 2008). In addition, uncoated, not positively charged TiO_2 NPs can cross the intestinal epithelium by the transcellular route (Koeneman et al. 2010).

There are three pathways by which nanomaterials could cross the epithelial layer as illustrated in Figure 13.4a. Nanomaterials could disrupt the junctional complexes without killing cells in the epithelial sheet and pass across the epithelial layer (arrow 1). Nanomaterials could kill cells within the epithelial sheet and pass through holes formed by dead cells (arrow 2). It is also possible that nanomaterials could make use of the transport function of epithelial cells used in nutrient uptake and pass through individual epithelial cells by transcytosis (arrow 3). All three of these possible routes of transport have been investigated.

One function of epithelial cells of the intestine is to undergo transcytosis. Transcytosis is a process whereby extracellular components are endocytosed on one side of the cell; in this case, the apical surface is exposed to the lumen of the gut, and exocytosed from the other surface of the cell, the basolateral surface where the circulatory system is present, without undergoing metabolism inside the cell. TiO_2 may be transported across the epithelium without disrupting epithelial integrity if it is transcytosed (Koeneman et al. 2010).

13.4 BEHAVIOR AND FATE OF NANOMATERIALS IN THE GI TRACT

Whether it exists as nanostructure food ingredient, nanosized particles, or nanocarriers incorporated in food packaging, human exposure to nanomaterials present in the food or food contact materials occurs during ingestion. The entire cascade of events together with absorption, distribution, metabolism, and excretion/elimination occur following ingestion and determine the internal exposure and toxicity of these substances. Nevertheless, due to the interactions of nanomaterials with the surrounding matrix and unexpected effects resulting from this, little is known regarding the behavior and fate of nanomaterials in the GI tract (EFSA 2009).

13.4.1 GI Absorption of NPs

NPs can be distributed to the same organ by several routes of exposure. For example, NPs found in the GI tract can originate from ingested products. Alternatively, they can also reach the GI tract via an indirect route. For example, particles can be absorbed in systemic circulation via the respiratory tract or skin. From there, particles can be distributed to the liver, taken up by hepatocytes, and excreted in bile to the GI tract.

In this section, the absorption of NPs into the human body via different exposure pathways such as inhalation, oral, skin, and parenteral is addressed. In pharmacokinetics, absorption represents the process by which unchanged compounds (e.g., NPs) proceed from the site of administration to the central blood circulation (site of measurement) (Werner et al. 2007).

The GI tract may represent an important port of entry for NPs since food products may eventually contain NPs (Lomer et al. 2002; Maynard and Michelson 2005). In addition, inhaled particles can be excreted via the mucociliary escalator and can be subsequently ingested into the GI tract. The contribution of the oral exposure following inhalation depends on the particles' physicochemical characteristics and size. The computer-based multiple-path particle dosimetry model (Price et al. 2002) can predict the amount of poorly soluble, nontoxic, solid NPs (10 nm and larger) transported from the lungs to the GI tract in rats. A validated extrapolation to the human situation is not possible yet.

Micronized particles can enter the body by a process called persorption, the paracellular uptake of particles from the GI tract into lymphatic and blood circulation (Volkheimer et al. 1968; Volkheimer 1974). Several studies in rats have shown that nano- and microsized particles (50 nm–20 μm) are mainly absorbed through Peyer's patches of the small intestine, while Peyer's patches only comprise a small percentage of the total surface of the small intestine. In addition, absorption via intestinal enterocytes has been demonstrated (Carr et al. 1996; Florence 2005; Hillery et al. 1994; Jani et al. 1990). Charge appears to be an important determinant of the extent of absorption. Positively charged particles seem to be absorbed more effectively through the GI tract than neutral and negatively charged particles (Florence 1997, 2005; Hussain et al. 2001; Janes et al. 2001). Additionally, the size of the particles determines the extent of absorption. Polystyrene NPs of 50 and 100 nm were found to be absorbed to the extent of 34% and 26%, respectively (Jani et al. 1990). The efficiency in absorption of 100 nm polystyrene particles (4 mg/mL) was found to be up to 250-fold higher compared to larger-sized (500 nm, 1 and 10 μm) polystyrene microparticles (4 mg/mL) (Desai et al. 1996).

Nevertheless, no data are available that link the GI absorption of polystyrene NPs to negative effects in rats. In addition, the variety of NPs tested with different physicochemical properties, the heterogeneity in experimental protocols, and the higher amount of M-cells in Peyer's patches of rodents compared to humans shows the need for proper transport studies in humans (Des Rieux et al. 2006). Unfortunately, clinical transport studies with NPs are currently still missing.

Earlier reports by Jani et al. indicated that orally administered NPs can be absorbed across the GI tract via the lymph nodes to the liver and spleen (Jani et al. 1990). Reports by Yoshifumi showed that NP substances are easily taken up by reticuloendothelial cells during drug transfer. The uptake of these variably sized particles can lead to different toxicological effects (Yoshifumi 2002). Studies on polystyrene latex NPs in the range of 50 nm–3 μm have shown that maximal absorption can occur with particles sized approximately 50–100 nm in diameter (Hussain et al. 2001). On the other hand, further studies by Hussain et al. found that even latex particles above 1 mm can be retained in

Peyer's patches. The ingestion of ultrafine particles by the GI tract can stimulate phagocytosis at the GI mucosa and cause antigen–antibody-mediated reactions and inflammatory responses and, from there onward, systematic transportation to other organs of the body (Hussain et al. 2001).

Nanomaterials present in food may be readily absorbed from the GI tract. Nanomaterial translocation takes place through the epithelium of the intestinal wall depending on the physiochemical properties, for example, size, surface charge, presence/absence of a ligand, lipophilicity/hydrophilicity, and physiology of the intestinal tract (Des Rieux et al. 2006). Oral administration of gold NPs to mice showed that the GI uptake of gold NPs increased with diminishing size (Hillyer and Albrecht 2001) and smaller particles are absorbed more readily and faster than larger particles (Szentkuri 1997).

On the other hand, it is also possible that the ingested nanomaterials may not remain in a free form in the lumen due to transformations such as agglomeration, adsorption, aggregation, or binding with other food components and, hence, are not readily available for translocation through the intestinal wall. Currently, only limited information is available on the absorption of nanomaterials after ingestion (FAO and WHO 2009). The translocation of nanosized particles potentially used as food components through the GI tract remains to be explored (EFSA 2009).

13.4.2 Distribution

Ingested nanomaterials can enter the capillaries, which will carry nanomaterials through the portal circulation to the liver, or nanomaterials enter the lymphatic system by way of the thoracic ducts upon contact with the intestinal submucosal tissue. Data obtained from experiments demonstrated that the distribution of NPs after oral administration is dependent on particle size. NPs with a smaller size have a more widespread tissue distribution to organs such as the brain, liver, lungs, and kidneys, while bigger particles (28 and 58 nm) remain almost solely inside the GI tract (Hillyer and Albrecht 2001). Studies have been carried out on the ability of NPs to penetrate the placental barrier. Moreover, certain nanomaterials (C_{60} fullerene) can pass across the placenta (Tsuchiya et al. 1996). Nevertheless, due to the conflicting results of some *in vitro* (Myllynen et al. 2008) and animal studies (Tsuchiya et al. 1996), no general conclusions on the penetrating power of NPs across the placental barrier can be made. No information is available on whether nanomaterials are transferred into milk (EFSA 2009).

13.4.3 Excretion/Elimination

There is very limited information on the excretion of absorbed nanomaterials. In animal studies, feeding rats with radioactive iridium NPs (192Ir) showed that the ingested NPs were not substantially taken up through the GI tract and were rapidly eliminated via feces within 2–3 days. No major NP translocations from the GI tract to other organs through the blood were observed (Kreyling et al. 2002). A positive surface charge was also found to increase both urinary and fecal excretion (Balogh et al. 2007).

NPs may primarily target respiratory organs; however, on the other hand, other organs, for example, the GI tract, also need to be considered, because NPs could get into the GI tract by many ways, such as indirectly via mucociliary movement, directly via the oral intake of water, cosmetics, food, drugs, and nanoscale drug-delivery systems (Meng et al. 2007).

13.5 TOXICITY OF NANOMATERIALS ON GI TRACT

13.5.1 Toxicity of TiO_2 NPs

Koeneman et al. investigated the possible pathways by which TiO_2 NPs could cross the epithelium layer by employing both toxicity and mechanistic studies. Microvillar organization was investigated

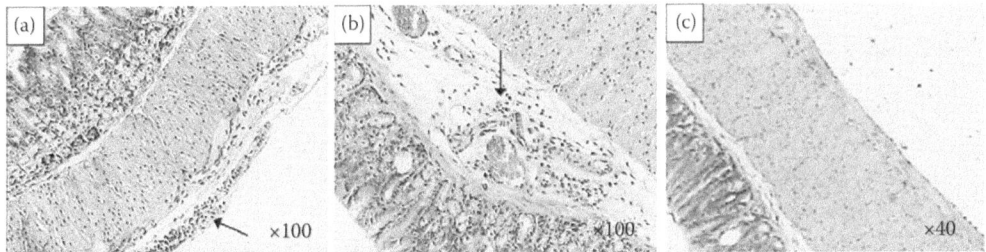

FIGURE 13.5 Histopathology of the stomach tissue (100× for A and B; 40× for C) in female mice 2 weeks postexposure to different sized TiO$_2$ particles by a single oral administration of (a) control group (only exposure to 0.5% HPMC), (b) 80 nm group, and (c) fine group. Arrows indicate the inflammation cells. (Reproduced with permission from Wang, J. et al. 2007b. *Toxicol Lett* 168:176–185.)

as a central structural change, as the microvilli present on the apical surface of the epithelial sheets are seminal to the cellular function of these intestinal cells. In addition, cytoplasmic signal transduction underlies all changes that cells make over the short term to environmental stimuli, and calcium signaling is a major form of signal transduction. They found that at 10 μg/mL and above, TiO$_2$ NPs cross the epithelial lining of the intestinal model by transcytosis, albeit at low levels (Koeneman et al. 2010).

A previous investigation reports that TiO$_2$ accumulates in the intestine of fish (Zhang et al. 2007) and rats (Jani et al. 1994) where TiO$_2$ is translocated to systemic organs throughout the body.

Wang et al. evaluated the toxicity of fine and nanoparticulate TiO$_2$: The acute toxicity of nanosized TiO$_2$ particles (25 and 80 nm) on adult mice and compared it with fine TiO$_2$ particles (155 nm). The histological photomicrographs of the stomach sections are shown in Figure 13.5a–c. As depicted from the photomicrograph, there were some inflammatory cells in the chorion layer of the stomach in mice of the 80 nm group (Figure 13.5), which was ascribed to the overload of particles in the stomach after a single oral administration of TiO$_2$ particles. In one of the control mice, inflammatory cells in the mucosal layer of the stomach were also observed (Figure 13.5a), but it was not representative. This effect may have been induced by the immune self-deficiency of this mouse (Wang et al. 2007b).

13.5.2 Toxicity of Nanoscale Zinc Powder

Wang et al. evaluated the acute toxicity in the oral exposure to nanoscale zinc powder in mice. The adult male and female healthy mice were gastrointestinally administered a dose of 5 g/kg body weight with two particle sizes, nanoscale zinc (N-Zn) and microscale zinc (M-Zn) powder; one group of mice was used as the control and treated with sodium carboxymethyl cellulose. The symptoms and mortality after zinc powder treatment were recorded. The effects of different sized particles on the blood element, blood coagulation, and serum biochemical levels were studied after 2 weeks of administration. Organs were collected for the histopathological examination. The N-Zn-treated mice showed more severe symptoms of vomiting, diarrhea, and lethargy in the beginning days than the M-Zn mice. Furthermore, during the initial 3 days, a 22% reduction in the weight gain of mice exposed to NPs was observed when compared to the control group. The deaths of two mice occurred in the N-Zn group after the first week of treatment. The mortalities after zinc powder treatment were confirmed by the intestinal obstruction of N-Zn aggregations. The histopathological examination found slight stomach and intestinal inflammation in almost all the nano and micro Zn-administered mice (Wang et al. 2006).

When mice were orally administered with 20 and 120 nm zinc oxide nanoparticles (ZnO nanoparticles) at different doses, it was found that the damaged target organs showed different dose response relationships. The 120 nm ZnO-treated mice had a positive dose-dependent pathological damage in the stomach (inflammation in gastric lamina propria, submucosa, and serosa layer), whereas 20 nm ZnO displayed a negative dose-dependent damage in the stomach (Wang et al. 2008a).

13.5.3 Toxicity of Copper Nanoparticles (Copper NPs)

Chen et al. investigated the oral toxicity of several types of NPs with GI tract exposure. Studies have shown that copper NPs can cytotoxically trigger injuries on the lymph, Payer's patches, liver, spleen, and kidney of experimental animals. In studies by Chen et al. after gavaging mice with copper NPs, they revealed that GI tract toxicity belongs to the Hodge and Sterner scale (three classes of moderate toxicity). Further studies revealed that the toxicity of nanosized copper particles was highly correlated with the particle size and specific surface area. Compared to microcopper (17 μm), nanocopper (23.5 nm) can rapidly interact with artificial gastric acid juice and can be transformed into ionic copper with an ultrahigh reactivity. They compared the toxicity of nanocopper with microcopper in mice. A few microcopper particle-treated mice exhibited symptoms of poisoning. However, all nanocopper-treated mice appeared to exhibit symptoms of alimentary canal disorder, such as loss of appetite, vomiting, diarrhea, and so on. The LD50 for the nano- and microcopper particles and cupric ions exposed to mice via oral gavage were 413, >5000, and 110 mg/kg body weight, respectively. The toxicity class for both nano and ionic copper particles was class 3 (moderately toxic) and for microcopper was class 5 (practically nontoxic) from the Hodge and Sterner scale. They also noticed tremors or hypopnea and arching of the back in some mice that received NPs. Parameters such as blood urea nitrogen, creatinine, total bile acid, and ALP (alkaline phosphatase) were significantly higher than in the controls. Results indicate a gender-dependent feature of nanotoxicity. Moreover, nanotoxicity depends on several factors, such as a huge specific surface area, ultrahigh reactivity, and so on (Chen et al. 2006).

It has been suggested that once inside an organism's stomach, nanocopper particles could react with protons (H^+) from gastric juice and become quickly ionized, resulting in an overload of ionic copper. In this case, the depletion of H^+ would then lead to a massive formation of HCO_3^- and the induction of metabolic alkalosis (Meng et al. 2007). Nano- and microcopper exhibit different biological behaviors *in vivo* via the oral exposure route. In terms of nanocopper particles, both copper overload and metabolic alkalosis contribute to their grave toxicity. Dissimilarly, microcopper does not stagnate in the stomach, and the velocity of ionization is much slower than with NPs. After the particles propelled into the small intestine by gastric emptying, the ionization reaction is prohibited because of a basic condition and is excreted as feces. For the direct intake of copper ions, transitory glomerulonephritis and alimentary canal disorders happen in experimental animals. These toxicological responses can be partially corrected within 72 h.

The toxicity of copper particles is highly correlated to particle size and specific surface areas because ultrahigh reactivity provokes nanocopper's *in vivo* toxicity. Nanocopper NPs may not compromise the mice directly; nevertheless, they lead to the accumulation of excessive alkalescent substances and heavy metal ions (copper ions), culminating in metabolic alkalosis and copper ion overloads. When nanocopper reacts with acid substances in the stomach, a great amount of proton ions is consumed. Metabolic alkalosis, as well as the poisonous copper ions, cause higher mortality than microcopper of the same dose (Meng et al. 2007).

13.5.4 Toxicity of Single-Walled Carbon Nanotubes

Smith et al. studied the toxicity of single-walled carbon nanotubes (SWCNTs) to rainbow trout. SWCNT exposure caused a dose-dependent rise in the rate of ventilation, gill pathologies (altered mucocytes, edema, and hyperplasia), and mucus secretion, along with SWCNT precipitation on the gill mucus. SWCNT exposures caused a statistically significant increase in Na^+ K^+ adenosine triphosphatase (ATP) activity in the gills and intestine but not in the brain (Smith et al. 2007).

Toxicity of Nanomaterials on the Gastrointestinal Tract

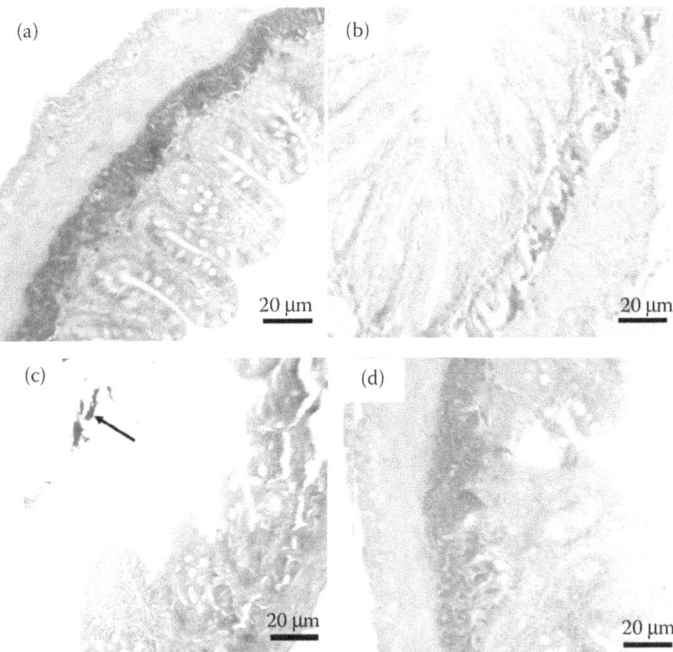

FIGURE 13.6 Histology of the intestine at the end of the experiment. (a) Fresh water control showing the normal intestine, (b) solvent control, (c) intestine from a fish in 0.1 mg L^{-1} SWCNT showing erosion of the epithelium and precipitated SWCNT in the gut lumen (arrow), and (d) intestine from a fish in 0.5 mg L^{-1} SWCNT showing fusion of the intestinal villi. Scale bar = 20 μm, sections were 8 μm thick and stained with Mallory's trichrome. (Reproduced with permission from Smith, C.J., Shaw, B.J. and Handy, R.D. 2007. *Aquat Toxicol* 82:94–109.)

Mammalian studies have raised concerns about the toxicity of CNTs; however, there are very limited data on the ecotoxicity to aquatic life. Stock solutions of dispersed SWCNTs were prepared using a combination of solvents (sodium dodecyl sulfate) and sonification. A semistatic test system was used to expose rainbow trout to a solvent control, freshwater control, 0.1, 0.25, or 0.5 mg L^{-1} SWCNTs for up to 10 days. Fish ingested water containing SWCNTs during exposure (presumably stress-induced drinking) that resulted in precipitated SWCNTs in the gut lumen and intestinal pathology (Smith et al. 2007).

The histology of the intestine is shown in Figure 13.6. Gross observations during dissection showed clear black deposits in the gut lumen, which indicated that the fish had been drinking the SWCNT-contaminated water. Subsequent histological examinations of the intestine showed no effects from the control solvent, but demonstrated some clear intestinal pathology associated with SWCNT exposure (Figure 13.6). All fishes from the 0.1 mg L^{-1} SWCNT treatment showed some areas of intestinal villi fusion, areas of inflammation and erosion, or total atrophy of the mucosa. Precipitated SWCNT could be seen in the gut lumen (Figure 13.6c, arrow). The injuries were observed to a lesser extent at 0.5 mg L^{-1} (in four out of six of the fishes examined) and 1.0 mg L^{-1} SWCNT (half of the fishes examined). There was no evidence of major bleeding from the blood vessels of the submucosa, but the tissue layer with the associated nerve plexus appeared more granular in the SWCNT-treated specimens than the controls (Smith et al. 2007).

13.5.5 Toxicity of Metal NPs

The zebrafish (*Danio rerio*) has become an important model species for the study of microbial communities in vertebrate intestines (Rawls et al. 2004, 2007), and this model has also been useful for assessing the toxicity of NPs (Henry et al. 2007; Park et al. 2010, 2011). Therefore, the primary objective of the study was to make an initial assessment on the effects of dietary copper nanoparticles or silver nanoparticles (Cu-NPs or Ag-NPs) on gut microbial communities in adult zebrafish. In addition, to evaluate gut microbiota, the presence of lesions in the intestinal epithelial mucosa after exposure to NPs was assessed by electron microscopy. While the uses of Cu-NPs are emerging, they do not currently present substantial exposure risks to organisms; on the other hand, Cu is an important environmental toxicant and the selection of Cu-NPs for this study was based on the interest to compare results with Ag-NPs, evaluate differences between nano- and bulk forms of Cu, and to enable comparisons with the previous studies of Cu-NPs (Griffitt et al. 2007; Shaw et al. 2012).

NPs can be ingested by organisms, and those with antimicrobial properties may disrupt beneficial endogenous microbial communities and affect the health of the organism. Zebrafish was fed with diets containing Cu-NPs, Ag-NPs (500 mg/kgfood), or an appropriate control for 14 days. By transmission electron microscopy (TEM), the intestinal epithelium integrity was examined and the structure of the microbial community within the intestine was assessed by the denaturing gradient gel electrophoresis (DGGE) of partial 16S rRNA. No lesions were observed in the intestinal epithelia; however, the presence of NPs in diets changed the structure of the intestinal microbial community. Some beneficial bacterial strains (e.g., *Cetobacterium somerae*) were suppressed to nondetectable levels by Cu-NPs exposure, and two unidentified bacterial clones from the *Firmicutes phylum* were sensitive to Cu, but were present in Ag-treated and control fish. In zebrafish, unique microbiome changes caused by the exposure to Ag- and Cu-NPs indicate that the ingestion of NPs could affect the function of the digestive system and the organism's health (Daniel et al. 2013).

13.5.6 Toxicity of Cadmium–Selenium Quantum Dot

Wang et al. evaluated the possible toxicity due to the exposure of cadmium–selenium quantum dots (CdSe QDs) via ingestion on enterocyte-like Caco-2 cells as a small intestine epithelial model. Cells were incubated in Cd^{+2} (2–200 nmol/mL) containing a medium for 24 h. A Cd^{+2} concentration of 200 nmol/mL resulted in a drop of the relative viability of Caco-2 to 0.62, which is considerably lower than the control. However, cytotoxicity largely depended on the coating and treatment of QDs (e.g., acid treatment and dialysis). This concluded that Caco-2 cell viability correlated with the concentration of free Cd^{+2} ions present in cell culture medium. CdSe QDs exposure to low gastric pH affected cytotoxicity, indicating that the route of exposure may be an important factor in QDs cytotoxicity (Wang et al. 2008b).

13.5.7 Toxicity of Si and SiO Particles

The GI route represents the most common and patient-compliant route of drug administration. However, some drugs have poor pharmacokinetic profiles when administered orally (Bimbo et al. 2011). The major barriers in oral delivery are the enzymes in the upper parts of the GI system, the pH changes in the GI tract, the absorption and efficient permeability across the intestinal wall (Ramesan and Sharma 2009), the presence of bile salts (Qian and Bogner 2012), and liver-mediated, first-pass metabolism (Bimbo et al. 2011). pH levels change throughout the GI tract: The pH in the stomach is 1–3, while in the small intestine, the pH is 6.5–7.0 and in the colon, it is 7.0–8.0 (Cheng et al. 2011). Drug absorption occurs through paracellular transport along the epithelial lumen of the small intestine. However, the tight junctions among adjacent enterocytes form a barrier for drugs (Foraker et al. 2003). The presence of bile salts can affect the structure of the drug molecule and, therefore, its function. The absorbed drugs then pass through entero-hepatic circulation and the active agent is cleared by the "first-pass metabolism" in the liver before reaching systemic circulation and the intended site.

In this complicated environment, Si and SiO particulates can have clear advantages for overcoming the various hurdles of oral delivery and protecting the encapsulated molecule. Si- and SiO-based materials are hydrophilic, which increase the wettability of water-insoluble drugs in the GI tract (Tan et al. 2010). Moreover, Si and SiO particles are low-pH resistant; this is a clear rationale for using them as a drug-delivery system for orally administered drugs to protect them from the low pH in the stomach.

In one *in vitro* study, the toxicity induced in human esophageal epithelial cells (NE083) was studied with crystalline and amorphous SiO NPs. The crystalline SiO NPs showed a dose dependency on Caco-2 cell viability. Compared to crystalline SiO, amorphous SiO NPs were less toxic at doses ranging from 0.156 to 10 μg/mL. TEM analyses have shown that the morphology of esophageal epithelial cells did not change following the uptake of amorphous SiO NPs. For crystalline SiO NPs, however, it was observed that the organelle membrane ruptured and there was direct contact between the NP and cytoplasm, which may lead to direct chemical exchanges and toxicities in esophageal epithelial cells (Chu et al. 2011).

Consistent with the idea that toxicity is dependent on surface area, Bimbo et al. observed that porous silicon (pSi) microparticles (10–25 μm) induced more toxicity by decreasing cell viability than small, pSi particles (97, 126, and 164 nm) in human colon carcinoma cells (CaCo-2). Unlike nonporous SiO particles, porous silica (pSiO) exhibit large surface areas that may cause toxicity in cells. In nontoxic concentrations, microparticles were not internalized by the CaCo-2 monolayers but were in close proximity to cells (Bimbo et al. 2011). At higher concentrations (2–14 mg/mL) of mesoporous pSiO microparticles, Caco-2 cell membrane integrity weakened along with a diminished cell metabolism and increased apoptotic signalings (Heikkila et al. 2007). Smaller porous particles (50 nm) also exhibited insignificant toxicity after treating at various concentrations from 1 to 500 μg/mL in human colon cancer cell line (HT-29) (Cheng et al. 2011). It was observed that cell viability was particle size dependent, while the production of intracellular ROS was particle concentration dependent.

Heikkila et al. carried out a cytotoxicity study of ordered mesoporous silica MCM-41 and SBA-15 microparticles on Caco-2 cells. Cytotoxicity of ordered mesoporous silica MCM-41 and SBA-15 microparticles (fractions between 1 and 160 μm) was determined *in vitro* on an undifferentiated human colon carcinoma, that is, Caco-2 cell line, taking into account the feasibility of using these silica-based materials in oral drug formulations. For assessing the effects of the MCM-41 and SBA-15 microparticles on Caco-2, the cellular endpoints employed were (1) cell membrane integrity, by monitoring live-cell protease activity (AFC) and by employing the flow cytometry method; (2) metabolic activity, by monitoring total ATP content via luminescence assay; and (3) activity of apoptotic effectors, by caspase-3/7 activity assay. The generation of ROS, specifically, the hydrogen peroxide (H_2O_2) and the superoxide radical (O_2^-); was also followed. MCM-41 and SBA-15 microparticles caused cytotoxic effects on Caco-2 cells at most of the tested concentrations (0.2–14 mg/mL) and incubation times (3 and 24 h). The effects on the cells included attenuated cell membrane integrity, reduced cell metabolism, and increased apoptotic signaling. The major cause for cytotoxicity was the heightened production of ROS, especially, the generation of the superoxide radical (O_2^-), after 3 h of incubation with the threshold dose of 1 mg/mL, evidently overwhelming the antioxidant defenses and causing mitochondrial dysfunction, therefore increasing apoptotic signaling (Heikkilä et al. 2010).

Jaganathan et al. assessed the biocompatibility of Si-based nano- and microparticles (Jaganathan and Godin 2012). Nonporous SiONPs, pSi, and pSiO were investigated as oral drug carriers due to their large pore volumes and, thus, the ability to load drugs in the pores. Possessing a high surface free energy due to their large surface area, drug molecules can be absorbed into the pores to reach a low state of free energy. The stability of pSi particles is important in the GI tract. It was shown that bare pSi exhibited high surface oxidation after 18 h of incubation in simulated intestinal fluid (Albrecht et al. 2009). After functionalizations with alkyl groups, Albrecht et al. demonstrated that the porous particles had high resistance to oxidation in the gastric and intestinal fluids (Albrecht et al. 2009). Surface oxidation and other chemical/physical exchanges between the particle and GI environment can produce unpredictable toxicities by ROS.

13.5.8 TOXICITY OF CHITOSAN NPS

Nevertheless, while the chitosan bulk polymer is biocompatible, the polymer when presented as NPs may not be quite as innocuous. Using lung and intestinal epithelial cell models, different mechanisms of the cellular uptake and distribution of dissolved or nanoparticulate chitosan were established (Huang et al. 2004; Loh et al. 2010; Ma and Lim 2003).

13.5.9 TOXICITY OF GOLD NPS

Pompa et al. investigated the effects of citrate-capped gold nanoparticles (Au NPs) upon ingestion by the model system, *Drosophila melanogaster*. The significant *in vivo* toxicity of Au NPs was observed, which elicited clear adverse effects in treated organisms, such as strong diminution of their life span and fertility, DNA fragmentation, as well as a substantial overexpression of the stress proteins.

The GI tissue was further investigated to assess any occurrences of DNA damage in 15 nm Au NP-treated individuals (100 pmol/L). For this purpose, a specific fluorescence kit based on terminal deoxynucleotidyl transferase (TdT)-mediated terminal transferase dUTP nick-end-labeling (TUNEL) was used. Importantly, among the numerous nuclei displaying no DNA damage, few cells were observed in which DNA fragmentation was evident (Figure 13.7). Such DNA damage was found to be distributed throughout the whole GI tissue with no apparent localization into specific regions.

Several midgut samples were analyzed with this technique and the observation of DNA nicks in enterocytes was highly reproducible. On the contrary, in supernatant (SN)-treated samples as well as in control flies, weak fluorescent signals related to DNA strand nicks were typically detected. A quantitative analysis of the TUNEL assay revealed an occurrence of DNA damage of around 8% (8.15% ± 2.46%) in Au NP-treated flies, while in the control and SN-treated samples, DNA fragmentation was <1% (0.84% ± 0.43%) (student's t-test $p < 0.001$). This experimental evidence indicates a clear toxic effect of Au NPs on GI tissue. The localization of the 15 nm Au NPs in the enterocytes, evidenced by TEM analyses (Figure 13.7), strongly suggests an indirect effect of the nanomaterials in causing DNA damage. Further analyses are required to understand this point better, but the observed fragmentation is very likely to be mediated by oxidative stress and/or related to early-stage apoptosis (Pompa et al. 2011).

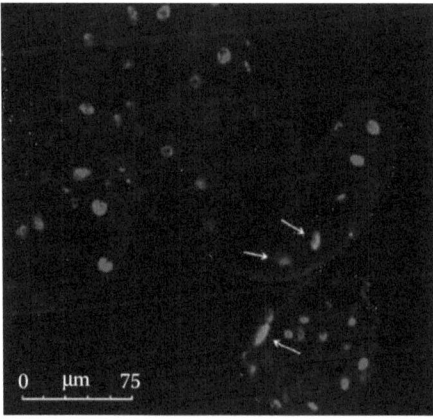

FIGURE 13.7 (See color insert.) Representative confocal microscopy image of *Drosophila* midgut in flies treated with 15 nm Au NPs (100 pmol/L). Nuclei are stained with Hoechst 33 342 (blue) while cells containing DNA strand nicks are detected by TUNEL assay and fluorescent red (highlighted by the white arrows).

REFERENCES

Albrecht, D.S., Lee, J.T., Molby, N. et al. 2009. Functionalized porous silicon in a simulated gastrointestinal tract: Modeling the biocompability of a monolayer protected nanostructured material. In *Materials Research Society Symposia Proceedings*, Zaragoza, Spain, vol. 1063, pp. 39–44.

Atuma, C., Strugala, V., Allen, A. and Holm, L. 2001. The adherent gastrointestinal mucus gel layer: Thickness and physical state *in vivo*. *Am J Physiol Gastrointest Liver Physiol* 280:G922–G929.

Avdeef, A. and Testa, B. 2002. Physicochemical profiling in drug research: A brief survey of the state-of-the-art of experimental techniques. *Cell Mol Life Sci* 59:1681–1689.

Ballestri, M., Baraldi, A., Gatti, A.M. et al. 2001. Liver and kidney foreign bodies granulomatosis in a patient with malocclusion, bruxism, and worn dental prostheses. *Gastroenterology* 121:1234–1238.

Balogh, L., Nigavekar, S.S., Nair, B.M. et al. 2007. Significant effect of size on the *in vivo* biodistribution of gold composite nanodevices in mouse tumor models. *Nanomedicine* 3(4):281–296.

Becker, S., Mundandhara, S., Devlin, R.B. and Madden, M. 2005. Regulation of cytokine production in human alveolar macrophages and airway epithelial cells in response to ambient air pollution particles: Further mechanistic studies. *Toxicol Appl Pharmacol* 207:269–275.

Bhaskar, K.R., Gong, D.H., Bansil, R. et al. 1991. Profound increase in viscosity and aggregation of pig gastric mucin at low pH. *Am J Physiol* 261:G827–G832.

Bimbo, L.M., Mäkilä, E., Laaksonen, T. et al. 2011. Drug permeation across intestinal epithelial cells using porous silicon nanoparticles. *Biomaterials* 32:2625–2633.

Borm, P.J., Robbins, D., Haubold, S. et al. 2006. The potential risks of nanomaterials: A review carried out for ECETOC. *Part Fibre Toxicol* 3:11.

Bykov, V.L. 1996. The tissue and cell defense mechanisms of the oral mucosa. *Morfologiia* 110:14–24.

Bykov, V.L. 1997. The functional morphology of the epithelial barrier of the oral mucosa. *Stomatologiia* (Mosk) 76:12–17.

Carr, K.E., Hazzard, R.A., Reid, S. and Hodges, G.M. 1996. The effect of size on uptake of orally administered latex microparticles in the small intestine and transport to mesenteric lymph nodes. *Pharm Res* 13(8):1205–1209.

Carrero-Sanchez, J.C., Elias, A.L., Mancilla, R. et al. 2006. Biocompatibility and toxicological studies of carbon nanotubes doped with nitrogen. *Nano Lett* 6:1609–1616.

Chen, E.Y., Garnica, M., Wang, Y.C., Chen, C.S. and Chin, W.C. 2011. Mucin secretion induced by titanium dioxide nanoparticles. *PLoS One* 6:e16198.

Cheng, S.-H., Liao, W.-N., Chen, L.-M. and Lee, C.-H. 2011. pH-controllable release using functionalized mesoporous silica nanoparticles as an oral drug delivery system. *J Mater Chem* 21:7130–7137.

Chen, Z., Meng, H., Xing, G.M. et al. 2006. Acute toxicological effect of copper nanoparticle *in vivo*. *Toxicol Lett* 163:109–120.

Chen, E.Y., Wang, Y.C., Chen, C.S. and Chin, W.C. 2010. Functionalized positive nanoparticles reduce mucin swelling and dispersion. *PLoS One* 5:e15434.

Chu, Z., Huang, Y., Tao, Q. and Li, Q. 2011. Cellular uptake, evolution, and excretion of silica nanoparticles in human cells. *Nanoscale* 3:3291–3299.

Chung, T.H., Wu, S.H., Yao, M. et al. 2007. The effect of surface charge on the uptake and biological function of mesoporous silica nanoparticles in 3T3-L1 cells and human mesenchymal stem cells. *Biomaterials* 28:2959–2966.

Clift, M.J., Rothen-Rutishauser, B., Brown, D.M. et al. 2008. The impact of different nanoparticle surface chemistry and size on uptake and toxicity in a murine macrophage cell line. *Toxicol Appl Pharmacol* 232:418–427.

Collins, L.M. and Dawes, C. 1987. The surface area of the adult human mouth and thickness of the salivary film covering the teeth and oral mucosa. *J Dent Res* 66:1300–1302.

Cone, R.A. 2009. Barrier properties of mucus. *Adv Drug Deliv Rev* 61:75–85.

Corfield, A.P., Wagner, S.A., Clamp, J.R., Kriaris, M.S. and Hoskins, L.C. 1992. Mucin degradation in the human colon: Production of sialidase, sialate O-acetylesterase, N-acetylneuraminate lyase, arylesterase, and glycosulfatase activities by strains of fecal bacteria. *Infect Immun* 60:3971–3978.

Daniel, L.M., Benjamin, J.S., Glenn, M.H. et al. 2013. Ingestion of metal–nanoparticle contaminated food disrupts endogenous microbiota in zebrafish (*Danio rerio*). *Environ Pollut* 174:157e163.

Dawson, M., Wirtz, D. and Hanes, J. 2003. Enhanced viscoelasticity of human cystic fibrotic sputum correlates with increasing microheterogeneity in particle transport. *J Biol Chem* 278:50393–50401.

Des Rieux, A., Fievez, V., Garinot, M., Schneider, Y.J. and Preat, V. 2006. Nanoparticles as potential oral delivery systems of proteins and vaccines: A mechanistic approach. *J Control Release* 1:1–27.

Desai, M.P., Labhasetwar, V., Amidon, G.L. and Levy, R.J. 1996. Gastrointestinal uptake of biodegradable microparticles: Effect of particle size. *Pharm Res* 13:1838–1845.

Diaz del Consuelo, I., Falson, F., Guy, R.H. and Jacques, Y. 2005. Transport of fentanyl through pig buccal and esophageal epithelia *in vitro*. Influence of concentration and vehicle pH. *Pharm Res* 22: 1525–1529.

Doherty, G.J. and McMahon, H.T. 2009. Mechanisms of endocytosis. *Ann Rev Biochem* 78:857–902.

Donaldson, K., Stone, V., Tran, C.L., Kreyling, W. and Borm, P.J. 2004. Nanotoxicology. *Occup Environ Med* 61:727–728.

Eom, H.J. and Choi, J. 2009. Oxidative stress of silica nanoparticles in human bronchial epithelial cell, Beas-2B. *Toxicol in Vitro* 23:1326–1332.

European Food Safety Authority (EFSA). 2009. Scientific opinion of the scientific committee on the potential risks arising from nanoscience and nanotechnologies on food and feed safety. http://www.efsa.europa.eu/EFSA/Scientific_Opinion/sc_op_ej958_nano_en,0.pdf?ssbinary=true.

Florence, A.T. 1997. The oral absorption of micro- and nanoparticulates: Neither exceptional nor unusual. *Pharm Res* 14(3):259–266.

Florence, A.T. 2005. Nanoparticle uptake by the oral route: Fulfilling its potential? *Drug Discov Today Technol* 2 (1):75–81.

Food and Agriculture Organization of the United Nations (FAO) and World Health Organization (WHO). 2009. Report of joint FAO/WHO expert meeting on the application of nanotechnologies in the food and agriculture sectors: Potential food safety implications. http://www.fao.org/ag/agn/agns/files/FAO_WHO_Nano_Expert_Meeting_Report_Final.pdf

Foraker, A.B., Walczak, R.J., Cohen, M.H., Boiarski, T.A., Grove, C.F. and Swaan, P.W. 2003. Microfabricated porous silicon particles enhance paracellular delivery of insulin across intestinal Caco-2 cell monolayers. *Pharm Res* 20:110–116.

Gatti, A.M. 2004. Biocompatibility of micro- and nano-particles in the colon. Part II. *Biomaterials* 25:385–392.

Gaumet, M., Gurny, R. and Delie, F. 2009. Localization and quantification of biodegradable particles in an intestinal cell model: The influence of particle size. *Eur J Pharm Sci* 36:465–473.

Geiser, M., Rothen-Rutishauser, B., Kapp, N., Schurch, S., Kreyling, W. and Schulz, H. 2005. Ultrafine particles cross cellular membranes by nonphagocytic mechanisms in lungs and in cultured cells. *Environ Health Perspect* 113(11):1555–1560.

Gerbert, A., Rothkotter, H.J. and Pabst, R. 1996. M-cells in Peyer's patches of the intestine. *Int Rev Cytol* 167:91–159.

Greulich, C., Diendorf, J., Simon, T., Eggeler, G., Epple, M. and Koller, M. 2011. Uptake and intracellular distribution of silver nanoparticles in human mesenchymal stem cells. *Acta Biomater* 7:347–354.

Griffitt, R.J., Weil, R., Hyndman, K.A. et al. 2007. Exposure to copper nanoparticles causes gill injury and acute lethality in zebrafish (*Danio rerio*). *Environ Sci Technol* 41 (23):8178–8186.

Harris, D. and Robinson, J.R. 1992. Drug delivery via the mucous membranes of the oral cavity. *J Pharm Sci* 81:1–10.

Harush-Frenkel, O., Rozentur, E., Benita, S., and Altschuler, Y. 2008. Surface charge of nanoparticles determines their endocytic and transcytotic pathway in polarized MDCK cells. *Biomacromolecules* 9:435–443.

He, Q., Zhang, Z., Gao, Y., Shi, J. and Li, Y. 2009. Intracellular localization and cytotoxicity of spherical mesoporous silica nano- and microparticles. *Small* 5:2722–2729.

Heikkila, T., Salonen, J., Tuura, J. et al. 2007. Evaluation of mesoporous TCPSi, MCM-41, SBA-15, and TUD-1 materials as API carriers for oral drug delivery. *Drug Deliv* 14:337–347.

Heikkilä, T., Santos, H.A., Kumar, N. et al. 2010. Cytotoxicity study of ordered mesoporous silica MCM-41 and SBA-15 microparticles on Caco-2 cells. *Eur J Pharm Biopharm* 74:483–494.

Henry, T.B., Menn, F., Fleming, J.T., Wilgus, J., Compton, R.L. and Sayler, G.S. 2007. Attributing the toxicity of aqueous C_{60} nano-aggregates to tetrahydrofuran decomposition products in larval zebrafish by assessment of gene expression. *Environ Health Perspect* 115(7):1059–1065.

Hillery, A.M., Jani, P.U. and Florence, A.T. 1994. Comparative, quantitative study of lymphoid and non-lymphoid uptake of 60 nm polystyrene particles. *J Drug Target* 2(2):151–156.

Hillyer, J.F. and Albrecht, R.M. 2001. Gastrointestinal persorption and tissue distribution of differently sized colloidal gold nanoparticles. *J Pharm Sci* 90(12):1927–1936.

Hoet, P.H.M., Bruske-Hohlfeld, I. and Salata, O.V. 2004. Nanoparticles—Known and unknown health risks. *J Nanobiotechnol* 2:12–27.

Hoskins, L.C. and Boulding, E.T. 1981. Mucin degradation in human colon ecosystems. Evidence for the existence and role of bacterial subpopulations producing glycosidases as extracellular enzymes. *J Clin Invest* 67:163–172.

Huang, D.M., Hung, Y., Ko, B.S. et al. 2005. Highly efficient cellular labeling of mesoporous nanoparticles in human mesenchymal stem cells: Implication for stem cell tracking. *FASEB J* 19:2014–2016.

Huang, M., Khor, E. and Lim, L.Y. 2004. Uptake and cytotoxicity of chitosan molecules and nanoparticles: Effects of molecular weight and degree of deacetylation. *Pharm Res* 21:344–353.

Huh, C.H., Bhutani, M.S., Farfan, E.B. and Bolch, W.E. 2003. Individual variations in mucosa and total wall thickness in the stomach and rectum assessed via endoscopic ultrasound. *Physiol Meas* 24: N15–N22.

Hussain, N., Jaitley, V. and Florence, A.T. 2001. Recent advances in the understanding of uptake of microparticulate across the gastrointestinal lymphatics. *Adv Drug Deliv Rev* 50:107–142.

Jaganathan, H. and Godin, B. 2012. Biocompatibility assessment of Si-based nano- and micro-particles. *Adv Drug Deliv Rev* 64:1800–1819.

Janes, K.A., Fresneau, M.P., Marazuela, A., Fabra, A. and Alonso, M.J. 2001. Chitosan nanoparticles as delivery systems for doxorubicin. *J Control Release* 73 (2–3):255–267.

Jani, P., Halbert, G.W., Langridge, J. and Florence, A.T. 1990. Nanoparticle uptake by the rat gastrointestinal mucosa: Quantitation and particle size dependency. *J Pharm Pharmacol* 42 (12):821–826.

Jani, P.U., Mccarthy, D.E. and Florence, A.T. 1994. Titanium dioxide (rutile) particle uptake from the rat GI tract and translocation to systemic organs after oral-administration. *Int J Pharm* 105:157–168.

Jeong, G.N., Jo, U.B., Ryu, H.Y., Kim, Y.S., Song, K.S. and Yu, I.J. 2010. Histochemical study of intestinal mucins after administration of silver nanoparticles in Sprague–Dawley rats. *Arch Toxicol* 84:63–69.

Jiang, X., Rocker, C., Hafner, M., Brandholt, S., Dorlich, R.M. and Nienhaus, G.U. 2010. Endo- and exocytosis of zwitterionic quantum dot nanoparticles by live HeLa cells. *ACS Nano* 4:6787–6797.

Knowles, M.R. and Boucher, R.C. 2002. Mucus clearance as a primary innate defense mechanism for mammalian airways. *J Clin Invest* 109:571–577.

Koeneman, B.A., Zhang, Y., Westerhoff, P., Chen, Y., Crittenden, J.C. and Capco, D.G. 2010. Toxicity and cellular responses of intestinal cells exposed to titanium dioxide. *Cell Biol Toxicol* 26:225–238.

Kreyling, W.G., Semmler, M., Erbe, F. et al. 2002. Translocation of ultrafine insoluble iridium particles from lung epithelium to extrapulmonary organs is size dependent but very low. *J Toxicol Environ Health Part A* 65:1513–1530.

Lagerlof, F. and Dawes, C. 1984. The volume of saliva in the mouth before and after swallowing. *J Dent Res* 63:618–621.

Lai, S.K., O'Hanlon, D.E., Harrold, S. et al. 2007. Rapid transport of large polymeric nanoparticles in fresh undiluted human mucus. *Proc Nat Acad Sci USA* 104:1482–1487.

Lanone, S. and Boczkowski, J. 2006. Biomedical applications and potential health risks of nanomaterials: Molecular mechanisms. *Curr Mol Med* 6:651–663.

Loh, J.W., Yeoh, G., Saunders, M. and Lim, L.-Y. 2010. Uptake and cytotoxicity of chitosan nanoparticles in human liver cells. *Toxicol Appl Pharmacol* 249:148–157.

Lomer, M.C.E., Hutchinson, C., Volkert, S. et al. 2004. Dietary sources of inorganic microparticles and their intake in healthy subjects and patients with Crohn's disease. *Brit J Nutr* 92:947–955.

Lomer, M.C., Thompson, R.P. and Powell, J.J. 2002. Fine and ultrafine particles of the diet: Influence on the mucosal immune response and association with Crohn's disease. *Proc Nutr Soc* 61(1):123–130.

Long, J.D. and Orlando, R.C. 1999. Esophageal submucosal glands. Structure and function. *Am J Gastroenterol* 94:2818–2824.

Ma, Z. and Lim, L.Y. 2003. Uptake of chitosan and associated insulin in Caco-2 cell monolayers: A comparison between chitosan molecules and chitosan nanoparticles. *Pharm Res* 20:1812–1819.

Matsuo, K., Ota, H., Akamatsu, T., Sugiyama, A. and Katsuyama, T. 1997. Histochemistry of the surface mucous gel layer of the human colon. *Gut* 40:782–789.

Maynard, A.D. and Michelson, E. 2005. The nanotechnology consumer products inventory. *Woodrow Wilson Int Center Scholars* 1–8.

Meng, H., Chen, Z., Xing, G. et al. 2007. Ultrahigh reactivity provokes nanotoxicity: Explanation of oral toxicity of nano-copper particles. *Toxicol Lett* 175(1–3):102–110.

Myllynen, P.K., Loughran, M.J., Vyvyan, H.C., Sormunen, R., Walsh, A.A. and Vähäkangas, K.H. 2008. Kinetics of gold nanoparticles in the human placenta. *Reprod Toxicol* 26(2):130–137.

Nel, A., Xia, T., Mädler, L. and Li, N. 2006. Toxic potential of materials at the nanolevel. *Science* 311:622–627.

Oberdörster, E. 2004. Manufactured nanomaterials (fullerenes, C_{60}) induce oxidative stress in the brain of juvenile largemouth bass. *Environ Health Persp* 112:1058–1062.

Oberdorster, G. 2000. Toxicology of ultrafine particles: *In vivo* studies. *Philos Trans Roy Soc Lond Series* 358 (1775):2719–2739.

Oberdörster, G., Ferin, J. and Lehnert, B.E. 1994. Correlation between particle-size, *in-vivo* particle persistence, and lung injury. *Environ Health Perspect* 102:173–179.

Oberdorster, G., Oberdorster, E. and Oberdorster, J. 2005. Nanotoxicology: An emerging discipline evolving from studies of ultrafine particles. *Environ Health Perspect* 113:823–839.

Olmsted, S.S., Padgett, J.L., Yudin, A.I., Whaley, K.J., Moench, T.R. and Cone, R.A. 2001. Diffusion of macromolecules and virus-like particles in human cervical mucus. *Biophys J* 81:1930–1937.

Park, J.W., Henry, T.B., Ard, S., Menn, F.M., Compton, R.N. and Sayler, G.S. 2011. Mixtures of 17α-ethinylestradiol (EE2) and aqueous C_{60} aggregates decrease bioavailability of EE2 and change C_{60} aggregate characteristics. *Nanotoxicology* 5(3):406–416.

Park, J.W., Henry, T.B., Menn, F.M., Compton, R.N. and Sayler, G.S. 2010. No bioavailability of 17α-ethinylestradiol (EE2) when associated with nC_{60} aggregates during dietary exposure in adult male zebrafish *Danio rerio*. *Chemosphere* 81:1227–1232.

Paul, J.A.B., David, R., Stephan, H. et al. 2006. The potential risks of nanomaterials: A review carried out for ECETOC. *Part Fibre Toxicol* 3:11

Penn, A., Murphy, G., Barker, S., Henk, W. and Penn, L. 2005. Combustion-derived ultrafine particles transport organic toxicants to target respiratory cells. *Environ Health Perspect* 113:956–963.

Pietzonka, P., Rothen-Rutishauser, B., Langguth, P., Wunderli-Allenspach, H., Walter, E. and Merkle, H.P. 2002. Transfer of lipophilic markers from PLGA and polystyrene nanoparticles to Caco-2 monolayers mimics particle uptake. *Pharm Res* 19:595–601.

Pompa, P.P., Vecchio, G., Galeone, A. et al. 2011. *In vivo* toxicity assessment of gold nanoparticles in *Drosophila melanogaster*. *Nano Res* 4(4):405–413.

Powell, J.J., Ainley, C.C. and Harvey, R.S. 1996. Characterization of inorganic microparticles in pigment cells of human gut associated lymphoid tissue. *Gut* 38:390–395.

Price, O.T., Asgharian, B., Miller, F.J., Cassee, F.R. and de Winter-Sorkina, R. 2002. Multiple Path Particle Dosimetry Model (MPPD v1.0): A Model for Human and Rat Airway Particle Dosimetry (Report No. 650010030). [CD-ROM]. Bilthoven, The Netherlands: National Institute of Public Health and the Environment (RIVM).

Qian, K.K. and Bogner, R.H. 2012. Application of mesoporous silicon dioxide and silicate in oral amorphous drug delivery systems. *J Pharm Sci* 101:444–463.

Rae, C.S., Khor, I.W. and Wang, Q. 2005. Systemic trafficking of plant virus nanoparticles in mice via the oral route. *Virology* 343(2):224–235.

Ramesan, R.M. and Sharma, C.P. 2009. Challenges and advances in nanoparticle-based oral insulin delivery. *Expert Rev Med Dev* 6.6:665.

Rawls, J.F., Mahowald, M.A., Goodman, A.L., Trent, C.M. and Gordon, J.I. 2007. *In vivo* imaging and genetic analysis link bacterial motility and symbiosis in the zebrafish gut. *Proc Nat Acad Sci USA* 104(18):7622–7627.

Rawls, J.F., Samuel, B.S. and Gordon, J.I. 2004. Gnotobiotic zebrafish reveal evolutionarily conserved responses to the gut microbiota. *Proc Nat Acad Sci USA* 101(13):4596–4601.

Sahay, G., Alakhova, D.Y. and Kabanov, A.V. 2010. Endocytosis of nanomedicines. *J Control Release* 145:182–195.

Saltzman, W.M., Radomsky, M.L., Whaley, K.J. and Cone, R.A. 1994. Antibody diffusion in human cervical mucus. *Biophys J* 66:508–515.

Sayes, C.M., Fortner, J.D., Guo, W. et al. 2004. The differential cytotoxicity of water soluble fullerenes. *Nano Lett* 4:1881–1887.

Schmidt, G., Decker, M., Ernst, H. et al. 2003. Small dimensions and material properties. *Europaische Akademie Graue Reihe in a Definition of Nanotechnology Bad Neuenahr* 134.

Schreiber, S. and Scheid, P. 1997. Gastric mucus of the guinea pig: Proton carrier and diffusion barrier. *Am J Physiol* 272:G63–G70.

Shaw, B.J., Al-Bairuty, G. and Handy, R.D. 2012. Effects of waterborne copper nanoparticles and copper sulfate on rainbow trout (*Oncorhynchus mykiss*): Physiology and accumulation. *Aquat Toxicol* 116–117:90–101.

Singh, S., Shi, T., Duffin, R. et al. 2007. Endocytosis, oxidative stress and IL-8 expression in human lung epithelial cells upon treatment with fine and ultrafine TiO_2: Role of the specific surface area and of surface methylation of the particles. *Toxicol Appl Pharmacol* 222:141–151.

Smith, C.J., Shaw, B.J. and Handy, R.D. 2007. Toxicity of single walled carbon nanotubes to rainbow trout, (*Oncorhynchus mykiss*): Respiratory toxicity, organ pathologies, and other physiological effects. *Aquat Toxicol* 82:94–109.

Squier, C.A. and Kremer, M.J. 2001. Biology of oral mucosa and esophagus. *Oxf J: JNCI Monogr* 2001(29):7–15.

Stoeger, T., Reinhard, C., Takenaka, S. et al. 2006. Instillation of six different ultrafine carbon particles indicates a surface area threshold dose for acute lung inflammation in mice. *Environ Health Perspect* 114:328–333.

Sudhakar, Y., Kuotsu, K. and Bandyopadhyay, A.K. 2006. Buccal bioadhesive drug delivery—A promising option for orally less efficient drugs. *J Control Release* 114:15–40.

Sun, W., Fang, N., Trewyn, B.G. et al. 2008. Endocytosis of a single mesoporous silica nanoparticle into a human lung cancer cell observed by differential interference contrast microscopy. *Anal Bioanal Chem* 391:2119–2125.

Szentkuri, L. 1997. Light microscopical observations on luminally administered dyes, dextrans, nanospheres and microspheres in the pre-epithelial mucus gel layer of the rat distal colon. *J Control Release* 46(3): 233–242.

Takenaka, S., Karg, E., Roth, C. et al. 2001. Pulmonary and systemic distribution of inhaled ultrafine silver particles in rats. *Environ Health Persp* 109(4):547–551.

Takizawa, H., Ohtoshi, T., Ohta, K. et al. 1992. Interleukin 6/B cell stimulatory factor-II is expressed and released by normal and transformed human bronchial epithelial cells. *Biochem Biophys Res Commun* 187:596–602.

Takubo, K. 2009. *Pathology of the Esophagus: An Atlas and Textbook*. Springer: Tokyo, Berlin, Heidelberg. .

Tan, A., Simovic, S., Davey, A.K., Rades, T., Boyd, B.J. and Prestidge, C.A. 2010. Silica nanoparticles to control the lipase-mediated digestion of lipid-based oral delivery systems. *Mol Pharm* 7:522–532.

Tran, C.L., Buchanan, D., Cullen, R.T., Searl, A., Jones, A.D. and Donaldson, K. 2000. Inhalation of poorly soluble particles. Influence of particle surface area on inflammation and clearance. *Inhal Toxicol* 12(12): 1113–1126.

Tsuchiya, T., Oguri, I., Yamakoshi, Y.N. and Miyata, N. 1996. Novel harmful effects of fullerene on mouse embryos *in vitro* and *in vivo*. *FEBS Lett* 393(1):139–145.

Vallhov, H., Qin, J., Johansson, S.M. et al. 2006. The importance of an endotoxin-free environment during the production of nanoparticles used in medical applications. *Nano Lett* 6:1682–1686.

Volkheimer, G. 1974. Passage of particles through the wall of the gastrointestinal tract. *Environ Health Perspect* 9:215–225.

Volkheimer, G., Schulz, F.H., Aurich, I., Strauch, S., Beuthin, K. and Wendlandt, H. 1968. Persorption of particles. *Digestion* 1(2):78–80.

Wang, B., Feng, W.-Y., Wang, T.-C. et al. 2006. Acute toxicity of nano- and micro-scale zinc powder in healthy adult mice. *Toxicol Lett* 161:115–123.

Wang, B., Feng, W.Y., Wang, M. et al. 2008a. Acute toxicological impact of nano- and submicron-scaled zinc oxide powder on adult mice. *J Nanopart Res* 10(2):263–276.

Wang, L., Nagesha, D.K., Selvarasah, S., Dokmeci, M.R. and Carrier, R.L. 2008b. Toxicity of CdSe nanoparticles in Caco-2 cell cultures. *J Nanobiotechnol* 6:11.

Wang, J.J., Sanderson, B.J. and Wang, H. 2007a. Cyto- and genotoxicity of ultrafine TiO_2 particles in cultured human lymphoblastoid cells. *Mutat Res* 628:99–106.

Wang, Z., Tiruppathi, C., Minshall, R.D. and Malik, A.B. 2009. Size and dynamics of caveolae studied using nanoparticles in living endothelial cells. *ACS Nano* 3:4110–4116.

Wang, J., Zhou, G., Chen, C. et al. 2007b. Acute toxicity and biodistribution of different sized titanium dioxide particles in mice after oral administration. *Toxicol Lett* 168:176–185.

Werner, I.H., Agnes, G.O., Wim, H.J., Flemming, R.C. and Adrienne, J.A.M.S. 2007. What do we (need to) know about the kinetic properties of nanoparticles in the body? *Regul Toxicol Pharmacol* 49:217–229.

Xia, T., Kovochich, M., Brant, J. et al. 2006. Comparison of the abilities of ambient and manufactured nanoparticles to induce cellular toxicity according to an oxidative stress paradigm. *Nano Lett* 6:1794–1807.

Yacobi, N.R., Demaio, L., Xie, J. et al. 2008. Polystyrene nanoparticle trafficking across alveolar epithelium. *Nanomed Nanotechnol Biol Med* 4:139–145.

Yoshifumi, T. 2002. Lipid formulation as a drug carrier for drug delivery. *Curr Pharm Des* 8(6):467–474 .

Zhang, X., Sun, H., Zhang, Z., Niu, Q., Chen, Y. and Crittenden, J.C. 2007. Enhanced bioaccumulation of cadmium in carp in the presence of titanium dioxide nanoparticles. *Chemosphere* 67:160–166.

Zhao, Y., Wang, B., Feng, W. and Bai, C. 2010. Nanosciences and nanotechnologies—Nanotoxicology: Toxicology and biological activities of nanomaterials, *Encyclopedia of Life Support System (EOLSS)*, UNESCO-EOLSS Publishers, USA:1–68.

14 Toxicity of Nanomaterials on the Liver, Kidney, and Spleen

Jayvadan Patel and Anita Patel

CONTENTS

14.1 Introduction 286
    14.1.1 Mechanisms of Nanomaterial Toxicity...286
    14.1.2 Liver, Spleen, Kidneys Uptake of Nanoparticles..287
    14.1.3 Adverse Health Effects of Liver and Kidney Uptake 288
14.2 Toxicity of Nanomaterials on the Liver...288
    14.2.1 Liver Morphology...288
    14.2.2 Phagocytosis in Kupffer Cells 288
    14.2.3 Potential Risk of Nanomaterials on the Liver 290
        14.2.3.1 Toxicity of Quantum Dots 290
        14.2.3.2 Toxicity of Silver Nanoparticles 290
        14.2.3.3 Toxicity of Copper Nanoparticles 291
        14.2.3.4 Toxicity of Dendrimer... 291
        14.2.3.5 Toxicity of Carbon Nanotubes 292
        14.2.3.6 Toxicity of Europium Hydroxide Nanorods..292
        14.2.3.7 Toxicity of Carbon Black and Polystyrene 293
        14.2.3.8 Toxicity of ZnO Nanoparticles 293
14.3 Toxicity of Nanomaterials on the Kidney..293
    14.3.1 Kidney Morphology..293
    14.3.2 Glomerular Filtration..294
    14.3.3 Tubular Reabsorption..295
    14.3.4 Potential Risk of Nanomaterials on the Kidney 295
        14.3.4.1 Toxicity of Metals and Heavy Metals 296
        14.3.4.2 Toxicity of Copper Nanoparticles 296
        14.3.4.3 Toxicity of Gold Nanoparticles...297
        14.3.4.4 Toxicity of Carbon Nanotubes 297
        14.3.4.5 Toxicity of Titanium Dioxide Nanoparticles 298
        14.3.4.6 Toxicity of Silica Particles 298
        14.3.4.7 Toxicity of QDs...300
        14.3.4.8 Toxicity of Dendrimers... 301
        14.3.4.9 Toxicity of Fullerenes 301
14.4 Toxicity of Nanomaterials on Spleen... 301
    14.4.1 Spleen Morphology... 301
    14.4.2 ABC Phenomenon 302
    14.4.3 Potential Risk of Nanomaterials on the Spleen 303
        14.4.3.1 Toxicity of Metal Nanoparticles 303
        14.4.3.2 Toxicity of Nanorods...303
        14.4.3.3 Toxicity of Silica Nanoparticles..304
        14.4.3.4 Toxicity of TiO_2 Nanoparticles 304

    14.4.3.5 Toxicity of Carbon Nanotubes 305
    14.4.3.6 Toxicity of Engineered Nanomaterials 306
14.5 Summary 306
References 306

14.1 INTRODUCTION

Human skin, lungs, and the gastrointestinal tract (GIT) are in constant contact with the environment. Although the skin is generally an effective barrier to foreign substances, the lungs and the GIT are more susceptible. These three ways are the most likely points of entry for natural or anthropogenic nanoparticles. Injections and implants are other potential routes of exposure, majorly limited to engineered materials.

Owing to their small size, nanoparticles can translocate from these entry portals into the circulatory and lymphatic systems, and finally to body tissues and organs. Few nanoparticles, depending on their composition and size, can produce permanent damage to cells by oxidative stress and/or organelle injuries.

Emphasizing that not all nanoparticles produce adverse health effects—the toxicity of nanoparticles depends on several factors, including their size, aggregation, composition, crystallinity, surface functionalization, and so on. In addition, the toxicity of any nanoparticle to an organism is determined by the individual's genetic complement, which provides the biochemical toolbox by which it can adapt to and fight toxic substances. Summarized here are the most extreme adverse health effects produced by nanoparticles in order to immediately increase the awareness of potential toxicity of some nanoparticles. Nanoparticles that are introduced into the circulatory system are associated with the occurrences of arteriosclerosis and blood clots, arrhythmia, heart diseases, and may finally lead to cardiac death. The translocation of these nanoparticles to other organs, like the liver and spleen, may lead to the likewise diseases of these organs. A vulnerability to some nanoparticles is related to the occurrence of autoimmune diseases such as systemic lupus erythematosus, rheumatoid arthritis, and scleroderma (Cristina et al. 2007).

Nanoparticles have been found to be distributed to the liver and spleen after intravenous injections (El-Ansarty and Al-Daihan 2009). Their distribution is followed by a rapid clearance from systemic circulation, predominately by the action of liver and splenic macrophages. Processes like clearance and opsonization, that prepare foreign materials to be more efficiently engulfed by macrophages, occur under certain conditions for nanoparticles depending on their size and surface characteristics. When inhaled, nanoparticles are found to be distributed to the liver and spleen. Nanoparticles are cleared from the alveolar region via phagocytosis by macrophages facilitated by the chemotactic attraction of alveolar macrophages to the deposition site. After oral exposure, nanoparticles distribute to the kidneys, liver, and spleen. However, some nanoparticles can accumulate in the liver during first-pass metabolism.

14.1.1 Mechanisms of Nanomaterial Toxicity

Nanomaterials can modify the physicochemical properties of materials as well as create the opportunity to increase their uptake and interactions with biological tissues through inhalation, ingestion, and injection. A combination of these effects can generate adverse biological effects in living cells. Nanomaterial toxicity can occur through several different mechanisms in the body. The main molecular mechanism of *in vivo* nanotoxicity is the induction of oxidative stress by the formation of free radicals (Lanone and Boczkowski 2006). In excess, free radicals cause damages to biological components through the oxidation of lipids, proteins, and DNA. The role of oxidative stress has been predominant in the induction or the enhancement of inflammation through the upregulation of redox-sensitive transcription factors such as NF-κB, activator protein-1, and kinases involved in inflammation (Rahman 2000; Rahman et al. 2005; Lanone and Boczkowski 2006). Free radicals can develop from various sources, including phagocytic cell response to foreign materials, the presence of transition metals, insufficient amounts of antioxidants, and the physicochemical properties of some nanomaterials and environmental factors (Lanone and Boczkowski 2006). The slow clearance and tissue accumulation (storage) of potential free radical-producing nanomaterials as well as the prevalence of numerous phagocytic cells in the organs of the reticuloendothelial system (RES) makes organs such as the liver and spleen the main targets of oxidative stress. In addition, organs of high blood flow that are exposed to nanomaterials, such as the kidneys and lungs, may also be affected.

Intracellularly, the interaction of nanomaterials with cellular components may occur, which can potentially disrupt or modify cell functions or lead to the production of reactive oxygen species (ROS). It may result in oxidative stress, inflammation, and the consequential damage to proteins, membranes, and DNA. The smaller a particle is, the greater is its surface area-to-volume ratio and the higher its chemical reactivity and biological activity. The greater chemical activity of nanomaterials results in the increased production of ROS, including free radicals. The production of ROS has been found in a diverse range of nanomaterials, including carbon fullerenes, carbon nanotubes (CNTs), and metal oxide nanoparticles. The extremely small size of nanomaterials also means that they much more readily gain entry into the human body than larger-sized particles (Yang et al. 2010, Chakraborty et al. 2011). Interactions of nanomaterials with the mitochondria and cell nucleus are being considered as main sources of toxicity. Unfried et al. reviewed that nanomaterials such as silver-coated gold nanoparticles block copolymer micelles; CNTs and fullerenes may be capable of localizing to mitochondria and inducing apoptosis and ROS formation, and this nuclear DNA damage, mutagenesis, cell-cycle arrest, and apoptosis induced by nanomaterials is a possible source of toxicity (Unfried et al. 2007). Although still under disputation, nanomaterials may be involved in the upregulation of xanthine oxidase and NADPH oxidase (nicotinamide adenine dinucleotide phosphate oxidase), which are free radical sources in macrophages and neutrophils (Lanone and Boczkowski 2006). Other mechanisms of toxicity from nanomaterials should be considered since nanomaterials immediately interact with their surrounding environments. Once introduced or absorbed into systemic circulation, their interaction with blood components can result in hemolysis and thrombosis. Additionally, the interactions of nanomaterials with the immune system have been known to increase immunotoxicity (Dobrovolskaia and McNeil 2007). In the liver, further metabolic modifications of nanomaterials, for example, by cytochrome P450, may cause hepatotoxicity by reactive intermediates (Lanone and Boczkowski 2006).

14.1.2 Liver, Spleen, Kidneys Uptake of Nanoparticles

Endothelial cells (cells that line the vascular system), having very tight junctions typically smaller than 2 nm, form a physical barrier for particles (Schwab and Pang 2000). Nevertheless, larger values, from 50 nm (Hussain et al. 2001) up to 100 nm (Schwab and Pang 2000), have been reported depending on the organ or tissue. A very tight endothelial junction is present in the brain, often called the blood–brain barrier. However, experiments performed on rats injected with ferritin macromolecules (with sizes around 10 nm) into the cerebrospinal fluid, demonstrated the passage of ferritin deep into the brain tissue. In certain organs such as the liver, the endothelium is fenestrated with pores of up to 100 nm, allowing the easier passage of larger particles. In the presence of inflammation, the permeability of the endothelium is increased, allowing a larger passage of particles.

Micro- and nanoparticle debris was detected by scanning electron microscopy in the organs and blood of patients with orthopedic implants, drug addiction (Gatti and Rivasi 2002), worn dental prostheses (Ballestri et al. 2001), blood diseases (Gatti et al. 2004), colon cancer, Crohn's disease, ulcerative colitis (Gatti 2004), and with diseases of unknown etiologies (Gatti and Rivasi 2002). Coal workers' autopsies revealed an increased amount of particles in the liver and spleen compared to noncoal workers (Donaldson et al. 2005). The pathway of exposure most likely involves the translocation of the inhaled nanoparticles from the lungs to systemic circulation, followed by organular uptake.

Rat inhalation studies with stainless-steel welding fumes showed that manganese accumulates in the blood and liver (Donaldson et al. 2005). Rat inhalation studies with 4–10 nm silver nanoparticles show that the nanoparticles enter the circulatory system within 30 min, and can be found in the liver, kidney, and heart after a day until being subsequently cleared from these organs after a week (Takenaka et al. 2001). Clearances from the liver can occur via biliary secretion into the small intestine (Cooper et al. 1979).

A case study shows that wearing dental bridges leads to the accumulation of nanoparticles in the liver and kidneys. The most probable absorption pathway was assumed to be via intestinal absorption (Ballestri et al. 2001). Scanning electron microscopy and energy-dispersive microanalytical techniques identified the chemical compositions of particles in liver and kidney biopsies as well in the stool as the same as porcelain from dental prostheses. The maximum size of particles found in the liver (20 μm) was larger than in the kidneys (below 6 μm), suggesting that particles are absorbed by the intestinal mucosa and translocate to the liver before reaching the circulatory system and kidneys. After the removal of dental bridges, particles in stool are no longer observed.

14.1.3 Adverse Health Effects of Liver and Kidney Uptake

Up to now, there is little knowledge on the effect of nanoparticles on organs such as the liver, kidneys, and spleen. However, one can speculate that as long as there is the translocation to and accumulation of nanoparticles in these organs, potentially adverse reactions and cytotoxicity may lead to disease.

Diseases with unknown origins have been correlated with the presence of micro- and nanoparticles in the kidneys and liver. For comparison, the liver and kidneys of healthy subjects did not show any debris. Particle debris has been found also in the liver of patients with worn orthopedic prosthesis (Gatti and Rivasi 2002).

Dental prosthesis debris internalized by intestinal absorption can lead to severe health conditions, including fever, enlarged spleen and liver, suppression of bile flow, and acute renal failure. These symptoms appeared about a year after the application of dental porcelain bridges. After the removal of dental bridges and subsequent treatment with steroids, the clinical symptoms declined (Ballestri et al. 2001).

14.2 TOXICITY OF NANOMATERIALS ON THE LIVER

14.2.1 Liver Morphology

The liver engages in numerous metabolic, immunological, and endocrine functions. It receives blood from the gut and the heart via the portal vein and hepatic artery, respectively (Figure 14.1a). Blood circulates through a permeable, discontinuous capillary network (the sinusoids) to reach the central and hepatic veins (Figure 14.1b). The sinusoids are 5–10 μm-wide blood vessels with a fenestrated epithelium without any basal membrane (Figure 14.1c). The size of the fenestrations (100–150 nm depending on the animal species) allows the almost unrestricted passage of plasma components to the perisinusoidal space, where the cords of parenchymal cells (hepatocytes) are situated (Wisse et al. 2008; Jacob et al. 2010). Inside the sinusoid capillaries, Kupffer cells are responsible for the phagocytic activity of the liver. These nonparenchymal cells belong to the mononuclear phagocyte system and represent 80–90% of the body's total macrophage population. This defense system was formerly known as the reticuloendothelial system, but the name became obsolete when it was understood that endothelial cells are not macrophages responsible for the clearance of pathogens.

14.2.2 Phagocytosis in Kupffer Cells

Phagocytosis occurs after the multivalent contacts of the colloid with macrophages and spreading of the cell membrane around the particle for engulftion. After ingestion, phagocytic vesicles

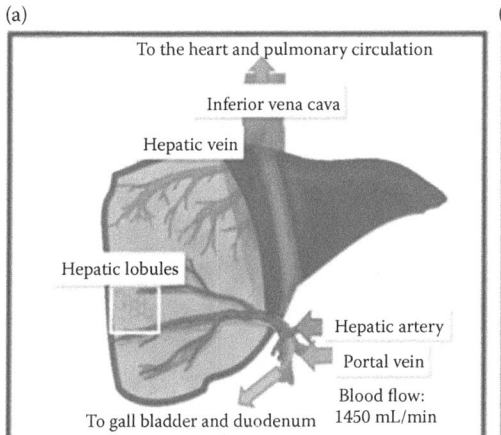
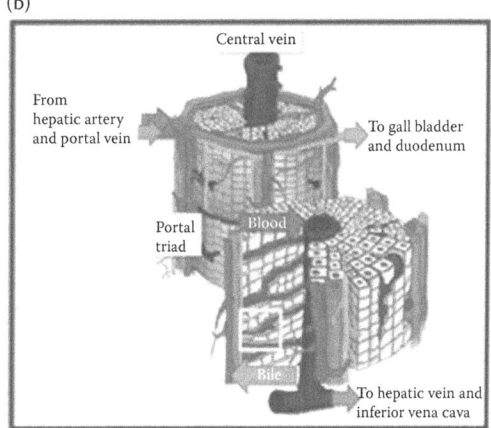
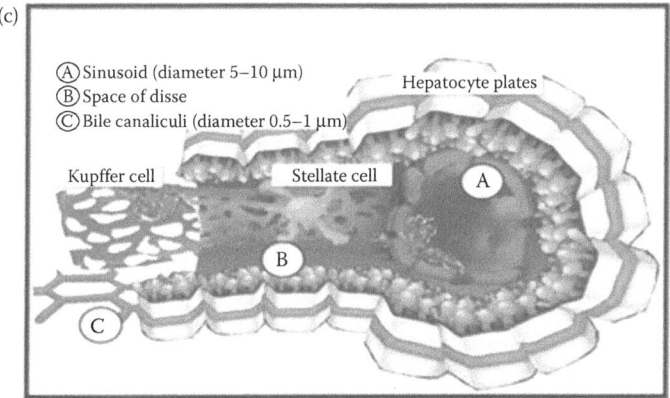

FIGURE 14.1 Structure of the liver. (a) The liver. The liver is perfused by the hepatic artery and portal vein. Blood exits through the hepatic vein into the inferior vena cava, and bile drains into the gall bladder. (b) The liver lobules. The liver parenchyma is composed of hexagonal lobules containing hepatocyte plates and the sinusoids. The blood in each lobule comes from the portal vein and hepatic arteries; it drains out through the central vein and into the inferior vena cava. (c) The liver sinusoid. Fenestrated hepatic sinusoids where the arterial and portal veins merge. (From Wisse, E. et al. 2008. *Gene Ther* 15:1193–1199. With permission; Adapted from Jacobs, F., Wisse, E. and De Geest, B. 2010. *Am J Pathol* 176:14–21.)

(phagosomes) coalesce with intracellular organelles containing digestive proteins and an acidic internal pH to mature into phagolysosomes and degrade the internalized colloid. The colloid is then eliminated by exocytosis after degradation or sequestered in residual bodies within the cell if it cannot be digested. Contacts between colloidal drug carriers (CDCs) and macrophages occur via the recognition of opsonins on the CDC surface or through interactions with scavenger receptors on Kupffer cells. The size and radius of the curvature of colloids highly influence both CDC-cell contacts and internalizations (Harashima et al. 1994; Champion et al. 2008; Jiang et al. 2008). Because Kupffer cells have a ruffled surface, privileged interactions and optimal phagocytosis occur when the colloids have a diameter between 1 and 3 μm (Champion et al. 2008; Doshi and Mitragotri 2010). Smaller particles offer less cooperative contacts with the cell membrane (Jiang et al. 2008), and exploit other entry ways into cells (e.g., fluid phase pinocytosis or endocytosis) (Petros and DeSimone 2010). Oppositely, larger particles cannot maximize contacts with the cell surface and require extensive cytoskeleton remodeling to be internalized. *In vitro*, the upper size limit for phagocytosis has been determined to be around 20 μm in diameter or with a volume at most 3 times that of the macrophage. Frustrated phagocytosis occurs when the membrane is stretched to the maximum without engulfing the whole system (Cannon and Swanson 1992).

The shape of CDCs also highly affects internalization (Gratton et al. 2008; Champion and Mitragotri 2009, 2006; Doshi and Mitragotri 2010). For large particles (size >1 µm), *in vitro* studies have shown that while elongated shapes promote interactions with the surface of macrophages (Gratton et al. 2008; Sharma et al. 2010), high-aspect ratios (i.e., the ratio of the larger over the smaller dimension) hinder the spreading of the optimal membrane and complete phagocytosis (Champion and Mitragotri 2006, 2009; Doshi and Mitragotri 2010). For smaller particles (size <1 µm), the shape influences the speed of internalization and the different pathways used to enter the cells (Gratton et al. 2008).

14.2.3 Potential Risk of Nanomaterials on the Liver

Nanoparticles taken up in the body and circulation are largely recovered in the liver, followed by the spleen. Studies of these organs are therefore highly relevant with regard to nanomaterials. As these organs have not been previously regarded as target organs, only limited information is available on the effects of nanomaterials. Nanoparticles are stated as possibly being capable of leading to thrombosis without any inflammation in the liver when they are injected into mice (Khandoga et al. 2004). This is supported by studies by Schwartz who showed that nanoparticles are potent with regards to modifying clotting factors in the blood (Schwartz 2001).

14.2.3.1 Toxicity of Quantum Dots

Possibly, the most important aspect of quantum dot (QD) toxicity is the stability of the core/shell/coating complex. The toxicity of core metals on the body can be prevented by stable complexes. Various studies suggested that under oxidative and photolytic conditions, quantum dots (QDs) are labile and subject to degradation, hence exposing the potentially toxic shell material, core metal components (e.g., cadmium, selenium), and intact core metalloid complex (Hardman 2006). QDs of cadmium–selenium (CdSe) were found to be acutely toxic under certain conditions for primary hepatocytes. Derfus et al. found that CdSe QDs were toxic to liver hepatocytes if exposed to air or UV light, due to oxygen combining with Se and releasing free Cd^{2+} under oxidative attack, which can bind to the sulfhydryl groups of critical mitochondria proteins leading to a subsequent mitochondria dysfunction and, ultimately, cell poisoning (Derfus et al. 2004; Zhang et al. 2006). They found that coating the CdSe QDs with ZnS, polyethylene glycol, or other coatings prevented toxicity during a 2-week incubation with hepatocytes.

The addition of the secondary surface coatings renders QDs biocompatible and can help protect against their degradation, but the method of functionalization (e.g., electrostatic, adsorption, covalent bonding, or multivalent chelation) and the composition of coating are important when considering QD durability, stability, and *in vivo* reactivity (Hardman 2006). Interestingly, a recent communication by Mancini et al. suggests an oxidative mechanism wherein hypochlorous acid and hydrogen peroxide, produced by phagocytic cells, may diffuse across secondary polymeric coatings and cause QD lattice defects and the solubility of the core (Mancini et al. 2008). Therefore, there may be *in vivo* oxidative mechanisms of QD degradation regardless of the presence of stable secondary coatings. The potential toxic effects of QDs have become a hot issue that must be further addressed before clinical applications would be possible. Most studies recommend that not all QDs are similar in their toxicity, and the toxicity of differing QDs must be considered individually. The adverse effects of QDs can be mitigated or eliminated by the proper choice of coating materials and modification techniques that reduce QD instability.

14.2.3.2 Toxicity of Silver Nanoparticles

Recently, numerous *in vitro* studies have shown that silver (Ag) nanoparticles have the potential to induce toxicity in cells derived from a variety of organs. The use of Ag nanoparticles in textiles and cosmetics has substantially increased the potential for human skin exposure. Ag nanoparticle crystals released from a commercial dressing were found to be toxic to both keratinocytes and fibroblasts (Poon and Burd 2004). It is interesting to see that fibroblasts appeared to be more sensitive to Ag nanoparticles than keratinocytes. Ag nanoparticles are found to induce cell death and oxidative stress in human fibrosarcoma and skin carcinoma cells (Arora et al. 2008), and it was demonstrated that Ag nanoparticles could enter cells and cause DNA damage and apoptosis in fibroblasts and liver cells (Arora et al. 2009). The possible mechanism of Ag nanoparticle toxicity in fibroblasts has been elaborated upon (Hsin et al. 2008) and Ag nanoparticles have been found to induce ROS and release of cytochrome-c into the cytosol and the translocation of Bax proteins to the mitochondria. These observations suggest that Ag nanoparticle-mediated apoptosis was mitochondria-dependent in fibroblasts. Some further observations have suggested that Ag nanoparticles induce a p53-mediated apoptotic pathway through which most of the chemotherapeutic drugs trigger apoptosis (Gopinath et al. 2010). It has been demonstrated by different studies that the lungs and liver are major target tissues for prolonged Ag nanoparticle exposure (Takenaka et al. 2001; Sung et al. 2008). The studies of Ag nanoparticles on rat liver cells have shown that a significant depletion of the antioxidant, glutathione reduced the mitochondrial membrane potential and increased ROS (Hussain et al. 2005). These findings suggested that Ag nanoparticle cytotoxicity is likely mediated through oxidative stress in liver cells. The expression of genes associated with cell cycle progression and apoptosis in human hepatoma cells was induced by a noncytotoxic dose of Ag nanoparticles (Kawata et al. 2009). Choi et al. studied the liver toxicity of Ag nanoparticles in adult zebrafish, where a number of cellular alterations, including the disruption of hepatic cell cords and apoptotic changes, were observed. The DNA double-strand break marker γ-H2AX and the expression of the p53 protein implied that Ag nanoparticles induced DNA damage. In addition, the p53-related pro-apoptotic genes Bax, Noxa, and p21 were upregulated (Choi et al. 2010a,b).

Experimental results showed that Ag nanoparticles were more toxic to mitochondria than Mn and Mn^{2+}. These mitochondria seem to be sensitive targets of cytotoxicity and deposition in the cytoplasm. In environmental aspects, silver leaks into the aquatic environment. 150,000 kg of silver enters into the aquatic system every year from industries (Helinor et al. 2010).

14.2.3.3 Toxicity of Copper Nanoparticles

Despite an increasing application of copper nanoparticles, there is a serious deficiency of information regarding their impact on human health and the environment. Copper is an essential trace element capable of producing toxic effects in animals or humans when ingested acutely or chronically (Bremner 1998). There are numerous data regarding the toxicity of copper compounds. Upon oral exposure, the liver and kidneys remain sensitive targets of copper toxicity. The manifestation of copper poisoning mainly includes drowsiness and anorexia in the early stages (Winge and Mehra 1990; Barceloux 1999) as well as acute tubular necrosis in the kidney and hepatocellular necrosis (Liu et al. 2004). Moreover, metabolic alkalosis and copper accumulation in the kidneys were detected in mice that were orally exposed to nanocopper particles (Galhardi et al. 2004). Although the potential risks of nanocopper particles on human health have been identified, its subacute toxicity has not been depicted (Lei et al. 2008).

14.2.3.4 Toxicity of Dendrimer

The reported hepatotoxicity has been consistent with biodistribution studies that revealed high liver accumulation for G3 (Generation 3) polyamidoamine dendrimers (IP) (Roberts et al. 1996), G3 and G4 polyamidoamine dendrimers (IV or IP) (Malik et al. 2000), biotinylated-polyamidoamine dendrimers (G0–4) (IV) (Wilbur et al. 1998), and polyethylene glycol–polyester dendritic hybrids (IV) (DeJesus et al. 2002). Roberts et al. observed the liver cell vacuolization of the cytoplasm in a 6-month toxicity study group after administration of G7 polyamidoamine dendrimers by the I.P. route once a week for 10 weeks (Roberts et al. 1996). A third-generation melamine dendrimer (MW 8067, 24 amine termini) was evaluated for *in vivo* acute toxicities (single dose, 48 h) and subchronic toxicities (three IP injections every 3 weeks over 6 weeks) (Neerman et al. 2004). The acute study indicated that the lethal dose was 160 mg/kg, giving rise to 100% mortality 6–12 h post injection. In both acute and subchronic studies, the hepatic function was normal at up to 10 mg/kg dendrimer as evaluated by changes in serum alanine transaminase activity. However, a significant increase in serum alanine transaminase activity was noticed for both acute and subchronic groups at the dose level of 40 mg/kg. In addition, the histopathological investigation showed extensive liver necrosis. *In vivo* dendrimer toxicity profiles are closely related to the chemical structure of the dendrimer; size and generation; exposure duration; biodistribution; and the rate, location, and mechanism of metabolism. Dendrimer toxicity is influenced by the nature of the terminal groups. Full generation polyamidoamine dendrimers with cationic amine terminal groups are more toxic than half-generation dendrimers with anionic carboxylic acid terminal groups. High generation dendrimers and high doses of cationic dendrimers usually lead to a greater *in vivo* toxicity.

14.2.3.5 Toxicity of Carbon Nanotubes

The size of aggregated CNTs is thought to be a primary concern for toxicity. Pristine (nonfunctionalized) CNTs are inherently hydrophobic; as a result, aggregation is expected and observed *in vivo*. For injection, pristine CNTs are suspended in biocompatible surfactants such as Tween 80 or Pluronic F108 (Cherukuri et al. 2006; Yang et al. 2008). Several studies have been conducted on the *in vivo* distribution of IV injected pristine single-walled carbon nanotubes (SWCNTs). The accumulation of CNTs was primarily determined to be in the liver, but is also found in the spleen and lungs. Acute toxicity was not observed in any tissue up to 24 h (Cherukuri et al. 2006; Yang et al. 2007, 2008). The accumulation in the liver was suggested to be due to a rapid surfactant displacement followed by the opsonization of serum proteins (Cherukuri et al. 2006). Yang et al. followed up distribution studies by looking at serum biomarkers of damage. Furthermore, they looked at markers for oxidative stress (glutathione and malondialdehyde) in liver samples post dose. Elevated levels of lactose dehydrogenase and alanine aminotransferase were concluded to be due to hepatic injury from accumulation in the liver. They also observed an increase in malondialdehyde and a decrease in glutathione in liver samples at the dose of 1.0 mg/mouse, which was indicative of increased levels of oxidative stress. Though no acute toxicity was determined histologically up to 90 days post dose, biomarkers representing hepatic injury due to oxidative stress should be further investigated (Yang et al. 2008). Recent studies have shown that once in the bloodstream, the slow clearance, a tendency to aggregate, and intrinsic properties can lead to oxidative stress especially in the liver, lungs, and spleen, eventually resulting in inflammation.

Histopathological evaluations of livers exposed to multi-walled carbon nanotubes (MWCNTs) showed remarkable morphological alterations such as individual cell necrosis indicated by karyorhxis as well as sinusoid dilation, hepatocyte disruption, vacuolation and swelling, fatty changes, hemorrhagic clots, and necrotic changes when compared to controls. Kupffer cells are the resident macrophages of the liver and play an important role in its normal physiology and homeostasis. These cells participate in acute and chronic responses of the liver to toxic compounds. Activation of Kupffer cells by toxic agents, both directly and indirectly, results in the release of an array of inflammatory responses, growth factors, and ROS. This activation appears to control acute hepatocyte injury along with chronic liver responses. The role of Kupffer cells in these diverse responses is the key to understanding the mechanisms involved in liver injury (Roberts et al. 2007).

14.2.3.6 Toxicity of Europium Hydroxide Nanorods

Chitta and coworkers synthesized europium hydroxide nanorods by microwave irradiation and characterized them using several analytical tools. They found that synthesized europium hydroxide nanorods are nontoxic to endothelial cells observed by the apoptosis assay. The IP injection of europium hydroxide nanorods at several doses in mice showed normal blood hematology and serum clinical chemistry except the slight elevation of liver enzymes. The histopathological examination of the liver, kidney, spleen, and lungs from every mouse, assayed on day 8 or day 60 after nanorod injection, showed none or only mild histological changes that indicate mild toxicities at highest dose of nanorods (Chitta et al. 2009).

14.2.3.7 Toxicity of Carbon Black and Polystyrene

The RES in the liver is exposed to all nanoparticles absorbed from the GIT into the cardiovascular system, in view of the fact that all blood exiting the GIT goes into the hepatic portal vein that directly diffuses through the liver. Nanoparticles with low toxicities, such as carbon black and polystyrene, stimulate macrophages *via* calcium signaling and ROS to make proinflammatory cytokines such as tumor necrosis factor alpha. Oxidative stress inhibited hepatocyte function and bile formation, while proinflammatory cytokines are also associated with the pathology of liver diseases (Brown et al. 2004).

14.2.3.8 Toxicity of ZnO Nanoparticles

Sharma et al. performed *in vivo* studies in mice to investigate the oral toxicity of ZnO nanoparticles. Results showed that a significant accumulation of nanoparticles in the liver leads to cellular injuries after the sub-acute oral exposure of ZnO nanoparticles (300 mg/kg) for 14 consecutive days. Orally administered ZnO nanoparticles led to liver damage, as revealed by histopathological examination which showed hepatocellular necrosis. This was also supported by elevated alanine aminotransferase (ALT) and alkaline phosphatase (ALP) levels in the serum. The levels of these enzymes rise in serum when the liver is damaged. The unaltered levels of creatinine, blood urea nitrogen, and bilirubin in this study indicated normal renal functions in the ZnO nanoparticle exposed group (Sharma et al. 2012). These results support the findings of Wang et al. (2008), who observed liver damage in mice after the oral exposure of ZnO nanoparticles, although at a higher dose of 5 g/kg.

14.3 TOXICITY OF NANOMATERIALS ON THE KIDNEY

14.3.1 KIDNEY MORPHOLOGY

The kidneys are responsible for the filtration of the blood. Their parenchyma is separated into two sections, the cortex and the medulla (Figure 14.2a), which comprise different parts of the nephron, the basic functional unit of the kidney (Figure 14.2b). Blood filtration occurs in the cortex through the glomerulus, a structure formed of the glomerular capillary network (Figure 14.2c). The filtering apparatus in the glomerular capillary network is composed of three consecutive elements that are essential for physiological functions (Figure 14.2d). The first component is the highly fenestrated endothelium. The fenestrations, 60–80 nm in size, are covered by a 200–300 nm thick glycocalyx layer adsorbed onto the luminal side of the endothelium (Haraldsson et al. 2008). This flexible glycoprotein barrier is anchored on the surface of endothelial cells and restricts the passage of blood components. The second structure that is essential in controlling the passage of solutes into the urine is the glomerular basement membrane (GBM). It is a 240–370 nm hydrated, fibrous, network layer of type IV collagen containing laminin, entactin, and proteoglycan, with high heparin and chondroitin sulfate contents (Haraldsson et al. 2008). The fibrous mesh of collagen and the abundant anionic charges dispersed in the GBM could together contribute to restrict fluid flux and the passage of negatively charged solutes. However, it is now believed that the GBM alone cannot be responsible for the high permeation selectivity of the glomerular filtration process (Haraldsson et al. 2008). The third component of the glomerulus is the visceral layer of Bowman's capsule that is composed of podocytes. These highly differentiated cells outline the glomerular capillaries.

Podocytes interact with the GBM via numerous inter-digitated foot processes that form filtration slits (Haraldsson et al. 2008). Finally, the mesangial cells and their extracellular matrix form the renal mesangium, which offers structural support to glomerular capillaries. They are derived from smooth muscle cells and possess phagocytic properties to keep the GBM filter free of debris as well as secretory functions solicited in case of glomerular injuries (Michael et al. 1980).

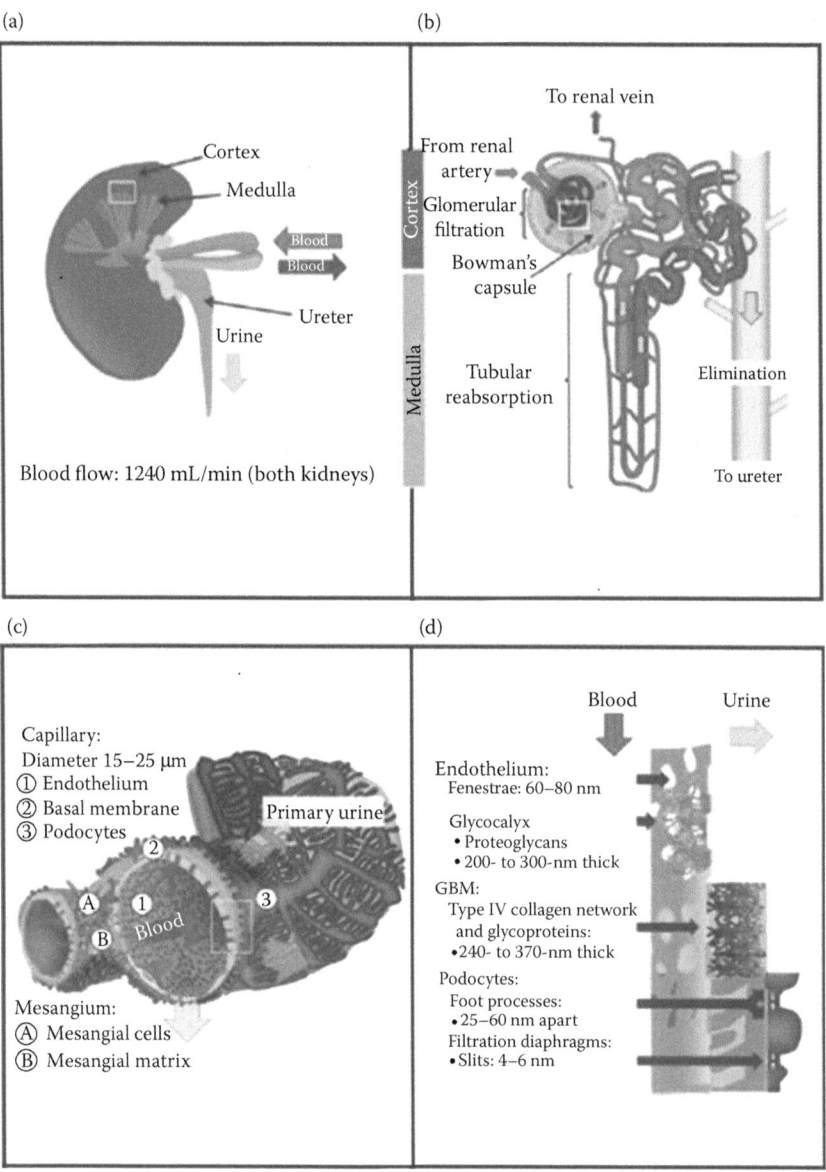

FIGURE 14.2 Blood filtration in the kidney. (a) The kidney. The kidney parenchyma is separated between the cortex and the medulla. (b) The nephron. Approximately 1 million nephrons span both regions and are responsible for blood filtration. Blood filtration occurs in the glomerulus. Bowman's capsule collects the filtrate before its transit through the tubules where it can be concentrated. The collecting duct gathers urine from multiple nephrons for exit to the ureter. (c) The glomerular capillary. The glomerular capillaries are composed of a highly fenestrated endothelium, the glomerular basal membrane (GBM) and the podocytes. The mesangium offers structural support. (d) The filtration apparatus. Three distinctive structures of the component together exert qualitative and quantitative restrictions over the filtrate.

14.3.2 Glomerular Filtration

Convective and diffusive forces in the glomerulus capillaries compel the filtration of blood solutes from the lumen of capillaries to Bowman's capsule. Structures of the glomerulus exert qualitative and quantitative controls on what is filtered. Proteins with hydrodynamic diameters (D_H) smaller than

5–6 nm are freely filtered by the glomerulus, while larger solutes are retained. For globular proteins, this size is equivalent to a molecular weight around 60 kDa. The molecular weight limit can be lower for nonglobular proteins (Maack et al. 1979). Grafting polymer chains to drugs is now a common method of circumventing renal filtration and increasing the biological half-life (Kopecek et al. 2001). Until today, PEG remains the most frequently utilized polymer to augment the D_H of therapeutic molecules (Harris and Chess 2003; Gao et al. 2010), but other hydrosoluble polymers (Seymour et al. 1987; Li et al. 1999, Gauthier and Klok 2010) as well as synthetic linear polypeptides (Schellenberger et al. 2009, Zalevsky et al. 2010) have also been employed for this purpose. Several soluble CDCs based on non-PEG polymers are currently in the preclinical and clinical stages of development.

Macromolecules with sizes around the filtration threshold must distort through the pores in order to be filtered. Factors affecting their deformability, such as hydration, flexibility, as well as intra- and intermolecular architecture, also highly influence their glomerular filtration (Bertrand et al. 2009; Fox et al. 2009; Nasongkla et al. 2009). Most particulate CDCs are too large to be filtered without prior biodegradation. Exceptions include QDs, shown to be filtered for diameters up to 5.5 nm (Choi et al. 2007, 2010a,b), and CNTs (Liu et al. 2008; Schipper et al. 2008). The unique shape of CNTs might promote their renal clearance by a favorable blood flow orientation (Lacerda et al. 2008; Ruggiero et al. 2010) or distinctive transcellular transport mechanisms (Kostarelos et al. 2007).

Nevertheless, additional studies are still needed to confirm that their renal filtration is sufficiently fast to prevent accumulations and toxicities in a clinical context. Recently, an interesting study has shown that the renal mesangium could be targeted by particulate CDCs with a defined size of ~75 nm (Choi et al. 2011). Compared to similar particles of larger and smaller sizes, colloids of this diameter are small enough to permeate the endothelial fenestrae and sufficiently large to interact with mesangial cells. The particles were trapped in the collagen network of the GBM for at least 24 h and were cleared by the residing macrophages.

However, compared to the organs of the mononuclear phagocyte system (MPS), the total amount found in the kidneys remained low (≤5% ID), and complementary investigations are required to determine whether this size-dependent passive targeting can be exploited for imaging or drug delivery purposes.

14.3.3 Tubular Reabsorption

Once a solute is filtered in Bowman's capsule and reaches the proximal and distal tubules, it can be reabsorbed by the epithelial cells. Tubular reabsorption regulates the elimination of ions and proteins. It occurs through receptor-mediated and fluid-phase endocytosis. After internalization, protein-containing vesicles can transcytose back to peritubular circulation (e.g., via FcRn) (Anderson and Sandlie 2009) or can be catabolized in lysosomes. In the latter case, catabolic products are subsequently released into systemic circulation (Maack et al. 1979). Soluble CDCs filtered through the glomerulus can also be reabsorbed. The accumulation of PEG–protein conjugates in intracellular vacuoles inside tubular epithelial cells has been reported (Bendele et al. 1998). In this case, the protein moiety seems essential for endocytosis by tubular cells. It is believed that the incomplete catabolism of nonbiodegradable polymers leads to a sequestration in intracellular organelles. Although the vacuolization process appears to be transient and no signs of toxicity were witnessed, one should be aware that data remain limited and the risk could increase upon chronic exposure.

14.3.4 Potential Risk of Nanomaterials on the Kidney

The most common route for the excretion of xenobiotics is via the kidneys into the urine. This process involves the filtration of the blood through a complex filtration membrane in the glomerulus of kidney nephrons. The size of the fenestrae within the glomerular filter is limited to allow the passage of small molecular weight substances while preventing the filtration of larger molecules. It is conceivable that particles of an appropriate size could easily block the renal fenestrae leading to rapid kidney failure and death.

14.3.4.1 Toxicity of Metals and Heavy Metals

The safety of metals and their *in vivo* use is an ongoing issue of contest. Divalent cations are recognized to be toxic even at low concentrations in the body. Because of the possible reabsorption and accumulation in the kidneys, the main problem with heavy metals is nephrotoxicity (Babier et al. 2005). For example, gadolinium-based contrast agents for clinical MRI analysis have been reported to be associated with acute renal failure (Akgun et al. 2006).

Wang et al. studied acute toxicity of nano- and micro-scale zinc powder in healthy adult mice. Renal histopathological examinations revealed that there were alterations of proteinaceous casts in the tubules as well as renal tubular dilatations in nano-scale zinc powder-treated mice. The histopathological finding demonstrated that the oral exposure of nano-scale zinc powder could cause more severe renal damage than micro-scale zinc powder, though the serum indicators did not show obvious changes (Figure 14.3). It was concluded that the oral exposure of high-dose nano-scale zinc powder could induce heavier renal damage and anemia in mice (Wang et al. 2006).

Free cadmium is accumulated in the kidneys of Sprague–Dawley rats after intraperitoneal injections, but intact nanoparticles were also detected, organular levels of which increased with the dose (Arslan et al. 2011).

14.3.4.2 Toxicity of Copper Nanoparticles

Sil et al. (2011) performed a dose-dependent study by estimating blood urea nitrogen (BUN) and creatinine levels; the different doses taken for this study were well correlated with the doses taken by Chen et al. (2006). After the dose-dependent study, 200, 413, and 600 mg/kg body weight copper nanoparticles were chosen for further studies. Serum BUN and creatinine levels were significantly increased in the group exposed to copper nanoparticles in comparison to the control group, suggesting renal dysfunction. Nanocopper exposure increased the production of ROS, reactive nitrogen species (RNS), and altered the levels of oxidative stress-related biomarkers in kidney tissue.

Exposure to nano-sized copper particles caused alterations in the levels of intracellular ROS as well as the production of NO (measured by nitrite level). It was also clear from this study that

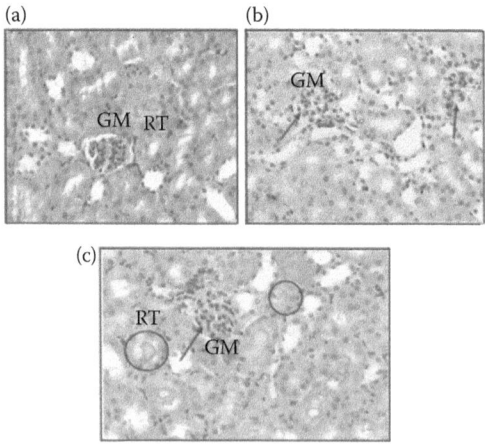

FIGURE 14.3 Kidney tissue from mice exposed to zinc powder at an acute toxic dose of 5 g/kg body weight on 14 days post-oral administration (magnification = 200). (a) Control group (instilled 1% sodium carboxy methyl cellulose). GM: glomerulus; RT: renal tubular. (b) Microscale group. Arrows show the glomerulus swelling. (c) Nanoscale group. Arrows show the glomerulus swelling. Circle area shows the proteinaceous casts in renal tubular. (Reprinted from Wang, B. et al. 2006. *Toxicol Lett* 161:115–123. With permission.)

the copper nanoparticles induced kidney toxicity and cell death via the activation of oxidative and nitrosative-stress responsive cell signaling. They observed that the exposure to copper nanoparticles increased the intracellular ROS and NO production in a dose-dependent manner. Both oxidative and nitrosative stress altered enzymatic and nonenzymatic antioxidant defenses. They assayed the activities of the antioxidant enzymes SOD (superoxide dismutase), CAT (catalase), GST (glutathione S-transferase), GR (glutathione reductase), GPx (glutathione peroxidase), and the levels of the nonenzymatic antioxidant molecules GSH (glutathione) and its metabolite GSSG (glutathione disulfide) and observed that the overload of intracellular copper nanoparticles decreased the activities of antioxidant enzymes and levels of GSH, along with increased GSSG, lipid peroxidation, and protein carbonylation. Since vitamin E treatments can prevent all these alterations related to oxidative stress induced by copper nanoparticles, it was concluded that ROS play a major role in copper nanoparticle-induced renal toxicity (Sil et al. 2011).

14.3.4.3 Toxicity of Gold Nanoparticles

Studies have shown that gold is heavily taken up by the kidneys, causing nephrotoxicity and also initiating eryptosis (erythrocyte suicidal death) (Sereemaspun et al. 2008; Sopjani et al. 2008). Studies have suggested that the size, shape, and surface charge are key factors related to the potential toxicity of medicinal gold complexes.

Gold nanoparticles produced occasional glomerular congestion in rats exposed to 10 or 20 nm particles for 7 days, but not in the glomeruli of rats exposed to 50 nm particles. The occasional dilatation of glomerular tuft blood capillaries was observed (Mohamed and Bashir 2011). Terentyuk et al. reported the proliferation of Bowman's capsule epithelial cells by gold nanoparticles, where 15 nm particles showed more effect than larger ones (Terentyuk et al. 2009). Renal tubules' epithelial lining exhibited cloudy swellings with pale cytoplasms and poorly delineated and displaced nuclei in all gold nanoparticle-treated rats. This alteration was more prominent in the proximal convoluted tubules than the distal ones, with more swelling induced by the 100 µL dose than the 50 µL dose, and with 10 and 20 nm-sized particles than larger ones. Cytoplasmic swelling might be exhibited as a result of disturbances in membrane functions that lead to the massive influx of water and Na$^+$ due to the effects of gold nanoparticles. This alteration might be accompanied by the leakage of liposome hydrolytic enzymes that lead to cytoplasmic degeneration and macromolecular crowding (Del 2005; Mohamed and Bashir 2011). Anisokaryosis (variable nuclei sizes) were observed in some renal cells. This change became noticeable 7 days after the administration of 50 nm gold nanoparticles. Several studies indicate that nuclear polymorphisms are seen in dysplasia and carcinomatous lesions (Zusman et al. 1991; Mohamed and Bashir 2011). Sections of gold nanoparticle-treated kidneys developed pyknosis in some epithelial cells of the proximal tubules with a lesser extent in the distal ones. This alteration was seen in all 10 and 20 nm gold nanoparticle-treated rats. Pyknotic nuclei exhibited the clumping and condensation of chromatin material in the periphery of the nuclei, together with the irregularity of nuclear membranes (Kumar et al. 2007; Mohamed and Bashir 2011). Histological alterations through the exposure of gold nanoparticles could be an indication of injured renal tissue due to toxicity that becomes unable to deal with as well as the accumulated residues resulting from metabolic and structural disturbances caused by gold particles. One might conclude that these alterations are size dependent, as smaller ones induced more damage to renal tissue with relation to gold nanoparticle exposure time. This might be due to the earlier accumulation of larger nanoparticles in the tissue while the smaller ones stay in the blood stream much longer due to recirculation (Mohamed and Bashir 2011).

14.3.4.4 Toxicity of Carbon Nanotubes

Reddy et al. found that the exposure to carbon nanoparticles at dosage levels of 3–300 µg/mL caused a dose-dependent cytotoxicity as revealed by mitochondrial function (MTT assay). It was found that the higher cytotoxicity of both nanoparticles was comparable with known cytotoxic agent and quartz (positive control) against all cell type tested. The IC_{50} values of both carbon nanoparticles and quartz were almost equal (39.85 ± 2.58 µg/mL), indicating the comparable cytotoxicity of carbon nanoparticles with quartz particles.

The damage to the cell membrane induced by carbon nanoparticles was also monitored by the lactate dehydrogenase enzyme (LDH) leakage assay, as LDH, a stable cytosolic enzyme in normal cells, can leak into the extracellular fluid only after membrane damage. Exposure to MWCNTs and quartz particles (10–100 µg/mL) for 48 h gave rise to a greater LDH release from human embryonic kidney (HEK) cells. The analysis of the particle exposure media for LDH demonstrates that carbon nanomaterials increase LDH leakage in a dose-dependent manner at a 48 h exposure period.

Concomitant cellular oxidative stress was manifested by reduced glutathione (GSH) levels and an increased lipid peroxidation. The contrary linear relationship between the exposure concentration and the GSH level indicated that the exposure to carbon nanoparticles reduced intracellular glutathione levels ($p < 0.01$). Moreover, the increased levels of thiobarbituric acid reactive substance content resulted in the production of malondialdehyde, an indication of lipid peroxidation. MWCNT exposures to HEK cells produced a concentration-dependent cytotoxicity. The exposure of MWCNTs to HEK cells resulted in cell membrane damage, increased lipid peroxidation, and decreased intracellular glutathione levels, indicating that oxidative stress contributes to MWCNTs induced cytotoxicity in HEK cells (Reddy et al. 2010).

14.3.4.5 Toxicity of Titanium Dioxide Nanoparticles

Titanium dioxide (TiO_2) as an inert, nontoxic pigment product, is biologically inert and used in cosmetics, paint, and building materials (Gillian et al. 2007). *In vivo* studies of the oral uptake of TiO_2 have shown it to enter blood circulation, leading to kidney and liver damage, severe pulmonary inflammation (Oberdörster et al. 2005), and emphysema.

Suxing et al. carried out a study to investigate the molecular mechanism of kidney injury in mice caused by the intragastric administration of TiO_2 nanoparticles (TiO_2 NPs). The results showed that TiO_2 NPs were accumulated in the kidneys, resulting in cell necrosis, nephric inflammation, and dysfunction. Nucleic factor-κB was activated by TiO_2 NPs exposure; promoting the expression of macrophage migration inhibitory factor; tumor necrosis factor-α; interferon-γ; interleukins-1β, -2, -4, -6, -8, -10, and -18; cross-reaction protein; transforming growth factor-β; and CYP1A1; while the expression of heat-shock protein 70 was inhibited. These findings implied that the TiO_2 NP-induced nephric injury in mice might be associated with the alteration in the expression of inflammatory cytokines and reduced detoxification of TiO_2 NPs (Suxing et al. 2011). Previous studies indicated that abnormal pathological changes in the mouse kidney and nephric dysfunction were not able to be triggered by the intraperitoneal injection of 5 mg/kg body weight TiO_2 NPs for 14 days, but with 50, 100, and 150 mg/kg body weight TiO_2 NPs exposure, impairment of kidney functions, and severe inflammatory response in the kidneys were observed (Zhao et al. 2010).

Numerous studies have demonstrated that damage to the kidneys of mice can be caused by exposure TiO_2 NPs. Wang et al. observed that the exposure to TiO_2 NPs to mice resulted in higher blood urea nitrogen and creatinine levels and the renal tubule was filled with proteinic liquids (Wang et al. 2007). Chen et al. also observed the dilatation of the renal glomerulus and proteinic liquid-filled renal tubule, but no kidney dysfunction was found in TiO_2 NP-treated mice (Chen et al. 2009). Furthermore, TiO_2 NPs were also suggested to induce nephric inflammation and impair nephric functions, which exerted its toxicity through ROS accumulation (Zhao et al. 2010). However, the molecular mechanism of TiO_2 NP-induced nephric injury remains unclear.

Acute toxicity with kidney injuries is also obtained with TiO_2 NPs in mice after oral or intraperitoneal administration (Wang et al. 2007; Chen et al. 2009).

14.3.4.6 Toxicity of Silica Particles

Omar et al. studied the effects of orally administered silica nanoparticles (nano-SiO_2) in a rat model in terms of biodistribution, toxicity, and changes to the elemental composition of feces and organs. A response to the administration of silica was observed in the rats as a change in the concentration of some elements in the excreted feces with respect to the control group. This response to silica was divided into a short-term and a long-term response. The short-term response was observed by the increase of excreted potassium (K) (up to 60%) from acute administration. No significant changes in K were found in subacute administration, but it was observed that fecal calcium (Ca), phosphorous (P), and magnesium (Mg) excretion was time-dependent and increased (up to 29%, 18%, and 39% at day 28 for Ca, P, and Mg, respectively), probably due to a decreased absorption of these elements as a result of silica administration. These increments present an increasing statistical significance, especially at day 28 (end of the subacute study) with respect to other days (Omar et al. 2012). A decrease in the absorption of Ca and P is of special importance because both Ca^{2+} and PO_4^{3-} ions control many essential cellular processes (Clapham 2007). Such a decrease in Ca and P, although not inducing an immediate short-term toxicity effect or perceptible damage on either the GI tract or liver and kidneys, may have some long-term repercussions such as changes in mitochondrial function, apoptotic cell death, or bone metabolism (Clapham 2007).

Isabelle et al. studied the cytotoxicity of different sizes of nano-SiO_2, investigated on two renal proximal tubular cell lines, human HK-2 and porcine LLC-PK1. The molecular pathways involved were studied with a focus on the involvement of oxidative stress. Specific inhibitors of endocytic pathways showed an internalization process by macropinocytosis and clathrin-mediated endocytosis for 100 nm nano-SiO_2 nanoparticles. Silica particles are known to induce nephropathy in workers by direct and indirect toxic effects via the deposition of particles in the renal parenchyma (Steenland et al. 2002; Isabelle et al. 2012).

In this study, proximal tubular (human HK-2, porcine LLC-PK1) cell lines were used to assess the potential toxicity of nano-SiO_2 on kidney cells. The analysis of the mechanism involved was based on the study of oxidative stress. A study demonstrated that nano-SiO_2 are internalized into the cell and especially localized in the cytoplasm. Studies also indicate that nano-SiO_2 is toxic on kidney cells and that their toxicity depends mainly on their size. Cell mortality increased significantly in LLC-PK1 as their size decreased, with a CI_{50} value sevenfold lower than that at 100 nm (Isabelle et al. 2012). Wang et al. reported that cytotoxicity was size- and time-dependent, with 20 nm silica nanoparticles being more cytotoxic than larger ones (50 nm). It appears that below 30 nm, changes in the structure increase the toxic potential due to variations in surface reactivities (Auffan et al. 2009a; Wang et al. 2009a).

For a mechanistic point of view, the production of ROS in kidney cells was significantly increased after exposure to nano-SiO_2, notably at 20 nm. As previously described on other cell types, the generation of ROS is size-dependent (Wang et al. 2009b; Napierska et al. 2010; Ye et al. 2010; Nabeshi et al. 2011; Gong et al. 2012). This effect is also dependent on the surface area of nano-SiO_2. Indeed, 20 nm nano-SiO_2 has a surface area which is found to be about fivefold more important as compared to 100 nm and produced significant oxidative stress in tubular cells compared to 100 nm particles, suggesting great surface reactivity at 20 nm. A high reactivity could lead to toxicity due to important nanoparticles interactions with biological systems and the cellular environment. Moreover, some authors suggest that the size-dependent toxic effect is caused by a different mechanism of ROS generation. In fact, the amount of hydroxyl radical generated in cellular models is highly dependent on the size, with high ROS production in 14 nm particles as compared to 100 and 500 nm silica nanoparticles (Shang et al. 2009).

Therefore, the toxic effects seem to be triggered by oxidative stress, as evidenced by great anion superoxide productions with 20 nm nano-SiO_2. This generation of primary ROS seems to occur through a mechanism that involves NADPH oxidase (Ushio-Fukai 2006; Nabeshi et al. 2011). After the ingestion of the xenobiotic into endosomes, NADPH oxidase is activated and generates ROS (Nabeshi et al. 2011). The anion superoxide formed could lead to the formation of radical hydroxyl, which is more reactive and destructive for cells. This radical hydroxyl induces the destruction of the membrane structure via the peroxidation of unsaturated lipids. This leads to a loss of physiological cellular integrity, resulting in cell death. The increase in ROS levels was associated with a concomitant elevation of malondialdehyde with 20 nm nano-SiO$_2$, which reflects lipid peroxidation. It has been reported by Wang et al. that oxidative stress plays an important role in the toxicity of silica nanoparticles on HEK cells, related to particle size (Wang et al. 2009a).

Furthermore, a strong correlation ($R^2 = 0.956$) was demonstrated between the decrease of cell viability and the increase of ROS production after exposure to 20 nm nano-SiO$_2$. In addition, oxygen radicals also resulted in the production of malondialdehyde after the exposure of 20 nm particles to HEK cells (Wang et al. 2009a). By identifying the mechanisms of internalization, it was found that silica nanoparticles predominantly penetrate the cells by macropinocytosis and clathrin-dependent endocytosis, resulting in the intracellular localization by endocytic vesicles as observed by transmission electron microscopy (Isabelle et al. 2012).

14.3.4.7 Toxicity of QDs

QDs, currently under development for a broad range of biomedical imaging applications, have also been exposed to induce autophagy in a variety of cell lines, including porcine kidney cells (Figure 14.4) and human mesenchymal stem cells (Seleverstov et al. 2006; Stern et al. 2008). As the underlying composition (e.g., CdSe, InGaP) and surface chemistries (e.g., protein-coated, PEGylated, silica coated) of the QDs used in these studies varied significantly, it would suggest that the unity of the nanoscale size was a significant factor in eliciting this common autophagic response. Consistent with this hypothesis, autophagy was not induced by QDs that had a tendency to aggregate to micro-scale particles intracellularly (Seleverstov et al. 2006). The dependence on the nanoscale size was also noted for neodymium oxide nanoparticle autophagy induction (Chen et al. 2005), with larger particles having reduced activities.

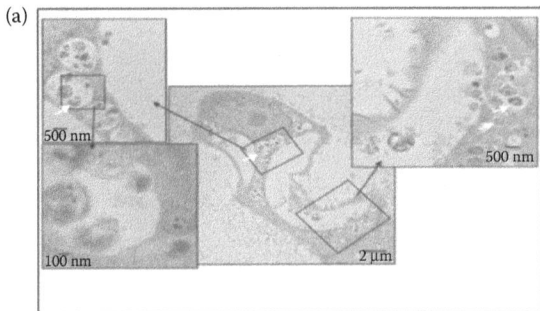

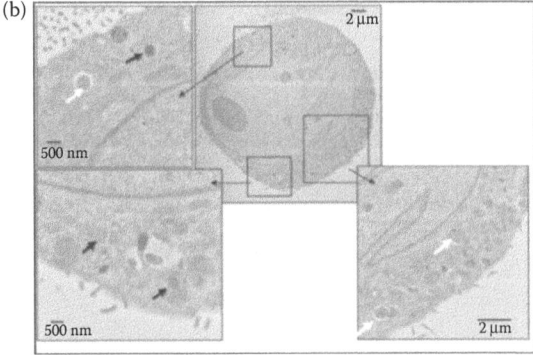

FIGURE 14.4 Transmission electron micrographs detailing autophagic vacuoles. Porcine kidney cells (LLC-PK1) were treated for 24 h with either (a) 10 nM CdSe or (b) 100 nM InGaP QDs. The white arrows represent autophagic vacuoles containing cellular debris. The black arrows represent lysosomal remnants, which consist of multilamellar vacuoles and electron-dense deposits. (Reproduced from Stern, S.T. et al. 2008. *Toxicol Sci* 106:140–152. With permission.)

Gold nanoparticles have recently been shown to increase the lysosomal pH in rat kidney cells, causing lysosomal dysfunction; this fact supports the alkalization mechanism of toxicity (Ma et al. 2011).

14.3.4.8 Toxicity of Dendrimers

Nephrotoxicity consistent with biodistribution studies has been reported, revealing a high kidney accumulation for G3 PAMAM dendrimers (IP) (Roberts et al. 1996), G3 and G4 PAMAM dendrimers (IV or IP) (Malik et al. 2000), biotinylated-PAMAM dendrimers (G0–4) (IV) (Wilbur et al. 1998), and PEG–polyester dendritic hybrids (IV) (DeJesus et al. 2002).

14.3.4.9 Toxicity of Fullerenes

Fullerenes are another category of carbon-based nanoparticles. The most common type of fullerenes have a molecular structure of C_{60} which takes the shape of a ball-shaped cage of carbon particles arranged in pentagons and hexagons. In an *in vivo* study, Chen et al. administered a water-soluble, polyalkylsulfonated C_{60} dispersion orally, intraperitoneally, and intravenously. No lethal damage was observed through oral administration, but the median lethal dose (LD_{50}) was estimated to be 600 mg/kg in intraperitoneal administration. C_{60} injected intraperitoneally or intravenously accumulated in the kidney and induced nephropathy (Chen et al. 1998).

14.4 TOXICITY OF NANOMATERIALS ON SPLEEN

14.4.1 Spleen Morphology

The spleen is a large, highly irrigated, lymphatic organ. It participates in the storage of spent blood components as well as in the maturation and recycling of lymphocytes. Anatomically, the spleen is confined and protected by a fibrous capsule, and fibrous trabeculae shape its inner structure (Figure 14.5a). Splenic blood flow is divided into two circulation pathways: one fraction transits through closed vasculature, as seen in other organs, and the remainder reaches the open circulation, where small arteries empty directly into the parenchyma (Figure 14.5b) (Moghimi 1995; Mebius and Kraal 2005). Spleen parenchyma is divided between red and white pulps. The red pulp is formed by a network of reticular fibers containing macrophages and senescent RBCs. It is engaged in the filtration of pathogens and old erythrocytes from the blood (Figure 14.5c). White pulp, situated in the vicinity of the arteries, is involved in the proliferation of lymphocytes (both B and T types). The marginal zone, delimiting the outer border of the white pulp, is composed of specialized macrophages. Efferent lymph vessels leave the spleen to reach lymphatic circulation. The spleen architecture varies between species (Demoy et al. 1999; Mebius and Kraal 2005). The most evident distinction is the existence of sinusoidal (human, rat) and nonsinusoidal (mice) splenic circulations. Judicious interpretation must be applied when comparing results between animal models. For example, distinct regions of the spleen contribute differently to particle retention in sinusoidal and nonsinusoidal spleens (Demoy et al. 1999).

CDC retention in the spleen is usually undesired because it can lead to immunogenic reactions (Dobrovolskaia and McNeil 2007). Because of the differences in the blood flow between the liver and spleen, splenic delivery is inversely related to hepatic uptake (Allen and Hansen 1991). Therefore, PEGylated CDCs that avoid uptake in the liver are delivered to the spleen in higher amounts compared to non-PEGylated colloids (Moghimi 1995; Peracchia et al. 1999). The architecture of the splenic parenchyma is highly tortuous and its reticular fiber meshwork decrease blood shear rates, maximizing interactions between blood components and splenocytes (Figure 14.5c). Moreover, extravasated colloids must squeeze through the elongated fenestrations of the sinusoids to reenter venous circulation. Physiologically, this process results in the splenic retention (and subsequent phagocytosis) of senescent erythrocytes (Mebius and Kraal 2005).

For CDCs, the physicochemical factors that limit passage through the sinusoidal sieve (high rigidity, large size (N200 nm) and elongated or irregular shapes) will therefore contribute to their

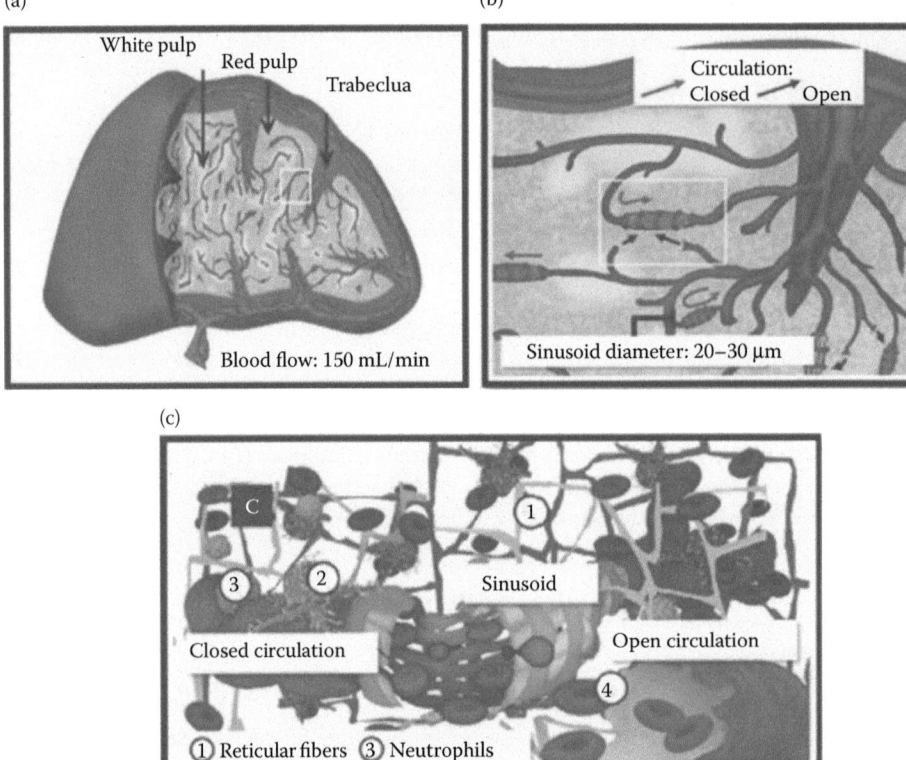

FIGURE 14.5 Structure of the spleen. (a) The spleen parenchyma is protected by a fibrous capsule and is composed of the white and red pulp. (b) From the trabecular circulation, the arteries empty into the splenic sinusoids (closed circulation) or directly into the parenchyma (open circulation). (c) Reticular fibers confer a highly tortuous architecture to the red pulp, and blood components must squeeze through the 200-nm wide fenestration of the sinusoids.

sequestration in the spleen (Demoy et al. 1999; Liu et al. 2008; Cho et al. 2009a). Sufficiently flexible particles can be stored in the spleen upon initial sieving and, if they avoid phagocytosis, are gradually released into the bloodstream within a few days (Merkel et al. 2011). Further developments and understandings of this strategy are needed to evaluate its potential for the delivery of active compounds.

14.4.2 ABC Phenomenon

Preclinical pharmacokinetic and biodistribution studies generally involve the administration of a single dose of CDCs. However, experiments have revealed that the blood clearance of a second CDC dose highly increases after an initial sensitization. This phenomenon, referred to as the accelerated blood clearance (ABC) effect, has been observed for various CDCs (e.g., liposomes) (Laverman et al. 2001), lipid complexes (Tagami et al. 2009), and polymeric nanoparticles (Ishihara et al. 2009). The ABC effect is divided into two phases, initial sensitization (induction) and effectuation. During induction, direct CDC interactions with B cells in the spleen induce the secretion of antibodies (Ishihara et al. 2009; Judge et al. 2006). Within 2–4 days of the first injection, the antibody titer in blood increases and the effectuation phase begins. CDCs injected during this period are opsonised by circulating antibodies and rapidly cleared by the liver. The blood circulation times of subsequent doses are much shortened compared to the values observed with the first injection. The ABC effect gradually decreases over 2–4 weeks. The ABC effect is more evident when CDCs are PEGylated. Even without prior sensitization, non-PEGylated CDCs are rapidly opsonized and captured by the liver. The induction of circulating antibodies, therefore, does not affect their clearance. Conversely, PEGylated CDCs are designed to slow down nonspecific capture by the MPS. The presence of specific interacting antibodies greatly enhances their clearance.

The secreted antibodies are mainly immunoglobulin-M (IgM) (Ishihara et al. 2009), but -G has also been reported (Judge et al. 2006). The specific epitopes targeted by the Ig's are still unclear, because high doses of non-PEGylated CDCs (N5 mg/kg) can induce an ABC effect on subsequently administered PEGylated CDCs (Laverman et al. 2001; Ishida et al. 2004). As with most immunization processes, the factors affecting the ABC induction phase rely on a subtle balance between immune responses and tolerance. Particulate CDCs possess characteristics, submicrometer sizes, and repetitive surface architectures similar to those of pathogens, which can stimulate immunogenic reactions (Bachmann and Jennings 2010). Nucleic acids further enhance immunogenicity (Judge et al. 2006; Tagami et al. 2009) while cytotoxic drugs have a deleterious effect on splenocytes and prevent the initiation of ABC (Laverman et al. 2001; Ishida et al. 2006). Interestingly, high doses (N5 mg/kg) of PEGylated CDCs seem to induce tolerance (Ishida et al. 2005, 2006; Tagami et al. 2009).

14.4.3 Potential Risk of Nanomaterials on the Spleen

The spleen is an important organ in the immune system and a constituting part of the RES (Xiaoyong et al. 2009). The slow clearance and tissue accumulation of potential free radical producing nanomaterials as well as the prevalence of numerous phagocytic cells in the organs of the RES make organs such as the liver and spleen to be the main targets of oxidative stress (Sumit et al. 2012).

14.4.3.1 Toxicity of Metal Nanoparticles

The *in vivo* renal clearance, biodistribution, and toxicity of gold nanoclusters were studied by Xiao-Dong et al. They concluded that gold nanoparticles have a high level of accumulation in the liver and spleen, and these accumulations can induce genetic changes and liver necrosis (Cho et al. 2010; Lipka et al. 2010). QDs larger than 15 nm can bypass renal excretion and can be accumulated in the liver and the spleen (Fischer et al. 2006; Cho et al. 2010; Zhou et al. 2011). It was found that gold nanoclusters (Au25 NCs) caused an acute immune response with a decreasing thymus index. Subsequently, the continuous Au accumulation in the liver and spleen destroys the immune system and causes irreparable damage. Furthermore, the liver and kidneys cannot obtain protection by the immune system. These damages can be reflected by clinical hematology and biochemistry (Xiao-Dong et al. 2012).

Jong et al. demonstrated the size-dependent tissue distribution of gold nanoparticles with the smallest (10 nm) nanoparticles showing the most widespread distribution (blood, liver, kidney, spleen, testis, heart, lung, thymus, and brain) whereas the larger particles (50, 100, and 250 nm) were detected only in the blood, liver, and spleen (Jong et al. 2008, Sumit et al. 2012).

Lankveld et al. carried out a kinetics study of the tissue distribution of different-sized silver nanoparticles and found that following a single exposure, the highest silver concentrations per gram organ were found in the spleen followed by the liver for both, 80 and 110 nm particles. They found that the accumulation of silver nanoparticles was present in all organs, but most significantly in the liver, lungs, and spleen and, thus, these organs may be potential target organs for toxicity after repeated exposures (Lankveld et al. 2010).

14.4.3.2 Toxicity of Nanorods

Yang et al. (2013) studied the long-term *in vivo* biodistribution and toxicity of $Gd(OH)_3$ nanorods. Biodistribution results by both, *in vivo* SPECT imaging and l-counter experiments showed that most of the $Gd(OH)_3$ nanorods were quickly cleared from the bodies of mice. However, the excretion from the spleen was slower than from other organs. Histology and hematology results showed that the spleen was slightly affected by the injection of the Gd(OH)$_3$ nanorods; there was a slight hyperplasia in the periarteriolar lymphoid sheath of the white pulp which may be caused by a nanotoxicity effect of the Gd(OH)$_3$ nanorods. This phenomenon has previously been observed in other nanoparticle-treated spleen tissues (Xiong et al. 2010).

14.4.3.3 Toxicity of Silica Nanoparticles

Sergey et al. carried out an experiment to study the *in vivo* toxicity of intravenously administered silica and silicon nanoparticles (nano-SiO$_2$). Most of the *in vivo* studies demonstrated the accumulation and persistence of nano-SiO$_2$ in macrophages of the liver and spleen, which was associated with a variable degree of inflammatory responses. The determinants of the observed toxicity are complex and include dose, surface area, particle size, and treatment regimen. The single intravenous administration of 11–15 nm nano-SiO$_2$ to rats at a dose of 7 mg/kg resulted in granuloma formation and mononuclear infiltration in the liver and spleen that persisted for 60 days post-injection. Silica and silicon nanoparticles were shown to cause granuloma formation in the organs of the RES, such as the liver and spleen (Sergey et al. 2012).

14.4.3.4 Toxicity of TiO$_2$ Nanoparticles

Li et al. studied splenic injury in mice induced by the intraperitoneal injection of TiO$_2$ NPs for 45 consecutive days and investigated the pathological changes in the spleen, the expression levels of the apoptotic genes, apoptosis and their proteins, and oxidative stress in the spleen. It was demonstrated that TiO$_2$ NPs had an obvious accumulation in the mouse spleen, resulting in the congestion and lymph nodule proliferation of splenic tissue and splenocyte apoptosis. TiO$_2$ NPs effectively activated caspase-3 and -9, decreased the Bcl-2 levels of genes and proteins, and increased the levels of cytochrome c genes and Bax and their protein expression, and promoted the accumulation of ROS. TiO$_2$ NPs induced apoptosis in mouse splenocytes via mitochondrial-mediated pathways leading to the reduction of immunity (Li et al. 2010).

14.4.3.4.1 The pathway suggested

Intraperitoneal injections of various doses of TiO$_2$ NPs increased coefficients of the spleen, and its significant accumulation in the mouse spleen induced histopathological changes of spleen, including congestion and lymph nodule proliferation. The swelling of splenocyte mitochondria was observed, splenocyte nuclei exhibited the classical morphology characteristics of apoptosis or necrosis: A reduction in nuclear size, chromatin condensation, and the appearance of a nucleolus cap. The swelling of the mitochondria suggested that TiO$_2$ NPs enter the cell, bind to the mitochondria, and lead to structural changes and subsequent effects (Li et al. 2010).

Apoptosis is an important way to maintain cellular homeostasis between cell division and cell death. It is well known that apoptosis can be triggered via two principal signaling pathways: The death receptor-mediated extrinsic apoptotic pathway and the recruitment of adaptor proteins, followed by activation of caspase-8 and the mitochondrion-mediated intrinsic apoptotic pathway. Previous investigations have revealed that TiO$_2$ NPs induced the apoptosis of the hepatocytes (Ma et al. 2009; Li et al. 2010), and speculated that the induction of apoptosis may be through the mitochondria-mediated pathway, in which mitochondrial permeability transition is promoted, followed by the release of apoptogenic factors such as apoptosis-inducing factor, cytochrome c, and caspases-3 and -9. The Bcl-2 family of proteins have also been considered as critical regulators of mitochondria-mediated apoptosis by functioning as either promoters (e.g., Bax and Bid) or inhibitors of cell death process (Suliman et al. 2001; Kandasamy et al. 2003; Yao et al. 2008).

As per the study, TiO$_2$ NPs were potent inducers of the expression of caspase-3 and -9 genes and their proteins in the mouse spleen, but did not alter caspase-8 expression. The mice treated with TiO$_2$ NPs showed obviously enhanced levels of Bax, the release of cytochrome c from the mitochondria to the cytosol, and a concomitant reduction of Bcl-2 levels. The extent of the alterations of the caspase-3, -9, and Bcl-2 family genes and their proteins varied with the doses of TiO_2 NPs. Nanoparticles can interact readily with biomolecules either on the surface or within the cells because of the small sizes of the nanoparticles. Cellular and subcellular distributions of the particles have great influences on protein aggregation, gene expression, and cell cytotoxicity (Chen and Gerion 2004; Jiang et al. 2008).

Oxidative stress is known to induce cellular death by apoptosis or necrosis (Hockenbery et al. 1993). The significant production of ROS and malondialdehyde (MDA) occurred in the mouse spleen treated with various doses of TiO_2 NPs, indicating that these TiO_2 NP-treated mouse spleens underwent severe oxidative stress. It has been reported that TiO_2 NPs could be phagocytized by neurons and microglia, which then released ROS and cause apoptosis (Dunford et al. 1995; Long et al. 2006, 2007; Wang et al. 2008). Interaction between H_2O_2 and $O_2\bullet-$ can create $\bullet OH$ and $1O_2$, which are far more destructive and can peroxidate the unsaturated lipids of the cell membrane (Fridovich 1978; Li et al. 2010).

14.4.3.5 Toxicity of Carbon Nanotubes

Many *in vivo* studies have shown that CNTs delivered to specific areas of the body are not confined to that area (Deng et al. 2007). For example, intravenously injected CNTs were shown to be taken up both by the liver and the spleen and then excreted rapidly through the kidneys (Deng et al. 2007; Schipper et al. 2008). In contrast, SWCNTs injected into the bloodstream of mice persisted within liver and spleen macrophages (Kupffer cells) (Yang et al. 2008). Inhaled MWCNTs were also shown to be able to suppress the immune function of the spleen through the signals coming from the lungs of exposed mice (Mitchell et al. 2007, 2009).

Xiaoyong et al. carried out a study to establish the splenic toxicity of water-soluble MWCNTs (SMWCNTs) in mice (Xiaoyong et al. 2009). CNTs *in vivo* can be readily scavenged from the blood and mainly entrapped by the liver, spleen, and lungs. SMWCNTs were used as a model to investigate the possible toxicity of CNTs to mouse spleens. When CNTs were injected into the blood, they could be quickly scavenged by the RES and the majority accumulated in mouse liver and spleen for a long time without being metabolized (Wang et al. 2004; Deng et al. 2007, 2008; Yang et al. 2007; Liu et al. 2008), which might affect the normal function of these tissues.

There may be two ways that the spleen scavenges SMWCNTs from blood circulation. One way is through opsonization, which makes them easily detectable and eliminated by macrophages in the spleen (Owens and Peppas 2006). The other way is that SMWCNTs can be trapped by splenic filtration. The size and deformability of foreign particles play a critical role in their clearance by the sinusoidal spleens of mice. Particles must be either small or deformable enough to avoid the splenic filtration process at the inter-endothelial cell slits in the walls of venous sinuses. The slit size rarely exceeds 200–500 nm in width and thus particles larger than 200 nm probably can be filtered by the spleen (Moghimi et al. 2001). Since SMWCNTs have dimensions of 270 nm in length and 13 nm in diameter, the single SMWCNT is hard to be trapped by splenic filtration; however, the aggregated SMWCNTs can be captured via this mechanism.

Many SMWCNTs accumulate in the spleen and no obvious decrease is observed over 2 months. It is interesting to note that accumulated SMWCNTs gradually transfer from the red pulp to the white pulp with increasing exposure times. The spleen is an exclusive immune organ, joining in the innate and adaptive immune systems. It is organized as a "tree" of branching arterial vessels, in which the small arterioles end in a venous sinusoidal system where the blood is collected from cords in the red pulp and transported to vena lienalis. The wall structure of sinuses allows the removal of foreign particles from circulation by macrophages and monocytes. Thus, it is easy to understand that SMWCNTs are enriched in the sinuses of red pulp early-on in post exposure. However, white pulp is organized as lymphoid sheaths with T- and B-cell compartments in the region of the branching arterial vessels, and is the center of the adaptive immune system (Mebius and Kraal 2005). How SMWCNTs transfer to the white pulp is currently unclear, but it may come from the translocation of antigen-presenting cells. The increasing SMWCNTs accumulated in the white pulp probably initiate the adaptive immune response to SMWCNTs (Xiaoyong et al. 2009).

14.4.3.6 Toxicity of Engineered Nanomaterials

Yokel et al. carried out an investigation to study the distribution and clearance of ceria engineered nanomaterial (ceria ENM) in blood. The presence of ceria ENM was verified by electron microscopy. The highest concentrations were seen in macrophages of RES tissues, as has been observed for other insoluble ENMs (Yokel et al. 2009). These results illustrate the prolonged retention of nondegradable ENM, consistent with reports of titanium (Fabian et al. 2008; Van et al. 2009), gold (Cho et al. 2009b), and functionalized QDs (Ballou et al. 2004; Fischer et al. 2006; Yang et al. 2007; Lin et al. 2008). Within cells of RES tissues, ceria ENM were often found in agglomerations (collections of particles loosely bound by relatively weak forces, such as van der Waals, electrostatic, physical entanglement, and surface tension). The prolonged determination of these ceria agglomerates was associated with histopathological changes, including granulomatous formations in the spleen (William et al. 2012).

No significant decreases of the amount (mass) of cerium was seen in the liver or spleen of rats up to 30 days after IV administration of 5 or 30 nm ceria ENM. Hepatic granulomas and giant cells containing agglomerates in the cytoplasm of the red pulp and thickened arterioles in white pulp were seen in the spleen (Dan et al. 2010).

14.5 SUMMARY

Nanomaterials have become an inherent part of our daily life, as they are being increasingly used in multifaceted areas such as medicines, prosthetics, engineering materials, articles, paints, clothes, and countless other household products. Ecotoxicological issues owing to nanomaterials have received even less attention, which is appalling since the production of nanomaterials has grown over the past few years.

In the last few years, research on toxicologically relevant properties of engineered nanoparticles has increased enormously. A number of international research projects and additional activities are ongoing in the European Union and the United States, nurturing the anticipation that more relevant technical and toxicological data will be published. Their widespread use allows for the likelihood of exposures for a variety of engineered nanoparticles during the whole lifecycle.

With recent developments in nanotechnology, the utilization of nanomaterials in various industrial applications such as medicine and cosmetics has enhanced successfully. Nanomaterials demonstrate useful properties such as electronic reactivity and tissue permeability that are absent in micromaterials. Hence, it is predicted that nanomaterials will develop as innovative materials in medicine and the cosmetics industry. Nevertheless, these innovative properties may be accompanied by unknown biological responses that could not have been detected by conventional toxicity assays.

Nanotoxicity is defined as the study of nature and mechanism of toxic effect of nanoscale materials on living organism and other biological system and is intended to address the toxicological activities of nanoparticles and their products to determine whether and to what extent they may pose a threat to the environment and to human health. It also deals with quantitative assessment of the severity and frequency of nanotoxic effects in relation to the exposure of the organism. The knowledge from nanotoxicological studies will be the basis for designing safe nanomaterials and nanoproducts, and also directs uses in nanomedical sciences. As various tissue and organs have different compositions, structures, and functions, toxic responses are most different once nanoparticles enter an organ.

REFERENCES

Akgun, H., Gonlusen, G., Cartwright, J., Suki, W.N., and Truong, L.D. 2006. Are gadolinium-based contrast media nephrotoxic? *Arch Pathol Lab Med* 130:1354–1357.

Allen, T.M. and Hansen, C. 1991. Pharmacokinetics of stealth versus conventional liposomes: Effect of dose. *Biochim Biophys Acta* 1068:133–141.

Anderson, J.T. and Sandlie, I. 2009. The versatile MHC Class 1-related FcRn protects IgG and albumin from degradation: Implications for the development of new diagnostics and therapeutics. *Drug Metab Pharmacokinet* 24:318–332.

Arora, S., Jain, J., Rajwade, J.M., and Paknikar, K.M. 2008. Cellular responses induced by silver nanoparticles: *In vitro* studies. *Toxicol Lett* 179(2):93–100.

Arora, S., Jain, J., Rajwade, J.M., and Paknikar, K.M. 2009. Interactions of silver nanoparticles with primary mouse fibroblasts and liver cells. *Toxicol Appl Pharmacol* 236(3):310–318.

Arslan, Z., Ates, M., McDuffy, W. et al. 2011. Probing metabolic stability of CdSe nanoparticles: Alkaline extraction of free cadmium from liver and kidney samples of rats exposed to CdSe nanoparticles. *J Hazard Mater* 192:192–199.

Auffan, M., Rose, J., Bottero, J.Y., Lowry, G.V., Jolivet, J.P., and Wiesner, M.R. 2009a. Towards a definition of inorganic nanoparticles from an environmental, health and safety perspective. *Nat Nanotechnol* 4:634–641.

Babier, O., Jacquillet, G., Tauc, M., Cougnon, M., and Poujeol, P. 2005. Effect of heavy metals on, and handling by, the kidney. *Nephron Physiol* 99:105–110.

Bachmann, M.F. and Jennings, G.T. 2010. Vaccine delivery: A matter of size, geometry, kinetics and molecular patterns. *Nat Rev Immunol* 10:787–796.

Ballestri, M., Baraldi, A., Gatti, A.M. et al. 2001. Liver and kidney foreign bodies granulomatosis in a patient with malocclusion, bruxism, and worn dental prostheses. *Gastroenterology* 121:1234–1238.

Ballou, B., Lagerholm, B.C., Ernst, L.A., Bruchez, M.P., and Waggoner, A.S. 2004. Noninvasive imaging of quantum dots in mice. *Bioconjug Chem* 15:79–86.

Barceloux, D.G. 1999. Copper. *J Toxicol Clin Toxicol* 37(2):217–230.

Bendele, A., Seely, J., Richey, C., Sennello, G., and Shopp, G. 1998. Renal tubular vacuolation in animals treated with polyethylene-glycol-conjugated proteins. *Toxicol Sci* 42:152–157.

Bertrand, N., Fleischer, J.G., Wasan, K.M., and Leroux, J.C. 2009. Pharmacokinetics and biodistribution of N-isopropylacrylamide copolymers for the design of pH-sensitive liposomes. *Biomaterials* 30:2598–2605.

Bremner, I. 1998. Manifestations of copper excess. *Am J Clin Nutr* 67(5):1069S–1073S.

Brown, D.M., Donaldson, K., Borm, P.J. et al. 2004. Calcium and ROS-mediated activation of transcription factors and TNF-α cytokine gene expression in macrophages exposed to ultrafine particles. *Am J Physiol* 286(2):L344–L353.

Cannon, G.J. and Swanson, J.A. 1992. The macrophage capacity for phagocytosis. *J Cell Sci* 101:907–913.

Chakraborty, S.P., Mahapatra, S.K., Sahu, S.K. et al. 2011. Antioxidative effect of folate-modified chitosan nanoparticles. *Asian Pac J Trop Biomed* 1(1):29–38.

Champion, J.A. and Mitragotri, S. 2006. Role of target geometry in phagocytosis. *Proc Natl Acad Sci USA* 103:4930–4934.

Champion, J.A. and Mitragotri, S. 2009. Shape induced inhibition of phagocytosis of polymer particles. *Pharm Res* 26:244–249.

Champion, J.A., Walker, A., and Mitragotri, S. 2008. Role of particle size in phagocytosis of polymeric microspheres. *Pharm Res* 25:1815–1821.

Chen, J., Dong, X., Zhao, J. et al. 2009. *In vivo* acute toxicity of titanium dioxide nanoparticles to mice after intraperitoneal injection. *J Appl Toxicol* 29:330–337.

Chen, F.Q. and Gerion, D. 2004. Fluorescent CdSe/ZnS nanocrystal-peptide conjugates for long-term, nontoxic imaging and nuclear targeting in living cells. *Nano Letters* 4:1827–1832.

Chen, Z., Meng, H., Xing, G. et al. 2006. Acute toxicological effects of copper nanoparticles *in vivo*. *Toxicol Lett* 163:109–120.

Chen, Y., Yang, L., Feng, C., and Wen, L.P. 2005. Nano neodymium oxide induces massive vacuolization and autophagic cell death in non-small cell lung cancer NCI-H460 cells. *Biochem Biophys Res Commun* 337:52–60.

Chen, H.H., Yu, C., Ueng, T.H., Chen, S. et al. 1998. Acute and subacute toxicity study of water-soluble polyalkylsulfonated C60 in rats. *Toxicol Pathol* 26:143–151.

Cherukuri, P., Gannon, C.J., Leeuw, T.K. et al. 2006. Mammalian pharmacokinetics of carbon nanotubes using intrinsic near-infrared fluorescence. *Proc Natl Acad Sci USA* 103:18882–18886.

Chitta, R.P., Soha, S.A.M., Enfeng, W. et al. 2009. *In vivo* toxicity studies of europium hydroxide nanorods in mice. *Toxicol Appl Pharmacol* 240:88–98.

Cho, M., Cho, W.S., Choi, M. et al. 2009a. The impact of size on tissue distribution and elimination by single intravenous injection of silica nanoparticles. *Toxicol Lett* 189:177–183.

Cho, W.S., Cho, M., Jeong, J., Choi, M., Cho, H.Y., Han, B.S. et al. 2009b. Acute toxicity and pharmacokinetics of 13 nm-sized PEG-coated gold nanoparticles. *Toxicol Appl Pharmacol* 236:16–24.

Cho, W.S., Cho, M., Jeong, J., Choi, M., Han, B.S., Shin, H.S. et al. 2010. Size-dependent tissue kinetics of PEG-coated gold nanoparticles. *Toxicol Appl Pharmacol* 245:116–123.

Choi, J.E., Kim, S., Ahn, J.H. et al. 2010b. Induction of oxidative stress and apoptosis by silver nanoparticles in the liver of adult zebrafish. *Aquatic Toxicol* 100:151–159.

Choi, H.S., Liu, W., Misra, P. et al. 2007. Renal clearance of nanoparticles. *Nat Biotechnol* 25:1165–1170.

Choi, H.S., Liu, W., Liu, F. et al. 2010a. Design considerations for tumour-targeted nanoparticles. *Nat Nanotechnol* 5:42–47.

Choi, C.H.J., Zuckerman, J.E., Webster, P., and Davis, M.E. 2011. Targeting kidney mesangium by nanoparticles of defined size. *Proc Natl Acad Sci USA* 108:6656–6661.

Clapham, D.E., 2007. Calcium signaling. *Cell* 131:1047–1058.

Cooper, J.R., Stradling, G.N., Smith, H., and Breadmore, S.E. 1979. The reactions of 1.0 nanometre diameter plutonium-238 dioxide particles with rat lung fluid. *Int J Radiat Biol Relat Stud Phys Chem Med* 36:453–466.

Cristina, B., Ivan, I.P.B., and Kevin, R. 2007. Nanomaterials and nanoparticles: Sources and toxicity. *Biointerphases* 2(4):MR17–MR172.

Dan, M., Tseng, M.T., Florence, R.L. et al. 2010. Short- and long-term biodistribution and oxidative stress effects of a systemically-introduced 5 nm ceria engineered nanomaterial. 49th Annual Meeting (March 7–11, 2010) of the Society of Toxicology, Salt Lake City, UT.

DeJesus, O.L.P., Ihre, H.R., Gagne, L., Frechet, J.M.J., and Szoka, F.C. 2002. Polyester dendritic systems for drug delivery applications: *In vitro* and *in vivo* evaluation. *Bioconjug Chem* 13:453–461.

Del, M.U. 2005. Swelling of hepatocytes injured by oxidative stress suggests pathological changes related to macromolecular crowding. *Med Hypotheses* 64(4):818–825.

Demoy, M., Andreux, J.P., Weingarten, C., Gouritin, B., Guilloux, V., and Couvreur, P. 1999. Spleen capture of nanoparticles: Influence of animal species and surface characteristics. *Pharm Res* 16:37–41.

Deng, X., Jia, G., Wang, H. et al. 2007. Translocation and fate of multi-walled carbon nanotubes *in vivo*. *Carbon* 45(7):1419–1424.

Deng, X.Y., Yang, S.T., Nie, H.Y., Wang, H.F., and Liu, Y.F. 2008. A generally adoptable radiotracing method for tracking carbon nanotubes in animals. *Nanotechnology* 19:075101.

Derfus, A.M., Chan, W.C.W., and Bhatia, S.N. 2004. Probing the cytotoxicity of semiconductor quantum dots. *Nano Letters* 4(1):11–18.

Dobrovolskaia, M.A., and McNeil, S.E. 2007. Immunological properties of engineered nanomaterials. *Nat Nanotechnol* 2:469–478.

Donaldson, K., Tran, L., Jimenez, L.A. et al. 2005. Combustion-derived nanoparticles: A review of their toxicology following inhalation exposure part. *Fibre Toxicol* 2:10.

Doshi, N. and Mitragotri, S. 2010. Macrophages recognize size and shape of their targets. *PLoS One* 5:e10051.

Dunford, R., Salinaro, A., Cai, L. et al. 1995. Chemical oxidation and DNA damage catalyzed by inorganic sunscreen ingredients. *Toxicol Lett* 80: 61–67.

El-Ansarty, A. and Al-Daihan, S. 2009. On the toxicity of therapeutically used nanoparticles: An overview. *J Toxicol* 1–9, Article ID 754810.

Fabian, E., Landsiedel, R., Ma-Hock, L., Wiench, K., Wohlleben, W., and Van, R.B. 2008. Tissue distribution and toxicity of intravenously administered titanium dioxide nanoparticles in rats. *Arch Toxicol* 82:151–157.

Fischer, H.C., Liu, L., Pang, K.S., and Chan, W.C.W. 2006. Pharmacokinetics of nanoscale quantum dots: *In vivo* distribution, sequestration, and clearance in the rat. *Adv Funct Mater* 16:1299–1305.

Fox, M.E., Szoka, F.C., and Fréchet, J.M.J. 2009. Soluble polymer carriers for the treatment of cancer: The importance of the molecular architecture. *Acc Chem Res* 42:1141–1151.

Fridovich, I. 1978. The biology in oxygen radical. *Science* 201:875–888.

Galhardi, C.M., Diniz, Y.S., Faine, L.A. et al. 2004. Toxicity of copper intake: Lipid profile, oxidative stress and susceptibility to renal dysfunction. *Food Chem Toxicol* 42(12):2053–2060.

Gao, W., Liu, W., Christensen, T., Zalutsky, M.R., and Chilkoti, A. 2010. *In situ* growth of a PEG like polymer from the C terminus of an intein fusion protein improves pharmacokinetics and tumor accumulation. *Proc Natl Acad Sci USA* 107:16432–16437.

Gatti, A.M. 2004. Biocompatibility of micro- and nano-particles in the colon. Part II. *Biomaterials* 25:385–392.

Gatti, A.M. and Rivasi, F. 2002. Biocompatibility of micro- and nanoparticles. Part I: In liver and kidney. *Biomaterials* 23:2381–2387.

Gatti, A.M., Montanari, S., Monari, E. et al. 2004. Detection of micro- and nano-sized biocompatible particles in the blood. *J Mater Sci Mater Med* 15:469–472.

Gauthier, M.A. and Klok, H.A. 2010. Polymer–protein conjugates: An enzymatic activity perspective. *Polym Chem* 1:1352–1373.

Gillian, F., Benjamin, J.S., and Richard, D.H. 2007. *Aquatic Toxicol* 84:415–430.

Gong, C., Tao, G., Yang, L., Liu, J., He, H., and Zhuang, Z. 2012. The role of reactive oxygen species in silicon dioxide nanoparticle-induced cytotoxicity and DNA damage in HaCaT cells. *Mol Biol Rep* 39:4915–4925.

Gopinath, P., Gogoi, S.K., Sanpui, P., Paul, A., Chattopadhyay, A., and Ghosh, S.S. 2010. Signaling gene cascade in silver nanoparticle induced apoptosis. *Colloids Surf B* 77(2):240–245.

Gratton, S.E., Ropp, P.A., Pohlhaus, P.D. et al. 2008. The effect of particle design on cellular internalization pathways. *Proc Natl Acad Sci USA* 105:11613–11618.

Haraldsson, B., Nyström, J., and Deen, W.M. 2008. Properties of the glomerular barrier and mechanisms of proteinuria. *Physiol Rev* 88:451–487.

Harashima, H., Sakata, K., Funato, K., and Kiwada, H. 1994. Enhanced hepatic uptake of liposomes through complement activation depending on the size of the liposomes. *Pharm Res* 11:402–406.

Hardman, R. 2006. A toxicologic review of quantum dots: Toxicity depends on physicochemical and environmental factors. *Environ Health Perspect* 114:165–172.

Harris, M.J. and Chess, R.B. 2003. Effect of pegylation of pharmaceuticals. *Nat Rev Drug Discov* 2:214–221.

Helinor, J.J., Gary, R.H., Frans, M.C., Karin, A., and Vicki, S. 2010. The biological mechanisms and physicochemical characteristics responsible for driving fullerene toxicity. *Toxicol Sci* 114:162–182.

Hockenbery, D.M., Oltvai, Z.N., Yin, X.M., Milliman, C.L., and Korsmeyer, S.J. 1993. Bcl-2 functions in an antioxidant pathway to prevent apoptosis. *Cell* 75:241–251.

Hsin, Y.H., Chen, C.F., Huang, S., Shih, T.S., Lai, P.S., and Chueh, P.J. 2008. The apoptotic effect of nanosilver is mediated by a ROS- and JNK-dependent mechanism involving the mitochondrial pathway in NIH3T3 cells. *Toxicol Lett* 179(3):130–139.

Hussain, N., Jaitley, V., and Florence, A.T. 2001. Recent advances in the understanding of uptake of microparticulate across the gastrointestinal lymphatics. *Adv Drug Delivery Rev* 50:107–142.

Hussain, S. M., Hess, K. L., Gearhart, J. M., Geiss, K. T., and Schlager, J. J. 2005. *In vitro* toxicity of nanoparticles in BRL 3A rat liver cells. *Toxicol In Vitro* 19(7):975–983.

Isabelle, P., Marie, M., Marine, R., Igor, P., and Béatrice, L.A. 2012. Implication of oxidative stress in size-dependent toxicity of silica nanoparticles in kidney cells. *Toxicology* 299:112–124.

Ishida, T., Atobe, K., Wang, X.Y., and Kiwada, H. 2006. Accelerated blood clearance of PEGylated liposomes upon repeated injections: Effect of doxorubicin encapsulation and high dose first injection. *J Control Release* 115:251–258.

Ishida, T., Harada, M., Wang, X.Y., Ichihara, M., Irimura, K., and Kiwada, H. 2005. Accelerated blood clearance of PEGylated liposomes following preceding liposome injection: Effects of lipid dose and PEG surface-density and chain length of the first-dose liposomes. *J Control Release* 105:305–317.

Ishida, T., Ichikawa, T., Ichihara, M., Sadsuka, Y., and Kiwada, H. 2004. Effect of physicochemical properties of initially injected liposomes on the clearance of subsequently injected PEGylated liposomes in mice. *J Control Release* 95:403–412.

Ishihara, T., Takeda, M., and Sakamoto, H. 2009. Accelerated blood clearance phenomenon upon repeated injection of PEG-modified PLA nanoparticles, *Pharm Res* 26:2270–2279.

Jacobs, F., Wisse, E., and De Geest, B. 2010. The role of the sinusoidal cells in hepatocyte directed gene transfer. *Am J Pathol* 176:14–21.

Jiang, W., Kim, B.Y., Rutka, J.T., and Chan, W.C. 2008. Nanoparticle-mediated cellular response is size-dependent. *Nat Nanotechnol* 3:145–150.

Jong, W.H.D., Hagens, W.I., Krystck, P., Burger, M.C., Sips, A.J.A.M., and Geertsma, R.E. 2008. Particle size-dependent organ distribution of gold nanoparticles after intravenous administration. *Biomaterials* 29:1912–1919.

Judge, A., McClintock, K., Phelps, J.R., and MacLachlan, I. 2006. Hypersensitivity and loss of disease site targeting caused by antibody responses to PEGylated liposomes. *Mol Ther* 13:328–337.

Kandasamy, K., Srinivasula, S.M., Alnemri, E.S. et al. 2003. Involvement of proapoptotic molecules Bax and Bak in tumor necrosis factor-related apoptosis-inducing ligand (TRAIL)—Induced mitochondrial disruption and apoptosis: Differential regulation of cytochrome c and Smac/DIABLO release. *Cancer Res* 63:1712–1721.

Kawata, K., Osawa, M., and Okabe, S. 2009. *In vitro* toxicity of silver nanoparticles at noncytotoxic doses to HepG2 human hepatoma cells. *Environ Sci Technol* 43(15):6046–6051.

Khandoga, A., Stampfl, A., and Takenaka, S. 2004. Ultrafine particles exert prothrombotic but not inflammatory effects on the hepatic microcirculation in healthy mice *in vivo*. *Circulation* 109:1320–1325.

Kopecek, J., Kopeckova-Rejmanova, P., Minko, T., Lu, Z.R., and Peterson, C.M. 2001. Water soluble polymers in tumor targeted delivery. *J Control Release* 74:147–158.

Kostarelos, K., Lacerda, L., and Pastorin, G. 2007. Functionalized carbon nanotube cellular uptake and internalisation mechanism is independent of functional group and cell type. *Nat Nanotechnol* 2:108–113.

Kumar, V., Abbas, A., Nelson, F., and Mitchell, R. 2007. Robbins basic pathology. *Robbins Basic Pathol* 6:9–10.

Lacerda, L., Herrero, M.A., Venner, K., Bianco, A., Prato, M., and Kostarelos, K. 2008. Carbon nanotube shape and individualization critical for renal excretion. *Small* 4:1130–1132.

Lankveld, D.P.K., Oomen, A.G., Krystek, P. et al. 2010. The kinetics of the tissue distribution of silver nanoparticles of different sizes. *Biomaterials* 31:8350–8361.

Lanone, S. and Boczkowski, J. 2006. Biomedical applications and potential health risks of nanomaterials: Molecular mechanisms. *Curr Mol Med* 6:651–663.

Laverman, P., Carstens, M.G., Boerman, O.C. et al. 2001. Factors affecting the accelerated blood clearance of polyethylene glycol-liposomes upon repeated injection. *J Pharmacol Exp Ther* 298:607–612.

Lei, R., Wu, C., and Yang, B. 2008. Integrated metabolomic analysis of the nano-sized copper particle-induced hepatotoxicity and nephrotoxicity in rats: A rapid *in vivo* screening method for nanotoxicity. *Toxicol Appl Pharmacol* 232(2):292–301.

Li, N., Ma, L.L., Wang, J. et al. 2010. Interaction between nano-anatase TiO_2 and liver DNA from mice *in vivo*. *Nanoscale Res Lett* 5:108–115.

Li, C., Price, J.E., Milas, L. et al. 1999. Antitumor activity of poly (L-glutamic acid)-paclitaxel on syngeneic and xenografted tumors. *Clin Cancer Res* 5:891–897.

Li, N., Yanmei, D., Mengmeng, H. et al. 2010. Spleen injury and apoptotic pathway in mice caused by titanium dioxide nanoparticles. *Toxicol Lett* 195:161–168.

Lin, P., Chen, J.W., Chang, L.W. et al. 2008. Computational and ultra structural toxicology of a nanoparticle, Quantum Dot 705, in mice. *Environ Sci Technol* 42:6264–6270.

Lipka, J., Semmler-Behnke, M., Sperling, R.A. et al. 2010. Biodistribution of PEG-modified gold nanoparticles following intratracheal instillation and intravenous injection. *Biomaterials* 31:6574–6581.

Liu, Z., Davis, C., Cai, W., He, L., Chen, X., and Dai, H. 2008. Circulation and long-term fate of functionalized, biocompatible single-walled carbon nanotubes in mice probed by Raman spectroscopy. *Proc Natl Acad Sci USA* 105:1410–1415.

Liu, G., Li, X., Qin, B., Xing, D., Guo, Y., and Fan, R. 2004. Investigation of the mending effect and mechanism of copper nano-particles on a tribologically stressed surface. *Tribol Lett* 17(4):961–966.

Long, T.C., Saleh, N., Tilton, R.D., Lowry, G., and Veronesi, B. 2006. Titanium dioxide (P25) produces reactive oxygen species in immortalized brain microglia (BV2): Implications for nanoparticle neurotoxicity. *Environ Sci Technol* 40:4346–4352.

Long, T.C., Tajuba, J., Sama, P. et al. 2007. Nano-TiO_2 stimulates ROS in brain microglia and damages neurons *in vitro*. *Environ Health Perspect* 115:1631–1637.

Ma, X., Wu, Y., Jin, S. et al. 2011. Gold nanoparticles induce autophagosome accumulation through size-dependent nanoparticle uptake and lysosome impairment. *ACS Nano* 5: 8629–8639.

Ma, L.L., Zhao, J.F., Wang, J. et al. 2009. The acute liver injury in mice caused by nano-anatase TiO_2. *Nanoscale Res Lett* 4:1275–2128.

Maack, T., Johnson, V., Kau, S.T., Figueiredo, J., and Sigulem, D. 1979. Renal filtration, transport, and metabolism of low molecular-weight proteins: A review. *Kidney Int* 16:251–270.

Malik, N., Wiwattanapatapee, R., Klopsch, R. et al. 2000. Dendrimers: Relationship between structure and biocompatibility *in vitro*, and preliminary studies on the biodistribution of I-125-labelled polyamidoamine dendrimers *in vivo*. *J Control Release* 65:133–148.

Mancini, M.C., Kairdolf, B.A., Smith, A.M., and Nie, S. 2008. Oxidative quenching and degradation of polymer encapsulated quantum dots: New insights into the long-term fate and toxicity of nanocrystals *in vivo*. *J Am Chem Soc* 130:10836–10837.

Mebius, R.E. and Kraal, G. 2005. Structure and function of the spleen. *Nat Rev Immunol* 5:606–616.

Merkel, T.J., Jones, S.W., Herlihy, K.P. et al. 2011. Using mechanobiological mimicry of red blood cells to extend circulation times of hydrogel microparticles. *Proc Natl Acad Sci USA* 108:586–591.

Michael, A.F., Keane, W.F., Raij, L., Vernier, R.L., and Mauer, S.M. 1980. Glomerular mesangium. *Kidney Int* 17:141–154.

Mitchell, L.A., Gao, J., Wal, R.V., Gigliotti, A., Burchiel, S.W., and McDonald, J.D. 2007. Pulmonary and systemic immune response to inhaled multiwalled carbon nanotubes. *Toxicol Sci* 100(1):203–214.

Mitchell, L.A., Lauer, F.T., Burchiel, S.W., and McDonald, J.D. 2009. Mechanisms for how inhaled multi-walled carbon nanotubes suppress systemic immune function in mice. *Nat Nanotechnol* 4(7):451–456.

Moghimi, S.M. 1995. Mechanisms of splenic clearance of blood cells and particles: Towards development of new splenotropic agents. *Adv Drug Delivery Rev* 17:103–115.

Moghimi, S.M., Hunter, A.C., and Murray, J.C. 2001. Long-circulating and target-specific nanoparticles: Theory to practice. *Pharmacol Rev* 53:283–318.

Mohamed, A.K.A. and Bashir, M.J. 2011. Renal tissue alterations were size-dependent with smaller ones induced more effects and related with time exposure of gold nanoparticles. *Lipids Health Dis* 10:163.

Nabeshi, H., Yoshikawa, T., Matsuyama, K. et al. 2011. Amorphous nanosilica induce endocytosis-dependent ROS generation and DNA damage in human keratinocytes. *Part Fibre Toxicol* 8:1.

Napierska, D., Thomassen, L.C., Lison, D., Martens, J.A., and Hoet, P.H. 2010. The nanosilic hazard: Another variable entity. *Part Fibre Toxicol* 7:39.

Nasongkla, N., Chen, B., Macaraeg, N., Fox, M.E., Fréchet, J.M.J., and Szoka, F.C. 2009. Dependence of pharmacokinetics and biodistribution on polymer architecture: Effect of cyclic versus linear polymers. *J Am Chem Soc* 131:3842–3843.

Neerman, M.F., Zhang, W., Parrish, A.R., and Simanek, E.E. 2004. *In vitro* and *in vivo* evaluation of a melamine dendrimer as a vehicle for drug delivery. *Int J Pharm* 281:129–132.

Oberdörster, G., Maynard, A., Donaldson, K. et al. 2005. Principles for characterizing the potential human health effects from exposure to nanomaterials: Elements of a screening strategy. *Part Fibre Toxicol* 2:1–35.

Omar, L., Julie, L., Lütfiye, A. et al. 2012. Effects of SiC nanoparticles orally administered in a rat model: Biodistribution, toxicity and elemental composition changes in faeces and organs. *Toxicol Appl Pharmacol* 264:232–245.

Owens III, D.E. and Peppas, N.A. 2006. Opsonization, biodistribution, and pharmacokinetics of polymeric nanoparticles. *Int J Pharm* 307:93–102.

Peracchia, M.T., Fattal, E., and Desmaële, D. 1999. Stealth(R) PEGylated polycyanoacrylate nanoparticles for intravenous administration and splenic targeting. *J Control Release* 60:121–128.

Petros, R.A. and DeSimone, J.M. 2010. Strategies in the design of nanoparticles for therapeutic applications. *Nat Rev Drug Discov* 9:615–627.

Poon, V.K.M. and Burd, A. 2004. *In vitro* cytotoxity of silver: Implication for clinical wound care. *Burns* 30(2):140–147.

Rahman, I. 2000. Regulation of nuclear factor-κB, activator protein-1, and glutathione levels by tumor necrosis factor-α and dexamethasone in alveolar epithelial cells. *Biochem Pharmacol* 60:1041–1049.

Rahman, I., Biswas, S.K., Jimenez, L.A. et al. 2005 Glutathione, stress responses, and redox signaling in lung inflammation. *Antioxid Redox Signaling* 7:42–59.

Reddy, Y.N., Reddy, A.R.N., Krishnaa, D.R., and Himabindu, V. 2010. Multi wall carbon nanotubes induce oxidative stress and cytotoxicity in human embryonic kidney (HEK293) cells. *Toxicology* 272:11–16.

Roberts, J.C., Bhalgat, M.K., and Zera, R.T. 1996. Preliminary biological evaluation of polyamidoamine (PAMAM) Starburst (TM) dendrimers. *J Biomed Mater Res A* 30:53–65.

Roberts, R.A., Ganey, P.E., Ju, C., Kamendulis, L.M., Rusyn, I., and Klaunig, J.E. 2007. Role of Kupffer cell in mediating hepatic toxicity and carcinogenesis. *Toxicol Sci* 96 (1):2–15.

Ruggiero, A., Villa, C.H., Bander, E. et al. 2010. Paradoxical glomerular filtration of carbon nanotubes. *Proc Natl Acad Sci USA* 107:12369–12374.

Schellenberger, V., Wang, C.-W., Geething, N.C. et al. 2009. A recombinant polypeptide extends the *in vivo* half-life of peptides and proteins in a tunable manner. *Nat Biotechnol* 27:1186–1191.

Schipper, M.L., Nakayama-Ratchford, N., Davis, C.R. et al. 2008. A pilot toxicology study of single-walled carbon nanotubes in a small sample of mice. *Nat Nanotechnol* 3:216–221.

Schwab, A.J. and Pang, K.S. 2000. The multiple indicator dilution method and its utility in risk assessment. *Environ Health Perspect* 108:861–872.

Schwartz, J. 2001. Air pollution and blood markers of cardiovascular risks. *Environ Health Perspects* 109(suppl. 3):405–409.

Seleverstov, O., Zabirnyk, O., Zscharnack, M. et al. 2006. Quantum dots for human mesenchymal stem cells labeling. A size-dependent autophagy activation. *Nano Letters* 6:2826–2832.

Sereemaspun, A., Rojanathanes, R., and Wiwanitkit, V. 2008. Effect of gold nanoparticle on renal cell: An implication for exposure risk. *Ren Fail* 30:323–325.

Sergey, I., Sergey, Z., Galina, Y., Vladimir, T., Dmitry, K., and Michael, G. 2012. *In vivo* toxicity of intravenously administered silica and silicon nanoparticles. *Materials* 5:1873–1889.

Seymour, L.W., Duncan, R., Strohalm, J., and Kopecek, J. 1987. Effect of molecular weight (Mw) on N (2-hydroxypropyl) methacrylamide copolymers on body distribution and rate of excretion after subcutaneous, intraperitoneal, and intravenous administration to rats. *J Biomed Mater Res* 21:1341–1358.

Shang, Y., Zhu, T., Li, T., and Zhao, J. 2009. Size-dependent hydroxyl radicals generation induced by SiO_2 ultra-fine particles: The role of surface iron. *Sci China Ser B Chem* 52:1033–1041.

Sharma, G., Valenta, D.T., and Altman, Y. 2010. Polymer particle shape independently influences binding and internalization by macrophages. *J Control Release* 147:408–412.

Sharma, V., Singh, P., Pandey, A.K., and Dhawan, A. 2012. Induction of oxidative stress, DNA damage and apoptosis in mouse liver after sub-acute oral exposure to zinc oxide nanoparticles. *Mutat Res* 745:84–91.

Sil, P.C., Sarkar, A., Das, J., and Manna, P. 2011. Nano-copper induces oxidative stress and apoptosis in kidney via both extrinsic and intrinsic pathways. *Toxicology* 290:208–217.

Sopjani, M., Föller, M., and Lang, F. 2008. Gold stimulates Ca^{2+} entry into and subsequent suicidal death of erythrocytes. *Toxicology* 244:271–279.

Steenland, K., Rosenman, K., Socie, E., and Valiante, D. 2002. Silicosis and end-stage renal disease. *Scand J Work Environ Health* 28:439–442.

Stern, S.T., Zolnik, B.S., McLeland, C.B., Clogston, J., Zheng, J., and McNeil, S.E. 2008. Induction of autophagy in porcine kidney cells by quantum dots: A common cellular response to nanomaterials? *Toxicol Sci* 106:140–152.

Suliman, A., Lam, A., Datta, R., and Srivastava, R.K. 2001. Intracellular mechanisms of TRAIL: Apoptosis through mitochondrial-dependent and -independent pathways. *Oncogene* 20:2122–2133.

Sumit, A., Jyutika, M.R., and Kishore, M.P. 2012. Nanotoxicology and *in vitro* studies: The need of the hour. *Toxicol Appl Pharmacol* 258:151–165.

Sung, J.H., Ji, J.H., Yoon, J.U. et al. 2008. Lung function changes in Sprague–Dawley rats after prolonged inhalation exposure to silver nanoparticles. *Inhalation Toxicol* 20(6):567–574.

Suxing, G., Zengli, Z., Lei, Z. et al. 2011. Molecular mechanism of kidney injury of mice caused by exposure to titanium dioxide nanoparticles. *J Hazardous Mater* 195:365–370.

Tagami, T., Nakamura, K., Shimizu, T., Ishida, T., and Kiwada, H. 2009. Effect of siRNA in PEGcoated siRNA-lipoplex on anti-PEG IgM production. *J Control Release* 137:234–240.

Takenaka, S., Karg, E., Roth, C. et al. 2001. Pulmonary and systemic distribution of inhaled ultrafine silver particles in rats. *Environ Health Perspect* 109(4):547–551.

Terentyuk, G., Maslyyakova, G., Suleymanova, L. et al. 2009. Tracking gold nanoparticles in the body. *J Biomed Opt* 14:19–16.

Unfried, K., Albrecht, C., Klotz, L-O. et al. 2007. Cellular response to nanoparticles: Target structures and mechanisms. *Nanotoxicology* 1:52–71.

Ushio-Fukai, M. 2006. Localizing NADPH oxidase-derived ROS. *Sci STKE* 349:re8.

Van, R.B., Landsiedel, R., Fabian, E., Burkhardt, S., Strauss, V., and Ma-Hock, L. 2009. Comparing fate and effects of three particles of different surface properties: Nano-TiO_2, pigmentary TiO_2 and quartz. *Toxicol Lett* 186:152–159.

Wang, B., Feng, W., Wang, M. et al. 2008. Acute toxicological impact of nano- and submicro-scaled zinc oxide powder on healthy adult mice. *J Nanopart Res* 10:263–276.

Wang, B., Feng, W.-Y., Wang, T.-C. et al. 2006. Acute toxicity of nano- and micro-scale zinc powder in healthy adult mice. *Toxicol Lett* 161:115–123.

Wang, F., Gao, F., Lan, M., Yuan, H., Huang, Y., and Liu, J. 2009a. Oxidative stress contributes to silica nanoparticle-induced cytotoxicity in human embryonic kidney cells. *Toxicol In Vitro* 23:808–815.

Wang, F., Jiao, C., Liu, J., Yuan, H., Lan, M., and Gao, F. 2009b. Oxidative mechanisms contribute to nanosize silican dioxide-induced developmental neurotoxicity in PC12 cells. *Toxicol In Vitro* 23:808–815.

Wang, J.X., Liu, Y., Jiao, F. et al. 2008. Time-dependent translocation and potential impairment on central nervous system by intranasally instilled TiO_2 nanoparticles. *Toxicology* 254:82–90.

Wang, H.F., Wang, J., Deng, X.Y. et al. 2004. Biodistribution of carbon single-wall nanotubes in mice. *J Nanosci Nanotechnol* 4:1019–1024.

Wang, J.X., Zhou, G.Q., Chen, C.Y. et al. 2007. Acute toxicity and biodistribution of different sized titanium dioxide particles in mice after oral administration. *Toxicol Lett* 168:176–185.

Wilbur, D.S., Pathare, P.M., Hamlin, D.K., Buhler, K.R., and Vessella, R.L. 1998. Biotin reagents for antibody pretargeting. 3. Synthesis, radioiodination, and evaluation of biotinylated starburst dendrimers. *Bioconjug Chem* 9:813–825.

William, K.B., Rui, C., Chunying, C., and Robert, A.Y. 2012. The neurotoxic potential of engineered nanomaterials. *NeuroToxicology* 33:902–910.

Winge, D.R. and Mehra, R.K. 1990. Host defenses against copper toxicity. *Int Rev Exp Pathol* 31:47–83.

Wisse, E., Jacobs, F., Topal, B., Frederik, P., and De Geest, B. 2008. The size of endothelial fenestrae in human liver sinusoids: Implications for hepatocyte-directed gene transfer. *Gene Ther* 15:1193–1199.

Xiao-Dong, Z., Di, W., Xiu, S., Pei-Xun, L., Fei-Yue, F., and Sai-Jun, F. 2012. *In vivo* renal clearance, biodistribution, toxicity of gold nanoclusters. *Biomaterials* 33:4628–4638.

Xiaoyong, D., Fei, W., Zhen, L. et al. 2009. The splenic toxicity of water soluble multi-walled carbon nanotubes in mice. *Carbon* 47:1421–1428.

Xiong, L.Q., Yang, T.S., Yang, Y., Xu, C.J., and 2Li, F.Y. 2010. Long-term *in vivo* biodistribution imaging and toxicity of polyacrylic acid-coated upconversion nanophosphors. *Biomaterials* 31:7078–7085.

Yang, R.S., Chang, L.W., Wu, J.P. et al. 2007. Persistent tissue kinetics and redistribution of nanoparticles, quantum dot 705, in mice: ICP-MS quantitative assessment. *Environ Health Perspect* 115:1339–1343.

Yang, S., Guo, W., Lin, Y. et al. 2007. Biodistribution of pristine single-walled carbon nanotubes *in vivo*. *J Phys Chem C* 111:17761–17764.

Yang, Z., Liu, Z.W., Allaker, R.P. et al. 2010. A review of nanoparticles functionality and toxicity on the central nervous system. *J R Soc Interface* S411–S422.

Yang, S., Wang, X., Jia, G. et al. 2008. Long-term accumulation and low toxicity of single-walled carbon nanotubes in intravenously exposed mice. *Toxicol Lett* 181:182–189.

Yang, Y., Yun, S., Ying, L. et al. 2013. Long-term *in vivo* biodistribution and toxicity of Gd (OH)$_3$ nanorods. *Biomaterials* 34:508–515.

Yao, J.C., Jiang, Z.Z., Duan, W.G. et al. 2008. Involvement of mitochondrial pathway in triptolide induced cytotoxicity in human normal liver L-02 cells. *Biol Pharm Bull* 31:592–597.

Ye, Y., Liu, J., Xu, J., Sun, L., Chen, M., and Lan, M. 2010. Nano-SiO$_2$ induces apoptosis via activation of p53 and Bax mediated by oxidative stress in human hepatic cell line. *Toxicol In Vitro* 24:751–758.

Yokel, R.A., Florence, R.M.D., Unrine, J. et al. 2009. Safety/toxicity assessment of ceria (a model engineered NP) to the brain. In Presentation: Interagency Nanotechnology Implications Grantees Workshop-EPA, NSF, NIEHS, NIOSH, and DOE; November 9–11. Las Vegas, NV.

Zalevsky, J., Chamberlain, A.K., Horton, H.M. et al. 2010. Enhanced antibody half-life improves *in vivo* activity. *Nat Biotechnol* 28:157–159.

Zhang, T., Stilwell, J.L., Gerion, D. et al. 2006. Cellular effect of high doses of silica-coated quantum dot profiled with high throughput gene expression analysis and high content cellomics measurements. *Nano Letters* 6:800–808.

Zhao, J.F., Wang, J., Wang, S.S. et al. 2010. The mechanism of oxidative damage in nephrotoxicity of mice caused by nano-anatase TiO$_2$. *J Exp Nanosci* 5:447–462.

Zhou, C., Long, M., Qin, Y., Sun, X., and Zheng, J. 2011. Luminescent gold nanoparticles with efficient renal clearance. *Angew Chem Int* 123:3226–3230.

Zusman, I., Kozlenko, M., and Zimber, A. 1991. Nuclear polymorphism and nuclear size in precarcinomatous and carcinomatous lesions in rat colon and liver. *Cytometry* 12(4):302–307.

15 Regulatory Implications of Nanotechnology

Lynn L. Bergeson and Michael F. Cole

CONTENTS

15.1 Introduction 315
15.2 The Beginnings of Regulation 315
15.3 Regulatory Challenges 316
15.4 Lack of Reliable Data 318
15.5 Lack of Reliable Tools 318
15.6 U.S. Food and Drug Administration 320
15.7 U.S. Environmental Protection Agency 326
15.8 Work Place Concerns 329
15.9 The European Union 330
    15.9.1 Cosmetic Directive 333
    15.9.2 Food Labeling 334
    15.9.3 Medical Device and Pharmaceutical 335
15.10 Nanomaterials in Products 336
    15.10.1 Classification, Labeling, and Packaging 336
15.11 Other Countries 338
    15.11.1 Germany 338
    15.11.2 United Kingdom 340
    15.11.3 France 341
    15.11.4 Denmark 341
    15.11.5 Netherlands 342
    15.11.6 Australia 342
    15.11.7 Canada 344
15.12 Conclusion 345

15.1 INTRODUCTION

This chapter examines the effect regulation, albeit nascent, is having or may have on the commercialization of nanomaterials in the pharmaceutical and medical fields in the United States and abroad. It will also discuss how authorities are addressing the effects the application of nanotechnology are believed to be having on the environment. It will examine the types of regulatory schemes being enacted or considered, and whether different systems are likely to have differing effects on nanomedicine, including the effect on competition among companies and countries.

15.2 THE BEGINNINGS OF REGULATION

In recent years, there has been an increase in the incorporation of nanomaterials in products and components utilized in the pharmaceutical and medical fields. There are a number of products containing nanomaterials already in commercial distribution and many more in the development and concept stages, respectively. Products are heralded as providing more targeted drug delivery that may be the key to greater success in treating cancer and other illnesses. At the same time, there are those who sound a note of caution due to the perceived potential for serious human adverse health effects because of the different characteristics of materials at the nanoscale level that come into contact with biological systems. It is postulated that generally smaller particles may be more toxic, with doses increasing as particle size decreases. It is claimed that present risk assessment methodologies are not sufficient to address this toxicity because they generally consider mass alone, whereas with nanotechnology the assessment must be of the number of particles and surface area. The critics claim that there are little or no data on the physiological response to nanoparticles. A recurrent concern, for example, is the mobility of nanoparticles, such as the potential for particles to pass freely in the bloodstream and cross the blood–brain barrier to affect neurological function. Because of the severe consequences of these and other anticipated biological interactions, some call for strict regulation of medical products utilizing nanotechnology, or even the ban of such products, until such time as controlled research establishes the safety of nanoparticles in specific applications. Conversely, proponents of the use of nanotechnology point to the lack of substantiated reports of harm and argue against overregulation that could stifle the development of products and material that could greatly benefit society.

Governments reviewing the need to regulate the use of nanotechnology to respond to questions of safety are faced with the most difficult sort of decision. There are at least identifiable theoretical risks that must be considered, but there are equally compelling arguments that most of the materials being developed on the nanoscale, such as nanosilver and titanium dioxide, pose no more risk of adverse health consequences than do their macro counterparts. There is strong pressure on a national level in many countries to support the development of products using nanomaterials, or fall behind the relentless press of competition for commercial strength throughout the rest of the world. Nanotech companies are also under pressure to justify the extensive investment already made and that needs to continue to be made into the future in advance of successful commercialization. Companies state repeatedly that they cannot survive and prosper in the face of stifling regulation. Regulatory bodies must take all these factors into account and develop a rational system that will signal to the public that its health and safety interests are being protected, while assuring developers and supporters that their effort will not be compromised if certain safeguards are observed.

15.3 REGULATORY CHALLENGES

A major problem confronting any governmental body considering the need to regulate nanotechnology is how new the technology is, the lack of evidence regarding the nature of the materials being used, and the effect they might have, individually and in the aggregate, on biological systems, and the short- and long-term effects on the environment. There is little agreement in the scientific community on the protocols for studying nanomaterials; little commonality in terms of the materials being tested; little in the way of long-term results; considerable inconsistencies in results, making comparisons difficult; and virtually nothing in the way of clinical testing.

The governing bodies of nations in the global nanotechnology race do not have the option of doing nothing when dealing with medicines or hardware used in serious situations, and the release of nanomaterials in the environment. The specter of some potential disaster resulting from the perceived lack of regulatory attention lurks in the background, and critics are unrelenting in their push to get attention for regulation. Evidence of potential harm, such as alleged harm resulting from exposure to some types of carbon nanotubes (CNTs), cannot be ignored. At the same time, as mentioned above, there is little hard evidence of any adverse effects being reported from the use of nanomaterials in humans, and that cautions against overreacting—particularly given the national and company competitive concerns.

A tempting choice for a regulatory body is the application of existing regulatory provisions for pharmaceuticals and medical products to ones produced by, and/or containing, nanomaterials. Doing so would circumvent the delays caused by the rulemaking process that developing an entirely new system would invite. Invariably, however, when the use of existing rules is raised for comment, the question arises as to whether that framework is relevant and/or adequate to address issues suggested by nano products. Here, the most pertinent question is often whether the proposed products fit into the existing scheme, or are so "new" that the original benefit-to-risk equations do not adequately address the risks posed by the products.

In some countries, there may not be a comprehensive body of regulations available for possible adaptation. While authorities consider what to do, they often times encourage voluntary action by the regulated industry.

At the other end of the spectrum, governing bodies may consider enactment of mandatory regulation of the entire class of products, or they may propose regulation of products on a case-by-case basis, as needed. Typically, such action so early in the life cycle of the products involved only follow-up reports of health hazards where regulatory action is mandated. Absent is the need to address pressing public health issues, the aim of any of these mechanisms will likely be to collect relevant information first. The government can be expected to seek information on products that are in distribution, or proposed for sale, that contain nanomaterials and whether they are labeled as such, who is producing them, the type of manufacturing safeguards that are utilized, and related questions.

A regulator can seek to obtain this information by asking manufacturers and developers to submit it voluntarily. It could be done by formulating a variety of data requests and publishing them widely. Voluntary action could be as simple as advising anyone proposing to sell a pharmaceutical or medical product utilizing nanotechnology to contact the regulator and discuss what, if any, action should be undertaken prior to going to market. The ultimate effectiveness of voluntary action is, of course, quantitative information that is forthcoming and the development of information on product performance establishes that the products generally present little or no risk. Also, information obtained can be useful, but only if everyone subject to regulation responds. Otherwise, an incomplete, and perhaps false, picture can be created of an industry.

Mandatory regulations, at least as to notification and identification of the use of nanomaterials, is more likely to be enacted, while at the same time deferring any premarket approval until more is known about what should be required in any such applications. The initial activity is often the agreement on a definition of the substance or product to be regulated, so the affected industry and the public will know what is being addressed. The next objective is to establish an inventory of products being sold, so that the regulator and the public are not "blind." Such regulations usually require registration of companies and locations of manufacture and distribution, and notification of products sold or to be sold. An integral part of the notification about products being sold is to know the intended uses and the product labeling meant to reflect those uses. Hand-in-hand with that is information on any adverse effect in the marketplace from use, and from clinical studies done on the product or similar products for the intended use involved.

At that point, the regulator will begin to accumulate the information likely to forestall a major product failure, and then go on to formulate proposals for dealing with the risks presented by premarket submissions or conformance to standards. Given the paucity of hard information from studies involving nanomaterials, the regulator may also seek to encourage meaningful testing. Eventually, the shape of the data requirements may be best governed by regulations detailing the information necessary to demonstrate safety and, perhaps, effectiveness.

While it is still in the early days with respect to the commercialization of nanomaterials for pharmaceuticals and medical products, many nations and multinational entities are beginning to confront the situation in earnest. In the space provided, it is not possible to address in detail each such global effort. Instead, the following sections will address in some detail efforts in the United States, the European Union (EU), and countries such as Canada and Australia, since collectively each has progressed the farthest in dealing with the challenge presented by the marketing of pharmaceuticals and medical products incorporating or produced using nanomaterials, and the effect that the testing, production, and use of those materials in products might have on the environment. The regulations and guidances put forward to date will be presented in the words of the regulator, so the reader can understand the thinking behind the actions being taken. The rest of the chapter will address how each of the above governing bodies have dealt with the issues of learning about inventory through registration and notification, about intended uses through labeling, product performance by adverse reaction reporting, and the encouragement of directed research to demonstrate safety.

15.4 LACK OF RELIABLE DATA

As referenced above, an equally daunting task for regulators is the real and potential effect of the release of engineered nanoparticles (ENPs) into the environment. As discussed below, the same sort of information problems exist. Reliable data are inadequate to address the core question of whether toxic and/or hazardous materials can be released into the environment during the manufacture, use of, or end-of-life stage of nano-enabled products. Questions thus abound as to whether nano applications will lead to environmental degradation, particularly from bioaccumulation of nanomaterials in biological systems.*

Stander and Theodore note that concerns arise because following exposure, chemical substances, including nanomaterials, can elude defense mechanisms and enter the body via various pathways, including inhalation, absorption through the skin, and ingestion.† There is also the risk of the release and absorption or inhalation from equipment malfunctions or unanticipated problems during use of products.‡

The evaluation process to determine the possibility of a risk resulting from nanomaterials exposure is difficult.§ A typical comment is that "…determining the hazard potentials and possible risks posed by the special physical and chemical properties of ENPs requires more detailed study. On one hand, there is currently no evidence that ENPs pose a significant threat to the environment; on the other, many gaps in our knowledge remain with regard to ENP ecotoxicity. The lack of evidence should by no means be interpreted to imply that environmental damage cannot occur."¶

15.5 LACK OF RELIABLE TOOLS

The tools used to assess risk are still under development. Presently, ecotoxicological research focuses primarily on controlled laboratory studies involving cell cultures or model organisms often using unrealistically maximum dosages to trigger effects. In laboratory studies, however, they can lead to analytical artifacts because some ENPs form large aggregates that can significantly alter the bioavailability and thus the toxicity of a material. Also, laboratory studies follow protocols originally developed for conventional organic chemicals and pesticides and do not in all cases consider the specific properties of nanomaterials.

Moreover, natural ecosystems are considerably more complex than a Petri dish, limiting the interpretability of laboratory results. Fortunately, no reported accidents involving major releases of ENPs into the environment have been documented to date. Ecological research on the behavior of ENPs must look to numerous studies from the geosciences that have examined the behavior of naturally occurring nanoparticles in the environment, but ENPs differ in certain respects from those occurring naturally. While natural nanoparticles are randomly structured and diffusely distributed in the environment, industrially produced suspensions or powders contain pure nanomaterials of very uniform size, shape, and structure. In the environment, nanomaterials can undergo a range of chemical processes that depend on many factors (e.g., pH value, salinity, concentration

---

* Stander, L. and L. Theodore. Environmental implications of nanotechnology—An update. *International Journal of Environmental Research and Public Health* 8(2): 470–479. http://www.mdpi.com/1660-4601/8/2/470.
† *Id.*
‡ *Id.*
§ NanoWerk, LLC. 2012. Nanotechnology and the environment—Hazard potentials and risks. http://www.nanowerk.com/spotlight/spotid=25937.php.
¶ *Id.*

differences, and the presence of organic or inorganic material). The characteristics and properties of a nanomaterial also play a major role. Bioavailability is decisive in determining potential toxicity. This depends strongly on whether nanoparticles remain stable in an environmental medium or are removed from the respective medium through agglomeration and deposition, or are transformed into a form that organisms cannot take up.

The current lack of data prevents a comprehensive and accurate picture of the fate and behavior of nanomaterials in the environment. It is difficult to compare results because different nanomaterials with different properties are used, and because both the methodology and the duration of the studies also often differ considerably. As interest in nanoscale materials matures, regulators are challenged to find ways to gather hazard information while continuing to assure the public that nano-enabled products do not pose risks that the government is tasked with addressing.

The issues to address are numerous. There is the effect on the air. As nanoparticles enter the atmosphere, they move from zones of higher concentration to zones of lower concentration (diffusion). Air currents distribute the particles rapidly; these can migrate great distances from their original source. Nonetheless, nanoparticles tend to aggregate into larger structures (agglomeration). Detecting nanoparticles in the air is very difficult because simple measurements of size distributions can hardly distinguish such agglomerates from natural particulates. The speed with which particles in the air are deposited on the ground, in the water, or onto plants (deposition) depends on particle diameter. Nanoparticles from the air are deposited much slower than larger particles due to their smaller diameters. In water, nanoparticles are relatively unstable because they rapidly adhere to one another due to electrostatic attractive forces and then sink as a result of gravity. Natural water bodies typically contain dissolved or distributed materials, including natural nanomaterials. As expected, synthetic nanomaterials that enter a natural water body bind themselves to such natural materials. The fate and behavior of nanomaterials in the water, however, are also influenced by factors such as pH, salinity (ionic strength), and the presence of organic material, which can lead to the decomposition of substances or of their aggregates and thus alter particle size and shape.

With regard to soil and sediment, the data are insufficient to draw any general conclusions. Considerably fewer studies are available for this sector than for water or air.

There are also potential environmental processes that can influence the behavior and the properties of nanomaterials. These include the aforementioned dissolution, precipitation and sedimentation, transformation, and agglomeration. As for toxicity, nanoparticles have been naturally present in the environment since the origin of earth in the form of combustion processes such as forest fires, in volcanic ash, in most natural waters, or as dust in the air due to weathering and erosion. Organisms produce various substances in nanoform in their cells (e.g., proteins, DNA) or are themselves only several nanometers large, such as viruses. During their evolution, all living organisms have adapted to an environment that contains nanoparticles, some of which can also be toxic (e.g., volcanic ash). This adaptation is a function of exposure, dose, and the speed with which habitats change. These natural nanoparticles in the environment are now accompanied by those that have been released unintentionally due to human activities such as household heating, industry, slash-and-burn clearance, transport and, most recently, through the industrial application of various extremely polymorphic synthetic nanoparticles in unknown amounts. This additional burden on humans and the environment has taken place over a very short period, at least from an evolutionary standpoint.

To date, no ecotoxicological studies are available that could explain in detail the mechanisms of uptake, distribution, metabolization, and excretion of nanoparticles. The few studies on the effects of ENPs on ecological communities failed to detect significant increases in mortality rates or changes in their compositions. Whether ENPs pose a risk to the environment depends not only on the toxicity of the respective material but also on exposure, that is, on the amount released into the environment. Unfortunately, little quantitative data are available and few studies have addressed the environmental exposure to nanomaterials. These are based on rough estimates of production volumes and releases, as well as on model calculations, which do not allow comprehensive risk assessments. Knowledge about production volumes alone is insufficient to estimate potential environmental risk: the actually released amounts must be known. To date, the assumption has been that nanomaterials that are tightly embedded in a matrix pose no or only minimal environmental risk. Nonetheless, only very few studies have examined the release of ENPs from consumer products. The most likely entry pathways of nanomaterials into the environment are sewage water and wastes. Wastes that contain nanomaterials can arise either during the production of the raw materials, the manufacture of products with nanomaterials, as well as at the end of the products' lifecycles.* The current domestic and international legal frameworks contain no specific regulations for treating wastes containing nanomaterials. A release of ENPs into the environment from wastes is possible, although virtually no studies have been conducted on this aspect. The assumption is that ENPs are efficiently removed by filters during waste incineration. Given the state of knowledge and the difficulty of extrapolating from studies involving naturally occurring material to ENPs, regulators must exercise caution. As noted, there is little hard evidence that ENPs pose a serious threat, but no real information to establish that they do not. So regulators have embarked on accumulating the needed information, analyzing it, and taking steps to avoid being blindsided by unanticipated problems.

In the following sections, the efforts of various national and international regulatory bodies in addressing the dual issue of products containing nanomaterials and the consequences of their manufacture on the environment will be examined.

15.6 U.S. FOOD AND DRUG ADMINISTRATION

The U.S. Food and Drug Administration (FDA) within the Department of Health and Human Services is the federal agency responsible for the implementation of regulations, guidances, and the like for pharmaceuticals and medical products such as devices, biologics, and diagnostics, while the U.S. Environmental Protection Agency (EPA) is responsible for addressing the environmental effects of nanomaterials. The FDA must consider the risks from the use of nanomaterials. Its efforts are made more difficult because various components within the agency are responsible for food and food packaging, human and animal drugs, devices and diagnostic products, cosmetics, vaccines, blood products and radiation emitting devices, and they do not act at the same speed or necessarily according to the same script. In part, this difference is due to the fact that the statutory provisions that the agency must administer for each group of products varies, with some products requiring pre-market approval, others needing to comply with standards or monographs, and still others regulated mainly by various guidance documents and voluntary submissions.

The criteria for marketing also varies, from requirements for proof of safety and effectiveness, safety alone, FDA required proof of lack of safety, and so forth. The agency has been discussing nanotechnology and how it might regulate products containing nanomaterials in earnest since its internal task force published recommendations in a report issued in 2007.† The result has been a variety of initiatives, depending on the class of products involved and the appropriate FDA center taking action. Throughout, the agency has moved cautiously, repeatedly stating that the existing regulatory framework should suffice, with some adjustments, and that the manufacturers intending to market products using nanomaterials in one way or another should meet with the agency early in the development cycle to come to agreement on the pathway to market and the data that should be generated. As will be seen below, there has been no effort yet to label products as containing nanomaterial, and no activity toward the mandated generation of an inventory of products employing nanotechnology. The FDA has been candid in the description of its efforts and intent in the several product areas it regulates, and since it is the dawning of regulation in the country with the most advanced regulatory system, it is useful to quote from the materials the FDA has supplied.

---

* *Id.* at 4–13.
† U.S. Food and Drug Administration. 2007. Nanotechnology Task Force Report. http://www.fda.gov/ScienceResearch/SpecialTopics/Nanotechnology/UCM2006659.htm.

As might be expected, the agency has developed its own definition of nanomaterial, which is contained in a Guidance issued in 2011.* In that document, the Agency highlighted the following:

1. *Engineered material or end product*
   This term is used to distinguish between products that have been engineered to contain nanoscale materials or involve the application of nanotechnology from those products that contain incidental or background levels of nanomaterials or those that contain materials that naturally occur in the nanoscale range. The FDA is particularly interested in the *deliberate* manipulation and control of particle size to produce specific properties, because the emergence of these new properties or phenomena may warrant further evaluation. (emphasis added)
2. *At least one dimension in the nanoscale range (~1–100 nm)*
   A size range of approximately 1–100 nm is commonly used in various working definitions or descriptions proposed by the regulatory and scientific community. In this size range, materials can exhibit new or altered physicochemical properties that enable novel applications. Accordingly, a range of approximately 1–100 nm should be applied as a first reference point in considering whether an FDA-regulated product contains nanomaterials or otherwise involves application of nanotechnology.
3. *Exhibits properties or phenomena ... that are attributable to its dimension(s)*
   These terms are used because properties and phenomena of materials at the nanoscale enable applications that can affect safety, effectiveness, performance, quality and, where applicable, public health impact of FDA-regulated products. For example, dimension-dependent properties or phenomena may be used for functional effects such as increased bioavailability, decreased dosage, or increased potency of a drug product, decreased toxicity of a drug product better detection of pathogens, enhanced protection offered by improved food packaging materials, or improved delivery of a functional ingredient or a nutrient in food. The properties and phenomena may be due to altered chemical, biological, or magnetic properties, altered electrical or optical activity, increased structural integrity, or other unique characteristics of nanoscale materials not normally observed in their larger counterparts (emphasis added). These changes may raise questions about the safety, effectiveness, performance, quality, or public health impact of the products. In addition, considerations such as routes of exposure, dosage, and behavior in various biological systems (including specific tissues and organs) are critical for evaluating the wide array of products under the FDA's jurisdiction.† (emphasis added)

In a Questions and Answers document accompanying the 2011 Guidance, the Agency stated how the Guidance should be applied:

> For products subject to premarket notice or review, FDA intends to incorporate attention to nanomaterials into its product-specific review procedures and apply certain considerations to better understand the properties and behavior of engineered nanomaterials. For products not subject to premarket review, manufacturers are encouraged to consult with FDA to reduce the risk of unintended harm to human or animal health. [NOTE: No product sold in the United States requires a premarket submission simply because it contains nanomaterials, and the incorporation of nanomaterials would be one element to cover in a submission otherwise required for a product.]
> Industry is encouraged to consult with the agency early in the product development process to address questions related to the regulatory status, safety, effectiveness, or public health impact of products that

---

* U.S. Food and Drug Administration. 2011. Draft guidance for industry: Considering whether an FDA-regulated product involves the application of nanotechnology. http://www.fda.gov/ScienceResearch/SpecialTopics/Nanotechnology/ucm257926.htm.
† *Id.* (citations omitted).

involve nanotechnology. FDA will offer technical advice and guidance to manufacturers, as needed, so that they can improve pre-market product development and safety assessments.*

In the absence of binding regulations, the agency is putting great stock in the invitation to consult. No one knows how successful that effort will be.

On many of the web pages of the several FDA centers, there are detailed guidances, and even references to regulations, to assist in the generation of data and the preparation and submission of a marketing application for products containing macro materials. A good example is the wealth of material on the Center for Food Safety and Applied Nutrition (CFSAN) site addressing food ingredients and packaging.[†] That degree of specificity is not available for products that employ deliberately manipulated nanoscale material. There are, however, a number of documents that provide insight into how the FDA would like to regulate products using deliberately manipulated nanomaterials. It is useful to consider those documents, center by center.

The Center for Drug Evaluation and Research (CDER) has posted on its website nanomaterial-related information in its *Manual of Policies and Procedures* (MAPP), effective June 3, 2010. The Manual provides chemistry, manufacturing, and controls (CMC) reviewers in CDER with "the framework by which relevant information about nanomaterial-containing drugs will now be captured in CMC reviews of CDER drug application submissions."[‡] CDER states that this is important, because to this point, "much of the information that was necessary to populate the fields of the database were not being captured consistently."[§] The information to be entered by the reviewer includes whether any nanoscale materials are included, and if so, what material, and its source. Next, the reviewer is to indicate whether the material is a reformulation of a previously approved product. That is followed by a description of the nanomaterial functionality (i.e., carrier, API, excipient, packaging, or other) and whether the material is soluble or insoluble in an aqueous environment. Then the reviewer is to note whether the particle size was reported in the application and, if so, the size reported. If the particle size was not included, the reviewer is to put down the reason for that, and then go on to discuss other disclosed properties, and the methods used to characterize the nanomaterial. Collection of this sort of information will assist in addressing the need to develop an inventory in order for the regulator not to be "blind," as discussed earlier.

A product review flow chart is included to make it easier to envision the steps of the review, and then there is a list of the common techniques used as the basis to characterize nanomaterials, such as morphology, surface, and chemical composition. More than 25 techniques are listed. A manufacturer can expect that its application will be considered incomplete if it does not address all of the points in the reporting format, or if the manufacturer does not use the listed techniques for characterization. It will make the review go more smoothly if the manufacturer explains why all the points were not addressed, if that is the case. The characterization is particularly relevant, since the reporting format states that reviewers can use their scientific judgment to determine the adequacy of the techniques used by the sponsor. The procedure can take the place of "hard" regulation at this juncture in the development of a regulatory scheme.

The Center for Veterinary Medicine (CVM) has a procedure similar to the CDER CMC policy described above.[¶] In the section on the purpose of the procedure, CVM writes that the procedure is intended to identify "points to consider for technical sections for products containing

---

* *Id.*, Questions and answers about the draft guidance.
† U.S. Food and Drug Administration. 2013. Ingredients, packaging, labeling. http://www.fda.gov/Food/IngredientsPackagingLabeling/default.htm.
‡ U.S. Food and Drug Administration. 2010. *CDER Manual of Policies & Procedures*, Chapter 5015.9, Reporting format for nanotechnology-related information in CMC review. http://www.fda.gov/AboutFDA/CentersOffices/OfficeofMedicalProductsandTobacco/CDER/ManualofPoliciesProcedures/default.htm.
§ *Id.*
¶ U.S. Food and Drug Administration. 2011. Center for Veterinary Medicine, *Program and Policy Procedure Manual* 1234.2600, Review of ONADE regulated products that contain nanomaterials or otherwise involve the use of nanotechnology. http://www.fda.gov/.../AnimalVeterinary/GuidanceComplianceEnforcement/PoliciesProceduresManual/ucm270271.pdf.

nanomaterials(s) or otherwise involve the application of nanotechnology that might require addition data or special steps to address potential safety or quality issues."*

The CVM procedure addresses the investigational stage, as well as the post-clinical stage. At the early phase, the reviewer is to note whether the applicant has addressed any unique safety concerns related to the use of nanomaterials in the formulation and how the applicant has attempted to characterize those hazards. The reviewer is to address the issue of whether additional separate safety studies of the measurement of additional safety parameters in the margin of safety studies are needed. The review catalog also addresses a review of formulation bridging studies, human food safety reviews, toxicology, residue chemistry, microbial food safety, and the CMC. The CVM document is useful because it not only lists the technical information to include, but what those studies and data should address. This effort is a good example of the FDA using its guidances to attempt to generate meaningful scientific information regarding a product.

There is an interesting article by Dr. Jan Simak† at the Biologics Research Projects link on the Center for Biologics Evaluation and Research (CBER) website. Dr. Simak describes the purpose of his research as a study of nanomaterials made of carbon atoms:

> Because of their unique and useful properties, carbon nanomaterials are under investigation for various applications in biomedical nanotechnologies, usually as structures called carbon fullerenes and carbon nanotubes (CNTs). The superior mechanical characteristics of CNTs make them very attractive materials for making devices for collecting, processing, and storing blood transfusion products, diagnostic biosensors (devices that use biological material such as enzymes, cells, or antibodies to detect a substance), drug delivery systems, and imaging devices for use inside blood vessels. Several studies, however, are raising safety concerns by showing toxic effects of fullerenes and CNTs. Therefore, one part of our research is focused on determining the mechanism of toxicity in these studies, by evaluating whether these nanomaterials are toxic to blood vessels and blood cells.

One important class of nanomaterials is made of carbon atoms. Because of their unique and useful properties, carbon nanomaterials are under investigation for various applications in biomedical nanotechnologies, usually as structures called carbon fullerenes and CNTs.

The paper goes on to describe how the CBER staff will explore the potential toxic effects of the nanotubes and other materials, and the paper gives clues to a potential applicant on what CBER deems important in studying the toxicity of nanomaterials:

> Thus the investigation of effects of fullerenes and CNTs on blood cells, platelets and endothelial cells is a critical safety issue. The results of our Critical Path research project on blood and vascular biocompatibility of carbon nanomaterials will help FDA to create regulatory policy for evaluating the safety and effectiveness of different biomedical nanomaterials.
>
> Our investigation is focused on effects of various commercially available carbon nanomaterials and their structural and chemical derivatives on plasma membranes of cultured endothelial cells, blood cells, and platelets isolated from blood. We use a variety of cell biology assays to study *in vitro* cytotoxicity of these materials, such as apoptosis, necrosis, autophagy, effects on the cell cycle, cell surface activation markers, and intracellular calcium concentration.‡

In a particular guidance section, issued about when to file a 510(k) for changes to an existing device, another center, the Center for Devices and Radiological Health (CDRH) stated that:

> *Important Note on 510(k) Devices that Contain Nanomaterials or Otherwise Involve the Application of Nanotechnology:* Nanotechnology is a new and evolving field for both the medical device industry

---

* *Id.* at 1.
† Simak, J. 2011. Investigation of potential toxic effects of engineered nanoparticles and biologic microparticles in blood and their biomarker applications. http://www.fda.gov/BiologicsBloodVaccines/ScienceResearch/BiologicsResearchAreas/ucm127045.htm.
‡ *Id.*

and the Agency. At this time, FDA has not adopted nanotechnology-specific criteria to assist manufacturers in determining when a change to a device that contains nanomaterials or otherwise involves the application of nanotechnology rises to the level of significance that requires submission of a new 510(k). For this reason, FDA recommends that manufacturers consult with the agency for any nanotechnology-related changes to devices to determine whether and how the change may affect the safety or effectiveness of the device. FDA plans on developing additional guidance to further explain the Agency's thinking on this matter. Contact the appropriate review division with any questions on devices that contain nanomaterials or otherwise involve the application of nanotechnology.[*]

The note demonstrates that the CDRH, like the other FDA centers, believes that the use of nanomaterials at least raises questions about whether there has been a fundamental change in the risk associated with use of the device, and the proof of whether that is so or not will be the centerpiece of any application involving a change to engineered nanomaterials in a regulated product. What the CDRH says also raises the issue of whether the center believes that the manufacturer has the option to make its own determination about whether the change in materials affects safety. The CDRH has no guidances at present. It did, however, host a public workshop to discuss nanomaterials in devices products. In advance of the meeting, the CDRH posted a document asking for comment on a variety of issues.[†]

Finally, the Center for Food Safety and Applied Nutrition (CFSAN) has issued two guidelines, one for cosmetics and one for food, that contain important information about the agency's thinking on the regulation of nanotechnology. The first document discusses the proof of safety for cosmetics incorporating engineered nanomaterials.[‡] There has been concerted opposition to the use of nanomaterials in cosmetics, especially fullerenes, because they pose a significant theoretical risk to users without a perceived benefit that would be worth any risk. Manufacturers do not have to submit premarket applications for cosmetics, but they are responsible for the generation and preservation of data demonstrating safety. Critics think that is not sufficient, but that is the way it is unless the United States Congress amends the statute. In the meantime, the CFSAN has issued the guidance, stating at the outset that:

> At the nanoscale, properties of materials might change resulting in changes to the product's performance, quality, safety, and/or effectiveness (Ref. 2). Nanomaterials may have chemical, physical, or biological properties that are different from their larger counterparts. The use of nanomaterials may alter the bioavailability of the cosmetic formulation (Ref. 5). In some of these cases, the traditional safety tests that have been used to determine the safety of cosmetic ingredients and finished products may not be fully applicable. As noted in the 2007 FDA Nanotechnology Task Force report, there may be a higher degree of uncertainty associated with nanoscale materials compared to conventional chemicals, both with respect to knowledge about them and the way that testing is performed. In Section III.B of this document, we highlight key scientific considerations relevant to the assessment of the safety of nanomaterials used in cosmetic products.
>
> If you wish to use a nanomaterial in a cosmetic product, either a new material or an altered version of an already marketed ingredient, FDA encourages you to meet with us to discuss the test methods and data needed to substantiate the product's safety, including chronic toxicity and other long-term toxicity data as appropriate.[§]

---

[*] U.S. Food and Drug Administration. Guidance for Industry and FDA Staff - 510(k) Device Modifications: Deciding When to Submit a 510(k) for a Change to an Existing Device. http://www.fda.gov/MedicalDevices/DeviceRegulationandGuidance/GuidanceDocuments/ucm265274.htm.

[†] U.S. Food and Drug Administration. 2010. Public workshop—Medical devices and nanotechnology: Manufacturing, characterization, and biocompatability considerations, questions for discussion. http://www.fda.gov/MedicalDevices/NewsEvents/WorkshopsConferences/ucm222591.htm.

[‡] U.S. Food and Drug Administration. 2012. Draft guidance for industry: Safety of nanomaterials in cosmetic products. http://www.fda.gov/Cosmetics/GuidanceComplianceRegulatoryInformation/ GuidanceDocuments/ucm300886.htm.

[§] *Id*. at 8.

The Center office begins by conceding that "FDA believes the current general framework for safety assessment which includes hazard identification, dose-response assessment, exposure assessment, and risk characterization is generally robust and flexible enough to be considered appropriate for nanomaterials, even though they can have properties that may be different from conventional ones."[*] That does not hold true for safety tests:

> However, standard safety tests may need to be modified or new methods developed to address (1) the key chemical and physical properties that may affect the toxicity profile of nanomaterials and (2) the effects of those properties on the function of the cosmetic formulation. The safety assessment for cosmetic products using nanomaterials should address the physico-chemical characteristics of the nanomaterials, impurities, if present, and the potential product and ingredient exposure levels to help determine what other testing may be needed. The safety assessment should include consideration of the toxicity of both the ingredients and their impurities; dosimetry for *in vitro* and *in vivo* toxicology studies, if needed; and clinical testing, if warranted. The safety assessment should also address the issues of toxicokinetics and toxicodynamics. The overall package of data and information should substantiate the safety of the product under the intended conditions of use.[†]

Each of the points mentioned is discussed in some detail in the document. The guidance concludes with a summary that is worth quoting at length:

> Summary of Recommendations—In summary, inclusion of nanomaterials in an FDA-regulated product or a change in the nanomaterials used might affect the quality, safety, effectiveness, and/or public health impact of the product. Therefore, as with any cosmetic product that has new or altered properties, data needs and testing methods should be evaluated accordingly to address the unique properties and function of the nanomaterials used in the cosmetic products as well as the questions that continue to remain about the applicability of traditional safety testing methods to products that involve nanotechnology. FDA recommends that the safety assessment for cosmetic products using nanomaterials should address several important factors such as
>
> - The physico-chemical characteristics
> - Agglomeration and size distribution of nanomaterials at the toxicity testing conditions which should correspond to those of a final product
> - Impurities
> - Potential product exposure levels, and the potential for agglomeration of nanoparticles in the final product
> - Dosimetry for *in vitro* and *in vivo* toxicology studies
> - *In vitro* and *in vivo* toxicological data on ingredients and their impurities, dermal penetration, irritation (skin and eye) and sensitization studies, mutagenicity/genotoxicity studies
> - Clinical studies to test the ingredient, or finished product, in human volunteers under controlled conditions
>
> FDA expects that the science surrounding nanomaterials will continue to evolve and be used in the development of new testing methods.
>
> In conclusion, the safety of a cosmetic product should be evaluated by analyzing the physico-chemical properties and the relevant toxicological endpoints of each ingredient in relation to the expected exposure levels resulting from the intended use of the finished product. If you wish to use a nanomaterial in a cosmetic product, either a new material or an altered version of an already marketed ingredient, FDA encourages you to meet with us to discuss the test methods and data needed to substantiate the product's safety, including short-term toxicity and other long-term toxicity data as appropriate. We welcome your contacting us with other questions relating to the use of nanomaterials in cosmetic products.[‡]

---

[*] *Id.* at 9.
[†] *Id.*
[‡] *Id.* at 18–19.

The other guidance document posted by the CFSAN relates to food and food packaging.[*]

The CFSAN identifies the purpose of the document "to describe the factors you should consider when determining whether a significant change in manufacturing process for a food substance already in the market:

- Affects the identity of the food substance
- Affects the safety of the use of the food substance
- Affects the regulatory status of the use of the food substance
- Warrants a regulatory submission to FDA"

As with all food substances, this guidance also is intended to recommend that manufacturers consult with FDA regarding a significant change in manufacturing process for a food substance already in the market, irrespective of any conclusion about whether that change affects the safety or regulatory status of the food substance.[†]

Over the past several years, the agency has engaged in a dialog with the regulated industry, opting to encourage contact prior to the submissions for products where one issue may be the presence of nanomaterials and providing forms of guidance for the generation of data where no submission is required. It has emphasized its expertise in dealing with small molecules, and declined to push too hard in the absence of hard information on product problems. There has been no concerted effort to mandate inventory information or require notification before marketing a product solely because it employs nanomaterials. There are no nanomaterial product registration requirements and no mandated report of adverse effects. The lynchpin has been to encourage dialogue with the centers, and it is an open question whether the FDA is acquiring information relating to regulatory aspects of nanotechnology quickly enough to be in a position to address issues developing in the foreseeable future.

15.7 U.S. ENVIRONMENTAL PROTECTION AGENCY

The FDA's strategy of the EPA, on the other hand, is somewhat more directed to implementing specific regulations based on two major statutes it administers. The law of broader application is the Toxic Substances Control Act (TSCA) that gives the EPA authority over "chemical substances," including nanoscale materials. To ensure that nanoscale materials are manufactured and used in a manner that protects against unreasonable risks to human health and the environment, the EPA is pursuing a comprehensive four-prong regulatory approach under TSCA. This approach includes: the Premanufacture Notifications (PMN), Significant New Use Rule (SNUR), information gathering authority, and test authority.[‡] These tools facilitate both the acquisition of information needed to make regulatory decisions and actions to attempt to prevent risk in the interim. Regarding the Premanufacture Notifications (PMN), TSCA requires manufacturers of new chemical substances to provide specific information to the EPA for review prior to manufacturing chemicals or introducing them into commerce. The EPA can take action to ensure that chemicals that pose a risk to human health or the environment are effectively controlled. Since 2005, the EPA has received and reviewed over 130 new chemical notices under TSCA for nanoscale materials, including CNTs. The agency has taken a number of actions to control and limit exposures to these chemicals, including:

---

[*] U.S. Food and Drug Administration. 2012. Draft guidance for industry: Assessing the effects of significant manufacturing process changes, including emerging technologies, on the safety and regulatory status of food ingredients and food contact substances, including food ingredients that are color additives. http://www.fda.gov/Food/GuidanceRegulation/GuidanceDocumentsRegulatoryInformation/IngredientsAdditivesGRASPackaging/ucm300661.htm.

[†] Id. at 4.

[‡] U.S. Environmental Protection Agency. 2010. Control of nanoscale materials under toxic substances control act. http://www.epa.gov/oppt/nano/.

- Limiting the uses of the nanoscale materials
- Requiring the use of personal protective equipment, such as impervious gloves and NIOSH approved respirators
- Limiting environmental releases
- Requiring testing to generate health and environmental effects data

The EPA has permitted limited manufacture of new chemical nanoscale materials through the use of administrative orders or SNURS. Under TSCA Section 5(a)(2), the EPA determines that a use of a chemical substance is a significant new use after considering all relevant factors, including:

- The projected volume of manufacturing and processing of a chemical substance
- The extent to which a use changes the type or form of exposure of human beings or the environment to a chemical substance
- The extent to which a use increases the magnitude and duration of exposure of human beings or the environment to a chemical substance
- The reasonably anticipated manner and methods of manufacturing, processing, distribution in commerce, and disposal of a chemical substance

The EPA has also allowed the manufacture of nanoscale chemical substances under the terms of certain regulatory exemptions, but only in circumstances where exposures are tightly controlled to protect against unreasonable risks (e.g., the protective equipment and environmental release limitations discussed above).[*]

The EPA has used its Significant New Use Rule (SNUR) authority to ensure that nanoscale materials receive appropriate regulatory review. The SNUR requires persons who intend to manufacture, import, or process new nanoscale materials that are identical to chemical substances listed on the TSCA Inventory, but in a way that diverges from the SNUR provisions, to submit a Significant New Use Notice (SNUN) at least 90 days before commencing that activity.

The EPA reportedly is also developing a proposed rule under TSCA to require the submission of additional information, something the FDA has yet to undertake. This rule reportedly would propose that persons who manufacture defined nanoscale materials notify the EPA of certain information, including production volume, methods of manufacture and processing, exposure and release information, and available health and safety data. Such information would assist the agency in evaluating the intended uses of the nanoscale materials and take action to prohibit or limit activities that may present an unreasonable risk to human health or the environment. To obtain more of the information needed for nanoscale materials, the agency may also seek to regulate a rule to require testing for certain nanoscale materials that are already in commerce. The EPA has said it is particularly interested in classes of nanoscale materials not already being tested by other United States and international organizations.[†]

The second statute EPA administers is the Federal Insecticide, Fungicide, and Rodenticide Act (FIFRA). According to the EPA, FIFRA and implementing regulations provide an effective framework for regulating pesticide products that contain a nanoscale material, a product of concern from an environmental standpoint.[‡] The EPA has indicated that the special properties that make nanoscale materials of potentially great benefit also can present new challenges for risk assessment and decision-making. Given the potential for nanoscale materials to pose different risks than their larger-sized counterparts, the EPA issued a notice in the *Federal Register* on June 10, 2011, outlining what it was considering to address the situation. This document describes several possible approaches for obtaining certain additional information on the composition of pesticide products.

---

[*] *Id.*
[†] *See* fn.8.
[‡] U.S. Environmental Protection Agency. 2011. Regulating pesticides that use nanotechnology. http://www.epa.gov/pesticides/regulating/nanotechnology.html.

The notice focuses particularly on information about what nanoscale materials are present in registered pesticide products. Under one approach, the EPA would use Section 6(a)(2) of FIFRA to obtain information regarding what nanoscale material is present in a registered pesticide product and its potential effects on humans or the environment. If the EPA adopted this approach, 40 C.F.R. 152.50(f)(3) would also require the inclusion of such information with any application for registration of a pesticide product that contains a nanoscale material.

Under an alternative approach, the EPA would obtain such information using Data Call-In notices (DCIs) under FIFRA Section 3(c)(2)(B). The EPA is reviewing whether this could be done under existing regulations, or whether the EPA would need to amend existing regulations to clarify that the information is required with any application for registration.

This document also proposes a new approach for how the EPA will determine on a case-by-case basis whether a nanoscale active or inert ingredient is a "new" active or inert ingredient for purposes of FIFRA and the Pesticide Registration Improvement Act (PRIA), even when an identical, nonnanoscale form of the nanoscale ingredient is already registered. Using the Section 6(a)(2) approach, the EPA would rely on FIFRA requirements for a registrant to inform EPA of relevant information relating to their products. Specifically, if at any time after the initial registration of a pesticide the registrant has additional factual information regarding unreasonable adverse effects on the environment of the pesticide, the registrant must provide such information to the agency. In the more recent past, EPA seems to have abandoned its efforts to integrate FIFRA Section 6(a)(2) in this way.

The EPA is mirroring the FDA informal mechanism of strongly recommending that companies contact the EPA's pesticide registration ombudsmen to arrange a pre-application conference as early as possible in the development of any pesticide that would be a product of nanotechnology or that would contain nanoscale material. During pre-application conferences, the EPA would expect that a company should provide information on: how the pesticide is made, how it is proposed to be used, and how people and the environment may be exposed to the product.*

Because nanoscale materials may have special properties, EPA's data requirements may need to be tailored to the specific characteristics of the product under consideration. That point is presently being argued in a case in federal court. The case arose after the EPA announced on December 1, 2011, that it was conditionally registering a pesticide product containing nanosilver as a new active ingredient. HeiQ AGS-20 is a silver-based antimicrobial pesticide product approved for use as a preservative for a plethora of textiles. As a condition of FIFRA registration, EPA stated that it was requiring additional data to confirm its assessment that HeiQ AGS-20 would not cause unreasonable adverse effects on human health or the environment. The agency proposed granting the nanosilver a four-year conditional registration.

On January 26, 2012, the Natural Resources Defense Council (NRDC) filed a lawsuit in the U.S. Court of Appeals for the Ninth Circuit against the EPA concerning its conditional registration. The NRDC sought to block the EPA from allowing nanosilver on the market without the legally required data about its suspected harmful effects on humans and wildlife. In a brief dated April 16, 2012, the NRDC argued that the EPA's decision that HeiQ AGS-20 will not cause unreasonable adverse effects on human health was not supported by substantial evidence. According to NRDC, in calculating the risks to human health, the EPA failed to evaluate the risks to infants, even though they have an especially high likelihood of exposure to AGS-20 because they are more likely than other age groups to chew on textiles coated with it. The NRDC stated that had the EPA properly taken infants into account, application of its own risk criteria "would have shown that AGS-20 poses unacceptable risks, and thus may have unreasonable adverse effects." The NRDC further argued that the EPA likewise failed to consider the risk of aggregate exposures from other nanosilver on the market. Had it done so, the NRDC stated, the EPA's analysis would have shown that registering AGS-20 created unacceptable risks. NRDC concluded that, because the EPA's finding of no

---

* Pesticides: Policies Concerning Products Containing Nanoscale Materials, 76 Fed. Reg. 35383 (June 17, 2011) (to be codified at 40 C.F.R. Chap. 1), https://federalregister.gov/a/2011-14943.

"unreasonable adverse effects" rested on "significantly understated risk assessments," its decision was not supported by substantial evidence and must be vacated.

On June 14, 2012, the EPA filed its answering brief, arguing that despite data gaps concerning HeiQ AGS-20, it had determined it had sufficient evidence to conclude "that even a three-year-old chewing and wearing a new AGS-20 treated textile every day for six months could potentially be exposed to no more than 1/1000th of the quantity of nanosilver which did not cause any adverse health effects in relevant scientific studies." Given the low risk, the EPA stated that it reasonably concluded that HeiQ AGS-20 will not cause unreasonable adverse effects to consumers, and its risk assessment "warrants substantial deference from the Court." On November 7, 2013, the Court largely ruled in EPA's favor. The Court found objectionable EPA's decision to decline to decide to require risk mitigation after calculating a margin of exposure for aggregate exposure to AGS-20 exactly of 1000. EPA and HeiQ subsequently revised the pesticide label to address this and the case is now resolved

15.8 WORK PLACE CONCERNS

In addition to products and the environment, the federal government has also displayed interest in regulating the workplaces where the products are manufactured. The National Institute for Occupational Safety and Health (NIOSH) has been involved with nanoscale materials as they relate to the issue of workplace exposure mentioned earlier. An early NIOSH effort is to request information and comment on silver nanoparticles. According to the notice, the NIOSH has initiated an evaluation of the scientific data on silver nanoparticles "to ascertain the potential health risks to workers and to identify gaps in knowledge so that appropriate laboratory and field research studies can be conducted." Giving an indication of the sort of data and information that might be used to regulate the workplace, the NIOSH wants to obtain additional information concerning:

1. Published and unpublished reports and findings from *in vitro* and *in vivo* toxicity studies with silver nanoparticles
2. Information on possible health effects observed in workers exposed to silver nanoparticles
3. Information on workplaces and products in which silver nanoparticles can be found
4. Description of work tasks and scenarios with a potential for exposure
5. Information on measurement methods and workplace exposure data
6. Information on control measures (e.g., engineering controls, work practices, personal protective equipment) that are being used in workplaces where potential exposures to silver nanoparticles occurs

NIOSH has also posted a document titled *General Safe Practices for Working with Engineered Nanomaterials in Research Laboratories*, which contains recommendations on engineering controls and safe practices for handling engineered nanomaterials in laboratories and some pilot-scale operations.[*] The document is designed "to be used in tandem with well-established practices and the laboratory's chemical hygiene plan." The guidance notes that experimental animal studies indicate that potentially adverse health effects may result from exposure to nanomaterials, and that the routes of exposure include inhalation, dermal exposure, and ingestion. The guidance concludes that "[t]he full range of occupational hygiene controls will be necessary to limit exposures to nanomaterials as a means to prevent adverse health outcomes in the research community. Engineering and administrative controls can eliminate or minimize the amount of nanomaterials that will be present in workplace air or settled on surfaces. Personal protective equipment can be used where other types of controls are not available or practical."

---

[*] Centers for Disease Control and Prevention, National Institute for Occupational Safety and Health. 2012. General safe practices for working with engineered nanomaterials in research laboratories. http://www.cdc.gov/niosh/docs/2012-147.

Finally, on September 5, 2012, the National Research Council released a pre-publication version of a report titled *Science for Environmental Protection: The Road Ahead.*[*] The EPA asked NRC to assess the EPA's overall capabilities "to develop, obtain, and use the best available scientific and technologic information and tools to meet persistent, emerging, and future mission challenges and opportunities." The report discusses nanotechnology as an example of using emerging science to address regulatory issues and support decision making. The Committee states that to have the capacity to address emerging tools, technologies, and challenges, the EPA "will need to have enough internal expertise to identify and collaborate with the expertise of all of its stakeholders in order to ask the right questions; determine what existing tools and strategies can be applied to answer those questions; determine the needs for new tools and strategies; develop, apply, and refine the new tools and strategies; and use the science to make recommendations based on hazards, exposures, and monitoring." According to the report, the example of engineered nanomaterials "illustrates some of the problems and pitfalls of current approaches to emerging technologies." While the EPA provided early funding regarding the use of nanotechnology in remediation, the report states that it missed the opportunity to support research addressing the environmental health and safety of nanomaterials, pollution prevention in the production of nanomaterials, and the use of nanotechnology to prevent pollution. The reasons for the delay in early intervention include "insufficient federal agency leadership, emphasis, and policy regarding *proactive* rather than *reactive* approaches to safer design." If the EPA intends to promote and guide early intervention in the design and production of emerging chemicals, materials, and products, the report states, "it will need to commit to this effort beyond its regulatory role."

15.9 THE EUROPEAN UNION

The EU and its member nations have been active in consideration of the same sorts of issues relating to nanomaterials, the environment, and the possible regulation of pharmaceuticals, foods, cosmetics, and other medical products containing nanomaterials. There are a variety of reports and descriptions of those activities readily available on the Internet. In this chapter, we will concentrate on present activity and how it compares to efforts underway on the United States.

A significant step in EU regulation was the adoption in 2011 of a definition of nanomaterial. The definition was adopted in the same year as the FDA acted, and the reasons for adoption were similar. Nanomaterial was stated to mean:

> ... a natural, incidental or manufactured material containing particles, in an unbound state or as an aggregate or as an agglomerate and where, for 50% or more of the particles in the number size distribution, one or more external dimensions is in the size range 1 nm–100 nm.
>
> In specific cases and where warranted by concerns for the environment, health, safety or competitiveness the number size distribution threshold of 50% may be replaced by a threshold between 1 and 50%.
>
> By derogation from point 2, fullerenes, graphene flakes and single wall carbon nanotubes with one or more external dimensions below 1 nm should be considered as nanomaterials.[†]

An accompanying explanation stated the case for the definition in much the same terms used by the FDA:

> Nanomaterials are currently governed by a variety of legislative instruments at EU and national level. However, definitions have been developed on a case-by-case basis and vary across sectors, creating unnecessary burdens for industry and hampering public debate about risks and benefits of these substances. This recommendation gives EU legislators a legal reference for nanomaterials, when adopting new or implementing existing legislation.

---

[*] National Research Council. 2012. Science for environmental protection: The road ahead. http://www.nap.edu/catalog.php?record_id=13510.

[†] Mantovani, E., A. Porcari, M.D. Morrison, and R.E. Geertsma. 2012. Developments in nanotechnologies regulations and standards 2012—Report of the observatory nano at 8. www.observatorynano.eu.

The experience of the first registration deadline (30 November 2010) under REACH, the EU's overarching chemicals policy, showed that companies needed more clarity about their obligations with regard to nanomaterials. REACH has a key role to play in generating information about the properties of nanomaterials as chemical substances. With the adopted definition it will be easier for companies to assess their registration dossiers and determine exactly when they should consider their products as nanomaterials.

European Environment Commissioner Janez Potočnik stated:

"I am happy to say that the EU is the first to come forward with a cross-cutting designation of nano-materials to be used for all regulatory purposes. We have come up with a solid definition based on scientific input and a broad consultation. Industry needs a clear coherent regulatory framework in this important economic sector, and consumers deserve accurate information about these substances. It is an important step towards addressing any possible risks for the environment and human health, while ensuring that this new technology can live up to its potential."

Nanomaterials are already being used in hundreds of applications and consumer products ranging from toothpaste to batteries, paints and clothing. Developing these innovative substances is an important driver for European competitiveness, and they have significant potential for progress in areas like medicine, environmental protection and energy efficiency. But as uncertainties remain about the risks they pose, a clear definition is needed to ensure that the appropriate chemical safety rules apply. The definition will help all stakeholders including industry associations, as it brings coherence to the variety of definitions that are currently in use in different sectors. The definition will be reviewed in 2014 in the light of technical and scientific progress.[*]

As referenced in the above explanation, the REACH regulation is perhaps the most important document to surface in the EU regarding nanomaterials and the regulation thereof. The most recent document pertaining to REACH is the communication from the European Commission to the European Parliament and others as "Secondary Regulatory Review on Nanomaterials," issued as SWD(2012) 288 Final on October 3, 2012. In the Review, the Commission states that the materials attracting attention in the EU presently are nano-titanium dioxide, nano zinc oxide, fullerenes, CNTs and nano silver—a list very similar to the most talked about materials in the United States. Benefits of the materials and the products containing them are listed as contributions to growth and jobs, innovation and competitiveness, the saving of lives, improved function of consumer products and the reduction of environmental impacts. As to the balance the Parliament should strive for, the Commission states that "[t]he applicable legislation must ensure a high level of health, safety and environmental protection. At the same time, it should permit access to innovative products and promote innovation and competitiveness."[†]

The Commission goes on to discuss safety, risk assessment and risk/benefit assessment, starting with language from the 2009 Report by the Scientific Committee on Emerging and Newly Identified Health Risks (SCENIHR). SCENIHR had stated then that widely used methodologies are generally applicable to nanomaterials, "specific aspects related to nanomaterials still require further development. This will remain so until there is sufficient scientific information available to characterise the harmful effects of nanomaterials on humans ...."[‡] The several guidances issued in the United States sound variations of this theme as well, for products as opposed to substances. It is a more cautious approach to regulation than would be immediate comprehensive regulation—that would

---

[*] European Commission, Joint Research Centre, Institute for Health and Consumer Protection. 2011. What is a nanomaterial? European Commission breaks new ground with a common definition. http://ihcp.jrc.ec.europa.eu/our_activities/nanotechnology/what-is-a-nanomaterial-european-commission-breaks-new-ground-with-a-common-definition.

[†] European Commission. 2012. Communication from the Commission to the European Parliament, the Council, and the European Economic and Social Committee. Second regulatory review on nanomaterials, at 3. http://ec.europa.eu/nanotechnology/policies_en.html.

[‡] *Id.* at 4.

likely regulate some nanomaterials and nano-containing products too much, and others too little. As the SCENIHR further pointed out, however, it is not an approach without risks:

> ... health and environmental hazards have been demonstrated for a variety of manufactured nanomaterials. The identified hazards indicate potential toxic effects of nanomaterials for man and the environment. However, it should be noted that not all nanomaterials induce toxic effects. Some manufactured nanomaterials have already been in use for a long time (e.g., carbon black, TiO2) showing low toxicity. Therefore, the hypothesis that smaller means more reactive, and thus more toxic, cannot be substantiated by the published data. In this respect nanomaterials are similar to normal chemicals/substances in that some may be toxic and some may not. As there is not yet a generally applicable paradigm for nanomaterial hazard identification, a case-by-case approach for the risk assessment of nanomaterials is still warranted.[*]

The Commission points out that a single REACH registration can include different forms of a substance, and the registrant must ensure the safety of all included forms and provide adequate information to address the different forms in the registration. According to the Second Regulatory Review, only seven substance registrations under REACH and 18 Notifications under the Classification, Labeling, and Packaging regulation (CLP) have listed nanomaterial as the substance in the voluntary field. The Commission knows of registrations for substances known to have nanomaterials that "do not mention clearly which forms are covered or how information relates to the nanoform. Only little information is specifically addressing safe use of the nanomaterials supposed to be covered by the specific dossiers."[†] The solution to this problem is that the Commission will review relevant regulatory options in its upcoming REACH review and discuss possible amendments regarding how to address nanomaterials. In addressing the matter this way, the Commission might be missing an opportunity to gather information on the "newness" of the nano form of a substance by requiring a side-by-side comparison of the nano and bulk versions of the substance in the registration. Further study would not be needed for such a step, and the information could be very relevant to the key issue in the regulation of nanoforms, namely whether information on the bulk substance is transferrable in the risk assessment of the nanoform. It is laudable that the Commission looks to develop best practices for assessing and reporting nanomaterials, but it is likely that the technology already exists for side-by-side comparison of effects on health. Indeed, the review points to a number of guidances developed on the subject for food, cosmetics, medicinal products, and medical devices.[‡]

The conclusion in the report of the Commission is worth considering at length:

> In the light of current knowledge and opinions of the EU Scientific and Advisory Committees and independent risk assessors, nanomaterials are similar to normal substances/substances in that some may be toxic and some may not. Possible risks are related to specific nanomaterials and specific uses. Therefore, nanomaterials require a risk assessment, which should be performed on a case-by-case basis using pertinent information Current risk assessment methods are applicable, even if work on specific aspects of risk assessment is still required.
>
> The definition of nanomaterials will be integrated in EU legislation, where appropriate. The Commission is currently working on detection, measurement and monitoring methods for nanomaterials and their validation to ensure the proper implementation of the definition.
>
> Important challenges relate primarily to establishing validated methods and instrumentation for detection, characterization and analysis, completing information on hazards of nanomaterials and developing methods to assess exposure to nanomaterials.
>
> Overall the Commission remains that REACH sets the best possible framework for the risk management of nanomaterials when they occur as substances or mixtures but more specific requirements for nanomaterials within the framework have proven necessary. The Commission envisages modifications in some of the REACH Annexes and encourages ECHA to further develop guidance for registrations after 2013.

---

[*] Id.
[†] Id. at 6.
[‡] Id. at 12–14.

The Commission will carefully follow developments, and report back to the Parliament, the Council and the European Economic and Social Committee within 3 years.[*]

Much of this language is of the same substance and tone as that quoted in the various FDA guidances earlier in this chapter. The key difference is whether the comments from the FDA relate to products incorporating nanomaterials rather than to the materials themselves.

15.9.1 Cosmetic Directive

There have been specific product class developments within the EU in this time period relating to cosmetics, food and related items, biocides, and medical devices and pharmaceuticals. The major development regarding cosmetics is Regulation (EC) No. 1223/2009, passed by the Parliament and Council on November 30, 2009. The regulation became effective on July 11, 2013, and replaced the "Cosmetics Directive," which in the words of the announcement "ensured that products circulate freely whilst guaranteeing a high level of protection for consumers."[†] The provisions of No. 1223/2009 "aim at ensuring that consumers' health is protected and that they are well informed by monitoring the composition and labeling of products. The Regulation also provides for the assessment of product safety and the prohibition of animal testing."[‡] The regulation includes specific provisions for nanomaterials (definition, requirements for notification, labeling and reporting of nanomaterials). There is also a Cosmetics Product Notification Portal, a central system where a distributor will have to submit information on the presence of nanomaterials in cosmetics, as well as plans for a guidance for safety assessment of nanomaterials in cosmetics.[§] The provisions of the regulation go a considerable way to satisfying the objectives of identification, registration (notification), safety labeling, and data collection and assessment. It is an advance on the somewhat tepid guidance on cosmetics issued by CFSAN in the United States.

The regulation continues to provide that free movement of cosmetics in the internal market is permitted if they comply with the provisions of No. 1223/2009. Compliance rests in the first instance with the entity distributing the cosmetic. A responsible person must be designated for each product placed on the market, and the responsible person is charged with ensuring compliance of the cosmetic products with the rules set out in the Regulation, especially those relating to human health, safety, and consumer information. The responsible person is obliged to maintain a product information file accessible to the public authorities, and furthermore, he or she has to identify the distributors to whom they supply the cosmetic product for a period of three years following the date on which the batch of the cosmetic product was made available to the distributor. The same applies to all other persons involved in the supply chain.

If a product does not comply, a responsible person is directed to take one of the two actions to render it compliant: (1) withdraw it from the market; or (2) recall it to the manufacturing company in all member states where the product is available. Where the responsible person does not take all appropriate measures, the competent national authorities are empowered to take the necessary corrective measures.

If a product which complies with the requirements of the Regulation presents or could present a serious risk to human health, the competent national authority shall take all necessary provisional measures to withdraw, recall, or restrict the availability of the product on the market.

The Annexes of the Regulation provide lists of prohibited or restricted substances with respect to use in cosmetic products. Certain colorants, preservatives, and UV-filters are also prohibited.

---

[*] Id. at 11.
[†] Europa. Summaries of EU Legislation. Cosmetic Products (from 2013). http://europa.eu/legislation_summaries/consumers/product_labelling_and_packaging/co0013_en.htm.
[‡] Id.
[§] Developments in Nanotechnologies Regulations and Standards 2012—Report of the Observatory Nano at 6.

No. 1223/2009 prohibits the use of substances recognized as carcinogenic, mutagenic, or toxic for reproduction (classified as CMR), apart from exceptional cases. It provides for a high level of protection of human health where nanomaterials are used in cosmetic products.

To improve consumer protection, containers or packaging must bear written information in indelible, easily legible, and visible lettering. This information must include:

- The name or registered name and the address of the responsible person
- The country of origin for imported products
- The weight or volume of the content at the time of packaging
- A use-by date for products kept in appropriate conditions
- Precautions for use, including for cosmetics for professional use
- The batch number of manufacture or the reference for identifying the cosmetic product
- The list of ingredients, that is, any substance or mixture intentionally used in the product during the process of manufacturing

Generally, the new Regulation applies in 2013. Some of its provisions, however, have options since December 1, 2010. These concern substances that are carcinogenic, mutagenic, or toxic for reproduction (classified as CMR).

15.9.2 Food Labeling

Regarding foods, the new EU Regulation 1169/2011 on the provision of food information to consumers considerably changes existing legislation on food labeling including:

- Nutrition information on processed foods
- Origin labeling of fresh meat from pigs, sheep, goats, and poultry
- Highlighting allergens, for example, peanuts or milk in the list of ingredients
- Better legibility, that is, minimum size of text
- Requirements on information on allergens also cover non-pre-packed foods including those sold in restaurants and cafés

The new rules will apply beginning December 13, 2014. The obligation to provide nutrition information will apply from December 13, 2016. The rules combine two directives into one legislation: labeling, presentation, and advertising of foodstuffs (2000/13/EC) and nutrition labeling for foodstuffs (90/496/EEC). These directives provide for a requirement for the labeling of ingredients in the form of nanomaterials, that is, "material plus the word 'nano' in brackets."[*] Similar labeling requirements are in place for biocides, and the said products must also have a separate evaluation of risks from nanomaterials used.[†]

In addition to the above, since 2006, the European Food Safety Authority (EFSA) has been following developments in nanotechnology within its remit—providing independent scientific advice and technical support to risk managers, including reviewing the current state of knowledge and latest developments in nanotechnology with regard to food and feed.[‡] In March 2009, the EFSA's Scientific Committee, which includes the chairs of all of EFSA's Panels, published a scientific opinion on nanoscience and nanotechnologies in relation to food and feed safety. A guidance document on how to assess potential risks related to certain food-related uses of nanotechnology followed in May 2011. It provides practical recommendations on how to assess applications from industry to use engineered nanomaterials (ENMs) in food additives, enzymes, flavorings, food contact materials, novel foods, food supplements, feed additives, and pesticides.

---

[*] Id.
[†] Id.
[‡] EFSA JOURNAL. 2009. The potential risks arising from nanoscience and nanotechnologies on food and feed safety. http://www.efsa.europa.eu/en/efsajournal/pub/958.htm.

Both publications followed requests for advice from the European Commission. While finalizing its outputs, EFSA took into account feedback received from public consultations. EFSA's Panels also consider the safety of specific nanomaterials, for instance in the areas of food additives and food contact materials.

15.9.3 Medical Device and Pharmaceutical

Finally, the European Medicines Agency addresses issues regarding nanomaterials in medical devices and pharmaceuticals. On its website, the Agency states that "nanotechnology has only partially been exploited in medicine" and "is being investigated as a way to improve the properties of medicines, such as their solubility or stability, and to develop medicines that may provide new ways to:

- Deliver medicines to the body
- Target medicines in the body more accurately
- Diagnose and treat diseases
- Support the regeneration of cells and tissues"*

Going forward, the Medicines Agency describes its activities in the area of nanotechnology as:

The European Medicines Agency follows the latest developments in nanotechnology that are relevant to the development of medicines. Recommendations from the Agency's Committee for Medicinal Products for Human Use (CHMP) have already led to the approval of a number of medicines based on nanotechnology. These include medicines containing:

- Liposomes (microscopic fatty structures containing the active substance), such as Caelyx (doxorubicin); Mepact (mifamurtide), and Myocet (doxorubicin)
- Nano-scale particles of the active substance, such as Abraxane (paclitaxel), Emend (aprepitant), and Rapamune (sirolimus)

The development of medicines using newer, innovative nanotechnology techniques may raise new challenges for the Agency in the future. These include discussions on whether the current regulatory framework is appropriate for these medicines and whether existing guidelines and requirements on the way the medicines are assessed and monitored are adequate.†

The Agency also needs to consider the acceptability of new testing methods and the availability of experts to guide the Agency's opinion-making.‡

Again, the language on the website echoes what the FDA is saying about its activities.

The CHMP mentioned by the European Medicines Agency is the committee responsible for preparing opinions on questions concerning medicines for human use. The members and alternates of the CHMP are nominated by European Union Member States in consultation with the Agency's Management Board. They are chosen on the strength of their qualifications and expertise with regard to the evaluation of medicines and serve for a renewable period of three years. In 2009, the CHMP established the *ad-hoc* expert group meeting on nanomedicines. This group includes selected experts from academia and the European regulatory network, which support the Agency's activities by providing specialist input on new scientific knowledge and help with the review of guidelines on nanomedicines. The group also helps the Agency's discussions with international partners on issues concerning nanomedicines.

---

* European Medicines Agency—Medicines and emerging science—Nanotechnology. http://www.ema.europa.eu/ema/index.jsp?curl=pages/special_topics/general/general_content_000345.jsp&mid=WC0b01ac05800baed9.
† *Id.*
‡ *Id.*

The European Medicines Agency has addressed products incorporating nanotechnology on a case-by-case basis, as detailed above. Observatory Nano notes that the "detailed authorization procedures required" for medical devices and pharmaceuticals in general "are considered adequate for nano-related products, although a case-by-case approach in the evaluation and authorization procedures is envisaged to take into account the peculiar properties of nanotechnologies."*

15.10 NANOMATERIALS IN PRODUCTS

The EU is also in the forefront of considering the effect of nanomaterials mentioned for use in products and processes on the environment. As is the case with the United States and with all forms of product and environmental regulation, the first task confronting the EU is obtaining information about potential risks to the environment and health and safety concerns. These risks, and to what extent they can be tackled by the existing risk assessment measures in the EU, have been the subject of several opinions of the Scientific Committee on Emerging and Newly Identified Health Risks (SCENIHR). The overall conclusion to date mirrors that in the United States. The SCENIHR opines that "even though nanomaterials are not *per se* dangerous, there still is scientific uncertainty about the safety of nanomaterials in many aspects and therefore the safety assessment of the substances must be done on a case-by-case basis."†

As discussed earlier, in the EU, "REACH is the over-arching legislation applicable to the manufacture, placing on the market and use of substances on their own, in preparations or in articles. Nanomaterials are covered by the definition of a 'substance' in REACH, even though there is no explicit reference to nanomaterials. The general obligations in REACH, such as registration of substances manufactured at 1 tonne or more and providing information in the supply chain apply as for any other substance."‡ The first registration deadline under REACH on November 30, 2010, applied to substances manufactured or imported at 1000 tonnes or more per year. The registrations of nanomaterials in this tonnage band will help to generate more information useful for the assessment of risks. The next registration deadline is on May 31, 2013, and applies to substances manufactured or imported at or above 100 tonnes per year. The European Chemicals Agency (ECHA) receives the registrations and the Agency plays a central role in the collection, evaluation and dissemination of information on substances and preparations, including nanomaterials.§

15.10.1 CLASSIFICATION, LABELING, AND PACKAGING

Nanomaterials that fulfill the criteria for classification as hazardous under Regulation 1272/2008 on CLP of substances and mixtures must be classified and labeled. This applies to nanomaterials as substances in their own right, or nanomaterials as special forms of the substance. Many of the related provisions, including safety data sheets and classification and labeling apply already today, independently of the tonnage in which the substances are manufactured or imported. Substances, including nanomaterials, meeting the classification criteria as hazardous should have been notified to the ECHA by January 3, 2011. Any further update to the classification must also be notified without undue delay. The ECHA has made a classification and labeling inventory.

There has been concern expressed that many nanomaterials will not be reviewed under REACH because of the tonnage limits for regulations to apply. The Center for International Environmental Law (CIEL), ClientEarth, and Friends of the Earth Germany (BUND) released a proposal on November 13, 2012, for EU legislation to address the risks of nanomaterials. This proposal was

---

* Developments in Nanotechnologies Regulations and Standards 2012—Report of the Observatory Nano at 7.
† European Commission—Environment—Nanomaterials. http://ec.europa.eu/environment/chemicals/nanotech/index.htm.
‡ *Id.*
§ *Id.*

prompted by the EC's October 3, 2012, *Communication on the Second Regulatory Review on Nanomaterials*. The Communication describes the REACH program as "the best possible framework for the risk management of nanomaterials when they occur as substances or mixtures" and notes that "more specific requirements for nanomaterials within the framework have proven necessary." The CIEL proposal maintains that further regulatory action is necessary, and recommends a "nano patch" for REACH, including an obligation for all nanomaterials to be considered distinct from their nonnanoscale counterparts and substantially lower volume thresholds for registration of nanoscale substances. It also calls for an EU-wide registry for all nanomaterials and products on the market.[*] The EC's Environment and Enterprise Directorate-Generals (DG) issued separate statements in response to the NGOs' proposal. The DG Environment stated: "We regard the NGO proposal ... as a reaction to the Commission's recent regulatory review on nanomaterials. We are looking forward to discussing our review with all the stakeholders." DG Enterprise released a more detailed statement, noting that the EC "does not consider appropriate at present to change the basic registration rules under REACH and the rules for when a chemical safety assessment is required under REACH." In addition, the DG Enterprise stated: "[T]he highest volume substances such as carbon black and synthetic amorphous silica, as well as the most debated substances such as titanium dioxide, zinc oxide and carbon nanotubes, have already been registered under REACH. Together, they represent the vast majority of nanomaterials on the market in terms of tonnage and sales value."

ECHA's REACH Technical Guidance Documents, until recently, gave no specific guidance concerning nanomaterials.[†] There was a technical manual on how to include information on nanomaterial in an IUCLID dossier, which is an integral part of every REACH registration.[‡] The manual describes possibilities and best practices to include nanomaterials and to structure the available nanomaterial information.[§] The latter is particularly important when nanomaterials are additional forms of a substance rather than substances in their own right.[¶] Further guidance on the clarification and elaboration of the REACH information requirements and the Chemical Safety Assessment in case of nanomaterials is needed as registration dossiers for nanomaterials must be prepared or updated by companies and reviewed by the ECHA.[**] Work in the scientific community as well as in international organizations has been ongoing for almost a decade on methodologies for assessing risks associated with nanomaterials.[††]

On this basis the Commission launched a comprehensive REACH Implementation Project on Nanomaterials (RIPoN) in 2009 to provide advice on key aspects of the implementation of REACH with regard to nanomaterials concerning Information Requirements and Chemical Safety Assessment.[‡‡] Based on the scientific and technical state-of-the-art recommendations made in reports to date, ECHA on April 30, 2012 published three new appendices, updating the *Guidance on Information Requirements and Chemical Safety Assessment*.[§§] These three new appendices are recommendations for registering nanomaterials.[¶¶]

Another comment of note was issued on October 3, 2012, when the EC announced its adoption of the *Communication on the Second Regulatory Review on Nanomaterials*, that assesses the

---

[*] Center for International Environmental Law (CIEL). 2012. High time to act on nanomaterials: A proposal for a "Nano Patch" for EU regulation. http://www.ciel.org/Chem/Nano_EU_13Nov2012.html.
[†] European Commission—Environment—Nanomaterials. http://ec.europa.eu/environment/chemicals/nanotech/index.htm.
[‡] *Id.*
[§] *Id.*
[¶] *Id.*
[**] *Id.*
[††] *Id.*
[‡‡] *Id.*
[§§] *Id.*
[¶¶] *Id. See also* ECHA. 2012. *Updated Guidance on Information Requirements and Chemical Safety Assessment for Nanomaterials.* http://echa.europa.eu/web/guest/guidance-documents/guidance-on-information-requirements-and-chemical-safety-assessment.

adequacy and implementation of EU legislation for nanomaterials, indicates follow-up actions, and responds to issues raised by the European Parliament (EP), EU Council, and the European Economic and Social Committee. The Communication concludes that "nanomaterials are similar to normal chemicals/substances in that some may be toxic and some may not."* Since possible risks are related to specific nanomaterials and specific uses, nanomaterials should be assessed on a case-by-case basis. The Communication states "Current risk assessment methods are applicable, even if work on particular aspects of risk assessment is still required."† According to the Communication, the EC "remains convinced that REACH sets the best possible framework for the risk management of nanomaterials when they occur as substances or mixtures but more specific requirements for nanomaterials within the framework have proven necessary.‡ The Commission envisages modifications in some of the REACH Annexes and encourages the ECHA to further develop guidance for registrations after 2013."§ To improve the availability of information, the EC states that it "… will create a web platform with references to all relevant information sources, including registries on a national or sector level, where they exist. In parallel, the Commission will launch an impact assessment to identify and develop the most adequate means to increase transparency and ensure regulatory oversight, including an in-depth analysis of the data gathering needs for such purpose. This analysis will include those nanomaterials currently falling outside existing notification, registration or authorization schemes."¶

The Communication is accompanied by an EC Staff Working Paper Document on types and uses of nanomaterials, including safety aspects, that responds to the EP's concern that the EC's approach to nanomaterials is jeopardized by a lack of information on the use and safety of nanomaterials that are already on the market. The Staff Working Paper Document provides detailed information on the definition of nanomaterials, nanomaterial markets, uses, benefits, health and safety aspects, risk assessment, and information and databases on nanomaterials. According to the Staff Working Paper Document, in response to the EP's call on the EC to compile a public inventory of the different types and uses of nanomaterials on the European market, the EC has compiled information on existing databases and intends to create an EC web platform on nanomaterial types and uses, including safety aspects.**

15.11 OTHER COUNTRIES

The initial effort at regulation has been concentrated in the United States and in the EU. To a lesser degree, however, there has been activity in other nations, and in the concluding sections of the chapter, the highlights of those efforts will be discussed.

15.11.1 Germany

Germany has encouraged a dialogue on the effects on the environment from the production and use of such materials. There have been recommendations regarding regulations and labeling, and there has been consideration of the need for further research. The German Federal Institute for Risk Assessment has posted its view on the potential for risk in the use of nanomaterials:

---

* Communication at 4.
† *Id.* at 11.
‡ *Id.*
§ *Id.*
¶ *Id.*
** European Commission. 2012. Communication from the Commission. Second regulatory review on nanomaterials, at 2. *See also* Commission Staff Working Document. 2008. Summary of legislation in relation to health, safety and environment aspects of nanomaterials, regulatory research needs and related measures. *COM* (2008)366 final. http://eur-lex.europa.eu/lexUriServ/LexUriServ.do?=uri=CELEX:52008SC2036:EN:NOT.

The scientific risk assessment of BfR (German Federal Institute for Risk Assessment) focuses on selectively manufactured nanomaterials. Whether these new nanomaterials or products containing such materials can involve unknown risks for consumers has not yet been definitively clarified in scientific terms. In a risk appraisal the dangerous properties on the one hand and the actual exposure on the other must be examined. This means that risks might result, more particularly, from nanoproducts in which dangerous nanomaterials occur in a nonbound form or can be easily released from them. Reasons for nanomaterials possibly involving risks are:

- The particular (physical–chemical) properties of a nanomaterial, for example, large reactive (reaction promoting) surfaces
- The special behavior in the human body, for example, a long retention time and the overcoming of natural biological barriers
- The burdening to be expected from a release
- BfR also deals with the risk assessment of nanotechnological applications in many different consumer articles such as chemicals, foods and their packaging, cosmetic agents, articles of daily use but also pesticides and biocides[*]

A 2011 German report has indicated that there will be the need to perform, "where necessary in individual cases," a "risk assessment based on internationally validated OECD methods and testing guidelines for nanomaterials."[†]

According to a January 2012 article on the Nanotechnology Industries Association website, there is organized political opposition to nanotechnology regulation as it stands now in Germany.[‡] The list of demands mirrors the points discussed at the beginning of this chapter, and already being implemented to varying degrees in the EU and in the United States:

On the 14th of December 2011, the German Social Democratic Party (SPD) published a press release outlining a motion (17/8158) to the German Bundestag (Parliament). The motion entitled 'Seize opportunities and reduce risks of nanotechnologies for consumers' ('*Chancender NanotechnologiennutzenundRisikenfürVerbraucherreduzieren*' orig.) requests action to be taken on the regulation of nanomaterials in regard to the following:

- Implementation of a EU nano-product register.
- Development of a national product register to be published online by 2013.
- Labeling requirement with the addition of "nano" on all consumer products containing nanoscale ingredients, in which nanomaterials are not embedded.
- Commitment to the drafting of an internationally accepted definition and examination of the EU definition on nanomaterials as a first step.
- Verification and possible application of the precautionary principle for consumer products containing nanomaterials in existing relevant legislations.
- Continuation of EU negotiations on the EU Novel Foods Regulation.
- Admission procedure and labeling of food products containing nanomaterials.
- Development of safe methods for disposing of waste products containing synthetic nanomaterials from the environment.
- Increase in funding for safety and risk research on nanomaterials by 10%.
- Development of strategies to address the skill shortage by fostering interest among young people in engineering careers and nanotechnology.

---

[*] German Federal Institute for Risk Assessment (BfR). 2013. Health assessment of nanotechnology. http://www.bfr.bund.de/en/health_assessment_of_nanotechnology-30439.html.
[†] Developments in Nanotechnologies Regulations and Standards 2012—Report of the Observatory Nano at 10.
[‡] NIA—Nanotechnology Industries Association. German fraction calls for tightening of nanotechnology regulation. 2012. http://nanotechia.org/global-new/german-faction-calls-for-tightening-of-nanotechnology-regulation.

15.11.2 UNITED KINGDOM

The United Kingdom supports the EU efforts and also is moving forward with case-by-case risk product evaluation, at least with regard to food and food contact materials, and an environmental assessment. There are ongoing studies of several of the most prominently mentioned nanomaterials: nanosilver, CNTs, and iron nanomaterials.

The Department for Environment, Food and Rural Affairs (Defra) is the government department in the United Kingdom that deals with nanotechnology and the environment. It makes policy and legislation, and works with others to deliver policies in areas such as the natural environment, biodiversity, plants and animals; sustainable development and the green economy; food, farming, and fisheries; animal health and welfare; and environmental protection and pollution control.[*]

Defra has stated that nanotechnologies have the potential to bring significant benefits to consumers, society, the environment, and the economy through a range of applications. Defra's primary interest in nanotechnologies is in managing any potential risks to the environment and human health via the environment. This area was addressed by the Royal Society and Royal Academy of Engineering (RS/RAEng) in a report titled "Nanoscience and nanotechnologies: opportunities and uncertainties."[†]

The Government's response to the recommendations was published in February 2005, and its implementation is being coordinated through the Nanotechnology Issues Dialogue Group (NIDG). Like the United States and the EU, Defra states that there is currently little evidence on which to determine the potential risks posed by engineered nanoscale materials. It is therefore difficult to assess the extent to which current controls and regulations cover these materials, or the type of additional measures that may be necessary to control potential risks.[‡] To address this, the U.K. government has developed a comprehensive program of research on potential risks and a Voluntary Reporting Scheme for engineered nanoscale materials.[§] The latter is of interest since it addresses the question of how to protect the public while information is being gathered on the effects of nanoscale materials. The UK's Voluntary Reporting Scheme (VRS) for Manufactured Nanomaterials was launched on September 22, 2006.[¶] The VRS was initially set up as a 2-year trial initiative for industry and research organizations to provide the Government with information relevant to understanding the potential risks posed by free engineered nanoscale materials.[**] During the 2-year trial thirteen data submissions were received, eleven from industry and 2 from academia.[††]

Following the conclusion of the trial, Defra stated that it has to be mindful of how to incorporate its reporting scheme or "build a bridge" to any future European reporting requirements for nanomaterials, including those of the REACH Regulation.[‡‡] Defra indicates that REACH is currently the subject of its own review, which aims to examine how the Regulation can be adapted to better address issues of specific relevance to nanomaterials (the previously described nanotechnology gap).[§§] Through its nanotechnologies research program, Defra has also commissioned research into the policy implications of nanotechnologies that will benefit the environment.[¶¶] It is investigating the opportunities and potential obstacles to adoption of a number applications of nanotechnology which could be used to cut use of nonrenewable energy sources and reduce greenhouse gas emissions,

---

[*] United Kingdom. Department for Environment, Food and Rural Affairs (Defra). http://www.defra.gov.uk/corporate/.
[†] Nanotechnology: Policy. http://archive.defra.gov.uk./environment/quality/nanotech/policy.htm.
[‡] *See, e.g.*, United Kingdom, Department for Environment, Food and Rural Affairs (Defra). "Nanotechnology: Policy." http://archive.defra.gov.uk/environment/quality/nanotech/policy.htm;.Chemicals, pesticides and nanotechnology." http://www.defra.gov.uk/environment/quality/chemicals/.
[§] Nanotechnology: Policy.
[¶] *Id.*
[**] *Id.*
[††] *Id.*
[‡‡] *Id.*
[§§] *Id.*
[¶¶] *Id.*

and Defra is also exploring the application of nanoscience in the areas of insulation, photovoltaics, electricity storage, engine efficiency and the hydrogen economy.* It is looking into applications of nanotechnologies that will benefit the environment. Potential applications include nanomaterials that can remove more efficiently pollutants from contaminated sites, and nano-based sensors that improve our ability to detect and eliminate pollution.†

15.11.3 FRANCE

The French Ministry of Ecology, Sustainable Development and Energy initiated a mandatory reporting program for nanomaterials, which took effect January 1, 2013, to establish an inventory of available nanomaterials to help with both information gathering and trying to guard against unanticipated accidents or malfunctions. Under Decree No. 2012-232, companies that manufacture, import, and/or distribute a "substance with nanoparticle status" in an amount of at least 100 grams per year must submit an annual report with substance identity, quantity, and use information. The report is due by May 1 of each year for information about nanoparticle substances produced/imported/distributed during the prior year.‡

15.11.4 DENMARK

Denmark is also considering creating a National Inventory. The Danish Consumer Council and the Danish Ecological Council, in cooperation with the Technical University of Denmark (DTU) Environment, have developed a database intended to help consumers identify more than 1200 products that may contain nanomaterials.§ The database is an inventory of products that contain nanomaterials or are marketed with the word "nano." The website asks users to help "improve and expand" the database "by telling us about products that claim to be 'nano' or you think may contain nano materials." Users are instructed to:

- Search shop shelves and find products, where the word "Nano" appears on the packaging or on the product itself
- Note down product name, manufacturer, address, and website
- Take a picture of the product
- Check the Nano Database to see if the product you have found has already been reported to the Danish Consumer Council
- Fill out the form on the website

After a report is received, it will be forwarded to DTU Environment, "who will examine the reported product and in turn authorize its publication in the database."¶

This initiative follows a draft amendment to the Danish Chemicals Act whereby the Minister of the Environment would have the authority to write a detailed order establishing the rules for a national database of mixtures and articles containing or releasing nanomaterials. The order would also require producers and importers to report products containing or releasing nanomaterials. The information in the database is intended to form the basis of an evaluation of whether the content of nanomaterials in products on the Danish market poses a risk for consumers and the environment. The Ministry plans for the first reports to be due in early 2014.

---

* *Id.*
† *Id.*
‡ Ministère de l'Écologie, Du Développement Durable et de L'Énergie. 2013. Prévention des Risques. La déclaration des nanomatériaux devient obligatoire. http://www.developpement-durable.gouv.fr/spip.php?page=article&id=30578.
§ Where's the nano? http://nano.taenk.dk/wheress-the-nan.
¶ The Nanodatabase. 2012. Welcome to the database. http://nano.taenk.dk/welcome-to-the-database; *see also* http://www.mim.dk/Nyheder/20120917_nano.htm.

15.11.5 Netherlands

In the Netherlands, two initiatives are noteworthy. First, the Netherlands National Institute for Public Health and the Environment (RIVM) published a report on June 29, 2012, titled *Interpretation and implications of the European Commission's definition on nanomaterials*. The Dutch ministries requested RIVM to interpret the meaning and implications of the European Commission's (EC) recommendation from a scientific perspective and to consider the implications for use in legislation. RIVM intends the report to provide the basis for discussions by policy makers and stakeholders on the use and further implementation of the recommended definition in national and international legal frameworks. RIVM concludes that, while the EC's recommendation contains the relevant aspects, further guidance is necessary to ensure the definition is interpreted consistently. According to RIVM, the next step is to incorporate the definition into legal frameworks regarding the interpretation of the term nanomaterial. RIVM states that this will lead to the collection of "nano-specific" data, contributing to further insight into the "nano-specific" properties and the fate, kinetics, and effects of nanomaterials. This will help focus on the specific needs for risk assessment and risk management of nanomaterials.[*]

The second initiative announced on May 22, 2013, was when the Health Council of the Netherlands noted the availability of a draft report proposing the implementation of an exposure registry and a system of health monitoring when working with ENPs. The draft report states that, due to the concerns and lack of knowledge, the Health Council "considers it prudent" to create an exposure registry. The Health Council recommends that the exposure registry be created for "insoluble and poorly in water-soluble nanoparticles in any composition or physical structure, including nanoparticles that are present in solid materials." The draft report acknowledges that, if solid materials are in good condition, "scarcely any nanoparticles will be released, but due to wear and tear and handling, such as drilling and sanding, it cannot be excluded that such particles can be released with all the associated risks." The draft report concludes that, "[f]rom the point of view of health, it is best to also register the solid materials." Data submitted to the registry would need to include chemical and physical properties, determinants of emission and exposure, and exposure concentrations. Regarding medical surveillance, the draft report concludes that implementation of a passive system is the best option. While a passive system would not provide answers quickly on whether health risks exist when working with nanoparticles, and if so, which type of health effects, when combined with other activities, such as targeted scientific research, it "may give a valuable contribution in the future to providing insight in the potential health risks due to exposure to nanoparticles." According to the Health Council, it will consider comments when preparing the final report, which will be presented to the State.[†]

15.11.6 Australia

In Australia, the Therapeutic Goods Administration (TGA) is responsible for the regulation of medical products, pharmaceuticals and cosmetics. TGA has taken an active interest in regulation of nanotechnology, spurred on by continuing controversy regarding the use of nanomaterials in sunscreen products, the source of great interest given the high rate of skin cancer in Australia. TGA describes its plans on its website:

> To date, the existing regulatory framework of the TGA has proved more than adequate to identify, assess and manage the risks associated with therapeutic products that incorporate nanotechnologies.

---

[*] National Institute for Public Health and the Environment. Ministry of Health, Welfare and Sport. 2012. Interpretation and implications of the European Commission's definition on nanomaterials. RIVM Report 601358001. http://www.rivm.nl/en/Library/Scientific/Reports/2012/juni/Interpretation_and_implications-of_the_European_Commission_s_definition_on_nanomaterials.

[†] *See* http://www.gezondheidsraad.nl/en/publications/working-engineered-nanoparticles-exposure-registry-and-system-health-monitoring.

The TGA, however, recognizes that the development of nanotechnologies, as well as other technologies, will continue to pose challenges to regulators into the future.

> The best response to such challenges is the maintenance and continued development of high quality scientific expertise within the agency together with ongoing interaction with sponsors, researchers, regulators and policy makers throughout Australia and internationally, to which the TGA is committed.
>
> As such, the TGA is conducting a comprehensive review of current regulatory arrangements to ensure that those arrangements remain adequate to assess and manage the risks that may be associated with products manufactured using increasingly sophisticated nanotechnologies.
>
> The TGA is also continuing to closely monitor developments around nanotechnology internationally to ensure a rapid response to any new issues identified.[*]

What Australia is doing is in the same vein as the action in the United States and the EU. There is no rush to expand the regulatory framework because the existing rules seem adequate, with the reservation that amendments can always be made if needed. The attitude in Australia, as in other jurisdictions, stems from a lack of any hard proof of a present health hazard, thus permitting more time to accumulate information to assist with a meaningful risk assessment.

One review that TGA has made is of the safety of two primary sunscreen ingredients: nanoparticle titanium dioxide and zinc oxide. TGA stated as follows, about the nanomaterials and sunscreens:

> Nanotechnology is used in some sunscreen products containing zinc oxide and titanium dioxide. These two chemicals are particularly valuable in sunscreens because they give broad protection from damaging sunlight. In conventional applications zinc oxide and titanium dioxide leave a white layer on the skin, but when they are reduced to nanoparticles this white layer does not appear while still providing the same level of protection from the sun. As such, this has proved a popular option with consumers. Recently, concern has been raised as to whether nanoparticles in sunscreen might be absorbed into viable cells below the skin's surface, risking damage to these cells. The Therapeutic Goods Administration actively monitors matters relating to the regulation of nanotechnologies and has found no current evidence to suggest that sunscreen nanoparticles pose greater safety risks than conventional products.[†]

In support of its position, TGA states that the potential for the two nanomaterials to cause adverse effects depends primarily on the ability of the nanoparticles to reach viable skin cells. On that score, TGA notes: "To date, the current weight of evidence suggests that TiO2 and ZnO nanoparticulates do not reach viable skin cells, rather, they remain on the surface of the skin and in the outer layer (stratum corneum) of the skin that is composed of non-viable, keratinized cells." The situation with sunscreen ingredients illustrates the situation facing regulatory agencies. There is a scientific and public clamor to take action to thwart the risks, but there is no real evidence of risk. This is the reason why there is an increased desire to require the submission of any information on adverse events in the use of the products incorporating nanomaterials. (Note: In Australia, all active ingredients, such as zinc oxide and titanium dioxide, must be declared on sunscreen labels, to help consumers make informed choices. It is not a requirement, however, for sunscreen labels to declare the particle size of the active ingredients.)

On October 22, 2012, Safe Work Australia announced the availability of a report titled *Human Health Hazard Assessment and Classification of Carbon Nanotubes*, as well as an information sheet on the report. The report recommends that MWCNTs should be classified as hazardous unless toxicological or other data for specific types imply otherwise. The National Industrial Chemicals Notification and Assessment Scheme (NICNAS) prepared the report in support of Safe Work Australia's Nanotechnology Work Health and Safety Program. NICNAS extensively reviewed the

---

[*] Australian Government. Department of Health and Ageing, Therapeutic Goods Administration. 2009. Nanotechnology and therapeutic products. http://www.tga.gov.au/industry/nanotechnology-qa.htm.

[†] Department of Health and Ageing. 2012. Current Issues. Sunscreen containing Nanoparticles. http://www.health.gov.au/internet/main/publishing.nsf/Content/currentissue-P10000045.

published literature on the toxicity of CNTs and concluded that multi-walled CNTs may cause damage to lungs through prolonged or repeated inhalation exposure. The report recommends classification as hazardous for repeated or prolonged inhalation exposure and for carcinogenicity. For all other endpoints, NICNAS found that CNTs either were not classified as hazardous, or cannot be classified as insufficient data are available. The information sheet summarizes the key findings from the report and discusses implications for manufacturers, importers, persons in control of a business or undertaking, and workers manufacturing or using products containing CNTs.[*]

15.11.7 Canada

The Canadian Government regulates medical and pharmaceutical products through Health Canada. Canada indicates in its official materials that nanotechnology and products derived from nanotechnology have a wide range of applications and the potential to impact many sectors, including the health and food sectors. In the health sector, the applications of nanotechnology impact new natural health products, medical devices, drugs, drug delivery systems, regenerative medicines, and diagnostic devices for improved detection and treatment of illnesses. In the food sector, nanomaterials could be used to preserve food, improve nutritional values, and enhance flavors. One of the most active components of Health Canada as it relates to nanotechnology is the Health Products and Food Branch (HPFB). HPFB participates in an interdepartmental Health Portfolio Nanotechnology Working Group, which gathers information and acts as a discussion forum for issues related to nanotechnology. This working group contains members from Health Canada, the Public Health Agency of Canada (PHAC), and the Canadian Institutes of Health Research (CIHR). Additionally, HPFB participates in the interdepartmental network chaired by Industry Canada.

Health Canada participates in a number of international initiatives, such as the Working Party on Manufactured Nanomaterials of the Organization for Economic Co-operation and Development (OECD) and the Technical Committee 229 of the International Organization for Standardization (ISO), and collaborates with international counterparts.

The authority to regulate nanotechnology in health-related products comes in part from the adoption of a broad working definition for nanomaterials to provide a consistent approach across several diverse regulatory program areas to identify regulated products and substances that may contain nanomaterials. The working definition enables the Department to establish internal inventories, to ask for additional information, and to integrate that new knowledge into regulatory decision-making processes. The first step to assuring adequate risk assessment and risk management is to identify potential nanomaterials using the working definition as a tool. The Canadian regulatory track is in line with what other countries are doing.

Currently, there are no regulations specific to nanotechnology-based health and food products. Health Canada relies on authorities within existing legislative and regulatory frameworks, which require the assessment of potential risks and benefits of products to the health and safety of Canadians before they can be authorized for sale.

According to Health Canada's working definition for nanomaterial, the term "nanoscale" means 1–100 nm inclusive. However, individual regulatory programs may request information above the 100 nm size range to an upper limit of 1000 nm in order to maintain flexibility to assess potential nanomaterials, including suspected nanoscale properties and phenomena. The 1000 nm cut-off attempts to separate characteristics attributable to macro-scaled materials from those of nanomaterials. In addition, for any regulated product or substance that contains nanomaterial and measures beyond 1 μm in size (e.g., bundles of CNTs that are very long), information may be requested for risk assessment purposes, regardless of the size.

---

[*] Safe Work Australia. 2012. Classification of carbon nanotubes as hazardous chemicals. http://www.safeworkaustralia.gov.au/sites/swa/about/publications/pages/classification-of-carbon-nanotubes-as-hazardous-chemicals.

To identify a nano-based product/material, the sponsor will be asked to self-identify when their application concerns a nanomaterial or "nanoproduct." This is a relatively new requirement worldwide, but one that is essential for the development of the inventory previously described.

Recently, the Drug Submission Application Form for Human, Veterinary, Disinfectant Drugs and Clinical Trial Application/Attestation (HC/SC 3011) was revised to facilitate this process. Section 59 of the revised form allows the sponsor to identify Medicinal (Active) Ingredient(s) or Non-medicinal Ingredient(s) listed under Section 56 or 57 that are a nanomaterial. A similar approach has been adapted for natural health products. It is planned that the Medical Devices Licence Application Form will also be revised to request the manufacturer to state whether their devices contain nanomaterials.

Health Canada encourages sponsors and other stakeholders to communicate with the responsible regulatory authority early in the development process, especially for combination products that are, contain, or make use of nanomaterials. In order to identify and assess potential risks and benefits of nanotechnology-based health and food products, the Department encourages manufacturers to request a pre-submission meeting with the responsible regulatory authority to discuss the type of information that may be required for their product's safety assessment.

In discussion with the sponsor, the Department may require the following types of information, including but not limited to: intended use of the nanomaterial, including any end product in which it will be used; manufacturing methods; characterization and physico-chemical properties of the nanomaterial, including identity, composition and purity; toxicological, eco-toxicological, metabolism and environmental fate data that may be both generic and specific to the nanomaterial if applicable; and risk assessment and risk management strategies, if considered or implemented.*

This language could well be part of the FDA information described early in the chapter. As in the United States, Canada's working definition was developed to be intentionally broad and will be applied more specifically in each regulatory program area. Future guidance specific to program areas and legislative and regulatory authorities will be developed in a manner that promotes a consistent set of approaches

15.12 CONCLUSION

Global governments are proceeding cautiously. The United States, EU, and other nations are only now commencing the first steps of regulatory development, the identification of what is to be regulated, and the requirement or request to notify authorities of the identifiable nanoproducts being sold and under what conditions.

As regulations are amended, applied, and interpreted, they may have a much greater effect on the market for pharmaceuticals and medicinal products containing nanomaterials, and the rate at which this is accomplished may cause a gap between countries that is not now apparent. The first possible effect to monitor is notification/registration. If registration is pursued vigorously, the public and the market could well learn of presently undisclosed applications of nanotechnology and other presence of nanomaterials in products. It has been widely reported that many companies are utilizing nanomaterials in products, but being very circumspect in what they disclose because of the fear of litigation or adverse reaction from consumers, patients, and professionals. Full disclosure and, of course, reports of real or perceived negative consequences from use could slow market development.

A more important factor will be if regulators ratchet up attempts to review products with nanomaterials prior to marketing. In the United States, the FDA has stated that the effect of the inclusion of nanomaterials will be considered in products requiring preclearance because of their intended uses. This could be an issue with prescription drugs, veterinary drugs, certain medical devices, and food packaging materials, because of premarket approval requirements for many products. There are no

---

* *See* Health Canada. Drugs and Health Products. Nanotechnology-based health products and food. http://www.hc-sc.gc.ca/dhp-mps/nano-eng.php.

generally agreed-upon methods of risk assessment for nanomaterials, so delays can be expected as regulators and industry seek to reach agreement on what testing will suffice to support approval for targeted applications. Such testing will likely increase the cost of product development—all considerations that have affected drug and device companies having to cope with premarket approval for products containing macro versions of substances. It can be postulated that the delay and cost will at least initially be greater for nanomaterials, until a great deal more information is generated and analyzed. Similar results can be anticipated for drugs and devices in the EU and elsewhere.

Another issue is whether nanomaterials, because of their characteristics and effects on biological systems, are "new" when compared to the macro version of substances, thereby requiring premarket review prior to marketing where it might otherwise not be required. A positive resolution to this question is important to the nanotechnology industry. In terms of generating support for the products under development, and those proposed, comparison to the macro form of the substance involved that has a solid safety record can be a great benefit, saving costly testing for the most part, and increasing consumer and patient confidence in the known.

Conversely, if a nanosubstance is found to be different from the macro form, considerable effort will be needed to overcome the finding. The "new" debate has been around as long as there have been products in development that could provide risk to the public—such as sunscreens. The statements made by the FDA for devices, drugs, cosmetics, and food packaging leave open the likelihood of "newness" as a topic, as do the REACH registration provisions and TSCA chemical substances debate involving different forms of a substance. Any situation where the similarity of a nanomaterial to its macro counterpart becomes a contested issue must be watched because it could change the curve for the success of the industry.

Another area to monitor is reporting of adverse events from products in use. Reports, especially involving some serious adverse effect, could discourage confidence, raise the specter of product liability litigation and tougher regulation, and slow down funding/investment. In an industry trying to gain a foothold, serious product problems could be crippling, given the manner in which the substances' safety has been touted.

Somewhat mundane by comparison to the above, but important to the growth of a new industry, are requirements that might be imposed on the design, choice of materials, manufacturing, and quality control provisions prior to release of a product for sale. This would involve the promulgation of some form of good manufacturing practices (GMPs) or quality systems regulation. Given the wide variety of procedures for the manufacture of nanoparticles, the lack of general agreement on acceptable methods, the concern about confidentiality in the process, and the relative lack of expertise on the part of regulators, the institution of GMPs could command a great deal of time on the part of the small companies presently making up the bulk of the industry.

The next problem would be inspections of the procedures put into place in reaction to the regulations and the expense and possible adverse publicity if the procedures are claimed to be inadequate or not being followed. This is another area of regulation that should be kept under close observation.

One of the most important functions of product and substance regulation is the mandatory generation of information regarding risk/benefit, or safety and effectiveness. As regulators are challenged to get more involved in the workings of the nanotechnology industry, the lack of hard information or procedures for addressing issues previously identified for all products will become increasingly apparent. To date, regulators around the world have been able to address the theoretical concerns about products by pointing to the absence of reports of adverse effects, or instances of product failures due to lack of controls. They have also fielded safety or risk concerns by opining that the nanoparticle forms have not been shown to present any additional documented risks to consumers and patients. That could change as more information-gathering mechanisms are implemented, and the manner in which regulatory bodies react could foretell the continued development or slowdown of commercial development of pharmaceuticals and medicinal products. How regulators react, and to what, will likely signal the emergence of any effects on the industry.

16 Ocular Toxicity of Nanoparticles

*Aditya Grover, Anjali Hirani, Yong Woo Lee,
Vijaykumar B. Sutariya, and Yashwant Pathak*[*]

CONTENTS

16.1 Use of Nanoparticles in Ocular Therapy 347
    16.1.1 Nanoceria 348
    16.1.2 CK30PEG 349
    16.1.3 Magnetic Nanoparticles 349
    16.1.4 Chitosan 350
    16.1.5 Polylactic-Co-Glycolic Acid 350
    16.1.6 Other Nanoparticles 351
        16.1.6.1 Poly(Alkyl-Cyanoacrylate) 351
        16.1.6.2 Poly-ε-Caprolactone Nanoparticles and Nanocapsules 351
        16.1.6.3 Nanomicelles 351
        16.1.6.4 Poly[(Cholesteryl Oxocarbonylamido Ethyl) Methyl Bis(Ethylene) Ammonium Iodide] 351
        16.1.6.5 Acrylate Polymers (Eudragit®) 351
        16.1.6.6 Solid Lipid Nanoparticles 351
References 352

16.1 USE OF NANOPARTICLES IN OCULAR THERAPY

The application of biodegradable nanoparticles in ocular therapies is of utmost importance to the field of ocular medicine. The enhancement of the current methods of therapy could mean a drastic change in the quality of life of people suffering from a number of degenerative posterior eye disorders. Currently, the most common form of ocular therapy involves topical treatment in the form of eyedrops due to its ease of use, minimal risk of infection, and patient compliability [1]. This method is limited in its effect, as natural processes in the eye flush the drug out of the tissue within the first minute of application; lacrimation is one such example [1,2]. The structural barriers in the ocular tissue combined with the difficulty in drug delivery make the posterior eye chamber a potentially neglected site of therapy [1].

To directly target the posterior eye chamber, a common method of therapy is intravitreal injection (IVT) to deliver drugs to the retina [1,2]. This method is not without any adverse side effects; the common ones include tissue damage and infections [1]. Nonbiodegradable forms of treatment are also used, by which a nanosized device is surgically implanted at the site of therapy [3]. The drawbacks of such a method are the relative large-sized incision required to implant a device; the repeated implantations of a new device, once the previous device has exhausted its drug supply; and neglecting removal of the device may cause it to be encapsulated by the fibrous tissue [3]. The possible complications with this type of therapy include retinal detachment, vitreous hemorrhage, and dissolution of the device, among others [3].

---

[*] Aditya Grover and Anjali Hirani are equal contributors to the work.

It behooves pharmaceutical researchers to develop nanotechnologies that bypass these invasive methods. Such particles improve tissue penetration, bypassing IVT, and provide sustained drug or gene therapy, a possible advantage that would bypass the need for viral gene therapy. However, this method is not without its own risk of infection and hemorrhagic complications [2,4,5]. By being able to control the dramatic increase in surface-area-to-volume ratio of nanoparticles as compared to comparable macroscopic devices, researchers could enhance tissue penetration and drug-delivery systems directly to the affected tissue [4,6].

The accessibility of the eye makes it a great target for nanoparticle therapy [2]. However, such technologies are not without their own drawbacks. Nanoparticles may be coated with toxic chemicals that could be released into the body during therapies or may build up into tissues and cause blockages, leading to an increase in interocular pressure [3,4]. The retina and optic neural tissue are very sensitive to toxic materials, which may cause unforeseen complications due to seepage of toxic materials during therapy [2,5,6].

The most common forms of nanoparticle toxicity that are expressed in the ocular tissue are oxidative stress, counteractions with cell membranes, and inflammation [2]. These parameters were tested in the trials conducted with the following nanoparticles outlined in this section of the chapter. Since the rabbit is the most common animal model used for ocular toxicity studies, most of the studies cited used this animal to obtain toxicity data [2]. The large lens of the mouse and rat makes them poor models, as administrating intraocular injections proves difficult [2]. In addition, a number of ocular complications arise in the mouse or rat by even the slightest touch of the eye by the needle, including inflammation or the development of cataracts [2]. Monkeys' eyes provide the best model for studying ocular toxicity of nanoparticles, but none of the following studies cited used the monkey as a model for the respective toxicity studies; it is unsure why not [2].

16.1.1 Nanoceria

One of the causes of retinal degenerative diseases, such as diabetic retinopathy, is oxidative damage due to reactive oxygen radical species [6]. Cerium oxide nanoparticles, also known as nanoceria, have been developed as antioxidants and free radical scavengers as a potential therapy for such neurodegenerative diseases [6]. When these nanoparticles are synthesized in the 3–5 nm range, they mimic the effects of the antioxidant enzymes, such as superoxide dismutase and catalase, which neutralize superoxide anions and hydrogen peroxides, respectively [6–8].

The primary method of administration for the nanoceria particles was via IVT. After injection, the retinal tissue accumulated the greatest concentration of nanoparticles and 70% of the particles were retained 120 days postinjection [6]. Nanoceria was not actively eliminated from the eye, as 90% of the injected nanoceria was retained in the eye 120 days postinjection [6]. The experimental half-life yield of the nanoceria in the retina was 414 days with a half-life of 525 days in the eye. Studies by Asati et al. [9] showed that the polymer coating of the nanoceria particles may induce a charge on the particles, varying the rates of uptake in different tissues and inducing localization in certain tissues. They also found that positively charged nanoceria particles could be taken up by more cell types than negatively charged particles [9].

Previous studies showed that weekly injected nanoceria did not have cytotoxic effects in the heart, kidney, brain, lung, spleen, and liver; cytotoxic studies in the ocular tissue also showed that nanoceria did not have toxic effects on healthy retinal tissue over a range of doses [6]. Investigations of four types of retinal tissue—superior and inferior central retina and superior and inferior peripheral retina—9 days post-IVT injection showed that there was no reduction in the thickness of these layers. This finding along with the lack of observable change in retinal function postnanoceria IVT injection as compared to saline-injected eyes further suggests that nanoceria does not have any short- or long-term cytotoxic effects on retinal tissue [6]. In addition, nanoceria particles were shown to be nontoxic to the optical neural tissue up to 120 days postinjection [6].

However, there may be some drawbacks to using nanoceria particles. Synthesis of nanoceria particles through the use of hexamethylenetetramine (HMT) may induce cytotoxic effects in the ocular tissue [6]. In addition, some cell culture studies with nanoceria have yielded results of particle aggregation, which may negatively affect the ocular tissue [10]. Given these possible negative effects of nanoceria particles, there are no negative effects on healthy retinal cells with enhanced redox nanoceria capacity [6]. The minimal cytotoxic effects of nanoceria particles on the retinal tissue make it a great candidate for the possible widespread of ocular drug therapies.

16.1.2 CK30PEG

Recombinant adeno-associated viruses (AAVs) have been extensively used in the ocular tissue as a gene therapeutic method by which long-term gene expression can be induced, thereby offering therapeutic intervention for defects in large genes [11,12]. Plasmid DNA (deoxyribonucleic acid) compacted with polyethylene glycol (PEG)-substituted polylysine (CK30PEG) nanoparticles offers a promising alternative to AAV gene therapies [13,14]. Subretinal injections of CK30PEG nanoparticles were able to induce persistent gene expression in mice for up to a year, opening up doors to efficiently deliver up to 20 kbp plasmid vectors efficiently to dividing and postmitotic cells [11,13,15]. These nanoparticles have been shown to be safe and effective in human clinical trials, thereby allowing for the direct targeting of molecular markers in photoreceptors and retinal pigment epithelium cells for gene therapies [11]. Similar therapeutic results have been shown in mouse models expressing the retinitis pigmentosa phenotype [16].

Histological examination showed that the subretinal delivery of CK30PEG nanoparticles did not induce infiltration of inflammatory cells in the eye [13]. Injected eyes did not show the proliferation of polymorphonuclear leukocytes (PMN), an early response to toxicity [13]. Myeloperoxidase (MPO) immunoreactivity, an activator of inflammatory signaling cascades, was not detected in injected eyes, nor was F4/80 microglia/macrophage marker, a response to ischemia-induced retinopathy [13]. PMN, MPO, and F4/80 markers were histologically expressed in positive experimental controls [13]. The lack of such expression in CK30PEG-injected eyes suggests that these nanoparticles do not cause an inflammatory cascade in injected eyes [13].

Enzyme-linked immunosorbent assay (ELISA) and real-time reverse-transcription PCR (qRT-PCR) were used to detect the presence of interleukin-8 (IL-8), monocyte chemotactic protein-1 (MCP-1), and tumor necrosis factor-α (TNF-α) proteins and mRNA, respectively, in CK30PEG-injected eyes, saline-injected eyes, and *Bacillus cereus* endophthalmitis eyes as a positive control [13]. IL-8 can be produced in response to inflammatory stimuli and is involved in the initiation and amplification of acute inflammatory response processes [13]. *B. cereus* eyes expressed elevated levels of IL-8 protein and mRNA, with lack of elevation of either, expressed by the nanoparticle and saline-injected eyes [13]. MCP-1 is a member of the chemokine family and recruits monocytes to sites of injury and infection [13]. Eyes injected with CK30PEG nanoparticles showed transient elevations in MCP-1 mRNA and protein that returned to saline-injected control levels after 1 day [13]. *B. cereus* positive control eye samples markedly expressed elevated levels of MCP-1 protein and mRNA [13]. TNF-α is produced by macrophages and monocytes as a part of the inflammation cascade pathway and apoptotic cell death [13]. There was no detectable increase in TNF-α levels following subretinal injection of CK30PEG, compared to the significantly increased TNF-α levels in endophthalmitis positive control eyes.

The lack in expression of inflammatory cascade proteins in eyes with subretinally injected CK30PEG suggests that this gene therapy model is nontoxic in the ocular tissue, thereby safely inducing sustained gene expression in necessary tissues [13].

16.1.3 MAGNETIC NANOPARTICLES

DNA-tethered magnetic nanoparticles (MNPs) are Food and Drug Administration (FDA)-approved MRI (magnetic resonance imaging) contrast agents able to successfully deliver genes to targeted ocular tissues [2,4,5]. These nanoparticles are nontoxic to the retinal tissue; apart from successfully transfecting ocular cells, eyes treated with MNP did not show signs of inflammation nor did they induce white blood cell infiltration, both intravitreally and subretinally [5]. Intraocular pressure in MNP-treated eyes remained the same as phosphate buffered saline (PBS)-treated eyes, suggesting that the particles did not disturb any intraocular meshwork [4].

Histological analysis of MNP-treated ocular tissue did not yield any signs of iron oxide toxicity by the nanoparticles [4]. MTT 3-(4,5-dimethylthiazol-2-yl)-2,5-diphenyltetrazolium bromide (MTT) cytotoxicity assays showed the biocompatibility of MNP coated with polyethylenoxide copolymers, suggesting that these nanoparticles are safe for intraocular injection [4]. In addition, the iron oxide was not shown to cause oxidative stress *in vivo* [2].

The use of uncoated MNP could lead to aggregation and oxidation *in vivo*; thus, natural, biocompatible, and biodegradable polymers are used to coat the nanoparticles [4]. Adaptation of the surface of the particle could enhance product delivery to tissues and could target specific tissues by the polymer coat used on the MNP [4]. Since iron oxide is a component of the MNP, exposure to external magnetic fields could be harmful to body tissues; however, these data were not provided [4]. In addition, iron could damage photoreceptors in the eye, thereby causing a serious side effect [4,17]. However, the low iron load in the MNP and the protective polymer coating prevented iron from leaking out into the tissue and showed no great amplitude difference in electroretinography (ERG) waveforms as compared to tissues injected with PBS [4]. Since MNP causes no major cytotoxicity issues *in vivo* for up to 5 months postinjection, MNP is one of the safest nanoparticle gene delivery mechanisms currently available [2,4].

16.1.4 CHITOSAN

Chitosan, a biocompatible and nontoxic deacetylated form of chitin derived from crustacean shells, is one of the least expensive and most widely used nanomaterial in ocular therapies [2,5]. The positively charged surface of the nanoparticles helps them interact well with the negatively charged corneal surface and has been successful in delivering drugs and genes to the ocular tissue [2,5].

Given these benefits, chitosan is a biomaterial that has different effects in different ocular tissues, making it compatible in some and incompatible in others [2,5]. Topically, chitosan shows biocompatibility and efficient gene delivery with little-to-none tolerance issues nor any tissue necrosis up to 24 h posttreatment [2]. However, chitosan induces acute inflammatory responses when injected intravitreally [2,5]. The severe inflammatory response caused a vitreous haze and membranous opacities caused by infiltration of a large number of monocytes to phagocytize the foreign polysaccharide-based chitosan nanoparticles, suggesting that the immunomodulatory hyalocytes in the vitreous humor are particularly sensitive to chitosan [2,5]. There were signs of retinal degradation at the sites of the most severe inflammation [5]. The dichotomy in the biological interaction of chitosan with different tissues suggests that chitosan is a promising nanoparticle for topical ocular therapy but a poor intraocular therapeutic nanoparticle.

16.1.5 POLYLACTIC-CO-GLYCOLIC ACID

Polylactic-co-glycolic acid (PLGA) is a copolymer of polylactic acid (PLA) and polyglycolic acid (PGA) [3]. This biodegradable, biocompatible copolymer is one of the most studied polymers and is a popular choice for the treatment of choroidal neovascularization [2,3]. The ratio of PLA and PGA in the synthesis of PLGA can be altered to change the therapeutic effects of the copolymer in biological systems by changing the total surface area, rate of drug release, and rate of polymer degradation [3]. Since PGA is synthesized using toxic solvents, improper formulations may cause toxicity in biological systems [3]. Furthermore, PLGA can be synthesized by emulsification in acetone and methylene chloride, also chemicals that may induce cytotoxicity [2]. However, dose-dependent studies of PLGA in biological systems did not yield any signs of cytotoxic effects on ocular tissues [2]. With no reports of cytotoxicity in the eye, PLGA remains one of the most widely used, FDA-approved nanoparticles for experimental nanotherapies [2].

16.1.6 OTHER NANOPARTICLES

16.1.6.1 Poly(Alkyl-Cyanoacrylate)

Poly (acyl-cyanocrylate) (PACA) is a colloidal suspension of nanoparticles shown to prolong the corneal penetration of hydrophilic and lipophilic drugs in the eye [18]. These compounds have a strong shelf life, as they maintained their mean size and appeared unchanged when stored at room temperature for 6 months [18]. The corneal penetration may be due to their colloidal nature; however, its therapeutic application may be hindered by the disruption caused to the corneal cell membrane [18].

16.1.6.2 Poly-ε-Caprolactone Nanoparticles and Nanocapsules

Poly-ε-caprolactone (PECL) is a hydrophobic, biodegradable, and biocompatible polymer of ε-caprolactone [3]. PECL nanoparticles are slowly broken down in biological systems by the hydrolysis of ester linkages but the nanoparticles can be mixed with more hydrophilic polymers to form copolymers that can be broken down at faster rates [3,19]. Experimental studies in rabbit eyes showed that the nanoparticles are well tolerated without any signs of inflammation in the anterior and posterior segments of the eye [20].

PECL nanocapsules enhance the penetration of lipophilic drugs in ocular tissues without damaging the cell membranes, with penetration rates more favorable than PECL nanoparticles [18]. The sizes of nanocapsules did not change after 6 months at room temperature, suggesting that the polymer coating imparts stability to the nanocapsules [18].

16.1.6.3 Nanomicelles

Nanomicelles, up to 100 nm in size, are a low-toxicity colloidal dispersion of molecules with a hydrophobic core and hydrophilic shell [1,21]. These molecules provide an excellent method by which to solubilize hydrophobic drugs with therapeutic concentrations and administer them to hydrophilic tissues, thereby lowering drug degradation and enhancing permeability [1]. Since the hydrophilic sclera is an efficient therapeutic pathway to the posterior eye, nanomicelles make it easier for hydrophobic drugs to efficiently diffuse to the posterior eye with minimal degradation [1]. In addition, the hydrophilicity of the nanomicelle shell may confer resistance against systemic circulation washout *via* ocular blood and lymphatic vessels [1]. This relatively new nanotherapeutic molecule has shown little toxicity in biological systems, but more toxicological studies are warranted before its widespread use [1].

16.1.6.4 Poly[(Cholesteryl Oxocarbonylamido Ethyl) Methyl Bis(Ethylene) Ammonium Iodide]

Poly[(cholesteryl oxocarbonylamido ethyl) methyl bis(ethylene) ammonium iodide] (PCEP) is a DNA-condensing agent with gene therapy potential in the eye [5]. Similar to MNP, PCEP did not induce inflammation when injected intravitreally or subretinally [5]. These relatively harmless molecules did not attract white blood cells to the site of injection [5]. Because of its inert biodegradability, and low toxicity, it serves as a potential ocular nanoparticle transfection agent [5].

16.1.6.5 Acrylate Polymers (Eudragit®)

Eudragit is a nanoparticle that improves drug stability and maximizes drug dosages and effects [2]. The biological activity of Eudragit can be directly manipulated by the choice of polymer used to make the Eudragit copolymer [2]. There were no reports of ocular toxicity after 10 min of application, with mild irritation in the first 10 min reported by 20–30% of subjects [2]. Because of its low- to nontoxicity, Eudragit serves as an ocular topically therapeutic nanoparticle.

16.1.6.6 Solid Lipid Nanoparticles

Solid lipid nanoparticles (SLNs), ranging in size up to 400 nm, are biodegradable and biocompatible, and have been used as nanoparticles since the 1990s [2]. The advantage of these nanoparticles is their advanced drug load that they can carry [2]. However, due to the nature of its synthesis, a potential source of toxicity may be the presence of excipients on the nanoparticle [2]. In addition, some surfactants dissociate from the nanoparticle during sterilization due to the high temperatures, another potential source of toxicity; there is no dissociation of surfactants at body temperature [2].

REFERENCES

1. Vadlapudi AD, Mitra AK. Nanomicelles: An emerging platform for drug delivery to the eye. *Therapeutic Delivery*. 2013;4(1):1–3.
2. Prow TW. Toxicity of nanomaterials to the eye. *Wiley Interdisciplinary Reviews: Nanomedicine and Nanobiotechnology*. 2010;2(4):317–33.
3. Christoforidis JB, Chang S, Jiang A, Wang J, Cebulla CM. Intravitreal devices for the treatment of vitreous inflammation. *Mediators of Inflammation*. 2012.
4. Raju HB, Hu Y, Vedula A, Dubovy SR, Goldberg JL. Evaluation of magnetic micro- and nanoparticle toxicity to ocular tissues. *PLoS One*. 2011;6(5):e17452.
5. Prow TW, Bhutto I, Kim SY, Grebe R, Merges C, McLeod DS et al. Ocular nanoparticle toxicity and transfection of the retina and retinal pigment epithelium. *Nanomedicine: Nanotechnology, Biology and Medicine*. 2008;4(4):340–9.
6. Wong LL, Hirst SM, Pye QN, Reilly CM, Seal S, McGinnis JF. Catalytic nanoceria are preferentially retained in the rat retina and are not cytotoxic after intravitreal injection. *PLoS One*. 2013;8(3):e58431.
7. Jessica E. Nanoceria exhibit redox state-dependent catalase mimetic activity. *Chemical Communications*. 2010;46(16):2736–8.
8. Self WT, Seal S. Nanoparticles of cerium oxide having superoxide dismutase activity. Google Patents; 2009.
9. Asati A, Santra S, Kaittanis C, Perez JM. Surface-charge-dependent cell localization and cytotoxicity of cerium oxide nanoparticles. *ACS Nano*. 2010;4(9):5321–31.
10. Verma A, Stellacci F. Effect of surface properties on nanoparticle–cell interactions. *Small*. 2010;6(1):12–21.
11. Han Z, Conley SM, Makkia R, Guo J, Cooper MJ, Naash MI. Comparative analysis of DNA nanoparticles and AAVs for ocular gene delivery. *PLoS One*. 2012;7(12):e52189.
12. Amado D, Mingozzi F, Hui D, Bennicelli JL, Wei Z, Chen Y et al. Safety and efficacy of subretinal readministration of a viral vector in large animals to treat congenital blindness. *Science Translational Medicine*. 2010;2(21):21ra16.
13. Ding X-Q, Quiambao AB, Fitzgerald JB, Cooper MJ, Conley SM, Naash MI. Ocular delivery of compacted DNA-nanoparticles does not elicit toxicity in the mouse retina. *PLoS One*. 2009;4(10):e7410.
14. Liu G, Li D, Pasumarthy MK, Kowalczyk TH, Gedeon CR, Hyatt SL et al. Nanoparticles of compacted DNA transfect postmitotic cells. *Journal of Biological Chemistry*. 2003;278(35):32578–86.
15. Fink T, Klepcyk P, Oette S, Gedeon C, Hyatt S, Kowalczyk T et al. Plasmid size up to 20 kbp does not limit effective *in vivo* lung gene transfer using compacted DNA nanoparticles. *Gene Therapy*. 2006;13(13):1048–51.
16. Cai X, Conley SM, Nash Z, Fliesler SJ, Cooper MJ, Naash MI. Gene delivery to mitotic and postmitotic photoreceptors via compacted DNA nanoparticles results in improved phenotype in a mouse model of retinitis pigmentosa. *The FASEB Journal*. 2010;24(4):1178–91.
17. Declercq SS, Meredith P, Rosenthal AR. Experimental siderosis in the rabbit: Correlation between electroretinography and histopathology. *Archives of Ophthalmology*. 1977;95(6):1051–8.
18. Calvo P, Vila-Jato JL, Alonso MJ. Comparative *in vitro* evaluation of several colloidal systems, nanoparticles, nanocapsules, and nanoemulsions, as ocular drug carriers. *Journal of Pharmaceutical Sciences*. 1996;85(5):530–6.
19. Pitt C. Poly-*e*-caprolactone and its copolymers. In: Chasin M, Langer R, eds. *Biodegradable Polymers as Drug Delivery Systems*. New York, NY: Marcel Dekker; 1990. p. 71.
20. Silva-Cunha A, Fialho SL, Naud MC, Behar-Cohen F. Poly-*e*-caprolactone intravitreous devices: An *in vivo* study. *Investigative Ophthalmology Visual Science*. 2009;50(5):2312–8.
21. Trivedi R, Kompella UB. Nanomicellar formulations for sustained drug delivery: Strategies and underlying principles. *Nanomedicine*. 2010;5(3):485–505.

17 Genotoxicity of Nanoparticles

Amaya Azqueta, Leire Arbillaga, and Adela López de Cerain

CONTENTS

17.1 Genotoxicity Testing of Nanoparticles: Nanogenotoxicology 353
17.2 Mechanism of Action 355
17.3 *In Vitro* versus *In Vivo* Assays, What Do They Tell Us? 355
17.4 State of the Art: Study of Methods Used in Nanogenotoxicology 355
17.5 Tests Used in Nanogenotoxicology 356
    17.5.1 Comet Assay 357
    17.5.2 Micronucleus Test 357
    17.5.3 Bacterial Reverse Mutation Test: Ames Test 358
    17.5.4 Previous *In Vitro* Cytotoxicity Studies 358
17.6 Test Conditions That Influence the Genotoxicity of NPs 358
17.7 Characteristics of NPs That Influence Potential Genotoxicity 359
17.8 Conclusion 360
References 360

17.1 GENOTOXICITY TESTING OF NANOPARTICLES: NANOGENOTOXICOLOGY

In the last two decades, the production and use of nanoparticles (NPs) have impressively increased. NPs have been defined as particles in a nanometer scale with at least one dimension of 100 nm or less, although, in the case of pharmaceutical NPs, the dimension should be in the nanometer range (Bawa et al. 2005).

NPs possess different physical, chemical, and biological properties compared with bulk materials of the same composition. The small size, keeping the mass unchanged with respect to the same bulk material, entails an increase in the surface area and, consequently, in the number of atoms that can react to produce a certain activity. Owing to their special characteristics, NPs are applied in several areas, including biomedicine. There are enormous potential advantages of nanotechnology in medicine and its wide use has given rise to a new area of medicine and research called nanomedicine. NPs are used in preventing diseases, as well as diagnosing, monitoring, treating (e.g., drug carriers), and relieving pain.

Nevertheless, the small size and the large surface area improve the cellular uptake, but may lead to cellular accumulations with poor clearances and subsequent chronic toxicities (Landsiedel et al. 2009; Singh et al. 2009). Biological nanomaterials are normally biodegradable and biocompatible and are considered to be less toxic or nontoxic to the human body (Kim et al. 2010b). In contrast, insoluble nanomaterials may accumulate in human tissues and organs and exert toxic effects over long-term administrations. Nanotoxicology is a special branch of toxicology that studies the adverse effects of NPs on living systems. The main problem of this new area is the fact that the term NP is too broad, as it covers particles with very different physical, chemical, and biological properties.

Genotoxicity studies are crucial for assessing the safety of NPs for the development of possible medical applications. The cell nucleus is one of the desired targets, implying a possible interaction with DNA and a possible induction of genetic damage. NPs can also reach the nucleus unintentionally or provoke the release of reactive oxygen and nitrogen species that can interact with DNA. They can even interact with DNA replication proteins or interfere with the DNA repair mechanism. These interactions with DNA may lead to other responses such as mutagenesis and carcinogenesis (Landsiedel et al. 2009; Singh et al. 2009). Nanogenotoxicology is the new area of research that studies the deleterious effects of NPs on DNA. Nowadays, despite the high development of NPs through medical or pharmaceutical applications, there are still very few studies regarding their genotoxicity.

Genetic lesions can be produced at different levels: primary and secondary DNA structure (e.g., alteration of bases, cross-links, and adducts) or chromosomes (changes in number—aneugenic effects—or structure—clastogenic effects) (Friedberg et al. 2006). Some of these lesions are always produced by a direct interaction between a chemical compound and DNA (i.e., chemical adducts) while others can also be produced by an indirect mode of action (e.g., producing ROS that interact with the DNA, or decreasing the DNA repair capacity of the cell). Special attention is given to the chemicals that produce mutations, since there is a clear relation between this type of lesion and carcinogenesis.

Lesions at the level of the primary and secondary structures of DNA can easily be induced and produce mutations if not repaired well, and so their presence can also be a risk factor for the occurrence of cancer. Numerical chromosomal changes have also been associated with tumorigenesis.

The importance of testing the genotoxicity of chemicals is obvious. In 2011, the International Conference on Harmonisation of Technical Requirements for Registration of Pharmaceuticals for Human Use (ICH) published a new version of the "Guidance on Genotoxicity Testing and Data Interpretation for Pharmaceuticals Intended for Human Use" (ICH S2(R1) 2011), which has been recently implemented. This guidance suggests a strategy to determine the genotoxicity of pharmaceutical compounds (Figure 17.1) and the use of OECD (Organisation for Economic Co-operation and Development) guidelines to carry out different assays. The OECD guidelines cover the testing of chemicals, including different protocols to carry out *in vitro* and *in vivo* genotoxicity tests (http://www.oecd-ilibrary.org/environment/oecd-guidelines-for-the-testing-of-chemicals-section-4-health-effects_20745788). Not all of the genotoxicity assays included in the ICH's strategy have an OECD guideline at the moment, but efforts are being made to solve this gap and for the *in vivo* comet assay a draft guideline already exists.

Until now, there have been no regulatory requirements to test the genetic safety of NPs and it is not clear if current strategies and standard assays for testing chemicals, ICH S2(R1) and OECD guidelines, are suitable to characterize the potential genotoxic risk of NPs. A lot of effort is being concentrated in developing a strategy and a battery of assays to check the genotoxicity of these materials in a reliable way.

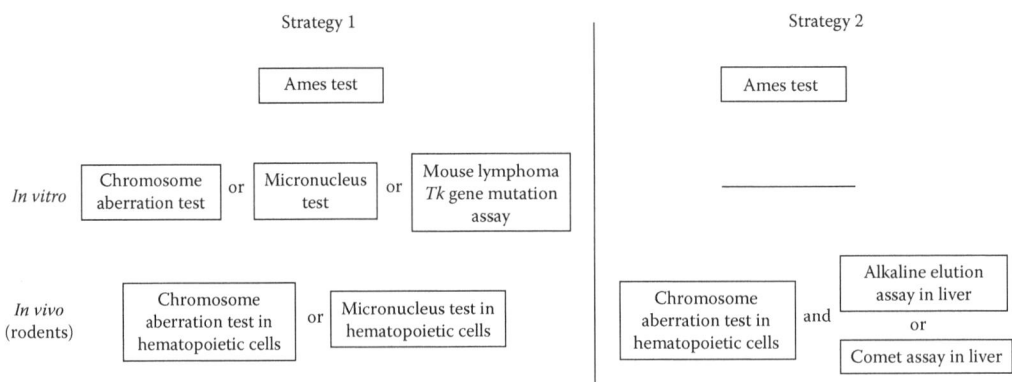

FIGURE 17.1 ICH's strategy for genotoxicity testing of pharmaceutical compounds.

17.2 MECHANISM OF ACTION

NPs enter the cell via endocytosis and end up in lysosomes where they are exposed to hydrolytic enzymes. Small NPs can also cross cell membranes and accumulate in specific subcellular structures and even the nucleus. Tsoli et al. (2005) described how small NPs (1–2 nm diameter) associate with DNA strands. But even if free NPs in cytoplasm cannot go through the nuclear membrane, they can have eventual access to genetic material during mitosis. Thus, some NPs can penetrate the nucleus and interfere directly with the structure and function of genomic DNA or the nuclear matrix.

The genotoxicity of NPs might also result from indirect DNA damage by the cellular production of reactive oxygen species (ROS), the depletion of the antioxidant defense, or the altered synthesis of DNA repair enzymes (Singh et al. 2009). ROS can be generated directly by the interaction between the NP surface and cell membranes, alteration of mitochondrial function, or the activation of the membrane-bound NADPH oxidase during phagocytosis (Konczol et al. 2011; Nabeshi et al. 2011; Norppa et al. 2011; Wang et al. 2012). They are also produced in inflammatory processes induced by the activation of neutrophils and macrophages by NPs (Greim et al. 2001; Schins 2002; Schins and Knaapen 2007; Donaldson et al. 2010). ROS can interfere with DNA, producing different lesions (e.g., 8-oxo-7,8-dihydro-2′-deoxyguanosine (8-oxo-gua), single-strand breaks) or stalled replication forks can lead to double strand breaks after replication, which can result in recombination, thus producing permanent genome rearrangements.

NPs can exert their genotoxic effects in a variety of organs due to their capability to enter systemic blood circulation. Moreover, they can cross epithelial and endothelial cell layers by transcytosis and subsequently travel along the dendrites, axons, and lymphatic vessels (Durnev 2008).

A decreased antioxidant defense or DNA repair activity are two other mechanisms of action by which an NP can induce genotoxicity, which have not yet been fully studied.

17.3 *IN VITRO* VERSUS *IN VIVO* ASSAYS, WHAT DO THEY TELL US?

The strategy to test the genotoxicity of pharmaceuticals for human use suggested in ICH S2(R1) includes a battery of *in vitro* and *in vivo* studies (Figure 17.1); both types of assays are needed to check different endpoints. Results obtained in *in vitro* studies require the eventual confirmation with an *in vivo* approach since *in vitro* assays lack the complexity of an organism.

In nanogenotoxicology, *in vitro* assays are widely used and can detect different mechanisms of action and endpoints. The localization of the NPs inside the nucleus as well as the production of ROS will give clues to distinguishing between different mechanisms of action (Schins 2002). ROS scavengers to check for a possible decrease in the DNA damage produced (Konczol et al. 2011; Toduka et al. 2012; Toyooka et al. 2012) as well as inhibitors of endocytosis, nitric oxide synthase, and cyclooxygenase-2 (Xu et al. 2009) have been used to elucidate the mechanism of oxidative stress induced by NPs *in vitro*.

Secondary inflammatory effects on DNA have to be studied *in vivo*. Relevant oxidative and inflammatory endpoints would help in the interpretation of *in vivo* genotoxicity outcomes (Nielsen et al. 2008; Borm et al. 2011; Downs et al. 2012).

17.4 STATE OF THE ART: STUDY OF METHODS USED IN NANOGENOTOXICOLOGY

About 240 papers from different searches (using the terms nanogenotoxicology, genotoxicology and NPs, genetic toxicology and NPs, and genotoxicity and NPs) in the PubMed database have been analyzed to check the different methodologies used to evaluate the genotoxic effects of NPs until now. Papers that involve genotoxicity studies of NPs with a potential application to nanomedicine, as well as ones that are currently in use, have been selected to be included for the analysis. Gene expression studies have not been included. One hundred and two papers met the criteria. Table 17.1

TABLE 17.1
Summary of the Different Genotoxicity Assays Used in the 102 Papers Analyzed

	Number of Papers[a]
In vitro	81
Comet assay	52
Micronucleus test	30
Ames test	9
Chromosome aberration test	9
Detection of γ-H2AX by immunostaining	9
Standard DNA gel electrophoresis	4
In vivo	16
Comet assay	6
Micronucleus test	11
In vitro + *in vivo*	5

Note: *In vitro* assays used in a marginal number of studies: sister chromatid exchange, alkaline unwinding assay, eukaryotic gene mutation assays, and chemical detection of different DNA lesions. *In vivo* assays used in a marginal number of studies: chromosome aberration test (bone marrow), detection of γ-H2AX by immunostaining (bone marrow), and DNA fragmentation (liver).

[a] The same study can use different assays.

summarizes some of the results of the analysis; there is a clear lack of *in vivo* studies in the characterization of the genotoxic effects of NPs.

The detection of 8-oxo-gua was performed in 22% of the *in vitro* studies, mainly by the comet assay in combination with formamidopyrimidine DNA glycosylase (FPG) or 8-oxoguanine DNA glycosylase (OGG1), and in 31% of the *in vivo* studies. In this case, the lesion was detected by different techniques and in different specimens (liver, lung, colon, bone marrow, and urine).

The *in vivo* micronucleus (MN) test was performed in the blood or bone marrow of rodents (rats and mice) and the comet assay in the liver, lung, brain, bone marrow, and blood.

The nuclear uptake of NPs, important in interpreting the results, was assessed in approximately 50% of the papers that use *in vitro* approaches; 65% of these studies localize the NPs in different cellular compartments. The presence of the NPs in different tissues was checked in 31% of *in vivo* studies, but a proper bioavailability study was not carried.

The generation of ROS and inflammatory markers were assayed in 25% of *in vivo* studies.

The most frequent NPs studied in the analyzed papers were TiO_2-, silver-, zinc-, and fullerenes, followed by gold-, silica-, and ferric NPs.

17.5 TESTS USED IN NANOGENOTOXICOLOGY

A Working Party on Manufactured Nanomaterials was established by the OECD in 2006. One of the aims of this organization is to review the OECD guidelines for testing the genotoxicity of chemicals in order to decide their applicability to testing NPs. In a preliminary review (Test Guidelines for their Applicability to Manufactured Nanomaterials, 2009), the working party recommended the bacterial reverse mutation test (OECD TG 471), mammalian chromosome aberration test (OECD TG 473), and mammalian cell gene mutation test (OECD TG 476) for *in vitro* testing; and the mammalian bone marrow chromosome aberration test (OECD TG 475), mammalian erythrocyte micronucleus test (OECD TG 474), and unscheduled DNA synthesis (UDS) test with mammalian liver cells for *in vivo* testing. Nevertheless, according to the literature, these are not the most widely used tests to check the genotoxicity of nanomaterials. Instead, the comet assay and the MN test, both in *in vitro* and in *in vivo*, are the most used techniques. Both assays are well-established methods for detecting the genotoxic effects of NPs, but some concerns have to be taken into account. The Ames test has also been used, but its suitability is still under debate.

17.5.1 COMET ASSAY

Compared with other genotoxicity assays, the comet assay appears to find positive results for NPs most consistently, which may be attributed to its high sensitivity toward capturing DNA damages as triggered by ROS (Collins 2009).

The standard comet assay or single-cell gel electrophoresis is a sensitive method to measure single- and double-strand breaks and abasic sites (Collins et al. 2008). The use of the bacterial enzymes, FPG or endonuclease III or the human enzyme, OGG1 allows the detection of oxidized bases, a more relevant DNA lesion (Azqueta et al. 2009). In the revised bibliography, 33% of the *in vitro* studies using the comet assay include the detection of oxidized bases by the inclusion of at least one of these enzymes.

The comet assay detects reversible DNA lesions, since strand breaks and oxidized bases can be repaired. The rejoining of single-strand breaks is very simple, but rejoining double-strand breaks and the repair of oxidized lesions are more complicated and prone to errors.

Although there are some guidelines and protocols in the literature (Tice et al. 2000; Hartmann et al. 2003; Collins and Azqueta 2012; Vasquez 2012), as yet there is no OECD guideline for the *in vitro* assay. Nevertheless, a draft OECD guideline for the *in vivo* assay has been published.

The comet assay can be applied to virtually any type of eukaryotic cell if a cell suspension can be obtained. This gives a big advantage in testing the genotoxicity of NPs *in vivo*, as many tissues can be checked with small modifications to the assays (e.g., for sperm). The protocol of the *in vitro* and *in vivo* comet assays used nowadays in nanogenotoxicity tests does not differ from the protocol used to test the genotoxicity of chemical compounds.

Pfaller et al. (2010) reported the presence of aggregates in the tails of comets when high concentrations of some NPs were tested *in vitro*; washing steps are important to remove residual NPs. If still present during the process, they can come in contact with naked DNA during the final processing steps of the comet assay and induce artificial damages (Stone et al. 2009).

17.5.2 MICRONUCLEUS TEST

The MN assay is a genotoxicity test for the detection of subnuclear DNA-containing fragments in the cytoplasm of interphase cells. MNs are distinct from main nuclei, and may originate from acentric chromosome fragments (i.e., lacking a centromere) or whole chromosomes that are unable to migrate to the poles during the anaphase stage of cell division. The assay detects the activity of clastogenic (with potential to break chromosomes) and aneugenic chemicals (with potential for lagging entire chromosomes) in cells that have undergone cell division during or after exposure. The *in vitro* test is described in OECD TG 487 and the *in vivo* test in OECD TG 474.

Several cell lines or human lymphocytes can be used in the *in vitro* version; bone marrow and/or peripheral blood cells of animals, usually rodents, are used in the *in vivo* version. MN tests can only be applied to dividing cells. Lesions detected with this assays are not reversible.

The *in vitro* MN assay is performed using different approaches (Doak et al. 2009; Gonzalez et al. 2011). Some of them use cytochalasin B to inhibit cytokinesis and generate binucleated cells, and it is in these cells that the MN frequency is scored, increasing the likelihood that the MNs have been formed during or since treatment rather than as preexisting ones. This compound inhibits endocytosis, an important cell uptake mechanism, which can interfere with the evaluation of the genotoxicity of some NPs (Doak et al. 2009; Gonzalez et al. 2011); it is important to treat the cells with NPs but without cytochalasin B during some time.

Washing steps after *in vitro* treatments are also important since the presence of NPs in the preparations being scored has been reported by Pfaller et al. (2010) when high concentrations of NPs are used.

17.5.3 Bacterial Reverse Mutation Test: Ames Test

The bacterial reverse mutation test is commonly employed as an initial screen for genotoxic activity and, in particular, for point mutation-inducing activity. The principle and the description of the test method are in OECD TG 471.

The bacterial reverse mutation test uses amino acid-requiring strains of *Salmonella typhimurium* and *Escherichia coli* to detect point mutations, which involve the substitution, addition, or deletion of one or a few DNA base pairs. The principle of this test is that it detects reversions, or back-mutations, of a mutated gene present in the test strains, restoring the functional capability of the bacteria to synthesize an essential amino acid. The revertant bacteria are detected by their ability to grow in the absence of the amino acid required by the parent test strain.

Surprisingly, test results have been negative whereas *in vitro* mammalian cell tests often give positive genotoxic responses (Doak et al. 2012; Jomini et al. 2012; Nam et al. 2013). The most likely hypothesis is that NPs cannot penetrate the walls of bacterial cells, so the Ames test is not suitable for testing the genotoxicity of NPs (Landsiedel et al. 2009; Singh et al. 2009; Nabeshi et al. 2011; Woodruff et al. 2012). However, some studies have shown that NPs can cross the bacterial wall of *E. coli* (Brayner et al. 2006) and *Pseudomonas aeruginosa* (Xu et al. 2004).

Nevertheless, Jomini et al. (2012) showed that the medium used during the exposure of bacterial cells to NPs prevents electrostatic interactions between them, leading to false-negative responses. They showed that the exposure of bacteria to TiO_2 NPs in a low ionic strength solution at a pH below the NP isoelectric point makes the assay suitable. More investigations should be done in order to validate this last hypothesis.

17.5.4 Previous *In Vitro* Cytotoxicity Studies

A preliminary cytotoxicity test must always be carried out in order to choose the adequate concentration range for *in vitro* genotoxicity studies.

NPs are optically active due to their high absorption and scattering effects, so standard assays for testing cytotoxicity may not be suitable. In the revised publications, close to 87% of the studies used these types of assays while close to 13% used the direct counting of cells or a proliferation assay. The last option is more complete since it conflates cell killing, growth, and cytostatic effects.

The issue of appropriate concentrations for testing the genotoxicity of chemicals is currently under discussion. Nowadays, concentrations causing 50% or less cytotoxicity are recommended; chemicals that exert cytotoxic effects are likely to give false-positive results in genotoxicity tests, the DNA damage being secondary to cytotoxicity.

17.6 TEST CONDITIONS THAT INFLUENCE THE GENOTOXICITY OF NPs

The differences in the responses between studies may be due to the test being used and to the different endpoints being measured, but also to the different test conditions employed.

The dosing obviously influences the genotoxic potential of NPs, but a dose-dependent relationship is not always clear. Several studies showed an inverse dose-dependent relationship between dose and cell genotoxicity using different NPs (Ghosh et al. 2010; Sergent et al. 2012). Calarco et al. (2013) showed that polyethylenimine-NPs uptake by cells was higher at lower concentrations, which indicates a saturation and limited cellular uptake capability when a higher concentration is used. The aggregation of NPs at higher doses can explain the inverse dose-dependent relationship between the dose and genotoxicity. Tang et al. (2010) concluded that the instillation of TiO_2 NPs in rats at low doses could reversibly affect metabolic function because NPs can migrate from the lungs to the liver and kidneys, but at higher doses, the particles will aggregate and deposit in the lungs without migration.

The movement of NPs through cells may probably be slower than chemical diffusion, so longer periods of treatment may be adequate; nowadays, there are very few *in vitro* studies with a treatment longer than 24 h (Doak et al. 2009). Moreover, treatments should be long enough to ensure the access of the NPs to the nucleus during mitosis (Gonzalez et al. 2011).

The susceptibility of different cell lines and different animal models to NPs has not yet been thoroughly studied. Paino et al. (2012) found that peripheral blood mononuclear cells were less sensitive to Au NP-induced DNA damage than were HepG2 (human cells derived from hepatocellular carcinoma). Nielsen et al. (2008) found differences in the sensitivity of two strains of mice while Kim et al. (2008) found differences in the accumulation of silver NPs in female and male kidneys. The susceptibility of organs can be due to differences in their metabolic rate, antioxidant defense, and DNA repair capability.

The different exposure routes used in *in vivo* experiments affects the NPs biodistribution and would give different genotoxic profiles in different organs. The clearance also depends on the route of administration (Nielsen et al. 2008).

Dispersion methods used to prepare the NPs (e.g., sonication or vortexing) can influence the results of genotoxicity studies; NP aggregates and agglomerates may complicate the interpretation of the results (Pfaller et al. 2010). Magdolenova et al. (2012) suggested that at least two different dispersion procedures should be used. Moreover, the solvent can affect the stability of the NPs (Pfaller et al. 2010).

Merhi et al. (2012) confirm the importance of serum and proteins in cell culture medium; the presence of serum partially protects the cells from the genotoxicity induced by positively charged NPs by decreasing their cellular uptake.

17.7 CHARACTERISTICS OF NPs THAT INFLUENCE POTENTIAL GENOTOXICITY

It has been demonstrated that some characteristics of NPs have a great impact on their genotoxicity, mainly because they modify biokinetics (Nel et al. 2006; Singh and Nalwa 2007; Gratton et al. 2008). Chemical composition, particle size, shape, surface charge, and coating can vary, among other characteristics.

The different chemical composition obviously influences the genotoxicity of NPs. Moreover, it has been demonstrated that different crystalline structures also change the genotoxic potential of NPs with the same chemical composition (Petkovic et al. 2011).

Many studies have demonstrated that small NPs exert more DNA damage than bigger ones, both *in vitro* and *in vivo* (Konczol et al. 2011; Park et al. 2011a,b; Downs et al. 2012; Jugan et al. 2012). It seems clear that small NPs can be taken up by the cells more easily, but some authors claim that larger NPs would penetrate the cell membrane more easily by endocytosis than smaller NPs (Hong et al. 2011). These large NPs would be present in vacuoles inside the cell, reducing their accessibility to the nucleus.

The surface properties of NPs give them the possibility to form aggregates. The stability of these agglomerates, or the NP *per se*, in the cell culture medium or body fluids and tissues will affect the biokinetics of the NPs (Doak et al. 2009). A positive correlation between DNA damage and positively charged NPs has been observed (Hong et al. 2011; Shah et al. 2012). Hong et al. (2011) found that positively charged, coated NPs seemed to be more concentrated inside the cell than NPs with negatively charged coatings. Coatings can have different functions depending on the coating material; its effects can cause or prevent extensive agglomeration of the NPs, or can protect or induce the interaction of NPs with cells or biomolecules. All these options have an impact on the genotoxic potential (Yin et al. 2010; Toyooka et al. 2012). Moreover, NPs can absorb organic molecules and macromolecules onto their surface, influencing the results of *in vitro* studies (Doak et al. 2009).

The shape can also affect the specific physicochemical and transport properties and, thus, the genotoxicity (Yang et al. 2009). It seems that materials with a high aspect ratio have a higher internalization rate and more of an impact on cellular functions than their low-aspect-ratio counterparts

(Gratton et al. 2008; Kim et al. 2010a; Stella 2011; Zhang et al. 2011). However, Liu et al. (2012) has shown the opposite.

The partial dissolution of metal NPs can release toxic metal ions, and this fact should be taken into account when interpreting results obtained with these types of NPs (Doak et al. 2009). Moreover, some transition metal ions released from certain NPs can induce the formation of ROS. The redox state of iron also modifies the cellular uptake of iron NPs and, thus, their genotoxicity (Singh et al. 2012).

17.8 CONCLUSION

The term NP is too broad, covering a wide range of particles with very distinct physical and chemical properties. Genotoxicological studies with NPs give conflicting results. This is probably due to their different properties, which have a great impact on their potential interactions with living cells or tissues; the different genotoxicity assays used; measuring different endpoints; and the different experimental conditions applied. Even the dose metrics vary between studies. Taking all of these variables into account, it is very difficult to compare different studies and to achieve general conclusions. Moreover, there is a lack of *in vivo* genotoxicity studies.

A thorough physiochemical characterization of NPs before testing and a battery approach including *in vitro* and *in vivo* genotoxicity studies are necessary to get reliable and comparable results. NPs should be characterized and tested in a representative physiological environment; this implies the development or adaptation of technologies to enable genotoxicity tests and physicochemical characterizations in such conditions.

Reliable experiments with well-characterized NPs will help researchers to develop new nongenotoxic NPs. Some genotoxic effects, specifically those induced by the production of ROS, can be mitigated by the use of antioxidants (Toduka et al. 2012). However, it should not be forgotten that genotoxic NPs could be useful for some medical purposes.

NPs per se can be used as a therapeutic agent, but they can also be a carrier of them. In the latter case, the cargo can play a critical role in the genotoxicity response (Lewis et al. 2010); a new genotoxicity test should be performed.

Taking into account the increasing production of different NPs in terms of the physicochemical parameters described, there will be a large number of samples to be tested in the near future. A proper screening method to test their genotoxicity would also be needed.

REFERENCES

Azqueta, A., Shaposhnikov, S., and A.R. Collins. 2009. DNA oxidation: Investigating its key role in environmental mutagenesis with the comet assay. *Mutat. Res.* 674:101–8.
Bawa, R., Bawa, S.R., Maebius, S.B., Flynn, T., and C. Wei. 2005. Protecting new ideas and inventions in nanomedicine with patents. *Nanomedicine* 1:150–8.
Borm, P.J., Tran, L., and K. Donaldson. 2011. The carcinogenic action of crystalline silica: A review of the evidence supporting secondary inflammation-driven genotoxicity as a principal mechanism. *Crit. Rev. Toxicol.* 41:756–70.
Brayner, R., Ferrari-Iliou, R., Brivois, N., Djediat, S., Benedetti, M.F., and F. Fievet. 2006. Toxicological impact studies based on *Escherichia coli* bacteria in ultrafine ZnO nanoparticles colloidal medium. *Nano Lett.* 6:866–70.
Calarco, A., Bosetti, M., Margarucci, S. et al. 2013. The genotoxicity of PEI-based nanoparticles is reduced by acetylation of polyethylenimine amines in human primary cells. *Toxicol. Lett.* 218:10–7.
Collins, A.R. 2009. Investigating oxidative DNA damage and its repair using the comet assay. *Mutat. Res.* 681:24–32.
Collins, A.R. and A. Azqueta. 2012. Single cell gel electrophoresis combined with lesion-specific enzymes to measure oxidative damage to DNA. *Methods Cell Biol.* 112:69–92.
Collins, A.R., Oscoz, A.A., Brunborg, G. et al. 2008. The comet assay: Topical issues. *Mutagenesis* 23:143–51.

Doak, S.H., Griffiths, S.M., Manshian, B. et al. 2009. Confounding experimental considerations in nanogenotoxicology. *Mutagenesis* 24:285–93.

Doak, S.H., Manshian, B., Jenkins, G.J., and N. Singh. 2012. *In vitro* genotoxicity testing strategy for nanomaterials and the adaptation of current OECD guidelines. *Mutat. Res.* 745:104–11.

Donaldson, K., Poland, C.A., and R.P. Schins. 2010. Possible genotoxic mechanisms of nanoparticles: Criteria for improved test strategies. *Nanotoxicology* 4:414–20.

Downs, T.R., Crosby, M.E., Hu, T. et al. 2012. Silica nanoparticles administered at the maximum tolerated dose induce genotoxic effects through an inflammatory reaction while gold nanoparticles do not. *Mutat. Res.* 745:38–50.

Durnev, A.D. 2008. Toxicology of nanoparticles. *Bull. Exp. Biol. Med.* 145:72–4.

Friedberg, E.C., Walker, G.C., Siede, W., Wood, R.D., Schultz, R.A., and T. Ellenberger. 2006. *DNA Repair and Mutagenesis*. ASM Press, Washington.

Ghosh, M., Bandyopadhyay, M., and A. Mukherjee. 2010. Genotoxicity of titanium dioxide (TiO_2) nanoparticles at two trophic levels: Plant and human lymphocytes. *Chemosphere* 81:1253–62.

Gonzalez, L., Sanderson, B.J., and M. Kirsch-Volders. 2011. Adaptations of the *in vitro* MN assay for the genotoxicity assessment of nanomaterials. *Mutagenesis* 26:185–91.

Gratton, S.E., Ropp, P.A., Pohlhaus, P.D. et al. 2008. The effect of particle design on cellular internalization pathways. *Proc. Natl. Acad. Sci. U.S.A.* 105:11613–8.

Greim, H., Borm, P., Schins, R. et al. 2001. Toxicity of fibers and particles. Report of the workshop held in Munich, Germany, October 26–27, 2000. *Inhal. Toxicol.* 13:737–54.

Hartmann, A., Agurell, E., Beevers, C. et al. 2003. Recommendations for conducting the *in vivo* alkaline comet assay. 4th International Comet Assay Workshop. *Mutagenesis* 18:45–51.

Hong, S.C., Lee, J.H., Lee, J. et al. 2011. Subtle cytotoxicity and genotoxicity differences in superparamagnetic iron oxide nanoparticles coated with various functional groups. *Int. J. Nanomedicine* 6:3219–31.

International Conference on Harmonisation (ICH). 2011. Harmonised Tripartite guideline S2 (R1). Guidance on genotoxicity testing and data interpretation for pharmaceuticals intended for human use.

Jomini, S., Lebille, J., Bauda, P., and C. Pagnout. 2012. Modifications of the bacterial reverse mutation test reveals mutagenicity of TiO_2 nanoparticles and byproducts from a sunscreen TiO_2-based nanocomposite. *Toxicol. Lett.* 215:54–61.

Jugan, M.-L., Barillet, S., Simon-Deckers, A. et al. 2012. Titanium dioxide nanoparticles exhibit genotoxicity and impair DNA repair activity in A549 cells. *Nanotoxicology* 6:501–13.

Kim, J.S., Song, K.S., Joo, H.J., Lee, J.H., and I.J. Yu. 2010a. Determination of cytotoxicity attributed to multiwall carbon nanotubes (MWCNT) in normal human embryonic lung cell (WI-38) line. *J. Toxicol. Environ. Health.* 73:1521–9.

Kim, S., Seong, K., Kim, O. et al. 2010b. Polyoxalate nanoparticles as a biodegradable and biocompatible drug delivery vehicle. *Biomacromolecules* 11:555–60.

Kim, Y.S., Kim, J.S., Cho, H.S. et al. 2008. Twenty-eight-day oral toxicity, genotoxicity, and gender-related tissue distribution of silver nanoparticles in Sprague-Dawley rats. *Inhal. Toxicol.* 20:575–83.

Konczol, M., Ebeling, S., Goldenberg, E. et al. 2011. Cytotoxicity and genotoxicity of size-fractionated iron oxide (magnetite) in A549 human lung epithelial cells: Role of ROS, JNK, and NF-jB. *Chem. Res. Toxicol.* 24:1460–75.

Landsiedel, R., Kapp, M.D., Schulz, M., Wiench, K., and F. Oesch. 2009. Genotoxicity investigations on nanomaterials: Methods, preparation and characterization of test material, potential artifacts and limitations—Many questions, some answers. *Mutat. Res.* 681:241–58.

Lewis, D.J., Bruce, C., Bohic, S. et al. 2010. Intracellular synchrotron nanoimaging and DNA damage/genotoxicity screening of novel lanthanide &endash;coated nanovectors. *Nanomedicine (Lond).* 5:1547–57.

Liu, W., Chaurand, P., Di Giorgio, C. et al. 2012. Influence of the length of imogolite-like nanotubes on their cytotoxicity and genotoxicity toward human dermal cells. *Chem. Res. Toxicol.* 25:2513–22.

Magdolenova, Z., Bilanicova, D., Pojana, G. et al. 2012. Impact of agglomeration and different dispersions of titanium dioxidenanoparticles on the human related *in vitro* cytotoxicity and genotoxicity. *J. Environ. Monit.* 14:455–64.

Merhi, M., Dombu, C.Y., Brient, A. et al. 2012. Study of serum interaction with a cationic nanoparticle: Implications for *in vitro* endocytosis, cytotoxicity and genotoxicity. *Int. J. Pharm.* 423:37–44.

Nabeshi, H., Yoshikawa, T., Matsuyama, K. et al. 2011. Amorphous nanosilica induce endocytosis-dependent ROS generation and DNA damage in human keratinocytes. *Part. Fibre Toxicol.* 8:1.

Nam, S.-H., Kim, S.W., and Y.-J. An. 2013. No evidence of the genotoxic potential of gold, silver, zinc oxide and titanium dioxide nanoparticles in the SOS chromotest. *J. Appl. Toxicol.* 33:1061–1069.

Nel, A., Xia, T., Madler, L., and N. Li. 2006. Toxic potential of materials at the nanolevel. *Science* 311:622–7.

Nielsen, G.D., Roursgaard, M., Jensen, K.A., Poulsen, S.S., and S.T. Larsen. 2008. *In vivo* biology and toxicology of fullerenes and their derivatives. *Basic Clin. Pharmacol. Toxicol.* 103:197–208.

Norppa, H., Catalán, J., Falck, G., Hannukainen, K., Siivola, K., and K. Savolainen. 2011. Nano-specific genotoxic effects. *J. Biomed. Nanotechnol.* 7:19.

Organisation for Economic Cooperation & Development (OECD). 1977. Test Guideline 471: Bacterial reverse mutation test. In: OECD Guidelines for testing of chemicals.

Organisation for Economic Cooperation & Development (OECD). 1977. Test Guideline 473. *In vitro* mammalian chromosome aberration test. In: OECD Guidelines for testing of chemicals.

Organisation for Economic Cooperation & Development (OECD). 1997. Test Guideline 474. Mammalian erythrocyte micronucleus. In: OECD Guidelines for testing of chemicals.

Organisation for Economic Cooperation & Development (OECD). 1997. Test Guideline 475. Mammalian bone marrow chromosome aberration test. In: OECD Guidelines for testing of chemicals.

Organisation for Economic Cooperation & Development (OECD). 1997. Test Guideline 476. *In vitro* mammalian cell gene mutation test. In: OECD Guidelines for testing of chemicals.

Organisation for Economic Cooperation & Development (OECD). 2009. Preliminary Review of OECD Test Guidelines for their Applicability to Manufactured Nanomaterials. Series of Safety of Manufactured Nanomaterials No. 15.

Organisation for Economic Cooperation & Development (OECD). 2010. Test Guideline 487. *In vitro* mammalian cell micronucleus test. In: OECD Guidelines for testing of chemicals.

Organisation for Economic Cooperation & Development (OECD). OECD guidelines for the testing of chemicals; http://www.oecd-ilibrary.org/environment/oecd-guidelines-for-the-testing-of-chemicals-section-4-health-effects_20745788.

Paino, I.M., Marangoni, V.S., de Oliveira R. de C., Antunes, L.M., and V. Zucolotto. 2012. Cyto and genotoxicity of gold nanoparticles in human hepatocellular carcinoma and peripheral blood mononuclear cells. *Toxicol. Lett.* 215:119–25.

Park, M.V., Neigh A.M., Vermeulen, J.P. et al. 2011a. The effect of particle size on the cytotoxicity, inflammation, developmental toxicity and genotoxicity of silver nanoparticles. *Biomaterials* 32:9810–7.

Park, M.V., Verharen, H.W., Zwart, E. et al. 2011b. Genotoxicity evaluation of amorphous silica nanoparticles of different sizes using the micronucleus and the plasmid lacZ gene mutation assay. *Nanotoxicology* 5:168–81.

Petkovic, J., Zegura, B., Stevanovic, M. et al. 2011. DNA damage and alterations in expression of DNA damage responsive genes induced by TiO_2 nanoparticles in human hepatoma HepG2 cells. *Nanotoxicology* 5:341–53.

Pfaller, T., Colognato, R., Nelissen, I. et al. 2010. The suitability of different cellular *in vitro* immunotoxicity and genotoxicity methods for the analysis of nanoparticle-induced events. *Nanotoxicology* 4:52–72.

PubMed database: www.ncbi.nlm.nih.gov/pubmed/

Schins, R.P. 2002. Mechanisms of genotoxicity of particles and fibers. *Inhal. Toxicol.* 14, 57–78.

Schins, R.P. and A.M. Knaapen. 2007. Genotoxicity of poorly soluble particles. *Inhal. Toxicol.* 19:189–98.

Sergent, J.A., Paget, V., and S. Chevillard. 2012. Toxicity and genotoxicity of nano-SiO_2 on human epithelial intestinal HT-29 cell line. *Ann. Occup. Hyg.* 56:622–30.

Shah, V., Taratula, O., Garbuzenco, O.B. et al. 2012. Genotoxicity of different nanocarriers: Possible modifications for the delivery of nucleic acids. *Curr. Drug Discov. Technol.* 10:8–15.

Singh, N., Jenkins G.J., Nelson, B.C. et al. 2012. The role of iron redox state in the genotoxicity of ultrafine superparamagnetic iron oxide nanoparticles. *Biomaterials* 33:163–70.

Singh, N., Manshian, B., Jenkins, G.J. et al. 2009. Nanogenotoxicology: The DNA damaging potential of engineered nanomaterials. *Biomaterials* 30, 3891–914.

Singh, S. and H.S. Nalwa. 2007. Nanotechnology and health safety- toxicity and risk assessments of nanostructured materials on human health. *J. Nanosci. Nanotechnol.* 7:3048–70.

Stella, G.M. 2011. Carbon nanotubes and pleural damage: Perspectives of nanosafety in the light of asbestos experience. *Biointerphases* 6:1–17.

Stone, V., Johnston, H., and R.P. Schins. 2009. Development of *in vitro* systems for nanotoxicology: Methodological considerations. *Crit. Rev. Toxicol.* 39:613–26.

Tang, M., Zhang, T., Xue, Y. et al. 2010. Dose dependent *in vivo* metabolic characteristics of titanium dioxide nanoparticles. *J. Nanosci. Nanotechnol.* 10:8575–83.

Tice, R.R., Agurell, E., Anderson, D. et al. 2000. Single cell gel/comet assay: Guidelines for *in vitro* and *in vivo* genetic toxicology testing. *Environ. Mol. Mutagen.* 35:206–21.

Toduka, Y., Toyooka, T., and Y. Ibuki. 2012. Flow cytometric evaluation of nanoparticles using side-scattered light and reactive oxygen species-mediated fluorescence–correlation with genotoxicity. *Environ. Sci. Technol.* 46:7629–36.

Toyooka, T., Amano, T., and Y. Ibuki. 2012. Titanium dioxide particles phosphorylate histone H2AX independent of ROS production. *Mutat. Res.* 742:84–91.

Tsoli, M., Kuhn, H., Brandau, W., Esche, H., and G. Schmid. 2005. Cellular uptake and toxicity of Au55 clusters. *Small* 1:841–4.

Vasquez, M.Z. 2012. Recommendations for safety testing with the *in vivo* comet assay. *Mutat. Res.* 747:142–56.

Wang, Z., Li, N., Zhao, J. et al. 2012. CuO nanoparticle interaction with human epithelial cells: Cellular uptake, location, export, and genotoxicity. *Chem. Res. Toxicol.* 25:1512–21.

Woodruff, R.S., Li, Y., Yan, J. et al. 2012. Genotoxicity evaluation of titanium dioxide nanoparticles using the Ames test and Comet assay. *J. Appl. Toxicol.* 32:934–43.

Xu, A., Chai, Y., Nohmi, T., and T.K. Hei. 2009. Genotoxic responses to titanium dioxide nanoparticles and fullerene in gpt delta transgenic MEF cells. *Part. Fibre Toxicol.* 6:3.

Xu, X.H., Brownlow, W.J., Kyriacou, S.V., Wan, Q., and J.J., Viola. 2004. Real-time probing of membrane transport in living microbial cells using single nanoparticles optics and living cell imaging. *Biochemistry* 43:10400–13.

Yang, H., Liu, C., Yang, D., Zhanga, H., and Z. Xi. 2009. Comparative study of cytotoxicity, oxidative stress and genotoxicity induced by four typical nanomaterials: The role of particle size, shape and composition. *J. Appl. Toxicol.* 29:69–78.

Yin, H., Casey, P.S., McCall, M.J., and M. Fenech. 2010. Effects of surface chemistry on cytotoxicity, genotoxicity, and the generation of reactive oxygen species induced by ZnO nanoparticles. *Langmuir* 26:15399–408.

Zhang, Y., Wang, B., Meng, X., Sun, G., and C. Gao. 2011. Influences of acid-treated multiwalled carbon nanotubes on fibroblasts: Proliferation, adhesion, migration, and wound healing. *Ann. Biomed. Eng.* 39:414–26.

18 Interactions of Polysaccharide-Coated Nanoparticles with Proteins

Christine Vauthier

CONTENTS

18.1 Introduction 365
18.2 Synthetic Identity of Polysaccharide-Coated Poly(alkylcyanoacrylate) Nanoparticles 368
18.3 Interactions of Polysaccharide-Coated Nanoparticles with Proteins 371
18.4 Conclusion 379
References 379

18.1 INTRODUCTION

Nanomaterials occurring as single particles within the nanosize range are small enough to diffuse in living tissues of the organisms and to penetrate into cells. The synthesis of artificial nanomaterials appeared reproducible enough at the end of the 1960s and beginning of the 1970s (Bangham et al., 1965, see for references Daniel, J.C. 2003). They appeared suitable to serve as drug carriers (Gregoriadis, 1976; Kreuter and Speiser, 1976; Kreuter, 2007) and attracted an immediate interest to realize the "magic bullet," a concept imagined by Paul Ehrlich, winner of the Nobel Prize in Physiology and Medicine in 1908. The rationale behind this idea was to develop a method of drug delivery by reducing the severe side effects of the drugs used in chemotherapies of cancer and infections, thanks to a better targeting of the drug to diseased tissues. The early stages of nanomedicine development have considered liposomal formulations of chemotherapeutic agents during the 1970s (Gregoriadis, 1976; Juliano, 1976; Gregoriadis et al., 1974). Since then, the field of nanomedicine has expanded considerably. Many types of nanomaterials have been proposed to serve as drug carriers for not only small molecules but also for biomacromolecules, including therapeutic peptides, proteins, and all kinds of nucleic acids that can be used to control the expression of a specific gene and can be applied to gene therapy. Proofs of concept for the delivery of most of these molecules, considering different modalities for their administration, have now been provided (Couvreur and Vauthier, 2006; Farokhzad and Langer, 2009; Etheridge et al., 2013; Lehner et al., 2013). Besides applications that overcome drug delivery challenges, several types of interesting nanomaterials were found to improve the performance of imaging techniques used in diagnostics (Liu et al., 2011). Theragnostic, a new nanomedicine field, uses a combination of drug delivery and diagnostics in a single nanomaterial (Mura and Couvreur, 2012). Finally, several types of nanomaterials can be used to potentiate the effect of radiotherapy (Bakht et al., 2012). They are used to focus the effects of radiations after implantation within tumors that enhances the efficacy of the radiotherapy. Several of these nanomaterials are bringing hope toward the development of noninvasive methods for the ablation of tumors with the ambition to displace classical surgery.

It is now established that many types of nanoparticles have the potential to revolutionize diagnostic and therapeutic methods (Etheridge et al., 2013; Lehner et al., 2013; Wang et al., 2013). However, success is closely conditioned by their biodistributions, as they are intended to be administered *in vivo* by a general route of administration, that is, oral or IV. In turn, the expected biodistribution is especially closely conditioned to how nanoparticles interact with biological media, those found on their way to their target sites from their sites of administration (Alexis et al., 2008; Owens and Peppas, 2006; Dobrovolskaia et al., 2008; Walczyk et al., 2010; Moyano and Rotello, 2011). Considering the administration of a nanomedicine by the IV route, which is a typical route considered, immediate interactions include the adsorption of proteins on the nanomaterial surface. A layer of proteins form at the surface of the nanomedicine changing their "synthetic identity" into a "biological identity" (Lynch et al., 2007; Dobrovolskaia et al., 2008; Norde, 2008; Walkey et al., 2012; Walkey and Chan, 2012). This event determines and regulates the subsequent biological response that will, in the end, control the *in vivo* fate of the nanomedicine, including its eventual toxicity. Understanding the interactions of nanomedicines with blood proteins is now considered to be a fundamental issue in predicting and controlling their biodistribution and the eventual toxicological reaction they may induce (Lynch et al., 2007; Walkey and Chan, 2012). While this is expected to provide for a better understanding on how the physicochemical properties of nanomedicines that define their "synthetic identity" influence the *in vivo* fate, it will enable a rational design of safe and efficient nanomedicines. Considering the broader range of applications offered by nanotechnologies that far exceed that of nanomedicines, the elucidation of interactions occurring between nanomaterials and proteins is expected to contribute to evaluating and elucidating health and environmental risks which may arise from their use in various domains of the industry and in consumer products distributed on a large scale.

Investigating the interactions of proteins with nanomaterials is a complex task. Nanomaterials occur with various compositions, sizes, shapes, and surface properties (Algar et al., 2011; Etheridge et al., 2013; Lehner et al., 2013). A glance at this complexity is provided by considering the nature of the components that compose nanomaterials developed as nanomedicines; these include lipids, polymers, carbon, and metal for the simpler systems. More complex systems are composites resulting from the association of metal colloids and lipids or polymers for instance. A second difficulty arises from the large numbers of proteins that are found in the blood and the broad range of concentrations in which they individually occur. There are over a thousand different proteins with differences in concentrations spread over 10 orders of magnitude (Anderson et al., 2004). Additionally, the adsorption of protein on a surface of a material is kinetically dependant and the composition of the adsorbed proteins evolves with time (Hirsh et al., 2013). Finally, investigating interactions of proteins with nanomaterials is not simply an extension of approaches applied to biomaterials. The range of nanomaterial sizes imposed the development of specific methods. Some were adapted from those used to evaluate biomaterials, but a lot needed to be developed specifically to be applied to nanomaterials (Lynch et al., 2007; Vauthier et al., 2009, 2011; Walkey and Chan, 2012; Welsch et al., 2013).

Research carried out on the interactions of nanoparticles with proteins has been intensified over the last two decades. They started soon after it was assumed that proteins play a fundamental role in defining the *in vivo* fate of nanomaterials, including those designed as drug carriers. Today, it is expected that the biodistribution of nanomaterials may be anticipated from the type of proteins that are adsorbed onto their surface. It is also assumed that it may be possible to target a nanomaterial to a specific site in the organism, taking advantage of a preferential adsorption of a certain plasma protein—if it exists—and to the specific design of the nanomaterial. While taking advantage of the preferential adsorption of proteins may be a strategy to conceive targeted nanoparticles for drug delivery, a systematic identification of proteins that adsorb onto nanomaterials could be used as a method to identify preferential sites of accumulation in the body for the purpose of anticipating the safety of a nanomaterial. Methods of camouflaging the surface of nanoparticles were developed in parallel to research carried out to decrypt the types of proteins adsorbing onto a nanoparticles surface. To this purpose, PEG was chosen because of its known antifouling properties from its uses to improve the resistance to the protein adsorption of biomaterials (Jeon and Andrade, 1991; Jeon et al., 1991; Gref et al., 1994; Papahadjopoulos et al., 1991; Banerjee et al., 2011). Together with observations of the occurrence of protein adsorption modifications, the addition of PEG at the surface of nanomaterials could modulate the pharmacokinetic and biodistribution of the corresponding nanoparticles (Gref et al., 1997, 2000; Peracchia et al., 1999; Vonarbourg et al., 2006, see for review Li and Huang, 2008). However, the use of PEG presents some limitations, following observations from clinicians, while the use of PEGylated nanomedicines was intensified in human medicine. The most serious concern is the appearance of a new type of toxicity that occurs as a hypersensitive reaction called C activation-related pseudoallergy (CARPA) (Jiskoot et al., 2009; Szebeni et al., 2011; Lehner et al., 2013). Only few works have considered the use of alternative polymers to PEG to modulate the surface properties of nanoparticles in the aim to modify protein adsorption. The few numbers of polymers used in this purpose included polyoxazolines, oligo- and polysaccharides, polyelectrolytes, and zwitterionic polymers (Passirani et al., 1999; Chauvierre et al., 2003; Labarre et al., 2005; Lemarchand et al., 2004; Estephan et al., 2010; Walkey and Chan, 2012; Welsch et al., 2013). Whatever the approach used to modify surface properties of nanoparticles, the control of the *in vivo* fate of nanomedicine from the protein adsorption pattern needs the understanding of the relationship between the synthetic identity of the nanomaterials, its biological identity, and the corresponding physiological response. Each can be described by many parameters that considerably increase the challenge. We should admit that we are still at the beginning. Nevertheless, a few general principles have emerged from the number of studies available in literature. For instance, Walkey and Chan (2012) have provided a comprehensive analysis of data collected from many of these works. They described an "adsorbome" of 125 plasma proteins for nanoparticles of various compositions and surface properties. From this analysis, they draw a relation between the physicochemical characteristics of the nanoparticles and the corresponding absorbome. The difficulty of this work came from the various sources of data, hence the wide range of nanoparticles. Unfortunately, this analysis could obviously not be based on standardized evaluations of nanomaterial properties and of protein adsorption. In agreement with the authors' findings, this would be the main restriction of this work. Nevertheless, such an analysis deserves respect and will be very useful in the future. The conclusion that came out of this work was that 2–6 proteins from the plasma predominantly adsorbed on the surface of a nanomaterial together with many more that adsorbed at a lower abundance. Another remarkable analysis was provided by Walkey et al. (2012) considering a series of PEGylated gold nanoparticles and investigating, in parallel, their "absorbome" and their uptake by macrophages. It also emerged from the work done that PEG chains grafted at the surface of nanoparticles control the adsorption of proteins by a steric effect (Gref et al., 2000; for other references, see Owens and Peppas, 2006; Alexis et al., 2008; Walkey et al., 2011). The use of polyelectrolytes as nanoparticle coating materials leads to a different mechanism of control of the adsorption of proteins on the nanoparticle surface. In this case, proteins were found to behave like multicounterions that can interact with the charge of polyelectrolytes via the patches of opposite charges they display on their surface (Ballauff and Borisov, 2006; Ballauff, 2007; Welsch et al., 2013). In our group, we have started investigating the interactions of proteins with polysaccharide-coated poly(alkylcyanoacrylate) (PACA) nanoparticles. These nanoparticles have the potential to improve the delivery of many types of drugs (Vauthier et al., 2007; Andrieux and Couvreur 2009; Nicolas and Couvreur, 2009). They can be obtained with various surface properties, thanks to their method of preparation and the large panel of polysaccharides that can be used as coating materials (Nicolas and Vauthier, 2011). The aim of this chapter was to propose a comprehensive review of our contribution to the understanding of the influences of nanoparticle properties on interactions with plasma proteins. In the first part, we summarize the characteristics of a series of polysaccharide-coated PACA nanoparticles that were synthesized to carry on our work. Results from the interactions of proteins with nanoparticles are presented in the second part of the chapter. As it will be explained, models drawn to describe the "synthetic identity" of several nanoparticles could be used to elucidate the mechanisms controlling the accessibility of proteins to the nanoparticle surface during protein adsorption. It also serves to understand the mechanisms beyond the capacity of the nanoparticles to trigger the activation of proteins involved in the immune system, which play a fundamental role to define the *in vivo* fate of IV injected nanoparticles.

18.2 SYNTHETIC IDENTITY OF POLYSACCHARIDE-COATED POLY(ALKYLCYANOACRYLATE) NANOPARTICLES

Polysaccharide-coated poly(alkylcyanoacrylate) (PACA) nanoparticles have been utilized as drug carriers for various purposes since 1979. Their design has evolved with time to adjust their properties with those required to fulfill different drug delivery strategies. These include functions required to associate various types of drugs and the functionalities needed to adjust their *in vivo* biodistribution after IV administration. It leads to various types of PACA nanoparticles that are all spherical in shape, but different in size, composition, and surface characteristics (Table 18.1).

In general, the size of polysaccharide-coated PACA nanoparticles ranges from a few tens of nanometers (45–50 nm in diameter for the smaller) to several hundreds of nanometers (350–400 nm in diameter for the larger). PACA nanoparticles are composed of copolymers, including at least one segment of PACA and one segment of a hydrophilic macromolecule (Douglas et al., 1984; Chauvierre et al., 2003; Bertholon et al., 2006a; Zandanel and Vauthier 2012). As depicted in Table 18.1, the copolymers take different structures depending on the conditions used for their synthesis. Thanks to the amphiphilic properties of the copolymers, nanoparticles produced in an aqueous medium occur as stable dispersions and take a core–corona structure. The hydrophobic segments of the copolymers precipitate to compose the core of the nanoparticles while the hydrophilic part is exposed at the nanoparticle surface (Figure 18.1).

This core–corona structure forms due to the thermodynamics of the system in the course of nanoparticle synthesis, which evolves to obtain the more stable aqueous dispersion of nanoparticles. A primary indication of the nature of the component included in the nanoparticle corona can be given by measuring the zeta potential of the nanoparticles. In general, the zeta potential shown by the nanoparticles is consistent with the value expected when considering the nature of the macromolecule (i.e., the polysaccharide) composing the hydrophilic part of the copolymer. Figure 18.2 gives examples of zeta potentials of PACA nanoparticles composed of copolymers including various types of polysaccharides.

Some of the nanoparticles included in Figure 18.2 were prepared with blends of polysaccharides and blends of polysaccharide and pluronic® F68. It is noteworthy that such nanoparticles display intermediate properties when compared with those of nanoparticles obtained with each hydrophilic macromolecule taken separately, indicating that they formed by the assembly of different copolymers produced during polymerization.

The determination of the "synthetic identity" of nanoparticles in aiming to understand their interactions with proteins requires a deeper description of their characteristics, especially of the structure of their corona. Such efforts were carried out on PEG-coated nanoparticles and nanoparticles coated with brushes of polyelectrolytes (Gref et al., 2000; Welsch et al., 2013). Models of the chains' organization of PEG or polyelectrolytes could be suggested, providing with a clear understanding of the nanoparticle surfaces' structures. In our work, we have characterized a series of polysaccharide-coated PACA nanoparticles to provide with a comprehensive view of the structure of the nanoparticle corona (Bertholon et al., 2006b; Vauthier et al., 2009, 2011; Zandanel and Vauthier, 2012). In contrast with the polymers composing PEGylated nanoparticles, polymers composing polysaccharide-coated PACA nanoparticles are synthesized at the time of the nanoparticles' preparation by emulsion polymerization. This implies that the structure of the copolymers composing the nanoparticles needs to be characterized from the obtained nanoparticles. As in the case of many amphiphilic copolymers, the polysaccharide–PACA copolymers composing the nanoparticles were difficult to characterize. The main difficulty arose from the fact that they were not soluble in solvents commonly used for polymer analysis. The only solvent in which the nanoparticle-copolymers could apparently dissolve was DMSO (dimethyl sulfoxide). However, obtaining a true solution in which polymer molecules occur as single and well-individualized chains could only be obtained at a low concentration. Above a concentration of 1.1 mg/mL, the copolymer molecules aggregated together, forming

TABLE 18.1
Conditions of Fabrication of a Series of Model Nanoparticles Coated with Dextran as Polysaccharides and Corresponding Models Deduced from the Characteristics of the Copolymers That Composed Them

Name	Method of Polymerization	Conditions of Polymerization	Structure of the Copolymer PIBCA*/Dextran	Model Proposed for the Nanoparticles
A1	Anionic polymerization	pH 1 Dextran 70 kDa 1.3%	(2)	Loose Loops D_H: 86 nm $\zeta = -8.7$ mV
A2	Anionic polymerization	pH 2.5 Dextran 70 kDa 1.3%	(4)	Tight loops D_H: 181 nm $\zeta = -1.1$ mV
R1	Radical polymerization	pH 1 Dextran 70 kDa 1.3%	(1)	Long hair D_H: 224 nm $\zeta = -11.6$ mV
R2	Radical polymerization	pH 1 Dextran 70 kDa 0.5%		Long hair D_H: 180 nm
R3	Radical polymerization	pH 1 Dextran 15 kDa 1.3%		Short hair D_H: 200 nm $\zeta = -19$ mV

*PIBCA: poly(isobutylcyanoacrylate).

larger objects that could correspond to micelles (Bertholon et al., 2006b). True solutions of copolymers occurred at lower concentrations than that required to perform analyses with the usual methods applied to characterize polymers from solutions. To turn around this analytical problem, methods of the selective degradation of two types of polymers included in the copolymer were set up (Figure 18.3).

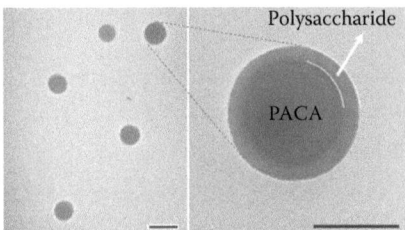

FIGURE 18.1 An electron micrograph of polysaccharide-coated PACA nanoparticles. The nanoparticles were prepared by radical polymerization of isobutylcyanoacrylate carried out with chitosan 20 kDa. The nanoparticles were composed of a core of PIBCA and a corona of chitosan as depicted on the right image. The nanoparticles deposited on a formvar-carbon-coated copper grid for electron microscopy were stained with phosphotungstic acid for 30 s. Scale bar = 100 nm. Cliché: H. de Martimprey. Service commun de Microscopie électronique, Orsay, France.

After purification, the polymer chains that were not degraded by the selective treatment of either the PACA chains or the polysaccharide segments could be characterized as a homopolymer. This methodology is time consuming and needs lots of effort in validating all steps. The selectivity of the degradation methods needed to be demonstrated. It was shown that the conditions of degradation applied to remove PACA did not modify the characteristics of the polysaccharides and vice versa. Nevertheless, this complex procedure provided with the structures of copolymers composing polysaccharide-coated PACA nanoparticles obtained in different conditions of synthesis. Depending on their structures, the copolymers can be classified into two categories that depend on the type of polymerization applied to synthesize the nanoparticles. They occurred as linear block copolymers when the nanoparticles were prepared by radical polymerization and as comb copolymers when the synthesis of the nanoparticles occurred by anionic polymerization (Table 18.1) (Bertholon et al., 2006b; Zandanel and Vauthier, 2012).

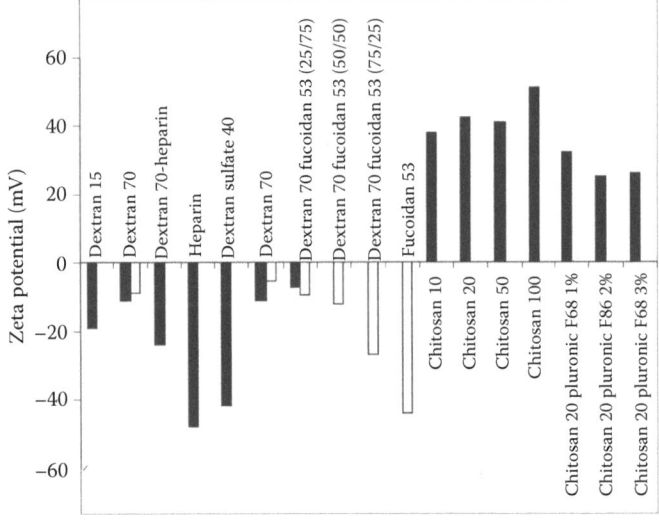

FIGURE 18.2 Zeta potential of nanoparticles produced with different polysaccharides. Numbers following the name of the polysaccharide indicates the molecular weight in kDa. White columns: nanoparticles obtained by anionic polymerization. Black columns: nanoparticles obtained by radical polymerization. (Data from Lira, M.C.B. et al., 2011. *Eur. J. Pharm. Biopharm.* 79:162–70; Bravo-Osuna, I. et al., 2007. *Biomaterials.* 28:2233–43; de Martimprey, H. et al., 2009. *Eur. J. Pharm. Biopharm.* 71(3):490–504; Labarre, D. et al., 2005. *Biomaterials.* 26:5075–84.)

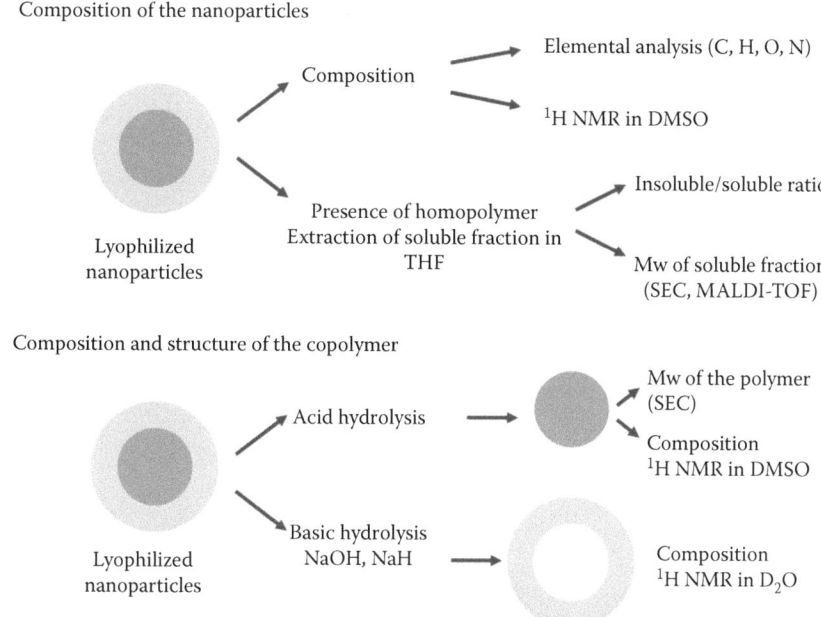

FIGURE 18.3 A scheme explaining the strategy developed to analyze the composition and the structure of copolymers composing polysaccharide-coated PACA nanoparticles.

This difference in the structure of the copolymer generates a fundamental difference in the characteristics of the nanoparticle corona, considering that the polysaccharide part of the copolymer is anchored in the core of the nanoparticles, thanks to the contribution of the PACA segments. Consequently, the number of PACA segments attached to the polysaccharide determines its spatial arrangement in the nanoparticle corona and its freedom to move (Table 18.1) (Chauvierre et al., 2004; Bertholon et al., 2006c). Simulations of the coronal structure of a few polysaccharide-coated PACA nanoparticles could be suggested. Models showing the configuration of polysaccharide chains in nanoparticle coronae could be drawn from the structure of the copolymers (Bertholon et al., 2006b; Zandanel and Vauthier 2012). Improved simulations were carried out from the copolymer's composition in polysaccharide and PACA, the structure of the copolymer, the diameter of the core of the nanoparticles, and their hydrodynamic diameter. Models drawn from the improved simulations can provide with information about the density of the polysaccharide chains within the nanoparticle corona (Vauthier et al., 2009, 2011). Table 18.2 summarizes the models established from the deep characterization of several PACA nanoparticles coated with dextran, which were then used to investigate their interactions with proteins.

18.3 INTERACTIONS OF POLYSACCHARIDE-COATED NANOPARTICLES WITH PROTEINS

When nanoparticles or, more generally, drug carrier nanotechnologies, are injected in the blood, proteins immediately interact with the foreign particles. At first, interactions consist of the adsorption or deposition of proteins on the material's surface. During this event, the particle's surface is decorated with blood proteins. The layer of proteins formed gives a "biological identity" to the nanoparticles that plays a fundamental role in subsequent events (Vonarbourg et al., 2006; Nilsson et al., 2007; Alexis et al., 2008; Dobrovolskaia et al., 2008). For instance, the nanoparticle surface's access to certain proteins can trigger the activation of the immune systems. When the immune system is activated, the particle is rapidly taken up by macrophages, in charge of the elimination of

TABLE 18.2
Fine Description of the Synthetic Identity of the Dextran-Coated Nanoparticles Prepared in Different Conditions

Nanoparticle A1	Nanoparticle R1	Nanoparticle R2

Size Characteristics

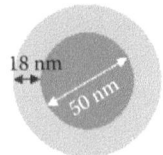 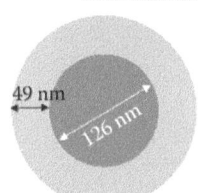 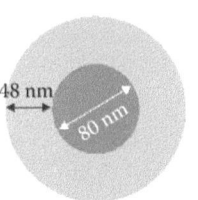

Arrangement of Dextran Chains in the Nanoparticle Corona

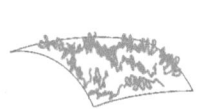 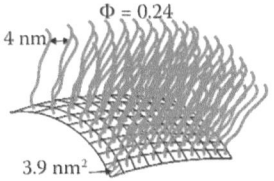 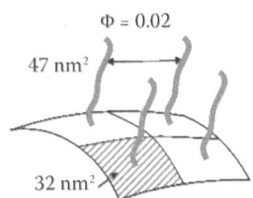

Fraction of the area of the core surface occupied by the anchorage of the polysaccharide chain: F
Mesh size of the polysaccharide chains in the corona: m
Maximal size of hydrophobic areas available on the nanoparticle core surface: A

$F = 50\%$ [a]	$F = 20\%$ [b]	$F = 7\%$ [b]
	$m = 4.0 \pm 0.2$ nm [b]	$m = 47 \pm 2$ nm [b]
	$A = 3.9 \pm 0.6$ nm² [b]	$A = 32 \pm 5$ nm² [b]

Note: See Table 18.1 for conditions of preparation. Φ gives the volume fraction occupied by dextran chains in the nanoparticle corona.

[a] Vauthier, C., Lindner, P., and Cabane, B., 2009. Configuration of bovine serum albumin adsorbed on polymer particles with grafted dextran corona. *Coll. Surf. B: Biointerfaces.* 69:207–15.

[b] Vauthier, C., Persson, B., Lindner, P., and Cabane, B. 2011. Protein adsorption and complement activation for di-block copolymer nanoparticles. *Biomaterials.* 32:1646–56.

---

foreign bodies. This strategy may be used to target drugs to the immune system, but it hampers the delivery of drugs to other organs and tissues. The approach used to escape the capture of drug carriers by macrophages is to mask the nanoparticle surface with hydrophilic polymers to modify the composition and amount of adsorbed proteins when nanoparticles are injected into the blood. The grafting of poly(ethylene glycol) (PEG) chains was a very efficient strategy to hamper the adsorption of proteins onto the surface of nanomaterials used as drug carriers (Gref et al., 2000; Owens and Peppas, 2006; Vonarbourg et al., 2006; Alexis et al., 2008; Li and Huang, 2008). It continues to be actively studied, as the understanding of how the particle's design influences the adsorption of proteins remains incomplete (Walkey et al., 2012; Walkey and Chang, 2012). Besides these works, we have started investigating the interactions of proteins with the model nanoparticles, PACA coated with polysaccharides, for which we could describe the synthetic identity of the hydrophilic corona. Our studies was aimed to answer several questions:

- Where were proteins adsorbed on the nanoparticle surface, assuming that they may remain on the top of the hydrophilic corona, become entrapped within the hydrophilic corona, or be adsorbed on the surface of the hydrophobic core?

- What are the parameters regulating the adsorption of proteins, assuming that the synthetic identity of the nanoparticle corona may greatly interfere in this process?
- What are the properties of the polysaccharide corona behind the activation or nonactivation of immune system proteins?

Answering the first question was an issue in estimating the role of the nanoparticle corona in controlling the interactions between nanoparticles and proteins. Obviously, this role was expected to differ greatly whether the proteins remained on top of the corona surface or penetrated through this layer to either remain or be retained in the corona (Figure 18.4a,b) or diffuse across to reach the nanoparticle's hydrophobic core (Figure 18.4c).

Results from a series of experiments designed to highlight proteins deposit on the surface of dextran-coated PIBCA nanoparticles led to the conclusion that proteins adsorbed onto the hydrophobic core of the nanoparticles (Figure 18.4c) (Vauthier et al., 2009). The obvious consequence drawn from this observation was that the properties of the corona are fundamental in controlling the accessibility of proteins to the nanoparticle core surface, hence controlling the whole phenomenon of protein deposition. Due to the anchorage of the dextran chains on the nanoparticle's core's surface, the space available to accommodate proteins adsorbing on the nanoparticle core is also a function of the corona characteristics. Using a rather small protein, bovine serum albumin (BSA), it could be shown that the access of proteins to the surface of the nanoparticle core greatly depended on both the density and conformation of the chains of dextran in the nanoparticle corona. The results of experiments performed with other proteins suggested that it is also greatly dependent on the protein's molecular weight, shape, and dimensions. A dense brush of dextran chains leaves spaces for the adsorption of proteins onto the core surface of nanoparticles of much smaller size than a loose brush. However, the maximum amount of BSA that can be accommodated on the nanoparticle core surface is almost equivalent (2.1 ± 0.4 mg/m² for nanoparticles R1 and 2.7 ± 0.2 mg/m² for nanoparticles R2) (Vauthier et al., 2011). This was explained by the rather small size of BSA. Comparing nanoparticles R1 and A1, the brush conformation of the chains of dextran in the corona of nanoparticles R1 allowed a lower amount of BSA to adsorb on the nanoparticle core than the loop conformation found on the surface of nanoparticles A1 (maximum amount of BSA adsorbed at the surface of nanoparticles A1 = 3.2 mg/m²) (Vauthier et al., 2009). Another important observation drawn from the experiment was that only proteins whose size suits that of the free spaces can adsorb onto the surface of the hydrophobic core. This is shown on Figure 18.5; nanoparticles with a dense brush of dextran (R1) adsorbed a majority of low-molecular weight proteins while the other nanoparticles (A1, A2, and R2) adsorbed larger amounts of immunoglobulin and fibrinogen, which are proteins with high molecular weights. The type of proteins that adsorbed on the nanoparticle surface can be selected by modulating both the conformation of the chains of dextran at the nanoparticle surface

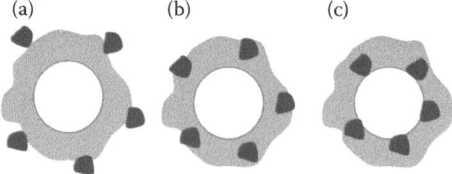

FIGURE 18.4 A scheme illustrating the different hypothesis drawn for protein adsorption on the nanoparticles. The particle is illustrated with a white core and a gray corona while the proteins are shown by the black spots. Adsorption of proteins taking place at the surface of the corona (a), within the polysaccharide corona (b) or on the surface of the hydrophobic core assuming that the proteins can diffuse across the polysaccharide corona (c). According to the experiments, proteins adsorbed on the surface of the hydrophobic core of dextran-coated PIBCA nanoparticles (hypothesis C). Considering nanoparticles coated with dextran sulfate or heparin, protein interactions may occur according to hypothesis A and/or B. (From Vauthier, C., Lindner, P., and Cabane, B., 2009. *Coll. Surf. B: Biointerfaces.* 69:207–15. With permission.)

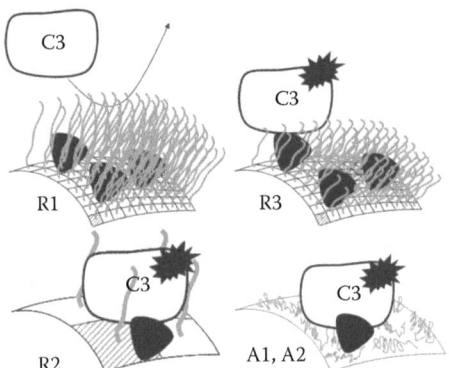

FIGURE 18.5 (**See color insert.**) A scheme illustrating the interactions of proteins with a series of nanoparticles having corona of dextran with different characteristics. The dark spot included in C3 indicates that the component C3 of the complement system is activated. Albumin adsorbed on the surface of the nanoparticle core appears as a dark triangle.

and the grafting density. A triage between proteins of high and low molecular weights applied at two levels. As explained above, the type of proteins that can adsorb onto the nanoparticle's surface depends on their size and the hydrophobic areas available on the nanoparticle core. It also depends on the characteristics of the corona, which functions as a molecular sieve, while proteins diffuse across.

Proteins whose size and geometry does not allow them to diffuse through the mesh of the polysaccharide layer that forms the corona are excluded from and are not able to reach the surface of the nanoparticle's hydrophobic core (Figure 18.6).

As it was found with PEG-coated nanomaterials, the density of the dextran chain stranded on the surface of nanoparticles, their molecular weight, and their conformation are factors that contribute to the selection of the amount and type of proteins that adsorb onto the nanoparticle's surface. These factors also control the activation of protein C3 of the complement system, which plays a central role in the activation of the immune system's complement cascade (Nilsson et al., 2007). A clear correlation was highlighted between the levels of activation produced by the nanoparticles and the mesh size characteristics of the polysaccharide brush composing their corona (Figure 18.7a). Considering this correlation, it clearly appeared that the steric exclusion of protein C3 from the nanoparticle corona could prevent the activation of the complement system. The correlation between the probability of insertion of protein C3 in the dextran corona and the level of activation measured was not as good (Figure 18.7b). This was interpreted by the fact that the activation of protein C3 may be triggered by a layer of already-adsorbed proteins which have changed conformation during adsorption onto the nanoparticle core surface, and that may be accessible to protein C3 from the external part of the nanoparticles.

As discussed above, the density, conformation, and concentration of dextran chains on the nanoparticle corona are factors controlling the adsorption of proteins and activation of complement. Activation of protein C3 of the complement system was also controlled by the molecular weight of the dextran chains in the brush (Figure 18.8).

Nanoparticles that activated protein C3 had a brush of dextran chains with a molecular weight lower than 60 kDa. With a brush formed by dextran of higher molecular weight, the nanoparticles hampered the activation of protein C3, which is mandatory in escaping the immune system in the vascular compartment. This result was not expected from the other characteristics of the nanoparticle corona because protein C3 was believed to be too large to diffuse through the dextran corona and reach the surface of the nanoparticle core to be activated. As illustrated in Figure 18.6 for nanoparticle R3, the activation of protein C3 could be triggered by a layer of adsorbed proteins protruding

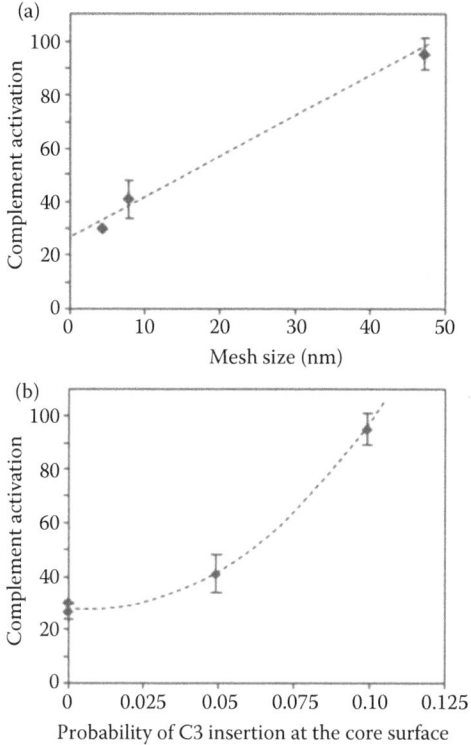

FIGURE 18.6 The relation between complement activation induced by the nanoparticles and the mesh size of the corona (a) and the insertion probability of the C3 protein at the core surface (b). (Reproduced from Vauthier, C. et al. 2011. *Biomaterials*. 32:1646–56. With permission.)

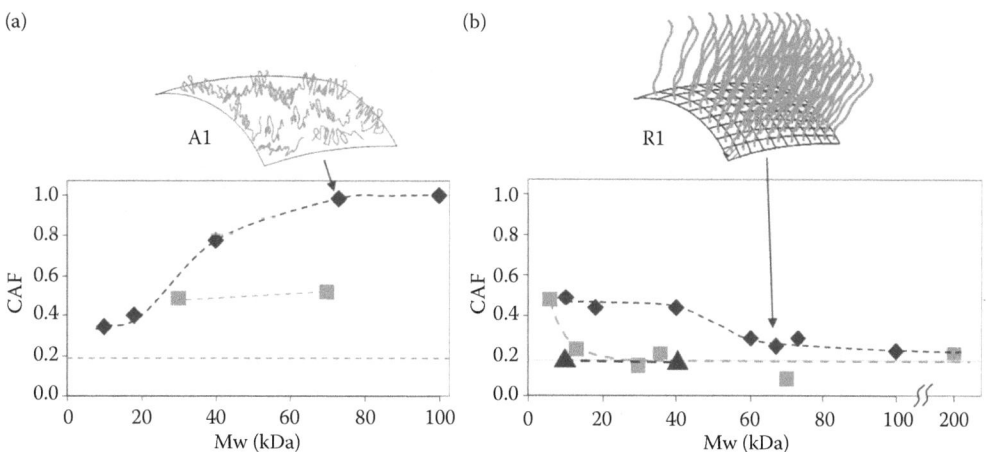

FIGURE 18.7 Complement activation factors (CAF) evaluated for nanoparticles prepared by anionic (a) and radical polymerization (b) with different polysaccharides (black diamond: dextran, gray squares: chitosan, black triangle: dextran sulfate) and varying their molecular weight. The basic activation of the complement system in the experimental conditions is indicated by the horizontal dotted line at a CAF of 0.195. On the top of the graphs, schemes illustrate the configuration of the polysaccharide chains in the corona of the nanoparticles according to the mechanism of polymerization applied to prepare the nanoparticles. (CAF values were taken from Bertholon, I., Vauthier, C., and Labarre, D. 2006a. *Pharm. Res.* 23:1313–1323; Model of the nanoparticle surface from Vauthier, C. et al. 2011. *Biomaterials*. 32:1646–56. With permission.)

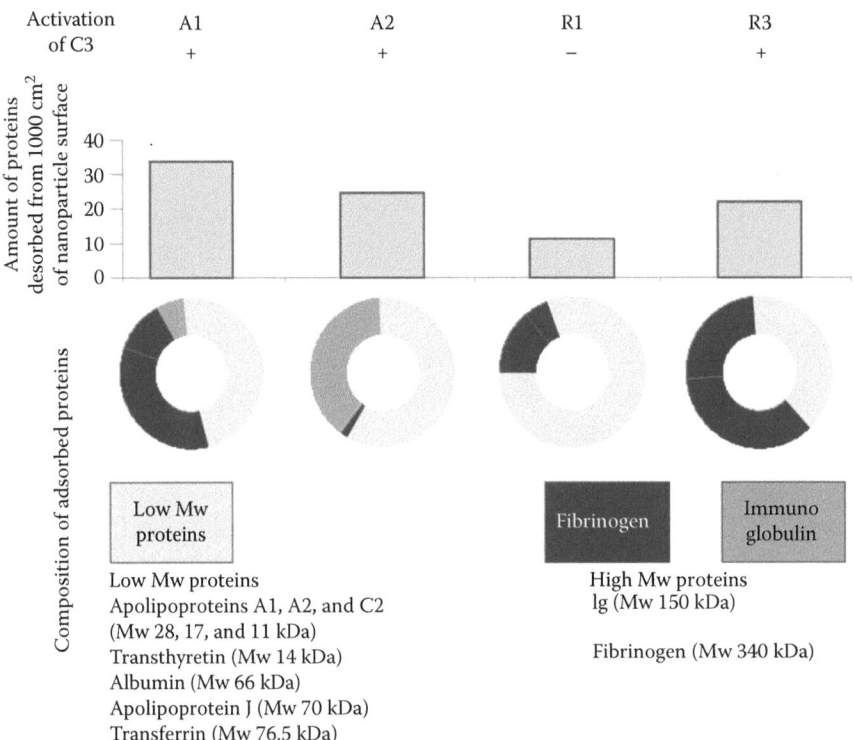

FIGURE 18.8 Protein adsorption on different dextran-coated nanoparticles and corresponding status of the activation of protein C3 of the complement system. The circles at the bottom represent the balance between proteins of high molecular weight (>100 kDa) (dark gray and black) and low molecular weight (<100 kDa) (light gray) that were found adsorbed at the nanoparticle surface after incubation in plasma. For proteins of high molecular weight a difference was made between fibrinogen (in black) and immunoglobulin (in dark gray).

out of the corona when the dextran chains are too short to hide them. On the contrary, when the dextran chains are long enough (at least 60 kDa), the layer of adsorbed proteins can be masked in an efficient way. The fact that protein C3 is excluded from the dextran brush and that the adsorbed proteins are masked by the dextran brush appeared as essential factors in hindering the activation of protein C3 of the complement system when considering dextran-coated nanoparticles.

A few conclusions can be drawn from the work done on dextran-coated PACA nanoparticles. While proteins were adsorbed to the hydrophobic core of the nanoparticles, the characteristics of the dextran corona controlled both the type of proteins that were adsorbed and subsequent events, including the activation of the complement system. A dense and thick dextran brush is needed to hinder the activation of the complement system. Besides the activation of the complement system, the composition of the adsorbed proteins can be modulated by varying the structure and characteristics of the corona. It is noteworthy that small changes can greatly influence the composition of the adsorbed proteins as illustrated by the comparison of protein adsorbed onto nanoparticles A1 and A2 (Figure 18.5). All the results obtained from the analysis of protein interactions with our nanoparticles coated with dextran chains suggested that the control of the adsorption of proteins is achieved by a similar mechanism to that described with PEG nanoparticles being mainly governed by a steric effect.

In contrast to dextran-coated nanoparticles, the nanoparticles coated with charged polysaccharides behaved very differently. Nanoparticles with brushes of chitosan, heparin, and dextran-sulfate worked well in preventing the activation of protein C3 even at low molecular weights. This effect was expected through designing nanoparticles coated with heparin, which is an inhibitor of complement system activation, thanks to an interference with factor H of the complement cascade (Nilsson et al., 2007). Interestingly, the inhibiting property of heparin was maintained in nanoparticles designed with a brush composed of a blend of dextran and heparin. Further analyses were pursued on the nanoparticles coated with dextran sulfate, with heparin and with a blend of heparin and dextran. The analysis of the proteins that adsorbed on these nanoparticles revealed a majority of high-molecular weight proteins, including fibrinogen and immunoglobulin (Figure 18.9) (Labarre et al., 2005).

The adsorption of these proteins onto the nanoparticle surface is in contradiction with their capacity to prevent the activation of the complement system. It is also in contradiction with the structure of a dense brush of polysaccharide chains forming the nanoparticle corona that should exclude large proteins from adsorption onto nanoparticles. Although these results disagreed with results obtained from previous studies carried out on almost neutral nanoparticles designed with a corona composed of PEG and dextran, they agreed with those reported by Ballauff et al. investigating protein adsorption on nanoparticles coated with polyelectrolyte brushes (Wittemann et al., 2003; Welsch et al., 2013). These authors suggested that the adsorption of proteins resulted from

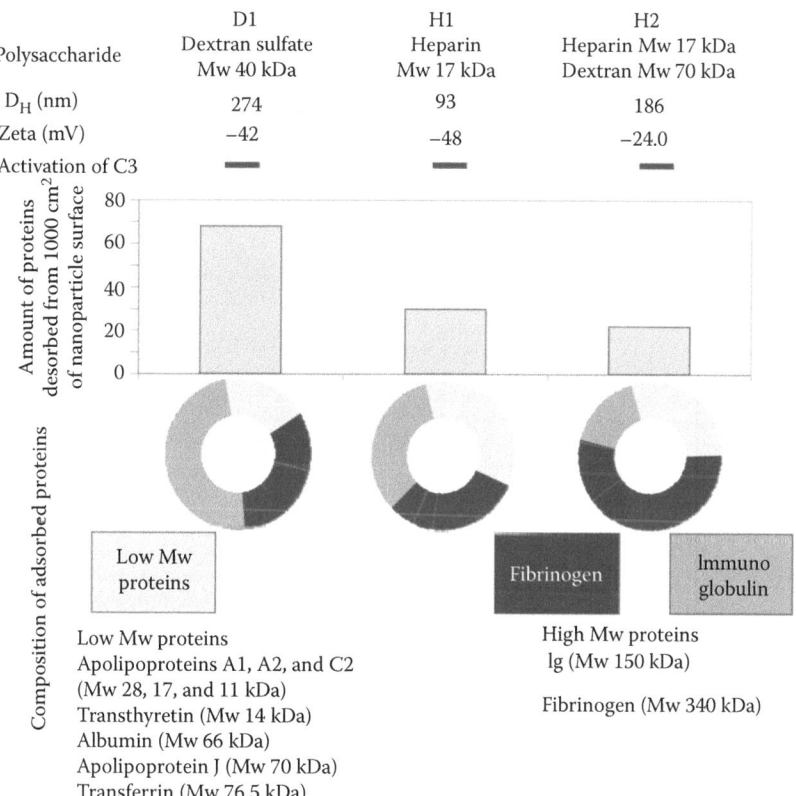

FIGURE 18.9 Protein adsorption on different negatively charged nanoparticles and corresponding status of the activation of protein C3 of the complement system. The circles at the bottom represent the balance between proteins of high molecular weight (>100 kDa) (dark gray and black) and low molecular weight (<100 kDa) (light gray) that were found adsorbed at the nanoparticle surface after incubation in plasma. For proteins of high molecular weight a difference was made between fibrinogen (in black) and immunoglobulins (in dark gray).

ionic interactions taking place between the negative charges of the polyelectrolyte grafted on the nanoparticle surface and patches of positive charges found on the proteins. It is believed that the proteins can then adsorb as a multivalent counter-ion after the displacement of the monovalent counter-ion found along the polyelectrolyte chains of the brush formed at the nanoparticle surface. This effect can take place even at the most external sites of the nanoparticle corona as illustrated by Figures 18.4a and b. The remarkable property of dextran sulfate to hinder the activation of the complement system despite the adsorption of immunoglobulin and fibrinogen needs to be elucidated. However, it can be assumed that the mechanism behind this effect is different from that depicted with dextran-coated nanoparticles and based on a steric exclusion effect.

At this stage, we can identify that polysaccharide-coated nanoparticles can prevent the activation of the complement system and control the adsorption of proteins by at least two mechanisms that were summarized by Table 18.3.

More work is now needed to understand how proteins interact with nanoparticles and what role the polysaccharide corona plays in this interaction. Nevertheless, the highly charged nanoparticles appeared as potential drug carriers that were able to bypass recognition mechanisms of the immune system followed by the activation of the complement cascade. Their capacity to adsorb high-molecular weight proteins without triggering the activation of the complement system is a new property. It can reasonably be assumed that they will show a different biodistribution compared to that of the nanoparticles with a corona that selects small molecular weight

TABLE 18.3
Summary on Mechanisms Controlling Activation of the Complement System and Adsorption of Proteins by Polysaccharide-Coated Nanoparticles and Major Type of Proteins Adsorbed on Corresponding Nanoparticles

Composition of the Nanoparticle Corona = Brush of Polysaccharide	Proposed Mechanism Behind the Control of the Adsorption of Proteins	Mechanism Behind the Control of the Activation of the Complement System	Type of Adsorbed Proteins
Dextran > 60 kDa	Adsorption on the hydrophobic core surface of the nanoparticles after diffusion through the polysaccharide corona. Selection of proteins that can adsorb arises by the characteristics of the nanoparticle corona which function as a molecular sieve and by the size and shape of proteins which should fit with that of the hydrophobic spaces available on the nanoparticle core surface	Steric exclusion	Majority of low molecular weight protein (apolipoprotein A1 (Mw 28.08 kDa) and C2 (Mw 11.3 kDa))
Heparin	Adsorption of proteins as a multivalent counterion by interactions of the charges of the polysaccharides composing the nanoparticle corona	Known activity against complement activation. Interference with factor H of the complement cascade	Majority of high molecular weight proteins. Immunoglobulin, fibrinogen
Dextran sulfate	Adsorption of proteins as a multivalent counterion by interactions of the charges of the polysaccharides composing the nanoparticle corona	To elucidate	Majority of high molecular weight proteins. Immunoglobulin (Mw 150 kDa), fibrinogen (Mw 340 kDa)
Chitosan	To elucidate	To elucidate	Unknown

proteins, thanks to a steric effect. We are still at the very beginning of understanding how polysaccharides can be used to tune nanoparticle properties to modulate protein adsorption in a perfectly controlled manner. Nevertheless, it already seems that polysaccharides occurring in various natures can offer much wider possibilities to finely tune interactions between proteins and nanoparticles and, in turn, provide a wider range of possibilities to control the *in vivo* fate of drug carriers other than PEG. Another very exciting perspective can emerge from the elucidation of the glyco-code that may provide keys to the design of nanoparticles having very defined interactions with proteins and, in turn, to obtain better controlled physiological responses and, finally, *in vivo* distributions.

18.4 CONCLUSION

Nanoparticles with a wide range of surface properties can be designed by choosing polysaccharides to create a hydrophilic corona on their surface. This modification strategy of surface properties greatly influences the amount and types of proteins that can adsorb onto the nanoparticle surface and the capacity of the nanoparticles to trigger the activation of the complement system. Interestingly, controlling the interactions of proteins with the nanoparticle surface can be achieved through different mechanisms by choosing the nature of the polysaccharide composing the nanoparticle corona and designing the structure of the nanoparticle corona. More investments are needed on systematic and fundamental works to explore the approach's full potential and to better understand how we can use polysaccharides to finely tune the interactions of proteins with nanoparticles. This will constitute a first step to the understanding of the physiological response and, in turn, the biodistribution after IV administration. The wide range of surface characteristics that can be generated in a controlled manner may also be an opportunity for having access to model nanoparticles that could be used to investigate sites of accumulation of nanomaterials depending on the profile of adsorbed proteins and interactions with the immune system. This would deserve a rationalization of the approaches developed to design more efficient drug carriers and the understanding of potential nanomaterial toxicological profiles from their interactions with proteins.

REFERENCES

Alexis, F., Pridgen, E., Molnar, L.K., and Farokhzad, O.C. 2008. Factors affecting the clearance and biodistribution of polymeric nanoparticles. *Mol. Pharm.* 5:505–15.

Algar, W.R., Prasuhn, D.E., Stewart, M.H., Jennings, T.L., Blanco-Canosa, J.B. et al. 2011. The controlled display of biomolecules on nanoparticles: A challenge suited to bioorthogonal chemistry. *Bioconjug. Chem.* 22:825–58.

Anderson, N.L., Polanski, M., Pieper, R., Gatlin, T., Tirumalai, R.S., Conrads, T.P. et al. 2004. The human plasma proteome: A nonredundant list developed by combination of four separate sources. *Mol. Cell Proteom.* 3:311–26.

Andrieux, K. and Couvreur, P. 2009. Polyalkylcyanoacrylate nanoparticles for delivery of drugs across the blood–brain barrier. *Wiley Interdiscip. Rev. Nanomed. Nanobiotechnol.* 1:463–74.

Ballauff, M. 2007. Spherical polyelectrolyte brushes. *Prog. Polym. Sci.* 32:1135–51.

Banerjee, I., Pangule, R.C., and Kane, R.S. 2011. Antifouling coatings: Recent developments in the design of surfaces that prevent fouling by proteins, bacteria, and marine organisms. *Adv. Mater.* 23:690–718.

Bangham, A.D., Standish, M.M., and Watkins, J.C. 1965. Diffusion of univalent ions across the lamellae of swollen phospholipids. *J. Mol. Biol.* 13:238–52.

Bakht, M.K., Sadeghi, M., Pourbaghi-Masouleh, M., and Tenreiro, C. 2012. Scope of nanotechnology-based radiation therapy and thermotherapy methods in cancer treatment. *Curr. Cancer Drug Targets* 12:998–1015.

Ballauff, M. and Borisov, O.V. 2006. Polyelectrolyte brushes. *Curr. Opin. Colloid Interf. Sci.* 11:316–23.

Bertholon, I., Vauthier, C., and Labarre, D. 2006a. Complement activation by core-shell poly(isobutylcyanoacrylate)-polysaccharide nanoparticles: Influences of surface morphology, length, and type of polysaccharide. *Pharm. Res.* 23:1313–1323.

Bertholon, I., Lesieur, S., Labarre, D., Besnard, M., and Vauthier C. 2006b. Characterization of dextran-poly(isobutylcyanoacrylate) copolymers obtained by redox radical and anionic emulsion polymerization. *Macromolecules*. 39:3559–67.

Bertholon, I., Hommel, H., Labarre, D., and Vauthier, C. 2006c. Properties of polysaccharides grafted on nanoparticles investigated by EPR. *Langmuir*. 22:5485–90.

Bravo-Osuna, I., Vauthier, C., Farabollini, A., Palmieri, G.F., and Ponchel, G. 2007. Mucoadhesion mechanism of chitosan and thiolated chitosan-poly(isobutyl cyanoacrylate) core-shell nanoparticles. *Biomaterials*. 28:2233–43.

Chauvierre, C., Labarre, D., Couvreur, P., and Vauthier, C. 2003. Novel polysaccharide-decorated poly(isobutyl cyanoacrylate) nanoparticles. *Pharm. Res.* 20:1786–93.

Chauvierre, C., Vauthier, C., Labarre, D., and Hommel, H. 2004. Evaluation of the surface properties of dextran coated poly(isobutylcyanoacrylate) nanoparticles by spin-labelling coupled with electron resonance spectroscopy. *Coll. Polym. Sci.* 282:1016–25.

Couvreur, P. and Vauthier, C. 2006. Nanotechnology: Intelligent design to treat complex disease. *Pharm. Res.* 23:1417–50.

Daniel, J.C. 2003. A long history with many future challenges to meet in the future—Free-radical emulsion polymerization and aqueous polymer dispersions. In: *Colloidal Polymers, Synthesis and Characterisation, Surfactant Science Series*. Ed. A. Elaissari, 115:1–22. New York: Marcel Dekker.

de Martimprey, H., Vauthier, C., Malvy, C., and Couvreur, P. 2009. Polymer nanocarriers for the delivery of small fragments of nucleic acids: Oligonucleotides and siRNA. *Eur. J. Pharm. Biopharm.* 71(3):490–504.

Dobrovolskaia, M.A., Aggarwal, P., Hall, J.B., and McNeil, S.E. 2008. Preclinical studies to understand nanoparticle interaction with the immune system and its potential effects on nanoparticle biodistribution. *Mol. Pharm.* 5:487–95.

Douglas, S.J., Illum, L., Davis, S.S., and Kreuter, J. 1984. Particle size and size distribution of poly(butyl 2 cyanoacrylate) nanoparticles. Influence of physicochemical factors. *J. Coll. Interface Sci.* 101:149–158.

Etheridge, M.L., Campbell, S.A., Erdman, A.G., Haynes, C.L., Wolf, S.M., and McCullough, J. 2013. The big picture on nanomedicine: The state of investigational and approved nanomedicine products. *Nanomedicine:NBM*. 9:1–14.

Estephan, Z.G., Jaber, J.A., and Schlenoff, J.B. 2010, Zwitterion-stabilized silica nanoparticles: Toward non-stick nano. *Langmuir*. 26:16884–9.

Farokhzad, O.C. and Langer, R. 2009. Impact of nanotechnology on drug delivery. *ACS Nano*. 1:16–20.

Gref, R., Lück, M., Quellec, P., Marchand, M., Dellacherie, E., Harnisch, S. et al. 2000. Stealth corona-core nanoparticles surface modified by polyethylene glycol (PEG): Influences of the corona (PEG chain length and surface density) and of the core composition on phagocytic uptake and plasma protein adsorption. *Colloids Surf B Biointerfaces*. 18:301–13.

Gref, R., Minamitake, Y., Peracchia, M.T., Domb, A., Trubetskoy, V., Torchilin, V. et al. 1997. Poly(ethylene glycol)-coated nanospheres: Potential carriers for intravenous drug administration. *Pharm. Biotechnol.* 10:167–98.

Gref, R., Minamitake, Y., Peracchia, M.T., Trubetskoy, V., Torchilin, V., and Langer, R. 1994. Biodegradable long-circulating polymeric nanospheres. *Science* 263:1600–3.

Gregoriadis, G. 1976. The carrier potential of liposomes in biology and medicine. *N. Eng. J. Med.* 295:704–10.

Gregoriadis, G., Wills, E.J., Swain, C.P., and Tavill, A.S. 1974. Drug-carrier potential of liposomes in cancer chemotherapy. *Lancet*. 1:1313–6.

Hirsh, S.L., McKenzie, D.R., Nosworthy, N.J., Denman, J.A., Sezerman, O.U., and Bilek, M.M.M. 2013. The Vroman effect: Competitive protein exchange with dynamic multilayer protein aggregates. *Coll. Surf. B: Biointerfaces* 103:395–404.

Jeon, S.I. and Andrade, J.D. 1991. Protein–surface interactions in the presence of polyethylene oxide: II. Effect of protein size. *J. Coll. Interf. Sci.* 142:159–166.

Jeon, S.I., Lee, J.H., Andrade, J.D., and De Gennes P.G. 1991. Protein–surface interactions in the presence of polyethylene oxide: I. Simplified theory. *J. Coll. Interf. Sci.* 142:149–158.

Jiskoot, W., Van Schie, R.M.F., Carstens, M.G., and Schellekens, H. 2009. Immunological risk of injectable drug delivery systems. *Pharm. Res.* 26:1303–14.

Juliano, R.L. 1976. The role of drug delivery systems in cancer chemotherapy. *Prog. Clin. Biol. Res.* 9:21–32.

Kreuter, J. 2007. Nanoparticles—A historical perspective. *Int. J. Pharm.* 331:1–10.

Kreuter, J. and Speiser, P.P. 1976. New adjuvants on a polymethylmethacrylate base. *Infect. Immun.* 13, 204–210.

Labarre, D., Vauthier, C., Chauvierre, C., Petri, B., Muller, R.H., and Chehimi, M.M. 2005. Interactions of blood proteins with poly (isobutylcyanoacrylate) nanoparticles decorated with a polysaccharidic brush. *Biomaterials*. 26:5075–84.

Lehner, R., Wang, X., Marsch, S., and Hunziker, P. 2013. Intelligent nanomaterials for medicine: Carrier platforms and targeting strategies in the context of clinical application. *Nanomedicine:NBM* 9:742–757.

Lemarchand, C., Gref, R., and Couvreur, P. 2004. Polysaccharide-decorated nanoparticles. *Eur. J. Pharm. Biopharm*. 58:327–41.

Li, S.D. and Huang, L. 2008. Pharmacokinetics and biodistribution of nanoparticles. *Mol. Pharm*. 5:496–504.

Lira, M.C.B., Santos-Magalhães, N.S., Nicolas, V., Marsaud, V., Silva, M.P., Ponchel, G. et al. 2011. Cytotoxicity and cellular uptake of newly synthesized fucoidan-coated nanoparticles. *Eur. J. Pharm. Biopharm*. 79:162–70.

Liu, F., Laurent, S., Fattahi, H., Vander Elst, L., and Muller, R.N. 2011. Superparamagnetic nanosystems based on iron oxide nanoparticles for biomedical imaging. *Nanomedicine (Lond)*. 6:519–28.

Lynch, I., Cedervall, T., Lundqvist, M., Cabaleiro-Lago, C., Linse, S. et al. 2007. The nanoparticle–protein complex as a biological entity; a complex fluids and surface science challenge for the 21st century. *Adv. Colloid Interface Sci*. 134–135:167–74.

Moyano, D.F. and Rotello, V.M. 2011. Nano meets biology: Structure and function at the nanoparticle interface. *Langmuir*. 27:10376–85.

Mura, S. and Couvreur, P. 2012. Nanotheranostics for personalized medicine. *Adv. Drug Deliv. Rev*. 64:1394–416.

Nicolas, J. and Couvreur, P. 2009. Synthesis of poly(alkyl cyanoacrylate)-based colloidal nanomedicines. *Wiley Interdiscip. Rev. Nanomed. Nanobiotechnol*. 1:111–27.

Nicolas, J. and Vauthier, C. 2011. Poly(alkylcyanoacrylate) nanosystems. In *Intracellular Delivery Fundamental Biomedical Technologies* ed. A. Prokov, pp. 225–251. New York: Springer, Volume 5, Part 2.

Nilsson, B., Ekdahl, K.N., Mollnes, T.E., and Lambris, J.D. 2007. The role of complement in biomaterial-induced inflammation. *Mol. Immunol*. 44:82–94.

Norde, W. 2008. My voyage of discovery to proteins in flatland…and beyond. *Coll. Surf. B. Biointerface*. 61:1–9.

Owens, D.E. and Peppas, N.A. 2006. Opsonization, biodistribution and pharmacokinetics of polymeric nanoparticles. *Int. J. Pharm*. 307:93–102.

Papahadjopoulos, D., Allen, T.M., Gabizon, A., Mayhew, E., Matthay, K., Huang, S.K. et al. 1991. Sterically stabilized liposomes: Improvements in pharmacokinetics and antitumor therapeutic efficacy. *Proc. Natl. Acad. Sci. U.S.A*. 88:11460–4.

Passirani, C., Ferrarini, L,. Barratt, G., Devissaguet, J.P., and Labarre, D. 1999. Preparation and characterization of nanoparticles bearing heparin or dextran covalently-linked to poly(methyl methacrylate*)*. *J. Biomater. Sci. Polym. Ed*. 10:47–62.

Peracchia, M.T., Fattal, E., Desmaële, D., Besnard, M., Noël, J.P., Gomis, J.M. et al. 1999. Stealth PEGylated polycyanoacrylate nanoparticles for intravenous administration and splenic targeting. *J. Control. Release*. 60:121–8.

Szebeni, J., Muggia, F., Gabizon, A., and Barenholz, Y. 2011. Activation of complement by therapeutic liposomes and other lipid excipient-based therapeutic products: Prediction and prevention. *Adv. Drug Deliv. Rev*. 63:1020–30.

Vauthier, C., Labarre, D., and Ponchel, G. 2007. Design aspects on poly(alkylcyanoacrylate) nanoparticles for targeted drug delivery. *J. Drug Targeting*. 15:641–63.

Vauthier, C., Lindner, P., and Cabane, B., 2009. Configuration of bovine serum albumin adsorbed on polymer particles with grafted dextran corona. *Coll. Surf. B: Biointerfaces*. 69:207–15.

Vauthier, C., Persson, B., Lindner, P., and Cabane, B. 2011. Protein adsorption and complement activation for di-block copolymer nanoparticles. *Biomaterials*. 32:1646–56.

Vonarbourg, A., Passirani, C., Saulnier, P., and Benoit, J.P. 2006. Parameters influencing the stealthiness of colloidal drug delivery systems. *Biomaterials*. 27:4356–73.

Walczyk, D., Baldelli Bombelli, F., Monopoli, M.P., Lynch, I., and Dawson, K.A. 2010. What the cell sees in bionanoscience. *J. Am. Chem. Soc*. 132:5761–68.

Walkey, C.D. and Chan, W.C.W. 2012. Understanding and controlling the interaction of nanomaterials with proteins in a physiological environment. *Chem. Soc. Chem*. 41, 2780–99.

Walkey, C.D., Olsen, J.B., Guo, H., Emili, A., and Chan, W.C. 2012. Nanoparticle size and surface chemistry determine serum protein adsorption and macrophage uptake. *J. Am. Chem. Soc*. 134:2139–47.

Wang, R., Billone, P.S., and Mullet, W.M. 2013. Nanomedicine in action: An overview of cancer nanomedicine on the market and in clinical trials. *J. Nanomed*. DOI:10.1155/2013/629681, Article ID: 629681, 12 pages.

Welsch, N., Lu, Y., Dzubiella, J., and Ballauff, M. 2013. Adsorption of proteins to dunctional polymeric nanoparticles. *Polymer.* 54:2835–49.

Wittemann, A., Haupt, B., and Ballauff, M. 2003. Adsorption of proteins on spherical polyelectrolyte brushes in aqueous solution. *Phys. Chem. Chem. Phys.* 5:1671–7.

Zandanel, C. and Vauthier, C. 2012. Poly(isobutylcyanoacrylate) nanoparticles decorated with chitosan: Effect of conformation of chitosan chains at the surface on complement activation properties. *J. Coll. Sci. Biotechnol.* 1:68–81.

19 Models for Risk Assessments of Nanoparticles

Sanjay Dey, Bhaskar Mazumder, and Yaswant Pathak

CONTENTS

19.1	Introduction	384
19.2	Interaction of NPs with Cells	385
	19.2.1 Interference of NPs to Cells Membrane	385
	19.2.2 Effect of NPs to Proteins and Macromolecules	385
	19.2.3 NP Interaction with DNA	386
19.3	Entry Routes of NPs into the Body	386
19.4	Conventional Risk Assessment	388
19.5	Structure for Risk Assessment	389
19.6	Issues Relevant to Risk Assessment	391
19.7	Stages of Risk Assessment	391
	19.7.1 Characterizations of Behavior of Engineered Nanomaterials	392
	19.7.2 An Assessment of Exposure to Engineered Nanomaterial	393
	19.7.3 Hazard Characterization	394
	19.7.3.1 Translocation of Engineered Nanomaterial into the Body	395
	19.7.3.2 Pulmonary Inflammation Induced by Engineered Nanomaterial	395
	19.7.3.3 Genotoxicity of Engineered Nanomaterial	396
	19.7.3.4 Carcinogenic Effects of Engineered Nanomaterial	396
	19.7.3.5 Effects of Engineered Nanomaterial on Circulation	397
	19.7.3.6 Other Remarks on the Effects of Engineered Nanomaterial	397
19.8	An Assessment of Particle Parameters Relevant for Biological Action	398
19.9	*In Vitro* Cytotoxicity Test	401
	19.9.1 Cell Viability Testing	401
	19.9.1.1 Detection of Mitochondrial Activity	402
	19.9.1.2 Detection of LDH Release upon Necrosis	402
	19.9.1.3 Fluorescein Diacetate Test	402
	19.9.1.4 WST-1 Cytotoxicity Assay	403
	19.9.1.5 Clonogenic Assay	403
	19.9.1.6 Hematoxylin and Eosin Assay	403
	19.9.1.7 Neutral Red Uptake Assay	403
	19.9.1.8 Trypan Blue Exclusion Assay	403
	19.9.1.9 Alamar Blue Reduction	404
	19.9.1.10 Commassie Blue Assay	404
	19.9.1.11 MetPLATE *E. coli* Bioassay	404
	19.9.1.12 CellTiter Assay	404
	19.9.1.13 Annexin V/Propidium Iodide Staining for Apoptotic and Necrotic Cells	404
	19.9.1.14 Detection of the Apoptosis Marker Caspase-3	405

	19.9.2	Stress Response	405
		19.9.2.1 Detection of ROS	405
		19.9.2.2 Glutathione Measurement	406
		19.9.2.3 Lipid Peroxidation Assay	406
		19.9.2.4 Nitrite Production	407
		19.9.2.5 Mitochondrial Membrane Potential Dissipation Assay	407
	19.9.3	Inflammatory Response	407
	19.9.4	Genotoxicity	407
		19.9.4.1 Comet Assay	408
		19.9.4.2 F_{pg}-Modified Comet Assay	408
		19.9.4.3 Ames Test	408
		19.9.4.4 Alkaline Single-Cell Microgel Electrophoresis Assay	408
		19.9.4.5 DNA Damage	409
	19.9.5	High-Throughput Screening Method	409
	19.9.6	Dose-Response Assessment	409
19.10	Application of Computational Approach for Risk Assessment		409
	19.10.1	Molecular Modeling Methods	409
	19.10.2	Quantitative Structure–Activity Relationships for Nanomaterials	410
19.11	Ecotoxicity of NPs		411
19.12	Limitations of *In Vitro* Assay in Toxicology		411
19.13	Challenges for NPs *In Vitro* Test Methods		412
19.14	European Union Approach to Nanomaterial Risk Assessment		413
References			413

19.1 INTRODUCTION

Nanomaterials have a wide range of emerging applications in biomedical, pharmaceutical, and biotechnological fields including biosensors, biomarkers, cancer therapy, deoxyribonucleic acid (DNA) delivery systems, drug-delivery systems, enzyme immobilization, gene delivery, tissue engineering, and as probes for confocal and electron microscopy [1].

Nowadays, nanotechnology-based products that claim the use of nanomaterials, are widely available in the market lists various products, including paint, cosmetics, personal-care products, and food supplements. However, the presence of nanoscale entities in these products has not been verified. The human exposure to nanomaterials may involve inhalation, ingestion, and dermal routes. These particles may also be directly injected into the human body for medical purposes. Once the nanoparticles (NPs) appear to the systemic circulation, the NPs may be capable of distributing to most organ systems and may even cross biological barriers, such as the blood–brain and blood–testis barriers [2]. Since the number of applications of nanomaterials is expected to rise even more in the future, long-term exposure and potential accumulation of these nanomaterials in the human body may result.

With the ongoing commercialization of nanotechnology products, human exposure to NPs will dramatically increase and an evaluation of their potential toxicity is essential. NPs applying for biomedical and pharmaceutical purposes must satisfy rigorous toxicity screening to obtain approval from regulatory authorities for use. The requirement to provide adequate biocompatibility profiles has led to renewed and increased interest in nanotoxicology research. Recently, a number of manufactured NPs have been shown to cause adverse effects *in vitro* and *in vivo* [3]. NPs have unique physicochemical properties attributed to their extremely small size, and possess extremely high surface-area-to-volume ratio that renders them highly reactive. High reactivity could potentially lead to toxicity due to harmful interactions of nanomaterials with biological systems and the environment [4]. It has been shown that the NPs of TiO_2 induced a much greater pulmonary–inflammatory response than larger particles of the same chemical content at equivalent mass doses [4]. Consequently, screening techniques commonly used for toxicity testing of macroscale substances may not be appropriate for NPs hazard characterization, but may have to be adapted or modified with regard to their nanospecific properties [5].

To facilitate a faster risk assessment of NPs, the use of *in vitro* studies has been suggested as a widely acceptable approach to evaluate the toxicity of nanomaterials [6]. Numerous *in vitro* studies investigating the cytotoxic, oxidative stress, and inflammation potential of nanomaterials are now published, although their value for predicting *in vivo* toxicity still remains to be demonstrated [7].

19.2 INTERACTION OF NPs WITH CELLS

19.2.1 Interference of NPs to Cells Membrane

Cell membranes, which are phospholipid bilayers, partition different intracellular compartments from extracellular compartments. They also encapsulate the different intracellular components. To facilitate exchanges of ions, molecules, and NPs between compartments and/or cells, membranes have to be permeable. The outer surface of the cell membrane that presents toward the extracellular compartment, allows selective transport of ions, molecules, and also NPs. Intracellular membranes separate distinct compartments such as mitochondria, vesicles, nucleus, and so on from the cytosol [8]. The stability of the membrane can be affected by NPs in two ways: either directly (e.g., physical damage) or indirectly (e.g., oxidation) that can lead to cell death. It is the ability of membranes to control intracellular homeostasis, through selective permeability and transport mechanisms, which makes them a vulnerable target for the possible damaging effects of NPs. Interactions of NPs with membranes largely depend on the NPs' surface properties. Therefore, the surface modifications of NPs are crucial in the design of drug-delivery systems to enhance uptake into cells [8]. NP size also plays an important role as it influences surface pressure and adhesion forces [9].

Research has shown that different nanomaterials can damage membranes by various processes (Figure 19.1) that lead to a compromise of membrane integrity, stability, and the formation of nanosized holes [10]. The physicochemical properties of NPs seem to be primarily responsible for changes in membrane morphology and stability [10]. Mitochondria emerge to be a major target for fullerenes [11] and carbon nanotubes (CNTs) [12]. However, other NPs (e.g., titanium dioxide, CNTs, polystyrene, silver, etc.) also seem to be able to alter mitochondrial function, leading to apoptosis [13,14]. Another preferential intracellular compartment is that of lysosomes, the cell's digestive system, where the NPs generally end up and the lysosomes try to either digest or excrete them [15]. The impact of size and shape of the nanomaterial on the ability of lysosomes to digest or excrete NPs is not fully understood.

19.2.2 Effect of NPs to Proteins and Macromolecules

The cell apparatus consists of a large amount of proteins and other macromolecules. These exist in the form of enzymes (e.g., gastrin), cell-signaling molecules (e.g., hormones), or structural proteins (e.g., tubulin). Their normal functioning is therefore essential for all vital cellular activities. Correct molecular conformation is essential for proteins to work as intended. The slight conformational changes of protein can alter or destroy its function. During their assembly process, chaperones play an important role in controlling the manner in which proteins fold [16], to obtain a certain conformation. NPs that can be of the same size of protein molecules are able to impede with cell-signaling processes [17] or interact with proteins [18], either by chaperone-like activity [19] or by changing the configuration of peptides in forms of aggregation and fibrillation [20]. Protein misfolding and peptide fibrillation leading to amyloid-like structures are associated with neurodegenerative diseases. Investigating the possible misformation and overproduction of proteins and macromolecules at the cellular level is important for nanotoxicological considerations [18].

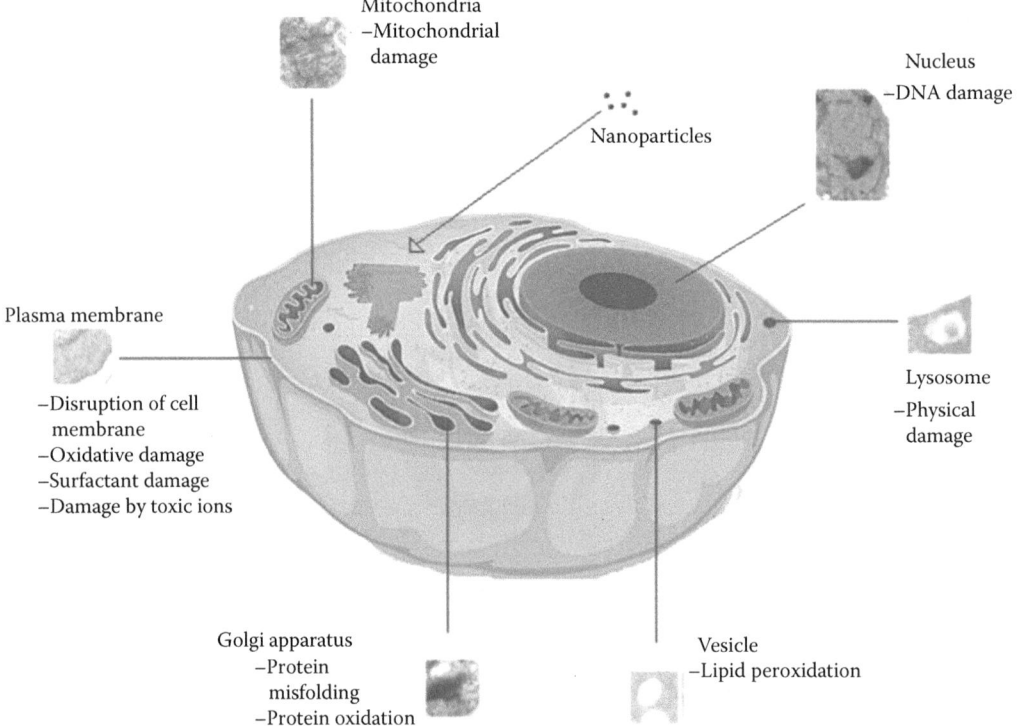

FIGURE 19.1 NP interaction with cells: intracellular targets and nanotoxicological mechanisms. (Reprinted with permission from Elsaesser A, Vyvyan Howard C. Toxicology of nanoparticles. *Advanced Drug Delivery Reviews*. 2012;64(2):12–137.)

19.2.3 NP Interaction with DNA

During the assessing of nanotoxicological risk caused by nanomaterials, the effect of NPs on DNA has attracted special attention. The researchers have reported that the NPs are able to enter the nuclear envelope and result in the possible genotoxic effects of NPs. The genotoxic effects of various types of NPs have been investigated [21]. However, these studies have not been able to give clear direction about the parameter of NPs that is mainly responsible for either positive or negative outcomes. Also, the mechanism of potential DNA damage is not fully understood. Apart from direct intercalation or the physical and/or electrochemical interaction with NPs [22], the generation of reactive oxygen species (ROS) plays a key role in DNA damage. This indicates that particles do not necessarily reach the nucleus but could induce genotoxicity via oxidative stress [23].

19.3 ENTRY ROUTES OF NPs INTO THE BODY

It has been demonstrated that NPs may gain access to the body via the airways, the skin, or via ingestion [24,25] (Figure 19.2). The micron-sized particles are largely trapped and cleared by the upper airway mucociliary escalator system, whereas particles <2.5 μm can get down to the alveoli. The deposition of inhaled ultrafine particles (UFPs) (aerodynamic diameter <100 nm) mainly deposit in the alveolar region [26]. After absorption across the lung epithelium, nanomaterials can enter the blood and lymph to reach cells in the bone marrow, lymph nodes, spleen, and heart [26]. It has been proven in cardiovascular events such as coagulation and cardiac rhythm disturbances due to the association of inhaled ambient UFPs and the heart [27,28]. The ultrafine silver particles were taken up by alveolar macrophages and aggregated silver particles persisted there for up to 7 days.

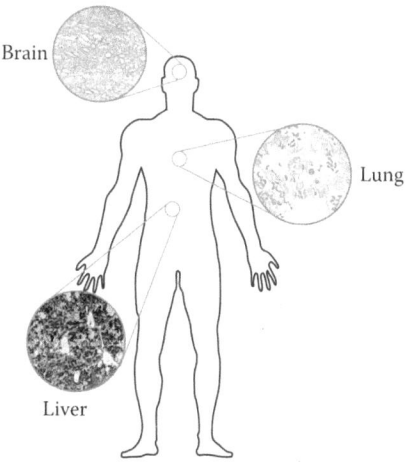

FIGURE 19.2 NP entry route into the body via the lung, particle accumulation in the liver, and the most vulnerable site: the brain. (Reprinted with permission from Elsaesser A, Vyvyan Howard C. Toxicology of nanoparticles. *Advanced Drug Delivery Reviews.* 2012;64(2):12–137.)

Aggregated silver NPs and some other nanomaterials have been shown to be cytotoxic to alveolar macrophage cells as well as epithelial lung cells [29].

Another potential exposure route in humans is via the skin [30]. The skin is a structured organ comprising three layers: the epidermis, the dermis, and the subcutaneous layer. The strongly keratinized stratum corneum acts as the primary protecting layer and may be the rate-limiting barrier to defend against the penetration of most micron-sized particles and harmful exogenetic toxicants. Skin exposure to nanomaterials can also occur during the intentional application of topical creams and other drug treatments [26]. Nanocrystalline magnesium oxide and titanium dioxide applied to dermatomed human skin (as dry powder, water suspension, and water/surfactant suspension) for 8 h did not show dermal absorption through human skin with intact functional stratum corneum. Whereas, titanium dioxide (TiO_2) NPs having a size range of 20–100 nm, when topically applied in porcine-, healthy human-, and human-grafted skin samples, get restricted to the topmost 3–5 corneocyte layers of the stratum corneum. However, TiO_2 particles could get through the human stratum corneum and reach the epidermis and even dermis. Flexing movement of normal skin was shown to facilitate the penetration of micrometer-sized fluorescent beads into the dermis [31]. The quantum dots (QDs) could penetrate the intact stratum corneum barrier and get localized within the epidermal and dermal layers [32]. In a clinical study, treatment of burns using nanosilver-coated dressings led to abnormal elevation of blood silver levels and argyria (blue or gray discoloration of the skin due to silver accumulation in the body over time which is a "cosmetic problem") [33]. The nanosilver-based dressings and surgical sutures have received approval for clinical application and good control of wound infection is achieved. However, their dermal toxicity is still a topic of scientific debate and concern. The nanocrystalline silver-coated dressing is more cytotoxic among the cultured keratinocyte extracts of several types of silver-containing dressings. Fullerene-based peptides are capable of penetrating into the intact skin [34]. Intradermally administered QDs could enter subcutaneous lymphatics [35] and regional lymph nodes [36]. Topically applied fine and ultrafine beryllium particles can be phagocytosed by macrophages and Langerhans cells, possibly leading to perturbations of the immune system [31]. Epidermal keratinocytes are capable of phagocytosing a variety of engineered NPs and setting off inflammatory responses [37]. Some other types of NPs, such as single-/multiwalled carbon nanotubes (SWCNTs, MWCNTs), QDs with surface coating and nanoscale titania, may produce toxic effects on epidermal keratinocytes and fibroblasts. They are capable of altering their gene/protein expression [38].

However, the major concern is that NPs could gain access to other organs, once having entered the body and reached the bloodstream [26]. Biodistribution studies of NPs revealed the low concentrations in liver, spleen, heart, and the brain [39,40]. Further concerns are the bioaccumulation of NPs in certain organs [41]. It is not yet clear to what extent the body is able to excrete NPs via urine [42] or whether residual NPs bioaccumulate in certain organs and may block the body's excretion systems. Certain types of NPs can pass through the gastrointestinal tract (GIT) and are rapidly eliminated in feces and urine. These indicate that the absorption occurs through the GIT barrier and entry into the systemic circulation [43]. However, some of the nanoparticulates can accumulate in the liver during first-pass metabolism [5]. After intravenous administration, NPs get distributed to the colon, lungs, bone marrow, liver, spleen, and the lymphatics [26,44,45]. Such distribution is followed by rapid clearance from the systemic circulation, predominantly by action of the liver and splenic macrophages [46]. Clearance and opsonization of NPs depends on size and surface characteristics [43]. Differential opsonization translates into variations in clearance rates and macrophage sequestration of NPs [46]. To increase the passive retention of nanomaterials in systemic circulation, the suppression of opsonization events is necessary at desired sites or anatomical compartments. For example, in the case of hydrophobic particles, a coating with poly(ethylene) glycol (PEG), would increase their hydrophilicity, hence increasing the systemic circulation time [47]. Whereas, PEGylated (polyethylene glycol coated) gold NPs (size range 10–30 nm) did not cross the perfused human placenta and are not detected in fetal circulation.

The inhaled NPs are distributed to the lungs, liver, heart, kidney, spleen, and brain [26,48,49]. The inhaled ultrafine silver NPs were distributed in the liver, lungs, and brain in rats. The NPs are cleared from the organs via phagocytosis in the alveolar region by macrophages [47].

The distribution of gold NPs into different organs depends on the size of them. The gold NPs having the size range <10 nm widely distributed in the blood, liver, spleen, kidney, testis, thymus, heart, lung, and brain whereas the larger particles (50, 100, and 250 nm) were detected only in the blood, liver, and spleen.

Owing to characteristic internalization and systemic distribution of inorganic and polymeric NPs, there is a growing interest in exploring their uses for imaging, systemic delivery of drugs, target-specific killing of cancerous cells, and so on. Understanding the relationship between the physicochemical properties (size, surface charge, hydrophilicity, etc.) of NPs and their ADME (absorption, distribution, metabolism, and elimination) characteristics is critical to achieve the desired biological effect. Muller et al. [88] have extensively reviewed the commonly studied nanomaterials, namely, iron oxide NPs, dendrimers, mesoporous silica particles, gold NPs, and CNTs with reference to their toxicity, biocompatibility, biodistribution, and biodegradation.

19.4 CONVENTIONAL RISK ASSESSMENT

The risk assessment is a very complex process. It involves the integration of information across a range of domains including source characterization, fate and transport, modeling, exposure assessment, and dose-response characteristics. It uses well-defined quantitative models to describe the relationships between the various elements of the paradigm shown in Figure 19.3. Here, we have briefly described how health risks have been traditionally identified and quantified based on information about exposure and dose-response relationships. Implicit in this process is the setting of "standards" or guidelines regarding "safe" or "acceptable" levels of exposure for a population. Figure 19.3 shows the relationship between the general environmental health framework (in the center) and the risk assessment framework.

Exposure is defined as the intensity of contact between the contaminant and the relevant biological sites of impact over a relevant time period. Exposure assessment includes assessing sources of pollutants and their strengths, measuring or modeling concentrations in environmental media, measuring or modeling human exposures through various pathways, and in some cases, even biological

Models for Risk Assessments of Nanoparticles

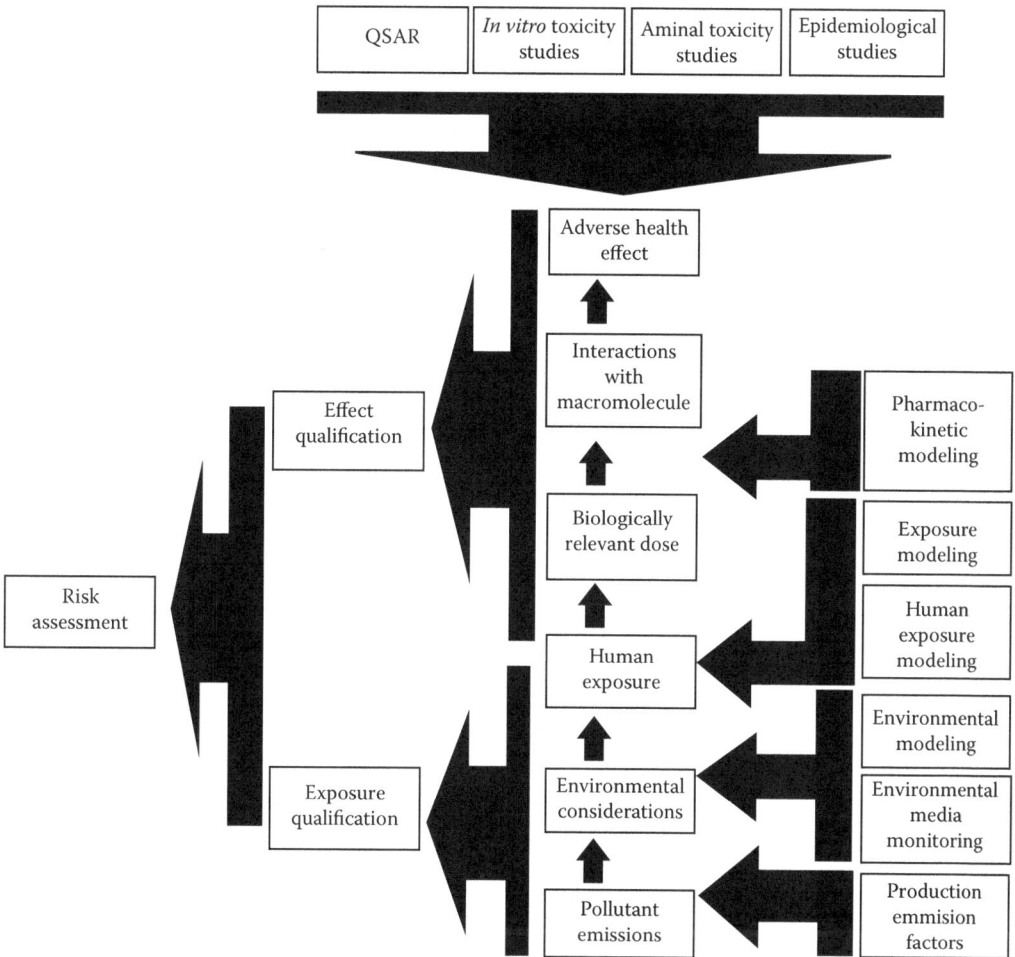

FIGURE 19.3 General environmental health framework for risk assessment. (From Gross MK et al. *Journal of Nanoparticle Research*. 2007;9:137–56. With permission.)

monitoring to measure tissue burden and thereby estimate the dose. The estimation of a biologically relevant dose from exposure information is, however, often very difficult and requires fairly detailed knowledge of the toxicokinetics of the pollutant in the human body. The lower portion of the diagram relates to the estimation of the health effects of the exposure and the biologically relevant dose. Information regarding the effects can come from *in vitro* and *in vivo* studies, quantitative structure–activity relationship (QSAR) modeling [50], and epidemiological studies. The quantification of exposures and effects allows the quantification of risk that, in turn, allows the proper allocation of resources to manage the risk. For example, risk assessment might allow the identification of populations or individuals at greater risk because their exposures are greater than some threshold identified in epidemiological studies.

19.5 STRUCTURE FOR RISK ASSESSMENT

Recent studies on nanoscale substances along with relevant research on UFPs from air pollution, metal fume, and mineral fibers provide an initial basis for evaluating the primary issues in a risk assessment framework of nanomaterials [5] (Figure 19.4).

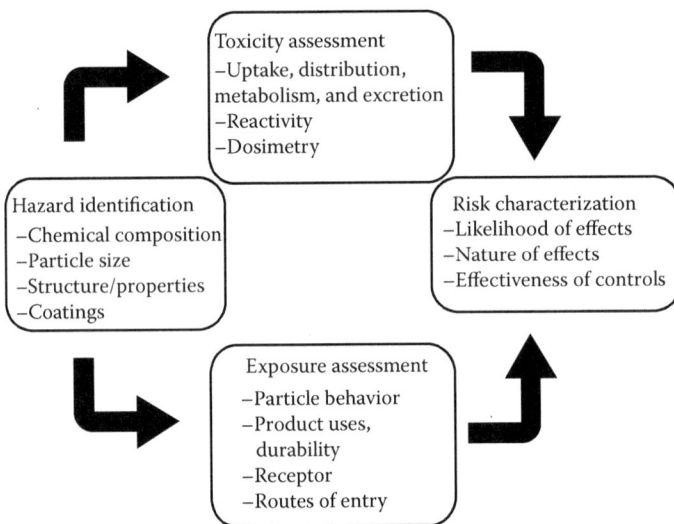

FIGURE 19.4 Risk assessment framework for nanomaterials. (From Tsuji JS et al. *Toxicology Science* 2006;89:42–50. With permission.)

Identification of hazards depends on the diverse characteristics of these particles. Hazards of novel structures will be less predictable than smaller-scale substances (e.g., metal oxides such as TiO_2 or ZnO). Encapsulation of the material either by surface coatings or within a matrix affects the reactivity and biological mobility of NPs [51], but the durability of such encapsulation needs to be considered.

Exposure assessment includes the entire life cycle of nanomaterials from synthesis to disposal. Exposure is likely the highest for workers, although product-specific evaluations should consider consumer uses, wear, disposal, and potential for environmental release and fate for products containing nanomaterials. Several nanomaterials (e.g., CNTs, metal oxides) tend to aggregate, which may reduce their ability to penetrate membranes, reach deep airways, or disperse. However, aggregation by fullerenes in water has been associated with increased solubility and antibacterial properties [52].

The inhaled nanocarbon particles in humans readily pass into systemic circulation, where they are a concern for cardiovascular toxicity. Iridium NPs administered to rats by the endotracheal tube were cleared by the airways to the GIT and eliminated, with <1% translocating to other organs [53]. For systemically absorbed particles, the liver may be a primary translocation site with spleen, bone marrow, heart, kidney, bladder, and brain as secondary sites [5].

The primary mechanism of action by inhalation or dermal routes appears to be free radical generation and oxidative stress associated with surface reactivity [5]. Oxidative stress associated with TiO_2 NPs, for example, results in early inflammatory responses such as an increase in polymorphonuclear cells, impaired macrophage phagocytosis, and/or fibroproliferative changes in rodents [54].

An inverse relationship has been observed between lung clearance rate and nano P25 TiO_2 toxicity in rodent species (i.e., order of increasing clearance and decreasing lung toxicity in a 90-day inhalation study: rats > mice > hamsters) [54]. Lung clearance of particles in humans is greater than in rats. Moreover, lung cancer in rats exposed to high levels of inert particles is thought to occur by a lung-overload mechanism that is not relevant for humans [55]. On the other hand, translocation to other organs may be a greater concern for humans. Such species differences are a concern in extrapolating results from animal models to humans [55].

Intratracheal inhalation studies in rats indicate that ultrafine or nano TiO_2 is less inflammatory than quartz, nickel or cobalt, or carbon black [56]. Studies in humans exposed to ZnO metal fumes (including nanoscale particles) have shown a relatively low order of toxicity associated with signs of pulmonary inflammation and metal fume fever that does not progress to the pulmonary disease [57]. In contrast, greater toxicity is expected for more reactive NPs such as fullerenes and CNTs.

Although most toxicological studies with nanomaterials have been *in vitro*, or short-term *in vivo* studies involving unnatural delivery (e.g., intratracheal instillation) in limited species and types of NPs, the National Toxicology Program is planning short- and long-term studies, including oral, dermal, and inhalation exposures for some NPs (http://ntp-server.niehs.nih.gov/files/nanoscale05.pdf). Nanomaterial research and risk assessments will ultimately need to address multiple potential health effects including cardiovascular, carcinogenicity, reproductive/developmental, immunological, and neurological.

Screening assessments of exposures to the more studied metal oxides could be conducted by developing toxicity benchmarks using the weight of evidence from studies of (1) nanoscale metal oxides in the toxicological and pharmacological literature, (2) fine-scale forms corrected for the proportionally greater surface area of nanoscale particles, (3) more toxic particles such as UFPs, and (4) the toxicology and epidemiology of metal fumes. Uncertainties in such assessments will have to be considered given data limitations; however, collectively, the available studies are beginning to reveal important features necessary for initial risk assessments of specific NPs.

19.6 ISSUES RELEVANT TO RISK ASSESSMENT

Critical steps in the risk assessment of nanomaterials remain, so far, the same as those used for the risk assessment of other types of chemicals, notably (1) hazard identification meaning identification of nanomaterial properties that may cause hazards to health; (2) hazard characterization, which requires defining of dose responses for critical target organ(s), cell(s), and mechanisms of toxicity [58]. Also, the potential of different engineered nanomaterials to react with constituents of cells at the port of entry and beyond, that is, lipids and proteins, should also be assessed [59]. To evaluate the translocation or distribution of these materials in the body, one should also assess their capability to cross internal barriers such as blood–brain barrier, blood–placental barrier, blood–testicular barrier, and many others [60].

The next and third step in risk assessment is the assessment of exposure. Nanomaterials are produced from many substances, in many forms and sizes, and with a variety of surface coatings. The health risk assessment of such diverse materials requires validated analytical methods both for their characterization in bulk samples, and for their detection and measurement in workplace air. This is because nanomaterial levels may be higher in occupational than in other environments, at least during certain operations; nanomaterials are handled in large quantities in workplaces and, hence, occupational settings carry the greatest potential for human exposure [61]. Risk assessment also requires an understanding of the transport processes between the source and human receptor and how they modify the characteristics of the nanomaterials [62]. Airborne nanomaterials ("aerosols") can be characterized by measuring several metrics, especially number concentration, surface area, and mass [61].

These four steps in risk assessment—hazard identification, hazard characterization, exposure assessment, and risk characterization are ultimately combined in the risk assessment process. Risk assessment integrates the results of the four steps of the risk assessment process and aims at assessing the likelihood of occurrence of a given hazard in a certain exposure situation. The ultimate goal of the current risk assessment paradigms is to be able to provide quantitative predictions of the given risks, enabling evidence-based risk management that is based on quantitative assessment of a given risk in a population.

19.7 STAGES OF RISK ASSESSMENT

Many of the most important health and safety concerns regarding nanomaterials and nanotechnologies are due to the lack of knowledge of levels of occupational and other types of exposure to nanomaterials during their production and use. Many nanomaterials are known to be much more reactive than their larger but chemically identical counterparts [63] and are having a potential to induce, for example, the formation of ROS, thereby rendering them *a priori* harmful to biological systems. The exposure of Degussa P25, a commercially available TiO_2 nanomaterial, at concentrations between 25 and 120 mg/L of cell culture medium to the cultured immortalized mouse microglia (BV_2), rat dopaminergic (DA), and primary cultures of embryonic rat striatum, found an immediate and prolonged release of ROS in BV_2 cells [64]. Naturally, this is an *in vitro* study without a relevance to risk assessment, but it provides insight into the possible mechanisms of the effects of nanosized TiO_2 on microglial cells. The agglomeration of particles though adds complexity to assessment of the true effect. However, the effective surface area of the agglomerates may not be that different from that of the primary particles forming the agglomerate. In an *in vivo* study, the exposure of commercial TiO_2 NPs, SiO_2 NPs, and nanosized TiO_2 to BALB/c mice, generated a gas-to-particle conversion process at 10 mg/m^3, and to SiO_2-coated TiO_2 NPs [65]. The inflammatory response induced by nanosized TiO_2 clearly exceeded that induced by the otherwise corresponding coarse particles [66]. Interestingly, the only particle that induced a dramatic inflammatory response in these animals was the SiO_2-coated TiO_2 NPs [65]. These findings indicate that in addition to the size of the particles, other features such as the surface chemistry potentially adds to the reactivity matters. Furthermore, recent observations on the translocation of engineered nanomaterials through the circulation due to their small size to any organ in the body have provided evidence that systemic effects due to exposure to the engineered nanomaterial are also possible [67]. The issue of size of particles and the possible harm to human health by nanoscale particles is further illustrated by a recent study that assessed risks of cancer in rats subsequent to exposure to different types of titanium dioxide (fine $TiO_2 < 2.5$ μm; ultrafine $TiO_2 < 0.1$ μm in diameter) [68]. They used the available rat dose-response data on these materials. The results of the modeling suggested that the maximum likelihood estimate (MLE) was much lower for the nanosized than for fine TiO_2 particles. At this stage, the reliability of any quantitative health risk estimates is limited due to the small amount of data on nanosized TiO_2. Nevertheless, it is quite obvious that nanosized particles seem to have more potential than chemically identical but larger particles to harm human health as the example of TiO_2 suggests.

19.7.1 CHARACTERIZATIONS OF BEHAVIOR OF ENGINEERED NANOMATERIALS

One major uncertainty in the safety and risk evaluation of the engineered nanomaterial arises from a lack of systematic knowledge about the physicochemical characteristics of the material arriving at the receptor (in this context the receiving organ or tissue), be it the nose, the skin, or the mouth of an exposed human. This is true for all forms of engineered nanomaterial, whether they exist in the form of macroscopic solid objects, as powders, emulsions or suspensions, or as aerosols (i.e., in the form of airborne particles). All these engineered nanomaterials are essentially constituted of NPs, or at least nanostructured building blocks such as agglomerates, with the potential of being released by some kind of mechanism into a transport chain from the "source" to the receptor (see the above definition of a receptor in this context) [69].

An issue requiring special and thorough consideration is the possible exposure to products containing the engineered nanomaterial during the entire life cycle of the products. Most products are not likely to cause exposure as long as the engineered nanomaterials are embedded in the polymers or other matrices in a given product [70]. The situation may change though, when a given material needs further processing, becomes waste, or is recycled. A detailed discussion of this issue is beyond the current presentation, but will require a thorough life-cycle analysis as the number and volume of products containing the engineered nanomaterial rapidly increases [71].

The tendency of airborne-engineered nanomaterial to the agglomerate that becomes attached to the ubiquitous background aerosol is of special importance because this may very rapidly change its specific size-related characteristics. Such transformations can be modeled for given scenarios provided enough information is available about the system [62]. The changes in airborne-engineered nanomaterial characteristics during transport have several consequences. For one, a change in airborne size will alter the deposition mechanisms for the engineered nanomaterial at the receptor and hence the effective dose. (It is not understood whether the agglomeration process affects the toxicological mechanisms as well.) Furthermore, growth and/or attachment of airborne-engineered nanomaterials to other particles challenges the established methods for their adequate characterization. Separation and identification of engineered nanomaterial against the submicron background aerosol originating from different sources is therefore a difficult or impossible task facing engineered nanomaterial monitoring and characterization in workplaces and other environments [72]. Currently, there are no rapid techniques available that would allow online distinction between background NPs and engineered nanomaterial. Instead, one has to collect the aerosol and do offline imaging or compositional analysis, for example, by transmission electron microscope (TEM). Therefore, TEM remains the gold standard for this type of analysis [72].

Novel materials introduce entirely new particle shapes, such as CNTs that, upon release into the air, can form larger fiber-like structures resembling, for example, crocidolite fibers with potentially serious health consequences [73]. The addition of fullerene structures to the backbone of SWCNTs produces nanobuds that may be associated with unexpected biological effects [74]. The effects of functionalization of engineered nanomaterials add to the complexity of their risk assessment and are beyond this discussion, and for the most part, are undone so far.

The thorough characterization of airborne-engineered nanomaterial is thus complicated by the dynamic behavior of the engineered nanomaterial as an aerosol as well as the structural complexity of the individual particles. The large set of parameters required for their complete characterization, the range of engineered nanomaterials already in use, and the multitude of biological responses create a special challenge for this undertaking. Currently, there is no robust set of devices that could be used for monitoring, measuring, and characterizing engineered nanoparticle (ENP) in workplace environments.

19.7.2 An Assessment of Exposure to Engineered Nanomaterial

Measurement and monitoring of engineered nanomaterial present in the air of workers' breathing zones means capturing all relevant information about the amount (number, surface area, or mass concentration) and size distribution, as well as shape, composition, and chemical reactivity of airborne-engineered nanomaterial in a given size class or a broad size range. Selection of the most relevant metric(s) for health-related sampling of the engineered nanomaterial is an important component in the development of the concepts, methods, and technology for engineered nanomaterial monitoring at workplaces [75]. For this purpose, understanding the relationship between engineered nanomaterial metrics and toxicological effects of engineered nanomaterial is necessary, but as yet it is unavailable. This is because a consensus on the correct metrics to be measured to assess the exposure to engineered nanomaterial has not yet been reached internationally.

The multitude of relevant engineered nanomaterial metrics, in combination with the different possible release mechanisms for engineered nanomaterial into workplace air as well as the poorly defined transport pathways between the source and receptor (which all affect the particle characteristics) make it imperative to establish and define the typical engineered nanomaterial exposure scenarios. For example, one should assess exposure to "fresh engineered nanomaterial" emitted from different production processes as well as for settings in which prevalent exposure occurs to "aged" and "attached" engineered nanomaterial [62]. These work processes may include, for example, handling and packing of partly agglomerated and aggregated engineered nanomaterial. It is important to obtain data from these types of exposure settings to be able to assess true exposure levels of engineered nanomaterial in workplaces [72].

Currently, aerosol technology disposes off a range of methods for monitoring of the engineered nanomaterial containing aerosols. These include online as well as off-line methods to capture a number of different metrics. The array of techniques and concepts for obtaining physical aerosol information related to concentration or size is relatively large, with a general trend away from mass toward number- or surface area-based techniques [76]. Some devices are stationary and capable of providing detailed information about size or composition, while others are "personal" at the expense of detailed information. The benefit of acquiring detailed size distribution data online (especially of number size distributions) is that such information can be easily and accurately converted into almost any other physical metric (except shape, which requires additional parameters). Keeping in mind the open-ended discussion about appropriate metrics; number size distribution data are thus the most valuable record that leaves the door open to *a posteriori* tests of new hypotheses concerning health risks of engineered nanomaterial.

In the past two decades, methods providing direct chemical, compositional, or biological information of airborne particles, usually on the basis of optical or mass spectroscopy, have received a major push in the field of atmospheric sciences and most recently, also in response to global security threats [77]. However, these devices are often quite expensive and complex, and the technology has not yet filtered down to affordable levels for health hazard evaluation.

Generally, one of the real challenges ahead for engineered nanomaterial monitoring and health risk assessment is (a) to redesign portable or personal and affordable "engineered nanomaterial-capable" instruments, (b) to expand the sensing technology available for engineered nanomaterial detection by adopting new options with realistic potential for real-time measurement and compact design, and (c) to extend the metrics into new areas such as CNT shape identification as well as surface chemical or catalytic properties. These latter concepts are especially interesting, since they have the potential of distinguishing engineered nanomaterial against background particles via special morphological features or function (not composition!).

The inability to separate engineered nanomaterial from the background NPs by straightforward concentration and size distribution measurements makes it impossible or at least highly problematic to set occupational exposure limits (OELs) for engineered nanomaterial. Harmful health effects, such as increased cardiac and pulmonary mortality, of ambient fine particles have been emphasized. Wide utilization of nanotechnologies globally is though a novel phenomenon, and is likely to have an impact on human health on a global scale in future. Besides, exposure to particle aerosols outdoors or indoors is in most cases exposure to a mixture of particles with a wide range of diameters and, hence, the importance of assessing the impact of the particle size range on effects on human health is a special challenge. There are thorough data, which suggest that exposure to UFPs, the ubiquitous background nanosized particles, may be especially harmful in inducing harmful health effects such as pulmonary inflammation, effects on circulation, and even increased cardiac mortality [78]. When assessing the health impact of engineered nanomaterial in occupational and other environments, the distinction between background ultrafine or NPs from engineered nanomaterial becomes especially important, because the ability of dissecting these effects is the prerequisite of setting of occupational exposure limits.

Each of the above avenues addresses an important demand: (i) making the current engineered nanomaterial monitoring technology more compact, more affordable, and more versatile will provide imminent short-term solutions required by toxicologists and the inhalation exposure community; (ii) new sensing technology will have a midterm effect by providing sophisticated measurement options for very small particles that can be adapted to the needs of aerosol-monitoring technology; and (iii) finally, the need of devices capable of capturing entirely new properties will provide new tools to characterize airborne-engineered nanomaterial. It will be important for these new devices to provide real-time and online data. However, the foregoing discussion also makes it clear than an ideal, all-purpose monitoring method will only become available (if it ever becomes available), once a clear link between health effects and engineered nanomaterial characteristics is well established for a majority of exposure scenarios, and this will only happen after sufficient data have been collected and analyzed.

19.7.3 Hazard Characterization

In this context, the discussion on health effects of engineered nanomaterial will be limited to the translocation of engineered nanomaterials in the body, and their effects on the lungs, on the circulation, genotoxic and possible carcinogenic effects, as well as some brief remarks on the effects of engineered nanomaterial on the brains. Other organs and their significance may only be briefly mentioned.

19.7.3.1 Translocation of Engineered Nanomaterial into the Body

The intravenously injected radioactive iridium191 NPs are rapidly excreted in the urine, but the fraction of dose is widely distributed to a number of organs in the body. The inhaled manganese oxide NPs enter into the olfactory bundle under the forebrain via the axons of the olfactory nerve in the nose, that is, in the olfactory epithelium, and they can reach other parts of the brain also through systemic inhalation. The inhalation of nanosized titanium dioxide NPs reached systemic circulation in rats. The inhaled engineered nanomaterial can reach systemic circulation in the body and via this route, can be distributed to a number of different target organs including the brain, liver, kidney, immunological system, and vessel walls. These findings emphasize the importance of exploring the ability of different engineered nanomaterials to reach the body via different routes of which the inhalation route is the most likely. Other routes such as the GIT or the skin will not be dealt with in this context. The former is less relevant in occupational safety and health or even in the consumer context, and penetration of the skin by engineered nanomaterial is not likely. However, titanium dioxide is increasingly being used, for example, in sunblock creams but its ability to penetrate beyond the stratum corneum is controversial [79]. In the future, when nanotechnology applications will become much more widespread and will be increasingly used in, for example, food items such as food additives or materials for food packaging, the significance of oral exposure and gastrointestinal absorption of engineered nanomaterial may considerably increase.

19.7.3.2 Pulmonary Inflammation Induced by Engineered Nanomaterial

Nanomaterial-induced pulmonary inflammation includes intratracheal installation or pharyngeal aspiration of engineered nanomaterial suspension. These models provide results of nanomaterial in the lungs that are similar or even identical to those induced through inhalational exposure, thereby providing evidence that the relevance of these models is quite good. It has been shown that CNTs, when introduced into the lungs, induce a strong pulmonary response in this organ at moderate doses of the material whether intratracheal installation, pharyngeal aspiration, or inhalational exposure have been used. It seems that these exposure models lead to relatively even distribution of the CNTs applied into the lung tissue. However, there seem to be slight differences when the effects of inhalational exposure with the other exposure models are compared. Intratracheal installation of SWCNTs suspension to rats at doses of 1 or 5 mg/kg, induced multifocal granulomas that were inconsistent with the lack of toxicity in bronchoalveolar lavage fluid (BAL), lack of lung toxicity based on proliferation parameters, and lack of dose effect relationship. Whereas, exposure of SWCNTs to mice induced pulmonary inflammation and granulomas in a dose-dependent fashion subsequent to pharyngeal aspiration of SWCNTs at lower or similar doses. It was also observed that oxidative stress played an important role in SWCNT-induced pulmonary toxicity by using reduced nicotinamide adenine dinucleotide phosphate (NADPH) oxidase deficient in C57BL/6 mice. The NADPH-oxidase null mice responded to SWCNTs with a marked exposure of polymorphonuclear leukocytes and elevated levels of apoptotic cells in the lungs, an increased production of proinflammatory cytokines, and lower levels of collagen deposition as compared to C57BL/6 control mice, providing new insight on the role of oxidative stress in SWCNT-induced pulmonary toxicity. The exposure of SWCNT in C57BL/6 mice by pharyngeal aspiration at doses of 0, 10, 20, and 40 μg/animal evoked a dose-dependent increase in the severity of acute pulmonary inflammation and progressive fibrosis and granulomas. Whereas, single long-term inhalational studies in Wistar rats exposed to MWCNTs at exposure levels of 0, 0.1, 0.5, or 2.5 mg/m^3 for 3 months, induced granulomatous-type inflammation.

It has been found that exposure of mice to titanium dioxide-engineered nanomaterial may cause pulmonary inflammation with elevated numbers of inflammatory cells together with increased levels of proinflammatory cytokine and chemokine proteins, especially when the surface of the particles has been modified by silica to increase the hydrophilicity of the titanium dioxide-engineered nanomaterial [65].

19.7.3.3 Genotoxicity of Engineered Nanomaterial

The number of studies exploring the genotoxic effect of engineered nanomaterial is small in view of the large variety of different engineered nanomaterials already in the market. The methods are being used for genotoxicity testing of engineered nanomaterials still inadequate for general conclusions. Still, it is unclear how well standard genotoxicity tests designed for soluble chemicals can be used to assess the genotoxicity of engineered nanomaterials. During the investigation, the possible mechanism of engineered nanomaterial genotoxicity may be linked with the inflammatory process; the question is whether *in vivo* tests are preferred over *in vitro* tests? It has been found that asbestos fibers, such as the *Salmonella* mutagenicity test does not appear to be responsive to insoluble engineered nanomaterial, probably because of the bacterial cell wall [80]. On the other hand, many engineered nanomaterials appear to be positive in tests of DNA damage and micronuclei [81].

Exposure of mice to SWCNTs by intratracheal installation induced aortic mtDNA damage [82] and DNA damage in bronchoalveolar lavage cells [83]. SWCNTs inhalation is more effective than pharyngeal aspiration in causing K-ras gene mutations in mouse lungs at moderate doses [84], and oral treatment of rats with SWCNT results in oxidative DNA adducts in the lungs and liver [85]. SWCNTs also increase DNA damage in Chinese hamster V79 lung fibroblasts and mouse embryo fibroblasts and oxidative DNA damage in FE1-Muta™ mouse lung epithelial cells, but have no significant effect on micronuclei in V79 cells or gene mutations in FE-1 cells [83,86]. As the doses in the *in vivo* studies are relatively high, therefore, the assessment of their relevance for risk is problematic. The results of the *in vitro* studies are in many cases inconsistent, and carry the most likely relevance to the risk assessment of these materials.

Xenopus laevis larvae did not show induction of micronuclei in blood erythrocytes when grown in the presence of double-walled CNTs [87]. A single intratracheal instillation of an MWCNT increased the frequency of micronucleated type II pneumocytes in rat lungs *in vivo* in association with a marked pulmonary inflammation [88]. *In vitro*, MWCNTs induced micronuclei in the rat lung epithelial RLE cells and human MCF-7 epithelial cells [88,89], DNA damage in RLE cells [89], human lung epithelial A549 cells [90], and mesothelial cells [91], and a slight increase in gene mutations in cultured mouse embryonic stem cells [92]. A mixture of SWCNTs and MWCNTs induce a dose-dependent elevation of DNA damage and an increase in micronuclei *in vitro* in human bronchial epithelial BEAS 2B cells [93].

In several research articles, the genotoxicity of nanosized TiO_2 has been described. However, due to the use of different types of TiO_2, various cell systems and variable assay conditions complicate the comparison of the existing studies [94]. There is some indication that especially anatase phase TiO_2 has genotoxic potential *in vitro* [94]. For example, the exposure of nanosized rutile (10×40 nm) or nanosized anatase (<25 nm) into BEAS 2B cells for the duration of 0, 24, 48, or 72 h, provides the evidence that in the comet assay, in both anatase and rutile dose-dependent effects were observed at 48 and 72 h. Micronuclei were only induced by nanosized rutile.

It indicates that more genotoxicity studies on engineered nanomaterial are immediately required, and the need for novel methods to assess engineered nanomaterial–genome interactions. The genotoxicity potential of different engineered nanomaterials has been identified but the possible mechanism is still unknown.

19.7.3.4 Carcinogenic Effects of Engineered Nanomaterial

The carcinogenic effects of asbestos may be due to the local generation of reactive oxygen and nitrogen species in association with emerging inflammation [95]. Studies with rats and mice have shown that the MWCNT induces oxidative stress, inflammation, granulomas, and fibrosis in the lungs [96]. The engineered nanomaterial with fibrogenic properties could induce cancers. Indeed, single intraperitoneal injection of high doses of crocidolite asbestos (1 × 1010 fibers, i.e., 3 mg in 1 mL of water) or MWCNT (1 × 109 fibers, i.e., 3 mg in 1 mL of water) to a sensitive p53 ± mouse strain for 120 days, the ability of MWCNTs to induce mesotheliomas exceeded the ability of crocidolite asbestos. In another study when Fischer 344 rats were given a single intrascrotal dose of 1 mg/kg MWCNT to induce exposure of the mesothelial lining of the intraperitoneal cavity, MWCNTs had a much higher potential to induce mesotheliomas in these animals than a comparable dose of crocidolite asbestos. In all studies in which MWCNTs have been given intraperitoneally [73], MWCNTs formed agglomerates or bundles, and single MWCNTs were not present. This type of agglomeration is also typical of MWCNTs and SWCNTs in occupational environments when the material exists in aerosol [97]. These observations suggest that agglomerates consisting of MWCNTs (or SWCNTs) may have the ability to induce mesotheliomas and asbestos-like morphological changes when in contact with the mesothelial lining of the abdominal cavity. This exposure mode has been criticized because the exposure is via the abdominal cavity rather than lungs [98]. It is also obvious that these data cannot be used to assess risks of exposure to CNT, but they may provide useful information for the identification of the possible hazard, that is, carcinogenic potential, of CNT. The issues to be considered when assessing the potential of MWCNTs to induce carcinogenicity include the route of exposure, the high dose given as a single bolus, fiber length, the possibility of removing the material from the peritoneal cavity, as well as the impact of the high concentrations at a given time in the abdominal cavity having a potential effect on agglomeration and aggregation of the material.

When MWCNTs were introduced into the abdominal cavity of mice, they induced asbestosis-like pathogenic changes in the mesothelial lining of the abdominal cavity. The typical changes were an increased number of inflammatory cells and protein exudates in the cavity as well as lesions in the mesothelium. These data provide evidence that MWCNTs, when in contact with mesothelial lining and mesothelial cells *in vivo*, have the capability to induce asbestos-like changes, even mesotheliomas, providing useful data for hazard assessment of these materials. High doses of MWCNTs do not induce mesotheliomas in Wistar rats after intraperitoneal injection.

19.7.3.5 Effects of Engineered Nanomaterial on Circulation

Urban nanosized aerosols have been shown to be associated with increased cardiac mortality in humans. The mechanisms of this effect have been delineated in several studies [67] and have been shown to be associated with oxidative stress in the blood vessel endothelial lining [99]. Engineered nanomaterials have also been assumed to have adverse effects on circulation, but so far, data on such effects have been scanty. The nanosized carbon black particles exert thrombogenic, that is, fibrinogen deposition and platelet adhesion, but not inflammatory effects in the microcirculation of healthy mice. The intratracheal exposure to particulate matter (diameter 0.1–2.5 μm) of rats impairs systemic microvascular endothelium-dependent dilatation and associated inflammation in the vessel wall. The inhalational exposure of rats to low concentrations of nanosized titanium dioxide-engineered nanomaterial augments particle-dependent microvascular dysfunction. Furthermore, the SWCNTs and MWCNTs induce platelet aggregation and vascular thrombosis. These findings provide evidence that at least some types of engineered nanomaterial can reach the systemic circulation through inhalation, and once in the bloodstream, they have the capacity to have a disturbing effect on microcirculation. This is an important observation because once in the bloodstream, the engineered nanomaterial can potentially induce its effects in any organ in the body as shown, for example, by some of the translocation studies [100].

19.7.3.6 Other Remarks on the Effects of Engineered Nanomaterial

Observations that engineered nanomaterials can reach the brain via the bloodstream once they have found an access to the systemic circulation have evoked much concern [100]. In an *in vitro* setting, it has been observed that the increased production of ROS in immortalized brain microglial cells and, in fact, convincing evidence on the effects of engineered nanomaterial in the neuronal cells is very limited. In another *in vivo* study with intranasal exposure to manganese oxide NPs [100] by using much lower doses provided evidence on translocation of these particles through olfactory axonal transport into the olfactory bundle and other parts of the brain with signs of neuronal effects. Brains are naturally of special concern due to the sensitivity of this target organ.

19.8 AN ASSESSMENT OF PARTICLE PARAMETERS RELEVANT FOR BIOLOGICAL ACTION

The unusual physicochemical properties of nanomaterials are attributable to their small size (surface area and size distribution), chemical composition (purity, crystallinity, electronic properties, etc.), surface structure (surface reactivity, surface groups, inorganic or organic coatings, etc.), solubility, shape, and aggregation. A question is being raised: "Do nanomaterials properties necessitate a new toxicological science?" The main characteristic of nanomaterials is their size in the transitional zone between individual atoms or molecules. This can modify the physicochemical properties of the material and increased uptake and interaction with biological tissues [101]. This combination of effects can generate adverse biological responses in living cells. The increase in surface area determines the potential number of reactive groups on the particle surface. The shape of the NPs has been shown to have a pronounced effect on the biological activity. It is reported that silver NPs undergo shape-dependent interactions with *Escherichia coli* [102]. In the case of anatase TiO_2 nanomaterial, it was shown that alteration to a fiber structure of >15 μm created a highly toxic particle that initiated an inflammatory response by alveolar macrophages and that length may be an important determinant of nanomaterial biocompatibility [103]. The water-soluble rosette nanotube structures display low pulmonary toxicity due to their biologically inspired design and self-assembled architecture. For metal oxide and carbon nanomaterials, the physicochemical characterization of nanomaterials and their interaction with biological media are essential for reliable studies. It was observed that for gold NPs (size 1.5 nm), the surface charge was a major determinant of their action on cellular processes; the charged NPs induce cell death through apoptosis and neutral NPs leading to necrosis in HaCaT cells [104]. Considering the physicochemical properties of various nanomaterials and their interactions with the biological environment, the challenges presented by simple nanoscale materials such as TiO_2, ZnO, Ag, CNTs, and CeO_2 are now beginning to be appreciated. But these simple materials are merely the vanguard of a new era of complex materials, where novel and dynamic functionality is engineered into multifaceted substances. If we are to meet the challenge of ensuring the safe use of this new generation of substances, it is time to move beyond "nano" toxicology and toward a new toxicology of sophisticated materials. Therefore, it is evident that physicochemical characteristics of the materials are very important with respect to the observed biological effects.

Physicochemical properties of engineered NPs are one of the most important factors that regulate the behavior of engineered NPs in the environment. Engineered NPs are synthesized for a particular application; therefore, the physicochemical properties of each NP vary considerably. However, universally agreed and essentially required properties for engineered NPs are chemical composition, mass, particle number and concentration, surface area concentration, size distribution, specific surface area, surface charge/zeta potential, stability, solubility, and nature of engineered NPs shell.

Variation in composition, size, or surface composition of engineered NPs, considerably changes their physicochemical properties. Properties such as solubility, transparency, color, conductivity, melting points, and catalytic behavior mainly depend on the particle size. Similarly, surface composition of the engineered NPs affects the dispersibility, optical properties, conductivity, and catalytic behavior of the particle. Metallic engineered NPs are usually coated with inorganic or organic compounds or surfactants to maintain their stability as colloidal solutions. Thus, surface properties of engineered NPs strongly depend on composition of these coatings also.

Furthermore, surface properties of engineered NPs decide their fate in the environment, for example, formation of colloidal solution or aggregation. In colloidal solution, engineered NPs remain dispersed and maintain their reactivity and catalytic behavior and thereby easily interact with various components of the environment. Whereas, in aggregation where engineered NPs tend to agglomerate with neighbors due to high surface energy, engineered NPs lose their reactivity as well as catalytic nature by forming large-sized particles. However, aggregation is a kinetic process; size may easily change with the passage of time or due to natural geothermal weathering processes. But reverting back to nanosized particles is a time-taking process and is also influenced by a variety of environmental factors, for example, pH, ionic strength, salinity, and so on. Engineered NPs also show the tendency to interact with the natural organic matter (NOM) or artificial organic compounds, which further direct their stability or aggregation. Aggregation of engineered NPs is enhanced in the presence of high-molecular-weight NOM compounds, whereas, low-molecular-weight NOM compounds increase the mobility of engineered NPs in colloidal solution.

Analytical procedures for assessing dissolution and structural transformation of metallic NPs include chemical assays and the use of techniques for structural characterization. Atomic arrangement can be studied in terms of the long-range (crystal parameters) and short-range order of the NPs' atomic structure. Of particular interest is the arrangement of atoms at the surface of metallic NPs. The various characterization techniques that are available for these purposes may be nonelement specific (x-ray diffraction, total scattering, Raman spectroscopy, etc.) or element specific (x-ray absorption spectroscopy, Mössbauer spectroscopy, nuclear magnetic resonance, etc.).

Particular attention has also been given to the interaction of the surface of metallic NPs with nutritive solutions such as cell growth media. Owing to their surface reactivity and affinity for ions in solution, metallic NPs have a high capacity to adsorb molecules (amino acids, proteins, sugar, or salts) within biological media. These strong interactions and the neutral pH (close to the point of zero charge of metallic NPs) can destabilize the colloidal suspension or passivate their surface. The agglomeration state must be determined using scattering techniques based on light or x-ray. During the toxicological study, the nature of culture broth plays an important role. A strong cytotoxicity is observed for CeO_2 NPs toward *E. coli* in 0.1 M KNO_3, whereas no significant effects occur in Luria–Bertani medium. Moreover, in particular applications, metallic NPs are functionalized on the surface with (in) organic compounds affecting their behavior in biological media. For instance, functionalization of magnetite NPs with dextran to enhance their blood circulation time induces an inhibition of BrdU incorporation with disruption of the F-actin and viculin filament of human cells. However, when magnetite NPs are functionalized with albumin to be recognized by specific cellular receptors, no significant biological effects are reported. The mechanisms of adsorption of the coating at the surface are crucial to understand NP modifications and ultimately reactivity. For instance, in neutral solutions, albumin is hydrolyzed and negatively charged that allows for chemisorptions at the surface of the positively charged iron oxide, whereas dextran is electrically neutral in similar conditions and the binding at the surface of iron oxide is more labile (physisorption). Consequently, once adsorbed, the albumin can persist at the surface of iron oxide and prevent a direct contact with the toxic magnetite NPs. On the contrary, dextran can be desorbed yielding to a direct contact between magnetite NPs and biological compounds.

Since NPs are typically engineered or postprocessed for specific applications, their physicochemical properties and reactivity can vary considerably. The two key physicochemical properties of ZnO NPs are relevant to or predictive of their ecotoxicity. The first is solubility and the second is photoreactivity of NPs. The solubility is determined by both intrinsic properties of the NPs (such as particle size, chemical composition) and environmental parameters of the exposure media (such as pH, temperature, organic matter, etc.). The second key property is photoreactivity of the NPs. This photoreactivity can cause photoinduced toxicity that is enhanced hundreds of times under environmentally relevant ultraviolet (UV) radiation (such as sunlight) as compared to laboratory fluorescent lighting. This phototoxicity has been demonstrated for both ZnO and TiO_2 NPs. It is suggested that photocatalytic activity of TiO_2 NPs, measured with a simple chemical ROS assay, may be a predictor for phototoxicity of TiO_2 NPs. Similar investigation on ZnO NPs is warranted to fully understand their potential hazard and risk to the environmental biota.

The most important tools in the study and characterization of inorganic nanomaterials are electron microscopy and related methods. Structural and morphological characteristics are not often accessible through x-ray crystallography methods, due to low scattering intensity of NPs, which leads to pronounced peak broadening as the NP size decreases. TEM helps to directly image the lattice structure of NPs in the order of a few nanometers as well as to obtain diffraction data, amplitudes, and phases of NP structures. TEM can also be used to determine the behavior and self-assembly of NPs under external influences such as magnetic fields. Further, elemental analysis of NPs can be made using energy-dispersive x-ray spectroscopy (EDS), and modern TEMs that are equipped with tools to perform elemental mapping and analysis using incident probe sizes in the order of a few nanometers in diameter.

From a toxicological point of view, TEM data can offer useful information on the surface properties of NPs. Since many NPs are being produced as mixed compositions, such as "core–shell" particles (e.g., silica-coated iron oxide NPs), the surface of the nanostructures and their properties become of considerable importance. Different structures may grow as NPs exposing different facets to the exterior depending on the mode of synthesis. For obtaining information on which facets are more prominent in a specific morphology or synthesis may very often only be attainable through TEM.

NPs are frequently being considered for diverse and important applications as a result of the high surface area resulting from the reduction in particle size in comparison to bulk materials of the same composition. The most typical method to determine the specific surface area of NPs is through the measurement of a nitrogen adsorption–desorption isotherm, which also gives information on the average pore size and pore volume.

When NPs are functionalized or tethered with organic groups, it becomes necessary to quantify the amount of functional groups that reside at the surface of their particle, their binding strength (e.g., covalent or ionic) to the particle, and their availability to perform their anticipated function. Spectroscopic techniques such as infrared spectroscopy, UV–visible spectroscopy, and Raman spectroscopy are invaluable both for the identification and to locate the presence and position of functional groups on NPs.

Particle size analysis may be conducted in a variety of ways. It is important to differentiate between the techniques. Electron microscopy-based methods such as scanning electron microscopy and TEM do not reflect the average particle sizes values that may be measured in solutions or biological media containing the dispersed particles. Therefore, one must distinguish between the hydrodynamic particle size measurements and those values obtained through electron microscopy observation.

Dynamic light scattering (DLS) offers a routine approach to measure average particle sizes in different media. This technique utilizes the time variation of scattered light from suspended particles to obtain their hydrodynamic size distribution. When the putative cytotoxicity of NPs is determined, it is prudent to consider whether cells do in fact encounter individual NPs, as opposed to aggregates or agglomorates of several NPs; this becomes particularly relevant when studying cellular recognition and internalization of NPs, as the actual size of the particle(s) will determine the route of cellular uptake (endocytosis, phagocytosis, etc.).

Data from some pulmonary toxicity studies in rats demonstrate that exposures to ultrafine/NPs may produce enhanced toxicity when compared to fine-sized (bulk) particle types of similar chemical composition [105]. Particle surface area and particle number determinations have been postulated to play significant roles in influencing the development of NP-related lung toxicity.

The assumptions made from these studies were that the only differences (i.e., variables) between the ultrafine and fine-sized particle types were the particle sizes. However, a closer analysis indicates that a number of other physicochemical characteristics including crystal structure, aggregation potential, and surface coatings were different in the various particle types that were being compared. Moreover, findings of other recent studies with nanoquartz and ultrafine titanium dioxide particle types demonstrate that the toxicity of some NP types may be related, in large part, to the surface reactivity of the particles, in influencing the development of inflammatory and cytotoxic responses in the lung [106].

Surfaces and interfaces of particles are particularly important components of nanoscale materials. As the particle size is decreased, the proportion of atoms found at the surface is magnified relative to the proportion inside its volume. This results in nanoscale particle types that are likely to become more reactive, thus generating more effective catalysts in a variety of applications. However, when considering the potential health implications, reactive groups on the surface of particles are also likely to influence the biological (potentially toxicological) effects when compared to nonreactive surfaces or surface coatings. As a consequence, modifications in surface chemistry forming the "shell" on a (core) NP type may be important and relevant for health effects following exposures. Moreover, surface coatings can be utilized to alter surface properties of NPs to prevent aggregation or agglomeration with different particle types. It is interesting to consider the surface coatings, as it may facilitate NP translocation from the respiratory tract to the systemic circulation and thereby significantly enhance NPs distribution and exposures to sites throughout the body [41]. It should be noted that two different NP types containing titanium dioxide as their "core" may not have the same or even similar hazard potentials. There can be differences in crystal structures (anatase vs. rutile), surface reactivity, aggregation status, particle size distribution, surface area, as well as surface coatings—including passivation and neutralization. These differences in physicochemical particle characteristics despite a similar "core," may result in comparative differences in the potencies of pulmonary inflammatory and cytotoxic endpoints [107].

It is necessary to evaluate physicochemical characteristics of NP prior to the initiation of toxicological experimentation. The point cannot be overemphasized that in the absence of an adequate description of the physicochemical characteristics of the NP type being studied (as well as the experimental conditions being employed), the results of toxicity experiments with nanoscale materials will have limited value or significance.

19.9 *IN VITRO* CYTOTOXICITY TEST

Currently, *in vitro* studies (using established cell lines and primary cells derived from target tissues) are widely adopted for risk characterization of NPs due to the following reasons: (1) *in vitro* studies give important information especially in terms of toxic mechanisms, (2) *in vitro* studies enable one to study the effects of NPs on individual genes, proteins, and other molecules, and (3) *in vitro* studies facilitate large-scale testing or NPs that are vastly produced to the increasing use of nanomaterials in consumer and industrial products.

The first step toward understanding how an agent will react in the body often involves cell culture studies. Compared to animal studies, *in vitro* studies are less ethically ambiguous, are easier to control, and are less expensive. In the case of cytotoxicity, it is important to recognize that in addition to the concentration of the potentially toxic agent being tested, cells in culture are sensitive to changes in their environment such as fluctuations in temperature, pH, nutrient, and waste concentrations. Therefore, controlling the experimental conditions is crucial to ensure that the measured cell death corresponds to the toxicity of the added NPs versus the unstable culturing conditions. In addition, as nanomaterials can adsorb dyes and can be redox active, it is important that the choice of the cytotoxicity assay is appropriate.

19.9.1 Cell Viability Testing

Cell viability is the most commonly investigated parameter in cytotoxicity testing. It is important to perform cell viability studies for each nanomaterial type because of their unique biological response. It was reported that *in vitro* toxicity of analyzed nanomaterials was not attributed to a defined physicochemical property and the accurate identification of nanomaterial cytotoxicity would require a matrix based on a set of sensitive cell lines and *in vitro* assays measuring different cytotoxicity endpoints. There is not a single method that is satisfactory for obtaining all the information on the toxicity. Since different NPs elicit different biological responses, to study mechanisms underlying toxicity, a combination of assays is often required.

19.9.1.1 Detection of Mitochondrial Activity

The colorimetric 3-(4,5-dimethylthiazol-2-yl)-2,5-diphenyltetrazolium bromide (MTT) assay is a widely used cell viability assay based on the reduction of the yellow tetrazolium salt MTT to a purple water-insoluble formazan by succinic dehydrogenase in cells bearing intact mitochondria [108]. It has been applied in numerous cytotoxicity studies and employed to validate other methods and to determine NP toxicity [108]. The MTT assays technique has been utilized for assessment of different NPs such as NPs composed of titanium dioxide, iron oxide, zinc oxide, chitosan, silica, for fullerenes, and for naked or chitosan-coated QD [109]. The absorption spectrum of reduced MTT is pH dependent, and metal ions such as Zn^{2+} interfere with the MTT reduction reaction [110].

19.9.1.1.1 Drawbacks of MTT Assay

1. The SWCNTs may interact with the substrate, thereby depleting free MTT and causing false-negative results [111].
2. Owing to optical properties of NPs, and their presence in the reaction mixture, in or on cell culture, cells may directly influence the readout by increasing the light absorption [112].
3. It has been found that the cytotoxic effects of carbonaceous nanomaterials on human epidermal keratinocytes using classical dye-based assays such as the MTT assay produce invalid results due to nanomaterial/dye interactions and/or nanomaterial adsorption of the dye/dye products [113].
4. MTT assay failed to report toxicity of certain porous silica microparticles due to spontaneous redox reactions where the MTT is reduced and the particle surfaces are oxidized simultaneously. However, for other completely oxidized particles, the assay yielded the expected results [114].

19.9.1.2 Detection of LDH Release upon Necrosis

The lactate dehydrogenase (LDH) release from damaged cells was measured using a colorimetric assay. LDH is a stable cytosolic enzyme that is released upon cell lysis. This method is based on the oxidation of the yellow tetrazolium salt of 2-(p-iodophenyl)-3-(p-nitrophenyl)-5-phenyltetrazolium chloride (INT) to a red formazan to evaluate the amounts of LDH released from the cytosol upon cellular necrosis [115]. The intensity of the color produced is proportional to the number of lysed cells. The LDH leakage (%) relative to the control without NPs is calculated as [test] − [control] × 100, where [test] is the absorbance of the test sample and [control] is the absorbance of the control sample [116].

LDH assays using INT as the substrate have been applied to assess the cytotoxic potential of various NPs. LDH is significantly deactivated under low pH conditions, whereas a high basic pH destabilizes the substrate [117]. Furthermore, metal ions (e.g., copper) have been shown to interfere with the LDH assay [118]. SWCNTs do not interact with the substrate INT [111].

19.9.1.3 Fluorescein Diacetate Test

Cell viability is also determined by the fluorescein diacetate (FDA) test [119]. In this method, the cells are treated with 1:1 solution of FDA and ethidium bromide and observed under fluorescence microscope with an excitation filter of 488 nm (blue light). Living cells are stained in green while dead cells exhibit their nucleus stained in orange. The survival percentage is obtained dividing the number of living cells by the total number of cells.

Models for Risk Assessments of Nanoparticles

19.9.1.4 WST-1 Cytotoxicity Assay

Measurement of cellular metabolic activity by 4-[3-(4-Iodophenyl)-2-(4-nitrophenyl)-3H-5-tetrazolio]-1,3-benzene disulfonate (WST-1) assay is based on the cleavage of the tetrazolium salt WST-1 to soluble formazan by succinate–tetrazolium reductase [120]. Succinate–tetrazolium reductase is a mitochondrial enzyme that is active only in viable cells. The amount of formazan is proportional to the mitochondrial activity. There are various ways in which the NPs may interfere with the WST-1 assay. Therefore, two different WST-1 interference tests are performed. The first interference test is designed to whether NPs (1) scatter or absorb light or (2) interference with the WST-1 reagent.

19.9.1.5 Clonogenic Assay

The clonogenic assay determines the ability of cells to form colonies after toxic treatment. Contrary to the WST-1 assay, it has the advantage of not using a colorimetric indicator dye. In fact, nanomaterials are known to frequently interact with the colorimetric indicator used in cytotoxicity assays [121]. Clonogenic assay has been applied to assess the cytotoxicity potential of various NPs, for instance, SWCNT [122], silica NPs [121], and so on.

19.9.1.6 Hematoxylin and Eosin Assay

In this method, the cells are treated with different concentrations of NPs and fixed with 4% polyoxymethylene. Then the cells are stained with hematoxylin and eosin and observed under a microscope. In general, the eosin imparts a pink-to-red color of the cytoplasm, and hematoxylin stains nucleus blue. This method has been applied for the assessment of cytotoxicity potential of silica NPs [123].

19.9.1.7 Neutral Red Uptake Assay

Since 1894, neutral red has been used as a viability stain and has since been implemented in numerous cytotoxicity, cell proliferation, and adhesion assays [124]. Neutral red (3-amino-7-dimethylamino-2-methylphenazine hydrochloride) is weakly cationic, and is thought to be taken up into the cytosol by nonionic diffusion through the cell membrane to then accumulate in the lysosomes of viable cells, while it is excluded from dead cells [125]. The uptake of neutral red may be detected via fluorescence or absorption measurement. So far, the neutral red uptake (NRU) in NIH_3T_3 mouse fibroblasts is the only validated *in vitro* method for toxicity testing [126]. A number of different NPs such as TiO_2 [127], SWCNTs and MWCNTs [128], and chitosan NPs [129] have been tested by NRU assays.

The color and intensity of light emission of neutral red is pH dependent [130]. It is known that neutral red has a high affinity to lipophilic structures (such as suberin or phenolic substances), the protonated dye interacts with negative charges, and covalently binds to cellular structures [130]. These properties were utilized to study the adsorption of neutral red to mercaptoethane sulfonate-protected gold NPs. The optical properties of neutral red were significantly influenced, and the NPs exerted static and dynamic energy transfer quenching of neutral red fluorescence [131]. Furthermore, SWCNTs have been shown to interact with neutral red and deplete the dye from the cell supernatant leading to false-positive results [132]. Hence, properties that are useful for technical applications may be deleterious in cytotoxicity testing.

19.9.1.8 Trypan Blue Exclusion Assay

Trypan blue is one of the many dye-recommended techniques of exclusion staining, counting and evaluation of cellular population, and acute cellular toxicity [133]. This method is based on the principle that the living cells do not incorporate the dye, whereas dead cells incorporate it owing to their damaged membrane. Cells are treated with different concentrations of NPs in well plates and after 24 h incubation, cells are trypsinized. Unexposed control cultures are maintained under the same conditions. The enumeration is carried out by immediate microscopic observation using the counting cell. At least six numerations per sample are performed to limit the risk of error. The percentage of viability corresponds to the percentage of living cells resistant to any adverse effect of the NPs.

19.9.1.9 Alamar Blue Reduction

Standard cytotoxicity assays that assess chemical toxicity may generate conflicting results with some nanomaterials (e.g., carbon nanomaterials) since dye-based assay may often provide false visibility and cytokine data [134]. The alamar blue reagent is added to the medium at a 10% concentration and incubated with cells for a few hours. The alamar blue reagent is converted from the nonfluorescent indicator dye resazurin into the highly red fluorescent metabolite resorufin *via* reduction reactions by metabolically active cells [134]. The fluorescence of each well is quantified on a spectrophotometer with excitation and emission wavelength. The resulting fluorescence is proportional to the number of viable cells per well. Fluorescence values are normalized to media controls and expressed as percent viability. Experimental controls are also conducted with the alamar blue to assess potential NPs/assay interaction [134].

19.9.1.10 Commassie Blue Assay

The cell viability of NPs is also determined by commassie blue assay [135]. In this method, the culture medium is treated with Bradford reagent. The optical density of the medium is determined at a particular wavelength using enzyme-linked immunosorbent assay (ELISA) plate reader.

19.9.1.11 MetPLATE *E. coli* Bioassay

This method is used to evaluate the potential toxicity of several metal-based NPs (e.g., nAg, nTiO$_2$, nZnO, and CdSe QDs) on *E. coli*. It is a β-galactosidase-based assay in which chlorophenol red β-galactopyranoside (CRPG) is used as a chromogenic substrate that is cleaved by the enzyme, forming galactopyranose and chlorophenol red as the by-products. During this event, yellow-colored CRPG is transformed into magneta-colored chlorophenol red, the concentration of which is proportional to the activity of β-galactosidase, which is quantified at a particular wavelength using the microplate reader. Moderately hard water is used as a negative control, while $Cu^{2+}L^{-1}$ (as $CuSO_4$) is used as a positive control with each set of analysis [136].

19.9.1.12 CellTiter Assay

Mitochondrial function was assessed using the CellTiter 96 AQueous one-solution assay [137]. The reagent contains tetrazolium compound [3-(4,5-dimethylthiazole-2-yl)-5-(3-carboxymethoxyphenyl)-2-(4-sulfophenyl)-2H-tetrazolium, inner salt; MST(a)] and electron-coupling reagent (phenazine ethosulfate). The absorbance is measured at a particular wavelength with a standard microplate reader. The quantity of formazan product as measured by the microplate reader is directly proportional to the number of living cells in culture. The relative cell viability (%) related to control containing cell culture medium without NPs is calculated by $[A]_{test}/[A]_{control} \times 100$ where $[A]_{test}$ is the absorbance of the test sample and $[A]_{control}$ is the absorbance of the control sample [137].

19.9.1.13 Annexin V/Propidium Iodide Staining for Apoptotic and Necrotic Cells

Annexin V (VAC alpha), which is regularly used to detect apoptotic cells [138], strongly binds to phosphatidyl serine in a calcium-dependent manner. Phosphatidyl serine is normally excluded from the extracellular side of the plasma membrane, but flips between the inner and the outer side upon the onset of apoptosis [138]. Fluorescently labeled Annexin V can therefore be used to detect apoptotic cells [138]. Necrotic cells will allow Annexin V to bind the inner part of the plasma membrane, resulting in false-negative results due to cell disintegration. Hence, cells have to be costained with propidium iodide (3,8-diamino-5-[3-(diethylmethylammonio) propyl]-6-phenylphenanthridinium diiodide) that will exclusively stain necrotic cells [139,140]. Investigation of apoptosis via Annexin

V staining and flow cytometry has been reported, for example, for TiO$_2$ NPs [141], for pure and polyhydroxylated fullerenes [142], for SWCNTs (111), and for QDs [143].

As apoptotic cells may easily detach from their substrate, it is essential to collect adherent and floating cells for Annexin V staining [144]. Consequently, NPs in the cell culture supernatant will be present in the cell suspension despite several washing steps. Gold NPs have been shown to bind propidium iodide and have to be taken up by intact cell culture cells. This process leads to false-positive results in the detection of necrotic cells [145].

19.9.1.14 Detection of the Apoptosis Marker Caspase-3

The detection of active Caspase-3 is one of the most commonly used apoptosis assays. Apoptosis may be triggered by different elicitors activating two main signaling cascades that converge in the activation of Caspase-3 [146]. The cysteine protease Caspase-3 is produced as a zymogen in the cytosol and is activated in the terminal apoptotic cascade by cleavage [113]. As soon as Caspase-3 is activated, cell death is inevitable. Activated Caspase-3 can be detected by measuring the cleavage of a Caspase-3 substrate (preferably the amino acids DEVD) linked to a chromophore (pNA) or fluorophore (AFC, AMC) that absorbs or emits light when separated from the substrate [114]. As yet, the Caspase-3 assay has been utilized to examine apoptosis in cell culture cells upon exposure, for example, to fullerenes [147], SWCNTs [148], silica NPs [149], QDs [143], and to TiO$_2$ NPs [64].

Caspase-3 is inhibited by trace metal ions, especially by Zn^{2+} ions [150]. On the other hand, Caspase-3 is relatively unsusceptible to changes in pH [150]. As described above, apoptotic cells may easily detach from their substrate, so that adherent and floating cells have to be used for Caspase-3 activity assays [144]. NPs in the cell culture supernatant will therefore remain in the cell suspension during measurement.

19.9.2 Stress Response

19.9.2.1 Detection of ROS

Cellular stress response is often investigated with H$_2$DCF-DA (20,70-dichlorodihydrofluoresc(e) in diacetate), which is a widely used probe for the *in vitro* detection of intracellular ROS [151].

The acetylated nonfluorescent molecule is taken up by cell culture cells, is presumably trapped in the cytosol by deacetylation, and becomes fluorescent upon intracellular oxidation [152]. Several studies on the exact mechanism responsible for H$_2$DCF oxidation offer varying conclusions; consequently, it has been suggested that DCF should be applied as a qualitative marker for cellular oxidative stress in general [152]. A possible increase in DCF fluorescence has been investigated after cell culture exposure [152].

Internalized, deacetylated H$_2$DCF does not exclusively remain in the cytosol but may accumulate in the extracellular space and react with catalytically active substances outside the cells [153]. DCF fluorescence is strongly pH dependent [154]. Additionally, NPs such as carbon-based materials may absorb light so that DCF fluorescence may be quenched. Controls including NPs and oxidized DCF have been suggested [155], but need to be verified for a linear relationship between NPs and dye concentration versus the decrease in fluorescence.

Exposure to NPs is known to cause an increase in ROS, which could lead to oxidative stress. ROS generation by NPs could be due to three factors [156]: (i) active redox cycling on the surface of NPs, particularly the metal-based NPs [157]; (ii) oxidative groups functionalized on NPs; and (iii) particle–cell interactions, especially in the lungs where there is a rich pool of ROS producers such as the inflammatory phagocytes, neutrophils, and macrophages (Figure 19.5). Overproduction of ROS activates a series of cytokine cascades, which include an upregulation of interleukins (ILs), kinases, and tumor necrosis factor (TNF-α) proinflammatory signaling processes as a counterreaction to oxidative stress [158]. Studies on TiO$_2$ NPs and C$_{60}$ fullerenes have shown that these NPs induce elevation of proinflammatory enzymes, such as IL-1, TNF-α, IL-6, macrophage inhibitory

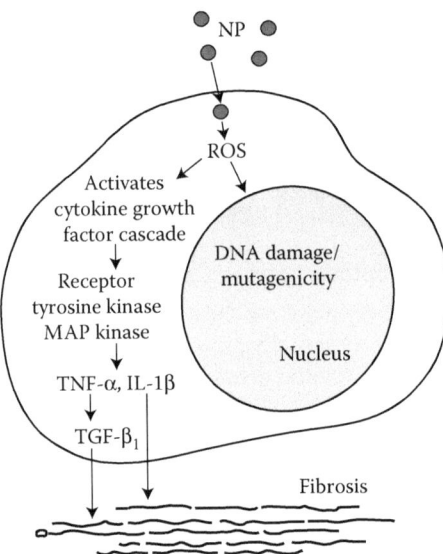

FIGURE 19.5 Possible mechanistic pathway for pulmonary toxicity induced by exposure to NPs. Exposure to NPs may lead to oxidative stress due to increased production of ROS and downstream signaling responses that promote fibrosis and produce genotoxicity. NP, nanoparticle; ROS, reactive oxygen species. (Reprinted with permission from Li JJ et al. Nanoparticle-induced pulmonary toxicity. *Experimental Biology and Medicine*. 2010;235:1025–33.)

protein, and monocyte chemotactic protein in rodent lungs [159]. When receptor tyrosine kinases, mitogen-activated protein kinases, and transcriptional factors, such as nuclear factor-kB and signal transducer and activator of transcription 1, are activated, the genes involved in inflammation and fibrosis are transcribed and expressed [160]. Stimulation of IL-1β and TNF-α heightens the expression of profibrotic proteins. More specifically, the latter is known to upregulate the production of transforming growth factor (TGF)-$β_1$, which potentiates collagen deposition by fibroblasts [161], while the former is associated with the expression of platelet-derived growth factor (PDGF)-AA and its receptor, PDGF receptor-α, which increases proliferation of myofibroblasts, promoting the formation of immature collagenous tissue within the lung [160].

19.9.2.2 Glutathione Measurement

Glutathione (GSH), an antioxidant, protects the cells from free radicals. GSH level is quantified using Ellman's reagent [162]. The level of GSH in the cells is used as a stress marker and is expressed as micromole per milligram of protein.

19.9.2.3 Lipid Peroxidation Assay

Lipid peroxidation level is estimated by extracting lipid hydroperoxide from cell lysate in chloroform. A solution containing ferrous ion is then added to the cell extract, which on reaction with lipid hydroperoxide, yields ferric ions. The resulting ferric ion is the determined colorimetric method using thiocyanate ion as chromogen and 13-hydroperoxy-octadecadienoic acid as standard [163].

Malondialdehyde (MDA), a product of lipid peroxide decomposition, serves as a reliable indicator of lipid peroxidation [123]. The extent of membrane lipid peroxidation is estimated by measuring the formation of MDA. MDA is one of the end products of membrane lipid peroxidation. MDA formed is evaluated using the thiobarbituric acid reactive species (TBARS) assay. This method is based on the quantification of the colored complex formed between thiobarbitutic acid (TBA) and MDA after acid hydrolysis reaction. This method was applied for the assessment of lipid peroxidation effects of NPs such as silica NPs [121] and CeO_2 NPs [164].

19.9.2.4 Nitrite Production

Nitrite production is determined by the Griess reagent (1:1 mixture of 1% sulfanilamide in 5% phosphoric acid and 0.1% N-1-naphthylethylenediamine dihydrochloride). In this method, the Griess reagent is mixed with culture supernatant and the intensity of color is determined at a particular wavelength using ELISA plate reader [135].

19.9.2.5 Mitochondrial Membrane Potential Dissipation Assay

The changes of mitochondrial membrane potential (MMP) after exposure of NPs are determined by the uptake of tetramethyl rhodamine ethyl ester (TMRE). TMRE is a red-orange fluorescent permeable cationic, lipophilic dye. It can be readily taken up by active mitochondria into negatively charged mitochondrial matrix. The intensity of fluorescent obtained is indicative of the MMP [165].

19.9.3 INFLAMMATORY RESPONSE

The *in vitro* method to evaluate the inflammatory response has been performed via ELISA. The ELISA method was first described in 1971, and enables simple and accurate quantification of inflammatory markers in cell culture supernatants through antibodies and enzymatic detection reactions [166]. ELISA results have been reported for NPs of different composition and origin, for example, for titanium dioxide [167], iron oxide [168], zinc oxide [169], carbon black [170], CNTs [171], fullerenes [172], silica [173], and for QDs [174]. The most commonly tested human and murine inflammatory markers are the chemokine interleukin-8 (IL-8), followed by TNF-α and IL-6. In some cases, IL-1β as well as a few other cytokine and stimulating factor concentrations are measured. The chemokine MIP-2 is usually quantified in rat model systems together with TNF-α and/or IL-6.

Recently, NPs have been reported to interfere with enzymatic immunoassays. Adsorption of cytokines by NPs *in vitro* was discussed for carbon NPs (IL-8, [175]) and for metal oxides (IL-6) [176]. Even depletion of trace nutrients or growth factors from cell culture media due to the high adhesive surface area of NPs was considered [175,177].

19.9.4 GENOTOXICITY

Nanogenotoxicology is yet another new term that was coined to represent the growing trend of research into NP-induced genotoxicity and carcinogenesis. Although there are still no conclusive links with NP-induced genotoxicity and lung cancer from past epidemiological studies and *in vivo* rodent experiments, some researchers have pointed out that long-term inflammation and oxidative stress present in tissue environment eventually induces DNA damage in cells and tissues [178]. This is of particular concern, especially if the NPs continue to generate an oxidative environment in the cell that causes gene mutations/deletions. This can lead to larger-scale mutagenesis and carcinogenicity, and subsequently, the development of tumors and cancer [156]. Already, more evidence has emerged regarding the DNA-damaging properties of certain classes of NPs, particularly, the metal-based NPs such as Ag NPs, Au NPs, and TiO_2 NPs [179–181]. One proposed mode of action for NP genotoxicity is the ability of signaling peptides functionalized on NPs such as CNTs that enable them to enter the nucleus via nuclear pores [182]. It is yet to be shown that such CNTs are able to cause genotoxicity, but it is believed that there is a greater potential of damage when NPs are able to get in close proximity to DNA. There are also other different mechanisms that may be specific to the elemental composition and shape of NPs, which could lead to DNA damage such as single-strand breaks, double-strand breaks, DNA deletions, and genomic instability in the form of increase in 8-hydroxy-20-deoxyguanosine levels [86]. While some researchers have found that exposure to TiO_2 NPs in rats could cause formation of lung granulomas [159], others have cautioned that appearance of granulomas does not necessarily mean that the tissue is cancerous as most tissues probably remain benign [183]. As most reports regarding NP toxicity were observed from experiments involving UV or irradiation exposure, the clinical relevance of these mechanistic experiments

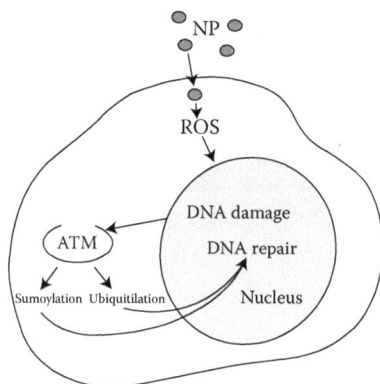

FIGURE 19.6 Postulated mechanism for the repair of damaged DNA on exposure to NP based on the ionizing radiation model. NP, nanoparticle; ATM, ataxia telangiectasia mutated. (Reprinted with permission from Li JJ et al. *Experimental Biology and Medicine*. 2010;235:1025–33.)

is questioned [184]. Nevertheless, a recent study has shown that TiO_2 NPs may be able to switch on regressive cancer cells. *In vivo* preimplantation of TiO_2 NPs, followed by coculturing of a regressive cancer cell line over the implantation site, was observed to induce tumorigenic characteristics such as an upregulation of TGF-β and prostaglandin E2 [185]. The long-term exposure to NPs, such as nanoparticulates in PM10, displayed genome instability under comet assay analysis, altered cell cycle kinetics in flow cytometry, and induced protein expression of p53 and DNA repair-related proteins, similar to that observed in irradiated cells. Hence, they postulate that these NPs could activate signaling pathways similar to ionizing radiation, resulting in carcinogenesis as a consequence of errors in DNA replication. DNA repair in ionizing radiation requires activation of ataxia telangiectasia-mutated protein, which is a serine-/threonine-specific kinase, and subsequently, the ubiquitylation-signaling cascade and sumoylation pathway (Figure 19.6).

19.9.4.1 Comet Assay

The comet assay is a widely used *in vitro* assay in fundamental research for DNA damage and repair, in genotoxicty testing of nanomaterials. This study has been employed to investigate the genotoxic potential of manufactured NPs that allowed DNA strand breaks or oxidative lesions. Considering the sensitivity of the assay, it can enable the assessment of its relative potency [186].

19.9.4.2 F_{pg}-Modified Comet Assay

The DNA formamidopyrimidine DNA glycosylase (F_{pg}) enzyme is used for detection of oxidative DNA base damage, in particular, 8-OH guanine. Therefore, F_{pg}-modified comet assay is used after treating with NPs [163].

19.9.4.3 Ames Test

Ames test (or bacterial reversion mutation test) is an *in vitro* assay used to assess the genotoxic potential of nanomaterials [187]. The test employs histidine-dependent (auxotrophic) mutant strains of *Salmonella typhimurium*. The test is usually employed as an adjunct technique because it is difficult to interpret that data generated in a prokaryotic system to a eukaryotic genotoxicity testing. Furthermore, results could be ambiguous in some instances when certain nanomaterials are not able to cross the bacterial wall or in situations where the nanomaterials are bactericidal.

19.9.4.4 Alkaline Single-Cell Microgel Electrophoresis Assay

The alkaline single-cell microgel electrophoresis assay is applied to detect DNA strand breaks and alkali labile as well as incomplete excision repair sites in single cells. The percentage of DNA in the tail is analyzed to quantify the induced DNA fragmentation. In this method, Dulbecco's modified Eagle medium (DMEM) is used as negative control and methyl methane sulfonate (MMS) is used as alkalyting agent and serves as a reliable positive control [186].

19.9.4.5 DNA Damage
DNA damage in cells is detected using γ-H2AX, a phosphorylated form of H2AX, which forms at the sites of DNA double-strand breaks [188].

19.9.5 High-Throughput Screening Method

Recent advances in cell-based assays allow for toxicity and/or efficacy screening of multiple nanomaterials at multiple concentrations with multiple cell lines, simultaneously. This expansion of the experimental design is practically enabled through the miniaturization and multiplexing of the experimental apparatus and method by utilization of either ultrasmall 384-well cell culture plates or nanodrop sample chambers on a chip. The nanodrop assay setup allows for different assays with suitable detection features (e.g., fluorescence, and luminescence) to be performed in a fraction of the volume without the cell activation or photometric effects of the culture plate since the cell culture is performed in a self-contained drop [189]. However, since cells are typically microns in size, nanodrops do not necessarily capture cells themselves, only fluidic cellular exudates for assay and analysis. By assaying numerous material types/functionalizations and material concentrations on numerous cell types, all in parallel, complex interactions between materials and cells may be ascertained through complex data analysis that correlates phenotypes with multiwell plates, cell culture, detection schemes, and recognition schemes [190].

19.9.6 Dose-Response Assessment

From the preceding section, it becomes obvious that without the use of standard operation procedures (SOPs), results from different studies will be difficult to compare. Dose indication for dose-response studies presents another problem. Dose-response relationships (relationship between dose, or level of exposure to a substance, and the incidence and severity of an effect) are more difficult to assess for NPs because mass per milliliter, which is commonly used for chemicals, may not be a suitable measure for NPs. It has been suggested to use either surface area or particle number because toxicity is correlated better with these parameters [191]. None of these measures proved to be ideal for all NP types and therefore, most studies use mass per volume for dosing. The correlation of dose and biological response is complicated by the fact that increases in the mass dose do neither increase the effect observed with lower concentrations nor do they act in the same way. Inhalation of low doses of TiO_2 NPs affects metabolites in the urine, whereas higher doses cause local effects in the lung [192].

19.10 Application of Computational Approach for Risk Assessment

19.10.1 Molecular Modeling Methods

The current computational methods show a great potential to predict properties, reactivity, and mechanisms of actions for various molecular systems, from small molecules up to large biomolecules. Quantum chemical calculations and molecular dynamics (MD) simulations can be useful to address the potential risks associated with nanomaterials. Results from such theoretical calculations might provide suggestions to experimentalists working in this research field. The knowledge gained from computational studies involving interactions of NPs with biological systems will be helpful to construct algorithms for assessing the likelihood of toxicity in a variety of natural environments [193]. Computer simulations could be valuable by examining the structure of surfaces and identifying even the smallest change in the position of atoms near edges, corners, surface steps, and defects. The mechanisms of the interatomic interactions between NPs and biological molecules are not well understood. Comprehension of the mechanisms of such interactions will aid safe production and utilization of the nanomaterials. Computational studies are helpful to understand the precise nature of interparticle interactions, the structure of the interface, and the packing of arrays and superstructures that are difficult to probe experimentally [194]. However, similar to experimentalists who face several issues, computational nanoscientists also have various challenges; for example, simulations involving many NPs are computationally too intensive and not feasible using advanced *ab initio* or density functional theory (DFT) approaches; convergence problems often occur in dealing with large molecules.

Experimental studies have concluded that the toxicity of carbon-based nanomaterials depends on a wide range of properties such as structures (single walled or multi walled), length and aspect ratios (in case of CNTs), surface area, surface charge, degree of aggregation, extent of oxidation, attached functional group(s), method of manufacturing, morphology, concentration, and dose [195]. Theoretical and computational nanoscientists should carefully consider all such physical parameters, which are intrinsically linked with toxicity of materials, before making predictions about risks of nanostructures.

19.10.2 Quantitative Structure–Activity Relationships for Nanomaterials

In fact, the toxicity properties of nanomaterials have been investigated by various groups of researchers, but considerable uncertainties in NPs' mechanism of toxicity still exist. Indeed, many studies proved much higher toxicity displayed by some of the NPs in comparison to bulk-size particles. For example, it is well known that metal oxide NPs possess higher toxicity than bulk-size counterparts of the same chemical composition [196].

The large number of NPs and the variety of their characteristics including sizes and coatings suggest that the only rational approach that avoids testing of every single NP is to find relationship between NP properties (i.e., physicochemical characteristics) and their toxicity. The current risk assessment paradigms, particularly in regulatory submissions for drugs and chemicals, generally depend on standardized methodologies. REACH legislation introduced in Europe allows computational tools in replacing experimental tests in some cases. Some computational tools, for example, QSAR [197], are essential for increasing throughput, reducing the burden of animal testing, providing details of the toxicity mechanisms, and generating novel hypotheses for risk assessment [198].

In the case of QSAR, the approach is useful not only in making predictions, but also in refining the existing risk assessment paradigms. As one of the examples, it is known that QSAR approaches applied to assessing risk may facilitate placement of chemicals with incomplete data sets in the appropriate risk categories. In addition, computational modeling includes physiologically-based pharmacokinetic (PBPK) models, as well as modeling dose response [198]. As a result, if a QSAR model is then developed, ideally, dose-response toxicity of untested NP can be predicted on the basis of its physicochemistry [197]. Actually, many investigations confirm that there is a strong need to extend the traditional QSAR paradigm to NPs. In connection to this, some recent studies have shown that QSAR can be effectively used for predicting the toxicity of NPs, other physicochemical properties, and therefore, one can assess the environmental risk of these materials [197].

The following three points need to be considered to extend QSAR approach to NPs:

1. QSAR methodology has been mainly developed for small organic compounds with diverse structural types, while NPs are large and structurally limited in diversity.
2. The experimental data accumulated for NPs are not yet sufficient to fully assess the toxicity; the empirical data are scarce and/or contradictory.
3. Regular QSAR descriptors, which are applicable for organic compounds, are generally not applicable for NPs. Although this problem is being rapidly addressed, some novel nanodescriptors are suggested.

To address all these issues, the following approaches are necessary. First of all, proper and reliable experimental data for the series of related but different size NPs should be available for QSAR analysis. It is vital to obtain consistent toxicological data, which will be used to develop QSAR models for toxicity assessment. In addition, it is quite important to develop the nanomaterials' property database-collected data on experimental physicochemical properties, biological activity, and toxicity endpoints for various sizes of nanomaterials. Second, the structural and size descriptors suitable for modeling NP reactivity have to be identified. These descriptors can be selected/adapted from the available pool of descriptors designed for conventional (small) compounds or calculated by applying quantum-mechanical (MD) methods, developed especially for NPs, or selected from known experimental physical properties. The close connection and interaction among all these areas—experimental toxicology, physicochemical NP characterization (nanodescriptor characterization), and computational nano-QSAR methods are essential for successfully understanding of the mechanisms of toxicity of nanomaterials and proper risk assessment.

19.11 ECOTOXICITY OF NPs

To ensure a "safe" nanotechnology industry, the need for hands-on research in the area ecotoxicology of nanomaterials has been emphasized [99]. Several assays for ecotoxicological testing of nanomaterials have been developed. Various mechanisms that govern toxicity as well as usefulness of bacterial systems to study toxicity of manufactured NPs have been explained. The suspensions of C_{60} have been shown to be toxic to bacteria [199], fathead minnows (*Pimephales promelas*) [200], and zebrafish embryos [200]. SWCNT-based nanomaterials to an estuarine copepod (*Amphiascus tenuiremis*), *Daphnia*, and rainbow trout have been reported [201–203]. The ecotoxicities of TiO_2, ZnO, and SiO_2 NPs suspended in water using *E. coli* and *Bacillus subtilis* as two model bacterial species were compared and reported that ZnO was toxic to *B. subtilis* [204]. Experiments on embryonic zebrafish demonstrated similar results; ZnO NPs were more toxic than TiO_2 or Al_2O_3 NPs [205]. In a comprehensive study on the 48-h acute toxicity of water suspensions of six manufactured nanomaterials (i.e., ZnO, TiO_2, Al_2O_3, C_{60}, SWCNTs, and MWCNTs) to *Daphnia magna*, using immobilization and mortality as toxicological endpoints, a dose dependence in acute toxicity was demonstrated [206]. The suitability of fish hepatocyte cultures as a model system for investigating the cellular uptake of engineered NPs have been illustrated [207]. Another model system for judging nanomaterials toxicity is zebrafish embryos; the model is also being useful for comparative biology because of the similarities between the zebrafish and human genomes, early-life development, and disease processes. In a study on ZnO toxicity in rodent lung and zebrafish embryos, data indicated reduced toxicity in the latter system upon doping of Fe in ZnO [208].

The release of nanomaterials to the environment during recycling and disposal is of particular concern for NPs incorporated into limited use and/or disposable products. Once released, these nanomaterials would readily undergo transformations via biotic and abiotic processes. Understanding environmental transformations and the fate of engineered nanomaterials will enable the design and development of environmentally benign nanomaterials, as well as their use as environmental tracers, in environmental sensing and in contaminant remediation. This was demonstrated in a biomimetic hydroquinone-based Fenton reaction that provides a new method to characterize transformations of nanoscale materials expected to occur under oxidative environmental conditions [209]. The current computational techniques are being used to study interactions of NPs with biological systems and these have been reviewed in Ref. [210]. Such studies could also be used to complement the experimental data on toxicity.

19.12 LIMITATIONS OF *IN VITRO* ASSAY IN TOXICOLOGY

In vitro toxicity assays are primarily utilized to investigate the generic cytotoxicity or genotoxicity of chemicals. While the traditional application of *in vitro* tests has been screening of chemicals, recent developments in molecular biology have helped to provide mechanistic information on toxicity [211]. The use of *in vitro* toxicity methods for the assessment of NPs was a subsequent development and the applicability of a prevalent *in vitro* test to NPs is reviewed below. Sutter [212] identified six general applications of *in vitro* toxicity assays, among which are the selection of the most appropriate animal model of humans and the rapid screening of a series of toxicants. In comparison to animal models, *in vitro* assays allow for a simpler, faster, and more cost-efficient assessment of defined toxicity endpoints [213]. However, *in vitro* test systems lack the complexity of animal models or the human body [211], and the metabolic activity of standardized cell lines has often not been comprehensively characterized [214]. According to Snodin [215], *in vitro* systems have no value for the prediction of biodistribution and target organ toxicity for the applied chemical and its metabolites. Besides these considerations, scientists have been striving to determine the correlation between the results obtained from *in vitro* and *in vivo* toxicity assessments. In the domain of particle toxicology, for example, Sayes et al. [169] found little correlation between *in vitro* and *in vivo* pulmonary toxicity of several fine NPs. On the other hand, Donaldson et al. [98] reported that the threshold for inflammation onset was identical *in vitro* and *in vivo* when the "particle surface area burden per unit of proximal alveolar region surface area" of different low-toxicity, low-solubility particles was used as a reference. Contradictory *in vitro* results on the same test substance accumulating in the literature as well as the continuous public concern for animal welfare and safe handling of substances have driven diverse international efforts to standardize and validate *in vitro* tests. The Economic Corporation & Development (OECD) validation harmonization report and programs such as the MEIC (multicenter evaluation of *in vitro* toxicity, e.g., Ref. [216]), the German Center for the Documentation and Validation of Alternative Methods (ZEBET), and standard protocols made available via the INVITTOX database run by the European Centre for the Validation of Alternative Methods (ECVAM), for example, are important aspects of an international standardization and validation process (reviewed, e.g., in Ref. [217]).

In vitro assays have general limitations for reliable risk assessment in the aspects of validation by *in vivo* experiments and a lack in standardized testing procedures, which are verified by reference materials or interlaboratory validation.

19.13 CHALLENGES FOR NPs *IN VITRO* TEST METHODS

Physicochemical properties of NPs often limit the use of established *in vitro* toxicity assays for risk assessment. Studies designed to determine NP toxicity should be ideally carried out using test systems that cannot be influenced by nanospecific properties [99]. Currently, however, NP risk analysis is impaired by the lack of standardized test systems that fulfill these criteria.

New test systems for toxicity screening address new endpoints or new toxicity biomarkers. For instance, cell culture systems reflecting better *in vivo* toxicity parameters have been recently developed [218]. New subdisciplines of toxicology, such as toxicogenomics, focus on studies of cellular products controlled by the genome (RNA, proteins, and metabolites) and provide new approaches to assess adverse biological effects of exogenous agents [219]. These technologies will enable a deeper understanding of biochemical pathways and cellular responses.

However, many of these strategies depend on conventional detection systems and may still be influenced by NP-specific properties. Novel technologies allowing marker-free cytotoxicity testing might overcome these obstacles. For example, cellular analysis can be performed using physical cell properties such as electrical resistance or refractive index. Digital holographic microscopy, for example, detects the integral refractive index of living single cells in cell culture medium, resulting in a phase shift of visible light [220]. Cell morphology reconstruction of the digitally captured holograms is performed by application of a spatial phase shifting, a nondiffractive reconstruction method [221]. This method can be applied to a marker-free online analysis of processes induced by drugs or toxic agents that lead to altered cell morphology, including apoptosis and cell swelling [222]. Electrochemical impedance spectroscopy measures cellular dielectric parameters such as membrane capacitance and conductivity that can change rapidly after exposure to toxic substances [223]. Tight cell layers grown on top of an electrode form an electrical resistance, which is altered after drug-induced cell morphology alteration such as cell rounding or cell–cell contact dissociation [224]. The online monitoring of cellular resistance can be integrated in conventional high-throughput plate formats [225]. Compared with a conventional MTT assay for cytotoxicity screening, electrochemical impedance spectroscopy was shown to be more sensitive [226].

After evaluation and validation, these new marker-free and live measurement technologies may define a new standard for a general *in vitro* toxicity testing of NPs unaffected by specific NP properties.

In conclusion, NPs exhibit size-specific properties limiting the application of established *in vitro* assays. NP *in vitro* toxicity testing therefore requires a careful characterization of particle properties and an extensive validation of assay systems when applying established methods for risk assessment. Future studies should only present data obtained with well-characterized particles and include reference materials when available. A major deficiency considering a wide range of NPs to be assessed is the current lack of nationally or internationally agreed reference NPs or nanomaterials and standardized test protocols.

Since the current *in vitro* test methods are likely to be influenced by NP-specific properties, NP risk assessment needs an adaptation of the existing cytotoxicity methods or the development of new test systems. Technologies based on marker-free detection of cellular endpoints have the potential to overcome nanomaterial-dependent assay limitations and will provide new standards for NP risk assessment.

19.14 EUROPEAN UNION APPROACH TO NANOMATERIAL RISK ASSESSMENT

The current risk assessment tools in regard to nanomaterial and nanotechnologies most likely, in many cases, lag behind the current understanding on nanomaterial and nanotechnologies [227], and this gap tends to increase. For this reason, intelligent and affordable testing strategies addressing the specific features of nanomaterial are required. Thus, there is an urgent need to update the REACH legislation to meet the challenges provided by the rapidly emerging novel nanomaterial and nanotechnologies.

REFERENCES

1. De Jong WH, Borm PJ. Drug delivery and nanoparticles: Applications and hazards. *International Journal of Nanomedicine*. 2008;3(2):133.
2. Kwon J-T, Hwang S-K, Jin H, Kim D-S, Minai-Tehrani A, Yoon H-J et al. Body distribution of inhaled fluorescent magnetic nanoparticles in the mice. *Journal of Occupational Health*. 2008;50(1):1–6.
3. Lewinski N, Colvin V, Drezek R. Cytotoxicity of nanoparticles. *Small*. 2008;4(1):26–49.
4. Oberdörster G, Maynard A, Donaldson K, Castranova V, Fitzpatrick J, Ausman K et al. Principles for characterizing the potential human health effects from exposure to nanomaterials: Elements of a screening strategy. *Particle and Fibre Toxicology*. 2005;2(1):8.
5. Oberdörster G, Oberdörster E, Oberdörster J. Nanotoxicology: An emerging discipline evolving from studies of ultrafine particles. *Environmental Health Perspectives*. 2005;113(7):823.
6. Service RF. *Nanotechnology—Can High-Speed Tests Sort out Which Nanomaterials Are Safe?*: American Association of Advancement Science 1200, New York Avenue, NW, Washington, DC 20005, USA; 2008.
7. Park MV, Lankveld DP, van Loveren H, de Jong WH. The status of *in vitro* toxicity studies in the risk assessment of nanomaterials. *Nanomedicine*. 2009;4(6):669–85.
8. Vasir JK, Labhasetwar V. Quantification of the force of nanoparticle–cell membrane interactions and its influence on intracellular trafficking of nanoparticles. *Biomaterials*. 2008;29(31):4244–52.
9. Peetla C, Labhasetwar V. Biophysical characterization of nanoparticle—Endothelial model cell membrane interactions. *Molecular Pharmaceutics*. 2008;5(3):418–29.
10. Ginzburg VV, Balijepalli S. Modeling the thermodynamics of the interaction of nanoparticles with cell membranes. *Nano Letters*. 2007;7(12):3716–22.

11. Foley S, Crowley C, Smaihi M, Bonfils C, Erlanger BF, Seta P et al. Cellular localisation of a water-soluble fullerene derivative. *Biochemical and Biophysical Research Communications*. 2002;294(1):116–9.
12. Zhu Y, Zhao Q, Li Y, Cai X, Li W. The interaction and toxicity of multi-walled carbon nanotubes with *Stylonychia mytilus*. *Journal of Nanoscience and Nanotechnology*. 2006;6(5):1357–64.
13. Xia T, Kovochich M, Brant J, Hotze M, Sempf J, Oberley T et al. Comparison of the abilities of ambient and manufactured nanoparticles to induce cellular toxicity according to an oxidative stress paradigm. *Nano Letters*. 2006;6(8):1794–807.
14. Jia G, Wang H, Yan L, Wang X, Pei R, Yan T et al. Cytotoxicity of carbon nanomaterials: Single-wall nanotube, multi-wall nanotube, and fullerene. *Environmental Science and Technology*. 2005;39(5):1378–83.
15. Al-Rawi M, Diabaté S, Weiss C. Uptake and intracellular localization of submicron and nano-sized SiO_2 particles in HeLa cells. *Archives of Toxicology*. 2011;85(7):813–26.
16. Dobson CM. Protein folding and misfolding. *Nature*. 2003;426(6968):884–90.
17. Marano F, Hussain S, Rodrigues-Lima F, Baeza-Squiban A, Boland S. Nanoparticles: Molecular targets and cell signalling. *Archives of Toxicology*. 2011;85(7):733–41.
18. Dawson KA, Salvati A, Lynch I. Nanotoxicology: Nanoparticles reconstruct lipids. *Nature Nanotechnology*. 2009;4:84–5.
19. Takahashi H, Sawada S-I, Akiyoshi K. Amphiphilic polysaccharide nanoballs: A new building block for nanogel biomedical engineering and artificial chaperones. *Acs Nano*. 2010;5(1):337–45.
20. Wagner SC, Roskamp M, Pallerla M, Araghi RR, Schlecht S, Koksch B. Nanoparticle-induced folding and fibril formation of coiled-coil-based model peptides. *Small*. 2010;6(12):1321–8.
21. Karlsson HL. The comet assay in nanotoxicology research. *Analytical and Bioanalytical Chemistry*. 2010;398(2):651–66.
22. Xie W, Wang L, Zhang Y, Su L, Shen A, Tan J et al. Nuclear targeted nanoprobe for single living cell detection by surface-enhanced Raman scattering. *Bioconjugate Chemistry*. 2009;20(4):768–73.
23. Myllynen P. Nanotoxicology: Damaging DNA from a distance. *Nature Nanotechnology*. 2009;4(12):795–6.
24. Stern ST, McNeil SE. Nanotechnology safety concerns revisited. *Toxicological Sciences*. 2008;101(1):4–21.
25. Elsaesser A, Vyvyan Howard C. Toxicology of nanoparticles. *Advanced Drug Delivery Reviews*. 2012;64(2):12–137.
26. Hagens WI, Oomen AG, de Jong WH, Cassee FR, Sips AJ. What do we (need to) know about the kinetic properties of nanoparticles in the body? *Regulatory Toxicology and Pharmacology*. 2007;49(3):217–29.
27. Nurkiewicz TR, Porter DW, Barger M, Millecchia L, Rao KMK, Marvar PJ et al. Systemic microvascular dysfunction and inflammation after pulmonary particulate matter exposure. *Environmental Health Perspectives*. 2006;114(3):412.
28. Yeates DB, Mauderly JL. Inhaled environmental/occupational irritants and allergens: Mechanisms of cardiovascular and systemic responses. Introduction. *Environmental Health Perspectives*. 2001;109(Suppl 4):479.
29. Soto K, Garza K, Murr L. Cytotoxic effects of aggregated nanomaterials. *Acta Biomaterialia*. 2007;3(3):351–8.
30. Crosera M, Bovenzi M, Maina G, Adami G, Zanette C, Florio C et al. Nanoparticle dermal absorption and toxicity: A review of the literature. *International Archives of Occupational and Environmental Health*. 2009;82(9):1043–55.
31. Tinkle SS, Antonini JM, Rich BA, Roberts JR, Salmen R, DePree K et al. Skin as a route of exposure and sensitization in chronic beryllium disease. *Environmental Health Perspectives*. 2003;111(9):1202.
32. Ryman-Rasmussen JP, Riviere JE, Monteiro-Riviere NA. Penetration of intact skin by quantum dots with diverse physicochemical properties. *Toxicological Sciences*. 2006;91(1):159–65.
33. Trop M, Novak M, Rodl S, Hellbom B, Kroell W, Goessler W. Silver-coated dressing acticoat caused raised liver enzymes and argyria-like symptoms in burn patient. *The Journal of Trauma and Acute Care Surgery*. 2006;60(3):648–52.
34. Rouse JG, Yang J, Ryman-Rasmussen JP, Barron AR, Monteiro-Riviere NA. Effects of mechanical flexion on the penetration of fullerene amino acid-derivatized peptide nanoparticles through skin. *Nano Letters*. 2007;7(1):155–60.
35. Gopee NV, Roberts DW, Webb P, Cozart CR, Siitonen PH, Warbritton AR et al. Migration of intradermally injected quantum dots to sentinel organs in mice. *Toxicological Sciences*. 2007;98(1):249–57.
36. Kim S, Lim YT, Soltesz EG, De Grand AM, Lee J, Nakayama A et al. Near-infrared fluorescent type II quantum dots for sentinel lymph node mapping. *Nature Biotechnology*. 2003;22(1):93–7.

37. Monteiro-Riviere NA, Nemanich RJ, Inman AO, Wang YY, Riviere JE. Multi-walled carbon nanotube interactions with human epidermal keratinocytes. *Toxicology Letters*. 2005;155(3):377–84.
38. Zhang LW, Zeng L, Barron AR, Monteiro-Riviere NA. Biological interactions of functionalized single-wall carbon nanotubes in human epidermal keratinocytes. *International Journal of Toxicology*. 2007;26(2):103–13.
39. Hillyer JF, Albrecht RM. Gastrointestinal persorption and tissue distribution of differently sized colloidal gold nanoparticles. *Journal of Pharmaceutical Sciences*. 2001;90(12):1927–36.
40. Nemmar A, Hoet PM, Vanquickenborne B, Dinsdale D, Thomeer M, Hoylaerts M et al. Passage of inhaled particles into the blood circulation in humans. *Circulation*. 2002;105(4):411–4.
41. Borm PJ, Robbins D, Haubold S, Kuhlbusch T, Fissan H, Donaldson K et al. The potential risks of nanomaterials: A review carried out for ECETOC. *Particle and Fibre Toxicology*. 2006;3(1):11.
42. Nigavekar SS, Sung LY, Llanes M, El-Jawahri A, Lawrence TS, Becker CW et al. 3H dendrimer nanoparticle organ/tumor distribution. *Pharmaceutical Research*. 2004;21(3):476–83.
43. Curtis J, Greenberg, M, Kester J, Phillips S, Krieger G. Nanotechnology and nanotoxicology: A primer for clinicians. *Toxicology Review*. 2006;25(4):245–60.
44. Fabian E, Landsiedel R, Ma-Hock L, Wiench K, Wohlleben W, van Ravenzwaay B. Tissue distribution and toxicity of intravenously administered titanium dioxide nanoparticles in rats. *Archives of Toxicology*. 2008;82(3):151–7.
45. Huang X-L, Zhang B, Ren L, Ye S-F, Sun L-P, Zhang Q-Q et al. *In vivo* toxic studies and biodistribution of near infrared sensitive Au–Au2S nanoparticles as potential drug delivery carriers. *Journal of Materials Science: Materials in Medicine*. 2008;19(7):2581–8.
46. Moghimi SM, Hunter AC, Murray JC. Nanomedicine: Current status and future prospects. *The FASEB Journal*. 2005;19(3):311–30.
47. Garnett M, Kallinteri P. Nanomedicines and nanotoxicology: Some physiological principles. *Occupational Medicine*. 2006;56(5):307–11.
48. Bérubé K, Balharry D, Sexton K, Koshy L, Jones T. Combustion-derived nanoparticles: Mechanisms of pulmonary toxicity. *Clinical and Experimental Pharmacology and Physiology*. 2007;34(10):1044–50.
49. Oberdörster G, Sharp Z, Atudorei V, Elder A, Gelein R, Lunts A et al. Extrapulmonary translocation of ultrafine carbon particles following whole-body inhalation exposure of rats. *Journal of Toxicology and Environmental Health Part A*. 2002;65(20):1531–43.
50. Coleman KP, Toscano WA, Wiese TE. QSAR models of the *in vitro* estrogen activity of bisphenol A analogs. *QSAR and Combinatorial Science*. 2003;22(1):78–88.
51. Warheit D, Reed K, Webb T. Pulmonary toxicity studies in rats with triethoxyoctylsilane (OTES)-coated, pigment-grade titanium dioxide particles: Bridging studies to predict inhalation hazard. *Experimental Lung Research*. 2003;29(8):593–606.
52. Fortner J, Lyon D, Sayes C, Boyd A, Falkner J, Hotze E et al. C_{60} in water: Nanocrystal formation and microbial response. *Environmental Science and Technology*. 2005;39(11):4307–16.
53. Kreyling W, Semmler M, Erbe F, Mayer P, Takenaka S, Schulz H et al. Translocation of ultrafine insoluble iridium particles from lung epithelium to extrapulmonary organs is size dependent but very low. *Journal of Toxicology and Environmental Health Part A*. 2002;65(20):1513–30.
54. Bermudez E, Mangum JB, Wong BA, Asgharian B, Hext PM, Warheit DB et al. Pulmonary responses of mice, rats, and hamsters to subchronic inhalation of ultrafine titanium dioxide particles. *Toxicological Sciences*. 2004;77(2):347–57.
55. Borm PJ, Schins RP, Albrecht C. Inhaled particles and lung cancer, part B: Paradigms and risk assessment. *International Journal of Cancer*. 2004;110(1):3–14.
56. Sato QZYKK, Donaldson KNNKK. Differences in the extent of inflammation caused by intratracheal exposure to three ultrafine metals: Role of free radicals. *Journal of Toxicology and Environmental Health Part A*. 1998;53(6):423–38.
57. Kuschner WG, D'Alessandro A, Wong H, Blanc PD. Early pulmonary cytokine responses to zinc oxide fume inhalation. *Environmental Research*. 1997;75(1):7–11.
58. Chou C-C, Hsiao H-Y, Hong Q-S, Chen C-H, Peng Y-W, Chen H-W et al. Single-walled carbon nanotubes can induce pulmonary injury in mouse model. *Nano Letters*. 2008;8(2):437–45.
59. Baciu CL, Becker J, Janshoff A, Sönnichsen C. Protein–membrane interaction probed by single plasmonic nanoparticles. *Nano Letters*. 2008;8(6):1724–8.
60. Elder A, Lynch I, Grieger K, Chan-Remillard S, Gatti A, Gnewuch H, Kenaway E et al. Human health risks of engineered nanomaterials: Critical knowledge gaps in nanomaterials risk assessment. In: Linkov I, Steevens J, eds. *Nanomaterials: Risks and Benefits*. Dordrecht: Springer; 2009. pp. 3–29.

61. Kuhlbusch T, Fissan H, Asbach C. Nanotechnologies and environmental risks: Measurement technologies and strategies. In: Linkov I, Steevens J, eds. *Nanomaterials: Risks and Benefits*. Dordrecht: Springer; 2009. pp. 233–43.
62. Seipenbusch M, Binder A, Kasper G. Temporal evolution of nanoparticle aerosols in workplace exposure. *Annals of Occupational Hygiene*. 2008;52(8):707–16.
63. Weber AP, Seipenbusch M, Kasper G. Size effects in the catalytic activity of unsupported metallic nanoparticles. *Journal of Nanoparticle Research*. 2003;5(3–4):293–8.
64. Long TC, Tajuba J, Sama P, Saleh N, Swartz C, Parker J et al. Nanosize titanium dioxide stimulates reactive oxygen species in brain microglia and damages neurons *in vitro*. *Environmental Health Perspectives*. 2007;115(11):1631.
65. Rossi EM, Pylkkänen L, Koivisto AJ, Vippola M, Jensen KA, Miettinen M et al. Airway exposure to silica-coated TiO_2 nanoparticles induces pulmonary neutrophilia in mice. *Toxicological Sciences*. 2010;113(2):422–33.
66. Pylkkänen L, Alenius H, Tuomi T, Savolainen K, eds. The effect of nanoparticles on the expression of mRNA and proteins of chemokines and cytokines in inflammatory cells in the lungs. *The Sixth Princess Chulabhorn International Science Congress*, Bangkok, Thailand; 2007.
67. Nurkiewicz TR, Porter DW, Hubbs AF, Cumpston JL, Chen BT, Frazer DG et al. Nanoparticle inhalation augments particle-dependent systemic microvascular dysfunction. *Part Fibre Toxicology*. 2008;5(1). doi:10.1186/1743-8977-5-1.
68. Dankovic D, Kuempel E, Wheeler M. An approach to risk assessment for TiO_2. *Inhalation Toxicology*. 2007;19(S1):205–12.
69. Nel AE, Mädler L, Velegol D, Xia T, Hoek EM, Somasundaran P et al. Understanding biophysicochemical interactions at the nano–bio interface. *Nature Materials*. 2009;8(7):543–57.
70. Nadagouda M, Varma RS. Risk reduction via greener synthesis of nober metal nanostructures. In: Linkov I, Steevens J, eds. *Nanomaterials: Risks and Benefits*. Dordrecht: Springer; 2009. pp. 209–17.
71. Owen R, Crane M, Grieger K, Handy R, Linkov I, Depkedge M. Strategic approaches for the management of environmental risk uncertainties posed by nanomaterials. In: Linkov I, Steevens J, eds. *Nanomaterials: Risks and Benefits*. Dordrecht: Springer; 2009. pp. 369–84.
72. Peters TM, Elzey S, Johnson R, Park H, Grassian VH, Maher T et al. Airborne monitoring to distinguish engineered nanomaterials from incidental particles for environmental health and safety. *Journal of Occupational and Environmental Hygiene*. 2008;6(2):73–81.
73. Sakamoto Y, Nakae D, Fukumori N, Tayama K, Maekawa A, Imai K et al. Induction of mesothelioma by a single intrascrotal administration of multi-wall carbon nanotube in intact male Fischer 344 rats. *The Journal of Toxicological Sciences*. 2009;34(1):65–76.
74. Nasibulin AG, Queipo P, Shandakov SD, Brown DP, Jiang H, Pikhitsa PV et al. Studies on mechanism of single-walled carbon nanotube formation. *Journal of Nanoscience and Nanotechnology*. 2006;6:1–14.
75. Maynard AD, Aitken RJ. Assessing exposure to airborne nanomaterials: Current abilities and future requirements. *Nanotoxicology*. 2007;1(1):26–41.
76. Vincent JH. *Aerosol Sampling: Science, Standards, Instrumentation and Applications*. Wiley.com; 2007.
77. Prather KA, Hatch CD, Grassian VH. Analysis of atmospheric aerosols. *Annual Review of Analytical Chemistry*. 2008;1:485–514.
78. Bergamaschi E. Occupational exposure to nanomaterials: Present knowledge and future development. *Nanotoxicology*. 2009;3(3):194–201.
79. Monteiro-Riviere NA, Tran CL. *Nanotoxicology: Characterization, Dosing and Health Effects*. CRC Press; 2007.
80. Yoshida R, Kitamura D, Maenosono S. Mutagenicity of water-soluble ZnO nanoparticles in Ames test. *The Journal of Toxicological Sciences*. 2009;34(1):119–22.
81. Landsiedel R, Kapp MD, Schulz M, Wiench K, Oesch F. Genotoxicity investigations on nanomaterials: Methods, preparation and characterization of test material, potential artifacts and limitations—Many questions, some answers. *Mutation Research/Reviews in Mutation Research*. 2009;681(2):241–58.
82. Li JG, Li WX, Xu JY, Cai XQ, Liu RL, Li YJ et al. Comparative study of pathological lesions induced by multiwalled carbon nanotubes in lungs of mice by intratracheal instillation and inhalation. *Environmental Toxicology*. 2007;22(4):415–21.
83. Jacobsen NR, Moller P, Jensen KA, Vogel U, Ladefoged O, Loft S et al. Lung inflammation and genotoxicity following pulmonary exposure to nanoparticles in ApoE$^{-/-}$ mice. *Part Fibre Toxicology*. 2009;6(2):2.

84. Shvedova AA, Kisin E, Murray AR, Johnson VJ, Gorelik O, Arepalli S et al. Inhalation vs. aspiration of single-walled carbon nanotubes in C57BL/6 mice: Inflammation, fibrosis, oxidative stress, and mutagenesis. *American Journal of Physiology—Lung Cellular and Molecular Physiology.* 2008; 295(4):L552–65.
85. Folkmann JK, Risom L, Jacobsen NR, Wallin H, Loft S, Møller P. Oxidatively damaged DNA in rats exposed by oral gavage to C_{60} fullerenes and single-walled carbon nanotubes. *Environmental Health Perspectives.* 2009;117(5):703.
86. Yang H, Liu C, Yang D, Zhang H, Xi Z. Comparative study of cytotoxicity, oxidative stress and genotoxicity induced by four typical nanomaterials: The role of particle size, shape and composition. *Journal of Applied Toxicology.* 2009;29(1):69–78.
87. Mouchet F, Landois P, Sarremejean E, Bernard G, Puech P, Pinelli E et al. Characterisation and *in vivo* ecotoxicity evaluation of double-wall carbon nanotubes in larvae of the amphibian *Xenopus laevis*. *Aquatic Toxicology.* 2008;87(2):127–37.
88. Muller J, Decordier I, Hoet PH, Lombaert N, Thomassen L, Huaux F et al. Clastogenic and aneugenic effects of multi-wall carbon nanotubes in epithelial cells. *Carcinogenesis.* 2008;29(2):427–33.
89. Muller J, Huaux FO, Fonseca A, Nagy JB, Moreau N, Delos M et al. Structural defects play a major role in the acute lung toxicity of multiwall carbon nanotubes: Toxicological aspects. *Chemical Research in Toxicology.* 2008;21(9):1698–705.
90. Karlsson HL, Cronholm P, Gustafsson J, Möller L. Copper oxide nanoparticles are highly toxic: A comparison between metal oxide nanoparticles and carbon nanotubes. *Chemical Research in Toxicology.* 2008;21(9):1726–32.
91. Pacurari M, Yin XJ, Ding M, Leonard SS, Schwegler-Berry D, Ducatman BS et al. Oxidative and molecular interactions of multi-wall carbon nanotubes (MWCNT) in normal and malignant human mesothelial cells. *Nanotoxicology.* 2008;2(3):155–70.
92. Zhu H, Yang H, Owen MR. Combined microarray analysis uncovers self-renewal related signaling in mouse embryonic stem cells. *Systems and Synthetic Biology.* 2007;1(4):171–81.
93. Lindberg HK, Falck GC-M, Suhonen S, Vippola M, Vanhala E, Catalán J et al. Genotoxicity of nanomaterials: DNA damage and micronuclei induced by carbon nanotubes and graphite nanofibres in human bronchial epithelial cells *in vitro*. *Toxicology Letters.* 2009;186(3):166–73.
94. Falck G, Lindberg H, Suhonen S, Vippola M, Vanhala E, Catalan J et al. Genotoxic effects of nanosized and fine TiO_2. *Human and Experimental Toxicology.* 2009;28(6–7):339–52.
95. Takagi A, Hirose A, Nishimura T, Fukumori N, Ogata A, Ohashi N et al. Induction of mesothelioma in p53± mouse by intraperitoneal application of multi-wall carbon nanotube. *The Journal of Toxicological Sciences.* 2008;33(1):105–16.
96. Lam C-W, James JT, McCluskey R, Arepalli S, Hunter RL. A review of carbon nanotube toxicity and assessment of potential occupational and environmental health risks. *CRC Critical Reviews in Toxicology.* 2006;36(3):189–217.
97. Maynard A. Experimental determination of ultrafine TiO_2 deagglomeration in a surrogate pulmonary surfactant: Preliminary results. *Annals of Occupational Hygiene.* 2002;46(Suppl 1):197–202.
98. Donaldson K, Borm P, Oberdorster G, Pinkerton K, Stone V, Tran C. Concordance between *in vitro* and *in vivo* dosimetry in the proinflammatory effects of low-toxicity, low-solubility particles: The key role of the proximal alveolar region. *Inhalation Toxicology.* 2008;20(1):53–62.
99. Nel A, Xia T, Mädler L, Li N. Toxic potential of materials at the nanolevel. *Science.* 2006;311(5761):622–7.
100. Elder A, Gelein R, Silva V, Feikert T, Opanashuk L, Carter J et al. Translocation of inhaled ultrafine manganese oxide particles to the central nervous system. *Environmental Health Perspectives.* 2006;114(8):1172.
101. Sonavane G, Tomoda K, Makino K. Biodistribution of colloidal gold nanoparticles after intravenous administration: Effect of particle size. *Colloids and Surfaces B: Biointerfaces.* 2008;66(2):274–80.
102. Pal S, Tak YK, Song JM. Does the antibacterial activity of silver nanoparticles depend on the shape of the nanoparticle? A study of the Gram-negative bacterium *Escherichia coli*. *Applied and Environmental Microbiology.* 2007;73(6):1712–20.
103. Jr RFH, Wu N, Porter D, Buford M, Wolfarth M, Holian A. Particle length-dependent titanium dioxide nanomaterials toxicity and bioactivity. *Particle and Fibre Toxicology.* 2009;6:35.
104. Schaeublin NM, Braydich-Stolle LK, Schrand AM, Miller JM, Hutchison J, Schlager JJ et al. Surface charge of gold nanoparticles mediates mechanism of toxicity. *Nanoscale.* 2011;3(2):410–20.
105. Donaldson K, Stone V, Clouter A, Renwick L, MacNee W. Ultrafine particles. *Occupational and Environmental Medicine.* 2001;58(3):211–6.

106. Warheit DB, Webb TR, Colvin VL, Reed KL, Sayes CM. Pulmonary bioassay studies with nanoscale and fine-quartz particles in rats: Toxicity is not dependent upon particle size but on surface characteristics. *Toxicological Sciences*. 2007;95(1):270–80.
107. Warheit DB, Hoke RA, Finlay C, Donner EM, Reed KL, Sayes CM. Development of a base set of toxicity tests using ultrafine TiO_2 particles as a component of nanoparticle risk management. *Toxicology Letters*. 2007;171(3):99–110.
108. Berridge MV, Herst PM, Tan AS. Tetrazolium dyes as tools in cell biology: New insights into their cellular reduction. *Biotechnology Annual Review*. 2005;11:127–52.
109. Hussain S, Hess K, Gearhart J, Geiss K, Schlager J. *In vitro* toxicity of nanoparticles in BRL 3A rat liver cells. *Toxicology in Vitro*. 2005;19(7):975–83.
110. Plumb JA, Milroy R, Kaye S. Effects of the pH dependence of 3-(4,5-dimethylthiazol-2-yl)-2,5-diphenyltetrazolium bromide–formazan absorption on chemosensitivity determined by a novel tetrazolium-based assay. *Cancer Research*. 1989;49(16):4435–40.
111. Wörle-Knirsch J, Pulskamp K, Krug H. Oops they did it again! Carbon nanotubes hoax scientists in viability assays. *Nano Letters*. 2006;6(6):1261–8.
112. Davis R, Lockwood P, Hobbs D, Messer R, Price R, Lewis J et al. *In vitro* biological effects of sodium titanate materials. *Journal of Biomedical Materials Research Part B: Applied Biomaterials*. 2007;83(2):505–11.
113. Yamin T-T, Ayala JM, Miller DK. Activation of the native 45-kDa precursor form of interleukin-1-converting enzyme. *Journal of Biological Chemistry*. 1996;271(22):13273–82.
114. Srinivasula SM, Saleh A, Ahmad M, Fernandes-Alnemri T, Alnemri ES. Isolation and assay of caspases. *Methods in Cell Biology*. 2001;66:1–27.
115. Wróblewski F, Ladue JS, eds. Lactic dehydrogenase activity in blood. *Proceedings of the Society for Experimental Biology and Medicine Society for Experimental Biology and Medicine* (New York, NY); 1955: Royal Society of Medicine.
116. Choi SY, Jeong S, Jang SH, Park J, Park JH, Ock KS et al. *In vitro* toxicity of serum protein-adsorbed citrate-reduced gold nanoparticles in human lung adenocarcinoma cells. *Toxicology in Vitro*. 2012;26(2):229–37.
117. Babson AL, Phillips G. A rapid colorimetric assay for serum lactic dehyurogenase. *Clinica Chimica Acta*. 1965;12(2):210–5.
118. Suska F, Gretzer C, Esposito M, Tengvall P, Thomsen P. Monocyte viability on titanium and copper coated titanium. *Biomaterials*. 2005;26(30):5942–50.
119. Strauss GH. Non-random cell killing in cryopreservation: Implications for performance of the battery of leukocyte tests (BLT), I. Toxic and immunotoxic effects. *Mutation Research/Environmental Mutagenesis and Related Subjects*. 1991;252(1):1–15.
120. Park MV, Annema W, Salvati A, Lesniak A, Elsaesser A, Barnes C et al. *In vitro* developmental toxicity test detects inhibition of stem cell differentiation by silica nanoparticles. *Toxicology and Applied Pharmacology*. 2009;240(1):108–16.
121. Passagne I, Morille M, Rousset M, Pujalté I, L'Azou B. Implication of oxidative stress in size-dependent toxicity of silica nanoparticles in kidney cells. *Toxicology*. 2012;299(2):112–24.
122. Herzog E, Casey A, Lyng FM, Chambers G, Byrne HJ, Davoren M. A new approach to the toxicity testing of carbon-based nanomaterials—The clonogenic assay. *Toxicology Letters*. 2007;174(1):49–60.
123. Ye Y, Liu J, Chen M, Sun L, Lan M. *In vitro* toxicity of silica nanoparticles in myocardial cells. *Environmental Toxicology and Pharmacology*. 2010;29(2):131–7.
124. Lee JK, Kim DB, Kim JI, Kim PY. *In vitro* cytotoxicity tests on cultured human skin fibroblasts to predict skin irritation potential of surfactants. *Toxicology in Vitro*. 2000;14(4):345–9.
125. Nemes Z, Dietz R, Lüth J, Gomba S, Hackenthal E, Gross F. The pharmacological relevance of vital staining with neutral red. *Experientia*. 1979;35(11):1475–6.
126. Spielmann H, Hoffmann S, Liebsch M, Botham P, Fentem JH, Eskes C et al. The ECVAM international validation study on *in vitro* tests for acute skin irritation: Report on the validity of the EPISKIN and EpiDerm assays and on the skin integrity function test. *ATLA—Alternatives to Laboratory Animals*. 2007;35(6):559.
127. Ramires P, Romito A, Cosentino F, Milella E. The influence of titania/hydroxyapatite composite coatings on *in vitro* osteoblasts behaviour. *Biomaterials*. 2001;22(12):1467–74.
128. Monteiro-Riviere NA, Inman AO, Zhang L. Limitations and relative utility of screening assays to assess engineered nanoparticle toxicity in a human cell line. *Toxicology and Applied Pharmacology*. 2009;234(2):222–35.
129. Huang M, Khor E, Lim L-Y. Uptake and cytotoxicity of chitosan molecules and nanoparticles: Effects of molecular weight and degree of deacetylation. *Pharmaceutical Research*. 2004;21(2):344–53.

130. Horobin RW, Kiernan JA. *Conn's Biological Stains. A Handbook of Dyes and Fluorochromes for Use in Biology and Medicine*, 10th edn. Oxford: BIOS Scientific Publishers; 2002.
131. Shang L, Zou X, Jiang X, Yang G, Dong S. Investigations on the adsorption behavior of neutral red on mercaptoethane sulfonate protected gold nanoparticles. *Journal of Photochemistry and Photobiology A: Chemistry*. 2007;187(2):152–9.
132. Casey A, Herzog E, Davoren M, Lyng F, Byrne H, Chambers G. Spectroscopic analysis confirms the interactions between single walled carbon nanotubes and various dyes commonly used to assess cytotoxicity. *Carbon*. 2007;45(7):1425–32.
133. Darolles C, Sage N, Armengaud J, Malard V. In vitro assessment of cobalt oxide particle toxicity: Identifying and circumventing interference. *Toxicology in Vitro*. 2013;27(6):1699–1710.
134. Saathoff JG, Inman AO, Xia XR, Riviere JE, Monteiro-Riviere NA. In vitro toxicity assessment of three hydroxylated fullerenes in human skin cells. *Toxicology in Vitro*. 2011;25(8):2105–12.
135. Singh RP, Ramarao P. Cellular uptake, intracellular trafficking and cytotoxicity of silver nanoparticles. *Toxicology Letters*. 2012;213(2):249–59.
136. Pokhrel LR, Silva T, Dubey B, El Badawy AM, Tolaymat TM, Scheuerman PR. Rapid screening of aquatic toxicity of several metal-based nanoparticles using the MetPLATE™ bioassay. *Science of the Total Environment*. 2012;426:414–22.
137. Braydich-Stolle L, Hussain S, Schlager JJ, Hofmann M-C. In vitro cytotoxicity of nanoparticles in mammalian germline stem cells. *Toxicological Sciences*. 2005;88(2):412–9.
138. Van Engeland M, Nieland LJ, Ramaekers FC, Schutte B, Reutelingsperger CP. Annexin V-affinity assay: A review on an apoptosis detection system based on phosphatidylserine exposure. *Cytometry*. 1998;31(1):1–9.
139. Aubry JP, Blaecke A, Lecoanet-Henchoz S, Jeannin P, Herbault N, Caron G et al. Annexin V used for measuring apoptosis in the early events of cellular cytotoxicity. *Cytometry*. 1999;37(3):197–204.
140. Bartkowiak D, Högner S, Baust H, Nothdurft W, Röttinger EM. Comparative analysis of apoptosis in HL60 detected by annexin-V and fluorescein-diacetate. *Cytometry*. 1999;37(3):191–6.
141. Aljandali A, Pollack H, Yeldandi A, Li Y, Weitzman SA, Kamp DW. Asbestos causes apoptosis in alveolar epithelial cells: Role of iron-induced free radicals. *Journal of Laboratory and Clinical Medicine*. 2001;137(5):330–9.
142. Isakovic A, Markovic Z, Todorovic-Markovic B, Nikolic N, Vranjes-Djuric S, Mirkovic M et al. Distinct cytotoxic mechanisms of pristine versus hydroxylated fullerene. *Toxicological Sciences*. 2006;91(1):173–83.
143. Chan W-H, Shiao N-H, Lu P-Z. CdSe quantum dots induce apoptosis in human neuroblastoma cells via mitochondrial-dependent pathways and inhibition of survival signals. *Toxicology Letters*. 2006;167(3):191–200.
144. Darzynkiewicz Z, Bedner E, Traganos F. Difficulties and pitfalls in analysis of apoptosis. *Methods in Cell Biology*. 2001;63:527–46.
145. Shukla S, Priscilla A, Banerjee M, Bhonde RR, Ghatak J, Satyam P et al. Porous gold nanospheres by controlled transmetalation reaction: A novel material for application in cell imaging. *Chemistry of Materials*. 2005;17(20):5000–5.
146. Waterhouse NJ, Green DR. Mitochondria and apoptosis: HQ or high-security prison? *Journal of Clinical Immunology*. 1999,19(6):378–87.
147. Harhaji L, Isakovic A, Vucicevic L, Janjetovic K, Misirkic M, Markovic Z et al. Modulation of tumor necrosis factor-mediated cell death by fullerenes. *Pharmaceutical Research*. 2008;25(6):1365–76.
148. Zeni O, Palumbo R, Bernini R, Zeni L, Sarti M, Scarfì MR. Cytotoxicity investigation on cultured human blood cells treated with single-wall carbon nanotubes. *Sensors*. 2008;8(1):488–99.
149. Thibodeau M, Giardina C, Hubbard AK. Silica-induced caspase activation in mouse alveolar macrophages is dependent upon mitochondrial integrity and aspartic proteolysis. *Toxicological Sciences*. 2003;76(1):91–101.
150. Segal MS, Beem E. Effect of pH, ionic charge, and osmolality on cytochromatic-mediated caspase-3 activity. *American Journal of Physiology—Cell Physiology*. 2001;281(4):C1196–204.
151. Jakubowski W, Bartosz G. 2, 7-dichlorofluorescein oxidation and reactive oxygen species: What does it measure? *Cell Biology International*. 2000;24(10):757–60.
152. Tarpey MM, Wink DA, Grisham MB. Methods for detection of reactive metabolites of oxygen and nitrogen: In vitro and in vivo considerations. *American Journal of Physiology—Regulatory, Integrative and Comparative Physiology*. 2004;286(3):R431–44.
153. Royall JA, Ischiropoulos H. Evaluation of 2′,7′-dichlorofluorescein and dihydrorhodamine 123 as fluorescent probes for intracellular H_2O_2 in cultured endothelial cells. *Archives of Biochemistry and Biophysics*. 1993;302(2):348–55.

154. Wrona M, Wardman P. Properties of the radical intermediate obtained on oxidation of 2′,7′-dichlorodihydrofluorescein, a probe for oxidative stress. *Free Radical Biology and Medicine*. 2006;41(4):657–67.
155. Aam BB, Fonnum F. Carbon black particles increase reactive oxygen species formation in rat alveolar macrophages *in vitro*. *Archives of Toxicology*. 2007;81(6):441–6.
156. Knaapen AM, Borm PJ, Albrecht C, Schins RP. Inhaled particles and lung cancer. Part A: Mechanisms. *International Journal of Cancer*. 2004;109(6):799–809.
157. Fahmy B, Cormier SA. Copper oxide nanoparticles induce oxidative stress and cytotoxicity in airway epithelial cells. *Toxicology in Vitro*. 2009;23(7):1365–71.
158. Fujii T, Hayashi S, Hogg JC, Vincent R, Van Eeden SF. Particulate matter induces cytokine expression in human bronchial epithelial cells. *American Journal of Respiratory Cell and Molecular Biology*. 2001;25(3):265–71.
159. Park E-J, Yoon J, Choi K, Yi J, Park K. Induction of chronic inflammation in mice treated with titanium dioxide nanoparticles by intratracheal instillation. *Toxicology*. 2009;260(1):37–46.
160. Bonner JC. The epidermal growth factor receptor at the crossroads of airway remodeling. *American Journal of Physiology—Lung Cellular and Molecular Physiology*. 2002;283(3):L528–30.
161. Sime P, Marr RA, Gauldie D, Xing Z, Hewlett BR, Graham FL, Gauldie J. Transfer of tumor necrosis factor-alpha to rat lung induces severe pulmonary inflammation and patchy interstitial fibrogenesis with induction of transforming growth factor-beta1 and myofibroblasts. *American Journal of Pathology*. 1998;153(3):825–32.
162. Ahamed M, Posgai R, Gorey TJ, Nielsen M, Hussain SM, Rowe JJ. Silver nanoparticles induced heat shock protein 70, oxidative stress and apoptosis in *Drosophila melanogaster*. *Toxicology and Applied Pharmacology*. 2010;242(3):263–9.
163. Shukla RK, Sharma V, Pandey AK, Singh S, Sultana S, Dhawan A. ROS-mediated genotoxity induced by titanium dioxide nanoparticles in human epidermal cells. *Toxicology in Vitro*. 2011;25(1):231–41.
164. Srinivas A, Rao PJ, Selvam G, Murthy PB, Reddy PN. Acute inhalation toxicity of cerium oxide nanoparticles in rats. *Toxicology Letters*. 2011;205(2):105–15.
165. Chairuangkitti P, Lawanprasert S, Roytrakul S, Aueviriyavit S, Phummiratch D, Kulthong K et al. Silver nanoparticles induce toxicity in A549 cells via ROS-dependent and ROS-independent pathways. *Toxicology in Vitro*. 2012;27:330–38.
166. Lequin RM. Enzyme immunoassay (EIA)/enzyme-linked immunosorbent assay (ELISA). *Clinical Chemistry*. 2005;51(12):2415–8.
167. Tao F, Kobzik L. Lung macrophage–epithelial cell interactions amplify particle-mediated cytokine release. *American Journal of Respiratory Cell and Molecular Biology*. 2002;26(4):499–505.
168. Wottrich R, Diabaté S, Krug HF. Biological effects of ultrafine model particles in human macrophages and epithelial cells in mono- and co-culture. *International Journal of Hygiene and Environmental Health*. 2004;207(4):353–61.
169. Sayes CM, Reed KL, Warheit DB. Assessing toxicity of fine and nanoparticles: Comparing *in vitro* measurements to *in vivo* pulmonary toxicity profiles. *Toxicological Sciences*. 2007;97(1):163–80.
170. Duffin R, Tran L, Brown D, Stone V, Donaldson K. Proinflammogenic effects of low-toxicity and metal nanoparticles *in vivo* and *in vitro*: Highlighting the role of particle surface area and surface reactivity. *Inhalation Toxicology*. 2007;19(10):849–56.
171. Davoren M, Herzog E, Casey A, Cottineau B, Chambers G, Byrne HJ et al. *In vitro* toxicity evaluation of single walled carbon nanotubes on human A549 lung cells. *Toxicology in Vitro*. 2007;21(3):438–48.
172. Sayes CM, Marchione AA, Reed KL, Warheit DB. Comparative pulmonary toxicity assessments of C_{60} water suspensions in rats: Few differences in fullerene toxicity *in vivo* in contrast to *in vitro* profiles. *Nano Letters*. 2007;7(8):2399–406.
173. Rao KMK, Porter DW, Meighan T, Castranova V. The sources of inflammatory mediators in the lung after silica exposure. *Environmental Health Perspectives*. 2004;112(17):1679.
174. Ryman-Rasmussen JP, Riviere JE, Monteiro-Riviere NA. Surface coatings determine cytotoxicity and irritation potential of quantum dot nanoparticles in epidermal keratinocytes. *Journal of Investigative Dermatology*. 2006;127(1):143–53.
175. Monteiro-Riviere NA, Inman AO. Challenges for assessing carbon nanomaterial toxicity to the skin. *Carbon*. 2006;44(6):1070–8.
176. Veranth JM, Kaser EG, Veranth MM, Koch M, Yost GS. Cytokine responses of human lung cells (BEAS-2B) treated with micron-sized and nanoparticles of metal oxides compared to soil dusts. *Part Fibre Toxicology*. 2007;4(2). doi:10.1186/1743-8977-4-2.
177. Guo L, Von Dem Bussche A, Buechner M, Yan A, Kane AB, Hurt RH. Adsorption of essential micronutrients by carbon nanotubes and the implications for nanotoxicity testing. *Small*. 2008;4(6):721–7.

178. Singh N, Manshian B, Jenkins GJ, Griffiths SM, Williams PM, Maffeis TG et al. Nanogenotoxicology: The DNA damaging potential of engineered nanomaterials. *Biomaterials.* 2009;30(23–24):3891–914.
179. AshaRani P, Low Kah Mun G, Hande MP, Valiyaveettil S. Cytotoxicity and genotoxicity of silver nanoparticles in human cells. *ACS Nano.* 2008;3(2):279–90.
180. Trouiller B, Reliene R, Westbrook A, Solaimani P, Schiestl RH. Titanium dioxide nanoparticles induce DNA damage and genetic instability *in vivo* in mice. *Cancer Research.* 2009;69(22):8784–9.
181. Li JJ, Zou L, Hartono D, Ong CN, Bay BH, Lanry Yung LY. Gold nanoparticles induce oxidative damage in lung fibroblasts *in vitro*. *Advanced Materials.* 2008;20(1):138–42.
182. Pantarotto D, Briand J-P, Prato M, Bianco A. Translocation of bioactive peptides across cell membranes by carbon nanotubes. *Chemical Communications.* 2004;10(1):16–7.
183. Muller J, Delos M, Panin N, Rabolli V, Huaux F, Lison D. Absence of carcinogenic response to multi-wall carbon nanotubes in a 2-year bioassay in the peritoneal cavity of the rat. *Toxicological Sciences.* 2009;110(2):442–8.
184. Roller M. Carcinogenicity of inhaled nanoparticles. *Inhalation Toxicology.* 2009;21(S1):144–57.
185. Onuma K, Sato Y, Ogawara S, Shirasawa N, Kobayashi M, Yoshitake J et al. Nano-scaled particles of titanium dioxide convert benign mouse fibrosarcoma cells into aggressive tumor cells. *The American Journal of Pathology.* 2009;175(5):2171–83.
186. Hackenberg S, Scherzed A, Kessler M, Hummel S, Technau A, Froelich K et al. Silver nanoparticles: Evaluation of DNA damage, toxicity and functional impairment in human mesenchymal stem cells. *Toxicology Letters.* 2011;201(1):27–33.
187. Arora S, Rajwade JM, Paknikar KM. Nanotoxicology and *in vitro* studies: The need of the hour. *Toxicology and Applied Pharmacology.* 2012;258(2):151–65.
188. Kim S, Choi JE, Choi J, Chung K-H, Park K, Yi J et al. Oxidative stress-dependent toxicity of silver nanoparticles in human hepatoma cells. *Toxicology in Vitro.* 2009;23(6):1076–84.
189. Lemaire F, Mandon CA, Reboud J, Papine A, Angulo J, Pointu H et al. Toxicity assays in nanodrops combining bioassay and morphometric endpoints. *PloS One.* 2007;2(1):e163.
190. Jan E, Byrne SJ, Cuddihy M, Davies AM, Volkov Y, Gun'ko YK et al. High-content screening as a universal tool for fingerprinting of cytotoxicity of nanoparticles. *ACS Nano.* 2008;2(5):928–38.
191. Wittmaack K. In search of the most relevant parameter for quantifying lung inflammatory response to nanoparticle exposure: Particle number, surface area, or what? *Environmental Health Perspectives.* 2007;115(2):187.
192. Tang M, Zhang T, Xue Y, Wang S, Huang M, Yang Y et al. Dose dependent *in vivo* metabolic characteristics of titanium dioxide nanoparticles. *Journal of Nanoscience and Nanotechnology.* 2010;10(12):8575–83.
193. Leszczynski J. Bionanoscience: Nano meets bio at the interface. *Nature Nanotechnology.* 2010;5(9):633–4.
194. Barnard AS. How can *ab initio* simulations address risks in nanotech? *Nature Nanotechnology.* 2009;4(6):332–5.
195. Lacerda L, Bianco A, Prato M, Kostarelos K. Carbon nanotubes as nanomedicines: From toxicology to pharmacology. *Advanced Drug Delivery Reviews.* 2006;58(14):1460–70.
196. Puzyn T, Gajewicz A, Leszczynska D, Leszczynski J. Nanomaterials—The next great challenge for QSAR modelers. In: Puzyn T, Leszczynski J, Cronin M, eds. *Recent Advances in QSAR Studies: Methods and Applications.* London, New York: M.T. Springer; 2010.
197. Puzyn T, Rasulev B, Gajewicz A, Hu X, Dasari TP, Michalkova A et al. Using nano-QSAR to predict the cytotoxicity of metal oxide nanoparticles. *Nature Nanotechnology.* 2011;6(3):175–8.
198. Bonnefoi MS, Belanger SE, Devlin DJ, Doerrer NG, Embry MR, Fukushima S et al. Human and environmental health challenges for the next decade (2010–2020). *Critical Reviews in Toxicology.* 2010;40(10):893–911.
199. Lyon DY, Adams LK, Falkner JC, Alvarez PJ. Antibacterial activity of fullerene water suspensions: Effects of preparation method and particle size. *Environmental Science and Technology.* 2006;40(14):4360–6.
200. Zhu S, Oberdörster E, Haasch ML. Toxicity of an engineered nanoparticle (fullerene, C_{60}) in two aquatic species, *Daphnia* and fathead minnow. *Marine Environmental Research.* 2006;62:S5–9.
201. Roberts AP, Mount AS, Seda B, Souther J, Qiao R, Lin S et al. *In vivo* biomodification of lipid-coated carbon nanotubes by *Daphnia magna*. *Environmental Science and Technology.* 2007;41(8):3025–9.
202. Smith CJ, Shaw BJ, Handy RD. Toxicity of single walled carbon nanotubes to rainbow trout, (*Oncorhynchus mykiss*): Respiratory toxicity, organ pathologies, and other physiological effects. *Aquatic Toxicology.* 2007;82(2):94–109.

203. Templeton RC, Ferguson PL, Washburn KM, Scrivens WA, Chandler GT. Life-cycle effects of single-walled carbon nanotubes (SWNTs) on an estuarine meiobenthic copepod. *Environmental Science and Technology*. 2006;40(23):7387–93.
204. Adams LK, Lyon DY, Alvarez PJ. Comparative eco-toxicity of nanoscale TiO_2, SiO_2, and ZnO water suspensions. *Water Research*. 2006;40(19):3527–32.
205. Zhu X, Zhu L, Duan Z, Qi R, Li Y, Lang Y. Comparative toxicity of several metal oxide nanoparticle aqueous suspensions to zebrafish (*Danio rerio*) early developmental stage. *Journal of Environmental Science and Health Part A*. 2008;43(3):278–84.
206. Zhu X, Zhu L, Chen Y, Tian S. Acute toxicities of six manufactured nanomaterial suspensions to *Daphnia magna*. *Journal of Nanoparticle Research*. 2009;11(1):67–75.
207. Scown TM, Goodhead RM, Johnston BD, Moger J, Baalousha M, Lead JR et al. Assessment of cultured fish hepatocytes for studying cellular uptake and (eco) toxicity of nanoparticles. *Environmental Chemistry*. 2010;7(1):36–49.
208. Xia T, Zhao Y, Sager T, George S, Pokhrel S, Li N et al. Decreased dissolution of ZnO by iron doping yields nanoparticles with reduced toxicity in the rodent lung and zebrafish embryos. *ACS Nano*. 2011;5(2):1223–35.
209. Metz KM, Mangham AN, Bierman MJ, Jin S, Hamers RJ, Pedersen JA. Engineered nanomaterial transformation under oxidative environmental conditions: Development of an *in vitro* biomimetic assay. *Environmental Science and Technology*. 2009;43(5):1598–604.
210. Makarucha A, Todorova N, Yarovsky I. Nanomaterials in biological environment: A review of computer modelling studies. *European Biophysics Journal*. 2011;40(2):103–15.
211. Hayashi Y. Designing *in vitro* assay systems for hazard characterization. Basic strategies and related technical issues. *Experimental and Toxicologic Pathology*. 2005;57:227–32.
212. Sutter TR. Molecular and cellular approaches to extrapolation for risk assessment. *Environmental Health Perspectives*. 1995;103(4):386.
213. Anderson D, Russell T, Savage Jr RE. The status of alternative methods in toxicology. *Toxicology Methods*. 1997;7(2):149–51.
214. Cimino MC. Comparative overview of current international strategies and guidelines for genetic toxicology testing for regulatory purposes. *Environmental and Molecular Mutagenesis*. 2006;47(5):362–90.
215. Snodin DJ. An EU perspective on the use of *in vitro* methods in regulatory pharmaceutical toxicology. *Toxicology Letters*. 2002;127(1):161–8.
216. Clemedson C, Ekwall B. Overview of the final MEIC results: I. The *in vitro–in vitro* evaluation. *Toxicology in Vitro*. 1999;13(4):657–63.
217. Liebsch M, Spielmann H. Currently available *in vitro* methods used in the regulatory toxicology. *Toxicology Letters*. 2002;127(1):127–34.
218. Lilienblum W, Dekant W, Foth H, Gebel T, Hengstler J, Kahl R et al. Alternative methods to safety studies in experimental animals: Role in the risk assessment of chemicals under the new European Chemicals Legislation (REACH). *Archives of Toxicology*. 2008;82(4):211–36.
219. Heijne WH, Kienhuis AS, van Ommen B, Stierum RH, Groten JP. Systems toxicology: Applications of toxicogenomics, transcriptomics, proteomics and metabolomics in toxicology. 2005;2(5):767–80.
220. Marquet P, Rappaz B, Magistretti PJ, Cuche E, Emery Y, Colomb T et al. Digital holographic microscopy: A noninvasive contrast imaging technique allowing quantitative visualization of living cells with subwavelength axial accuracy. *Optics Letters*. 2005;30(5):468–70.
221. Kemper B, Carl D, Höink A, Von Bally G, Bredebusch I, Schenekenburger J. Modular digital holographic microscopy system for marker free quantitative phase contrast imaging of living cells. *Proc. SPIE 6191, Biophotonics and New Therapy Frontiers*; April 2006;6191:61910T-1–61910T-8.
222. Schnekenburger J, Kemper B. Digital holography imaging—A 3D view on drug induced apoptosis. *Screening Trends of Drug Discovery*. 2008;9:24–6.
223. Ratanachoo K, Gascoyne PR, Ruchirawat M. Detection of cellular responses to toxicants by dielectrophoresis. *Biochimica et Biophysica Acta (BBA)—Biomembranes*. 2002;1564(2):449–58.
224. Ceriotti L, Ponti J, Colpo P, Sabbioni E, Rossi F. Assessment of cytotoxicity by impedance spectroscopy. *Biosensors and Bioelectronics*. 2007;22(12):3057–63.
225. Ressler J, Grothe H, Motrescu E, Wolf B, eds. New concepts for chip-supported multi-well-plates: Realization of a 24-well-plate with integrated impedance-sensors for functional cellular screening applications and automated microscope aided cell-based assays. *Engineering in Medicine and Biology Society, 2004 IEMBS'04 26th Annual International Conference of the IEEE*, San Francisco, CA; 2004: IEEE.

226. Ponti J, Ceriotti L, Munaro B, Farina M, Munari A, Whelan M et al. Comparison of impedance-based sensors for cell adhesion monitoring and *in vitro* methods for detecting cytotoxicity induced by chemicals. *ATLA Alternatives to Laboratory Animals*. 2006;34(5):515–25.
227. Linkov I, Satterstrom FK. Nanomaterial risk assessment and risk management: Review of regulatory frameworks. In: Linkov I, Ferguson E, Magar VS, eds. *Real Time and Deliberative Decision Making: Application for Risk Assessment for Non-chemical Stressors*. Amsterdam: Springer; 2008. pp. 129–58.
228. Kandlikar MG, Ramachandran AD, Maynard A, Murdock B, Toscano W. Health risk assessment for nanoparticles: A case for using expert judgment. *Journal of Nanoparticle Research*. 2007;9:137–56.
229. Tsuji JS, Maynard AD, Howard PC, James JT, Lam CW, Warheit DB et al. Research strategies for safety evaluation of nanomaterials, part IV: Risk assessment of nanoparticles. *Toxicology Science* 2006;89:42–50.
230. Li JJ, Muralikrishnan S, Ng C, Yung LL, Bay B. Nanoparticle-induced pulmonary toxicity. *Experimental Biology and Medicine*. 2010;235:1025–33.

20 Immunotoxicity of Carbon Nanoparticles

Paulami Pal, Bhaskar Mazumder, and Yaswant Pathak

CONTENTS

20.1	Introduction	425
20.2	Immunology	426
20.3	Carbon Nanoparticle and Its Routes of Entry within the Body	427
20.4	Physical Attribute of Carbon Nanoparticles Responsible for Toxicity	429
20.5	Nanoparticles Responsible for Toxicity	430
20.6	Cell Uptake of Carbon Nanoparticle	430
20.7	Mechanism of Carbon Nanoparticle Toxicity	430
	20.7.1 Free Radical Formation	431
	20.7.2 Reactive Oxygen Species	431
	20.7.3 Increased Inflammatory Responses	431
	20.7.4 Granuloma Formation	432
	20.7.5 Apoptosis	432
20.8	Carbon Nanoparticle-Induced Oxidative Stress	432
20.9	Inflammatory Response and Phagocytosis	433
20.10	Antigenicity	433
20.11	Causes of Carbon Nanoparticle Immunotoxicity and Control	433
20.12	Effects of Nanoparticle-Adsorbed Proteins (Peptides) on Immune Response	434
20.13	Lung Toxicity	435
	20.13.1 Effects on Lungs	436
	20.13.2 Lungs Disorder That May Rise from Exposure to Nanoparticles	437
	20.13.2.1 Pulmonary Fibrosis	437
	20.13.2.2 Pneumoconiosis	437
	20.13.2.3 Exacerbation of Asthma	438
20.14	Cytotoxicity	438
20.15	Dermal Toxicity	439
20.16	Genotoxicity	439
20.17	Conclusion	440
References		441

20.1 INTRODUCTION

Nanoscience and nanotechnology is the study of materials or particles having a size ranging between 0.1 and 100 nm. Advancements in nanoscience have thrown a new light of hope on biomedical science and therapy. The U.S. Environmental Protection Agency has classified nanoparticles (NPs) into four distinct groups: (1) carbon-based materials, (2) metal-based materials, (3) dendrimers, and (4) composites, including nanoclay [1].

In this chapter, we will focus on carbon-based NPs and their toxicity with special emphasis on immunotoxicity of carbon NPs. Carbon nanofibers or vapor-grown nanofibers are cylindrical nanostructures with graphene layers arranged as stacked cones, cups, or plates. These carbon nanofibers when wrapped into perfect cylinders with graphene layers are called carbon nanotubes (CNTs). The most common form of carbon NPs is CNTs. They consist of carbon atoms arranged in a series of condensed benzene rings wrapped in a tubular form. Depending on the number of sheets rolled, there are two types of CNTs: *single-walled carbon nanotubes* (SWCNTs) and *multiwalled carbon nanotubes* (MWCNTs). But regardless of its type, CNTs possess many significant properties such as high aspect ratio, tremendous strength, ultra-light weight, high thermal conductivity, and electronic properties ranging from metallic to semiconducting [2].

Recent research show a growing area of investigation on toxicity related to the exposure of NPs within the human body with the increased use of nanotechnology in drug delivery and environmental science. These NPs enter the body either accidentally from the environment or by the deliberate placing of NPs within the body through drug delivery or different biomedical applications. The consideration of the immunomodulating potential of NPs, including both immunosuppressive and immunostimulatory after binding with plasma protein, is a nascent field of research and hence much data have not been documented. Observation from different *in vitro* and *in vivo* toxicity studies showed inflammation, cytokine production, oxidative stress, cytoskeletal changes, apoptosis, and alteration in vesicular trafficking, in gene expression, and cell signaling [3].

20.2 IMMUNOLOGY

Immunology is the study of the relationship between body system, pathogens, and immunity. The science of immunology deals with organs and cellular components that comprise the immune system and their function and interaction. The immune system has been divided into two types: (1) innate immune system and (2) acquired or adaptive immune system. Again, an acquired immune system can be classified into humoral and cellular components [4] (Figure 20.1).

The various parts of the immune system are connected via the blood and lymphatic systems. Bone marrow, spleen, lymph nodes, thymus, and mucosa-associated lymphoid tissues are the main organs of the immune system and are involved in the manufacturing, maturation, differentiation, proliferation, and storage of immune cells. The blood is composed of red and white blood corpuscles together with other molecules such as various complement proteins and immunoglobulins. White blood corpuscles, also known as leukocytes, play a major role in the immune system. They are made up mainly of polymorphonuclear granulocytes (PMNs), in addition to monocytes, natural killer cells, B and T lymphocytes. PMNs are composed of neutrophils (phagocytic cells), eosinophils, and basophils. T and B lymphocytes, natural killer cells, and PMNs are the main cells of the immune system. In addition, dendritic cells and macrophages are essential components of the immune system that act as a chemical agent or antigen-presenting cells (APCs). Mononuclear phagocytic system (MPS) is the name given to the part of the immune system that consists of phagocytic cells, such as blood monocytes and macrophages accumulated in lymph nodes, liver, spleen, and other tissues [4] (Figure 20.2).

The humoral (antibody) response is characterized as the interaction between antibody and antigens. Antibodies are specific proteins released from a particular class of immune cells called B lymphocytes. Antigens are defined as anything that enkindles generation of antibodies; hence, they are

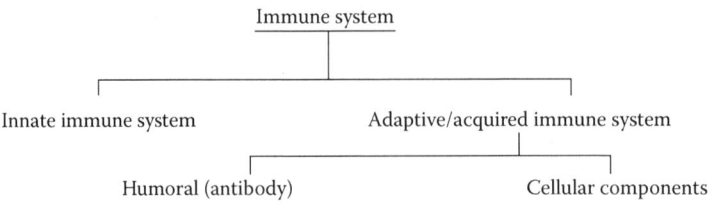

FIGURE 20.1 Classification of immune system.

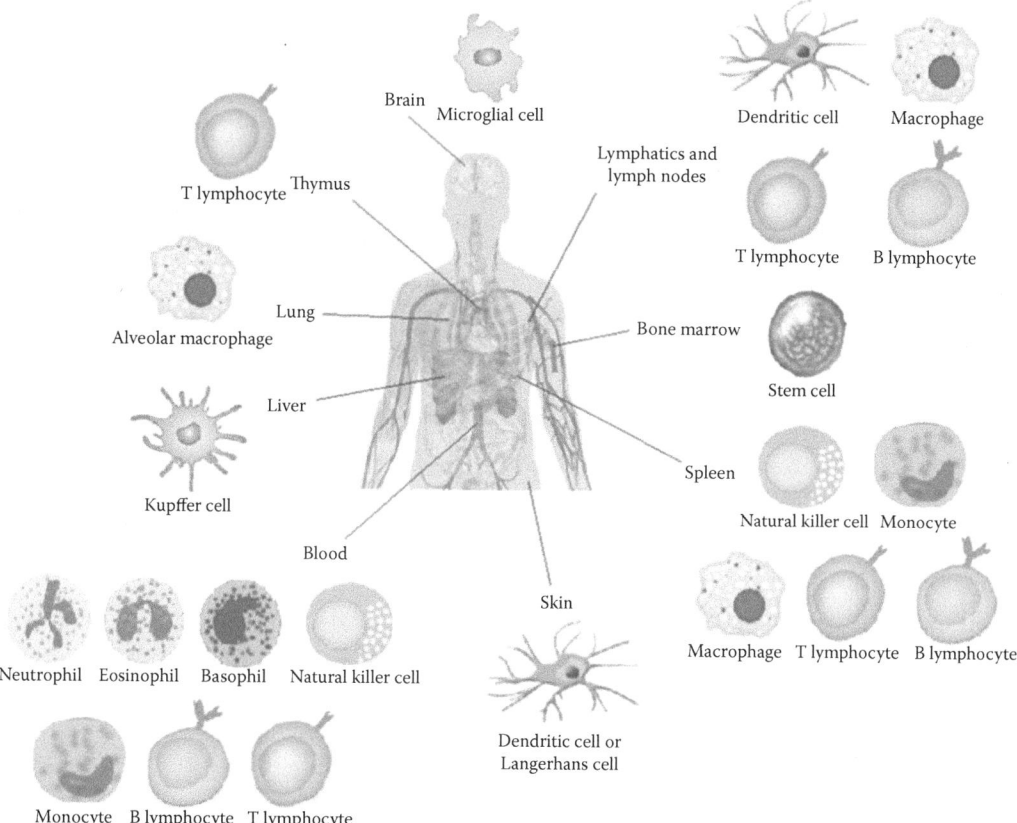

FIGURE 20.2 Various parts of the immune system and the presence of its cellular components in organs. (From Elsabahy M, Wooley KL. *Chemical Society Reviews*. 2013. With permission.)

also called antibody generators. Immunology rests on an understanding of the properties of these two biological entities. Cellular response, which can not only kill infected cells in its own right, is also crucial in controlling the antibody response. Put simply, both systems are highly interdependent [4].

Immunotoxicity can be defined as any adverse effect on the immune system that can result from exposure to foreign matter (Figure 20.3). To determine the effects of any foreign agent on the immune system of the human body, in general, five adverse event categories are studied. (1) Immunosuppression refers to impairment of any component of the immune system resulting in a decreased immune function. (2) Immunogenicity is an immune reaction evoked by a specific agent termed as a stressor and/or its metabolites, possibly resulting in an allergic response due to multiple exposure to the stressor. (3) Hypersensitivity is the immunological sensitization by a specific stressor and/or its metabolites, resulting in a strong adverse response. (4) Autoimmunity refers to a pathological process whereby the immune system responds to self-antigens. (5) Adverse immunostimulation refers to any antigen-nonspecific, inappropriate, or uncontrolled activation of some component of the immune system [4].

20.3 CARBON NANOPARTICLE AND ITS ROUTES OF ENTRY WITHIN THE BODY

CNTs are made up of carbon atoms arranged in a series of condensed benzene rings, wrapped in a tubular form. They represent a new allotropic form of carbon fabricated from the fullerene family that can be represented as rolled-up graphite sheets held together by van der Waal's force. The

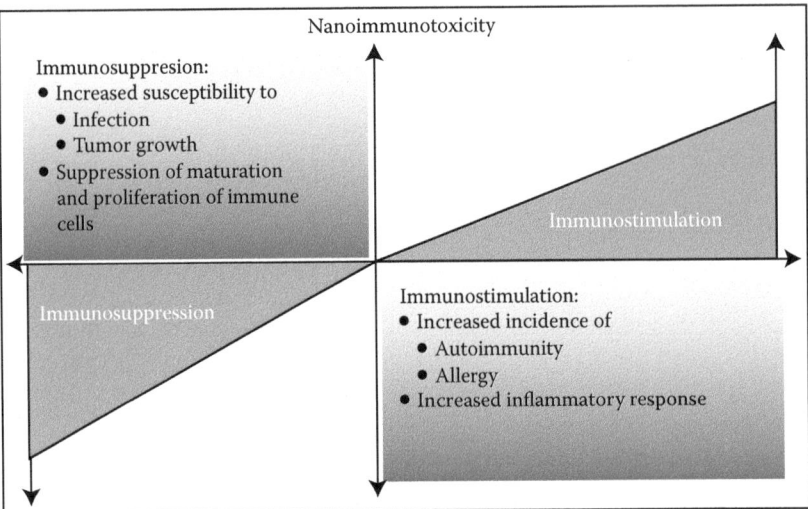

FIGURE 20.3 Schematic overview of nanoimmunotoxicology. Immune dysfunction resulting from exposure to nanomaterials or particles may take the form of specific stimulation of any components of the immune system leading to allergy, autoimminuty disorder, and so on. It might also take the form of immunosuppression leading to more infectious pathologies and tumor growth.

structure of CNTs can be imagined as the cylindrical roll-up of one or more graphene sheets held together with each carbon atom bonded to three neighboring carbon atoms forming only sp^2 hybridization in a honeycomb arrangement [5]. Owing to its high aspect ratio (length/diameter), these CNTs represent a nearly one-dimensional structure. Since sp^2 hybridization is stronger than the sp^3 hybridization found in diamonds, CNTs have a unique strength. From an atomic point of view, a nanotube can be divided into two parts, that is, the side wall and the end cap. The end caps can be considered as hemispherical fullerenes, curved in 2D and the side wall contains less distorted carbon atoms and is curved in 1D.

Depending on the method of synthesis, CNTs have lengths that vary from some hundred nanometers to several millimeters, but their diameters depend on their class: (a) SWCNTs are 0.4–3 nm in diameter and (b) MWCNTs are 2–500 nm in diameter. SWCNTs are formed when only one graphene sheet is rolled in a tubular fashion. The MWCNTs also consist of several cylinders of graphitic shells with a layer spacing of 0.3–0.4 nm, ideally closed at each end by half a fullerene [6]. CNTs can be classified as carbon nanohorns (CNHs), nanobuds, and nanotorus, depending upon their shape [7]. The CNTs are commercially manufactured using heavy metal as catalysts and substrates that include silica (SiO_2). The CNTs have low mass density, high electrical and magnetic properties, high thermal conductivity, and good mechanical properties. Hence, exposure to carbon NPs may be accidental, occupational, and also on biomedical grounds for diagnosis and therapeutics. In addition to the potential for occupational exposure with increasing production of carbon NPs, environmental exposures to low concentrations of MWCNTs but not SWCNTs, have been reported [8]. The presence of MWCNTs in the environment may be due to combustion emissions, as they have been found in the effluents from natural gas combustion.

The applications of carbon NPs are listed as follows:

1. CNTs in controlled and targeted drug delivery
2. CNTs as a nonviral gene delivery system
3. CNTs in diagnostics and imaging
4. Cancer therapy
5. Vaccine delivery

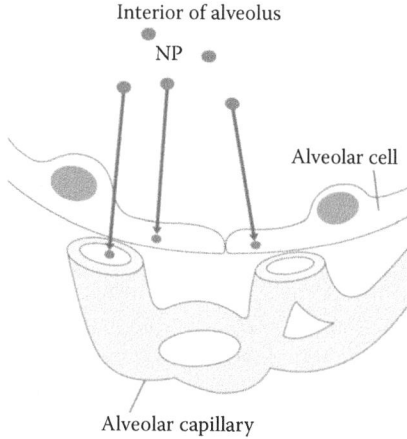

FIGURE 20.4 Entry of nanoparticles into alveolar cells and the pulmonary circulation. (From Jasmine Jia'en Li et al. *Experimental Biology and Medicine* 2010;235:1025–33.)

There are three main routes of entry of carbon nanoparticles (CNPs) into the human body. First, CNPs are inhaled into the body from the atmospheric air via the upper respiratory tract at the workplace and from the environment, which has become a common event. Second, CNPs enter the human body by oral ingestion. Third, CNP entry occurs via the dermal route, either by injection into the dermal layer or by absorption through the pores of the skin, which are mainly mediated by exposure to therapeutic or cosmetic applications. The lungs would be the first line of contact for particles that gain entry into the body through inhalation and hence the most likely organ for accumulation and long-term exposure (Figure 20.4).

20.4 PHYSICAL ATTRIBUTE OF CARBON NANOPARTICLES RESPONSIBLE FOR TOXICITY

Research suggests that the physicochemical properties such as size, shape, charge, and surface groups of the NPs determine their immunotoxicity. The physical attributes of CNTs such as fiber shape, length, and the aggregation status influences the immunological responses and their local deposition in tissues [9]. The shape and length of CNTs can determine the internalization of CNTs by macrophages and hence the immune response. Shorter CNTs are reported to be less toxic than the longer CNTs. Shorter-length CNTs when injected subcutaneously in rats were found in the cytosol of the macrophages after 4 weeks, but longer CNTs were found to be free floating and causing inflammation [10]. This confirms that the toxicity of MWCNTs was length dependent and comparable with that of asbestos toxicity. In a study, abdominal cavities of rats were injected with MWCNTs and asbestos fibers. After 24 h of exposure, an increased immune response was observed, and after 7 days of exposure, granuloma formation was observed in both cases. The increased immune response due to MWCNTs exposure was named "frustrated phagocytosis" wherein macrophages were unable to engulf long CNTs mainly because of their length.

Surface area and surface chemistry also influence the toxicity of a carbon NP. In one study, five carbon-based materials, graphite, SWCNTs, MWCNTs, active carbon, and carbon black were tested for their toxicity on fibroblast cells [11,12]. SWCNTs were found to be the most toxic because they had the lowest surface area. It was found that hydrophobic SWCNTs with low surface area induce enhanced toxic effects. CNTs influenced extracellular matrix protein signaling that resulted in the deformation of cell membranes, and displacement of cell organelles finally leading to cell death. When two NPs have comparable surface area, toxicity depends on their surface chemistry [13]. It was observed that unrefined SWCNTs were more toxic than the refined SWCNTs. This was because the unrefined SWCNTs agglomerated and their surface area decreased, and lower surface area resulted in higher toxicity, whereas refined SWCNTs remained dispersed and thus were less toxic [14].

20.5 NANOPARTICLES RESPONSIBLE FOR TOXICITY

It has been known for some time that NPs are not harmless but can harbor adverse effects. NPs are the main drivers of proinflammatory effects in cases of particulate matter (PM) toxicity because they are the main particulate type found in PM mixtures, thus implying that NPs may possess some intrinsic toxicity. In this context, we should discuss what properties of NP would lead to toxic effects. One would be the small size of PN as it would lead to a large surface area per unit mass, and particle toxicology can tell that this is often correlated with increased reactivity. Additionally, the larger surface area also leads to an increased possibility of the formation of free radicals (i.e., superoxide anion or hydroxyl radicals), which therefore lead to oxidative stress. Thus, this forms the underlying mechanism responsible for the inflammatory responses to carbon nano particles (CNP) exposure. While the size plays an important role in the toxicity of the particles, the shape of the NP, as well as surface modifications, can also affect absorption and toxic potential. Another effect of the high surface area per unit mass of the NP is that it may be responsible for the adsorption of various organic compounds from air and this phenomenon increases biological interactions within the body. It has been shown by a number of epidemiological studies that airborne PM from combustion sources such as motor vehicles or industrial exhaust contributes to morbidity and cardiovascular and respiratory mortality [9].

20.6 CELL UPTAKE OF CARBON NANOPARTICLE

The fact that CNPs can cross the cell membrane had been postulated and observed by several researchers. This property of CNPs has enabled their use in drug delivery and gene therapy. To deliver the drug to the target site, an active molecule must be attached to the delivery device by a covalent or a noncovalent bond. After reaching the target site, the drug has to enter the specified cell. The process of internalization can be achieved in two ways: either only the drug can enter the cell alone leaving the carrier outside or it can enter the cell along with the carrier to initiate the internalization process for cell uptake of CNPs. Uptake of a drug along with the carried molecule is more efficient as a drug delivery mechanism because after penetration into the cell, the drug–carrier complex will degrade releasing the drug inside the cell, whereas in the other internalization process, the extracellular environment causes degradation of a drug–carrier conjugate and then the drug penetrates the lipid membrane to enter the cell. The mechanism for cellular uptake of CNTs is not yet entirely understood. Nevertheless, two ways of internalization have been estimated: (1) through the endocytosis pathway, and (2) via the endocytosis-independent pathway through passive diffusion across the lipid bilayer in a needle-like manner. There are five methods to internalize macromolecules or NPs in mammalian cells: phagocytosis (via mannose receptor-, complement receptor-, Fcc receptor-, and scavenger receptor-mediated pathways), clathrin-mediated endocytosis, macro-pinocytosis, caveolin-mediated pathways, and clathrin/caveolin-independent endocytosis. Cellular uptake of CNPs is influenced by its dimensions and surface chemistry and the process is considered to be dose and time dependent [9]. The mechanism of SWCNT uptake by cell has been shown in Figure 20.5.

20.7 MECHANISM OF CARBON NANOPARTICLE TOXICITY

CNP-exposed cells are subjected to oxidative stress due to the induction of toxic oxidants and enzymes. An increased level of oxidative stress leads to inflammation and cytotoxicity. Protein kinase and nuclear factor kappa B (NF-κB) signaling pathways and proinflammatory cytokines regulate apoptosis in response to oxidative stress. Decreased cell viability and higher levels of the proinflammatory cytokines interleukin-8 (IL-8) and IL-1β indicate that MWCNTs can initiate an inflammatory response in human keratinocytes (Heks) at 0.4 mg/mL dose. Apoptosis may result

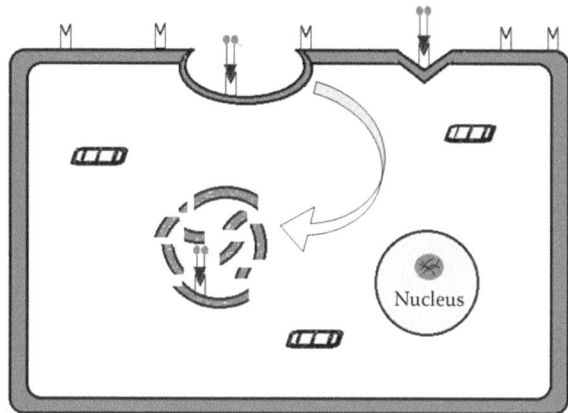

FIGURE 20.5 Mechanism of SWCNT uptake by cell via endocytosis pathway.

from mitochondrial disruption and the release of proapoptotic factors. Several mechanisms may be based on the toxicity of CNTs as described below [1].

20.7.1 Free Radical Formation

The main cause of the toxicity is related to the oxidative stress by free radical formation. These free radicals oxidize lipids, proteins, and DNA. Oxidative stress can upregulate transcription factors sensitive to redox activator protein-1, and kinases that cause inflammatory responses. Slow clearance caused by agglomeration or accumulation of these NPs can produce free radicals in the organs of the reticuloendothelial system (RES) such as the spleen, lungs, and kidneys, which are easy targets for oxidative stress [9].

20.7.2 Reactive Oxygen Species

Reactive oxygen species (ROS) are chemically reactive oxygen-containing molecules which are formed as by-products of normal oxygen metabolism. However, the level of ROS may increase due to environmental stress, such as exposure to radiation, foreign particles, and so on. ROS can lead to detrimental effects on the cells, such as apoptosis, DNA damage, oxidation of amino acids, and inactivity of enzymes. Studies have shown that damaged DNA SWCNTs cause changes in the cell cycle and apoptosis signal by the generation of ROS. Most of the cells cultured in media containing CNTs alter the G1 phase of their cell cycle. Wang et al. have found that apoptosis was induced in PC12 cells by 4–5 times higher concentrations of ROS in cells exposed to SWCNTs (200 g/mL) [15].

20.7.3 Increased Inflammatory Responses

Inflammatory responses produced by exposing mice to carbon black was compared to asbestos and MWCNTs. While carbon black initiated a normal foreign body response, where the immune system recognized and destroyed foreign particles, MWCNTs and asbestos exposure increased the release of polymorphonuclear leukocytes and protein exudation, indicating an increased inflammatory response. The same group also showed the difference in toxicity levels between short and long MWCNTs. They attributed the increased inflammatory response of long MWCNTs and asbestos to "frustrated phagocytosis" in which the macrophages are unable to engulf the long needle-shaped CNTs. By contrast, in another study done on a mouse macrophage RAW 264.7 cell line, CNTs induced ROS-related necrosis, apoptosis, and chromosomal damage, but did not induce an inflammatory response [16].

20.7.4 Granuloma Formation

The granuloma is a small nodule or a small collection of immune cells produced when the immune system attempts to remove foreign substances, but cannot eliminate them. CNTs and asbestos both were found to cause granulomas in mice exposed for 7 days. Intratracheal instillation of SWCNTs (diameter ranging from 0.7 to 1.5 nm) in the lungs of rats lead to the obstruction of the airway as a result of granuloma formation with a 15% mortality within a day [17].

20.7.5 Apoptosis

Apoptosis is the process of programmed cell death that may occur in multicellular organisms. SWCNTs caused apoptosis of five carbon-based nanomaterials, including SWCNTs and MWCNTs for toxicity in human fibroblast cells. Some researchers hypothesized that when dispersed with small surface hydrophobic materials, CNTs show increased toxicity and also proposed that the mechanism of toxicity due to the CNT was due to the extracellular matrix protein signaling resulting in changes in the cell skeleton and the subsequent displacement of organelles, resulting in membrane deformation and finally, apoptosis [11].

20.8 CARBON NANOPARTICLE-INDUCED OXIDATIVE STRESS

The oxidative stress results in cells exposed to CNPs. According to the hierarchical oxidative stress hypothesis, the lowest level of oxidative stress is associated with the induction of antioxidant and detoxification enzymes. The genes that encode the phase II enzymes are under the control of the transcription factor Nrf-2. Nrf-2 activates the promoters of phase II genes via an antioxidant response element. Defects or aberrancy of this protective response pathway may determine disease susceptibility during ambient particle exposure. At higher levels of oxidative stress, this protective response is overtaken by inflammation and cytotoxicity. Inflammation is initiated through the activation of proinflammatory signaling cascades (e.g., mitogen-activated protein kinase and NF-jB cascades), whereas programmed cell death could result from mitochondrial perturbation and the release of proapoptotic factors [18].

Exposure to NPs is a source of increasing ROS, leading to oxidative stress. Three factors mostly govern the process of ROS generation by NPs as outlined by Knaapen et al.: (i) active redox cycling on the surface of NPs, particularly, the metal-based NPs; (ii) oxidative groups functionalized on NPs; and (iii) particle–cell interactions, especially in the lungs where there is a rich pool of ROS producers such as the inflammatory phagocytes, neutrophils, and macrophages [18]. Owing to the overproduction of ROS, a series of cytokine cascades are activated, which include an upregulation of interleukins (IL), kinases, and tumor necrosis factor α (TNF-α) proinflammatory signaling processes as a counterreaction to oxidative stress. It has been observed in some studies on C60 fullerenes that these NPs induce elevation of proinflammatory enzymes, such as IL-1, TNF-α, IL-6, macrophage inhibitory protein, and monocyte chemotactic protein in rodent lungs. When receptor tyrosine kinases, mitogen-activated protein kinases, and transcriptional factors, such as nuclear factor-κB and signal transducer and activator of transcription 1, are activated, the genes involved in inflammation and fibrosis are transcribed and expressed. Stimulation of IL-1b and TNF-α complicates the expression of profibrotic proteins. More specifically, the latter is known to increase the production of the transforming growth factor (TGF)-b1, which potentiates collagen deposition by fibroblasts, while the former is associated with the expression of a platelet-derived growth factor (PDGF)-AA and its receptor, PDGF receptor-a, which increases proliferation of myofibroblasts, promoting the formation of immature collagenous tissue within the lung. Hence, oxidative stress in a cell is usually measured as a function of ROS assay due to the production of free radical species [18].

20.9 INFLAMMATORY RESPONSE AND PHAGOCYTOSIS

Inflammation is an integral part of the immune response. The inflammatory or *in vitro* proinflammatory effects of NPs are one of the most widely studied phenomena in nanotoxicological studies. However, most of these studies reported mainly changes in a few cytokines or chemokines but lacked mechanistic data. Moreover, NPs can reach distinct body organs after passing through diverse epithelial or endothelial barriers, which gives particular significance to the evaluation of their interaction with different types of macrophages and other immune cells. Particular notice must be paid in evaluating the inflammatory potentials of NPs because various types of NPs (e.g., CB and TiO_2) can adsorb proinflammatory mediators. The adsorption of GMCSF, interleukin 6 (IL-6), and TNF-α on CB (13 nm) and TiO_2 (15 nm) NPs under *in vitro* conditions has been confirmed [19].

Among all the adverse effects caused by NPs, inflammation (a biological reaction of tissues to harmful stimuli) appears to be the most common factor. Nitric oxide, TNF-α, and IL-8 are key inflammatory mediators when macrophages are activated. In an interaction between chemically modified SWCNTs and B lymphocytes, T lymphocytes, and macrophages, it was found that functionalized SWCNTs were taken up by cells without inducing toxicity [19]. It was observed that only the less soluble ones preserved lymphocytes' functionality while provoking the secretion of proinflammatory cytokines by macrophages. Strikingly, after CNTs were taken up by murine and rat macrophage cells, no inflammatory mediators such as NO, TNF-α, and IL-8 were found. However, a dose- and time-dependent increase of intracellular ROS and a decrease of the mitochondrial membrane potential occurred. Again, with the purified CNTs, no effect was reported upon incubation. Inflammatory responses were also observed when human epidermal keratinocytes or human skin fibroblast were exposed to CNTs. The underlying mechanism may be due to the production of ROS, leading to the activation of the NF-jB. 30-nm CNTs have the potential to penetrate skin tissue within 2–3 min during microimaging MRI experiments. Cell adhesion function can be altered by nanotubes [20].

20.10 ANTIGENICITY

Biotechnology-derived pharmaceuticals can cause specific antibody response (antigenicity). Antibodies are specialized proteins produced by plasma B cells in response to an antigen or foreign materials. The immune response to a composite NP-based drug potentially involves antibodies for both the particles and the surface groups. To date, there are very limited studies on the antigenicity of functionalized NPs and none of them report CNT-specific antibody generation. In one study, CNTs functionalized with a peptide antigen (B cell epitope from the foot-and-mouth disease virus, FMDV) was examined. The CNT-FMDV was recognized by antibodies equally well as the free peptide and the immunization of mice with the CNT-FMDV clearly enhanced anti-FMDV peptide antibody responses. Moreover, no immune response to CNTs was detected, which is an important issue in view of epitopic suppression when peptide antigen carriers are used [21,22]. A variety of factors, such as particle surface properties and functional groups, may ultimately affect the systemic antigenicity of CNTs when it was used as a drug carrier.

20.11 CAUSES OF CARBON NANOPARTICLE IMMUNOTOXICITY AND CONTROL

Size, shape, structure, and surface all play a role in defining nanotoxicity. The aggregation status and p–p electronic effects may also be significant in the case of CNTs. CNTs have an unusually large surface area/mass ratio. The large surface area gives the particles more opportunity to contact the cellular membrane and proteins, as well as a greater capacity for the absorption and transport of bioactive substances. The larger surface area also suggests that chemical modification may have a significant impact on the biological activities of CNTs.

TABLE 20.1
Immunotoxicity of Carbon-Based Nanomaterials

Sl. No.	Carbon Nanoparticle	Summary
1	Carbon black (<100 nm)	Induction of MCP-1, CCL2, IL-6, C-reactive protein, and exaggeration of atherosclerosis in animals
2	Carbon black (14 nm)	Induction of slight expression of CD80 and MHC class II and significant expression of CD86 and DEC205 in endothelial cells
3	Single-walled carbon nanotubes (PEG coating 1–5 nm in diameter, 50–200 nm in length)	Persistence of SWNT for several months in kidney and liver without obvious toxicity
4	Single-walled carbon nanotubes (1–4 nm in diameter)	Biodegradation of single-walled carbon nanotubes by hypochlorite and ROS mediated by human neutrophil myeloperoxidase
5	Single-walled carbon nanotubes (1–2 nm in diameter, 20 nm to several μm in length)	Induction of ROS, inflammatory cytokines, and expression of apoptosis-related genes in macrophages
6	Single-walled carbon nanotubes (800 nm length)	Inhibition of production of IL-8, 6, TNF-α, and MCP-1 in A549 cells
7	Multiwalled carbon nanotubes (10–30 nm in diameter, 30–50 nm in length)	Induction of fibrosis in asthma animal model and suggestion of the role of TGF-β and PDGF
8	Multiwalled carbon nanotubes (20–40 nm in diameter, 5–30 μm in length)	Induction of ROS, inflammatory cytokines, and activation of NF-κB, in A549 or BEAS-2B cells
9	C60 fullerene (0.7 nm in diameter)	No toxicity in animal lung

Source: From Jang J, Lim D-H and Choi I-H. *Immune Network* 2010;10(3):85–91. With permission.

Contamination is one of the reasons for CNTs' potential damage. The presence of such impurities interferes with experiments conducted on the inherent toxicity of CNTs. Transition metals are particularly effective as catalysts of oxidative stress in cells, tissues, and biofluids. In a particular study [23], interactions of two types of SWCNT (1) iron-rich (nonpurified) SWCNT (26% of iron) and (2) iron-stripped (purified) SWCNT (0.23 wt% of iron) with RAW264.7 macrophages was studied. Each type of SWCNT was able to generate intracellular production of superoxide radicals or nitric oxide in the cells. Less pure iron-rich SWCNT were more effective in generating hydroxyl radicals, and superoxide radicals, accumulating lipid hydroperoxides, and causing significant loss of intracellular low-molecular-weight thiols (GSH). Therefore, the inflammatory responses caused by nanotubes with metals can be particularly damaging. Oxidative species generated during inflammatory response can interact with transition metals to trigger redox-cycling cascades with a remarkable oxidizing potential to deplete endogenous reserves of antioxidants and induce oxidative damage to macromolecules.

Chemical modifications of NP surface have the potential to confer improved biocompatibility of CNPs. A nanocombinatorial chemistry approach was used to generate an MWCNT library containing 80 different surface modifications [24]. In addition to the successful regulation of protein binding and cytotoxicity, they also showed different roles in activating immune systems as measured by nitric oxide generation. Compared with the precursor, MWCNT-COOH, many modified MWCNTs exhibited lower immune responses. More biocompatible and immune-friendly nanomedicine carriers can be developed through iterative screening and optimization studies (Table 20.1).

20.12 EFFECTS OF NANOPARTICLE-ADSORBED PROTEINS (PEPTIDES) ON IMMUNE RESPONSE

The biological fate and (re)biodistribution of NPs strongly depend on the physicochemical characteristics of the particles and the proteins that NPs encounter in the body, particularly in the plasma.

Many serum proteins have been found to bind to carbon black (CB), TiO_2, or acrylamide NPs. Among the proteins identified, many, such as apolipoprotein E, granulocyte macrophage colony-stimulating factor (GM-CSF), or transferrin, are ligands for cellular receptors. These proteins adsorbed on the NP surfaces may contribute to the biological effects of NPs through activation/inactivation of receptor-dependent signaling.

The amount (along with the functional and structural properties) of the adsorbed proteins determines the interactions of these nanomaterials with the cells and contributes to their biological responses. These kinds of interactions mainly depend on the chemical nature, surface, and size of the NPs. Another important consideration in this case may be the possibility of conformational changes in the structure of adsorbed proteins. Such changes have been shown for a few NP types (e.g., Si NPs induced a helical structure, including a catalytic site, on unstructured peptides in solution). Under *in vivo* conditions, such interactions may cause a change or loss in function of the adsorbed proteins and may also result in the presentation of novel peptide motifs to the immune system. In a recent publication, the authors speculated that such interactions can also lead to autoreactivity against self-epitopes and may result in a persistent cell-mediated immune response, but further mechanistic studies are needed to confirm such hypotheses [25]. Mainly proteins such as immunoglobulins and components of the complement system are adsorbed from blood to the NP surface, which can act as signals for innate and/or adaptive immune responses. Purified SWCNTs and double-walled CNTs (DWCNTs) have been shown to activate the human serum complement system in a potent manner (comparable to equal weight of zymosan) by the classical pathway of human serum complement activation [26]. Moreover, DWCNTs can also activate alternative pathways of complement activation. This activation of complement may be due to selective binding of C1q CNTs (classical pathway activation), whereas C3b binding may be postulated as the mechanism of alternative pathway activation.

The phenomenon of protein binding is important in immune responses, but the method to study the phenomenon still remains unclear. A fundamental obstacle is the observation that the nature of adsorbed proteins on the NPs also depends on the cell culture media.

20.13 LUNG TOXICITY

The liberation of fine CNPs in the atmosphere is a highly energy-exhaustive process and current dosage and processes do not release significantly high amounts of airborne CNPs into the environment. However, there is a possible chance of health hazard upon cumulative accumulation effects, especially when CNTs are handled in higher amounts. NPs may enter the body because of inhalation, ingestion, cutaneous absorption, or through even the circulatory system. The respiratory system is one of the most important systems identified for quick absorption and deposition of NPs in the body. Owing to their small size, CNTs can be easily borne in air and be inhaled into the lungs [27]. Therefore, pulmonary toxicity is of prime importance.

The potential mechanism of NP-induced pulmonary toxicity tentatively is explained in Figure 20.6. The initial acute inflammatory reaction is probably caused by damage to pulmonary epithelial type I cells. The response includes robust neutrophilic pneumonia followed by recruitment and activation of macrophages. The unusual feature of the response is an early switch from the acute phase response to events resulting from fibrogen with significant lung deposition of collagen and elastin. This is accompanied by a characteristic change in the production and release of proinflammatory (TNF-α, interleukin-1 h) to anti-inflammatory profibrogenic cytokines (TGF-b, interleukin-10). Fibrogenic inflammatory responses were accompanied by a detrimental decrease in lung function and increased susceptibility to infection [27].

Oxidative stress is a consistent measure for the evaluation of the toxic response in general. While NPs enter into the lungs, oxidative stress tends to increase the formation of ROS, followed by oxidation of lipids, and malondialdehyde (MDA) and glutathione (GSH) formation. Lung damage is accompanied with an increase in MDA levels and a decrease in GSH (an antioxidant that helps protect cells and tissues from ROS) levels [28].

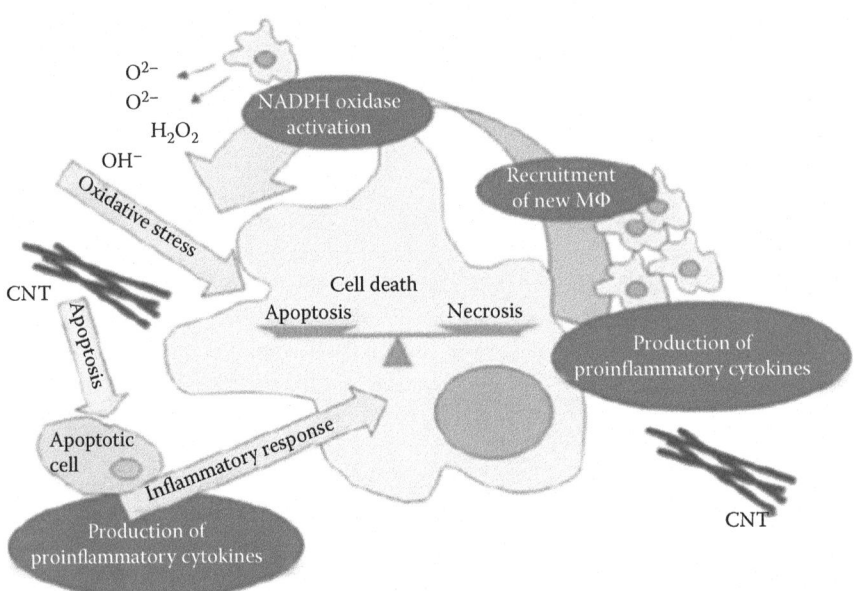

FIGURE 20.6 In the lung, the initial target for CNTs is probably type I epithelial cells whose necrotic death stimulates a proinflammatory response and recruitment of inflammatory cells. Interactions include oxidative burst due to activation of NADPH oxidase and possible interactions of nanoparticles with microbial pathogens. NADPH oxidase complex is activated in macrophages during inflammation and it acts as the major source for the generation of reactive oxygen species, such as superoxide O_2–d radicals that disproportionate to form hydrogen peroxide (H_2O_2). Transition metals, through their interactions with O_2–d and H_2O_2, act as catalysts for the formation of highly reactive hydroxyl (OH^-) radicals. Oxidatively modified lipids generated by cyclooxygenase (COX-2) and lipooxygenase (LOX) participate in amplification of the inflammatory response via recruitment of new inflammatory cells. (From Yu Y. et al. *Nanoscale Research Letters* 2008;3:271–7.)

There are several factors that regulate the distribution of nanomaterials through the respiratory tract, such as their chemical reactivity with body proteins, the size or surface characteristics, and so on. CNTs can escape clearance by macrophages and pass from the alveoli to the blood circulation followed by migration to different organs of the body. Physical attributes such as size, surface modifications, and so on play a crucial role in extrapulmonary translocation of CNPs. Following the deposition of CNPs, they are translocated to extrapulmonary sites, from where they reach the circulatory system via transcytosis. Removal of these deposited particles may take place by both physical and chemical translocation processes. The soluble components of NPs either in lipids or in intracellular fluid undergo absorption into protein or subcellular components, or into extracellular fluids. Owing to biodistribution of CNPs, they may reach the liver and even central nervous system after entering the body through inhalation. Alveolar macrophages form the first line of defense against NPs entering the lungs. A dose-dependent increase in inflammatory response has been observed with SWCNTs. Aggregated CNTs can form granulomatous inflammation, including discrete granulomas with hypertrophic epithelial cells in the lungs. CNPs are mostly eliminated from the body by macrophages through phagocytosis. But the ones that are not removed by phagocytosis penetrate epithelial cells and/or find their way into interstitial spaces, finally entering into systemic circulation and the lymphatic system.

20.13.1 Effects on Lungs

Different types of NPs can induce diverse inflammatory responses in the lung. For instance, owing to the toxicity of SWCNTs, epithelioid granulomas and interstitial inflammation were observed in mice lungs 7 and 90 days after intratracheal instillation, which was found to be higher as compared with other NPs, such as carbon black and quartz particles. Oxidative stress is most likely the major underlying mechanism for inflammation responses by NPs, which lead to the activation of different transcription factors with subsequent enhanced synthesis of proinflammatory proteins [27].

CNTs and NPs are also believed to cause adverse effects through inflammation and induction of proinflammatory molecules. However, it must be noted that some researchers have found metal contaminants (such as iron and nickel) in the nanotube production process to be the main causative agent of oxidative stress. Pulmonary inflammation may also result in changes in membrane permeability, which in turn can result in particle distribution extending beyond the lung and indirectly affecting cardiovascular performance. Moreover, NPs have the potential to enter the brain and blood circulation and subsequently other major organs, inciting inflammation in these places. Inflammation arising as a result of NP exposure could lead to pulmonary diseases or exacerbation of existing lung disorders [27].

20.13.2 Lungs Disorder That May Rise from Exposure to Nanoparticles

20.13.2.1 Pulmonary Fibrosis

Pulmonary fibrosis occurs as a result of increased tissue reactivity leading to the formation and accumulation of fibrous connective tissue. Fibrosis can take many forms, varying from severe forms that cause distortion of lung architecture, inducing bronchiectasis and chronic respiratory infection, to milder forms, which comprise of restrictive ventilatory defects causing hypoxemia, cor pulmonale, and pulmonary hypertension. The first step in pulmonary fibrosis is inflammatory response when immune cells comprising macrophages and neutrophils are excessively activated. These immune cells release toxic mediators, which result in the loss of epithelial integrity and promotion of tissue injury. When this happens, the cell normally employs a repair mechanism wherein mesenchymal cells are activated. These mesenchymal cells have a threefold function, which includes extracellular matrix deposition, reepithelialization, and restoration of normal lung architecture. However, certain patients show an abnormality in tissue remodeling and excessive matrix deposition, which leads to progressive scarring and fibrosis. The presence of MWCNTs in the subpleural region in the lungs of mice leads to fibrosis and scarring. This has become a matter of grave concern as nonclearance and persistence of MWCNTs could cause inflammation in the sensitive mesothelium, leading to mesothelioma formation. Another study also suggests that pulmonary fibrosis induced by MWCNTs may be exacerbated in people with existing lung inflammation. The irregularity in tissue remodeling and fibrosis, with reference to particle inhalation, may be due to an exaggerated inflammatory response that is driven by the inability to clear toxic particles from the lungs via the usual protective mechanisms. This whole cascade may be initiated by interactions of alveolar macrophages with lung epithelial cells, or even directly by interstitial fibroblasts. The extent of fibrotic response may also determine the severity of loss of tissue function. Generally, fibrosis occurs in the following sequence: (a) organization of the immature fibrinous tissue with the formation of new blood vessels and increased blood supply; (b) proliferation of myofibroblasts; (c) increased deposition of extracellular matrix; and finally (d) scar formation. Under conditions of normal lung function, immature intralumenal collagenous tissue may be eliminated by the fibrinolytic system with concomitant apoptosis of myofibroblasts, thus favoring reepithelialization. Depending on the degree of injury to the alveoli, removing the continual exposure to NPs may allow reepithelialization. In chronic cases of injury, however, lung function may be lost [29].

20.13.2.2 Pneumoconiosis

Pneumoconiosis, an occupational lung disease, is clinically classified into two categories, fibrotic and nonfibrotic. While the fibrotic process involves focal nodular or diffuse fibrosis, nonfibrotic lesions involve particle-laden macrophages, with minimal or no fibrosis. The former comprises silicosis, coal worker's pneumoconiosis, asbestosis, and berylliosis, which are caused by persistent inhalation of silica particles, washed coal particles, asbestos fibers, and beryllium particles, respectively. Nonfibrotic lesions include siderosis, stannosis, and baritosis that are caused by particles of iron oxide, tin oxide, and barium sulfate, respectively. Among these pulmonary lesions, silicosis, coal worker's pneumoconiosis, and asbestosis dominate the most common clinical cases. Over the last three decades, death rates due to asbestosis have increased tremendously, overwhelming the decrease in death rates due to the other two types of pneumoconiosis. Here, it is to be noted that these clinical conditions are influenced by a multitude of particle types, varying in size and concentration. These particulate clouds are mineral and combustion-derived, and are found most commonly in developing nations. NPs are the most toxic of the particles found in particulate clouds and are the most significant contributors to fibrogenicity. It is hypothesized that NPs could also behave like asbestos *in vivo* since some NPs, particularly the carbon rods, have similar shape, size, and properties. In one study, it has been shown that asbestos-like pathogenic behavior of MWCNTs in mice induces inflammation and formation of granulomas. Although there have not been any confirmed reported clinical cases of engineered NP-induced pulmonary fibrosis, it should be remembered that the rapidly increasing exposure levels may cause serious issues, considering the extent to which NPs are integrated into technology [27].

20.13.2.3 Exacerbation of Asthma

Asthma is a disease state of lung hypersensitivity caused by inflammation of the airways, making asthmatic individuals more vulnerable to NP-induced lung toxicity. Many early studies have shown that deposition of fine particles are most enhanced in the lungs of patients with chronic obstructive lung disease, including asthma. Since inhaled ultrafine particles (UFPs) have higher deposition efficiency in the pulmonary region, more UFPs are retained in the lung with each breath in comparison with larger particles. In cases of asthmatic patients, airway obstruction causes air trapping and thus an increase in alveolar volume, causing a net increase in UFP deposition through diffusion, although impairment of alveolar ventilation may prove to be inhibitory. Since alveolar volume increases during exercise, the deposition in healthy individuals is also higher during exercise than while at rest. However, this increase is not significant in asthmatic patients, perhaps because the increased alveolar volume and airway turbulence is inherently present. Dead space ventilation increases the minute respiration of patients with obstructive lung disease. This phenomenon along with hyperinflation, which is seen even in mild cases of asthma, is speculated to increase the diffusional deposition of UFPs in the distal airways and alveoli. The increase in particle numbers in the lungs has been reported to be 74% in asthmatic patients compared with healthy subjects. Therefore, it would appear that greater NP deposition in the lungs would exacerbate airway inflammation in susceptible individuals. Another concern would be how the use of steroids in asthmatic individuals would affect NP lung toxicity. Steroids, such as the various forms of corticosteroids, are used in the treatment of asthma as it helps to control and reduce inflammation in the airways by inhibiting cyclooxygenases and production of superoxides. But how effective steroids are in counteracting NP toxicity is still not known (30).

20.14 CYTOTOXICITY

Cells on exposure to CNTs may result in a variety of cell fates. CNTs are known to cause necrosis, where cells lose cell membranes and burst rapidly. There have been reports that cells stop growing or dividing actively, hence losing their viability on treatment with CNTs. Cells can also activate a genetic program of controlled cell death better known as apoptosis. The probable mechanism of photothermal killing of cancer cells with CNTs and graphene involved both necrotic and apoptotic cell death characterized by caspase activation or DNA fragmentation and cell membrane damage [31]. Apoptosis-associated genes can be upregulated and tyrosine kinase activities can be decreased, with downregulation of the expression of the related genes. The mechanism of SWCNTs cytotoxicity has been evaluated in terms of induced changes on cytoskeletons and cell morphology. It is known that certain proteins such as focal adhesion kinase (FAK) cadherin, collagen, and fibronectin play an important role in cell adhesion. CNTs introduce themselves into cell membranes and agitate the surface protein receptors. FAK leads to reduction in cell proliferation and adhesion. SWCNTs disturb the distribution of FAK in a human cell line (HEK293) along with a decrease in cell adhesion [20]. SWCNTs can indirectly cause toxicity by altering the composition of the culture media by reacting with them, hence reducing the availability of medium components to the cells. For example, when A549 cells were grown on media containing SWCNTs, significant cytotoxicity was observed [28].

Guinea pig alveolar macrophages, on exposure with SWCNTs and MWCNTs, are reported to demonstrate cytotoxicity in many cases. High concentration of pristine and oxidized MWCNTs have been shown to generate loss of viability of the human Jurkat T cells and human peripheral blood lymphocytes. A comparative study of the toxicity of pristine and oxidized MWCNT in human Jurkat T leukemia cells has shown that the latter were more toxic [32].

On the other hand, highly purified CNTs have reported to have low or no toxicity. In a report, purified SWCNTs were taken up slowly by human macrophage cells, which showed low toxicity [33] and CNTs found across the cell membrane of rat macrophages (NR8383) showed no cytotoxicity [34].

20.15 DERMAL TOXICITY

Skin is at a high potential for both occupational and environmental exposure to any kind of airborne NPs. If CNTs get to penetrate the stratum corneum cells and become stuck into the feasible epidermal cell layers of the skin, they may enter the keratinocytes directly or trigger the production of proinflammatory cytokines or initiate other consequence. Moreover, once the NPs are housed inside the avascular epidermis, they are difficult to be removed by phagocytosis. Studies on skin irritation by CNTs are extremely limited at this time. One recent study illustrated that the length of CNT modulates inflammation response.

Currently, not much evidence is available on whether nanomaterials can actually be absorbed across the skin's stratum corneum and can accumulate in dermal tissues. When some products containing SWCNTs and MWCNTs were tested for dermal toxicity and eye irritability using rabbits and skin sensitization using guinea pigs, none of the SWCNTs and MWCNTs was found to cause toxic skin sensitization effects. Only one of the products containing MWCNTs caused slight eye irritation [35]. In separate studies, SWCNTs that contain iron as an impurity showed dermal toxicity on murine epidermal cells (JB6 P+), EpiDerm FT engineered skin cells, and immunocompetent hairless SKH-1 mice [36].

The large surface area of the skin and small size of NPs make it complicated to determine the location of CNTs in the skin and within the systemic circulation. If they are systemically absorbed, CNTs would be distributed right through the entire body or may be deposited in major organs, hence making their detection and quantization even more difficult. Furthermore, CNTs are not absorbed by the skin in a similar fashion as that of chemical absorption. This may be because of the fact that they vary widely in their shape, size, and physiochemical properties, which could affect their dermal toxic potential and their ability to penetrate across the skin. Standardization becomes even more challenging for surface-modified CNTs. MWCNTs infiltrate through the stratum corneum, concentrate, and trigger an irritation response in dermal cells [37]. In some cases, proteomics studies were also conducted to determine the effect of MWCNTs on human epidermal keratinocytes. As compared to control cells, the CNT-treated cells show much variation in the expression of several proteins, which infer information about cell cycle inhibition, deregulation of intermediate filament expression, altered vesicular trafficking, as well as membrane protein downregulation [38].

20.16 GENOTOXICITY

Genotoxicity can be described as the lethal effect of NPs on a cell's genetic material. The first alarm regarding genotoxicity of nanomaterials was raised by the Royal Society and Royal Academy of Engineering in 2004. CNTs have the affinity as well as ability to interact with DNA, thus interpreting them potentially mutagenic or carcinogenic. By the virtue of CNTs being cohesive in nature, they have a tendency to form stable aggregates, causing inflammatory and oxidative stress at the sites of their accumulation. These effects, over the course of time, might lead to tissue/organ destruction and increase the risk of cancer.

Following almost the same fashion as that of asbestos, CNTs are also able to induce cancer and mesothelioma. Inside biological systems, they stay as stable aggregates in a micron size. Animal studies have shown that MWCNTs and SWCNTs can induce stress-related inflammatory responses, reactive oxygen and nitrogen species, and genotoxic effects associated with these effects [27].

Carcinogenic properties of CNTs are coupled with long-term genotoxic stress. CNTs can interact to cause genotoxicity mainly in two ways: (1) direct interaction with DNA or the mitotic apparatus and (2) indirectly via oxidative stress and inflammatory response. MWCNTs, due to its long and thin morphology, are capable of producing asbestos-like toxic responses. There have been interesting studies conducted on intra-abdominal injection of MWCNTs showing mesothelioma-inducing effects [39]. CNTs are also known to be cytotoxic and cause DNA damage.

20.17 CONCLUSION

Because of the fact that nanotechnological products and nanomedicine research are comparatively nascent, no standardized guidelines for assessing immunotoxicity due to CNPs are currently available. Many important issues need to be addressed in order to develop a new generation of nanomedicines. Available data (Table 20.2) strongly suggest that CNTs upon entering cells cause ROS

TABLE 20.2
Pathophysiology and Toxicity Effects of CNTs

Experimental NM Effects	Possible Pathophysiological Outcomes
ROS generation[a]	Protein, DNA, and membrane injury,[a] oxidative stress[b]
Oxidative stress[a]	Phase II enzyme induction, inflammation,[b] mitochondrial perturbation[a]
Mitochondrial perturbation[a]	Inner membrane damage,[a] permeability transition (PT) pore opening,[a] energy failure,[a] apoptosis,[a] aponecrosis, cytotoxicity
Inflammation[a]	Tissue infiltration with inflammatory cells,[b] fibrosis,[b] granulomas,[b] atherogenesis,[b] acute phase protein expression (e.g., C-reactive protein)
Uptake by reticuloendothelial system[a]	Asymptomatic sequestration and storage in liver,[a] spleen, lymph nodes,[b] possible organ enlargement and dysfunction
Protein denaturation, degradation[a]	Loss of enzyme activity,[a] auto-antigenicity
Nuclear uptake[a]	DNA damage, nucleoprotein clumping,[a] autoantigens
Uptake in neuronal tissue[a]	Brain and peripheral nervous system injury
Perturbation of phagocytic function,[a] "particle overload," mediator release[a]	Chronic inflammation,[b] fibrosis,[b] granulomas,[b] interference in clearance of infectious agents[b]
Endothelial dysfunction, effects on blood clotting[a]	Atherogenesis,[a] thrombosis,[a] stroke, myocardial infarction
Generation of neoantigens, breakdown in immune tolerance	Autoimmunity, adjuvant effects
Altered cell cycle regulation	Proliferation, cell cycle arrest, senescence
DNA damage	Mutagenesis, metaplasia, carcinogenesis

Source: Table cited at Yu Y. et al. *Nanoscale Research Letters* 2008;3:271–7; Table originally from Nel A. et al. Toxic potential of materials at the nanolevel. *Science* 2006;311:622–7. Reprinted with permission of AAAS.

[a] Effects supported by limited experimental evidence.
[b] Effects supported by limited clinical evidence.

production and interact with the immune systems. A better understanding of the mechanisms of CNTs' interaction with immune systems is still needed for developing and optimizing biocompatible nanomedicine carriers.

REFERENCES

1. Jain S, Singh SR, Pillai S. Toxicity issues related to biomedical applications of carbon nanotubes. *Journal of Nanomedicine & Nanotechnology*. 2013;3(5):140.
2. Dolatabadi JEN, Jamali AA, Hasanzadeh M, Omidi Y. Quercetin delivery into cancer cells with single walled carbon nanotubes. *International Journal of Bioscience, Biochemistry and Bioinformatics*. 2011;1(1):21–5.
3. Canesi L, Ciacci C, Betti M, Fabbri R, Canonico B, Fantinati A et al. Immunotoxicity of carbon black nanoparticles to blue mussel hemocytes. *Environment International*. 2008;34(8):1114–9.
4. Hussain S, Vanoirbeek JA, Hoet PH. Interactions of nanomaterials with the immune system. *Wiley Interdisciplinary Reviews: Nanomedicine and Nanobiotechnology*. 2012;4(2):169–83.
5. Ezzati Nazhad Dolatabadi J, Omidi Y, Losic D. Carbon nanotubes as an advanced drug and gene delivery nanosystem. *Current Nanoscience*. 2011;7(3):297–314.
6. Kim SN, Rusling JF, Papadimitrakopoulos F. Carbon nanotubes for electronic and electrochemical detection of biomolecules. *Advanced Materials*. 2007;19(20):3214–28.
7. Beg S, Rizwan M, Sheikh AM, Hasnain MS, Anwer K, Kohli K. Advancement in carbon nanotubes: Basics, biomedical applications and toxicity. *Journal of Pharmacy and Pharmacology*. 2011;63(2):141–63.
8. Murr L, Garza K, Soto K, Carrasco A, Powell T, Ramirez D et al. Cytotoxicity assessment of some carbon nanotubes and related carbon nanoparticle aggregates and the implications for anthropogenic carbon nanotube aggregates in the environment. *International Journal of Environmental Research and Public Health*. 2005;2(1):31–42.
9. Donaldson K, Stone V, Tran C, Kreyling W, Borm PJ. Nanotoxicology. *Occupational and Environmental Medicine*. 2004;61(9):727–8.
10. Poland CA, Duffin R, Kinloch I, Maynard A, Wallace WA, Seaton A et al. Carbon nanotubes introduced into the abdominal cavity of mice show asbestos-like pathogenicity in a pilot study. *Nature Nanotechnology*. 2008;3(7):423–8.
11. Tian F, Cui D, Schwarz H, Estrada GG, Kobayashi H. Cytotoxicity of single-wall carbon nanotubes on human fibroblasts. *Toxicology In Vitro*. 2006;20(7):1202–12.
12. Tran PA, Zhang L, Webster TJ. Carbon nanofibers and carbon nanotubes in regenerative medicine. *Advanced Drug Delivery Reviews*. 2009;61(12):1097–114.
13. Jiang J, Oberdörster G, Biswas P. Characterization of size, surface charge, and agglomeration state of nanoparticle dispersions for toxicological studies. *Journal of Nanoparticle Research*. 2009;11(1):77–89.
14. Aillon KL, Xie Y, El-Gendy N, Berkland CJ, Forrest ML. Effects of nanomaterial physicochemical properties on *in vivo* toxicity. *Advanced Drug Delivery Reviews*. 2009;61(6):457–66.
15. Wang J, Rahman MF, Duhart HM, Newport GD, Patterson TA, Murdock RC et al. Expression changes of dopaminergic system-related genes in PC12 cells induced by manganese, silver, or copper nanoparticles. *Neurotoxicology*. 2009;30(6):926–33.
16. Ye S, Jiang Y, Zhang H, Wang Y, Wu Y, Hou Z et al. Multi-walled carbon nanotubes induce apoptosis in RAW 264.7 cell-derived osteoclasts through mitochondria-mediated death pathway. *Journal of Nanoscience and Nanotechnology*. 2012;12(3):2101–12.
17. Sachar S, Saxena RK. Cytotoxic effect of poly-dispersed single walled carbon nanotubes on erythrocytes *in vitro* and *in vivo*. *PloS One*. 2011;6(7):e22032.
18. Knaapen AM, Borm PJ, Albrecht C, Schins RP. Inhaled particles and lung cancer. Part A: Mechanisms. *International Journal of Cancer*. 2004;109(6):799–809.
19. Dumortier H, Lacotte S, Pastorin G, Marega R, Wu W, Bonifazi D et al. Functionalized carbon nanotubes are non-cytotoxic and preserve the functionality of primary immune cells. *Nano Letters*. 2006;6(7):1522–8.
20. Cui D, Tian F, Ozkan CS, Wang M, Gao H. Effect of single wall carbon nanotubes on human HEK293 cells. *Toxicology Letters*. 2005;155(1):73–85.
21. Pantarotto D, Partidos CD, Graff R, Hoebeke J, Briand J-P, Prato M et al. Synthesis, structural characterization, and immunological properties of carbon nanotubes functionalized with peptides. *Journal of the American Chemical Society*. 2003;125(20):6160–4.

22. Pantarotto D, Partidos CD, Hoebeke J, Brown F, Kramer E, Briand J-P et al. Immunization with peptide-functionalized carbon nanotubes enhances virus-specific neutralizing antibody responses. *Chemistry & Biology.* 2003;10(10):961–6.
23. Kagan V, Tyurina Y, Tyurin V, Konduru N, Potapovich A, Osipov A et al. Direct and indirect effects of single walled carbon nanotubes on RAW 264.7 macrophages: Role of iron. *Toxicology Letters.* 2006;165(1):88–100.
24. Zhou H, Mu Q, Gao N, Liu A, Xing Y, Gao S et al. A nano-combinatorial library strategy for the discovery of nanotubes with reduced protein-binding, cytotoxicity, and immune response. *Nano Letters.* 2008;8(3):859–65.
25. Gustafsson Å, Lindstedt E, Elfsmark LS, Bucht A. Lung exposure of titanium dioxide nanoparticles induces innate immune activation and long-lasting lymphocyte response in the Dark Agouti rat. *Journal of Immunotoxicology.* 2011;8(2):111–21.
26. Salvador-Morales C, Flahaut E, Sim E, Sloan J H, Green ML, Sim RB. Complement activation and protein adsorption by carbon nanotubes. *Molecular Immunology.* 2006;43(3):193–201.
27. Lam C-w, James JT, McCluskey R, Arepalli S, Hunter RL. A review of carbon nanotube toxicity and assessment of potential occupational and environmental health risks. *CRC Critical Reviews in Toxicology.* 2006;36(3):189–217.
28. Casey A, Herzog E, Lyng F, Byrne H, Chambers G, Davoren M. Single walled carbon nanotubes induce indirect cytotoxicity by medium depletion in A549 lung cells. *Toxicology Letters.* 2008;179(2):78–84.
29. Byrne JD, Baugh JA. The significance of nanoparticles in particle-induced pulmonary fibrosis. *McGill Journal of Medicine: MJM.* 2008;11(1):43.
30. Hussain S, Vanoirbeek JA, Luyts K, De Vooght V, Verbeken E, Thomassen LC et al. Lung exposure to nanoparticles modulates an asthmatic response in a mouse model. *European Respiratory Journal.* 2011;37(2):299–309.
31. Hu X, Cook S, Wang P, Hwang H-m, Liu X, Williams QL. In vitro evaluation of cytotoxicity of engineered carbon nanotubes in selected human cell lines. *Science of the Total Environment.* 2010;408(8):1812–7.
32. Bottini M, Bruckner S, Nika K, Bottini N, Bellucci S, Magrini A et al. Multi-walled carbon nanotubes induce T lymphocyte apoptosis. *Toxicology Letters.* 2006;160(2):121–6.
33. Fiorito S, Serafino A, Andreola F, Bernier P. Effects of fullerenes and single-wall carbon nanotubes on murine and human macrophages. *Carbon.* 2006;44(6):1100–5.
34. Pulskamp K, Diabaté S, Krug HF. Carbon nanotubes show no sign of acute toxicity but induce intracellular reactive oxygen species in dependence on contaminants. *Toxicology Letters.* 2007;168(1):58–74.
35. Ema M, Matsuda A, Kobayashi N, Naya M, Nakanishi J. Evaluation of dermal and eye irritation and skin sensitization due to carbon nanotubes. *Regulatory Toxicology and Pharmacology.* 2011;61(3):276–81.
36. Murray A, Kisin E, Leonard S, Young S, Kommineni C, Kagan V et al. Oxidative stress and inflammatory response in dermal toxicity of single-walled carbon nanotubes. *Toxicology.* 2009;257(3):161–71.
37. Prow TW, Grice JE, Lin LL, Faye R, Butler M, Becker W et al. Nanoparticles and microparticles for skin drug delivery. *Advanced Drug Delivery Reviews.* 2011;63(6):470–91.
38. Witzmann FA, Monteiro-Riviere NA. Multi-walled carbon nanotube exposure alters protein expression in human keratinocytes. *Nanomedicine: Nanotechnology, Biology and Medicine.* 2006;2(3):158–68.
39. Takagi A, Hirose A, Nishimura T, Fukumori N, Ogata A, Ohashi N et al. Induction of mesothelioma in p53+/− mouse by intraperitoneal application of multi-wall carbon nanotube. *The Journal of Toxicological Sciences.* 2008;33(1):105–16.
40. Jia'en Li J, Muralikrishnan S, Ng C-T, Yung L-YL, Bay B-H. Nanoparticle induced pulmonary toxicity. *Experimental Biology and Medicine* 2010;235:1025–33.

Index

A

AA, *see* Atomic absorption spectrophotometer (AA)
AAVs, *see* Adeno-associated viruses (AAVs)
ABC effect, *see* Accelerated blood clearance effect (ABC effect)
Absorbome, 367
Absorption, digestion, metabolism, and elimination (ADME), 25, 195
    of nanoparticles, 216
AC, *see* Alternating current (AC)
Accelerated blood clearance effect (ABC effect), 302–303
ACH, *see* Air changes per hour (ACH)
Acid-functionalized MWCNTs (MWCNT-AT), 136
Acid-functionalized SWCNTs (SWCNT-AF), 136
Acinus, 227
Acridine orange/ethidium bromide double staining (AO/EB staining), 112
Acrylate polymers (Eudragit®), 351; *see also* Ocular toxicity
Activator protein 1 (AP-1), 233
Active pharmacological ingredient (API), 2, 176
Adeno-associated viruses (AAVs), 349
Adenosine triphosphatase (ATP), 274
ADME, *see* Absorption, digestion, metabolism, and elimination (ADME)
Adverse immunostimulation, 427
Aerogels, 16
Aerosols, 81, 392
    airborne nanomaterials as, 391
    human health and, 394, 397
    ion trap in MEAD, 55
    to monitor engineered nanomaterial, 393
AES, *see* Auger electron spectroscopy (AES)
AFM, *see* Atomic force microscopy (AFM)
Agglomerate, 162
Agglomeration, 35
Aggregate, 162
Airborne nanomaterials, 391
Air changes per hour (ACH), 61
Airway epithelium, 227
Alamar blue, 32; *see also* Cell viability testing; Nanotoxicity assessment
    assay, 122
    reduction, 404
Alanine aminotransferase (ALT), 33, 109, 293
Albumin, 237; *see also* Pulmonary drug delivery
    -bound paclitaxel, 238
    bovine serum, 238
    glutamate toxicity, 239
    human serum, 237–238
Alkaline phosphatase (ALP), 274, 293
Alkaline single-cell microgel electrophoresis assay, 408–409; *see also* Genotoxicity
ALP, *see* Alkaline phosphatase (ALP)
ALT, *see* Alanine aminotransferase (ALT)
Alternating current (AC), 120
Alveoli, 227, 229
Ames test, 215, 408; *see also* Bacterial reverse mutation test; Genotoxicity
Annexin V (VAC alpha), 404–405; *see also* Cell viability testing
    to determine membrane integrity, 32
Antibodies, 426
Antibody generators, *see* Antigens
Antigenicity, 433
Antigen-presenting cells (APCs), 426
Antigens, 426
Antioxidant enzymes, 297
AO/EB staining, *see* Acridine orange/ethidium bromide double staining (AO/EB staining)
AP-1, *see* Activator protein 1 (AP-1)
APCs, *see* Antigen-presenting cells (APCs)
APF, *see* Assigned protection factor (APF)
API, *see* Active pharmacological ingredient (API)
AP-MWCNTs, *see* As-prepared MWCNTs (AP-MWCNTs)
Apoptosis, 432
Aspartate aminotransferase (AST), 33, 93
Aspartate transaminase (AST), 251
As-prepared MWCNTs (AP-MWCNTs), 136
Assigned protection factor (APF), 64
AST, *see* Aspartate aminotransferase (AST); Aspartate transaminase (AST)
Asthma, 438
Atomic absorption spectrophotometer (AA), 54
Atomic force microscopy (AFM), 5, 121–122, 132, 163, 164
    analysis of AgNPs treated BHK21 and HT29 cells, 214
ATP, *see* Adenosine triphosphatase (ATP)
Auger electron spectroscopy (AES), 28
Autoimmunity, 427

B

B lymphocytes, 424
Bacterial reverse mutation test, 358
BAL, *see* Bronchoalveolar lavage fluid (BAL)
Bands, 67
BBB, *see* Blood–brain barrier (BBB)
BEGM, *see* Bronchial epithelial growth medium (BEGM)
BET method, *see* Brunauer–Emmett–Teller method (BET method)
Bhc, *see* 4-Bromo-7-hydroxycoumarin (Bhc)
BhcP, *see* 4-Bromo-7-hydroxycoumarin polymer (BhcP)
Bioelectronics, 117; *see also* Biosensors
Biological barriers, 384
Biological nanomaterials, 353
Biological safety cabinets (BSCs), 61
Biopersistence, 181
Biosensors, 12, 117, 124; *see also* Nanotoxicity assessment
    advantages and applications, 118

Biosensors (*Continued*)
  cellular-based, 120
  chip-based, 120
  inflammatory biomarkers detection, 123–124
  for nanotoxicity biomarker detection, 122
  paper-based biosensor, 124
  planar microelectrode, 118
  whole-cell impedance-based, 118
Blood–brain barrier (BBB), 74, 287
Blood compatibility, 108–109
Blood urea nitrogen (BUN), 296
Bottom-up nanofabrication technologies, 3, 4; *see also* Nanotechnology
Bovine serum albumin (BSA), 238
BrdU, *see* 5-Bromo-2-deoxyuridine (BrdU)
4-Bromo-7-hydroxycoumarin (Bhc), 104
4-Bromo-7-hydroxycoumarin polymer (BhcP), 112
5-Bromo-2-deoxyuridine (BrdU), 30
Bronchial epithelial growth medium (BEGM), 168
Bronchoalveolar lavage fluid (BAL), 395
Brunauer–Emmett–Teller method (BET method), 28
BSA, *see* Bovine serum albumin (BSA)
BSCs, *see* Biological safety cabinets (BSCs)
Buckminsterfullerene, *see* Fullerenes
Buckyballs, *see* Fullerenes
BUN, *see* Blood urea nitrogen (BUN)

C

C_{60}, *see* Fullerenes
C activation-related pseudoallergy (CARPA), 367
CAF, *see* Complement activation factors (CAF)
CAGR, *see* Compound annual growth rate (CAGR)
Canadian Institutes of Health Research (CIHR), 344
Carbonaceous nanomaterials, 249–250
Carbon black (CB), 435
Carbon fiber microelectrodes, 121
Carbon nanofibers (CNF), 64
Carbon nanohorns (CNHs), 428
Carbon nanoparticles (CNP)
  antigenicity, 433
  apoptosis, 432
  applications of, 428
  cell uptake of, 430
  effect of contaminants in, 434
  factors in distribution of, 436
  free radical formation, 431
  granuloma formation, 432
  immunotoxicity of, 434
  to improve biocompatibility, 434
  increased inflammatory responses, 431
  -induced oxidative stress, 432
  MWCNTs, 428
  NP responsible for toxicity, 430
  physical attribute for toxicity, 429–430
  pulmonary toxicity, 435–437
  reactive oxygen species, 431
  routes of entry, 429
  SWCNTs, 428
  target in lung, 436
  toxicity mechanism, 430–432
Carbon nanotubes (CNTs), 11, 64, 86, 131, 137, 231, 426, 427; *see also* Carbon nanoparticles (CNP); Multiwall carbon nanotubes (MWCNTs); Pulmonary drug delivery; Single-walled carbon nanotubes (SWCNTs)
  applications, 132, 133
  cardiovascular toxicity of, 250–251
  classification, 428
  cytotoxicity due to, 438–439
  dermal toxicity, 439
  diameter of, 428
  fibers, 131
  genotoxicity, 439–440
  health effects of, 191–192
  kidney toxicity, 297–298
  liver toxicity, 292
  pulmonary toxicity, 135–137, 232–234
  spleen toxicity, 305
  structure of, 428
  surface functionalization, 136
  toxicity, 132–134, 231–232, 287, 440
  types, 132
Carcinogenicity, 216
Cardiovascular events, 386
Cardiovascular system (CVS), 27
Cardiovascular toxicity of nanomaterials, 249
  carbonaceous nanomaterials, 249–250
  carbon nanotubes, 250–251
  cerium dioxide nanoparticles, 255
  gold nanomaterials, 252–253
  graphene, 251
  iron oxide nanomaterials, 253–254
  metallic nanomaterials, 252, 255
  perspectives, 255
  platinum nanoparticles, 255
  quantum dots, 254
  silica nanomaterials, 251–252
  silver nanoparticles, 255
  yttrium oxide nanoparticles, 255
  zinc oxide nanoparticles, 255
CARPA, *see* C activation-related pseudoallergy (CARPA)
Caspase-3, 405; *see also* Cell viability testing
CAT, *see* Catalase (CAT)
Catalase (CAT), 297
CB, *see* Carbon black (CB); Control banding (CB)
CBER, *see* Center for Biologics Evaluation and Research (CBER)
CCK-8 assay, *see* Cell counting kit-8 assay (CCK-8 assay)
CDCs, *see* Colloidal drug carriers (CDCs)
CDER, *see* Center for Drug Evaluation and Research (CDER)
CDRH, *see* Center for Devices and Radiological Health (CDRH)
Cell apparatus, 385
Cell counting kit-8 assay (CCK-8 assay), 112
Cell membranes, 385
CellTiter assay, 404; *see also* Cell viability testing
Cellular immune response, 427
Cellular stress response, 405; *see also* Nanoparticle risk assessment
  glutathione measurement, 406
  lipid peroxidation assay, 406
  mitochondrial membrane potential dissipation assay, 407
  nitrite production, 407
  pulmonary toxicity pathway, 406
  ROS detection, 405–406

Index

Cell viability testing, 401; *see also* Nanoparticle risk assessment
  alamar blue reduction, 404
  Annexin V/Propidium iodide staining, 404–405
  apoptosis marker Caspase-3 detection, 405
  CellTiter assay, 404
  clonogenic assay, 403
  commassie blue assay, 404
  fluorescein diacetate test, 402
  hematoxylin and eosin assay, 403
  LDH release detection, 402
  MetPLATE *Ecoli* bioassay, 404
  mitochondrial activity detection, 402
  neutral red uptake assay, 403
  trypan blue exclusion assay, 403–404
  WST-1 cytotoxicity assay, 403
Center for Biologics Evaluation and Research (CBER), 323
Center for Devices and Radiological Health (CDRH), 323–324
Center for Drug Evaluation and Research (CDER), 322
Center for Food Safety and Applied Nutrition (CFSAN), 322, 324
Center for International Environmental Law (CIEL), 336
Center for Veterinary Medicine (CVM), 322
Central nervous system (CNS), 77
Cerium dioxide nanoparticles, 255
Cerium oxide nanoparticles, *see* Nanoceria
CFE, *see* Colony-forming efficiency (CFE)
cfm/ft2, *see* Cubic feet per minute per square feet (cfm/ft2)
CFSAN, *see* Center for Food Safety and Applied Nutrition (CFSAN)
Chaperones, 385
Chemical vapor deposition (CVD), 131
Chemistry, manufacturing, and controls (CMC), 322
Chip-based biosensors, 120; *see also* Biosensors
Chit-AgNTs, *see* Chitosan-coated silver nanotriangles (Chit-AgNTs)
Chitosan, 238, 350; *see also* Ocular toxicity
Chitosan-coated silver nanotriangles (Chit-AgNTs), 112
Chlorophenol red β-galactopyranoside (CRPG), 404
CHMP, *see* Committee for Medicinal Products for Human Use (CHMP)
CIEL, *see* Center for International Environmental Law (CIEL)
CIHR, *see* Canadian Institutes of Health Research (CIHR)
CK30PEG, *see* PEG-substituted polylysine (CK30PEG)
Classification, Labeling, and Packaging (CLP), 332; *see also* Nanotechnology regulatory implications
  CIEL proposal, 336
  Communication on the Second Regulatory Review on Nanomaterials, 337–338
  DG Enterprise, 337
  IUCLID dossier, 337
Clathrin-dependent mechanism, 187–188
Clonogenic assay, 403; *see also* Cell viability testing
CLP, *see* Classification, Labeling, and Packaging (CLP)
CMC, *see* Chemistry, manufacturing, and controls (CMC)
CNF, *see* Carbon nanofibers (CNF)
CNHs, *see* Carbon nanohorns (CNHs)
CNS, *see* Central nervous system (CNS)
CNTs, *see* Carbon nanotubes (CNTs)
Codensation particle counter (CPC), 54
Colloidal drug carriers (CDCs), 289–290
Colony-forming efficiency (CFE), 31

Comb-like poly(ethylene glycol) (CPEG), 104
Comb-like polyethylene glycol-2-diazo-1, 2-naphthoquinone (CPEG-*g*-DNQ), 112
Comet assay, 31, 123, 357, 408; *see also* Genotoxicity; Nanotoxicity assessment
Commassie blue assay, 404; *see also* Cell viability testing
Committee for Medicinal Products for Human Use (CHMP), 335
Complement activation factors (CAF), 375
Composites, 16
Compound annual growth rate (CAGR), 77
Computed tomography (CT), 33
Condensed phase, 160
Control banding (CB), 67–68; *see also* Occupational nanoparticle exposures
Copper nanoparticles
  GIT toxicity, 274
  kidney toxicity, 296–297
  liver toxicity, 291
COX-2, *see* Cyclooxygenase (COX-2)
CPC, *see* Codensation particle counter (CPC)
CPEG, *see* Comb-like poly(ethylene glycol) (CPEG)
CPEG-*g*-DNQ, *see* Comb-like polyethylene glycol-2-diazo-1, 2-naphthoquinone (CPEG-*g*-DNQ)
CPK, *see* Creatine phosphokinase (CPK)
Creatine phosphokinase (CPK), 33
Crohn's disease, 265
CRPG, *see* Chlorophenol red β-galactopyranoside (CRPG)
CT, *see* Computed tomography (CT)
Cubic feet per minute per square feet (cfm/ft2), 61
CVD, *see* Chemical vapor deposition (CVD)
CVM, *see* Center for Veterinary Medicine (CVM)
CVS, *see* Cardiovascular system (CVS)
Cyclooxygenase (COX-2), 436
Cytokines, 123
Cytotoxicity, 438–439

D

Data Call-In notices (DCIs), 328
DCF, *see* 2′,7′-Dichlorofluorescein (DCF)
DCFDA, *see* 2,7-Dihydrodichlorofluorescein diacetate (DCFDA)
DCFH assay, *see* 2-,7-Dichlorofluorescein assay (DCFH assay)
DCIs, *see* Data Call-In notices (DCIs)
DCS, *see* Differential centrifugal sedimentation (DCS)
Defra, *see* Department for Environment, Food and Rural Affairs (Defra)
Degussa P25, 392
Denaturing gradient gel electrophoresis (DGGE), 276
Density functional theory (DFT), 409
Density of states (DOS), 164
Deoxyribonucleic acid (DNA), 250, 384; *see also* Genotoxicity
  to assess cell proliferation, 30
  cancer risk detection, 31
  -chips, 12
  damage, 18, 20, 124, 386, 406
  damage detection, 118, 123, 409
  detection, 13
  dimensions scale, 3
  double-strand break marker, 291
  in gene expression, 212

Deoxyribonucleic acid (*Continued*)
  in gene therapy, 11
  microarrays, 212
  mutation detection, 9
  in nanotechnology, 2
  plasmid, 148, 231
  -tethered magnetic nanoparticles, 349
Department for Environment, Food and Rural Affairs (Defra), 340
Department of Energy (DOE), 66
DEPs, *see* Diesel exhaust particles (DEPs)
Dermal delivery, 182
Dermal toxicity, 439
Design for the Environment (DfE), 60
Dextran, 238
Dextran-coated nanoparticles; *see also* Polysaccharide-coated poly(alkylcyanoacrylate) nanoparticles (PACA nanoparticles); Protein adsorption
  fabrication conditions, 369
  identity of, 372
  protein adsorption, 376
  protein interactions, 374
DfE, *see* Design for the Environment (DfE)
DFT, *see* Density functional theory (DFT)
DG, *see* Directorate-Generals (DG)
DGGE, *see* Denaturing gradient gel electrophoresis (DGGE)
DHE, *see* Dihydroethidium (DHE)
2-Diazo-1, 2-naphthoquinone (DNQ), 104
2′,7′-Dichlorodihydrofluorescein diacetate (H_2DCF-DA), 211, 405
2′,7′-Dichlorofluorescein (DCF), 211
2-,7-Dichlorofluorescein assay (DCFH assay), 123; *see also* Nanotoxicity assessment
Diesel exhaust particles (DEPs), 27
Differential centrifugal sedimentation (DCS), 165
Differential mobility analyzer (DMA), 54
Digestive tract, *see* Gastrointestinal tract (GIT)
2,7-Dihydrodichlorofluorescein diacetate (DCFDA), 31
Dihydroethidium (DHE), 31
Diketopiperazine derivative, 239
Dilution ventilation (DV), 61
3-(4,5-dimethylthiazol-2-yl)-2, 5-diphenyltetrazolium bromide (MTT), 104, 350; *see also* Cell viability testing
  assay, 113, 402
  drawbacks, 402
Directorate-Generals (DG), 336
Disk centrifugation, *see* Differential centrifugal sedimentation (DCS)
DL-lactide-co-glycolide, *see* Poly (DL-lactide-co-glycolide)
DLS, *see* Dynamic light scattering (DLS)
DMA, *see* Differential mobility analyzer (DMA)
DMEM, *see* Dulbecco's modified Eagle medium (DMEM)
DNQ, *see* 2-Diazo-1, 2-naphthoquinone (DNQ)
DOE, *see* Department of Energy (DOE)
DOS, *see* Density of states (DOS)
Dose, 87
Double-walled carbon nanotubes (DWCNTs), 132, 435; *see also* Carbon nanotubes (CNTs)
Drug delivery, 229, 238, 239; *see also* Pulmonary drug delivery
DTU, *see* Technical University of Denmark (DTU)
Dulbecco's modified Eagle medium (DMEM), 409
DV, *see* Dilution ventilation (DV)
DWCNTs, *see* Double-walled carbon nanotubes (DWCNTs)
Dynamic light scattering (DLS), 28, 136, 165, 400

E

EAD, *see* Electrical aerosol detector (EAD)
EBSD, *see* Electron backscatter diffraction (EBSD)
EC, *see* European Commission (EC)
ECG, *see* Electrocardiogram (ECG)
ECHA, *see* European Chemicals Agency (ECHA)
ECVAM, *see* European Centre for the Validation of Alternative Methods (ECVAM)
ED 50, 95
EDS, *see* Energy dispersive spectrometer (EDS)
EELS, *see* Electron energy loss spectroscopy (EELS)
EFSA, *see* European Food Safety Authority (EFSA)
EGFR, *see* Epidermal growth factor receptor (EGFR)
EIS biosensors, *see* Electrical impedance sensing biosensors (EIS biosensors)
Electrical aerosol detector (EAD), 54
Electrical impedance sensing biosensors (EIS biosensors), 118, 120
Electrocardiogram (ECG), 253
Electrochemical fuel cells, 15
Electron backscatter diffraction (EBSD), 163
Electron energy loss spectroscopy (EELS), 28
Electron paramagnetic resonance (EPR), 28, 212; *see also* Nanotoxicity assessment
  assay, 123
Electron spin resonance (ESR), 28, 217
Electroretinography (ERG), 350
ELISA, *see* Enzyme-linked immunosorbent assay (ELISA)
Embryonic stem cells (ESCs), 253
Endoplasmic reticulum (ER), 255
Endothelial cells, 287; *see also* Nanomaterial toxicity on kidney; Nanomaterial toxicity on liver; Nanomaterial toxicity on spleen
  uptake of nanoparticles, 287–288
Energy dispersive spectrometer (EDS), 163, 400
Engineered nanomaterials (ENMs), 161, 334; *see also* Nanoparticle risk assessment
  applications of, 209
  assessment of exposure to, 393–394
  behavior characterizations, 392–393
  carcinogenic effects of, 396–397
  effects of, 397–398
  effects on circulation, 397
  genotoxicity of, 396
  pulmonary inflammation induced by, 395–396
  risk assessment challenges, 394
  toxicity on spleen, 306
  translocation into body, 395
Engineered nanoparticles (ENPs), 318, 393; *see also* Nanoparticle risk assessment
  exposure and dose metrics, 261–262
  particle parameter assessment, 398–401
ENMs, *see* Engineered nanomaterials (ENMs)
ENPs, *see* Engineered nanoparticles (ENPs)
Enzyme-linked immunosorbent assay (ELISA), 13
  to detect IL-8, 349
  to determine cell viability of NPs, 404

measuring cytokine, 109, 215
  to quantify inflammatory markers, 123–124
Enzyme-responsive systems, 107 See also Stimuli-responsive nanomaterials
EP, see European Parliament (EP)
EPA, see U.S. Environmental Protection Agency (EPA)
Epidermal growth factor receptor (EGFR), 148
Epithelium, 267
EPR, see Electron paramagnetic resonance (EPR)
ER, see Endoplasmic reticulum (ER)
ERG, see Electroretinography (ERG)
ESCs, see Embryonic stem cells (ESCs)
ESR, see Electron spin resonance (ESR)
EU, see European Union (EU)
Eudragit®, see Acrylate polymers (Eudragit®)
European Centre for the Validation of Alternative Methods (ECVAM), 412
European Chemicals Agency (ECHA), 336
  REACH Technical Guidance Documents, 337
European Commission (EC), 342
European Food Safety Authority (EFSA), 334
European Parliament (EP), 338
European Union (EU), 317; see also Nanotechnology regulatory implications
  cosmetic directive, 333–334
  food labeling, 334–335
  medical device and pharmaceutical, 335–336
  nanomaterial definition, 330
  REACH regulation, 331
  SCENIHR, 331–332
  Second Regulatory Review, 332
Europium hydroxide nanorods, 292–293
Exposure, 388

F

FAE, see Follicle-associated epithelium (FAE)
FAK, see Focal adhesion kinase (FAK)
FCM, see Flow cytometry (FCM)
FDA, see U.S. Food and Drug Administration (FDA)
FDA test, see Fluorescein diacetate test (FDA test)
Federal Insecticide, Fungicide, and Rodenticide Act (FIFRA), 327
Feet per minute (fpm), 62
Fe_2O_3-loaded lecithin/PLGA, see Iron oxide-loaded lecithin/poly(lactic-co-glycolic acid) (Fe_2O_3-loaded lecithin/PLGA)
FFF, see Field flow fractionation (FFF)
Field flow fractionation (FFF), 28, 166
FIFRA, see Federal Insecticide, Fungicide, and Rodenticide Act (FIFRA)
Flow cytometry (FCM), 28
Fluorescein diacetate test (FDA test), 402; see also Cell viability testing
Fluorescein o-methacrylate (FMA), 251
Fluorescence resonance energy transfer (FRET), 11
FMA, see Fluorescein o-methacrylate (FMA)
Focal adhesion kinase (FAK), 438
Follicle-associated epithelium (FAE), 267
Formamidopyrimidine DNA glycosylase (FPG), 356
  Fpg-modified comet assay, 408; see also Genotoxicity
Formazan-based assays, 122; see also Nanotoxicity assessment
Fourier transform infrared spectroscopy (FTIR), 136
FPG, see Formamidopyrimidine DNA glycosylase (FPG)
fpm, see Feet per minute (fpm)
Free radical formation, 430
Free released ion concentration determination, 169–170
FRET, see Fluorescence resonance energy transfer (FRET)
Frustrated phagocytosis, 429
FTIR, see Fourier transform infrared spectroscopy (FTIR)
Fullerenes, 5, 57, 249
  health effects, 190–191
  kidney toxicity, 301
  mitochondrial localization, 188
  and MWCNT, 58
  pulmonary toxicity issues, 235–236
  structure of, 235
  toxicity, 146, 187, 287

G

Gas chromatography/mass spectrometry (GC/MS), 54
Gastrointestinal tract (GIT), 25, 181, 262, 286; see also Nanomaterial toxicity on GI tract
Gastrointestinal uptake of NPs, 262; see also Nanomaterial toxicity on GI tract
  acellular layers, 265–266
  active uptake mechanisms of nanomaterials, 269
  adverse health effects, 264
  associated diseases and treatment, 265
  behavior and fate of nanomaterials, 270
  Crohn's disease, 265
  in diseased subjects, 264–265
  distribution, 272
  endocytic routes, 269
  epithelial layers, 267
  excretion/elimination, 272
  exocytosis of NPs, 269
  exposure sources of NPs, 263
  GI absorption of NPs, 271–272
  through GI barrier, 265
  nanomaterial interaction with mucus layer, 266–267
  orogastrointestinal epithelia, 268
  permeation through orogastrointestinal barriers, 267–270
  reaction-reduced toxicity, 264
  routes of epithelial transport, 270
  size-and charge-dependent, 263–264
  translocation, 264
  uptake and clearance of NPs, 262
GBM, see Glomerular basement membrane (GBM)
GC/MS, see Gas chromatography/mass spectrometry (GC/MS)
GEEC, see GPI-anchored protein-enriched compartment (GEEC)
Gelatin, 238
Gene
  and drug delivery, 238, 239; see also Pulmonary drug delivery
  therapy, 11
Generally regarded as safe (GRAS), 105, 113
General ventilation, see Dilution ventilation (DV)
Genetic lesions, 354
Genotoxicity, 407, 439–440; see also Nanogenotoxicology; Nanoparticle risk assessment
  alkaline single-cell microgel electrophoresis assay, 408–409

Genotoxicity (*Continued*)
   Ames test, 408
   comet assay, 408
   DNA damage detection, 409
   DNA repair, 408
   Fpg-modified comet assay, 408
GFP, *see* Green fluorescence protein (GFP)
GIT, *see* Gastrointestinal tract (GIT)
Glomerular basement membrane (GBM), 293
Glutathione (GSH), 297, 406, 435; *see also* Nanotoxicity assessment
   assay, 123
Glutathione disulfide (GSSG), 297
Glutathione peroxidase (GPx), 297
Glutathione reductase (GR), 297
Glutathione *S*-transferase (GST), 297
Glycophosphatidylinositol (GPI), 269
Glycoprotein IIb/IIIa (GPIIb/IIIa), 252
GM-CSF, *see* Granulocyte macrophage colony-stimulating factor (GM-CSF)
GMPs, *see* Good manufacturing practices (GMPs)
GNPs, *see* Gold nanoparticles (GNPs)
GO, *see* Graphene oxide (GO)
Gold nanoparticles (GNPs), 252
   cardiovascular toxicities, 253
   distribution in human organs, 388
   GI tract toxicity, 278
   kidney toxicity, 297
   solution, 4
Good manufacturing practices (GMPs), 346
GPI, *see* Glycophosphatidylinositol (GPI)
GPI-anchored protein-enriched compartment (GEEC), 269
GPIIb/IIIa, *see* Glycoprotein IIb/IIIa (GPIIb/IIIa)
GPx, *see* Glutathione peroxidase (GPx)
GQD, *see* Graphene quantum dots (GQD)
GR, *see* Glutathione reductase (GR)
Granulocyte macrophage colony-stimulating factor (GM-CSF), 435
Granuloma formation, 432
Graphene, 132, 161; *see also* Single-walled carbon nanotubes (SWCNTs)
   cardiovascular toxicity of, 251
Graphene oxide (GO), 161
Graphene quantum dots (GQD), 194
GRAS, *see* Generally regarded as safe (GRAS)
Green fluorescence protein (GFP), 106
GSH, *see* Glutathione (GSH)
GSSG, *see* Glutathione disulfide (GSSG)
GST, *see* Glutathione *S*-transferase (GST)

H

H&E, *see* Hematoxylin and eosin (H&E)
HAECs, *see* Human aortic endothelial cells (HAECs)
HCMECs, *see* Human cardiac microvascular endothelial cells (HCMECs)
HCMs, *see* Human cardiac myocytes (HCMs)
H$_2$DCF-DA, *see* 2′,7′-Dichlorodihydrofluorescein diacetate (H$_2$DCF-DA)
HDMECs, *see* Human dermal microvascular endothelial cells (HDMECs)
Health Products and Food Branch (HPFB), 344
HeiQ AGS-20, 328
Hematoxylin and eosin (H&E), 109, 113; *see also* Cell viability testing
   assay, 403
Heme oxygenase-1 (HO-1), 251
HEPA, *see* High-efficiency particulate air (HEPA)
Hexamethylenetetramine (HMT), 349
High-efficiency particulate air (HEPA), 55
High-information-content data streams, 217
High-throughput screening (HTS), 218, 409; *see also* Nanoparticle risk assessment
HMT, *see* Hexamethylenetetramine (HMT)
HO-1, *see* Heme oxygenase-1 (HO-1)
HPFB, *see* Health Products and Food Branch (HPFB)
HPHSEP-*star*-PEP, *see* Hyperbranched multiarm copolyphosphates (HPHSEP-*star*-PEP)
HSA, *see* Human serum albumin (HSA)
HTS, *see* High-throughput screening (HTS)
Human aortic endothelial cells (HAECs), 254
Human cardiac microvascular endothelial cells (HCMECs), 254
Human cardiac myocytes (HCMs), 254
Human dermal microvascular endothelial cells (HDMECs), 253
Human respiratory system, 50
Human serum albumin (HSA), 108, 237
Human umbilical vein endothelial cells (HUVECs), 252
Humoral immune response, 426
HUVECs, *see* Human umbilical vein endothelial cells (HUVECs)
Hyaluronic acid, 238
8-Hydroxy deoxyguanosine (8-OHdG), 31, 124
Hyperbranched multiarm copolyphosphates (HPHSEP-*star*-PEP), 113
Hypersensitivity, 427

I

IA-SEM, *see* Ion abrasion SEM (IA-SEM)
ICAM-1, *see* Intracellular cell adhesion molecule-1 (ICAM-1)
ICH, *see* International Conference on Harmonization (ICH)
ICP–AES, *see* Inductively coupled plasma–atomic emission spectroscopy (ICP–AES)
ICP–MS, *see* Inductively coupled plasma–mass spectrometry (ICP–MS)
ICP-OES, *see* Inductively coupled plasma–optic emission spectrometry (ICP-OES)
ICR, *see* Imprinting control region (ICR)
ICRP, *see* International Commission on Radiological Protection (ICRP)
IgM, *see* Immunoglobulin-M (IgM)
IH, *see* Industrial hygiene (IH)
Immune system
   adverse immunostimulation, 427
   autoimmunity, 427
   cellular response, 427
   classification, 426
   humoral response, 426
   hypersensitivity, 427
   immunogenicity, 427
   immunosuppression, 427
   immunotoxicity, 427, 434
   inflammation, 433

parts of, 427
protein binding, 435
Immunoassays, 123; see also Nanotoxicity assessment
Immunochromatographic test strip, 124
Immunoglobulin-M (IgM), 303
Immunology, 426
Implantable devices, 9
Imprinting control region (ICR), 233
IM transistors, see Inversion-mode transistors (IM transistors)
Increased inflammatory responses, 431
Inductively coupled plasma–atomic emission spectroscopy (ICP–AES), 121
Inductively coupled plasma–mass spectrometry (ICP–MS), 28, 169
Inductively coupled plasma–optic emission spectrometry (ICP-OES), 28
Industrial hygiene (IH), 56
Inflammation markers, 215, 254
Inorganic nanoparticles, 192–193
Institute of Occupational Safety and Health (IOSH), 62
INT, see 2-(p-Iodophenyl)-3-(p-nitrophenyl)-5-phenyltetrazolium chloride (INT)
Interleukin, 405, 432
    IL-6, 123
    IL-8, 123, 349
    1β, 134
International Commission on Radiological Protection (ICRP), 51
International Conference on Harmonization (ICH), 34
International Organization for Standardization (ISO), 108, 160, 344
Intracellular cell adhesion molecule-1 (ICAM-1), 250
Intravenous (IV), 250
Intravitreal injection (IVT), 347
Inversion-mode transistors (IM transistors), 14
In vitro toxicity assays, 411; see also Nanoparticle risk assessment
    challenges for, 412–413
    limitations, 412
2-(p-Iodophenyl)-3-(p-nitrophenyl)-5-phenyltetrazolium chloride (INT), 402
Ion abrasion SEM (IA-SEM), 28
Ion corona, 166
IOSH, see Institute of Occupational Safety and Health (IOSH)
IR, see Ischemia/reperfusion (IR)
Iron oxide-loaded lecithin/poly(lactic-*co*-glycolic acid) (Fe_2O_4-loaded lecithin/PLGA), 112
Iron oxide nanomaterials, 253–254
Iron oxide/poly(*N*-isopropylacrylamide-*co*-acrylic acid) (Fe_2O_4/P(NIPPAAm-*co*-AAc)), 112
Ischemia/reperfusion (IR), 250
ISO, see International Organization for Standardization (ISO)
IUCLID dossier, 337
IV, see Intravenous (IV)
IVT, see Intravitreal injection (IVT)

K

Kidney, 293; see also Nanomaterial toxicity on kidney
    blood filtration in, 294
    tissue exposed to zinc, 296
    uptake of NPs, 287–288

L

Lactate dehydrogenase (LDH), 32–33, 109, 134, 402
    assay, 113
LAMS, see Light-activated mesoporous silica nanoparticles (LAMS)
Lateral flow immunoassay (LFIA), 124
LC-MS, see Liquid chromatography–mass spectrometry (LC-MS)
LDA, see Linear discriminant analyses (LDA)
LDH, see Lactate dehydrogenase (LDH)
LEV, see Local exhaust ventilation (LEV)
LFIA, see Lateral flow immunoassay (LFIA)
Ligand-coated NPs, 25
Light-activated mesoporous silica nanoparticles (LAMS), 113
Linear discriminant analyses (LDA), 219
Lipid peroxidation assay, 406; see also Cellular stress response
Lipooxygenase (LOX), 436
Lipopolysaccharide (LPS), 190
Liquid chromatography–mass spectrometry (LC-MS), 28
Liver, 288; see also Nanomaterial toxicity on liver
    structure of, 289
    uptake of NPs, 287–288
Local exhaust ventilation (LEV), 61
4-[3-(4-Lodophenyl)-2-(4-nitrophenyl)-3H-5-tetrazolio]-1,3-benzene disulfonate (WST-1), 403
LOX, see Lipooxygenase (LOX)
LPS, see Lipopolysaccharide (LPS)
Lung, 227
    -related disorders, 227
    structural organization of, 228

M

Magic bullet, 365
Magnetic nanoparticles (MNPs), 349–350; see also Ocular toxicity
Magnetic resonance imaging (MRI), 11, 349
MALDI-TOF, see Matrix-assisted laser desorption/ionization–time of flight (MALDI-TOF)
Malondialdehyde (MDA), 305, 406, 435
Manual of Policies and Procedures (MAPP), 322
MAP, see Mean arterial pressure (MAP)
MAPP, see Manual of Policies and Procedures (MAPP)
Matrix-assisted laser desorption/ionization–time of flight (MALDI-TOF), 28
Matrix metalloproteases (MMPs), 107, 113
Maximum likelihood estimate (MLE), 392
Maximum tolerated dose (MTD), 33
MCP-1, see Monocyte chemotactic protein-1 (MCP-1)
MD, see Molecular dynamics (MD)
MDA, see Malondialdehyde (MDA)
MEAD, see Modified EAD (MEAD)
Mean arterial pressure (MAP), 252
MEIC, see Multicenter evaluation of *in vitro* toxicity (MEIC)
Mercaptopropionic acid (MPA), 254
Mercaptoundecanoic acid (MUA), 254
Mesoporous silica nanoparticles (MSNs), 251
Mesothelioma, 181
Metallic nanomaterials, 252, 255
Methyl methane sulfonate (MMS), 409

MetPLATE *Ecoli* bioassay, 404; *see also* Cell viability testing
MFCs, *see* Microbial fuel cells (MFCs)
Microbial fuel cells (MFCs), 15–16
Micronuclei (MNi), 34
Micronucleus (MN), 215, 356
  test, 357
μ-XANES, *see* Micro-x-ray absorption near-edge structure (μ-XANES)
Micro-x-ray absorption near-edge structure (μ-XANES), 170
μ-XRF, *see* Synchrotron microfocused x-ray fluorescence (μ-XRF)
Mitochondrial membrane potential dissipation assay (MMP dissipation assay), 407; *see also* Cellular stress response
MLE, *see* Maximum likelihood estimate (MLE)
MMP dissipation assay, *see* Mitochondrial membrane potential dissipation assay (MMP dissipation assay)
MMPs, *see* Matrix metalloproteases (MMPs)
MMS, *see* Methyl methane sulfonate (MMS)
MN, *see* Micronucleus (MN)
MNi, *see* Micronuclei (MNi)
MNPs, *see* Magnetic nanoparticles (MNPs)
Modified EAD (MEAD), 55
Molecular dynamics (MD), 409
Molecular imaging, 11
Molecular modeling methods, 409–410; *see also* Nanoparticle risk assessment
Monocyte chemotactic protein-1 (MCP-1), 123, 250, 349
Mononuclear phagocyte system (MPS), 25, 295, 426
MPA, *see* Mercaptopropionic acid (MPA)
MPO, *see* Myeloperoxidase (MPO)
MPS, *see* Mononuclear phagocyte system (MPS)
MRI, *see* Magnetic resonance imaging (MRI)
MSNs, *see* Mesoporous silica nanoparticles (MSNs)
MTD, *see* Maximum tolerated dose (MTD)
MTT assay, *see* 3-(4,5-dimethylthiazol-2-yl)-2,5-diphenyltetrazolium bromide assay (MTT assy)
MUA, *see* Mercaptoundecanoic acid (MUA)
Mucin proteins, 265
Mucociliary escalator, 228
Mucus, 225, 265
  nanomaterial interaction with, 266
Multicenter evaluation of *in vitro* toxicity (MEIC), 412
Multiwall carbon nanotubes (MWCNTs), 20, 132, 426, 428; *see also* Carbon nanotubes (CNTs)
  acid functionalization, 136
  as-prepared, 136
  cardiovascular toxicity of, 250–251
  diameter of, 426
  health effects of, 191–192
  liver toxicity, 292
  pulmonary toxicity, 135, 233–234
  spleen toxicity, 305
MWCNT-AT, *see* Acid-functionalized MWCNTs (MWCNT-AT)
MWCNTs, *see* Multiwall carbon nanotubes (MWCNTs)
Myeloperoxidase (MPO), 349

N

NADPH, *see* Nicotinamide adenine dinucleotide phosphate (NADPH)
Nanobiosensor, 12; *see also* Nanomaterials (NMs)
Nanocapsules, 74
Nanocarriers, 10; *see also* Nanomaterials (NMs)
Nanoceria, 348; *see also* Ocular toxicity
Nanoclusters, 78
Nanocomposites, 16–17; *see also* Nanomaterials (NMs)
NanoEHS, 159
Nanoenabled technologies, 12
Nanofiber, 161
Nanogenotoxicology, 353, 360, 407; *see also* Genotoxicity
  action mechanism, 355
  Ames test, 358
  comet assay, 357
  cytotoxicity studies, 358
  factors influencing genotoxicity, 358–360
  genotoxicity testing, 353, 356
  ICH's strategy for genotoxicity testing, 354
  in vitro vs. *in vivo* assays, 355
  micronucleus test, 357
  study of methods used in, 355–356
Nanoimmunotoxicology, 428
Nanoinformatics, 221
Nanomaterial, 161
  manufactured, 161
  metallic, 252, 255
  size and toxicity, 164–165
Nanomaterial applications, 8, 9
  antimicrobial nanopowders and coatings, 12
  biosensor and biolabels, 12, 13
  cancer diagnosis and treatment, 10–11
  catalysis and pollutant elimination, 14
  coatings, 16
  in commonly used products, 16
  in computer technology, 13–14
  cosmetics, 16
  cutting tools, 17
  drug delivery systems, 10
  environmental applications, 14
  extraction and separation techniques, 13
  fuel cells, 15–16
  gene therapy and transfection, 11
  insulation materials, 16
  local anesthetic toxicity, 11
  lubricants, 17
  in medicine and pharmacy, 9
  molecular diagnostics and imaging, 11–12
  nanocomposites, 16–17
  nasal vaccination, 10
  nucleic acid sequence and protein detection, 13
  paint, 17
  scratch-resistant materials, 16
  self-cleaning windows, 16
  sensors, 15
  textiles, 16
  tissue engineering, 9–10
  water remediation, 15
Nanomaterial biointeractions, 19, 20; *see also* Nanotoxicity
  ecotoxicity, 20
  with environment, 19–23
  marine environment, 22–23
  NMs and water, 21
  in soil, 21
  transport in environment, 19
Nanomaterial characterization, 159, 170
  by electron microscopy, 166–167

Index

free released ion concentration determination, 169–170
physicochemical characterization, 163–165
primary NMs in dry state, 160
by Raman spectroscopy, 167–168
techniques, 163
Nanomaterial effects on organ systems, 26
cardiovascular system, 27
central nervous system, 27
gastrointestinal tract, 26
integumentary system, 27
pulmonary system, 26
reticuloendothelial systems, 27
Nanomaterials (NMs), 5, 260; *see also* Nanoparticles (NPs); Nanotechnology; Nanotoxicity; Occupational nanoparticle exposures; Pharmacokinetics; Protein adsorption; Pulmonary drug delivery
-based gene and drug delivery, 225, 240–241
biological fate of, 24
biological interactions of, 18
classification of, 6–8
dissolution in biological matrix, 168–170
as drug carriers, 365
future considerations, 36
-induced inflammation, 261, 395
modified, 5
nanostructured materials, 7
protein adsorption and, 366
spill kit, 66
toxicity, 18–19
Nanomaterial toxicity, 261
mechanisms of, 286–287
shape, 261
size, 261
surface, 261
Nanomaterial toxicity on GI tract, 272; *see also* Gastrointestinal uptake of NPs
cadmium–selenium quantum dot, 276
chitosan NPs, 278
copper nanoparticles, 274
gold NPs, 278
metal NPs, 275
nanoscale zinc powder, 273–274
Si and SiO particles, 276–278
single-walled carbon nanotubes, 274–275
TiO$_2$ NPs, 272–273
Nanomaterial toxicity on kidney, 293; *see also* Endothelial cells; Kidney
carbon nanotubes, 297–298
copper nanoparticles, 296–297
dendrimers, 301
fullerenes, 301
glomerular filtration, 294–295
gold nanoparticles, 297
metals and heavy metals, 296
QDs, 300
risk of on kidney, 295–296
silica particles, 298–300
titanium dioxide nanoparticles, 298
tubular reabsorption, 295
Nanomaterial toxicity on liver, 288; *see also* Endothelial cells; Liver
carbon black and polystyrene, 293
carbon nanotubes, 292
copper nanoparticles, 291
dendrimer, 291–292
europium hydroxide nanorods, 292–293
phagocytosis in Kupffer cells, 288–290
quantum dots, 290
risk on liver, 290
silver nanoparticles, 290–291
ZnO nanoparticles, 293
Nanomaterial toxicity on spleen, 301; *see also* Endothelial cells; Spleen
ABC phenomenon, 302–303
carbon nanotubes, 305
engineered nanomaterials, 306
metal nanoparticles, 303
nanorods, 303–304
risk on spleen, 303
silica nanoparticles, 304
TiO$_2$ nanoparticles, 304–305
Nanomedicine, 77; *see also* Nanoparticles; Protein adsorption
administration of, 366
global scenario, 77–78
marketed products, 78
PEGylated, 367
synthetic and biological identity, 366
Nanometer, 5
Nanomicelles, 193, 351; *see also* Ocular toxicity
Nano-MOUDI impactor, 54
Nano-object, 161
Nanoparticle risk assessment, 384; *see also* Cellular stress response; Cell viability testing; Engineered nanomaterial; Genotoxicity; *In vitro* toxicity assays
application of computational approach, 409
conventional risk assessment, 388–389
dose-response assessment, 409
ecotoxicological testing, 411
entry routes of NPs into body, 386–388
environmental health framework, 389
European Union approach, 413
exposure assessment, 388, 390
hazard characterization, 394
hazard identification, 390
high-throughput screening method, 409
inflammatory response, 407
in vitro cytotoxicity test, 401
issues relevant to risk assessment, 391
molecular modeling methods, 409–410
particle parameter assessment, 398–401
quantitative structure–activity relationships, 410–411
stages, 391
structure for risk assessment, 389–391
Nanoparticles (NPs), 2, 36, 141, 161, 210–211, 353; *see also* Carbon nanotubes (CNTs); Gastrointestinal uptake of NPs; Nanomaterials (NMs); Nanomedicine; Nanoparticle risk assessment; Nanosystems; Nanotechnology; Occupational nanoparticle exposures; Particulate matter (PM); Physicochemical characterization-dependent toxicity; Polysaccharide-coated poly(alkylcyanoacrylate) nanoparticles (PACA nanoparticles) ; Protein adsorption
-adsorbed protein effect, 434–435
advantages of, 75

Nanoparticles (Continued)
  aggregation of, 87
  anthropogenic, 50
  applications, 12, 74, 194
  and asthma, 438
  characteristics of, 75, 81, 83–84
  deposition curves for respiratory tract regions, 51
  before discovery of, 76
  diseases associated with inhaled, 85
  as drug carriers, 12, 230
  effect on proteins and macromolecules, 385
  effects on lungs, 436–437
  exposure and dose metrics, 261–262
  exposure pathways, 84
  exposures to, 50
  factors for toxicity of, 85
  gold, 4
  groups in, 425
  importance in today's era, 76
  inflammatory effects of, 433
  interaction with DNA, 386
  interference to cell membrane, 385
  ligand-coated, 25
  metallic, 163
  multidimensional issues, 80
  nanotoxicity by, 433
  pathway for pulmonary toxicity induced by exposure to, 406
  pharmaceutical, 3
  pharmaceutical NPs, 3
  physical properties of, 181
  and pneumoconiosis, 437–348
  polymeric, 11
  properties of, 81
  and protein interaction, 366
  and pulmonary fibrosis, 437
  radioactive iridium191, 395
  responsible for toxicity, 430
  ROS generation by, 405
  -sized carbon blacks, 260
  size of, 74
  sources of airborne, 50
  studies of inhaled, 147
  surface-functionalized, 11
  techniques to characterize, 29
  toxicity, 286
  toxicity comparison studies, 95
  toxicological effects of, 186–187
Nanoparticles and human health, 175, 194–195
  carbon nanotubes, 191
  cell signaling, 188
  charged nanoparticles, 183–184
  clathrin-dependent mechanism, 187–188
  delivery vehicle design, 179
  dermal delivery, 182
  design biosafety, 189
  endocytosis and wrapping time, 187
  exposure route in toxicity, 180–182
  fullerenes, 190–191
  inorganic nanoparticles, 192–193
  nanoparticle interactions, 178, 184, 188
  organic nanoparticles, 193
  protein corona formation, 184, 185, 187
  quantum dots, 194
  risk prediction, 176–179
  ROS and inflammation, 188–189
  safety, 175–176
  silver nanoparticles, 176–177, 178
  size and surface area correlation, 183
  skin applications, 177
  specific characteristics, 182–184
  topical application products, 177
  toxicity mechanisms, 184–189
  toxicokinetics in body, 180
  toxicological profile assessment, 178, 179
Nanoparticle surface area monitor (NSAM), 54
Nanoparticle toxicity *in silico* evaluation, 217
  artificial intelligence, 221
  dense vs. disrupted cell membrane, 220, 221
  global vs. local models, 219
  high-throughput screening, 218
  nanoinformatics, 221
  quantitative structure activity relationship, 218–219
  TiO_2 nanoparticles, 220
Nanoparticle toxicity *in vitro* evaluation, 211
  assays of ROS, 211–212
  cytotoxicity assays, 212
  cytotoxicity screening, 211
  gene expression analysis, 212–213
  genotoxicity testing, 214–215
  hemolysis test, 214
  inflammation assay, 215
  intracellular localization, 212
  in vitro assay, 215
Nanoparticle toxicity *in vivo* evaluation, 216
  ADME of nanoparticles, 216
  animal models, 216
  carcinogenicity studies, 216–217
  toxicity predication, 217
Nanoparticle tracking analysis (NTA), 165
Nanoparticulate drug delivery systems, 3; *see also* Nanotechnology
Nanoplate, 161
Nanopowders, 78
Nanorod, 161
Nanoscale, 161
Nanoscale Science, Engineering, and Technology (NSET), 49
Nanosensor, 12
Nanosheet, 161
Nano-sized formulations, 3; *see also* Nanotechnology
Nanosphere carriers, 10; *see also* Nanomaterials (NMs)
Nanostructure, 7, 161, 162; *see also* Nanomaterials (NMs)
Nanosystems; *see also* Nanoparticles; Nanotechnology
  advantages of, 75
  limitations of, 76
  types and applications, 78, 82
Nanotechnology, 2, 36, 73, 260; *see also* Nanomaterials (NMs); Nanoparticles (NPs); Nanosystems
  applications, 2
  bottom-up methods, 3, 4
  dimensions scale of, 3
  drug delivery systems, 3
  genesis of, 4–5
  marketed imaging/diagnostic and biomaterial, 81

Index

marketed nano-delivery products, 79–80
nano-sized formulations, 3
nanovehicles, 2–3
shift of healthcare system, 77
terms in, 160–162
top-down methods, 3, 4
Nanotechnology Issues Dialogue Group (NIDG), 340
Nanotechnology regulatory implications, 315, 345–346; *see also* Classification, Labeling, and Packaging (CLP); European Union (EU); U.S. Environmental Protection Agency (EPA); U.S. Food and Drug Administration (FDA)
  Australia, 342–344
  beginnings of regulation, 315–316
  Canada, 344–345
  Denmark, 341
  France, 341
  Germany, 338–339
  lack of reliable data, 318
  lack of reliable tools, 318–320
  nanomaterials in products, 336
  Netherlands, 342
  regulatory challenges, 316–318
  United Kingdom, 340–341
  work place concerns, 329–330
Nanotoxicity, 17, 37, 75, 119, 306; *see also* Nanomaterial biointeractions; Nanomaterials (NMs); Nanotoxicity assessment; Pharmacokinetics; Physicochemical characterization-dependent toxicity
  ADME/PK studies, 34
  in body, 23
  cell proliferation, 31–32
  cell viability, 32
  characterization, 28, 30
  crystallinity and, 19
  DNA synthesis and damage, 30–31
  exocytosis, 32
  future considerations, 36
  genotoxicity and carcinogenic studies, 34–35
  hemolysis, 32–33
  immunogenicity, 31
  mechanisms of, 97
  metallic NPs and bacteria, 20
  molecular mechanisms of, 23–25
  multidimensional issues affecting, 83
  oxidative stress, 31
  physicochemical characteristic induced, 85
  prevention, 35
  shape and, 18
  surfactant mediums, 19
Nanotoxicity assessment, 117, 124, 152; *see also* Biosensors; Nanotoxicity
  assays for gene expression alteration, 153
  assays on genomic level, 123
  atomic force microscopy, 121–122
  carbon fiber microelectrode, 121
  cell uptake studies of nanoparticles, 153
  cell viability/proliferation assay, 122–123, 152–153
  in vitro, 30, 152
  in vivo, 33–34
  methods, 119, 122
  physicochemical characterization, 152
  ROS measurement, 123
  techniques and devices for, 120–121
Nanotoxicity mechanism, 145; *see also* Polymeric nanoparticles (PNs); Solid lipid nanoparticle (SLN)
  based on different nanomaterial, 146
  nanoparticle size, 145–148
Nanotoxicology, 17, 84, 194, 353
  area of research, 210
  concepts in, 180
  nanoparticles, 210–211
Nanotube, 161
Nanovehicles, 2–3; *see also* Nanotechnology
Nanowire, 161
National Industrial Chemicals Notification and Assessment Scheme (NICNAS), 343
National Institute for Occupational Safety and Health (NIOSH), 329
National Nanotechnology Initiative (NNI), 5
National Research Council (NRC), 56
Natural organic matter (NOM), 399
Natural Resources Defense Council (NRDC), 328–329
NBT, *see* Nitroblue tetrazolium (NBT)
Near-infrared (NIR), 104
Near-infrared-emitting quantum dots (NIR QDs), 194
Neonatal rat ventricular myocytes (NRVMs), 251
Netherlands National Institute for Public Health and the Environment (RIVM), 342
Neurodegenerative diseases
Neutral red uptake (NRU), 403; *see also* Cell viability testing; Nanotoxicity assessment
  assay, 122
NF-κB, *see* Nuclear factor κB (NF-κB)
NICNAS, *see* National Industrial Chemicals Notification and Assessment Scheme (NICNAS)
Nicotinamide adenine dinucleotide phosphate (NADPH), 395
  oxidase, 287
NIDG, *see* Nanotechnology Issues Dialogue Group (NIDG)
NIOSH, *see* National Institute for Occupational Safety and Health (NIOSH)
NIR, *see* Near-infrared (NIR)
NIR QDs, *see* Near-infrared-emitting quantum dots (NIR QDs)
Nitric oxide (NO), 151
Nitroblue tetrazolium (NBT), 31
NMs, *see* Nanomaterials (NMs)
NNI, *see* National Nanotechnology Initiative (NNI)
NO, *see* Nitric oxide (NO)
NOEL, *see* No observable effect level (NOEL)
NOM, *see* Natural organic matter (NOM)
No observable effect level (NOEL), 33
NOS, *see* Reactive nitrogen species (RNS; NOS)
NPs, *see* Nanoparticles (NPs)
NRC, *see* National Research Council (NRC)
NRDC, *see* Natural Resources Defense Council (NRDC)
NRU, *see* Neutral red uptake (NRU)
NRVMs, *see* Neonatal rat ventricular myocytes (NRVMs)
NSAM, *see* Nanoparticle surface area monitor (NSAM)
NSET, *see* Nanoscale Science, Engineering, and Technology (NSET)
NTA, *see* Nanoparticle tracking analysis (NTA)
Nuclear factor κB (NF-κB), 233

O

Occupational exposure limits (OELs), 52, 394; see also Occupational nanoparticle exposures
Occupational nanoparticle exposures, 49, 68; see also Nanoparticle
  administrative controls, 62–63
  assessment, 52
  control and management practices, 56
  control measure selection, 52
  engineering control, 56, 60–62
  factors influencing exposure risk, 52
  fire and explosion control, 65
  health surveillance, 66–67
  in laboratories, 55–56
  measurement instruments, 53
  metrics, 52–53
  monitoring, 53
  number concentrations, 57–59
  package labeling, 60
  personal protective equipment, 64–65
  priority of control categories, 60
  risk-based management strategy, 67–68
  risk matrix, 63
  routes for, 50–52
  spills and waste management, 65–66
  time-integrated measurements, 54
  time-resolved measurements, 54–55
Ocular therapy, 347; see also Ocular toxicity
Ocular toxicity, 347
  acrylate polymers (Eudragit®), 351
  chitosan, 350
  CK30PEG, 349
  magnetic nanoparticles, 349–350
  nanoceria, 348
  nanomicelles, 351
  nanoparticles in ocular therapy, 347–348
  poly(alkyl-cyanoacrylate), 351
  poly[(cholesteryl oxocarbonylamido ethyl) methyl bis(ethylene) ammonium iodide], 351
  poly-ε-caprolactone, 351
  polylactic-co-glycolic acid, 350
  solid lipid nanoparticles, 351
OECD, see Organisation for Economic Co-operation and Development (OECD)
OELs, see Occupational exposure limits (OELs)
OGG1, see 8-Oxoguanine DNA glycosylase (OGG1)
8-OHdG, see 8-Hydroxy deoxyguanosine (8-OHdG)
ONBP, see O-nitrobenzyl polymer (ONBP)
O-nitrobenzyl polymer (ONBP), 113
Organic nanoparticles, 193
Organisation for Economic Co-operation and Development (OECD), 344, 354, 412
Orogastrointestinal epithelia, 268
8-Oxoguanine DNA glycosylase (OGG1), 356

P

PAA, see Polyacrylic acid (PAA)
PACA, see Poly(alkyl-cyanoacrylate) (PACA)
PACA nanoparticles, see Polysaccharide-coated poly(alkylcyanoacrylate) nanoparticles (PACA nanoparticles)
Packaging group I (PG I), 63
PAMAM, see Polyamidoamine (PAMAM)
Particle, 162
Particulate matter (PM), 430; see also Nanoparticles (NPs)
PBPK, see Physiologically-based pharmacokinetic (PBPK)
PBS, see Phosphate-buffered saline (PBS)
PCEP, see Poly[(cholesteryl oxocarbonylamido ethyl) methyl bis(ethylene) ammonium iodide] (PCEP)
PCL-b-PEEP, see Poly(ε-caprolactone)-b-poly(ethylethylene phosphate) (PCL-b-PEEP)
PCR, see Polymerase chain reaction (PCR)
PDGF, see Platelet-derived growth factor (PDGF)
PDGF receptor-a, 406
PECL, see Poly-ε-caprolactone (PECL)
PEG, see Polyethylene glycol (PEG)
PEG–anthracene crosslinker, see Polyethylene glycol–anthracene crosslinker (PEG–anthracene crosslinker)
PEGDA–MMP-sensitive peptide–platinum, see Poly(ethylene glycol) diacrylate hydrogel–matrix metalloprotease-sensitive peptide–platinum (PEGDA–MMP-sensitive peptide–platinum)
PEG-substituted polylysine (CK30PEG), 349; see also Ocular toxicity
PEGylation, 108
PENS, see Personal nanoparticle sampler (PENS)
Personal nanoparticle sampler (PENS), 53, 54
Personal protective equipment (PPE), 56, 64; see also Occupational nanoparticle exposures
Persorption, 271
Pesticide Registration Improvement Act (PRIA), 328
Peyer's patches, see Follicle-associated epithelium (FAE)
PGA, see Polyglycolic acid (PGA)
PG I, see Packaging group I (PG I)
PHAC, see Public Health Agency of Canada (PHAC)
Pharmaceutical nanoparticles, 3, 141–142; see also Polymeric nanoparticles (PNs); Solid lipid nanoparticle (SLN)
Pharmacokinetics (PK), 25, 216; see also Nanomaterials (NMs); Nanotoxicity
  absorption, 25
  distribution, 25
  elimination, 26
  metabolism, 25–26
Phosphate-buffered saline (PBS), 219, 350
Photoresponsive materials, 104; see also Stimuli-responsive nanomaterials
  chitosan-coated silver nanotriangles, 105
  NIR light effects on biological systems, 105–106
  photo-activated NP-based system, 104
pH-responsive systems, 106; see also Stimuli-responsive nanomaterials
Physicochemical characterization-dependent toxicity, 84; see also Nanoparticles; Nanotoxicity
  bio-persistence-dependent toxicity, 91, 94
  cellular uptake and binding-dependent toxicity, 94–95
  concentration and drug–loading-dependent toxicity, 87–88, 91
  dose-dependent toxicity, 87, 90
  future perspectives, 96
  material composition-dependent toxicity, 91, 93
  NMs and relative cytotoxicity index on macrophage cell, 97
  particle shape and aspect ratio, 86, 87

Index

size and size distribution dependent toxicity, 88, 92
structural arrangement, 86, 88
structural properties-dependent toxicity, 86
surface area-dependent toxicity, 89, 91, 92
surface chemistry and charge, 87, 89
surface coatings-dependent nanotoxicity, 95, 96
Physiologically-based pharmacokinetic (PBPK), 410
PK, see Pharmacokinetics (PK)
PLA, see Polylactic acid (PLA)
Planar microelectrode biosensors, 118; see also Biosensors
Platelet-derived growth factor (PDGF), 406, 432
Platinum nanoparticles, 255
PLGA, see Poly(lactic-co-glycolic acid) (PLGA)
PM, see Particulate matter (PM)
PMAA nanohydrogels, see Poly(methacrylic acid)-based nanohydrogels (PMAA nanohydrogels)
PMN, see Premanufacture Notifications (PMN)
PMNs, see Polymorphonucleocytes (PMNs)
Pneumoconiosis, 437–348
PNIPAM, PVCL, and PVCL-grafted Poly(ethylene oxide), 113
PNs, see Polymeric nanoparticles (PNs)
Polyacrylic acid (PAA), 251
Poly(alkyl-cyanoacrylate) (PACA), 239, 351; see also Ocular toxicity
Polyamidoamine (PAMAM), 148
Poly-β-aminoester ketal-2, 113
Polycaprolactone, 239
Poly[(cholesteryl oxocarbonylamido ethyl) methyl bis(ethylene) ammonium iodide] (PCEP), 351; see also Ocular toxicity
Poly (DL-lactide-co-glycolide), 238
Poly-ε-caprolactone (PECL), 351; see also Ocular toxicity
Poly(ε-caprolactone)-b-poly(ethylethylene phosphate) (PCL-b-PEEP), 113
Polyethylene glycol (PEG), 25, 108, 121, 349; see also Protein adsorption
in nanomedicine, 366–367
PEGylation, 108
Polyethylene glycol–anthracene crosslinker (PEG–anthracene crosslinker), 113
Poly (ethylene glycol) conjugates, 239
Poly(ethylene glycol) diacrylate hydrogel–matrix metalloprotease-sensitive peptide–platinum (PEGDA–MMP-sensitive peptide–platinum), 113
Polyglycolic acid (PGA), 350
Polylactic acid (PLA), 238, 350
Poly(lactic-co-glycolic acid) (PLGA), 18, 113, 147, 350; see also Ocular toxicity
Poly-L-lysine–neutravidin–PEG, see Poly-L-lysine–neutravidin–poly(ethylene glycol) (Poly-L-lysine–neutravidin–PEG)
Poly-L-lysine–neutravidin–poly(ethylene glycol) (Poly-L-lysine–neutravidin–PEG), 113
Polymerase chain reaction (PCR), 13
Polymeric nanoparticles (PNs), 142, 153; see also Nanotoxicity assessment; Nanotoxicity mechanism; Solid lipid nanoparticle (SLN)
biocompatibility of, 152
concentration of, 150–151
conversion of, 151
degradation of, 151
interaction with biological systems, 148
nanotoxicity of, 142, 143, 145
physicochemical characteristics, 149
preparation and characterization, 142, 143, 144
route of administration of, 151–152
safe and commonly used, 150
size, 145
solvents used in preparation of, 152
Polymerization, synthetic, 103
Poly(methacrylic acid)-based nanohydrogels (PMAA nanohydrogels), 113
Polymorphonucleocytes (PMNs), 349, 426
Poly[(N-isopropylacrylamide)-co-(methacrylic acid)] (P(NIPAM-co-MAA)), 113
Polysaccharide-coated poly(alkylcyanoacrylate) nanoparticles (PACA nanoparticles), 367; see also Dextran-coated nanoparticles; Protein adsorption
copolymers composition and structure analysis, 371
core–corona structure, 368
electron micrograph of, 370
interaction with proteins, 371
models from characterization of, 370, 371
size of, 368
synthetic identity, 368–371
Porous silica (pSiO), 277
Porous silicon (pSi), 277
PPE, see Personal protective equipment (PPE)
Premanufacture Notifications (PMN), 326
Prevention through-design strategy (PtD strategy), 56
PRIA, see Pesticide Registration Improvement Act (PRIA)
Protein adsorption, 366, 379; see also Dextran-coated nanoparticles; Polysaccharide-coated poly(alkylcyanoacrylate) nanoparticles (PACA nanoparticles)
absorbome, 367
activation levels and mesh size, 374, 375
complement system activation and, 378
on dextran-coated nanoparticles, 374
on different dextran-coated nanoparticles, 376
on different negatively charged nanoparticles, 377
hypothesis for, 373
nanoparticles and proteins, 366
PEG and, 366–367
proteins size and NPs hydrophobic core, 374, 375
Protein corona, 24, 25
pSi, see Porous silicon (pSi)
pSiO, see Porous silica (pSiO)
PtD strategy, see Prevention through-design strategy (PtD strategy)
Public Health Agency of Canada (PHAC), 344
Pulmonary disorders, 225
Pulmonary drug delivery, 225, 240–241; see also Albumin; Pulmonary system
advantages, 226
buckminsterfullerene, 235–236
carbon nanotubes, 231–234
gene and drug delivery agents, 238, 239
historical prospective, 226–227
nanomaterials used and toxicity concerns, 230–231
nanotechnology and pulmonary therapy, 230
NPs as drug delivery carriers, 230
possible pathways following CNTs inhalation, 232
silica, 239–240
system design, 226
titanium dioxide, 236–237

Pulmonary fibrosis, 437
Pulmonary system, 227; *see also* Pulmonary drug delivery
  acinus, 227
  airway epithelium, 227
  alveoli, 227, 229
  lungs, 227, 228
  mucociliary escalator, 228
  respiratory tract, 227
Pulmonary toxicity, 435–437

Q

QDs, *see* Quantum dots (QDs)
qRT-PCR, *see* Real-time reverse-transcription PCR (qRT-PCR)
QSAR, *see* Quantitative structure–activity relationships (QSAR)
Quantitative structure–activity relationships (QSAR), 218, 389, 410–411; *see also* Nanoparticle risk assessment
  workflow paradigm, 219
Quantum dots (QDs), 6, 151, 161, 194, 254
  exposure route in humans, 387
  kidney toxicity, 300
  liver toxicity, 290

R

RAECs, *see* Rat aortic endothelial cells (RAECs)
Raman spectroscopy, 167
  advantages and disadvantages, 168
  study of Au NPs, 169
Rat aortic endothelial cells (RAECs), 251
RBCs, *see* Red blood cells (RBCs)
REACH Implementation Project on Nanomaterials (RIPoN), 337
Reactive nitrogen species (RNS; NOS), 188, 296
Reactive oxygen species (ROS), 211, 431
  in apoptosis, 123, 211
  cell membrane and protein damage, 262
  DNA damage, 122, 124, 386
  effects, 287
  macrophages activation assessment, 109
  mitigation, 360
  NADPH oxidase, 299
  from nanomaterials, 146
  from nanoparticles, 186
  and neurodegenerative diseases, 27
  NP surface area and, 18, 75
  toxicity of, 190
Reactive species (RS), 75
Real-time reverse-transcription PCR (qRT-PCR), 349
Recommended exposure limits (RELs), 64; *see also* Occupational nanoparticle exposures
Red blood cells (RBCs), 112
Redox potential, 164
Redox-sensitive transcription factors, 286
Reduced graphene oxide (rGO), 161
RELs, *see* Recommended exposure limits (RELs)
RES, *see* Reticuloendothelial system (RES)
Resazurin, *see* Alamar blue
Respiratory protection program (RPP), 64
Respiratory tract, 227
Reticuloendothelial system (RES), 18, 287, 288, 431

Retinoid–phospholipid prodrug, 107
rGO, *see* Reduced graphene oxide (rGO)
RIPoN, *see* REACH Implementation Project on Nanomaterials (RIPoN)
Risk level (RL), 67
RIVM, *see* Netherlands National Institute for Public Health and the Environment (RIVM)
RL, *see* Risk level (RL)
RNS, *see* Reactive nitrogen species (RNS; NOS)
ROS, *see* Reactive oxygen species (ROS)
Royal Society and Royal Academy of Engineering (RS/RAEng), 340
RPP, *see* Respiratory protection program (RPP)
RS, *see* Reactive species (RS)
RS/RAEng, *see* Royal Society and Royal Academy of Engineering (RS/RAEng)

S

Saliva, 265
Scanning electron microscopy (SEM), 28
  live cells exposed to nanoparticles, 213
Scanning mobility particle sizer (SMPS), 54
SCENIHR, *see* Scientific Committee on Emerging and Newly Identified Health Risks (SCENIHR)
Scientific Committee on Emerging and Newly Identified Health Risks (SCENIHR), 331–332
SEC, *see* Size exclusion chromatography (SEC)
Secondary ion mass spectroscopy (SIMS), 28
Secretory phospholipase A_2 IIA (sPLA_2), 107, 113
Selectin P (SELP), 252
Self-organizing map (SOM), 218
SELP, *see* Selectin P (SELP)
SEM, *see* Scanning electron microscopy (SEM)
SERS, *see* Surface enhanced Raman scattering (SERS)
SF, *see* Superfect (SF)
Significant New Use Notice (SNUN), 327
Significant New Use Rule (SNUR), 326, 327
Silica, 239–240
  -based materials, 251
Silica nanoparticles (SNPs), 251
  biomedical applications of, 251
  GI tract toxicity, 276–278
  kidney toxicity, 298–300
  spleen toxicity, 304
Silicon nanowires, 14
Silver nanoparticles, 176–177, 178, 255
  liver toxicity, 290–291
SIMS, *see* Secondary ion mass spectroscopy (SIMS)
Single-cell gel electrophoresis assay, *see* Comet assay
Single-walled carbon nanotubes (SWCNTs), 19, 132, 426, 428; *see also* Carbon nanotubes (CNTs)
  acid functionalization, 136
  cardiovascular toxicity of, 250–251
  characterization by Raman spectroscopy, 168
  diameter of, 426
  GI tract toxicity, 274–275
  health effects of, 191–192
  liver toxicity, 292
  non induction of toxicity, 433
  and pulmonary toxicity issues, 233, 234
  pulmonary toxicity of, 134–135
  spleen toxicity, 305
  uptake, 431

Index

Size exclusion chromatography (SEC), 166
Skin, 27, 181–182, 387
SLN, *see* Solid lipid nanoparticle (SLN)
Small intestine epithelial cell, 263
SMPS, *see* Scanning mobility particle sizer (SMPS)
SN, *see* Supernatant (SN)
SNPs, *see* Silica nanoparticles (SNPs)
SNUN, *see* Significant New Use Notice (SNUN)
SNUR, *see* Significant New Use Rule (SNUR)
Social Democratic Party (SPD), 339
SOD, *see* Superoxide dismutase (SOD)
Soil solution chemistry parameters, 21
Solid lipid nanoparticle (SLN), 142, 153, 193; *see also* Nanotoxicity assessment; Nanotoxicity mechanism; Ocular toxicity; Polymeric nanoparticles (PNs)
    advantage of, 351–352
    solvents used in preparation of, 152
SOM, *see* Self-organizing map (SOM)
SOPs, *see* Standard operation procedures (SOPs)
SPD, *see* Social Democratic Party (SPD)
SPIONPs, *see* Superparamagnetic iron oxide nanoparticles (SPIONPs)
sPLA$_2$, *see* Secretory phospholipase A$_2$ IIA (sPLA$_2$)
Spleen, 301; *see also* Nanomaterial toxicity on spleen
    structure of, 302
SPR, *see* Surface plasmon resonance (SPR)
SRXRF, *see* Synchrotron radiation-induced x-ray fluorescence (SRXRF)
Standard operation procedures (SOPs), 409
Stimuli-responsive delivery systems, 103; *see also* Stimuli-responsive nanomaterials
    biocompatibility assessment methods for, 110, 111
    photo-activated NP-based system, 104
    properties of, 109
Stimuli-responsive nanomaterials, 103; *see also* Photoresponsive materials; Stimuli-responsive delivery systems
    biocompatibility assessment methods, 108–109
    enzyme-responsive systems, 107
    factors influencing biocompatibility, 107–108
    perspectives on field of, 109, 112
    pH-responsive systems, 106
Stressor, 427
Superfect (SF), 148
Supernatant (SN), 278
Superoxide dismutase (SOD), 31, 297
Superparamagnetic iron oxide nanoparticles (SPIONPs), 254
Surface enhanced Raman scattering (SERS), 167, 253
Surface plasmon resonance (SPR), 124, 163
SWCNT-AF, *see* Acid-functionalized SWCNTs (SWCNT-AF)
SWCNTs, *see* Single-walled carbon nanotubes (SWCNTs)
Synchrotron microfocused x-ray fluorescence (μ-XRF), 170
Synchrotron radiation-induced x-ray fluorescence (SRXRF), 28

T

TBA, *see* Thiobarbitutic acid (TBA)
TBARS, *see* Thiobarbituric acid reactive species (TBARS)
TdT, *see* Terminal deoxynucleotidyl transferase (TdT)
TEAC, *see* Trolox equivalent antioxidant capacity assay (TEAC)
Technical University of Denmark (DTU), 341
TEM, *see* Transmission electron microscopy (TEM)
Terminal deoxynucleotidyl transferase (TdT), 278
Terminal deoxynucleotidyl transferase-mediated terminal transferase dUTP nick-end-labeling (TUNEL), 278
Terrestrial ecosystems, 21
Tetramethyl rhodamine ethyl ester (TMRE), 407
Tetrazolium compound, 404
TGA, *see* Therapeutic Goods Administration (TGA); Thioglycolic acid (TGA)
TGF, *see* Transforming growth factor (TGF)
Therapeutic Goods Administration (TGA), 342
Thiobarbituric acid reactive species (TBARS), 406
Thiobarbitutic acid (TBA), 406
Thioglycolic acid (TGA), 254
Time-weighted average (TWA), 64
TIMP-4, *see* Tissue inhibitor of matrix metalloproteinase-4 (TIMP-4)
Tissue inhibitor of matrix metalloproteinase-4 (TIMP-4), 250
Titanium dioxide (TiO$_2$)
    exposure route in humans, 387
    GI tract toxicity, 272–273
    kidney toxicity, 298
    nanoparticles, 220
    pulmonary toxicity issues, 236–237
    spleen toxicity, 304–305
TMRE, *see* Tetramethyl rhodamine ethyl ester (TMRE)
TNF-a, *see* Tumor necrosis factor-a (TNF-a)
Top-down nanofabrication technologies, 3, 4; *see also* Nanotechnology
Toxicogenomics, 210, 221
Toxic Substances Control Act (TSCA), 326
Transcytosis, 270
Transforming growth factor (TGF), 432
    TGF-β$_1$, 406
Transmission electron microscopy (TEM), 28, 121
    NM characterization, 166–167
    sections of HepG2 cells, 213
    small intestine epithelial cell, 263
    TiO$_2$ uptake by, 169
Trolox equivalent antioxidant capacity assay (TEAC), 212
Trypan Blue assay, 122; *see also* Cell viability testing; Nanotoxicity assessment
    exclusion assay, 403–404
TSCA, *see* Toxic Substances Control Act (TSCA)
Tumor necrosis factor-a (TNF-a), 124, 134, 349, 432, 405
TUNEL, *see* Terminal deoxynucleotidyl transferase-mediated terminal transferase dUTP nick-end-labeling (TUNEL)
TWA, *see* Time-weighted average (TWA)

U

UDS, *see* Unscheduled DNA synthesis (UDS)
UFPs, *see* Ultrafine particles (UFPs)
Ultrafine particles (UFPs), 50, 386, 438
Ultraviolet (UV), 8, 104, 260
Unscheduled DNA synthesis (UDS), 356
U.S. Environmental Protection Agency (EPA), 320; *see also* Nanotechnology regulatory implications
    chemical substance use, 327
    to control chemical exposure, 326–327

U.S. Environmental Protection Agency (EPA) *(Continued)*
   FIFRA, 327–328
   NRDC vs., 328–329
   SNUR and SNUN, 327
   TSCA, 326
U.S. Food and Drug Administration (FDA), 149, 251, 320; *see also* Nanotechnology regulatory implications
   CBER, 323
   CDRH, 323–324
   CFSAN guidelines, 324–325
   CMC reviewers in CDER, 322
   CVM procedure, 322–323
   food and packaging guidelines, 326
   nanomaterial definition, 321
   2011 Guidance, 321–322
UV, *see* Ultraviolet (UV)

V

VAC alpha, *see* Annexin V (VAC alpha)
VEDICM, *see* Video-enhanced differential interference contrast microscopy (VEDICM)
Video-enhanced differential interference contrast microscopy (VEDICM), 212
Voluntary Reporting Scheme (VRS), 340

VRS, *see* Voluntary Reporting Scheme (VRS)

W

Whole-cell; *see also* Biosensors
   EIS-based sensors, 120
   impedance-based biosensors, 118
WST-1, *see* 4-[3-(4-Iodophenyl)-2-(4-nitrophenyl)-3H-5-tetrazolio]-1, 3-benzene disulfonate (WST-1)
WST-1 cytotoxicity assay, 403; *see also* Cell viability testing

X

XPS, *see* X-ray photoelectron spectroscopy (XPS)
X-ray diffraction (XRD), 54, 163
X-ray photoelectron spectroscopy (XPS), 28, 136, 163
XRD, *see* X-ray diffraction (XRD)

Y

Yttrium oxide nanoparticles, 255

Z

Zinc oxide nanoparticles, 255
   liver toxicity, 293